# Modern Quantum Mechanics

Modern Quantum Mechanics is a classic graduate level textbook, covering the main quantum mechanics concepts in a clear, organized, and engaging manner. The original author, J. J. Sakurai, was a renowned theorist in particle theory. The Third Edition, revised by Jim Napolitano, introduces topics that extend the text's usefulness into the twenty-first century such as advanced mathematical techniques associated with quantum-mechanical calculations, while at the same time retaining classic developments such as neutron interferometer experiments, Feynman path integrals, correlation measurements, and Bell's inequality. A solution manual for instructors using this textbook can be downloaded from www.cambridge.org/sakurai3.

**J. J. Sakurai** was a noted theorist in particle physics and Professor of Physics at UCLA (1970–1982) and University of Chicago (1964–1970). He received his Ph.D. from Cornell University in 1958. He contributed greatly to the field of particle physics before passing away at the age of 49 in 1982, while he was visiting CERN in Geneva.

In addition he held visiting staff appointments at the California Institute of Technology, Universities of Tokyo and Nagoya, University of Paris d'Orsay, Scuola Normale Superiore at Pisa, Stanford Linear Accelerator, CERN at Geneva, and Max Planck Institute at Munich. He was a Sloan Fellow (1962–1966), Fellow of the American Physical Society (1964–1982), a Guggenheim Fellow (1975–1976) and a von Humboldt Fellow (1981–1982).

**Jim Napolitano** is Professor of Physics and Department Chair at the College of Science and Technology, Temple University. He is an experimental nuclear physicist, with over 320 articles published in refereed journals and an h-index of 81. He shared in the 2016 Breakthrough Prize in Fundamental Physics and currently works on experiments using parity-violating electron scattering. An innovative educator, he has developed coursework and curricula at Rensselaer Polytechnic Institute and Temple University. In all cases, his teaching and instructional development make use of modern techniques. Professor Napolitano has also published textbooks on quantum mechanics, experimental physics, and using MATHEMATICA for physics.

# Modern Quantum Mechanics

## Third Edition

J. J. SAKURAI

JIM NAPOLITANO
Temple University, Philadelphia, PA

CAMBRIDGE
UNIVERSITY PRESS

Shaftesbury Road, Cambridge CB2 8EA, United Kingdom

One Liberty Plaza, 20th Floor, New York, NY 10006, USA

477 Williamstown Road, Port Melbourne, VIC 3207, Australia

314–321, 3rd Floor, Plot 3, Splendor Forum, Jasola District Centre, New Delhi – 110025, India

103 Penang Road, #05–06/07, Visioncrest Commercial, Singapore 238467

Cambridge University Press is part of Cambridge University Press & Assessment,
a department of the University of Cambridge.

We share the University's mission to contribute to society through the pursuit of
education, learning and research at the highest international levels of excellence.

www.cambridge.org
Information on this title: www.cambridge.org/9781108473224
DOI: 10.1017/9781108587280

First published 2021 (version 3, July 2022)

Printed in the United Kingdom by TJ Books Limited, Padstow Cornwall, July 2022

*A catalogue record for this publication is available from the British Library*

ISBN 978-1-108-47322-4 Hardback

Additional resources for this publication at www.cambridge.org/sakurai3.

# Contents

*Preface*                                                    *page* xiii
*Preface to the Revised First Edition*                              xvii
*In Memoriam to J. J. Sakurai*                                       xix
*Foreword from the First Edition*                                    xxi

**1  Fundamental Concepts**                                            1
1.1   The Stern–Gerlach Experiment                                     1
    1.1.1   Description of the Experiment          2
    1.1.2   Sequential Stern–Gerlach Experiments   4
    1.1.3   Analogy with Polarization of Light     6
1.2   Kets, Bras, and Operators                                      10
    1.2.1   Ket Space                             10
    1.2.2   Bra Space and Inner Products          12
    1.2.3   Operators                             13
    1.2.4   Multiplication                        14
    1.2.5   The Associative Axiom                 15
1.3   Base Kets and Matrix Representations                           16
    1.3.1   Eigenkets of an Observable            16
    1.3.2   Eigenkets as Base Kets                17
    1.3.3   Matrix Representations                18
    1.3.4   Spin $\frac{1}{2}$ Systems            21
1.4   Measurements, Observables, and the Uncertainty Relations       22
    1.4.1   Measurements                          22
    1.4.2   Spin $\frac{1}{2}$ Systems, Once Again  24
    1.4.3   Compatible Observables                27
    1.4.4   Incompatible Observables              29
    1.4.5   The Uncertainty Relation              31
1.5   Change of Basis                                                33
    1.5.1   Transformation Operator               33
    1.5.2   Transformation Matrix                 34
    1.5.3   Diagonalization                       35
    1.5.4   Unitary Equivalent Observables        36
1.6   Position, Momentum, and Translation                            37
    1.6.1   Continuous Spectra                    37
    1.6.2   Position Eigenkets and Position Measurements  38
    1.6.3   Translation                           40

|  |  | 1.6.4 | Momentum as a Generator of Translation | 42 |
|  |  | 1.6.5 | The Canonical Commutation Relations | 45 |
|  | 1.7 | Wave Functions in Position and Momentum Space |  | 47 |
|  |  | 1.7.1 | Position-Space Wave Function | 47 |
|  |  | 1.7.2 | Momentum Operator in the Position Basis | 49 |
|  |  | 1.7.3 | Momentum-Space Wave Function | 49 |
|  |  | 1.7.4 | Gaussian Wave Packets | 51 |
|  |  | 1.7.5 | Generalization to Three Dimensions | 53 |
|  | Problems |  |  | 54 |

**2  Quantum Dynamics**                                                                    62

|  | 2.1 | Time Evolution and the Schrödinger Equation |  | 62 |
|  |  | 2.1.1 | Time-Evolution Operator | 62 |
|  |  | 2.1.2 | The Schrödinger Equation | 65 |
|  |  | 2.1.3 | Energy Eigenkets | 67 |
|  |  | 2.1.4 | Time Dependence of Expectation Values | 68 |
|  |  | 2.1.5 | Spin Precession | 69 |
|  |  | 2.1.6 | Neutrino Oscillations | 71 |
|  |  | 2.1.7 | Correlation Amplitude and the Energy-Time Uncertainty Relation | 74 |
|  | 2.2 | The Schrödinger Versus the Heisenberg Picture |  | 75 |
|  |  | 2.2.1 | Unitary Operators | 75 |
|  |  | 2.2.2 | State Kets and Observables in the Schrödinger and the Heisenberg Pictures | 77 |
|  |  | 2.2.3 | The Heisenberg Equation of Motion | 78 |
|  |  | 2.2.4 | Free Particles: Ehrenfest's Theorem | 79 |
|  |  | 2.2.5 | Base Kets and Transition Amplitudes | 81 |
|  | 2.3 | Simple Harmonic Oscillator |  | 83 |
|  |  | 2.3.1 | Energy Eigenkets and Energy Eigenvalues | 83 |
|  |  | 2.3.2 | Time Development of the Oscillator | 88 |
|  | 2.4 | Schrödinger's Wave Equation |  | 91 |
|  |  | 2.4.1 | Time-Dependent Wave Equation | 91 |
|  |  | 2.4.2 | The Time-Independent Wave Equation | 92 |
|  |  | 2.4.3 | Interpretations of the Wave Function | 94 |
|  |  | 2.4.4 | The Classical Limit | 96 |
|  | 2.5 | Elementary Solutions to Schrödinger's Wave Equation |  | 97 |
|  |  | 2.5.1 | Free Particle in Three Dimensions | 97 |
|  |  | 2.5.2 | The Simple Harmonic Oscillator | 99 |
|  |  | 2.5.3 | The Linear Potential | 101 |
|  |  | 2.5.4 | The WKB (Semiclassical) Approximation | 104 |
|  | 2.6 | Propagators and Feynman Path Integrals |  | 108 |
|  |  | 2.6.1 | Propagators in Wave Mechanics | 108 |
|  |  | 2.6.2 | Propagator as a Transition Amplitude | 112 |
|  |  | 2.6.3 | Path Integrals as the Sum over Paths | 114 |

2.6.4   Feynman's Formulation                                    115
2.7   Potentials and Gauge Transformations                       120
    2.7.1   Constant Potentials                                  120
    2.7.2   Gravity in Quantum Mechanics                         122
    2.7.3   Gauge Transformations in Electromagetism             126
    2.7.4   The Aharonov–Bohm Effect                             131
    2.7.5   Magnetic Monopole                                    135
Problems                                                         138

3  Theory of Angular Momentum                                    149
3.1   Rotations and Angular Momentum Commutation Relations       149
    3.1.1   Finite Versus Infinitesimal Rotations                149
    3.1.2   Infinitesimal Rotations in Quantum Mechanics         152
    3.1.3   Finite Rotations in Quantum Mechanics                153
    3.1.4   Commutation Relations for Angular Momentum           154
3.2   Spin $\frac{1}{2}$ Systems and Finite Rotations            155
    3.2.1   Rotation Operator for Spin $\frac{1}{2}$             155
    3.2.2   Spin Precession Revisited                            157
    3.2.3   Neutron Interferometry Experiment to Study $2\pi$ Rotations  158
    3.2.4   Pauli Two-Component Formalism                        159
    3.2.5   Rotations in the Two-Component Formalism             161
3.3   SO(3), SU(2), and Euler Rotations                          163
    3.3.1   Orthogonal Group                                     163
    3.3.2   Unitary Unimodular Group                             164
    3.3.3   Euler Rotations                                      166
3.4   Density Operators and Pure Versus Mixed Ensembles          169
    3.4.1   Polarized Versus Unpolarized Beams                   169
    3.4.2   Ensemble Averages and Density Operator               170
    3.4.3   Time Evolution of Ensembles                          175
    3.4.4   Continuum Generalizations                            176
    3.4.5   Quantum Statistical Mechanics                        176
3.5   Eigenvalues and Eigenstates of Angular Momentum            180
    3.5.1   Commutation Relations and the Ladder Operators       180
    3.5.2   Eigenvalues of $J^2$ and $J_z$                       182
    3.5.3   Matrix Elements of Angular-Momentum Operators        184
    3.5.4   Representations of the Rotation Operator             185
3.6   Orbital Angular Momentum                                   188
    3.6.1   Orbital Angular Momentum as Rotation Generator       188
    3.6.2   Spherical Harmonics                                  191
    3.6.3   Spherical Harmonics as Rotation Matrices             194
3.7   Schrödinger's Equation for Central Potentials              195
    3.7.1   The Radial Equation                                  196
    3.7.2   The Free Particle and Infinite Spherical Well        198

|       | 3.7.3 | The Isotropic Harmonic Oscillator | 199 |
|       | 3.7.4 | The Coulomb Potential | 201 |
| 3.8 | Addition of Angular Momenta | | 205 |
|       | 3.8.1 | Simple Examples of Angular-Momentum Addition | 205 |
|       | 3.8.2 | Formal Theory of Angular-Momentum Addition | 208 |
|       | 3.8.3 | Recursion Relations for the Clebsch–Gordan Coefficients | 212 |
|       | 3.8.4 | Clebsch–Gordan Coefficients and Rotation Matrices | 216 |
| 3.9 | Schwinger's Oscillator Model of Angular Momentum | | 218 |
|       | 3.9.1 | Angular Momentum and Uncoupled Oscillators | 218 |
|       | 3.9.2 | Explicit Formula for Rotation Matrices | 222 |
| 3.10 | Spin Correlation Measurements and Bell's Inequality | | 224 |
|       | 3.10.1 | Correlations in Spin-Singlet States | 224 |
|       | 3.10.2 | Einstein's Locality Principle and Bell's Inequality | 226 |
|       | 3.10.3 | Quantum Mechanics and Bell's Inequality | 229 |
| 3.11 | Tensor Operators | | 231 |
|       | 3.11.1 | Vector Operator | 231 |
|       | 3.11.2 | Cartesian Tensors Versus Irreducible Tensors | 233 |
|       | 3.11.3 | Product of Tensors | 235 |
|       | 3.11.4 | Matrix Elements of Tensor Operators; the Wigner–Eckart Theorem | 236 |
|       | Problems | | 240 |

**4 Symmetry in Quantum Mechanics** — 249

| 4.1 | Symmetries, Conservation Laws, and Degeneracies | | 249 |
|       | 4.1.1 | Symmetries in Classical Physics | 249 |
|       | 4.1.2 | Symmetry in Quantum Mechanics | 250 |
|       | 4.1.3 | Degeneracies | 251 |
|       | 4.1.4 | SO(4) Symmetry in the Coulomb Potential | 252 |
| 4.2 | Discrete Symmetries, Parity, or Space Inversion | | 256 |
|       | 4.2.1 | Wave Functions under Parity | 258 |
|       | 4.2.2 | Symmetrical Double-Well Potential | 261 |
|       | 4.2.3 | Parity-Selection Rule | 263 |
|       | 4.2.4 | Parity Nonconservation | 264 |
| 4.3 | Lattice Translation as a Discrete Symmetry | | 265 |
| 4.4 | The Time-Reversal Discrete Symmetry | | 270 |
|       | 4.4.1 | Digression on Symmetry Operations | 272 |
|       | 4.4.2 | Time-Reversal Operator | 275 |
|       | 4.4.3 | Wave Function | 279 |
|       | 4.4.4 | Time Reversal for a Spin $\frac{1}{2}$ System | 280 |
|       | 4.4.5 | Interactions with Electric and Magnetic Fields; Kramers Degeneracy | 283 |
|       | Problems | | 285 |

**5 Approximation Methods**     288
   5.1   Time-Independent Perturbation Theory: Nondegenerate Case     288
      5.1.1   Statement of the Problem     288
      5.1.2   The Two-State Problem     289
      5.1.3   Formal Development of Perturbation Expansion     291
      5.1.4   Wave Function Renormalization     295
      5.1.5   Elementary Examples     296
   5.2   Time-Independent Perturbation Theory: The Degenerate Case     300
      5.2.1   Linear Stark Effect     303
   5.3   Hydrogenlike Atoms: Fine Structure and the Zeeman Effect     305
      5.3.1   The Relativistic Correction to the Kinetic Energy     305
      5.3.2   Spin-Orbit Interaction and Fine Structure     307
      5.3.3   The Zeeman Effect     311
      5.3.4   Van der Waals' Interaction     314
   5.4   Variational Methods     316
   5.5   Time-Dependent Potentials: The Interaction Picture     320
      5.5.1   Statement of the Problem     320
      5.5.2   The Interaction Picture     321
      5.5.3   Time-Dependent Two-State Problems: Nuclear Magnetic Resonance, Masers, and So Forth     323
      5.5.4   Spin Magnetic Resonance     325
      5.5.5   Maser     326
   5.6   Hamiltonians with Extreme Time Dependence     327
      5.6.1   Sudden Approximation     328
      5.6.2   Adiabatic Approximation     328
      5.6.3   Berry's Phase     331
      5.6.4   Example: Berry's Phase for Spin $\frac{1}{2}$     333
      5.6.5   Aharonov–Bohm and Magnetic Monopoles Revisited     335
   5.7   Time-Dependent Perturbation Theory     337
      5.7.1   Dyson Series     337
      5.7.2   Transition Probability     339
      5.7.3   Constant Perturbation     341
      5.7.4   Harmonic Perturbation     345
   5.8   Applications to Interactions with the Classical Radiation Field     347
      5.8.1   Absorption and Stimulated Emission     347
      5.8.2   Electric Dipole Approximation     348
      5.8.3   Photoelectric Effect     350
      5.8.4   Spontaneous Emission     352
   5.9   Energy Shift and Decay Width     355
   Problems     358

**6 Scattering Theory**     371
   6.1   Scattering as a Time-Dependent Perturbation     371
      6.1.1   Transition Rates and Cross Sections     373

6.1.2    Solving for the $T$ Matrix                                   374
6.1.3    Scattering from the Future to the Past                       376
6.2    The Scattering Amplitude                                       376
6.2.1    Wave Packet Description                                      381
6.2.2    The Optical Theorem                                          381
6.3    The Born Approximation                                         384
6.3.1    The Higher-Order Born Approximation                          387
6.4    Phase Shifts and Partial Waves                                 388
6.4.1    Free-Particle States                                         388
6.4.2    Partial-Wave Expansion                                       392
6.4.3    Unitarity and Phase Shifts                                   394
6.4.4    Determination of Phase Shifts                                397
6.4.5    Hard-Sphere Scattering                                       398
6.5    Eikonal Approximation                                          400
6.5.1    Partial Waves and the Eikonal Approximation                  403
6.6    Low-Energy Scattering and Bound States                         405
6.6.1    Rectangular Well or Barrier                                  406
6.6.2    Zero-Energy Scattering and Bound States                      408
6.6.3    Bound States as Poles of $S_l(k)$                            410
6.7    Resonance Scattering                                           412
6.8    Symmetry Considerations in Scattering                          416
6.9    Inelastic Electron-Atom Scattering                             419
6.9.1    Nuclear Form Factor                                          423
Problems                                                              424

7   Identical Particles                                               429
7.1    Permutation Symmetry                                           429
7.2    Symmetrization Postulate                                       433
7.3    Two-Electron System                                            434
7.4    The Helium Atom                                                437
7.5    Multiparticle States                                           441
7.6    Density Functional Theory                                      443
7.6.1    The Energy Functional for a Single Particle                  443
7.6.2    The Hohenberg–Kohn Theorem                                   445
7.6.3    The Kohn–Sham Equations                                      447
7.6.4    Models of the Exchange-Correlation Energy                    450
7.6.5    Application to the Helium Atom                               451
7.7    Quantum Fields                                                 454
7.7.1    Second Quantization                                          454
7.7.2    Dynamical Variables in Second Quantization                   456
7.7.3    Example: The Degenerate Electron Gas                         460
7.8    Quantization of the Electromagnetic Field                      464
7.8.1    Maxwell's Equations in Free Space                            465

| | | |
|---|---|---|
| | 7.8.2 Photons and Energy Quantization | 467 |
| | 7.8.3 The Casimir Effect | 468 |
| | 7.8.4 Concluding Remarks | 472 |
| | Problems | 474 |
| **8** | **Relativistic Quantum Mechanics** | **478** |
| | 8.1 Paths to Relativistic Quantum Mechanics | 478 |
| | 8.1.1 Natural Units | 479 |
| | 8.1.2 The Energy of a Free Relativistic Particle | 479 |
| | 8.1.3 The Klein–Gordon Equation | 480 |
| | 8.1.4 An Interpretation of Negative Energies | 484 |
| | 8.1.5 The Klein–Gordon Field | 485 |
| | 8.1.6 Summary: The Klein–Gordon Equation and the Scalar Field | 489 |
| | 8.2 The Dirac Equation | 490 |
| | 8.2.1 The Conserved Current | 491 |
| | 8.2.2 Free-Particle Solutions | 493 |
| | 8.2.3 Interpretation of Negative Energies | 494 |
| | 8.2.4 Electromagnetic Interactions | 495 |
| | 8.3 Symmetries of the Dirac Equation | 496 |
| | 8.3.1 Angular Momentum | 497 |
| | 8.3.2 Parity | 497 |
| | 8.3.3 Charge Conjugation | 498 |
| | 8.3.4 Time Reversal | 499 |
| | 8.3.5 CPT | 501 |
| | 8.4 Solving with a Central Potential | 502 |
| | 8.4.1 The One-Electron Atom | 505 |
| | 8.5 Relativistic Quantum Field Theory | 509 |
| | Problems | 510 |
| | **Appendix A Electromagnetic Units** | **514** |
| | **Appendix B Elementary Solutions to Schrödinger's Wave Equation** | **520** |
| | **Appendix C Hamiltonian for a Charge in an Electromagnetic Field** | **530** |
| | **Appendix D Proof of the Angular-Momentum Rule (3.358)** | **532** |
| | **Appendix E Finding Clebsch–Gordan Coefficients** | **534** |
| | **Appendix F Notes on Complex Variables** | **535** |
| | *Bibliography* | 541 |
| | *Index* | 544 |

# Preface

This book covers the material on quantum mechanics typically found in a first year graduate physics curriculum. The approach emphasizes states, operators, eigenvalues, and representations from the start. Building on these foundations, the reader sees, for example, how the Schrödinger representation is just one of several ways to realize quantum dynamics, and how classical physics emerges as an approximation. This approach also helps the reader gain an appreciation of purely quantum-mechanical phenomena, for example the magnetic moment and spin of an electron, that have no classical analogue.

The intended audience is the same as for earlier editions, that is, students having taken upper level undergraduate coursework in quantum physics, classical mechanics and electromagnetism, multivariable calculus, and ordinary and partial differential equations.

Professor Jun John Sakurai originally conceived the idea for this textbook, I think inspired by Dirac's monograph. Sakurai's life was cut short suddenly, as he was preparing the first manuscript. His colleague San Fu Tuan took over as Editor, completing a seven chapter manuscript for Addison-Wesley, who published the First Edition in 1985 and a Revised Edition in 1993. Some time later, I started work on the Second Edition for Pearson (who had since acquired Addison-Wesley). This volume contained a lot of new material, including an eighth chapter, and was published in 2010. The text was reissued by Cambridge University Press in 2017, which was also when I started work on the Third Edition.

Quantum mechanics has always fascinated me, but it was the First Edition of *Modern Quantum Mechanics* that finally explained to me the logical progression from fundamental assumptions to practical applications, with classical physics emerging as an approximation. When I first taught this material at Rensselaer Polytechnic Institute, I used the Revised Edition, but found myself supplementing with my own notes on solutions of the Schrödinger equation and other topics. I also tried to use my course to prepare students for quantum field theory, introducing second quantization and relativistic quantum mechanics, neither of which were included in Sakurai's book.

I was therefore pleased to be asked to take on the Second Edition. Sections were added to Chapters Two and Three on solutions to the Schrödinger equation. I reversed the order of Chapters Six and Seven, so that Scattering Theory came first, and I reworked the treatment so that it was based on the formal theory of time-dependent perturbations. The following chapter on Identical Particles was augmented to include second quantization and the quantization of the free electromagnetic field, and I added a new chapter on Relativistic Quantum Mechanics. I also included several connections throughout the book to experimental measurements, and worked to fix a number of idiosyncrasies that I found when I taught out of the book.

The result was a text that, I thought, achieved my goal of a high level treatment respecting Sakurai's vision, adding reference to additional modern concepts and experiment, and preparing the reader for quantum field theory and beyond. The first two chapters lay the mathematical and physical foundations for the rest of the book, and connect the reader to undergraduate topics in wave mechanics. Chapter Three covers angular momentum from the perspective of the rotation operator, with strong connections to important concepts such as the density operator, central potentials, and Bell's inequality. Groups are also introduced here, with further exposition in Chapter Four. Applications to "real world" problems are the focus of Chapters Five and Six, all the while keeping to the focus of building on the fundamentals. Chapters Seven and Eight move the discussion towards the "next" course in quantum mechanics, covering many-body formalism and the inclusion of special relativity.

The Third Edition keeps the same ordering of the eight chapters. Significant new material has been added, but I also worked to clarify some of the discussions and to fix various issues that I discovered after teaching out of the Second Edition. In fact, I compiled a long list of "Typographical Errors, Mistakes, and Comments" based on covering nearly the entire book in class, and working through all of the end-of-chapter problems. The Third Edition addresses all of the errors. It also addresses most of the comments, having to give up on some only for lack of time.

There are three new sections of new material. Despite its increasing use in condensed matter physics, I found no treatments of density functional theory in any quantum mechanics textbook. So, I added Section 7.6 to introduce the subject and take it through to its application in the helium atom. A reviewer's suggestion inspired me to add Section 8.1.5 to show how the Klein–Gordon field, built using second quantization, fixes the problems of negative energies and nonpositive definite probability currents in the Klein–Gordon wave equation. The Second Edition treated spontaneous emission only as an end-of-chapter problem, but Section 5.8.4 now goes through the derivation, with some details and numerical calculations left as problems.

I added new appendices on the Hamiltonian for a Charge in an Electromagnetic Field, Notes on Complex Analysis, and Calculating Clebsch–Gordan Coefficients. The appendix on Electromagnetic Units has been significantly revised, and I updated the appendix on Elementary Solutions to Schrödinger's Wave Equation to better connect to the discussions in the text.

Instructors may elect to pick and choose from topics in the book, and not necessarily in the order of presentation. Chapter One should be covered first, since it lays down the notation and fundamental assumptions. One could then, for example, take parts of Chapters Three and Four to expand on operators, observables, and symmetries, prior to discussing dynamics in Chapter Two. Many other combinations are possible. Indeed, throughout the book, I have tried to refer to other places in the text where relevant related material is covered or discussed.

As befits a graduate level textbook, the strategy here is to lay down the principles, following up with implications by deduction. Some example calculations are carried through in the text, but the end-of-chapter problems are generally meant to extend the discussion, and not simply practice what was covered. As such, I recommend that instructors choose problems, from the text or otherwise, that follow this idea, including connection to experimental measurements, where practical.

In several places in the book, either explicitly or implicitly, computer calculations are necessary to completely follow the arguments or to work the problems. I worked through these using MATHEMATICA, and am happy to share the code with anyone who would like to see it, but any other programming language or application can also be used, of course.

---

Producing the Second Edition was a long process that would not have been possible without help from many, many people. Colleagues in physics include John Cummings, Jack Fishburn, Joel Giedt, David Hertzog, Barry Holstein, Bob Jaffe, Matthew Kirby, Joe Levinger, Alan Litke, Kam-Biu Luk, Bob McKeown, Harry Nelson, Joe Paki, Murray Peshkin, Olivier Pfister, Mike Snow, John Townsend, San Fu Tuan, David Van Baak, Dirk Walecka, and Tony Zee. The people at Addison-Wesley/Pearson who guided me included Adam Black, Ashley Eklund, Deb Greco, Dyan Menezes, John Rogosich, and Jim Smith.

So many others were very helpful to me as I developed the Third Edition. This includes colleagues Kieron Burke, Mark Caprio, Carl Carlson, Benjamin Chandran, Chris Cocuzza, Martha Constantinou, Patrick Fasano, Jeremias Gonzalez, Aaron Kaplan (with special thanks for helping me learn DFT), Toh-Ming Lu, Carl Maes, Andreas Metz, Jerry Miller, Djordje Minic, Adilson Motter, Nick Murphy, Steve Naculich, Celso Nishi, John Perdew, Jon Rosner, and Roland Winkler. I am forever grateful to Simon Capelin at Cambridge University Press, for first bringing to me the possibility of republishing the Second Edition, and encouraging me to consider a Third Edition. Other key people at CUP include Jane Adams, Nick Gibbons, Lisa Pinto, and Ilaria Tassistro.

I can only offer my sincere apologies to people I should have listed, but whose name doesn't appear because I've been careless with note keeping. There are also the very many people who, over the past several years, offered comments, some of which I've not been able to incorporate.

Finally, I give a special acknowledgement for Stuart Freedman, my mentor, colleague, and friend. Stuart's Ph.D. thesis experiment was the first verification of the violation of Bell's inequality, and he used this to stoke my interest in quantum mechanics. His guidance during my years as a graduate student and young scientist shaped my career, and he remained my friend and counselor until his untimely passing.

*Jim Napolitano*
*Philadelphia, PA*

# Preface to the Revised First Edition

Since 1989 the Editor has enthusiastically pursued a revised edition of Modern Quantum Mechanics by his late great friend J. J. Sakurai, in order to extend this text's usefulness into the twenty-first century. Much consultation took place with the panel of Sakurai friends who helped with the original edition, but in particular with Professor Yasuo Hara of Tsukuba University and Professor Akio Sakurai of Kyoto Sangyo University in Japan.

This book is intended for the first year graduate student who has studied quantum mechanics at the junior or senior level. It does not provide an introduction to quantum mechanics for the beginner. The reader should have had some experience in solving time-dependent and time-independent wave equations. A familiarity with the time evolution of the Gaussian wave packet in a force-free region is assumed, as is the ability to solve one-dimensional transmission-reflection problems. Some of the general properties of the energy eigenfunctions and the energy eigenvalues should also be known to the student who uses this text.

The major motivation for this project is to revise the main text. There are three important additions and/or changes to the revised edition, which otherwise preserves the original version unchanged. These include a reworking of certain portions of Section 5.2 on time-independent perturbation theory for the degenerate case by Professor Kenneth Johnson of M.I.T., taking into account a subtle point that has not been properly treated by a number of texts on quantum mechanics in this country. Professor Roger Newton of Indiana University contributed refinements on lifetime broadening in Stark effect, additional explanations of phase shifts at resonances, the optical theorem, and on non-normalizable state. These appear as "remarks by the editor" or "editor's note" in the revised edition. Professor Thomas Fulton of the Johns Hopkins University reworked his Coulomb Scattering contribution (Section 7.13) so that it now appears as a shorter text portion emphasizing the physics, with the mathematical details relegated to Appendix C.

Though not a major part of the text, some additions were deemed necessary to take into account developments in quantum mechanics that have become prominent since November 1, 1982. To this end, two supplements are included at the end of the text. Supplement I is on adiabatic change and geometrical phase (popularized by M. V. Berry since 1983) and is actually an English translation of the supplement on this subject written by Professor Akio Sakurai for the Japanese version of *Modern Quantum Mechanics* (copyright © Yoshioka-Shoten Publishing of Kyoto). Supplement II is on non-exponential decays written by my colleague here, Professor Xerxes Tata, and read over by Professor E. C. G. Sudarshan of the University of Texas at Austin. Though non-exponential decays have a long history theoretically, experimental work on transition rates that tests indirectly such decays was done only in 1990. Introduction of additional material is of course a subjective matter on the part of the Editor; the readers will evaluate for themselves its appropriateness. Thanks to Professor Akio Sakurai, the revised edition has been "finely toothcombed" for misprint errors of the first ten printings of the original edition. My colleague, Professor Sandip Pakvasa, provided overall guidance and encouragement to me throughout this process of revision.

In addition to the acknowledgments above, my former students Li Ping, Shi Xiaohong, and Yasunaga Suzuki provided the sounding board for ideas on the revised edition when taking my graduate quantum mechanics course at the University of Hawaii during the spring of 1992. Suzuki provided the initial translation from Japanese of Supplement I as a course term paper. Dr. Andy Acker provided me with computer graphic assistance. The Department of Physics and Astronomy and particularly the High Energy Physics Group of the University of Hawaii at Manoa provided again both the facilities and a conducive atmosphere for me to carry out my editorial task. Finally I wish to express my gratitude to Physics (and sponsoring) Senior Editor, Stuart Johnson, and his Editorial Assistant, Jennifer Duggan, as well as Senior Production Coordinator Amy Willcutt, of Addison-Wesley for their encouragement and optimism that the revised edition will indeed materialize.

*San Fu TUAN*
*Honolulu, Hawaii*

# In Memoriam to J. J. Sakurai

Jun John Sakurai was born in 1933 in Tokyo and came to the United States as a high school student in 1949. He studied at Harvard and at Cornell, where he received his Ph.D. in 1958. He was then appointed assistant professor of Physics at the University of Chicago, and became a full professor in 1964. He stayed at Chicago until 1970 when he moved to the University of California at Los Angeles, where he remained until his death. During his lifetime he wrote 119 articles in theoretical physics of elementary particles as well as several books and monographs on both quantum and particle theory.

The discipline of theoretical physics has as its principal aim the formulation of theoretical descriptions of the physical world that are at once concise and comprehensive. Because nature is subtle and complex, the pursuit of theoretical physics requires bold and enthusiastic ventures to the frontiers of newly discovered phenomena. This is an area in which Sakurai reigned supreme with his uncanny physical insight and intuition and also his ability to explain these phenomena in illuminating physical terms to the unsophisticated. One has but to read his very lucid textbooks on *Invariance Principles and Elementary Particles* and *Advanced Quantum Mechanics* as well as his reviews and summer school lectures to appreciate this. Without exaggeration I could say that much of what I did understand in particle physics came from these and from his articles and private tutoring.

When Sakurai was still a graduate student, he proposed what is now known as the V-A theory of weak interactions, independently of (and simultaneously with) Richard Feynman, Murray Gell-Mann, Robert Marshak, and George Sudarshan. In 1960 he published in *Annals of Physics* a prophetic paper, probably his single most important one. It was concerned with the first serious attempt to construct a theory of strong interactions based on Abelian and non-Abelian (Yang–Mills) gauge invariance. This seminal work induced theorists to attempt an understanding of the mechanisms of mass generation for gauge (vector) fields, now realized as the Higgs mechanism. Above all it stimulated the search for a realistic unification of forces under the gauge principle, now crowned with success in the celebrated Glashow–Weinberg–Salam unification of weak and electromagnetic forces. On the phenomenological side, Sakurai pursued and vigorously advocated the vector mesons dominance model of hadron dynamics. He was the first to discuss the mixing of $\omega$ and $\phi$ meson states. Indeed, he made numerous important contributions to particle physics phenomenology in a much more general sense, as his heart was always close to experimental activities.

I knew Jun John for more than 25 years, and I had the greatest admiration not only for his immense powers as a theoretical physicist but also for the warmth and generosity of his spirit. Though a graduate student himself at Cornell during 1957–1958, he took time from his own pioneering research in K-nucleon dispersion relations to help me (via extensive correspondence) with my Ph.D. thesis on the same subject at Berkeley. Both Sandip Pakvasa and I were privileged to be associated with one of his last papers on weak couplings of heavy quarks, which displayed once more his infectious and intuitive style of doing physics. It is of course gratifying to us in retrospect that Jun John counted this paper among the score of his published works that he particularly enjoyed.

The physics community suffered a great loss at Jun John Sakurai's death. The personal sense of loss is a severe one for me. Hence I am profoundly thankful for the opportunity to edit and complete his manuscript on *Modern Quantum Mechanics* for publication. In my faith no greater gift can be given me than an opportunity to show my respect and love for Jun John through meaningful service.

*San Fu Tuan*
*(From the First Edition)*

# Foreword from the First Edition

J. J. Sakurai was always a very welcome guest here at CERN, for he was one of those rare theorists to whom the experimental facts are even more interesting than the theoretical game itself. Nevertheless, he delighted in theoretical physics and in its teaching, a subject on which he held strong opinions. He thought that much theoretical physics teaching was both too narrow and too remote from application: "...we see a number of sophisticated, yet uneducated, theoreticians who are conversant in the LSZ formalism of the Heisenberg field operators, but do not know why an excited atom radiates, or are ignorant of the quantum theoretic derivation of Rayleigh's law that accounts for the blueness of the sky." And he insisted that the student must be able to use what has been taught: "The reader who has read the book but cannot do the exercises has learned nothing."

He put these principles to work in his fine book *Advanced Quantum Mechanics* (1967) and in *Invariance Principles and Elementary Particles* (1964), both of which have been very much used in the CERN library. This new book, *Modern Quantum Mechanics*, should be used even more, by a larger and less specialized group. The book combines breadth of interest with a thorough practicality. Its readers will find here what they need to know, with a sustained and successful effort to make it intelligible.

J. J. Sakurai's sudden death on November 1, 1982 left this book unfinished. Reinhold Bertlmann and I helped Mrs. Sakurai sort out her husband's papers at CERN. Among them we found a rough, handwritten version of most of the book and a large collection of exercises. Though only three chapters had been completely finished, it was clear that the bulk of the creative work had been done. It was also clear that much work remained to fill in gaps, polish the writing, and put the manuscript in order.

That the book is now finished is due to the determination of Noriko Sakurai and the dedication of San Fu Tuan. Upon her husband's death, Mrs. Sakurai resolved immediately that his last effort should not go to waste. With great courage and dignity she became the driving force behind the project, overcoming all obstacles and setting the high standards to be maintained. San Fu Tuan willingly gave his time and energy to the editing and completion of Sakurai's work. Perhaps only others close to the hectic field of high-energy theoretical physics can fully appreciate the sacrifice involved.

For me personally, J. J. had long been far more than just a particularly distinguished colleague. It saddens me that we will never again laugh together at physics and physicists and life in general, and that he will not see the success of his last work. But I am happy that it has been brought to fruition.

*John S. Bell*
*CERN, Geneva*

# Fundamental Concepts

The revolutionary change in our understanding of microscopic phenomena that took place during the first 27 years of the twentieth century is unprecedented in the history of natural sciences. Not only did we witness severe limitations in the validity of classical physics, but we found the alternative theory that replaced the classical physical theories to be far richer in scope and far richer in its range of applicability.

The most traditional way to begin a study of quantum mechanics is to follow the historical developments – Planck's radiation law, the Einstein–Debye theory of specific heats, the Bohr atom, de Broglie's matter waves, and so forth – together with careful analyses of some key experiments such as the Compton effect, the Franck–Hertz experiment, and the Davisson–Germer–Thompson experiment. In that way we may come to appreciate how the physicists in the first quarter of the twentieth century were forced to abandon, little by little, the cherished concepts of classical physics and how, despite earlier false starts and wrong turns, the great masters – Heisenberg, Schrödinger, and Dirac, among others – finally succeeded in formulating quantum mechanics as we know it today.

However, we do not follow the historical approach in this book. Instead, we start with an example that illustrates, perhaps more than any other example, the inadequacy of classical concepts in a fundamental way. We hope that by exposing the reader to a "shock treatment" at the onset, he or she may be attuned to what we might call the "quantum-mechanical way of thinking" at a very early stage.

This different approach is not merely an academic exercise. Our knowledge of the physical world comes from making assumptions about nature, formulating these assumptions into postulates, deriving predictions from those postulates, and testing those predictions against experiment. If experiment does not agree with the prediction, then, presumably, the original assumptions were incorrect. Our approach emphasizes the fundamental assumptions we make about nature, upon which we have come to base all of our physical laws, and which aim to accommodate profoundly quantum-mechanical observations at the outset.

## 1.1 The Stern–Gerlach Experiment

The example we concentrate on in this section is the Stern–Gerlach experiment, originally conceived by O. Stern in 1921 and carried out in Frankfurt by him in collaboration with

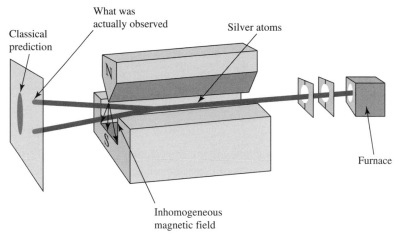

Classical prediction / What was actually observed / Silver atoms / Furnace / Inhomogeneous magnetic field

**Fig. 1.1**   The Stern–Gerlach experiment.

W. Gerlach in 1922.[1] This experiment illustrates in a dramatic manner the necessity for a radical departure from the concepts of classical mechanics. In the subsequent sections the basic formalism of quantum mechanics is presented in a somewhat axiomatic manner but always with the example of the Stern–Gerlach experiment in the back of our minds. In a certain sense, a two-state system of the Stern–Gerlach type is the least classical, most quantum-mechanical system. A solid understanding of problems involving two-state systems will turn out to be rewarding to any serious student of quantum mechanics. It is for this reason that we refer repeatedly to two-state problems throughout this book.

### 1.1.1  Description of the Experiment

We now present a brief discussion of the Stern–Gerlach experiment, which is discussed in almost any book on modern physics.[2] First, silver (Ag) atoms are heated in an oven. The oven has a small hole through which some of the silver atoms escape. As shown in Figure 1.1, the beam goes through a collimator and is then subjected to an inhomogeneous magnetic field produced by a pair of pole pieces, one of which has a very sharp edge.

We must now work out the effect of the magnetic field on the silver atoms. For our purpose the following oversimplified model of the silver atom suffices. The silver atom is made up of a nucleus and 47 electrons, where 46 out of the 47 electrons can be visualized as forming a spherically symmetrical electron cloud with no net angular momentum. If we ignore the nuclear spin, which is irrelevant to our discussion, we see that the atom as a whole does have an angular momentum, which is due solely to the spin – intrinsic as opposed to orbital – angular momentum of the single 47th (5$s$) electron. The 47 electrons

---

[1]  For an excellent historical discussion of the Stern–Gerlach experiment, see "Stern and Gerlach: how a bad cigar helped reorient atomic physics," by Friedrich and Herschbach, *Phys. Today*, **56** (2003) 53.

[2]  For an elementary but enlightening discussion of the Stern–Gerlach experiment, see French and Taylor (1978), pp. 432–438.

are attached to the nucleus, which is $\sim 2 \times 10^5$ times heavier than the electron; as a result, the heavy atom as a whole possesses a magnetic moment equal to the spin magnetic moment of the 47th electron. In other words, the magnetic moment $\boldsymbol{\mu}$ of the atom is proportional to the electron spin $\mathbf{S}$,

$$\boldsymbol{\mu} \propto \mathbf{S}, \tag{1.1}$$

where the precise proportionality factor turns out to be $e/m_e c$ ($e < 0$ in this book) to an accuracy of about 0.2%.

Because the interaction energy of the magnetic moment with the magnetic field is just $-\boldsymbol{\mu} \cdot \mathbf{B}$, the $z$-component of the force experienced by the atom is given by

$$F_z = \frac{\partial}{\partial z}(\boldsymbol{\mu} \cdot \mathbf{B}) \simeq \mu_z \frac{\partial B_z}{\partial z}, \tag{1.2}$$

where we have ignored the components of $\mathbf{B}$ in directions other than the $z$-direction. Because the atom as a whole is very heavy, we expect that the classical concept of trajectory can be legitimately applied, a point which can be justified using the Heisenberg uncertainty principle to be derived later. With the arrangement of Figure 1.1, the $\mu_z > 0$ ($S_z < 0$) atom experiences an upward force, while the $\mu_z < 0$ ($S_z > 0$) atom experiences a downward force. The beam is then expected to be split according to the values of $\mu_z$. In other words, the SG (Stern–Gerlach) apparatus "measures" the $z$-component of $\boldsymbol{\mu}$ or, equivalently, the $z$-component of $\mathbf{S}$ up to a proportionality factor.

The atoms in the oven are randomly oriented; there is no preferred direction for the orientation of $\boldsymbol{\mu}$. If the electron were like a classical spinning object, we would expect all values of $\mu_z$ to be realized between $|\boldsymbol{\mu}|$ and $-|\boldsymbol{\mu}|$. This would lead us to expect a continuous bundle of beams coming out of the SG apparatus, as indicated in Figure 1.1, spread more or less evenly over the expected range. Instead, what we experimentally observe is more like the situation also shown in Figure 1.1, where two "spots" are observed, corresponding to one "up" and one "down" orientation. In other words, the SG apparatus splits the original silver beam from the oven into *two distinct* components, a phenomenon referred to in the early days of quantum theory as "space quantization." To the extent that $\boldsymbol{\mu}$ can be identified within a proportionality factor with the electron spin $\mathbf{S}$, only two possible values of the $z$-component of $\mathbf{S}$ are observed to be possible, $S_z$ up and $S_z$ down, which we call $S_z+$ and $S_z-$. The two possible values of $S_z$ are multiples of some fundamental unit of angular momentum; numerically it turns out that $S_z = \hbar/2$ and $-\hbar/2$, where

$$\begin{aligned} \hbar &= 1.0546 \times 10^{-27} \,\text{erg-s} \\ &= 6.5822 \times 10^{-16} \,\text{eV-s}. \end{aligned} \tag{1.3}$$

This "quantization" of the electron spin angular momentum[3] is the first important feature we deduce from the Stern–Gerlach experiment.

Figure 1.2a shows the result one would have expected from the experiment. According to classical physics, the beam should have spread itself over a vertical distance corresponding

---

[3] An understanding of the roots of this quantization lies in the application of relativity to quantum mechanics. See Section 8.2 of this book for a discussion.

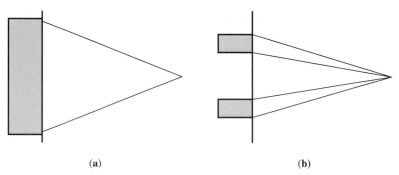

(a)                                                          (b)

**Fig. 1.2** (a) Classical physics prediction for results from the Stern–Gerlach experiment. The beam should have been spread out vertically, over a distance corresponding to the range of values of the magnetic moment times the cosine of the orientation angle. Stern and Gerlach, however, observed the result in (b), namely that only two orientations of the magnetic moment manifested themselves. These two orientations did not span the entire expected range.

to the (continuous) range of orientation of the magnetic moment. Instead, one observes Figure 1.2b which is completely at odds with classical physics. The beam mysteriously splits itself into two parts, one corresponding to spin "up" and the other to spin "down."

Of course, there is nothing sacred about the up-down direction or the $z$-axis. We could just as well have applied an inhomogeneous field in a horizontal direction, say in the $x$-direction, with the beam proceeding in the $y$-direction. In this manner we could have separated the beam from the oven into an $S_x+$ component and an $S_x-$ component.

## 1.1.2  Sequential Stern–Gerlach Experiments

Let us now consider a sequential Stern–Gerlach experiment. By this we mean that the atomic beam goes through two or more SG apparatuses in sequence. The first arrangement we consider is relatively straightforward. We subject the beam coming out of the oven to the arrangement shown in Figure 1.3a, where SG$\hat{z}$ stands for an apparatus with the inhomogeneous magnetic field in the $z$-direction, as usual. We then block the $S_z-$ component coming out of the first SG$\hat{z}$ apparatus and let the remaining $S_z+$ component be subjected to another SG$\hat{z}$ apparatus. This time there is only one beam component coming out of the second apparatus, just the $S_z+$ component. This is perhaps not so surprising; after all if the atom spins are up, they are expected to remain so, short of any external field that rotates the spins between the first and the second SG$\hat{z}$ apparatuses.

A little more interesting is the arrangement shown in Figure 1.3b. Here the first SG apparatus is the same as before but the second one (SG$\hat{x}$) has an inhomogeneous magnetic field in the $x$-direction. The $S_z+$ beam that enters the second apparatus (SG$\hat{x}$) is now split into two components, an $S_x+$ component and an $S_x-$ component, with equal intensities. How can we explain this? Does it mean that 50% of the atoms in the $S_z+$ beam coming out of the first apparatus (SG$\hat{z}$) are made up of atoms characterized by both $S_z+$ and $S_x+$, while the remaining 50% have both $S_z+$ and $S_x-$? It turns out that such a picture runs into difficulty, as will be shown below.

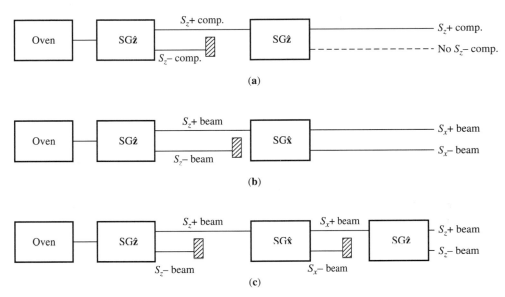

Sequential Stern–Gerlach experiments.

We now consider a third step, the arrangement shown in Figure 1.3c, which most dramatically illustrates the peculiarities of quantum-mechanical systems. This time we add to the arrangement of Figure 1.3b yet a third apparatus, of the SG$\hat{z}$ type. It is observed experimentally that *two* components emerge from the third apparatus, not one; the emerging beams are seen to have *both* an $S_z+$ component and an $S_z-$ component. This is a complete surprise because after the atoms emerged from the first apparatus, we made sure that the $S_z-$ component was completely blocked. How is it possible that the $S_z-$ component which, we thought, we eliminated earlier reappears? The model in which the atoms entering the third apparatus are visualized to have both $S_z+$ and $S_x+$ is clearly unsatisfactory.

This example is often used to illustrate that in quantum mechanics we cannot determine both $S_z$ and $S_x$ simultaneously. More precisely, we can say that the selection of the $S_x+$ beam by the second apparatus (SG$\hat{x}$) completely destroys any *previous* information about $S_z$.

It is amusing to compare this situation with that of a spinning top in classical mechanics, where the angular momentum

$$\mathbf{L} = I\boldsymbol{\omega} \tag{1.4}$$

can be measured by determining the components of the angular velocity vector $\boldsymbol{\omega}$. By observing how fast the object is spinning in which direction we can determine $\omega_x$, $\omega_y$, and $\omega_z$ simultaneously. The moment of inertia $I$ is computable if we know the mass density and the geometric shape of the spinning top, so there is no difficulty in specifying both $L_z$ and $L_x$ in this classical situation.

It is to be clearly understood that the limitation we have encountered in determining $S_z$ and $S_x$ is not due to the incompetence of the experimentalist. By improving the experimental techniques we cannot make the $S_z-$ component out of the third apparatus in Figure 1.3c disappear. The peculiarities of quantum mechanics are imposed upon us by the experiment itself. The limitation is, in fact, inherent in microscopic phenomena.

### 1.1.3  Analogy with Polarization of Light

Because this situation looks so novel, some analogy with a familiar classical situation may be helpful here. To this end we now digress to consider the polarization of light waves. This analogy will help us develop a mathematical framework for formulating the postulates of quantum mechanics.

Consider a monochromatic light wave propagating in the $z$-direction. A linearly polarized (or plane polarized) light with a polarization vector in the $x$-direction, which we call for short an *x-polarized light*, has a space-time dependent electric field oscillating in the $x$-direction

$$\mathbf{E} = E_0\hat{\mathbf{x}} \cos(kz - \omega t). \tag{1.5}$$

Likewise, we may consider a $y$-polarized light, also propagating in the $z$-direction,

$$\mathbf{E} = E_0\hat{\mathbf{y}} \cos(kz - \omega t). \tag{1.6}$$

Polarized light beams of type (1.5) or (1.6) can be obtained by letting an unpolarized light beam go through a Polaroid filter. We call a filter that selects only beams polarized in the $x$-direction an *x-filter*. An $x$-filter, of course, becomes a $y$-filter when rotated by 90° about the propagation ($z$) direction. It is well known that when we let a light beam go through an $x$-filter and subsequently let it impinge on a $y$-filter, no light beam comes out provided, of course, we are dealing with 100% efficient Polaroids; see Figure 1.4a.

The situation is even more interesting if we insert between the $x$-filter and the $y$-filter yet another Polaroid that selects only a beam polarized in the direction – which we call the $x'$-direction – that makes an angle of 45° with the $x$-direction in the $xy$ plane; see Figure 1.4b.

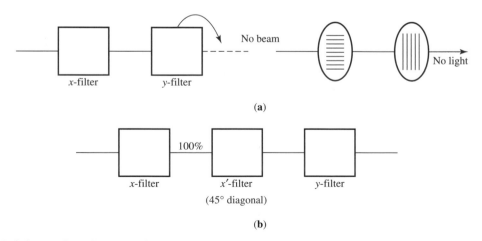

(a)

(b)

**Fig. 1.4**   Light beams subjected to Polaroid filters.

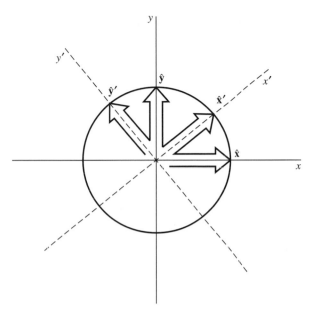

**Fig. 1.5**  Orientations of the $x'$- and $y'$-axes.

This time, there is a light beam coming out of the $y$-filter despite the fact that right after the beam went through the $x$-filter it did not have any polarization component in the $y$-direction. In other words, once the $x'$-filter intervenes and selects the $x'$-polarized beam, it is immaterial whether the beam was previously $x$-polarized. The selection of the $x'$-polarized beam by the second Polaroid destroys any previous information on light polarization. Notice that this situation is quite analogous to the situation that we encountered earlier with the SG arrangement of Figure 1.3b, provided that the following correspondence is made:

$$S_z \pm \text{atoms} \leftrightarrow x\text{-}, y\text{-polarized light}$$
$$S_x \pm \text{atoms} \leftrightarrow x'\text{-}, y'\text{-polarized light,}$$

(1.7)

where the $x'$- and the $y'$-axes are defined as in Figure 1.5.

Let us examine how we can quantitatively describe the behavior of $45°$-polarized beams ($x'$- and $y'$-polarized beams) within the framework of classical electrodynamics. Using Figure 1.5 we obtain

$$E_0 \hat{\mathbf{x}}' \cos(kz - \omega t) = E_0 \left[ \frac{1}{\sqrt{2}} \hat{\mathbf{x}} \cos(kz - \omega t) + \frac{1}{\sqrt{2}} \hat{\mathbf{y}} \cos(kz - \omega t) \right],$$

$$E_0 \hat{\mathbf{y}}' \cos(kz - \omega t) = E_0 \left[ -\frac{1}{\sqrt{2}} \hat{\mathbf{x}} \cos(kz - \omega t) + \frac{1}{\sqrt{2}} \hat{\mathbf{y}} \cos(kz - \omega t) \right].$$

(1.8)

In the triple-filter arrangement of Figure 1.4b the beam coming out of the first Polaroid is an $\hat{\mathbf{x}}$-polarized beam, which can be regarded as a linear combination of an $x'$-polarized beam and a $y'$-polarized beam. The second Polaroid selects the $x'$-polarized beam, which can in turn be regarded as a linear combination of an $x$-polarized and a $y$-polarized beam. And finally, the third Polaroid selects the $y$-polarized component.

Applying correspondence (1.7) from the sequential Stern–Gerlach experiment of Figure 1.3c, to the triple-filter experiment of Figure 1.4b suggests that we might be able to represent the spin state of a silver atom by some kind of vector in a new kind of two-dimensional vector space, an abstract vector space not to be confused with the usual two-dimensional $(xy)$ space. Just as $\hat{\mathbf{x}}$ and $\hat{\mathbf{y}}$ in (1.8) are the base vectors used to decompose the polarization vector $\hat{\mathbf{x}}'$ of the $\hat{\mathbf{x}}'$-polarized light, it is reasonable to represent the $S_x+$ state by a vector, which we call a *ket* in the Dirac notation to be developed fully in the next section. We denote this vector by $|S_x; +\rangle$ and write it as a linear combination of two base vectors, $|S_z; +\rangle$ and $|S_z; -\rangle$, which correspond to the $S_z+$ and the $S_z-$ states, respectively. So we may conjecture

$$|S_x; +\rangle \overset{?}{=} \frac{1}{\sqrt{2}}|S_z; +\rangle + \frac{1}{\sqrt{2}}|S_z; -\rangle \tag{1.9a}$$

$$|S_x; -\rangle \overset{?}{=} -\frac{1}{\sqrt{2}}|S_z; +\rangle + \frac{1}{\sqrt{2}}|S_z; -\rangle \tag{1.9b}$$

in analogy with (1.8). Later we will show how to obtain these expressions using the general formalism of quantum mechanics.

Thus the unblocked component coming out of the second (SG$\hat{\mathbf{x}}$) apparatus of Figure 1.3c is to be regarded as a superposition of $S_z+$ and $S_z-$ in the sense of (1.9a). It is for this reason that two components emerge from the third (SG$\hat{\mathbf{z}}$) apparatus.

The next question of immediate concern is: How are we going to represent the $S_y\pm$ states? Symmetry arguments suggest that if we observe an $S_z\pm$ beam going in the $x$-direction and subject it to an SG$\hat{\mathbf{y}}$ apparatus, the resulting situation will be very similar to the case where an $S_z\pm$ beam going in the $y$-direction is subjected to an SG$\hat{\mathbf{x}}$ apparatus. The kets for $S_y\pm$ should then be regarded as a linear combination of $|S_z; \pm\rangle$, but it appears from (1.9) that we have already used up the available possibilities in writing $|S_x; \pm\rangle$. How can our vector space formalism distinguish $S_y\pm$ states from $S_x\pm$ states?

An analogy with polarized light again rescues us here. This time we consider a circularly polarized beam of light, which can be obtained by letting a linearly polarized light pass through a quarter-wave plate. When we pass such a circularly polarized light through an $x$-filter or a $y$-filter, we again obtain either an $x$-polarized beam or a $y$-polarized beam of equal intensity. Yet everybody knows that the circularly polarized light is totally different from the 45°-linearly polarized ($x'$-polarized or $y'$-polarized) light.

Mathematically, how do we represent a circularly polarized light? A right circularly polarized light is nothing more than a linear combination of an $x$-polarized light and a $y$-polarized light, where the oscillation of the electric field for the $y$-polarized component is 90° out of phase with that of the $x$-polarized component:[4]

$$\mathbf{E} = E_0 \left[ \frac{1}{\sqrt{2}}\hat{\mathbf{x}}\cos(kz - \omega t) + \frac{1}{\sqrt{2}}\hat{\mathbf{y}}\cos\left(kz - \omega t + \frac{\pi}{2}\right) \right]. \tag{1.10}$$

---

[4] Unfortunately, there is no unanimity in the definition of right versus left circularly polarized light in the literature.

It is more elegant to use complex notation by introducing $\varepsilon$ as follows:

$$\mathrm{Re}(\varepsilon) = \mathbf{E}/E_0. \tag{1.11}$$

For a right circularly polarized light, we can then write

$$\varepsilon = \left[ \frac{1}{\sqrt{2}}\hat{\mathbf{x}}e^{i(kz-\omega t)} + \frac{i}{\sqrt{2}}\hat{\mathbf{y}}e^{i(kz-\omega t)} \right], \tag{1.12}$$

where we have used $i = e^{i\pi/2}$.

We can make the following analogy with the spin states of silver atoms:

$$\begin{aligned} S_y + \text{atom} &\leftrightarrow \text{right circularly polarized beam,} \\ S_y - \text{atom} &\leftrightarrow \text{left circularly polarized beam.} \end{aligned} \tag{1.13}$$

Applying this analogy to (1.12), we see that if we are allowed to make the coefficients preceding base kets complex, there is no difficulty in accommodating the $S_y\pm$ atoms in our vector space formalism:

$$|S_y; \pm\rangle \overset{?}{=} \frac{1}{\sqrt{2}}|S_z; +\rangle \pm \frac{i}{\sqrt{2}}|S_z; -\rangle, \tag{1.14}$$

which are obviously different from (1.9). We thus see that the two-dimensional vector space needed to describe the spin states of silver atoms must be a *complex* vector space; an arbitrary vector in the vector space is written as a linear combination of the base vectors $|S_z; \pm\rangle$ with, in general, complex coefficients. The fact that the necessity of complex numbers is already apparent in such an elementary example is rather remarkable.

The reader must have noted by this time that we have deliberately avoided talking about photons. In other words, we have completely ignored the quantum aspect of light; nowhere did we mention the polarization states of individual photons. The analogy we worked out is between kets in an abstract vector space that describes the spin states of individual atoms with the polarization vectors of the *classical electromagnetic field*. Actually we could have made the analogy even more vivid by introducing the photon concept and talking about the probability of finding a circularly polarized photon in a linearly polarized state, and so forth; however, that is not needed here. Without doing so, we have already accomplished the main goal of this section: to introduce the idea that quantum-mechanical states are to be represented by vectors in an abstract complex vector space.[5]

Finally, before outlining the mathematical formalism of quantum mechanics, we remark that the physics of a Stern–Gerlach apparatus is of far more than simply academic interest. The ability to separate spin states of atoms has tremendous practical interest as well. Figure 1.6 shows the use of the Stern–Gerlach technique to analyze the result of spin manipulation in an atomic beam of cesium atoms. The only stable isotope, $^{133}\mathrm{Cs}$, of this alkali atom has a nuclear spin $I = 7/2$, and the experiment sorts out the $F = 4$ hyperfine magnetic substate, giving nine spin orientations. This is only one of many examples where this once mysterious effect is used for practical devices. Of course, all of these uses only go

---

[5] The reader who is interested in grasping the basic concepts of quantum mechanics through a careful study of photon polarization may find Chapter 1 of Baym (1969) extremely illuminating.

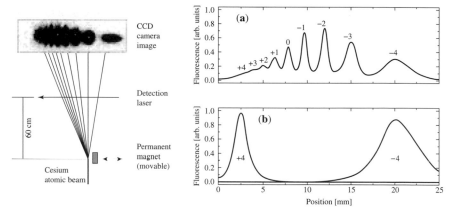

**Fig. 1.6**   A modern Stern–Gerlach apparatus, used to separate spin states of atomic cesium, taken from Lison et al., *Phys. Rev. A*, **61** (1999) 013405. The apparatus is shown on the left, while the data show the nine different projections for the spin-four atom, (a) before and (b) after optical pumping is used to populate only extreme spin projections. The spin quantum number $F = 4$ is a coupling between the outermost electron in the atom and the nuclear spin $I = 7/2$.

to firmly establish this effect, and the quantum-mechanical principles which we will now present and further develop.

# 1.2 Kets, Bras, and Operators

In the preceding section we showed how analyses of the Stern–Gerlach experiment led us to consider a complex vector space. In this and the following section we formulate the basic mathematics of vector spaces as used in quantum mechanics. Our notation throughout this book is the bra and ket notation developed by P. A. M. Dirac. The theory of linear vector spaces had, of course, been known to mathematicians prior to the birth of quantum mechanics, but Dirac's way of introducing vector spaces has many advantages, especially from the physicist's point of view.

## 1.2.1 Ket Space

We consider a complex vector space whose dimensionality is specified according to the nature of a physical system under consideration. In Stern–Gerlach type experiments where the only quantum-mechanical degree of freedom is the spin of an atom, the dimensionality is determined by the number of alternative paths the atoms can follow when subjected to an SG apparatus; in the case of the silver atoms of the previous section, the dimensionality is just two, corresponding to the two possible values $S_z$ can assume.[6] Later, in Section 1.6,

---

[6] For many physical systems the dimension of the state space is denumerably infinite. While we will usually indicate a finite number of dimensions, $N$, of the ket space, the results also hold for denumerably infinite dimensions.

we consider the case of continuous spectra, for example, the position (coordinate) or momentum of a particle, where the number of alternatives is nondenumerably infinite, in which case the vector space in question is known as a **Hilbert space** after D. Hilbert, who studied vector spaces in infinite dimensions.

In quantum mechanics a physical state, for example, a silver atom with a definite spin orientation, is represented by a **state vector** in a complex vector space. Following Dirac, we call such a vector a **ket** and denote it by $|\alpha\rangle$. This state ket is postulated to contain complete information about the physical state; everything we are allowed to ask about the state is contained in the ket. Two kets can be added:

$$|\alpha\rangle + |\beta\rangle = |\gamma\rangle. \tag{1.15}$$

The sum $|\gamma\rangle$ is just another ket. If we multiply $|\alpha\rangle$ by a complex number $c$, the resulting product $c|\alpha\rangle$ is another ket. The number $c$ can stand on the left or on the right of a ket; it makes no difference:

$$c|\alpha\rangle = |\alpha\rangle c. \tag{1.16}$$

In the particular case where $c$ is zero, the resulting ket is said to be a **null ket**.

One of the physics postulates is that $|\alpha\rangle$ and $c|\alpha\rangle$, with $c \neq 0$, represent the same physical state. In other words, only the "direction" in vector space is of significance. Mathematicians may prefer to say that we are here dealing with rays rather than vectors.

An **observable**, such as momentum and spin components, can be represented by an **operator**, such as $A$, in the vector space in question. Quite generally, an operator acts on a ket *from the left*,

$$A \cdot (|\alpha\rangle) = A|\alpha\rangle, \tag{1.17}$$

which is yet another ket. There will be more on multiplication operations later.

In general, $A|\alpha\rangle$ is *not* a constant times $|\alpha\rangle$. However, there are particular kets of importance, known as **eigenkets** of operator $A$, denoted by

$$|a'\rangle, |a''\rangle, |a'''\rangle, \dots \tag{1.18}$$

with the property

$$A|a'\rangle = a'|a'\rangle, \quad A|a''\rangle = a''|a''\rangle, \dots \tag{1.19}$$

where $a', a'', \dots$ are just numbers. Notice that applying $A$ to an eigenket just reproduces the same ket apart from a multiplicative number. The set of numbers $\{a', a'', a''', \dots\}$, more compactly denoted by $\{a'\}$, is called the set of **eigenvalues** of operator $A$. When it becomes necessary to order eigenvalues in a specific manner, $\{a^{(1)}, a^{(2)}, a^{(3)}, \dots\}$ may be used in place of $\{a', a'', a''', \dots\}$.

The physical state corresponding to an eigenket is called an **eigenstate**. In the simplest case of spin $\frac{1}{2}$ systems, the eigenvalue-eigenket relation (1.19) is expressed as

$$S_z|S_z; +\rangle = \frac{\hbar}{2}|S_z; +\rangle, \quad S_z|S_z; -\rangle = -\frac{\hbar}{2}|S_z; -\rangle, \tag{1.20}$$

where $|S_z; \pm\rangle$ are eigenkets of operator $S_z$ with eigenvalues $\pm\hbar/2$. Here we could have used just $|\hbar/2\rangle$ for $|S_z; +\rangle$ in conformity with the notation $|a'\rangle$, where an eigenket is labeled by its eigenvalue, but the notation $|S_z; \pm\rangle$, already used in the previous section, is more convenient here because we also consider eigenkets of $S_x$:

$$S_x|S_x; \pm\rangle = \pm\frac{\hbar}{2}|S_x; \pm\rangle. \tag{1.21}$$

We remarked earlier that the dimensionality of the vector space is determined by the number of alternatives in Stern–Gerlach type experiments. More formally, we are concerned with an $N$-dimensional vector space spanned by the $N$ eigenkets of observable $A$. Any arbitrary ket $|\alpha\rangle$ can be written as

$$|\alpha\rangle = \sum_{a'} c_{a'}|a'\rangle, \tag{1.22}$$

with $a', a'', \ldots$ up to $a^{(N)}$, where $c_{a'}$ is a complex coefficient. The question of the uniqueness of such an expansion will be postponed until we prove the orthogonality of eigenkets.

### 1.2.2 Bra Space and Inner Products

The vector space we have been dealing with is a ket space. We now introduce the notion of a **bra space**, a vector space "dual to" the ket space. We postulate that corresponding to every ket $|\alpha\rangle$ there exists a bra, denoted by $\langle\alpha|$, in this dual, or bra, space. The bra space is spanned by eigenbras $\{\langle a'|\}$ which correspond to the eigenkets $\{|a'\rangle\}$. There is a one-to-one correspondence between a ket space and a bra space:

$$|\alpha\rangle \overset{DC}{\leftrightarrow} \langle\alpha|$$
$$|a'\rangle, |a''\rangle, \ldots \overset{DC}{\leftrightarrow} \langle a'|, \langle a''|, \ldots \tag{1.23}$$
$$|\alpha\rangle + |\beta\rangle \overset{DC}{\leftrightarrow} \langle\alpha| + \langle\beta|$$

where DC stands for **dual correspondence**. Roughly speaking, we can regard the bra space as some kind of mirror image of the ket space.

The bra dual to $c|\alpha\rangle$ is postulated to be $c^*\langle\alpha|$, *not* $c\langle\alpha|$, which is a very important point. More generally, we have

$$c_\alpha|\alpha\rangle + c_\beta|\beta\rangle \overset{DC}{\leftrightarrow} c_\alpha^*\langle\alpha| + c_\beta^*\langle\beta|. \tag{1.24}$$

We now define the **inner product** of a bra and a ket.[7] The product is written as a bra standing on the left and a ket standing on the right, for example,

$$\langle\beta|\alpha\rangle = (\underset{\text{bra}(c)\,\text{ket}}{(\langle\beta|) \cdot (|\alpha\rangle)}). \tag{1.25}$$

This product is, in general, a complex number. Notice that in forming an inner product we always take one vector from the bra space and one vector from the ket space.

We postulate two fundamental properties of inner products. First,

$$\langle\beta|\alpha\rangle = \langle\alpha|\beta\rangle^*. \tag{1.26}$$

---

[7] In the literature an inner product is often referred to as a *scalar product* because it is analogous to $\mathbf{a}\cdot\mathbf{b}$ in Euclidean space; in this book, however, we reserve the term *scalar* for a quantity invariant under rotations in the usual three-dimensional space.

In other words, $\langle\beta|\alpha\rangle$ and $\langle\alpha|\beta\rangle$ are complex conjugates of each other. Notice that even though the inner product is, in some sense, analogous to the familiar scalar product $\mathbf{a}\cdot\mathbf{b}$, $\langle\beta|\alpha\rangle$ must be clearly distinguished from $\langle\alpha|\beta\rangle$; the analogous distinction is not needed in real vector space because $\mathbf{a}\cdot\mathbf{b}$ is equal to $\mathbf{b}\cdot\mathbf{a}$. Using (1.26) we can immediately deduce that $\langle\alpha|\alpha\rangle$ must be a real number. To prove this just let $\langle\beta|\to\langle\alpha|$.

The second postulate on inner products is

$$\langle\alpha|\alpha\rangle \geq 0, \tag{1.27}$$

where the equality sign holds only if $|\alpha\rangle$ is a *null ket*. This is sometimes known as the postulate of **positive definite metric**. From a physicist's point of view, this postulate is essential for the probabilistic interpretation of quantum mechanics, as will become apparent later.[8]

Two kets $|\alpha\rangle$ and $|\beta\rangle$ are said to be **orthogonal** if

$$\langle\alpha|\beta\rangle = 0, \tag{1.28}$$

even though in the definition of the inner product the bra $\langle\alpha|$ appears. The orthogonality relation (1.28) also implies, via (1.26),

$$\langle\beta|\alpha\rangle = 0. \tag{1.29}$$

Given a ket which is not a null ket, we can form a **normalized ket** $|\tilde{\alpha}\rangle$, where

$$|\tilde{\alpha}\rangle = \left(\frac{1}{\sqrt{\langle\alpha|\alpha\rangle}}\right)|\alpha\rangle, \tag{1.30}$$

with the property

$$\langle\tilde{\alpha}|\tilde{\alpha}\rangle = 1. \tag{1.31}$$

Quite generally, $\sqrt{\langle\alpha|\alpha\rangle}$ is known as the **norm** of $|\alpha\rangle$, analogous to the magnitude of vector $\sqrt{\mathbf{a}\cdot\mathbf{a}} = |\bar{\mathbf{a}}|$ in Euclidean vector space. Because $|\alpha\rangle$ and $c|\alpha\rangle$ represent the same physical state, we might as well require that the kets we use for physical states be normalized in the sense of (1.31).[9]

### 1.2.3 Operators

As we remarked earlier, observables like momentum and spin components are to be represented by operators that can act on kets. We can consider a more general class of operators that act on kets; they will be denoted by $X$, $Y$, and so forth, while $A$, $B$, and so on will be used for a restrictive class of operators that correspond to observables.

An operator acts on a ket from the left side,

$$X\cdot(|\alpha\rangle) = X|\alpha\rangle, \tag{1.32}$$

---

[8] Attempts to abandon this postulate led to physical theories with "indefinite metric." We shall not be concerned with such theories in this book.

[9] For eigenkets of observables with continuous spectra, different normalization conventions will be used; see Section 1.6.

and the resulting product is another ket. Operators $X$ and $Y$ are said to be **equal**,

$$X = Y, \tag{1.33}$$

if

$$X|\alpha\rangle = Y|\alpha\rangle \tag{1.34}$$

for an *arbitrary* ket in the ket space in question. Operator $X$ is said to be the **null operator** if, for any *arbitrary* ket $|\alpha\rangle$, we have

$$X|\alpha\rangle = 0. \tag{1.35}$$

Operators can be added; addition operations are commutative and associative:

$$X + Y = Y + X, \tag{1.36a}$$

$$X + (Y + Z) = (X + Y) + Z. \tag{1.36b}$$

With the single exception of the time-reversal operator to be considered in Chapter 4, the operators that appear in this book are all linear, that is,

$$X(c_\alpha|\alpha\rangle + c_\beta|\beta\rangle) = c_\alpha X|\alpha\rangle + c_\beta X|\beta\rangle. \tag{1.37}$$

An operator $X$ always acts on a bra from the *right* side

$$(\langle\alpha|) \cdot X = \langle\alpha|X, \tag{1.38}$$

and the resulting product is another bra. The ket $X|\alpha\rangle$ and the bra $\langle\alpha|X$ are, in general, *not* dual to each other. We define the symbol $X^\dagger$ as

$$X|\alpha\rangle \overset{\text{DC}}{\leftrightarrow} \langle\alpha|X^\dagger. \tag{1.39}$$

The operator $X^\dagger$ is called the **Hermitian adjoint**, or simply the adjoint, of $X$. An operator $X$ is said to be Hermitian if

$$X = X^\dagger. \tag{1.40}$$

## 1.2.4 Multiplication

Operators $X$ and $Y$ can be multiplied. Multiplication operations are, in general, *noncommutative*, that is,

$$XY \neq YX. \tag{1.41}$$

Multiplication operations are, however, associative:

$$X(YZ) = (XY)Z = XYZ. \tag{1.42}$$

We also have

$$X(Y|\alpha\rangle) = (XY)|\alpha\rangle = XY|\alpha\rangle, \quad (\langle\beta|X)Y = \langle\beta|(XY) = \langle\beta|XY. \tag{1.43}$$

Notice that

$$(XY)^\dagger = Y^\dagger X^\dagger \tag{1.44}$$

because

$$XY|\alpha\rangle = X(Y|\alpha\rangle) \overset{DC}{\leftrightarrow} (\langle\alpha|Y^\dagger)X^\dagger = \langle\alpha|Y^\dagger X^\dagger. \tag{1.45}$$

So far, we have considered the following products: $\langle\beta|\alpha\rangle, X|\alpha\rangle, \langle\alpha|X$, and $XY$. Are there other products we are allowed to form? Let us multiply $|\beta\rangle$ and $\langle\alpha|$, in that order. The resulting product

$$(|\beta\rangle) \cdot (\langle\alpha|) = |\beta\rangle\langle\alpha| \tag{1.46}$$

is known as the **outer product** of $|\beta\rangle$ and $\langle\alpha|$. We will emphasize in a moment that $|\beta\rangle\langle\alpha|$ is to be regarded as an operator; hence it is fundamentally different from the inner product $\langle\beta|\alpha\rangle$, which is just a number.

There are also "illegal products." We have already mentioned that an operator must stand on the left of a ket or on the right of a bra. In other words, $|\alpha\rangle X$ and $X\langle\alpha|$ are examples of illegal products. They are neither kets, nor bras, nor operators; they are simply nonsensical. Products like $|\alpha\rangle|\beta\rangle$ and $\langle\alpha|\langle\beta|$ are also illegal when $|\alpha\rangle$ and $|\beta\rangle$ ($\langle\alpha|$ and $\langle\beta|$) are ket (bra) vectors belonging to the same ket (bra) space.[10]

### 1.2.5  The Associative Axiom

As is clear from (1.42), multiplication operations among operators are associative. Actually the associative property is postulated to hold quite generally as long as we are dealing with "legal" multiplications among kets, bras, and operators. Dirac calls this important postulate the **associative axiom of multiplication**.

To illustrate the power of this axiom let us first consider an outer product acting on a ket:

$$(|\beta\rangle\langle\alpha|) \cdot |\gamma\rangle. \tag{1.47}$$

Because of the associative axiom, we can regard this equally well as

$$|\beta\rangle \cdot (\langle\alpha|\gamma\rangle), \tag{1.48}$$

where $\langle\alpha|\gamma\rangle$ is just a number. So the outer product acting on a ket is just another ket; in other words, $|\beta\rangle\langle\alpha|$ can be regarded as an operator. Because (1.47) and (1.48) are equal, we may as well omit the dots and let $|\beta\rangle\langle\alpha|\gamma\rangle$ stand for the operator $|\beta\rangle\langle\alpha|$ acting on $|\gamma\rangle$ or, equivalently, the number $\langle\alpha|\gamma\rangle$ multiplying $|\beta\rangle$. (On the other hand, if (1.48) is written as $(\langle\alpha|\gamma\rangle) \cdot |\beta\rangle$, we cannot afford to omit the dot and brackets because the resulting expression would look illegal.) Notice that the operator $|\beta\rangle\langle\alpha|$ rotates $|\gamma\rangle$ into the direction of $|\beta\rangle$. It is easy to see that if

$$X = |\beta\rangle\langle\alpha|, \tag{1.49}$$

[10] Later in the book we will encounter products like $|\alpha\rangle|\beta\rangle$, which are more appropriately written as $|\alpha\rangle \otimes |\beta\rangle$, but in such cases $|\alpha\rangle$ and $|\beta\rangle$ always refer to kets from *different* vector spaces. For instance, the first ket belongs to the vector space for electron spin, the second ket to the vector space for electron orbital angular momentum; or the first ket lies in the vector space of particle 1, the second ket in the vector space of particle 2, and so forth.

then

$$X^\dagger = |\alpha\rangle\langle\beta|, \tag{1.50}$$

which is left as an exercise.

In a second important illustration of the associative axiom, we note that

$$(\underbrace{\langle\beta|}_{\text{bra}}) \cdot (\underbrace{X|\alpha\rangle}_{\text{ket}}) = (\underbrace{\langle\beta|X}_{\text{bra}}) \cdot (\underbrace{|\alpha\rangle}_{\text{ket}}). \tag{1.51}$$

Because the two sides are equal, we might as well use the more compact notation

$$\langle\beta|X|\alpha\rangle \tag{1.52}$$

to stand for either side of (1.51). Recall now that $\langle\alpha|X^\dagger$ is the bra that is dual to $X|\alpha\rangle$, so

$$\begin{aligned}
\langle\beta|X|\alpha\rangle &= \langle\beta| \cdot (X|\alpha\rangle) \\
&= \{(\langle\alpha|X^\dagger) \cdot |\beta\rangle\}^* \\
&= \langle\alpha|X^\dagger|\beta\rangle^*,
\end{aligned} \tag{1.53}$$

where, in addition to the associative axiom, we used the fundamental property of the inner product (1.26). For a *Hermitian X* we have

$$\langle\beta|X|\alpha\rangle = \langle\alpha|X|\beta\rangle^*. \tag{1.54}$$

## 1.3 Base Kets and Matrix Representations

### 1.3.1 Eigenkets of an Observable

Let us consider the eigenkets and eigenvalues of a Hermitian operator $A$. We use the symbol $A$, reserved earlier for an observable, because in quantum mechanics Hermitian operators of interest quite often turn out to be the operators representing some physical observables. We begin by stating an important theorem.

**Theorem 1** *The eigenvalues of a Hermitian operator $A$ are real; the eigenkets of $A$ corresponding to different eigenvalues are orthogonal.*

**Proof** First, recall that

$$A|a'\rangle = a'|a'\rangle. \tag{1.55}$$

Because $A$ is Hermitian, we also have

$$\langle a''|A = a''^*\langle a''|, \tag{1.56}$$

where, $a', a'', \ldots$ are eigenvalues of $A$. If we multiply both sides of (1.55) by $\langle a''|$ on the left, both sides of (1.56) by $|a'\rangle$ on the right, and subtract, we obtain

$$(a' - a''^*)\langle a''|a'\rangle = 0. \tag{1.57}$$

Now $a'$ and $a''$ can be taken to be either the same or different. Let us first choose them to be the same; we then deduce the reality condition (the first half of the theorem)

$$a' = a'^*, \tag{1.58}$$

where we have used the fact that $|a'\rangle$ is not a null ket. Let us now assume $a'$ and $a''$ to be different. Because of the just proved reality condition, the difference $a' - a''^*$ that appears in (1.57) is equal to $a' - a''$, which cannot vanish, by assumption. The inner product $\langle a''|a'\rangle$ must then vanish:

$$\langle a''|a'\rangle = 0 \quad (a' \neq a''), \tag{1.59}$$

which proves the orthogonality property (the second half of the theorem).          □

We expect on physical grounds that an observable has real eigenvalues, a point that will become clearer in the next section, where measurements in quantum mechanics will be discussed. The theorem just proved guarantees the reality of eigenvalues whenever the operator is Hermitian. That is why we talk about Hermitian observables in quantum mechanics.

It is conventional to normalize $|a'\rangle$ so the $\{|a'\rangle\}$ form an **orthonormal** set:

$$\langle a''|a'\rangle = \delta_{a''a'}. \tag{1.60}$$

We may logically ask: Is this set of eigenkets complete? Since we started our discussion by asserting that the whole ket space is spanned by the eigenkets of $A$, the eigenkets of $A$ must therefore form a complete set by *construction* of our ket space.[11]

## 1.3.2  Eigenkets as Base Kets

We have seen that the normalized eigenkets of $A$ form a complete orthonormal set. An arbitrary ket in the ket space can be expanded in terms of the eigenkets of $A$. In other words, the eigenkets of $A$ are to be used as base kets in much the same way as a set of mutually orthogonal unit vectors is used as base vectors in Euclidean space.

Given an arbitrary ket $|\alpha\rangle$ in the ket space spanned by the eigenkets of $A$, let us attempt to expand it as follows:

$$|\alpha\rangle = \sum_{a'} c_{a'}|a'\rangle. \tag{1.61}$$

Multiplying $\langle a''|$ on the left and using the orthonormality property (1.60), we can immediately find the expansion coefficient,

$$c_{a'} = \langle a'|\alpha\rangle. \tag{1.62}$$

In other words, we have

$$|\alpha\rangle = \sum_{a'} |a'\rangle\langle a'|\alpha\rangle, \tag{1.63}$$

---

[11] The astute reader, already familiar with wave mechanics, may point out that the completeness of eigenfunctions we use can be proved by applying the Sturm–Liouville theory to the Schrödinger wave equation. But to "derive" the Schrödinger wave equation from our fundamental postulates, the completeness of the position eigenkets must be assumed.

which is analogous to an expansion of a vector $\mathbf{V}$ in (real) Euclidean space:

$$\mathbf{V} = \sum_i \hat{\mathbf{e}}_i (\hat{\mathbf{e}}_i \cdot \mathbf{V}), \tag{1.64}$$

where $\{\hat{\mathbf{e}}_i\}$ form an orthogonal set of unit vectors. We now recall the associative axiom of multiplication: $|a'\rangle\langle a'|\alpha\rangle$ can be regarded either as the number $\langle a'|\alpha\rangle$ multiplying $|a'\rangle$ or, equivalently, as the operator $|a'\rangle\langle a'|$ acting on $|\alpha\rangle$. Because $|\alpha\rangle$ in (1.63) is an arbitrary ket, we must have

$$\sum_{a'} |a'\rangle\langle a'| = 1, \tag{1.65}$$

where the 1 on the right-hand side is to be understood as the identity *operator*. Equation (1.65) is known as the **completeness relation** or **closure**.

It is difficult to overestimate the usefulness of (1.65). Given a chain of kets, operators, or bras multiplied in legal orders, we can insert, in any place at our convenience, the identity operator written in form (1.65). Consider, for example $\langle\alpha|\alpha\rangle$; by inserting the identity operator between $\langle\alpha|$ and $|\alpha\rangle$, we obtain

$$\langle\alpha|\alpha\rangle = \langle\alpha| \cdot \left( \sum_{a'} |a'\rangle\langle a'| \right) \cdot |\alpha\rangle$$

$$= \sum_{a'} |\langle a'|\alpha\rangle|^2. \tag{1.66}$$

This, incidentally, shows that if $|\alpha\rangle$ is normalized, then the expansion coefficients in (1.61) must satisfy

$$\sum_{a'} |c_{a'}|^2 = \sum_{a'} |\langle a'|\alpha\rangle|^2 = 1. \tag{1.67}$$

Let us now look at $|a'\rangle\langle a'|$ that appears in (1.65). Since this is an outer product, it must be an operator. Let it operate on $|\alpha\rangle$:

$$(|a'\rangle\langle a'|) \cdot |\alpha\rangle = |a'\rangle\langle a'|\alpha\rangle = c_{a'} |a'\rangle. \tag{1.68}$$

We see that $|a'\rangle\langle a'|$ selects that portion of the ket $|\alpha\rangle$ parallel to $|a'\rangle$, so $|a'\rangle\langle a'|$ is known as the **projection operator** along the base ket $|a'\rangle$ and is denoted by $\Lambda_{a'}$:

$$\Lambda_{a'} \equiv |a'\rangle\langle a'|. \tag{1.69}$$

The completeness relation (1.65) can now be written as

$$\sum_{a'} \Lambda_{a'} = 1. \tag{1.70}$$

### 1.3.3 Matrix Representations

Having specified the base kets, we now show how to represent an operator, say $X$, by a square matrix. First, using (1.65) twice, we write the operator $X$ as

$$X = \sum_{a''} \sum_{a'} |a''\rangle\langle a''|X|a'\rangle\langle a'|. \tag{1.71}$$

There are altogether $N^2$ numbers of form $\langle a''|X|a'\rangle$, where $N$ is the dimensionality of the ket space. We may arrange them into an $N \times N$ square matrix such that the column and row indices appear as follows:

$$\underset{\text{row}\quad\text{column}}{\langle a''|\,X\,|a'\rangle} \ . \tag{1.72}$$

Explicitly we may write the matrix as

$$X \doteq \begin{pmatrix} \langle a^{(1)}|X|a^{(1)}\rangle & \langle a^{(1)}|X|a^{(2)}\rangle & \cdots \\ \langle a^{(2)}|X|a^{(1)}\rangle & \langle a^{(2)}|X|a^{(2)}\rangle & \cdots \\ \vdots & \vdots & \ddots \end{pmatrix}, \tag{1.73}$$

where the symbol $\doteq$ stands for "is represented by."[12]

Using (1.53), we can write

$$\langle a''|X|a'\rangle = \langle a'|X^\dagger|a''\rangle^*. \tag{1.74}$$

At last, the Hermitian adjoint operation, originally defined by (1.39), has been related to the (perhaps more familiar) concept of *complex conjugate transposed*. If an operator $B$ is Hermitian, we have

$$\langle a''|B|a'\rangle = \langle a'|B|a''\rangle^*. \tag{1.75}$$

The way we arranged $\langle a''|X|a'\rangle$ into a square matrix is in conformity with the usual rule of matrix multiplication. To see this just note that the matrix representation of the operator relation

$$Z = XY \tag{1.76}$$

reads

$$\begin{aligned} \langle a''|Z|a'\rangle &= \langle a''|XY|a'\rangle \\ &= \sum_{a'''}\langle a''|X|a'''\rangle\langle a'''|Y|a'\rangle. \end{aligned} \tag{1.77}$$

Again, all we have done is to insert the identity operator, written in form (1.65), between $X$ and $Y$!

Let us now examine how the ket relation

$$|\gamma\rangle = X|\alpha\rangle \tag{1.78}$$

can be represented using our base kets. The expansion coefficients of $|\gamma\rangle$ can be obtained by multiplying $\langle a'|$ on the left:

$$\begin{aligned} \langle a'|\gamma\rangle &= \langle a'|X|\alpha\rangle \\ &= \sum_{a''}\langle a'|X|a''\rangle\langle a''|\alpha\rangle. \end{aligned} \tag{1.79}$$

---

[12] We do not use the equality sign here because the particular form of a matrix representation depends on the particular choice of base kets used. The operator is different from a representation of the operator just as the actress is different from a poster of the actress.

But this can be seen as an application of the rule for multiplying a square matrix with a column matrix, once the expansion coefficients of $|\alpha\rangle$ and $|\gamma\rangle$ are themselves arranged to form column matrices as follows:

$$|\alpha\rangle \doteq \begin{pmatrix} \langle a^{(1)}|\alpha\rangle \\ \langle a^{(2)}|\alpha\rangle \\ \langle a^{(3)}|\alpha\rangle \\ \vdots \end{pmatrix}, \quad |\gamma\rangle \doteq \begin{pmatrix} \langle a^{(1)}|\gamma\rangle \\ \langle a^{(2)}|\gamma\rangle \\ \langle a^{(3)}|\gamma\rangle \\ \vdots \end{pmatrix}. \tag{1.80}$$

Likewise, given

$$\langle \gamma| = \langle \alpha|X, \tag{1.81}$$

we can regard

$$\langle \gamma|a'\rangle = \sum_{a''} \langle \alpha|a''\rangle \langle a''|X|a'\rangle. \tag{1.82}$$

So a bra is represented by a row matrix as follows:

$$\langle \gamma| \doteq (\langle \gamma|a^{(1)}\rangle, \langle \gamma|a^{(2)}\rangle, \langle \gamma|a^{(3)}\rangle, \dots) = (\langle a^{(1)}|\gamma\rangle^*, \langle a^{(2)}|\gamma\rangle^*, \langle a^{(3)}|\gamma\rangle^*, \dots). \tag{1.83}$$

Note the appearance of complex conjugation when the elements of the column matrix are written as in (1.83). The inner product $\langle \beta|\alpha\rangle$ can be written as the product of the row matrix representing $\langle \beta|$ with the column matrix representing $|\alpha\rangle$:

$$\langle \beta|\alpha\rangle = \sum_{a'} \langle \beta|a'\rangle \langle a'|\alpha\rangle$$

$$= (\langle a^{(1)}|\beta\rangle^*, \langle a^{(2)}|\beta\rangle^*, \dots) \begin{pmatrix} \langle a^{(1)}|\alpha\rangle \\ \langle a^{(2)}|\alpha\rangle \\ \vdots \end{pmatrix}. \tag{1.84}$$

If we multiply the row matrix representing $\langle \alpha|$ with the column matrix representing $|\beta\rangle$, then we obtain just the complex conjugate of the preceding expression, which is consistent with the fundamental property of the inner product (1.26). Finally, the matrix representation of the outer product $|\beta\rangle\langle \alpha|$ is easily seen to be

$$|\beta\rangle\langle \alpha| \doteq \begin{pmatrix} \langle a^{(1)}|\beta\rangle\langle a^{(1)}|\alpha\rangle^* & \langle a^{(1)}|\beta\rangle\langle a^{(2)}|\alpha\rangle^* & \dots \\ \langle a^{(2)}|\beta\rangle\langle a^{(1)}|\alpha\rangle^* & \langle a^{(2)}|\beta\rangle\langle a^{(2)}|\alpha\rangle^* & \dots \\ \vdots & \vdots & \ddots \end{pmatrix}. \tag{1.85}$$

The matrix representation of an observable $A$ becomes particularly simple if the eigenkets of $A$ themselves are used as the base kets. First, we have

$$A = \sum_{a''}\sum_{a'} |a''\rangle\langle a''|A|a'\rangle\langle a'|. \tag{1.86}$$

But the square matrix $\langle a''|A|a'\rangle$ is obviously diagonal,

$$\langle a''|A|a'\rangle = \langle a'|A|a'\rangle\delta_{a'a''} = a'\delta_{a'a''}, \tag{1.87}$$

so

$$A = \sum_{a'} a' |a'\rangle \langle a'|$$
$$= \sum_{a'} a' \Lambda_{a'}. \tag{1.88}$$

## 1.3.4  Spin $\frac{1}{2}$ Systems

It is here instructive to consider the special case of spin $\frac{1}{2}$ systems. The base kets used are $|S_z; \pm\rangle$, which we denote, for brevity, as $|\pm\rangle$. The simplest operator in the ket space spanned by $|\pm\rangle$ is the identity operator, which, according to (1.65), can be written as

$$1 = |+\rangle\langle+| + |-\rangle\langle-|. \tag{1.89}$$

According to (1.88), we must be able to write $S_z$ as

$$S_z = (\hbar/2)[(|+\rangle\langle+|) - (|-\rangle\langle-|)]. \tag{1.90}$$

The eigenket-eigenvalue relation

$$S_z|\pm\rangle = \pm(\hbar/2)|\pm\rangle \tag{1.91}$$

immediately follows from the orthonormality property of $|\pm\rangle$.

It is also instructive to look at two other operators,

$$S_+ \equiv \hbar|+\rangle\langle-|, \quad S_- \equiv \hbar|-\rangle\langle+|, \tag{1.92}$$

which are both seen to be *non*-Hermitian. The operator $S_+$, acting on the spin-down ket $|-\rangle$, turns $|-\rangle$ into the spin-up ket $|+\rangle$ multiplied by $\hbar$. On the other hand, the spin-up ket $|+\rangle$, when acted upon by $S_+$, becomes a null ket. So the physical interpretation of $S_+$ is that it raises the spin component by one unit of $\hbar$; if the spin component cannot be raised any further, we automatically get a null state. Likewise, $S_-$ can be interpreted as an operator that lowers the spin component by one unit of $\hbar$. Later we will show that $S_\pm$ can be written as $S_x \pm iS_y$.

In constructing the matrix representations of the angular-momentum operators, it is customary to label the column (row) indices in *descending* order of angular-momentum components, that is, the first entry corresponds to the maximum angular-momentum component, the second, the next highest, and so forth. In our particular case of spin $\frac{1}{2}$ systems, we have

$$|+\rangle \doteq \begin{pmatrix} 1 \\ 0 \end{pmatrix}, \quad |-\rangle \doteq \begin{pmatrix} 0 \\ 1 \end{pmatrix}, \tag{1.93a}$$

$$S_z \doteq \frac{\hbar}{2}\begin{pmatrix} 1 & 0 \\ 0 & -1 \end{pmatrix}, \quad S_+ \doteq \hbar \begin{pmatrix} 0 & 1 \\ 0 & 0 \end{pmatrix}, \quad S_- \doteq \hbar \begin{pmatrix} 0 & 0 \\ 1 & 0 \end{pmatrix}. \tag{1.93b}$$

We will come back to these explicit expressions when we discuss the Pauli two-component formalism in Chapter 3.

# 1.4 Measurements, Observables, and the Uncertainty Relations

## 1.4.1 Measurements

Having developed the mathematics of ket spaces, we are now in a position to discuss the quantum theory of measurement processes. This is not a particularly easy subject for beginners, so we first turn to the words of the great master, P. A. M. Dirac, for guidance (Dirac (1958), p. 36): "A measurement always causes the system to jump into an eigenstate of the dynamical variable that is being measured." What does all this mean? We interpret Dirac's words as follows: Before a measurement of observable $A$ is made, the system is assumed to be represented by some linear combination

$$|\alpha\rangle = \sum_{a'} c_{a'}|a'\rangle = \sum_{a'}|a'\rangle\langle a'|\alpha\rangle. \tag{1.94}$$

When the measurement is performed, the system is "thrown into" one of the eigenstates, say $|a'\rangle$ of observable $A$. In other words,

$$|\alpha\rangle \xrightarrow{A \text{ measurement}} |a'\rangle. \tag{1.95}$$

For example, a silver atom with an arbitrary spin orientation will change into either $|S_z; +\rangle$ or $|S_z; -\rangle$ when subjected to an SG apparatus of type SG$\hat{z}$. Thus a *measurement usually changes the state*. The only exception is when the state is already in one of the eigenstates of the observable being measured, in which case

$$|a'\rangle \xrightarrow{A \text{ measurement}} |a'\rangle \tag{1.96}$$

with certainty, as will be discussed further. When the measurement causes $|\alpha\rangle$ to change into $|a'\rangle$, it is said that $A$ is measured to be $a'$. It is in this sense that the result of a measurement yields one of the eigenvalues of the observable being measured.

Given (1.94), which is the state ket of a physical system before the measurement, we do not know in advance into which of the various $|a'\rangle$ the system will be thrown as the result of the measurement. We do postulate, however, that the probability for jumping into some particular $|a'\rangle$ is given by

$$\text{Probability for } a' = |\langle a'|\alpha\rangle|^2, \tag{1.97}$$

provided that $|\alpha\rangle$ is normalized.

Although we have been talking about a single physical system, to determine probability (1.97) empirically, we must consider a great number of measurements performed on an ensemble, that is, a collection, of identically prepared physical systems, all characterized by the same ket $|\alpha\rangle$. Such an ensemble is known as a **pure ensemble**. (We will say more about ensembles in Chapter 3.) As an example, a beam of silver atoms which survive the first SG$\hat{z}$ apparatus of Figure 1.3 with the $S_z-$ component blocked is an example of a pure ensemble because every member atom of the ensemble is characterized by $|S_z; +\rangle$.

The probabilistic interpretation (1.97) for the squared inner product $|\langle a'|\alpha\rangle|^2$ is one of the fundamental postulates of quantum mechanics, so it cannot be proven. Let us note, however, that it makes good sense in extreme cases. Suppose the state ket is $|a'\rangle$ itself even before a measurement is made; then according to (1.97), the probability for getting $a'$, or, more precisely, for being thrown into $|a'\rangle$, as the result of the measurement is predicted to be 1, which is just what we expect. By measuring $A$ once again, we, of course, get $|a'\rangle$ only; quite generally, repeated measurements of the same observable in succession yield the same result.[13] If, on the other hand, we are interested in the probability for the system initially characterized by $|a'\rangle$ to be thrown into some other eigenket $|a''\rangle$ with $a'' \neq a'$, then (1.97) gives zero because of the orthogonality between $|a'\rangle$ and $|a''\rangle$. From the point of view of measurement theory, orthogonal kets correspond to mutually exclusive alternatives; for example, if a spin $\frac{1}{2}$ system is in $|S_z; +\rangle$, it is not in $|S_z; -\rangle$ with certainty.

Quite generally, the probability for anything must be nonnegative. Furthermore, the probabilities for the various alternative possibilities must add up to unity. Both of these expectations are met by our probability postulate (1.97).

We define the **expectation value** of $A$ taken with respect to state $|\alpha\rangle$ as

$$\langle A \rangle \equiv \langle \alpha | A | \alpha \rangle. \tag{1.98}$$

To make sure that we are referring to state $|\alpha\rangle$, the notation $\langle A \rangle_\alpha$ is sometimes used. Equation (1.98) is a definition; however, it agrees with our intuitive notion of *average measured value* because it can be written as

$$\langle A \rangle = \sum_{a'} \sum_{a''} \langle \alpha | a'' \rangle \langle a'' | A | a' \rangle \langle a' | \alpha \rangle$$
$$= \sum_{a'} \underset{\substack{\uparrow \\ \text{measured value } a'}}{a'} \quad \underbrace{|\langle a' | \alpha \rangle|^2}_{\text{probability for obtaining } a'}. \tag{1.99}$$

It is very important not to confuse eigenvalues with expectation values. For example, the expectation value of $S_z$ for spin $\frac{1}{2}$ systems can assume *any* real value between $-\hbar/2$ and $+\hbar/2$, say $0.273\hbar$; in contrast, the eigenvalue of $S_z$ assumes only two values, $\hbar/2$ and $-\hbar/2$.

To clarify further the meaning of measurements in quantum mechanics, we introduce the notion of a **selective measurement**, or *filtration*. In Section 1.1 we considered a Stern–Gerlach arrangement where we let only one of the spin components pass out of the apparatus while we completely blocked the other component. More generally, we imagine a measurement process with a device that selects only one of the eigenkets of $A$, say $|a'\rangle$, and rejects all others; see Figure 1.7. This is what we mean by a selective measurement; it is also called filtration because only one of the $A$ eigenkets filters through the ordeal. Mathematically we can say that such a selective measurement amounts to applying the projection operator $\Lambda_{a'}$ to $|\alpha\rangle$:

$$\Lambda_{a'} |\alpha\rangle = |a'\rangle \langle a' | \alpha \rangle. \tag{1.100}$$

J. Schwinger has developed a formalism of quantum mechanics based on a thorough examination of selective measurements. He introduces a measurement symbol $M(a')$ in the beginning, which is identical to $\Lambda_{a'}$ or $|a'\rangle\langle a'|$ in our notation, and deduces a number

---

[13] Here successive measurements must be carried out immediately afterward. This point will become clear when we discuss the time evolution of a state ket in Chapter 2.

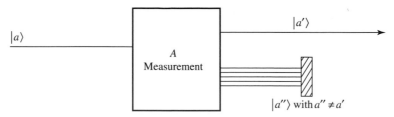

Fig. 1.7 Selective measurement.

of properties of $M(a')$ (and also of $M(b',a')$ which amount to $|b'\rangle\langle a'|$) by studying the outcome of various Stern–Gerlach type experiments. In this way he motivates the entire mathematics of kets, bras, and operators. In this book we do not follow Schwinger's path; the interested reader may consult Gottfried (1966).

## 1.4.2  Spin $\frac{1}{2}$ Systems, Once Again

Before proceeding with a general discussion of observables, we once again consider spin $\frac{1}{2}$ systems. This time we show that the results of sequential Stern–Gerlach experiments, when combined with the postulates of quantum mechanics discussed so far, are sufficient to determine not only the $S_{x,y}$ eigenkets, $|S_x; \pm\rangle$ and $|S_y; \pm\rangle$, but also the operators $S_x$ and $S_y$ themselves.

First, we recall that when the $S_x+$ beam is subjected to an apparatus of type SG$\hat{z}$, the beam splits into two components with equal intensities. This means that the probability for the $S_x+$ state to be thrown into $|S_z; \pm\rangle$, simply denoted as $|\pm\rangle$, is $\frac{1}{2}$ each; hence,

$$|\langle +|S_x; +\rangle| = |\langle -|S_x; +\rangle| = \frac{1}{\sqrt{2}}. \tag{1.101}$$

We can therefore construct the $S_x+$ ket as follows:

$$|S_x; +\rangle = \frac{1}{\sqrt{2}}|+\rangle + \frac{1}{\sqrt{2}}e^{i\delta_1}|-\rangle, \tag{1.102}$$

with $\delta_1$ real. In writing (1.102) we have used the fact that the *overall* phase (common to both $|+\rangle$ and $|-\rangle$) of a state ket is immaterial; the coefficient of $|+\rangle$ can be chosen to be real and positive by convention. The $S_x-$ ket must be orthogonal to the $S_x+$ ket because the $S_x+$ alternative and $S_x-$ alternative are mutually exclusive. This orthogonality requirement leads to

$$|S_x; -\rangle = \frac{1}{\sqrt{2}}|+\rangle - \frac{1}{\sqrt{2}}e^{i\delta_1}|-\rangle, \tag{1.103}$$

where we have, again, chosen the coefficient of $|+\rangle$ to be real and positive by convention. We can now construct the operator $S_x$ using (1.88) as follows:

$$\begin{aligned} S_x &= \frac{\hbar}{2}[(|S_x; +\rangle\langle S_x; +|) - (|S_x; -\rangle\langle S_x; -|)] \\ &= \frac{\hbar}{2}[e^{-i\delta_1}(|+\rangle\langle -|) + e^{i\delta_1}(|-\rangle\langle +|)]. \end{aligned} \tag{1.104}$$

Notice that the $S_x$ we have constructed is Hermitian, just as it must be. A similar argument with $S_x$ replaced by $S_y$ leads to

$$|S_y; \pm\rangle = \frac{1}{\sqrt{2}}|+\rangle \pm \frac{1}{\sqrt{2}}e^{i\delta_2}|-\rangle, \tag{1.105}$$

$$S_y = \frac{\hbar}{2}[e^{-i\delta_2}(|+\rangle\langle-|) + e^{i\delta_2}(|-\rangle\langle+|)]. \tag{1.106}$$

Is there any way of determining $\delta_1$ and $\delta_2$? Actually there is one piece of information we have not yet used. Suppose we have a beam of spin $\frac{1}{2}$ atoms moving in the $z$-direction. We can consider a sequential Stern–Gerlach experiment with SG$\hat{x}$ followed by SG$\hat{y}$. The results of such an experiment are completely analogous to the earlier case leading to (1.101):

$$|\langle S_y; \pm|S_x; +\rangle| = |\langle S_y; +|S_x; -\rangle| = \frac{1}{\sqrt{2}}, \tag{1.107}$$

which is not surprising in view of the invariance of physical systems under rotations. Inserting (1.103) and (1.105) into (1.107), we obtain

$$\frac{1}{2}|1 \pm e^{i(\delta_1-\delta_2)}| = \frac{1}{\sqrt{2}}, \tag{1.108}$$

which is satisfied only if

$$\delta_2 - \delta_1 = \frac{\pi}{2} \quad \text{or} \quad -\frac{\pi}{2}. \tag{1.109}$$

We thus see that the matrix elements of $S_x$ and $S_y$ cannot all be real. If the $S_x$ matrix elements are real, the $S_y$ matrix elements must be purely imaginary (and vice versa). Just from this extremely simple example, the introduction of complex numbers is seen to be an essential feature in quantum mechanics. It is convenient to take the $S_x$ matrix elements to be real[14] and set $\delta_1 = 0$; if we were to choose $\delta_1 = \pi$, the positive $x$-axis would be oriented in the opposite direction. The second phase angle $\delta_2$ must then be $-\pi/2$ or $\pi/2$. The fact that there is still an ambiguity of this kind is not surprising. We have not yet specified whether the coordinate system we are using is right-handed or left-handed; given the $x$- and the $z$-axes there is still a twofold ambiguity in the choice of the positive $y$-axis. Later we will discuss angular momentum as a generator of rotations using the right-handed coordinate system; it can then be shown that $\delta_2 = \pi/2$ is the correct choice.

To summarize, we have

$$|S_x; \pm\rangle = \frac{1}{\sqrt{2}}|+\rangle \pm \frac{1}{\sqrt{2}}|-\rangle, \tag{1.110a}$$

$$|S_y; \pm\rangle = \frac{1}{\sqrt{2}}|+\rangle \pm \frac{i}{\sqrt{2}}|-\rangle, \tag{1.110b}$$

---

[14] This can always be done by adjusting arbitrary phase factors in the definition of $|+\rangle$ and $|-\rangle$. This point will become clearer in Chapter 3, where the behavior of $|\pm\rangle$ under rotations will be discussed.

and

$$S_x = \frac{\hbar}{2}[(|+\rangle\langle-|) + (|-\rangle\langle+|)], \tag{1.111a}$$

$$S_y = \frac{\hbar}{2}[-i(|+\rangle\langle-|) + i(|-\rangle\langle+|)]. \tag{1.111b}$$

The $S_x\pm$ and $S_y\pm$ eigenkets given here are seen to be in agreement with our earlier guesses (1.9) and (1.14) based on an analogy with linearly and circularly polarized light. (Note, in this comparison, that only the relative phase between the $|+\rangle$ and $\langle-|$ components is of physical significance.) Furthermore, the non-Hermitian $S_\pm$ operators defined by (1.92) can now be written as

$$S_\pm = S_x \pm iS_y. \tag{1.112}$$

The operators $S_x$ and $S_y$, together with $S_z$ given earlier, can be readily shown to satisfy the commutation relations

$$[S_i, S_j] = i\varepsilon_{ijk}\hbar S_k, \tag{1.113}$$

and the anticommutation relations

$$\{S_i, S_j\} = \frac{1}{2}\hbar^2\delta_{ij}, \tag{1.114}$$

where the commutator [ , ] and the anticommutator { , } are defined by

$$[A,B] \equiv AB - BA, \tag{1.115a}$$

$$\{A,B\} \equiv AB + BA. \tag{1.115b}$$

(We make use of the totally antisymmetric symbol $\varepsilon_{ijk}$ which has the value $+1$ for $\varepsilon_{123}$ and any cyclic permutation of indices, the value $-1$ for $\varepsilon_{213}$ and any cyclic permutation of indices, and the value 0 when any two indices are the same. We also make use of the implied summation convention, that is the assumption that we perform a summation over any pair of repeated indices.) The commutation relations in (1.113) will be recognized as the simplest realization of the angular-momentum commutation relations, whose significance will be discussed in detail in Chapter 3. In contrast, the anticommutation relations in (1.114) turn out to be a *special* property of spin $\frac{1}{2}$ systems.

We can also define the operator $\mathbf{S}\cdot\mathbf{S}$, or $\mathbf{S}^2$ for short, as follows:

$$\mathbf{S}^2 \equiv S_x^2 + S_y^2 + S_z^2. \tag{1.116}$$

Because of (1.114), this operator turns out to be just a constant multiple of the identity operator

$$\mathbf{S}^2 = \left(\frac{3}{4}\right)\hbar^2. \tag{1.117}$$

We obviously have

$$[\mathbf{S}^2, S_i] = 0. \tag{1.118}$$

As will be shown in Chapter 3, for spins higher than $\frac{1}{2}$, $\mathbf{S}^2$ is no longer a multiple of the identity operator; however, (1.118) still holds.

## 1.4.3 Compatible Observables

Returning now to the general formalism, we will discuss compatible versus incompatible observables. Observables $A$ and $B$ are defined to be **compatible** when the corresponding operators commute,

$$[A, B] = 0, \qquad (1.119)$$

and **incompatible** when

$$[A, B] \neq 0. \qquad (1.120)$$

For example, $\mathbf{S}^2$ and $S_z$ are compatible observables, while $S_x$ and $S_z$ are incompatible observables.

Let us first consider the case of compatible observables $A$ and $B$. As usual, we assume that the ket space is spanned by the eigenkets of $A$. We may also regard the same ket space as being spanned by the eigenkets of $B$. We now ask: How are the $A$ eigenkets related to the $B$ eigenkets when $A$ and $B$ are compatible observables?

Before answering this question we must touch upon a very important point we have bypassed earlier, the concept of *degeneracy*. Suppose there are two (or more) linearly independent eigenkets of $A$ having the same eigenvalue; then the eigenvalues of the two eigenkets are said to be **degenerate**. In such a case the notation $|a'\rangle$ that labels the eigenket by its eigenvalue alone does not give a complete description; furthermore, we may recall that our earlier theorem on the orthogonality of different eigenkets was proved under the assumption of no degeneracy. Even worse, the whole concept that the ket space is spanned by $\{|a'\rangle\}$ appears to run into difficulty when the dimensionality of the ket space is larger than the number of distinct eigenvalues of $A$. Fortunately, in practical applications in quantum mechanics, it is usually the case that in such a situation the eigenvalues of *some other* commuting observable, say $B$, can be used to label the degenerate eigenkets.

Now we are ready to state an important theorem.

**Theorem 2**   *Suppose that $A$ and $B$ are compatible observables, and the eigenvalues of $A$ are nondegenerate. Then the matrix elements $\langle a''|B|a'\rangle$ are all diagonal. (Recall here that the matrix elements of $A$ are already diagonal if $\{|a'\rangle\}$ are used as the base kets.)*

**Proof**   The proof of this important theorem is extremely simple. Using the definition (1.119) of compatible observables, we observe that

$$\langle a''|[A, B]|a'\rangle = (a'' - a')\langle a''|B|a'\rangle = 0. \qquad (1.121)$$

So $\langle a''|B|a'\rangle$ must vanish unless $a' = a''$, which proves our assertion.     □

We can write the matrix elements of $B$ as

$$\langle a''|B|a'\rangle = \delta_{a'a''}\langle a'|B|a'\rangle. \qquad (1.122)$$

So both $A$ and $B$ can be represented by diagonal matrices with the *same* set of base kets. Using (1.71) and (1.122) we can write $B$ as

$$B = \sum_{a''} |a''\rangle\langle a''|B|a''\rangle\langle a''|. \qquad (1.123)$$

Suppose that this operator acts on an eigenket of $A$:

$$B|a'\rangle = \sum_{a''}|a''\rangle\langle a''|B|a''\rangle\langle a''|a'\rangle = (\langle a'|B|a'\rangle)|a'\rangle. \tag{1.124}$$

But this is nothing other than the eigenvalue equation for the operator $B$ with eigenvalue

$$b' \equiv \langle a'|B|a'\rangle. \tag{1.125}$$

The ket $|a'\rangle$ is therefore a **simultaneous eigenket** of $A$ and $B$. Just to be impartial to both operators, we may use $|a',b'\rangle$ to characterize this simultaneous eigenket.

We have seen that compatible observables have simultaneous eigenkets. Even though the proof given is for the case where the $A$ eigenkets are nondegenerate, the statement holds even if there is an $n$-fold degeneracy, that is,

$$A|a'^{(i)}\rangle = a'|a'^{(i)}\rangle \quad \text{for} \quad i = 1,2,\ldots,n \tag{1.126}$$

where $|a'^{(i)}\rangle$ are $n$ mutually orthonormal eigenkets of $A$, all with the same eigenvalue $a'$. To see this, all we need to do is construct appropriate linear combinations of $|a'^{(i)}\rangle$ that diagonalize the $B$ operator by following the diagonalization procedure to be discussed in Section 1.5.

A simultaneous eigenket of $A$ and $B$, denoted by $|a',b'\rangle$, has the property

$$A|a',b'\rangle = a'|a',b'\rangle, \tag{1.127a}$$

$$B|a',b'\rangle = b'|a',b'\rangle. \tag{1.127b}$$

When there is no degeneracy, this notation is somewhat superfluous because it is clear from (1.125) that if we specify $a'$, we necessarily know the $b'$ that appears in $|a',b'\rangle$. The notation $|a',b'\rangle$ is much more powerful when there are degeneracies. A simple example may be used to illustrate this point.

Even though a complete discussion of orbital angular momentum will not appear in this book until Chapter 3, the reader may be familiar from his or her earlier training in elementary wave mechanics that the eigenvalues of $\mathbf{L}^2$ (orbital angular momentum squared) and $L_z$ (the $z$-component of orbital angular momentum) are $\hbar^2 l(l+1)$ and $m_l\hbar$, respectively, with $l$ an integer and $m_l = -l, -l+1,\ldots,+l$. To characterize an orbital angular momentum state completely, it is necessary to specify *both* $l$ and $m_l$. For example, if we just say $l=1$, the $m_l$ value can still be 0, +1, or −1; if we just say $m_l = 1$, $l$ can be 1, 2, 3, 4, and so on. Only by specifying *both* $l$ and $m_l$ do we succeed in uniquely characterizing the orbital angular momentum state in question. Quite often a **collective index** $K'$ is used to stand for $(a',b')$, so that

$$|K'\rangle = |a',b'\rangle. \tag{1.128}$$

We can obviously generalize our considerations to a situation where there are several (more than two) mutually compatible observables, namely,

$$[A,B] = [B,C] = [A,C] = \cdots = 0. \tag{1.129}$$

Assume that we have found a **maximal** set of commuting observables; that is, we cannot add any more observables to our list without violating (1.129). The eigenvalues of individual operators $A$, $B$, $C$,... may have degeneracies, but if we specify a combination $(a', b', c', ...)$, then the corresponding simultaneous eigenket of $A$, $B$, $C$,... is uniquely specified. We can again use a collective index $K'$ to stand for $(a', b', c', ...)$. The orthonormality relation for

$$|K'\rangle = |a', b', c', ...\rangle \qquad (1.130)$$

reads

$$\langle K''|K'\rangle = \delta_{K'K''} = \delta_{aa'}\delta_{bb'}\delta_{cc'}..., \qquad (1.131)$$

while the completeness relation, or closure, can be written as

$$\sum_{K'}|K'\rangle\langle K'| = \sum_{a'}\sum_{b'}\sum_{c'}...|a', b', c', ...\rangle\langle a', b', c', ...| = 1. \qquad (1.132)$$

We now consider measurements of $A$ and $B$ when they are compatible observables. Suppose we measure $A$ first and obtain result $a'$. Subsequently, we may measure $B$ and get result $b'$. Finally we measure $A$ again. It follows from our measurement formalism that the third measurement always gives $a'$ with certainty, that is, the second ($B$) measurement does not destroy the previous information obtained in the first ($A$) measurement. This is rather obvious when the eigenvalues of $A$ are nondegenerate:

$$|\alpha\rangle \xrightarrow{A \text{ measurement}} |a', b'\rangle \xrightarrow{B \text{ measurement}} |a', b'\rangle \xrightarrow{A \text{ measurement}} |a', b'\rangle. \qquad (1.133)$$

When there is degeneracy, the argument goes as follows: After the first ($A$) measurement, which yields $a'$, the system is thrown into some linear combination

$$\sum_{i}^{n} c_{a'}^{(i)}|a', b^{(i)}\rangle, \qquad (1.134)$$

where $n$ is the degree of degeneracy and the kets $|a', b^{(i)}\rangle$ all have the same eigenvalue $a'$ as far as operator $A$ is concerned. The second ($B$) measurement may select just one of the terms in the linear combination (1.134), say, $|a', b^{(j)}\rangle$, but the third ($A$) measurement applied to it still yields $a'$. Whether or not there is degeneracy, $A$ measurements and $B$ measurements do not interfere. The term *compatible* is indeed deemed appropriate.

### 1.4.4  Incompatible Observables

We now turn to incompatible observables, which are more nontrivial. The first point to be emphasized is that incompatible observables do not have a complete set of simultaneous eigenkets. To show this let us assume the converse to be true. There would then exist a set of simultaneous eigenkets with property (1.127a) and (1.127b). Clearly,

$$AB|a', b'\rangle = Ab'|a', b'\rangle = a'b'|a', b'\rangle. \qquad (1.135)$$

Likewise,

$$BA|a', b'\rangle = Ba'|a', b'\rangle = a'b'|a', b'\rangle; \qquad (1.136)$$

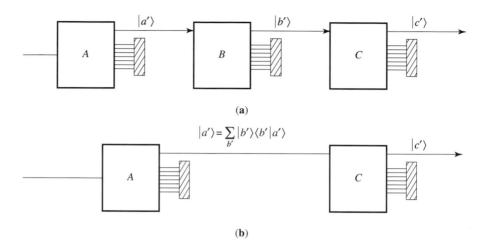

Fig. 1.8  Sequential selective measurements.

hence,

$$AB|a',b'\rangle = BA|a',b'\rangle, \tag{1.137}$$

and thus $[A,B] = 0$ in contradiction to the assumption. So in general, $|a',b'\rangle$ does not make sense for incompatible observables. There is, however, an interesting exception; it may happen that there exists a subspace of the ket space such that (1.137) holds for all elements of this subspace, even though $A$ and $B$ are incompatible. An example from the theory of orbital angular momentum may be helpful here. Suppose we consider an $l = 0$ state ($s$-state). Even though $L_x$ and $L_z$ do *not* commute, this state *is* a simultaneous eigenstate of $L_x$ and $L_z$ (with eigenvalue zero for both operators). The subspace in this case is one dimensional.

We already encountered some of the peculiarities associated with incompatible observables when we discussed sequential Stern–Gerlach experiments in Section 1.1. We now give a more general discussion of experiments of that type. Consider the sequence of selective measurements shown in Figure 1.8a. The first ($A$) filter selects some particular $|a'\rangle$ and rejects all others, the second ($B$) filter selects some particular $|b'\rangle$ and rejects all others, and the third ($C$) filter selects some particular $|c'\rangle$ and rejects all others. We are interested in the probability of obtaining $|c'\rangle$ when the beam coming out of the first filter is normalized to unity. Because the probabilities are multiplicative, we obviously have

$$|\langle c'|b'\rangle|^2 |\langle b'|a'\rangle|^2. \tag{1.138}$$

Now let us sum over $b'$ to consider the total probability for going through all possible $b'$ routes. Operationally this means that we first record the probability of obtaining $c'$ with all but the first $b'$ route blocked, then we repeat the procedure with all but the second $b'$ blocked, and so on; then we sum the probabilities at the end and obtain

$$\sum_{b'} |\langle c'|b'\rangle|^2 |\langle b'|a'\rangle|^2 = \sum_{b'} \langle c'|b'\rangle\langle b'|a'\rangle\langle a'|b'\rangle\langle b'|c'\rangle. \tag{1.139}$$

We now compare this with a different arrangement, where the $B$ filter is absent (or not operative); see Figure 1.8b. Clearly, the probability is just $|\langle c'|a'\rangle|^2$, which can also be written as follows:

$$|\langle c'|a'\rangle|^2 = |\sum_{b'}\langle c'|b'\rangle\langle b'|a'\rangle|^2 = \sum_{b'}\sum_{b''}\langle c'|b'\rangle\langle b'|a'\rangle\langle a'|b''\rangle\langle b''|c'\rangle. \qquad (1.140)$$

Notice that expressions (1.139) and (1.140) are different! This is remarkable because in both cases the pure $|a'\rangle$ beam coming out of the first ($A$) filter can be regarded as being made up of the $B$ eigenkets

$$|a'\rangle = \sum_{b'}|b'\rangle\langle b'|a'\rangle, \qquad (1.141)$$

where the sum is over all possible values of $b'$. The crucial point to be noted is that the result coming out of the $C$ filter depends on whether or not $B$ measurements have actually been carried out. In the first case we experimentally ascertain which of the $B$ eigenvalues are actually realized; in the second case, we merely imagine $|a'\rangle$ to be built up of the various $|b'\rangle$ in the sense of (1.141). Put in another way, actually recording the probabilities of going through the various $b'$ routes makes all the difference even though we sum over $b'$ afterwards. Here lies the heart of quantum mechanics.

Under what conditions do the two expressions become equal? It is left as an exercise for the reader to show that for this to happen, in the absence of degeneracy, it is sufficient that

$$[A,B] = 0 \quad \text{or} \quad [B,C] = 0. \qquad (1.142)$$

In other words, the peculiarity we have illustrated is characteristic of incompatible observables.

## 1.4.5 The Uncertainty Relation

The last topic to be discussed in this section is the uncertainty relation. Given an observable $A$, we define an **operator**

$$\Delta A \equiv A - \langle A\rangle, \qquad (1.143)$$

where the expectation value is to be taken for a certain physical state under consideration. The expectation value of $(\Delta A)^2$ is known as the **dispersion** of $A$. Because we have

$$\langle(\Delta A)^2\rangle = \langle(A^2 - 2A\langle A\rangle + \langle A\rangle^2)\rangle = \langle A^2\rangle - \langle A\rangle^2, \qquad (1.144)$$

the last equality of (1.144) may be taken as an alternative definition of dispersion. Sometimes the terms **variance** and **mean square deviation** are used for the same quantity. Clearly, the dispersion vanishes when the state in question is an eigenstate of $A$. Roughly speaking, the dispersion of an observable characterizes "fuzziness." For example, for the $S_z+$ state of a spin $\frac{1}{2}$ system, the dispersion of $S_x$ can be computed to be

$$\langle S_x^2\rangle - \langle S_x\rangle^2 = \hbar^2/4. \qquad (1.145)$$

In contrast the dispersion $\langle(\Delta S_z)^2\rangle$ obviously vanishes for the $S_z+$ state. So, for the $S_z+$ state, $S_z$ is "sharp," a vanishing dispersion for $S_z$, while $S_x$ is fuzzy.

We now state the uncertainty relation, which is the generalization of the well-known $x$-$p$ uncertainty relation to be discussed in Section 1.6. Let $A$ and $B$ be observables. Then for any state we must have the following inequality:

$$\langle (\Delta A)^2 \rangle \langle (\Delta B)^2 \rangle \geq \frac{1}{4} |\langle [A,B] \rangle|^2. \tag{1.146}$$

To prove this we first state three lemmas.

**Lemma 1**   *The Schwarz inequality*

$$\langle \alpha | \alpha \rangle \langle \beta | \beta \rangle \geq |\langle \alpha | \beta \rangle|^2, \tag{1.147}$$

*which is analogous to*

$$|\mathbf{a}|^2 |\mathbf{b}|^2 \geq |\mathbf{a} \cdot \mathbf{b}|^2 \tag{1.148}$$

*in real Euclidian space.*

**Proof**   First note

$$(\langle \alpha | + \lambda^* \langle \beta |) \cdot (|\alpha \rangle + \lambda |\beta \rangle) \geq 0, \tag{1.149}$$

where $\lambda$ can be any complex number. This inequality must hold when $\lambda$ is set equal to $-\langle \beta | \alpha \rangle / \langle \beta | \beta \rangle$:

$$\langle \alpha | \alpha \rangle \langle \beta | \beta \rangle - |\langle \alpha | \beta \rangle|^2 \geq 0, \tag{1.150}$$

which is the same as (1.147).   □

**Lemma 2**   *The expectation value of a Hermitian operator is purely real.*

**Proof**   The proof is trivial, just use (1.75).   □

**Lemma 3**   *The expectation value of an anti-Hermitian operator, defined by $C = -C^\dagger$, is purely imaginary.*

**Proof**   The proof is also trivial.   □

Armed with these lemmas, we are in a position to prove the uncertainty relation (1.146). Using Lemma 1 with

$$\begin{aligned} |\alpha\rangle &= \Delta A |\rangle, \\ |\beta\rangle &= \Delta B |\rangle, \end{aligned} \tag{1.151}$$

where the blank ket $|\rangle$ emphasizes the fact that our consideration may be applied to *any* ket, we obtain

$$\langle (\Delta A)^2 \rangle \langle (\Delta B)^2 \rangle \geq |\langle \Delta A \Delta B \rangle|^2, \tag{1.152}$$

where the Hermiticity of $\Delta A$ and $\Delta B$ has been used. To evaluate the right-hand side of (1.152), we note

$$\Delta A \Delta B = \frac{1}{2}[\Delta A, \Delta B] + \frac{1}{2}\{\Delta A, \Delta B\},\qquad(1.153)$$

where the commutator $[\Delta A, \Delta B]$, which is equal to $[A, B]$, is clearly anti-Hermitian

$$([A,B])^\dagger = (AB - BA)^\dagger = BA - AB = -[A,B].\qquad(1.154)$$

In contrast, the anticommutator $\{\Delta A, \Delta B\}$ is obviously Hermitian, so

$$\langle \Delta A \Delta B \rangle = \frac{1}{2}\underbrace{\langle [A,B] \rangle}_{\text{purely imaginary}} + \frac{1}{2}\underbrace{\langle \{\Delta A, \Delta B\} \rangle}_{\text{purely real}},\qquad(1.155)$$

where Lemmas 2 and 3 have been used. The right-hand side of (1.152) now becomes

$$|\langle \Delta A \Delta B \rangle|^2 = \frac{1}{4}|\langle [A,B] \rangle|^2 + \frac{1}{4}|\langle \{\Delta A, \Delta B\} \rangle|^2.\qquad(1.156)$$

The proof of (1.146) is now complete because the omission of the second (the anticommutator) term of (1.156) can only make the inequality relation stronger.[15]

Applications of the uncertainty relation to spin $\frac{1}{2}$ systems will be left as exercises. We come back to this topic when we discuss the fundamental $x$-$p$ commutation relation, that is, the Heisenberg uncertainty principle, in Section 1.6.

## 1.5 Change of Basis

### 1.5.1 Transformation Operator

Suppose we have two incompatible observables $A$ and $B$. The ket space in question can be viewed as being spanned either by the set $\{|a'\rangle\}$ or by the set $\{|b'\rangle\}$. For example, for spin $\frac{1}{2}$ systems $|S_x\pm\rangle$ may be used as our base kets; alternatively, $|S_z\pm\rangle$ may be used as our base kets. The two different sets of base kets, of course, span the same ket space. We are interested in finding out how the two descriptions are related. Changing the set of base kets is referred to as a **change of basis** or a **change of representation.** The basis in which the base eigenkets are given by $\{|a'\rangle\}$ is called the $A$ representation or, sometimes, the $A$ diagonal representation because the square matrix corresponding to $A$ is diagonal in this basis.

Our basic task is to construct a transformation operator that connects the old orthonormal set $\{|a'\rangle\}$ and the new orthonormal set $\{|b'\rangle\}$. To this end, we first show the following.

**Theorem 3**   *Given two sets of base kets, both satisfying orthonormality and completeness, there exists a unitary operator U such that*

$$|b^{(1)}\rangle = U|a^{(1)}\rangle, |b^{(2)}\rangle = U|a^{(2)}\rangle,\ldots,|b^{(N)}\rangle = U|a^{(N)}\rangle.\qquad(1.157)$$

---

[15] In the literature most authors use $\Delta A$ for our $\sqrt{\langle(\Delta A)^2\rangle}$ so the uncertainty relation is written as $\Delta A \Delta B \geq \frac{1}{2}|\langle [A,B] \rangle|$. In this book, however, $\Delta A$ and $\Delta B$ are to be understood as operators [see (1.143)], not numbers.

*By a* **unitary operator** *we mean an operator fulfilling the conditions*

$$U^\dagger U = 1 \qquad\qquad (1.158)$$

*as well as*

$$UU^\dagger = 1. \qquad\qquad (1.159)$$

**Proof**  We prove this theorem by explicit construction. We assert that the operator

$$U = \sum_k |b^{(k)}\rangle\langle a^{(k)}| \qquad\qquad (1.160)$$

will do the job and we apply this $U$ to $|a^{(l)}\rangle$. Clearly,

$$U|a^{(l)}\rangle = |b^{(l)}\rangle \qquad\qquad (1.161)$$

is guaranteed by the orthonormality of $\{|a'\rangle\}$. Furthermore, $U$ is unitary:

$$U^\dagger U = \sum_k \sum_l |a^{(l)}\rangle\langle b^{(l)}|b^{(k)}\rangle\langle a^{(k)}| = \sum_k |a^{(k)}\rangle\langle a^{(k)}| = 1, \qquad (1.162)$$

where we have used the orthonormality of $\{|b'\rangle\}$ and the completeness of $\{|a'\rangle\}$. We obtain relation (1.159) in an analogous manner.  $\square$

## 1.5.2 Transformation Matrix

It is instructive to study the matrix representation of the $U$ operator in the old $\{|a'\rangle\}$ basis. We have

$$\langle a^{(k)}|U|a^{(l)}\rangle = \langle a^{(k)}|b^{(l)}\rangle, \qquad\qquad (1.163)$$

which is obvious from (1.161). In other words, the matrix elements of the $U$ operator are built up of the inner products of old base bras and new base kets. We recall that the rotation matrix in three dimensions that changes one set of unit base vectors $(\hat{\mathbf{x}}, \hat{\mathbf{y}}, \hat{\mathbf{z}})$ into another set $(\hat{\mathbf{x}}', \hat{\mathbf{y}}', \hat{\mathbf{z}}')$ can be written as (Goldstein et al. (2002), pp. 134–144 for example)

$$R = \begin{pmatrix} \hat{\mathbf{x}}\cdot\hat{\mathbf{x}}' & \hat{\mathbf{x}}\cdot\hat{\mathbf{y}}' & \hat{\mathbf{x}}\cdot\hat{\mathbf{z}}' \\ \hat{\mathbf{y}}\cdot\hat{\mathbf{x}}' & \hat{\mathbf{y}}\cdot\hat{\mathbf{y}}' & \hat{\mathbf{y}}\cdot\hat{\mathbf{z}}' \\ \hat{\mathbf{z}}\cdot\hat{\mathbf{x}}' & \hat{\mathbf{z}}\cdot\hat{\mathbf{y}}' & \hat{\mathbf{z}}\cdot\hat{\mathbf{z}}' \end{pmatrix}. \qquad\qquad (1.164)$$

The square matrix made up of $\langle a^{(k)}|U|a^{(l)}\rangle$ is referred to as the **transformation matrix** from the $\{|a'\rangle\}$ basis to the $\{|b'\rangle\}$ basis.

Given an arbitrary ket $|\alpha\rangle$ whose expansion coefficients $\langle a'|\alpha\rangle$ are known in the old basis,

$$|\alpha\rangle = \sum_{a'} |a'\rangle\langle a'|\alpha\rangle, \qquad\qquad (1.165)$$

how can we obtain $\langle b'|\alpha\rangle$, the expansion coefficients in the new basis? The answer is very simple: Just multiply (1.165) (with $a'$ replaced by $a^{(l)}$ to avoid confusion) by $\langle b^{(k)}|$

$$\langle b^{(k)}|\alpha\rangle = \sum_l \langle b^{(k)}|a^{(l)}\rangle\langle a^{(l)}|\alpha\rangle = \sum_l \langle a^{(k)}|U^\dagger|a^{(l)}\rangle\langle a^{(l)}|\alpha\rangle. \qquad (1.166)$$

In matrix notation, (1.166) states that the column matrix for $|\alpha\rangle$ in the new basis can be obtained just by applying the square matrix $U^\dagger$ to the column matrix in the old basis:

$$(\text{new}) = (U^\dagger)(\text{old}). \tag{1.167}$$

The relationships between the old matrix elements and the new matrix elements are also easy to obtain:

$$
\begin{aligned}
\langle b^{(k)}|X|b^{(l)}\rangle &= \sum_m \sum_n \langle b^{(k)}|a^{(m)}\rangle\langle a^{(m)}|X|a^{(n)}\rangle\langle a^{(n)}|b^{(l)}\rangle \\
&= \sum_m \sum_n \langle a^{(k)}|U^\dagger|a^{(m)}\rangle\langle a^{(m)}|X|a^{(n)}\rangle\langle a^{(n)}|U|a^{(l)}\rangle.
\end{aligned} \tag{1.168}
$$

This is simply the well-known formula for a **similarity transformation** in matrix algebra,

$$X' = U^\dagger X U. \tag{1.169}$$

The **trace** of an operator $X$ is defined as the sum of diagonal elements:

$$\text{tr}(X) = \sum_{a'} \langle a'|X|a'\rangle. \tag{1.170}$$

Even though a particular set of base kets is used in the definition, $\text{tr}(X)$ turns out to be independent of representation, as shown:

$$
\begin{aligned}
\sum_{a'} \langle a'|X|a'\rangle &= \sum_{a'}\sum_{b'}\sum_{b''} \langle a'|b'\rangle\langle b'|X|b''\rangle\langle b''|a'\rangle \\
&= \sum_{b'}\sum_{b''} \langle b''|b'\rangle\langle b'|X|b''\rangle \\
&= \sum_{b'} \langle b'|X|b'\rangle.
\end{aligned} \tag{1.171}
$$

We can also prove

$$\text{tr}(XY) = \text{tr}(YX), \tag{1.172a}$$

$$\text{tr}(U^\dagger X U) = \text{tr}(X), \tag{1.172b}$$

$$\text{tr}(|a'\rangle\langle a''|) = \delta_{a'a''}, \tag{1.172c}$$

$$\text{tr}(|b'\rangle\langle a'|) = \langle a'|b'\rangle. \tag{1.172d}$$

## 1.5.3 Diagonalization

So far we have not discussed how to find the eigenvalues and eigenkets of an operator $B$ whose matrix elements in the old $\{|a'\rangle\}$ basis are assumed to be known. This problem turns out to be equivalent to that of finding the unitary matrix that diagonalizes $B$. Even though the reader may already be familiar with the diagonalization procedure in matrix algebra, it is worth working out this problem using the Dirac bra-ket notation.

We are interested in obtaining the eigenvalue $b'$ and the eigenket $|b'\rangle$ with the property

$$B|b'\rangle = b'|b'\rangle. \tag{1.173}$$

First, we rewrite this as

$$\sum_{a'} \langle a''|B|a'\rangle \langle a'|b'\rangle = b'\langle a''|b'\rangle. \tag{1.174}$$

When $|b'\rangle$ in (1.173) stands for the $l$th eigenket of operator $B$, we can write (1.174) in matrix notation as follows:

$$\begin{pmatrix} B_{11} & B_{12} & B_{13} & \cdots \\ B_{21} & B_{22} & B_{23} & \cdots \\ \vdots & \vdots & \vdots & \ddots \end{pmatrix} \begin{pmatrix} C_1^{(l)} \\ C_2^{(l)} \\ \vdots \end{pmatrix} = b^{(l)} \begin{pmatrix} C_1^{(l)} \\ C_2^{(l)} \\ \vdots \end{pmatrix}, \tag{1.175}$$

with

$$B_{ij} = \langle a^{(i)}|B|a^{(j)}\rangle, \tag{1.176a}$$

and

$$C_k^{(l)} = \langle a^{(k)}|b^{(l)}\rangle, \tag{1.176b}$$

where $i, j, k$ run up to $N$, the dimensionality of the ket space. As we know from linear algebra, nontrivial solutions for $C_k^{(l)}$ are possible only if the characteristic equation

$$\det(B - \lambda 1) = 0 \tag{1.177}$$

is satisfied. This is an $N$th order algebraic equation for $\lambda$, and the $N$ roots obtained are to be identified with the various $b^{(l)}$ we are trying to determine. Knowing $b^{(l)}$ we can solve for the corresponding $C_k^{(l)}$ up to an overall constant to be determined from the normalization condition. Comparing (1.176b) with (1.163), we see that the $C_k^{(l)}$ are just the elements of the unitary matrix involved in the change of basis $\{|a'\rangle\} \rightarrow \{|b'\rangle\}$.

For this procedure the Hermiticity of $B$ is important. For example, consider $S_+$ defined by (1.92) or (1.112). This operator is obviously non-Hermitian. The corresponding matrix, which reads in the $S_z$ basis as

$$S_+ \doteq \hbar \begin{pmatrix} 0 & 1 \\ 0 & 0 \end{pmatrix}, \tag{1.178}$$

cannot be diagonalized by any unitary matrix. In Chapter 2 we will encounter eigenkets of a non-Hermitian operator in connection with a coherent state of a simple harmonic oscillator. Such eigenkets, however, are known *not* to form a complete orthonormal set, and the formalism we have developed in this section cannot be immediately applied.

### 1.5.4 Unitary Equivalent Observables

We conclude this section by discussing a remarkable theorem on the unitary transform of an observable.

**Theorem 4**    *Consider again two sets of orthonormal basis $\{|a'\rangle\}$ and $\{|b'\rangle\}$ connected by the $U$ operator* (1.160). *Knowing $U$, we may construct a* **unitary transform** *of A,*

*UAU$^{-1}$; then A and UAU$^{-1}$ are said to be* **unitary equivalent observables**. *The eigenvalue equation for A,*

$$A|a^{(l)}\rangle = a^{(l)}|a^{(l)}\rangle, \tag{1.179}$$

*clearly implies that*

$$UAU^{-1}U|a^{(l)}\rangle = a^{(l)}U|a^{(l)}\rangle. \tag{1.180}$$

*But this can be rewritten as*

$$(UAU^{-1})|b^{(l)}\rangle = a^{(l)}|b^{(l)}\rangle. \tag{1.181}$$

This deceptively simple result is quite profound. It tells us that the $|b'\rangle$ are eigenkets of $UAU^{-1}$ with *exactly the same eigenvalues* as the $A$ eigenvalues. In other words, *unitary equivalent observables have identical spectra.*

The eigenket $|b^{(l)}\rangle$, by definition, satisfies the relationship

$$B|b^{(l)}\rangle = b^{(l)}|b^{(l)}\rangle. \tag{1.182}$$

Comparing (1.181) and (1.182), we infer that $B$ and $UAU^{-1}$ are simultaneously diagonalizable. A natural question is, is $UAU^{-1}$ the same as $B$ itself? The answer quite often is yes in cases of physical interest. Take, for example, $S_x$ and $S_z$. They are related by a unitary operator, which, as we will discuss in Chapter 3, is actually the rotation operator around the $y$-axis by angle $\pi/2$. In this case $S_x$ itself is the unitary transform of $S_z$. Because we know that $S_x$ and $S_z$ exhibit the same set of eigenvalues, namely, $+\hbar/2$ and $-\hbar/2$, we see that our theorem holds in this particular example.

## 1.6  Position, Momentum, and Translation

### 1.6.1  Continuous Spectra

The observables considered so far have all been assumed to exhibit discrete eigenvalue spectra. In quantum mechanics, however, there are observables with continuous eigenvalues. Take, for instance, $p_z$, the $z$-component of momentum. In quantum mechanics this is again represented by a Hermitian operator. In contrast to $S_z$, however, the eigenvalues of $p_z$ (in appropriate units) can assume any real value between $-\infty$ and $\infty$.

The rigorous mathematics of a vector space spanned by eigenkets that exhibit a continuous spectrum is rather treacherous. The dimensionality of such a space is obviously infinite. Fortunately, many of the results we worked out for a finite-dimensional vector space with discrete eigenvalues can immediately be generalized. In places where straightforward generalizations do not hold, we indicate danger signals.

We start with the analogue of eigenvalue equation (1.19), which, in the continuous spectrum case, is written as

$$\xi|\xi'\rangle = \xi'|\xi'\rangle, \tag{1.183}$$

where $\xi$ is an operator and $\xi'$ is simply a number. The ket $|\xi'\rangle$ is, in other words, an eigenket of operator $\xi$ with eigenvalue $\xi'$, just as $|a'\rangle$ is an eigenket of operator $A$ with eigenvalue $a'$.

In pursuing this analogy we replace the Kronecker symbol by Dirac's $\delta$-function, a discrete sum over the eigenvalues $\{a'\}$ by an integral over the *continuous variable* $\xi'$, so

$$\langle a'|a''\rangle = \delta_{a'a''} \to \langle \xi'|\xi''\rangle = \delta(\xi' - \xi''),\tag{1.184a}$$

$$\sum_{a'} |a'\rangle\langle a'| = 1 \to \int d\xi' |\xi'\rangle\langle \xi'| = 1,\tag{1.184b}$$

$$|\alpha\rangle = \sum_{a'} |a'\rangle\langle a'|\alpha\rangle \to |\alpha\rangle = \int d\xi' |\xi'\rangle\langle \xi'|\alpha\rangle,\tag{1.184c}$$

$$\sum_{a'} |\langle a'|\alpha\rangle|^2 = 1 \to \int d\xi' |\langle \xi'|\alpha\rangle|^2 = 1,\tag{1.184d}$$

$$\langle \beta|\alpha\rangle = \sum_{a'} \langle \beta|a'\rangle\langle a'|\alpha\rangle \to \langle \beta|\alpha\rangle = \int d\xi' \langle \beta|\xi'\rangle\langle \xi'|\alpha\rangle,\tag{1.184e}$$

$$\langle a''|A|a'\rangle = a'\delta_{a'a''} \to \langle \xi''|\xi|\xi'\rangle = \xi'\delta(\xi'' - \xi').\tag{1.184f}$$

Notice in particular how the completeness relation (1.184b) is used to obtain (1.184c) and (1.184e).

## 1.6.2  Position Eigenkets and Position Measurements

In Section 1.4 we emphasized that a measurement in quantum mechanics is essentially a filtering process. To extend this idea to measurements of observables exhibiting continuous spectra it is best to work with a specific example. To this end we consider the position (or coordinate) operator in one dimension.

The eigenkets $|x'\rangle$ of the position operator $x$ satisfying

$$x|x'\rangle = x'|x'\rangle\tag{1.185}$$

are postulated to form a complete set. Here $x'$ is just a number with the dimension of length 0.23 cm, for example, while $x$ is an operator. The state ket for an arbitrary physical state can be expanded in terms of $\{|x'\rangle\}$:

$$|\alpha\rangle = \int_{-\infty}^{\infty} dx' |x'\rangle\langle x'|\alpha\rangle.\tag{1.186}$$

We now consider a highly idealized selective measurement of the position observable. Suppose we place a very tiny detector that clicks only when the particle is precisely at $x'$ and nowhere else. Immediately after the detector clicks, we can say that the state in question is represented by $|x'\rangle$. In other words, when the detector clicks, $|\alpha\rangle$ abruptly "jumps into" $|x'\rangle$ in much the same way as an arbitrary spin state jumps into the $S_z+$ (or $S_z-$) state when subjected to an SG apparatus of the $S_z$ type.

In practice the best the detector can do is to locate the particle within a narrow interval around $x'$. A realistic detector clicks when a particle is observed to be located within some narrow range $(x' - \Delta/2, x' + \Delta/2)$. When a count is registered in such a detector, the state ket changes abruptly as follows:

$$|\alpha\rangle = \int_{-\infty}^{\infty} dx''|x''\rangle\langle x''|\alpha\rangle \xrightarrow{\text{measurement}} \int_{x'-\Delta/2}^{x'+\Delta/2} dx''|x''\rangle\langle x''|\alpha\rangle. \qquad (1.187)$$

Assuming that $\langle x''|\alpha\rangle$ does not change appreciably within the narrow interval, the probability for the detector to click is given by

$$|\langle x'|\alpha\rangle|^2 dx', \qquad (1.188)$$

where we have written $dx'$ for $\Delta$. This is analogous to $|\langle a'|\alpha\rangle|^2$ for the probability for $|\alpha\rangle$ to be thrown into $|a'\rangle$ when $A$ is measured. The probability of recording the particle *somewhere* between $-\infty$ and $\infty$ is given by

$$\int_{-\infty}^{\infty} dx'|\langle x'|\alpha\rangle|^2, \qquad (1.189)$$

which is normalized to unity if $|\alpha\rangle$ is normalized:

$$\langle\alpha|\alpha\rangle = 1 \Rightarrow \int_{-\infty}^{\infty} dx'\langle\alpha|x'\rangle\langle x'|\alpha\rangle = 1. \qquad (1.190)$$

The reader familiar with wave mechanics may have recognized by this time that $\langle x'|\alpha\rangle$ is the wave function for the physical state represented by $|\alpha\rangle$. We will say more about this identification of the expansion coefficient with the $x$-representation of the wave function in Section 1.7.

The notion of a position eigenket can be extended to three dimensions. It is assumed in nonrelativistic quantum mechanics that the position eigenkets $|\mathbf{x}'\rangle$ are complete. The state ket for a particle with internal degrees of freedom, such as spin, ignored can therefore be expanded in terms of $\{|\mathbf{x}'\rangle\}$ as follows:

$$|\alpha\rangle = \int d^3x'|\mathbf{x}'\rangle\langle\mathbf{x}'|\alpha\rangle, \qquad (1.191)$$

where $\mathbf{x}'$ stands for $x'$, $y'$, and $z'$; in other words, $|\mathbf{x}'\rangle$ is a *simultaneous* eigenket of the observables $x$, $y$, and $z$ in the sense of Section 1.4:

$$|\mathbf{x}'\rangle \equiv |x',y',z'\rangle, \qquad (1.192a)$$

$$x|\mathbf{x}'\rangle = x'|\mathbf{x}'\rangle, \quad y|\mathbf{x}'\rangle = y'|\mathbf{x}'\rangle, \quad z|\mathbf{x}'\rangle = z'|\mathbf{x}'\rangle. \qquad (1.192b)$$

To be able to consider such a simultaneous eigenket at all, we are implicitly assuming that the three components of the position vector can be measured simultaneously to arbitrary degrees of accuracy; hence, we must have

$$[x_i, x_j] = 0, \qquad (1.193)$$

where $x_1$, $x_2$, and $x_3$ stand for $x$, $y$, and $z$, respectively.

### 1.6.3  Translation

We now introduce the very important concept of translation, or spatial displacement. Suppose we start with a state that is well localized around $\mathbf{x}'$. Let us consider an operation that changes this state into another well-localized state, this time around $\mathbf{x}' + d\mathbf{x}'$ with everything else (for example, the spin direction) unchanged. Such an operation is defined to be an **infinitesimal translation** by $d\mathbf{x}'$, and the operator that does the job is denoted by $\mathscr{T}(d\mathbf{x}')$:

$$\mathscr{T}(d\mathbf{x}')|\mathbf{x}'\rangle = |\mathbf{x}' + d\mathbf{x}'\rangle, \tag{1.194}$$

where a possible arbitrary phase factor is set to unity by convention. Notice that the right-hand side of (1.194) is again a position eigenket, but this time with eigenvalue $\mathbf{x}' + d\mathbf{x}'$. Obviously $|\mathbf{x}'\rangle$ is *not* an eigenket of the infinitesimal translation operator.

By expanding an arbitrary state ket $|\alpha\rangle$ in terms of the position eigenkets we can examine the effect of infinitesimal translation on $|\alpha\rangle$:

$$|\alpha\rangle \to \mathscr{T}(d\mathbf{x}')|\alpha\rangle = \mathscr{T}(d\mathbf{x}') \int d^3x' |\mathbf{x}'\rangle\langle\mathbf{x}'|\alpha\rangle = \int d^3x' |\mathbf{x}' + d\mathbf{x}'\rangle\langle\mathbf{x}'|\alpha\rangle. \tag{1.195}$$

We also write the right-hand side of (1.195) as

$$\int d^3x' |\mathbf{x}' + d\mathbf{x}'\rangle\langle\mathbf{x}'|\alpha\rangle = \int d^3x' |\mathbf{x}'\rangle\langle\mathbf{x}' - d\mathbf{x}'|\alpha\rangle \tag{1.196}$$

because the integration is over all space and $\mathbf{x}'$ is just an integration variable. This shows that the wave function of the translated state $\mathscr{T}(d\mathbf{x}')|\alpha\rangle$ is obtained by substituting $\mathbf{x}' - d\mathbf{x}'$ for $\mathbf{x}'$ in $\langle\mathbf{x}'|\alpha\rangle$.

There is an equivalent approach to translation that is often treated in the literature. Instead of considering an infinitesimal translation of the physical system itself, we consider a change in the coordinate system being used such that the origin is shifted in the *opposite* direction, $-d\mathbf{x}'$. Physically, in this alternative approach we are asking how the *same* state ket would look to another observer whose coordinate system is shifted by $-d\mathbf{x}'$. In this book we try not to use this approach. Obviously it is important that we do not mix the two approaches!

We now list the properties of the infinitesimal translation operator $\mathscr{T}(-d\mathbf{x}')$. The first property we demand is the unitarity property imposed by probability conservation. It is reasonable to require that if the ket $|\alpha\rangle$ is normalized to unity, the translated ket $\mathscr{T}(d\mathbf{x}')|\alpha\rangle$ also be normalized to unity, so

$$\langle\alpha|\alpha\rangle = \langle\alpha|\mathscr{T}^\dagger(d\mathbf{x}')\mathscr{T}(d\mathbf{x}')|\alpha\rangle. \tag{1.197}$$

This condition is guaranteed by demanding that the infinitesimal translation be unitary:

$$\mathscr{T}^\dagger(d\mathbf{x}')\mathscr{T}(d\mathbf{x}') = 1. \tag{1.198}$$

Quite generally, the norm of a ket is preserved under unitary transformations. For the second property, suppose we consider two successive infinitesimal translations, first by $d\mathbf{x}'$ and subsequently by $d\mathbf{x}''$, where $d\mathbf{x}'$ and $d\mathbf{x}''$ need not be in the same direction. We expect the net result to be just a *single* translation operation by the vector sum $d\mathbf{x}' + d\mathbf{x}''$, so we demand that

$$\mathscr{J}(d\mathbf{x}'')\,\mathscr{J}(d\mathbf{x}') = \mathscr{J}(d\mathbf{x}' + d\mathbf{x}''). \qquad (1.199)$$

For the third property, suppose we consider a translation in the opposite direction; we expect the opposite-direction translation to be the same as the inverse of the original translation:

$$\mathscr{J}(-d\mathbf{x}') = \mathscr{J}^{-1}(d\mathbf{x}'). \qquad (1.200)$$

For the fourth property, we demand that as $d\mathbf{x}' \to 0$, the translation operation reduce to the identity operation

$$\lim_{d\mathbf{x}' \to 0} \mathscr{J}(d\mathbf{x}') = 1 \qquad (1.201)$$

and that the difference between $\mathscr{J}(d\mathbf{x}')$ and the identity operator be of first order in $d\mathbf{x}'$.

We now demonstrate that if we take the infinitesimal translation operator to be

$$\mathscr{J}(d\mathbf{x}') = 1 - i\mathbf{K}\cdot d\mathbf{x}', \qquad (1.202)$$

where the components of $\mathbf{K}$, $K_x$, $K_y$, and $K_z$, are **Hermitian operators**, then all the properties listed are satisfied. The first property, the unitarity of $\mathscr{J}(d\mathbf{x}')$, is checked as follows:

$$\begin{aligned}
\mathscr{J}^{\dagger}(d\mathbf{x}')\,\mathscr{J}(d\mathbf{x}') &= (1 + i\mathbf{K}^{\dagger}\cdot d\mathbf{x}')(1 - i\mathbf{K}\cdot d\mathbf{x}') \\
&- 1 - i(\mathbf{K} - \mathbf{K}^{\dagger})\cdot d\mathbf{x}' + 0[(d\mathbf{x}')^2] \\
&\simeq 1, \qquad (1.203)
\end{aligned}$$

where terms of second order in $d\mathbf{x}'$ have been ignored for an infinitesimal translation. The second property (1.199) can also be proved as follows:

$$\begin{aligned}
\mathscr{J}(d\mathbf{x}'')\,\mathscr{J}(d\mathbf{x}') &= (1 - i\mathbf{K}\cdot d\mathbf{x}'')(1 - i\mathbf{K}\cdot d\mathbf{x}') \\
&\simeq 1 - i\mathbf{K}\cdot(d\mathbf{x}' + d\mathbf{x}'') \\
&= \mathscr{J}(d\mathbf{x}' + d\mathbf{x}''). \qquad (1.204)
\end{aligned}$$

The third and fourth properties are obviously satisfied by (1.202).

Accepting (1.202) to be the correct form for $\mathscr{J}(d\mathbf{x}')$, we are in a position to derive an extremely fundamental relation between the $\mathbf{K}$ operator and the $\mathbf{x}$ operator. First, note that

$$\mathbf{x}\,\mathscr{J}(d\mathbf{x}')|\mathbf{x}'\rangle = \mathbf{x}|\mathbf{x}' + d\mathbf{x}'\rangle = (\mathbf{x}' + d\mathbf{x}')|\mathbf{x}' + d\mathbf{x}'\rangle \qquad (1.205a)$$

and

$$\mathscr{J}(d\mathbf{x}')\mathbf{x}|\mathbf{x}'\rangle = \mathbf{x}'\,\mathscr{J}(d\mathbf{x}')|\mathbf{x}'\rangle = \mathbf{x}'|\mathbf{x}' + d\mathbf{x}'\rangle; \qquad (1.205b)$$

hence,

$$[\mathbf{x}, \mathscr{J}(d\mathbf{x}')]\,|\mathbf{x}'\rangle = d\mathbf{x}'|\mathbf{x}' + d\mathbf{x}'\rangle \simeq d\mathbf{x}'|\mathbf{x}'\rangle, \qquad (1.206)$$

where the error made in writing the last part of (1.206) is of second order in $d\mathbf{x}'$. Now $|\mathbf{x}'\rangle$ can be *any* position eigenket, and the position eigenkets are known to form a complete set. We must therefore have an **operator identity**

$$[\mathbf{x}, \mathscr{J}(d\mathbf{x}')] = d\mathbf{x}', \tag{1.207}$$

or

$$-i\mathbf{x}\mathbf{K}\cdot d\mathbf{x}' + i\mathbf{K}\cdot d\mathbf{x}'\mathbf{x} = d\mathbf{x}', \tag{1.208}$$

where on the right-hand sides of (1.207) and (1.208) $d\mathbf{x}'$ is understood to be the number $d\mathbf{x}'$ multiplied by the identity operator in the ket space spanned by $|\mathbf{x}'\rangle$. By choosing $d\mathbf{x}'$ in the direction of $\hat{\mathbf{x}}_j$ and forming the scalar product with $\hat{\mathbf{x}}_i$, we obtain

$$[x_i, K_j] = i\delta_{ij}, \tag{1.209}$$

where again $\delta_{ij}$ is understood to be multiplied by the identity operator.

### 1.6.4  Momentum as a Generator of Translation

Equation (1.209) is the fundamental commutation relation between the position operators $x, y, z$ and the $K$ operators $K_x, K_y, K_z$. Remember that so far the $K$ operator is *defined* in terms of the infinitesimal translation operator by (1.202). What is the physical significance we can attach to $\mathbf{K}$?

J. Schwinger, lecturing on quantum mechanics, once remarked, "... for fundamental properties we will borrow only names from classical physics." In the present case we would like to borrow from classical mechanics the notion that momentum is the generator of an infinitesimal translation. An infinitesimal translation in classical mechanics can be regarded as a canonical transformation,

$$\mathbf{x}_{\text{new}} \equiv \mathbf{X} = \mathbf{x} + d\mathbf{x}, \quad \mathbf{p}_{\text{new}} \equiv \mathbf{P} = \mathbf{p}, \tag{1.210}$$

obtainable from the generating function (Goldstein et al. (2002), pp. 386 and 403)

$$F(\mathbf{x}, \mathbf{P}) = \mathbf{x}\cdot\mathbf{P} + \mathbf{p}\cdot d\mathbf{x}, \tag{1.211}$$

where $\mathbf{p}$ and $\mathbf{P}$ refer to the corresponding momenta.

This equation has a striking similarity to the infinitesimal translation operator (1.202) in quantum mechanics, particularly if we recall that $\mathbf{x}\cdot\mathbf{P}$ in (1.211) is the generating function for the identity transformation $(\mathbf{X} = \mathbf{x}, \mathbf{P} = \mathbf{p})$. We are therefore led to speculate that the operator $\mathbf{K}$ is in some sense related to the momentum operator in quantum mechanics.

Can the $K$ operator be identified with the momentum operator itself? Unfortunately the dimension is all wrong; the $K$ operator has the dimension of 1/length because $\mathbf{K}\cdot d\mathbf{x}'$ must be dimensionless. But it appears legitimate to set

$$\mathbf{K} = \frac{\mathbf{p}}{\text{universal constant with the dimension of action}}. \tag{1.212}$$

From the fundamental postulates of quantum mechanics there is no way to determine the actual numerical value of the universal constant. Rather, this constant is needed here because, historically, classical physics was developed before quantum mechanics using units convenient for describing macroscopic quantities – the circumference of the Earth, the mass of 1 cm$^3$ of water, the duration of a mean solar day, and so forth. Had microscopic physics been formulated before macroscopic physics, the physicists would have almost certainly chosen the basic units in such a way that the universal constant appearing in (1.212) would be unity.

An analogy from electrostatics may be helpful here. The interaction energy between two particles of charge $e$ separated at a distance $r$ is proportional to $e^2/r$; in unrationalized Gaussian units, the proportionality factor is just 1, but in rationalized mks units, which may be more convenient for electrical engineers, the proportionality factor is $1/4\pi\varepsilon_0$. (See Appendix A.)

The universal constant that appears in (1.212) turns out to be the same as the constant $\hbar$ that appears in L. de Broglie's relation, written in 1924,

$$\frac{2\pi}{\lambda} = \frac{p}{\hbar}, \tag{1.213}$$

where $\lambda$ is the wavelength of a "particle wave." In other words, the $K$ operator is the quantum-mechanical operator that corresponds to the wave number, that is, $2\pi$ times the reciprocal wavelength, usually denoted by $k$. With this identification the infinitesimal translation operator $\mathscr{T}(d\mathbf{x}')$ reads

$$\mathscr{T}(d\mathbf{x}') = 1 - i\mathbf{p}\cdot d\mathbf{x}'/\hbar, \tag{1.214}$$

where $\mathbf{p}$ is the momentum operator. The commutation relation (1.209) now becomes

$$[x_i, p_j] = i\hbar\delta_{ij}. \tag{1.215}$$

The commutation relations (1.215) imply, for example, that $x$ and $p_x$ (but not $x$ and $p_y$) are incompatible observables. It is therefore impossible to find simultaneous eigenkets of $x$ and $p_x$. The general formalism of Section 1.4 can be applied here to obtain the **position-momentum uncertainty relation** of W. Heisenberg:

$$\langle(\Delta x)^2\rangle\langle(\Delta p_x)^2\rangle \geq \hbar^2/4. \tag{1.216}$$

Some applications of (1.216) will appear in Section 1.7.

So far we have concerned ourselves with infinitesimal translations. A finite translation, that is, a spatial displacement by a finite amount, can be obtained by successively compounding infinitesimal translations. Let us consider a finite translation in the $x$-direction by an amount $\Delta x'$:

$$\mathscr{T}(\Delta x'\hat{\mathbf{x}})|\mathbf{x}'\rangle = |\mathbf{x}' + \Delta x'\hat{\mathbf{x}}\rangle. \tag{1.217}$$

By compounding $N$ infinitesimal translations, each of which is characterized by a spatial displacement $\Delta x'/N$ in the $x$-direction, and letting $N \to \infty$, we obtain

$$\mathscr{T}(\Delta x'\hat{\mathbf{x}}) = \lim_{N\to\infty}\left(1 - \frac{ip_x\Delta x'}{N\hbar}\right)^N$$

$$= \exp\left(-\frac{ip_x\Delta x'}{\hbar}\right). \tag{1.218}$$

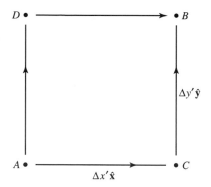

**Fig. 1.9**    Successive translations in different directions.

Here $\exp(-ip_x\Delta x'/\hbar)$ is understood to be a function of the *operator $p_x$*; generally, for any operator $X$ we have

$$\exp(X) \equiv 1 + X + \frac{X^2}{2!} + \cdots. \tag{1.219}$$

A fundamental property of translations is that successive translations in different directions, say in the $x$- and $y$-directions, commute. We see this clearly in Figure 1.9; in shifting from $A$ and $B$ it does not matter whether we go via $C$ or via $D$. Mathematically,

$$\begin{aligned}
\mathscr{T}(\Delta y'\hat{\mathbf{y}})\,\mathscr{T}(\Delta x'\hat{\mathbf{x}}) &= \mathscr{T}(\Delta x'\hat{\mathbf{x}} + \Delta y'\hat{\mathbf{y}}), \\
\mathscr{T}(\Delta x'\hat{\mathbf{x}})\,\mathscr{T}(\Delta y'\hat{\mathbf{y}}) &= \mathscr{T}(\Delta x'\hat{\mathbf{x}} + \Delta y'\hat{\mathbf{y}}).
\end{aligned} \tag{1.220}$$

This point is not so trivial as it may appear; we will show in Chapter 3 that rotations about different axes do *not* commute. Treating $\Delta x'$ and $\Delta y'$ up to second order, we obtain

$$[\mathscr{T}(\Delta y'\hat{\mathbf{y}}), \mathscr{T}(\Delta x'\hat{\mathbf{x}})] = \left[\left(1 - \frac{ip_y\Delta y'}{\hbar} - \frac{p_y^2(\Delta y')^2}{2\hbar^2} + \cdots\right),\right.$$

$$\left.\left(1 - \frac{ip_x\Delta x'}{\hbar} - \frac{p_x^2(\Delta x')^2}{2\hbar^2} + \cdots\right)\right]$$

$$\simeq -\frac{(\Delta x')(\Delta y')[p_y,p_x]}{\hbar^2}. \tag{1.221}$$

Because $\Delta x'$ and $\Delta y'$ are arbitrary, requirement (1.220), or

$$[\mathscr{T}(\Delta y'\hat{\mathbf{y}}), \mathscr{T}(\Delta x'\hat{\mathbf{x}})] = 0, \tag{1.222}$$

immediately leads to

$$[p_x,p_y] = 0, \tag{1.223}$$

or, more generally,

$$[p_i,p_j] = 0. \tag{1.224}$$

This commutation relation is a direct consequence of the fact that translations in different directions commute. Whenever the generators of transformations commute, the corresponding group is said to be **Abelian**. The translation group in three dimensions is Abelian.

Equation (1.224) implies that $p_x$, $p_y$, and $p_z$ are mutually compatible observables. We can therefore conceive of a simultaneous eigenket of $p_x, p_y, p_z$, namely,

$$|\mathbf{p}'\rangle \equiv |p_x', p_y', p_z'\rangle, \tag{1.225a}$$

$$p_x|\mathbf{p}'\rangle = p_x'|\mathbf{p}'\rangle, \qquad p_y|\mathbf{p}'\rangle = p_y'|\mathbf{p}'\rangle, \qquad p_z|\mathbf{p}'\rangle = p_z'|\mathbf{p}'\rangle. \tag{1.225b}$$

It is instructive to work out the effect of $\mathscr{T}(d\mathbf{x}')$ on such a momentum eigenket:

$$\mathscr{T}(d\mathbf{x}')|\mathbf{p}'\rangle = \left(1 - \frac{i\mathbf{p} \cdot d\mathbf{x}'}{\hbar}\right)|\mathbf{p}'\rangle = \left(1 - \frac{i\mathbf{p}' \cdot d\mathbf{x}'}{\hbar}\right)|\mathbf{p}'\rangle. \tag{1.226}$$

We see that the momentum eigenket remains the same even though it suffers a slight phase change, so unlike $|\mathbf{x}'\rangle$, $|\mathbf{p}'\rangle$ *is* an eigenket of $\mathscr{T}(d\mathbf{x}')$, which we anticipated because

$$[\mathbf{p}, \mathscr{T}(d\mathbf{x}')] = 0. \tag{1.227}$$

Notice, however, that the eigenvalue of $\mathscr{T}(d\mathbf{x}')$ is complex; we do not expect a real eigenvalue here because $\mathscr{T}(d\mathbf{x}')$, though unitary, is not Hermitian.

### 1.6.5  The Canonical Commutation Relations

We summarize the commutator relations we inferred by studying the properties of translation:

$$[x_i, x_j] = 0, \quad [p_i, p_j] = 0, \quad [x_i, p_j] = i\hbar\delta_{ij}. \tag{1.228}$$

These relations form the cornerstone of quantum mechanics; in his book, P. A. M. Dirac (1958) calls them the "fundamental quantum conditions." More often they are known as the **canonical commutation relations**, or the **fundamental commutation relations**.

Historically it was W. Heisenberg who, in 1925, showed that the combination rule for atomic transition lines known at that time could best be understood if one associated arrays of numbers obeying certain multiplication rules with these frequencies. Immediately afterward M. Born and P. Jordan pointed out that Heisenberg's multiplication rules are essentially those of matrix algebra, and a theory was developed based on the matrix analogues of (1.228), which is now known as **matrix mechanics**.[16]

Also in 1925, P. A. M. Dirac observed that the various quantum-mechanical relations can be obtained from the corresponding classical relations just by replacing classical Poisson brackets by commutators, as follows:

$$[ \quad , \quad ]_{\text{classical}} \rightarrow \frac{[ \quad , \quad ]}{i\hbar}, \tag{1.229}$$

---

[16] Appropriately, $pq - qp = h/2\pi i$ is inscribed on the gravestone of M. Born in Göttingen.

where we may recall that the classical Poisson brackets are defined for functions of $q$ and $p$ as

$$[A(q,p), B(q,p)]_{\text{classical}} \equiv \sum_s \left( \frac{\partial A}{\partial q_s} \frac{\partial B}{\partial p_s} - \frac{\partial A}{\partial p_s} \frac{\partial B}{\partial q_s} \right). \tag{1.230}$$

For example, in classical mechanics, we have

$$[x_i, p_j]_{\text{classical}} = \delta_{ij}, \tag{1.231}$$

which in quantum mechanics turns into (1.215).

Dirac's rule (1.229) is plausible because the classical Poisson brackets and quantum-mechanical commutators satisfy similar algebraic properties. In particular, the following relations can be proved regardless of whether [ , ] is understood as a classical Poisson bracket or as a quantum-mechanical commutator:

$$[A,A] = 0 \tag{1.232a}$$

$$[A,B] = -[B,A] \tag{1.232b}$$

$$[A,c] = 0 \quad (c \text{ is just a number}) \tag{1.232c}$$

$$[A+B,C] = [A,C] + [B,C] \tag{1.232d}$$

$$[A,BC] = [A,B]C + B[A,C] \tag{1.232e}$$

$$[A,[B,C]] + [B,[C,A]] + [C,[A,B]] = 0, \tag{1.232f}$$

where the last relation is known as the **Jacobi identity**.[17] However, there are important differences. First, the dimension of the classical Poisson bracket differs from that of the quantum-mechanical commutator because of the differentiations with respect to $q$ and $p$ appearing in (1.230). Second, the Poisson bracket of real functions of $q$ and $p$ is purely real, while the commutator of two Hermitian operators is anti-Hermitian (see Lemma 3 of Section 1.4). To take care of these differences the factor $i\hbar$ is inserted in (1.229).

We have deliberately avoided exploiting Dirac's analogy in obtaining the canonical commutation relations. Our approach to the commutation relations is based solely on (1) the properties of translations and (2) the identification of the generator of translation with the momentum operator modulo a universal constant with the dimension of action. We believe that this approach is more powerful because it can be generalized to situations where observables have no classical analogues. For example, the spin angular-momentum components we encountered in Section 1.4 have nothing to do with the $p$ and $q$ of classical mechanics; yet, as we will show in Chapter 3, the spin angular-momentum commutation relations can be derived using the properties of rotations just as we derived the canonical commutation relations using the properties of translations.

[17] It is amusing that the Jacobi identity in quantum mechanics is much easier to prove than its classical analogue.

# 1.7  Wave Functions in Position and Momentum Space

## 1.7.1  Position-Space Wave Function

In this section we present a systematic study of the properties of wave functions in both position and momentum space. For simplicity let us return to the one-dimensional case. The base kets used are the position kets satisfying

$$x|x'\rangle = x'|x'\rangle, \tag{1.233}$$

normalized in such a way that the orthogonality condition reads

$$\langle x''|x'\rangle = \delta(x'' - x'). \tag{1.234}$$

We have already remarked that the ket representing a physical state can be expanded in terms of $|x'\rangle$,

$$|\alpha\rangle = \int dx'|x'\rangle\langle x'|\alpha\rangle, \tag{1.235}$$

and that the expansion coefficient $\langle x'|\alpha\rangle$ is interpreted in such a way that

$$|\langle x'|\alpha\rangle|^2\, dx' \tag{1.236}$$

is the probability for the particle to be found in a narrow interval $dx'$ around $x'$. In our formalism the inner product $\langle x'|\alpha\rangle$ is what is usually referred to as the **wave function** $\psi_\alpha(x')$ for state $|\alpha\rangle$:

$$\langle x'|\alpha\rangle = \psi_\alpha(x'). \tag{1.237}$$

In elementary wave mechanics the probabilistic interpretations for the expansion coefficient $c_{a'}(=\langle a'|\alpha\rangle)$ and for the wave function $\psi_\alpha(x')(=\langle x'|\alpha\rangle)$ are often presented as separate postulates. One of the major advantages of our formalism, originally due to Dirac, is that the two kinds of probabilistic interpretations are unified; $\psi_\alpha(x')$ *is* an expansion coefficient [see (1.235)] in much the same way as $c_{a'}$ is. By following the footsteps of Dirac we come to appreciate the unity of quantum mechanics.

Consider the inner product $\langle\beta|\alpha\rangle$. Using the completeness of $|x'\rangle$, we have

$$\langle\beta|\alpha\rangle = \int dx'\langle\beta|x'\rangle\langle x'|\alpha\rangle$$
$$= \int dx'\psi_\beta^*(x')\psi_\alpha(x'), \tag{1.238}$$

so $\langle\beta|\alpha\rangle$ characterizes the overlap between the two wave functions. Note that we are not defining $\langle\beta|\alpha\rangle$ as the overlap integral; the identification of $\langle\beta|\alpha\rangle$ with the overlap integral *follows* from our completeness postulate for $|x'\rangle$. The more general interpretation of $\langle\beta|\alpha\rangle$, *independent of representations*, is that it represents the probability amplitude for state $|\alpha\rangle$ to be found in state $|\beta\rangle$.

This time let us interpret the expansion

$$|\alpha\rangle = \sum_{a'} |a'\rangle\langle a'|\alpha\rangle \tag{1.239}$$

using the language of wave functions. We just multiply both sides of (1.239) by the position eigenbra $\langle x'|$ on the left. Thus

$$\langle x'|\alpha\rangle = \sum_{a'} \langle x'|a'\rangle\langle a'|\alpha\rangle. \tag{1.240}$$

In the usual notation of wave mechanics this is recognized as

$$\psi_\alpha(x') = \sum_{a'} c_{a'} u_{a'}(x'),$$

where we have introduced an **eigenfunction** of operator $A$ with eigenvalue $a'$:

$$u_{a'}(x') = \langle x'|a'\rangle. \tag{1.241}$$

Let us now examine how $\langle\beta|A|\alpha\rangle$ can be written using the wave functions for $|\alpha\rangle$ and $|\beta\rangle$. Clearly, we have

$$\langle\beta|A|\alpha\rangle = \int dx' \int dx'' \langle\beta|x'\rangle\langle x'|A|x''\rangle\langle x''|\alpha\rangle$$

$$= \int dx' \int dx'' \psi_\beta^*(x')\langle x'|A|x''\rangle\psi_\alpha(x''). \tag{1.242}$$

So to be able to evaluate $\langle\beta|A|\alpha\rangle$, we must know the matrix element $\langle x'|A|x''\rangle$, which is, in general, a function of the two variables $x'$ and $x''$.

An enormous simplification takes place if observable $A$ is a function of the position operator $x$. In particular, consider

$$A = x^2, \tag{1.243}$$

which actually appears in the Hamiltonian for the simple harmonic oscillator problem to be discussed in Chapter 2. We have

$$\langle x'|x^2|x''\rangle = (\langle x'|)\cdot(x''^2|x''\rangle) = x'^2\delta(x'-x''), \tag{1.244}$$

where we have used (1.233) and (1.234). The double integral (1.242) is now reduced to a *single* integral:

$$\langle\beta|x^2|\alpha\rangle = \int dx' \langle\beta|x'\rangle x'^2 \langle x'|\alpha\rangle$$

$$= \int dx' \psi_\beta^*(x') x'^2 \psi_\alpha(x'). \tag{1.245}$$

In general,

$$\langle\beta|f(x)|\alpha\rangle = \int dx' \psi_\beta^*(x')f(x')\psi_\alpha(x'). \tag{1.246}$$

Note that the $f(x)$ on the left-hand side of (1.246) is an operator, while the $f(x')$ on the right-hand side is not an operator.

## 1.7.2  Momentum Operator in the Position Basis

We now examine how the momentum operator may look in the $x$-basis, that is, in the representation where the position eigenkets are used as base kets. Our starting point is the definition of momentum as the generator of infinitesimal translations:

$$\left(1 - \frac{ip\Delta x'}{\hbar}\right)|\alpha\rangle = \int dx' \, \mathscr{J}(\Delta x')|x'\rangle\langle x'|\alpha\rangle$$

$$= \int dx' \, |x' + \Delta x'\rangle\langle x'|\alpha\rangle$$

$$= \int dx' \, |x'\rangle\langle x' - \Delta x'|\alpha\rangle$$

$$= \int dx' \, |x'\rangle \left(\langle x'|\alpha\rangle - \Delta x' \frac{\partial}{\partial x'}\langle x'|\alpha\rangle\right). \qquad (1.247)$$

Comparison of both sides yields

$$p|\alpha\rangle = \int dx' \, |x'\rangle \left(-i\hbar \frac{\partial}{\partial x'}\langle x'|\alpha\rangle\right) \qquad (1.248)$$

or

$$\langle x'|p|\alpha\rangle = -i\hbar \frac{\partial}{\partial x'}\langle x'|\alpha\rangle, \qquad (1.249)$$

where we have used the orthogonality property (1.234). For the matrix element $p$ in the $x$-representation, we obtain

$$\langle x'|p|x''\rangle = -i\hbar \frac{\partial}{\partial x'}\delta(x' - x''). \qquad (1.250)$$

From (1.248) we get a very important identity:

$$\langle \beta|p|\alpha\rangle = \int dx' \langle \beta|x'\rangle \left(-i\hbar \frac{\partial}{\partial x'}\langle x'|\alpha\rangle\right)$$

$$= \int dx' \, \psi_\beta^*(x') \left(-i\hbar \frac{\partial}{\partial x'}\right)\psi_\alpha(x'). \qquad (1.251)$$

In our formalism (1.251) is not a postulate; rather, it has been *derived* using the basic properties of momentum. By repeatedly applying (1.249), we can also obtain

$$\langle x'|p^n|\alpha\rangle = (-i\hbar)^n \frac{\partial^n}{\partial x'^n}\langle x'|\alpha\rangle, \qquad (1.252)$$

$$\langle \beta|p^n|\alpha\rangle = \int dx' \, \psi_\beta^*(x')(-i\hbar)^n \frac{\partial^n}{\partial x'^n}\psi_\alpha(x'). \qquad (1.253)$$

## 1.7.3  Momentum-Space Wave Function

So far we have worked exclusively in the $x$-basis. There is actually a complete symmetry between $x$ and $p$, apart from occasional minus signs, which we can infer from the canonical commutation relations. Let us now work in the $p$-basis, that is, in the momentum representation.

For simplicity we continue working in one-space. The base eigenkets in the $p$-basis specify

$$p|p'\rangle = p'|p'\rangle \tag{1.254}$$

and

$$\langle p'|p''\rangle = \delta(p' - p''). \tag{1.255}$$

The momentum eigenkets $\{|p'\rangle\}$ span the ket space in much the same way as the position eigenkets $\{|x'\rangle\}$. An arbitrary state ket $|\alpha\rangle$ can therefore be expanded as follows:

$$|\alpha\rangle = \int dp'|p'\rangle\langle p'|\alpha\rangle. \tag{1.256}$$

We can give a probabilistic interpretation for the expansion coefficient $\langle p'|\alpha\rangle$; the probability that a measurement of $p$ gives eigenvalue $p'$ within a narrow interval $dp'$ is $|\langle p'|\alpha\rangle|^2 dp'$. It is customary to call $\langle p'|\alpha\rangle$ the **momentum-space wave function**; the notation $\phi_\alpha(p')$ is often used:

$$\langle p'|\alpha\rangle = \phi_\alpha(p'). \tag{1.257}$$

If $|\alpha\rangle$ is normalized, we obtain

$$\int dp'\langle\alpha|p'\rangle\langle p'|\alpha\rangle = \int dp'|\phi_\alpha(p')|^2 = 1. \tag{1.258}$$

Let us now establish the connection between the $x$-representation and the $p$-representation. We recall that in the case of the discrete spectra, the change of basis from the old set $\{|a'\rangle\}$ to the new set $\{|b'\rangle\}$ is characterized by the transformation matrix (1.163). Likewise, we expect that the desired information is contained in $\langle x'|p'\rangle$, which is a function of $x'$ and $p'$, usually called the **transformation function** from the $x$-representation to the $p$-representation. To derive the explicit form of $\langle x'|p'\rangle$, first recall (1.249); letting $|\alpha\rangle$ be the momentum eigenket $|p'\rangle$, we obtain

$$\langle x'|p|p'\rangle = -i\hbar\frac{\partial}{\partial x'}\langle x'|p'\rangle \tag{1.259}$$

or

$$p'\langle x'|p'\rangle = -i\hbar\frac{\partial}{\partial x'}\langle x'|p'\rangle. \tag{1.260}$$

The solution to this differential equation for $\langle x'|p'\rangle$ is

$$\langle x'|p'\rangle = N\exp\left(\frac{ip'x'}{\hbar}\right), \tag{1.261}$$

where $N$ is the normalization constant to be determined in a moment. Even though the transformation function $\langle x'|p'\rangle$ is a function of two variables, $x'$ and $p'$, we can temporarily regard it as a function of $x'$ with $p'$ fixed. It can then be viewed as the probability amplitude for the momentum eigenstate specified by $p'$ to be found at position $x'$; in other words, it is just the wave function for the momentum eigenstate $|p'\rangle$, often referred to as the momentum eigenfunction (still in the $x$-space). So (1.261) simply says that the wave function of a momentum eigenstate is a plane wave. It is amusing that we have obtained this plane wave solution without solving the Schrödinger equation (which we have not yet written down).

To get the normalization constant $N$ let us first consider

$$\langle x'|x''\rangle = \int dp' \langle x'|p'\rangle\langle p'|x''\rangle. \tag{1.262}$$

The left-hand side is just $\delta(x' - x'')$; the right-hand side can be evaluated using the explicit form of $\langle x'|p'\rangle$:

$$\delta(x' - x'') = |N|^2 \int dp' \exp\left[\frac{ip'(x' - x'')}{\hbar}\right]$$

$$= 2\pi\hbar|N|^2\delta(x' - x''). \tag{1.263}$$

Choosing $N$ to be purely real and positive by convention, we finally have

$$\langle x'|p'\rangle = \frac{1}{\sqrt{2\pi\hbar}} \exp\left(\frac{ip'x'}{\hbar}\right). \tag{1.264}$$

We can now demonstrate how the position-space wave function is related to the momentum-space wave function. All we have to do is rewrite

$$\langle x'|\alpha\rangle = \int dp' \langle x'|p'\rangle\langle p'|\alpha\rangle \tag{1.265a}$$

and

$$\langle p'|\alpha\rangle = \int dx' \langle p'|x'\rangle\langle x'|\alpha\rangle \tag{1.265b}$$

as

$$\psi_\alpha(x') = \left[\frac{1}{\sqrt{2\pi\hbar}}\right] \int dp' \exp\left(\frac{ip'x'}{\hbar}\right) \phi_\alpha(p') \tag{1.266a}$$

and

$$\phi_\alpha(p') = \left[\frac{1}{\sqrt{2\pi\hbar}}\right] \int dx' \exp\left(\frac{-ip'x'}{\hbar}\right) \psi_\alpha(x'). \tag{1.266b}$$

The pair of equations is just what one expects from Fourier's inversion theorem. Apparently the mathematics we have developed somehow "knows" Fourier's work on integral transforms.

## 1.7.4  Gaussian Wave Packets

It is instructive to look at a physical example to illustrate our basic formalism. We consider what is known as a **Gaussian wave packet**, whose $x$-space wave function is given by

$$\langle x'|\alpha\rangle = \left[\frac{1}{\pi^{1/4}\sqrt{d}}\right] \exp\left[ikx' - \frac{x'^2}{2d^2}\right]. \tag{1.267}$$

This is a plane wave with wave number $k$ modulated by a Gaussian profile centered on the origin. The probability of observing the particle vanishes very rapidly for $|x'| > d$; more quantitatively, the probability density $|\langle x'|\alpha\rangle|^2$ has a Gaussian shape with width $d$.

We now compute the expectation values of $x$, $x^2$, $p$, and $p^2$. The expectation value of $x$ is clearly zero by symmetry:

$$\langle x \rangle = \int_{-\infty}^{\infty} dx' \langle \alpha|x'\rangle x' \langle x'|\alpha\rangle = \int_{-\infty}^{\infty} dx' |\langle x'|\alpha\rangle|^2 x' = 0. \tag{1.268}$$

For $x^2$ we obtain

$$\begin{aligned}
\langle x^2 \rangle &= \int_{-\infty}^{\infty} dx' x'^2 |\langle x'|\alpha\rangle|^2 \\
&= \left(\frac{1}{\sqrt{\pi}d}\right) \int_{-\infty}^{\infty} dx' x'^2 \exp\left[\frac{-x'^2}{d^2}\right] \\
&= \frac{d^2}{2},
\end{aligned} \tag{1.269}$$

which leads to

$$\langle (\Delta x)^2 \rangle = \langle x^2 \rangle - \langle x \rangle^2 = \frac{d^2}{2} \tag{1.270}$$

for the dispersion of the position operator. The expectation values of $p$ and $p^2$ can also be computed as follows:

$$\langle p \rangle = \hbar k \tag{1.271a}$$

$$\langle p^2 \rangle = \frac{\hbar^2}{2d^2} + \hbar^2 k^2, \tag{1.271b}$$

which is left as an exercise. The momentum dispersion is therefore given by

$$\langle (\Delta p)^2 \rangle = \langle p^2 \rangle - \langle p \rangle^2 = \frac{\hbar^2}{2d^2}. \tag{1.272}$$

Armed with (1.270) and (1.272), we can check the Heisenberg uncertainty relation (1.216); in this case the uncertainty product is given by

$$\langle (\Delta x)^2 \rangle \langle (\Delta p)^2 \rangle = \frac{\hbar^2}{4}, \tag{1.273}$$

independent of $d$, so for a Gaussian wave packet we actually have an *equality* relation rather than the more general inequality relation (1.216). For this reason a Gaussian wave packet is often called a *minimum uncertainty wave packet*.

We now go to momentum space. By a straightforward integration, just completing the square in the exponent, we obtain

$$\begin{aligned}
\langle p'|\alpha\rangle &= \left(\frac{1}{\sqrt{2\pi\hbar}}\right)\left(\frac{1}{\pi^{1/4}\sqrt{d}}\right)\int_{-\infty}^{\infty} dx' \exp\left(\frac{-ip'x'}{\hbar} + ikx' - \frac{x'^2}{2d^2}\right) \\
&= \sqrt{\frac{d}{\hbar\sqrt{\pi}}} \exp\left[\frac{-(p'-\hbar k)^2 d^2}{2\hbar^2}\right].
\end{aligned} \tag{1.274}$$

This momentum-space wave function provides an alternative method for obtaining $\langle p \rangle$ and $\langle p^2 \rangle$, which is also left as an exercise.

The probability of finding the particle with momentum $p'$ is Gaussian (in momentum space) centered on $\hbar k$, just as the probability of finding the particle at $x'$ is Gaussian (in position space) centered on zero. Furthermore, the widths of the two Gaussians are inversely proportional to each other, which is just another way of expressing the constancy of the uncertainty product $\langle (\Delta x)^2 \rangle \langle (\Delta p)^2 \rangle$ explicitly computed in (1.273). The wider the spread in the $p$-space, the narrower the spread in the $x$-space, and vice versa.

As an extreme example, suppose we let $d \to \infty$. The position-space wave function (1.267) then becomes a plane wave extending over all space; the probability of finding the particle is just constant, independent of $x'$. In contrast, the momentum-space wave function is $\delta$-function-like and is sharply peaked at $\hbar k$. In the opposite extreme, by letting $d \to 0$, we obtain a position-space wave function localized like the $\delta$-function, but the momentum-space wave function (1.274) is just constant, independent of $p'$.

We have seen that an extremely well-localized (in the $x$-space) state is to be regarded as a superposition of momentum eigenstates with all possible values of momenta. Even those momentum eigenstates whose momenta are comparable to or exceed $mc$ must be included in the superposition. However, at such high values of momentum, a description based on nonrelativistic quantum mechanics is bound to break down.[18] Despite this limitation our formalism, based on the existence of the position eigenket $|x'\rangle$, has a wide domain of applicability.

### 1.7.5  Generalization to Three Dimensions

So far in this section we have worked exclusively in one-space for simplicity, but everything we have done can be generalized to three-space, if the necessary changes are made. The base kets to be used can be taken as either the position eigenkets satisfying

$$\mathbf{x}|\mathbf{x}'\rangle = \mathbf{x}'|\mathbf{x}'\rangle \tag{1.275}$$

or the momentum eigenkets satisfying

$$\mathbf{p}|\mathbf{p}'\rangle = \mathbf{p}'|\mathbf{p}'\rangle. \tag{1.276}$$

They obey the normalization conditions

$$\langle \mathbf{x}'|\mathbf{x}''\rangle = \delta^3(\mathbf{x}' - \mathbf{x}'') \tag{1.277a}$$

and

$$\langle \mathbf{p}'|\mathbf{p}''\rangle = \delta^3(\mathbf{p}' - \mathbf{p}''), \tag{1.277b}$$

where $\delta^3$ stands for the three-dimensional $\delta$-function

$$\delta^3(\mathbf{x}' - \mathbf{x}'') = \delta(x' - x'')\delta(y' - y'')\delta(z' - z''). \tag{1.278}$$

---

[18]  It turns out that the concept of a localized state in relativistic quantum mechanics is far more intricate because of the possibility of "negative energy states," or pair creation. See Chapter 8 of this textbook.

The completeness relations read

$$\int d^3x' |\mathbf{x}'\rangle\langle\mathbf{x}'| = 1 \tag{1.279a}$$

and

$$\int d^3p' |\mathbf{p}'\rangle\langle\mathbf{p}'| = 1, \tag{1.279b}$$

which can be used to expand an arbitrary state ket:

$$|\alpha\rangle = \int d^3x' |\mathbf{x}'\rangle\langle\mathbf{x}'|\alpha\rangle, \tag{1.280a}$$

$$|\alpha\rangle = \int d^3p' |\mathbf{p}'\rangle\langle\mathbf{p}'|\alpha\rangle. \tag{1.280b}$$

The expansion coefficients $\langle\mathbf{x}'|\alpha\rangle$ and $\langle\mathbf{p}'|\alpha\rangle$ are identified with the wave functions $\psi_\alpha(\mathbf{x}')$ and $\phi_\alpha(\mathbf{p}')$ in position and momentum space, respectively.

The momentum operator, when taken between $|\beta\rangle$ and $|\alpha\rangle$, becomes

$$\langle\beta|\mathbf{p}|\alpha\rangle = \int d^3x' \psi_\beta^*(\mathbf{x}')(-i\hbar\mathbf{\nabla}')\psi_\alpha(\mathbf{x}'). \tag{1.281}$$

The transformation function analogous to (1.264) is

$$\langle\mathbf{x}'|\mathbf{p}'\rangle = \left[\frac{1}{(2\pi\hbar)^{3/2}}\right]\exp\left(\frac{i\mathbf{p}'\cdot\mathbf{x}'}{\hbar}\right), \tag{1.282}$$

so that

$$\psi_\alpha(\mathbf{x}') = \left[\frac{1}{(2\pi\hbar)^{3/2}}\right]\int d^3p' \exp\left(\frac{i\mathbf{p}'\cdot\mathbf{x}'}{\hbar}\right)\phi_\alpha(\mathbf{p}') \tag{1.283a}$$

and

$$\phi_\alpha(\mathbf{p}') = \left[\frac{1}{(2\pi\hbar)^{3/2}}\right]\int d^3x' \exp\left(\frac{-i\mathbf{p}'\cdot\mathbf{x}'}{\hbar}\right)\psi_\alpha(\mathbf{x}'). \tag{1.283b}$$

It is interesting to check the dimension of the wave functions. In one-dimensional problems the normalization requirement (1.190) implies that $|\langle x'|\alpha\rangle|^2$ has the dimension of inverse length, so the wave function itself must have the dimension of $(\text{length})^{-1/2}$. In contrast, the wave function in three-dimensional problems must have the dimension of $(\text{length})^{-3/2}$ because $|\langle\mathbf{x}'|\alpha\rangle|^2$ integrated over all spatial volume must be unity (dimensionless).

# Problems

**1.1** A beam of silver atoms is created by heating a vapor in an oven to 1000°C, and selecting atoms with a velocity close to the mean of the thermal distribution. The beam moves through a one-meter long magnetic field with a vertical gradient 10 T/m, and impinges a screen one meter downstream of the end of the magnet. Assuming the silver atom has spin $\frac{1}{2}$ with a magnetic moment of one Bohr magneton, find the separation distance in millimeters of the two states on the screen.

**1.2**  Prove

$$[AB, CD] = -AC\{D, B\} + A\{C, B\}D - C\{D, A\}B + \{C, A\}DB.$$

**1.3**  For the spin $\frac{1}{2}$ state $|S_x; +\rangle$, evaluate both sides of the inequality (1.146), that is

$$\left\langle (\Delta A)^2 \right\rangle \left\langle (\Delta B)^2 \right\rangle \geq \frac{1}{4} |\langle [A, B] \rangle|^2$$

for the operators $A = S_x$ and $B = S_y$, and show that the inequality is satisfied. Repeat for the operators $A = S_z$ and $B = S_y$.

**1.4**  Suppose a $2 \times 2$ matrix $X$ (not necessarily Hermitian, nor unitary) is written as

$$X = a_0 + \boldsymbol{\sigma} \cdot \mathbf{a},$$

where the matrices $\boldsymbol{\sigma}$ are given in (3.50) and $a_0$ and $a_{1,2,3}$ are numbers.
a. How are $a_0$ and $a_k (k = 1, 2, 3)$ related to tr$(X)$ and tr$(\sigma_k X)$?
b. Obtain $a_0$ and $a_k$ in terms of the matrix elements $X_{ij}$.

**1.5**  Show that the determinant of a $2 \times 2$ matrix $\boldsymbol{\sigma} \cdot \mathbf{a}$ is invariant under

$$\boldsymbol{\sigma} \cdot \mathbf{a} \rightarrow \boldsymbol{\sigma} \cdot \mathbf{a}' \equiv \exp\left(\frac{i\boldsymbol{\sigma} \cdot \hat{\mathbf{n}}\phi}{2}\right) \boldsymbol{\sigma} \cdot \mathbf{a} \exp\left(\frac{-i\boldsymbol{\sigma} \cdot \hat{\mathbf{n}}\phi}{2}\right),$$

where the matrices $\boldsymbol{\sigma}$ are given in (3.50). Find $a_k'$ in terms of $a_k$ when $\hat{\mathbf{n}}$ is in the positive $z$-direction and interpret your result.

**1.6**  Using the rules of bra-ket algebra, prove or evaluate the following:
a. tr$(XY) = $ tr$(YX)$, where $X$ and $Y$ are operators;
b. $(XY)^\dagger = Y^\dagger X^\dagger$, where $X$ and $Y$ are operators;
c. $\exp[if(A)] = ?$ in ket-bra form, where $A$ is a Hermitian operator whose eigenvalues are known;
d. $\sum_{a'} \psi_{a'}^*(\mathbf{x}')\psi_{a'}(\mathbf{x}'')$, where $\psi_{a'}(\mathbf{x}') = \langle \mathbf{x}' | a' \rangle$.

**1.7**  a. Consider two kets $|\alpha\rangle$ and $|\beta\rangle$. Suppose $\langle a'|\alpha\rangle, \langle a''|\alpha\rangle, \ldots$ and $\langle a'|\beta\rangle, \langle a''|\beta\rangle, \ldots$ are all known, where $|a'\rangle, |a''\rangle, \ldots$ form a complete set of base kets. Find the matrix representation of the operator $|\alpha\rangle\langle\beta|$ in that basis.
b. We now consider a spin $\frac{1}{2}$ system and let $|\alpha\rangle$ and $|\beta\rangle$ be $|S_z; +\rangle$ and $|S_x; +\rangle$, respectively. Write down explicitly the square matrix that corresponds to $|\alpha\rangle\langle\beta|$ in the usual ($s_z$ diagonal) basis.

**1.8**  Suppose $|i\rangle$ and $|j\rangle$ are eigenkets of some Hermitian operator $A$. Under what condition can we conclude that $|i\rangle + |j\rangle$ is also an eigenket of $A$? Justify your answer.

**1.9**  Consider a ket space spanned by the eigenkets $\{|a'\rangle\}$ of a Hermitian operator $A$. There is no degeneracy.
a. Prove that

$$\prod_{a'}(A - a')$$

is the null operator.

b.  What is the significance of

$$\prod_{a'' \neq a'} \frac{(A - a'')}{(a' - a'')}?$$

c.  Illustrate (a) and (b) using $A$ set equal to $S_z$ of a spin $\frac{1}{2}$ system.

**1.10**  Using the orthonormality of $|+\rangle$ and $|-\rangle$, prove

$$[S_i, S_j] = i\varepsilon_{ijk}\hbar S_k, \quad \{S_i, S_j\} = \left(\frac{\hbar^2}{2}\right)\delta_{ij},$$

where

$$S_x = \frac{\hbar}{2}(|+\rangle\langle-| + |-\rangle\langle+|), \quad S_y = \frac{i\hbar}{2}(-|+\rangle\langle-| + |-\rangle\langle+|),$$

$$S_z = \frac{\hbar}{2}(|+\rangle\langle+| - |-\rangle\langle-|).$$

**1.11**  Construct $|\mathbf{S} \cdot \hat{\mathbf{n}}; +\rangle$ such that

$$\mathbf{S} \cdot \hat{\mathbf{n}}|\mathbf{S} \cdot \hat{\mathbf{n}}; +\rangle = \left(\frac{\hbar}{2}\right)|\mathbf{S} \cdot \hat{\mathbf{n}}; +\rangle$$

where $\hat{\mathbf{n}}$ is characterized by the angles shown in the figure. Express your answer as a linear combination of $|+\rangle$ and $|-\rangle$. [*Note:* The answer is

$$\cos\left(\frac{\beta}{2}\right)|+\rangle + \sin\left(\frac{\beta}{2}\right)e^{i\alpha}|-\rangle.$$

But do not just verify that this answer satisfies the above eigenvalue equation. Rather, treat the problem as a straightforward eigenvalue problem. Also do not use rotation operators, which we will introduce later in this book.]

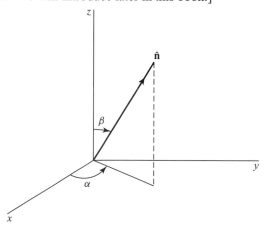

**1.12**  The Hamiltonian operator for a two-state system is given by

$$H = a(|1\rangle\langle1| - |2\rangle\langle2| + |1\rangle\langle2| + |2\rangle\langle1|),$$

where $a$ is a number with the dimension of energy. Find the energy eigenvalues and the corresponding energy eigenkets (as linear combinations of $|1\rangle$ and $|2\rangle$).

**1.13**  A two-state system is characterized by the Hamiltonian

$$H = H_{11}|1\rangle\langle 1| + H_{22}|2\rangle\langle 2| + H_{12}[|1\rangle\langle 2| + |2\rangle\langle 1|]$$

where $H_{11}, H_{22}$, and $H_{12}$ are real numbers with the dimension of energy, and $|1\rangle$ and $|2\rangle$ are eigenkets of some observable ($\neq H$). Find the energy eigenkets and corresponding energy eigenvalues. Make sure that your answer makes good sense for $H_{12} = 0$.

**1.14**  A spin $\frac{1}{2}$ system is known to be in an eigenstate of $\mathbf{S} \cdot \hat{\mathbf{n}}$ with eigenvalue $\hbar/2$, where $\hat{\mathbf{n}}$ is a unit vector lying in the $xz$-plane that makes an angle $\gamma$ with the positive $z$-axis.
a.  Suppose $S_x$ is measured. What is the probability of getting $+\hbar/2$?
b.  Evaluate the dispersion in $S_x$, that is,

$$\langle (S_x - \langle S_x \rangle)^2 \rangle.$$

(For your own peace of mind check your answers for the special cases $\gamma = 0, \pi/2$, and $\pi$.)

**1.15**  A beam of spin $\frac{1}{2}$ atoms goes through a series of Stern–Gerlach type measurements as follows.
a.  The first measurement accepts $s_z = \hbar/2$ atoms and rejects $s_z = \hbar/2$ atoms.
b.  The second measurement accepts $s_n = \hbar/2$ atoms and rejects $s_n = -\hbar/2$ atoms, where $s_n$ is the eigenvalue of the operator $\mathbf{S} \cdot \hat{\mathbf{n}}$, with $\hat{\mathbf{n}}$ making an angle $\beta$ in the $xz$-plane with respect to the $z$-axis.
c.  The third measurement accepts $s_z = -\hbar/2$ atoms and rejects $s_z = \hbar/2$ atoms.

What is the intensity of the final $s_z = -\hbar/2$ beam when the $s_z = \hbar/2$ beam surviving the first measurement is normalized to unity? How must we orient the second measuring apparatus if we are to maximize the intensity of the final $s_z = -\hbar/2$ beam?

**1.16**  A certain observable in quantum mechanics has a $3 \times 3$ matrix representation as follows:

$$\frac{1}{\sqrt{2}} \begin{pmatrix} 0 & 1 & 0 \\ 1 & 0 & 1 \\ 0 & 1 & 0 \end{pmatrix}.$$

a.  Find the normalized eigenvectors of this observable and the corresponding eigenvalues. Is there any degeneracy?
b.  Give a physical example where all this is relevant.

**1.17**  Let $A$ and $B$ be observables. Suppose the simultaneous eigenkets of $A$ and $B$ $\{|a',b'\rangle\}$ form a *complete* orthonormal set of base kets. Can we always conclude that

$$[A,B] = 0?$$

If your answer is yes, prove the assertion. If your answer is no, give a counterexample.

**1.18** Two Hermitian operators anticommute:

$$\{A, B\} = AB + BA = 0.$$

Is it possible to have a simultaneous (that is, common) eigenket of $A$ and $B$? Prove or illustrate your assertion.

**1.19** Two observables $A_1$ and $A_2$, which do not involve time explicitly, are known not to commute,

$$[A_1, A_2] \neq 0,$$

yet we also know that $A_1$ and $A_2$ both commute with the Hamiltonian:

$$[A_1, H] = 0, \quad [A_2, H] = 0.$$

Prove that the energy eigenstates are, in general, degenerate. Are there exceptions? As an example, you may think of the central-force problem $H = \mathbf{p}^2/2m + V(r)$, with $A_1 \rightarrow L_z$, $A_2 \rightarrow L_x$.

**1.20** a. The simplest way to derive the Schwarz inequality goes as follows. First, observe

$$((\langle \alpha| + \lambda^* \langle \beta|) \cdot (|\alpha\rangle + \lambda|\beta\rangle)) \geq 0$$

for any complex number $\lambda$; then choose $\lambda$ in such a way that the preceding inequality reduces to the Schwarz inequality.

   b. Show that the equality sign in the generalized uncertainty relation holds if the state in question satisfies

$$\Delta A |\alpha\rangle = \lambda \Delta B |\alpha\rangle$$

with $\lambda$ purely *imaginary*.

   c. Explicit calculations using the usual rules of wave mechanics show that the wave function for a Gaussian wave packet given by

$$\langle x'|\alpha\rangle = (2\pi d^2)^{-1/4} \exp\left[\frac{i\langle p\rangle x'}{\hbar} - \frac{(x' - \langle x\rangle)^2}{4d^2}\right]$$

satisfies the minimum uncertainty relation

$$\sqrt{\langle (\Delta x)^2\rangle}\sqrt{\langle (\Delta p)^2\rangle} = \frac{\hbar}{2}.$$

Prove that the requirement

$$\langle x'|\Delta x|\alpha\rangle = (\text{imaginary number})\langle x'|\Delta p|\alpha\rangle$$

is indeed satisfied for such a Gaussian wave packet, in agreement with (b).

**1.21** a. Compute

$$\langle (\Delta S_x)^2\rangle \equiv \langle S_x^2\rangle - \langle S_x\rangle^2,$$

where the expectation value is taken for the $S_z+$ state. Using your result, check the generalized uncertainty relation

$$\langle(\Delta A)^2\rangle\langle(\Delta B)^2\rangle \geq \frac{1}{4}|\langle[A,B]\rangle|^2,$$

with $A \rightarrow S_x$, $B \rightarrow S_y$.

b. Check the uncertainty relation with $A \rightarrow S_x$, $B \rightarrow S_y$ for the $S_x+$ state.

**1.22** Find the linear combination of $|+\rangle$ and $|-\rangle$ kets that maximizes the uncertainty product

$$\langle(\Delta S_x)^2\rangle\langle(\Delta S_y)^2\rangle.$$

Verify explicitly that for the linear combination you found, the uncertainty relation for $S_x$ and $S_y$ is not violated.

**1.23** Evaluate the $x$-$p$ uncertainty product $\langle(\Delta x)^2\rangle\langle(\Delta p)^2\rangle$ for a one-dimensional particle confined between two rigid walls

$$V = \begin{cases} 0 & \text{for } 0 < x < a, \\ \infty & \text{otherwise.} \end{cases}$$

Do this for both the ground and excited states.

**1.24** Estimate the rough order of magnitude of the length of time that an ice pick can be balanced on its point if the only limitation is that set by the Heisenberg uncertainty principle. Assume that the point is sharp and that the point and the surface on which it rests are hard. You may make approximations which do not alter the general order of magnitude of the result. Assume reasonable values for the dimensions and weight of the ice pick. Obtain an approximate numerical result and express it *in seconds*.

**1.25** Consider a three-dimensional ket space. If a certain set of orthonormal kets, say, $|1\rangle$, $|2\rangle$, and $|3\rangle$, are used as the base kets, the operators $A$ and $B$ are represented by

$$A \doteq \begin{pmatrix} a & 0 & 0 \\ 0 & -a & 0 \\ 0 & 0 & -a \end{pmatrix}, \quad B \doteq \begin{pmatrix} b & 0 & 0 \\ 0 & 0 & -ib \\ 0 & ib & 0 \end{pmatrix}$$

with $a$ and $b$ both real.

a. Obviously $A$ exhibits a degenerate spectrum. Does $B$ also exhibit a degenerate spectrum?

b. Show that $A$ and $B$ commute.

c. Find a new set of orthonormal kets which are simultaneous eigenkets of both $A$ and $B$. Specify the eigenvalues of $A$ and $B$ for each of the three eigenkets. Does your specification of eigenvalues completely characterize each eigenket?

**1.26** a. Prove that $(1/\sqrt{2})(1 + i\sigma_x)$, where the matrix $\sigma_x$ is given in (3.50), acting on a two-component spinor can be regarded as the matrix representation of the rotation operator about the $x$-axis by angle $-\pi/2$. (The minus sign signifies that the rotation is clockwise.)

b. Construct the matrix representation of $S_z$ when the eigenkets of $S_y$ are used as base vectors.

**1.27** Some authors define an *operator* to be real when every member of its matrix elements $\langle b'|A|b''\rangle$ is real in some representation ($\{|b'\rangle\}$ basis in this case). Is this concept representation independent, that is, do the matrix elements remain real even if some basis other than $\{|b'\rangle\}$ is used? Check your assertion using familiar operators such as $S_y$ and $S_z$ (see Problem 1.26) or $x$ and $p_x$.

**1.28** Construct the transformation matrix that connects the $S_z$ diagonal basis to the $S_x$ diagonal basis. Show that your result is consistent with the general relation

$$U = \sum_r |b^{(r)}\rangle\langle a^{(r)}|.$$

**1.29** a. Suppose that $f(A)$ is a function of a Hermitian operator $A$ with the property $A|a'\rangle = a'|a'\rangle$. Evaluate $\langle b''|f(A)|b'\rangle$ when the transformation matrix from the $a'$ basis to the $b'$ basis is known.

b. Using the continuum analogue of the result obtained in (a), evaluate

$$\langle \mathbf{p}''|F(r)|\mathbf{p}'\rangle.$$

Simplify your expression as far as you can. Note that $r$ is $\sqrt{x^2+y^2+z^2}$, where $x$, $y$, and $z$ are *operators*.

**1.30** a. Let $x$ and $p_x$ be the coordinate and linear momentum in one dimension. Evaluate the classical Poisson bracket

$$[x,F(p_x)]_{classical}.$$

b. Let $x$ and $p_x$ be the corresponding quantum-mechanical operators this time. Evaluate the commutator

$$\left[x,\exp\left(\frac{ip_x a}{\hbar}\right)\right].$$

c. Using the result obtained in (b), prove that

$$\exp\left(\frac{ip_x a}{\hbar}\right)|x'\rangle \quad (x|x'\rangle = x'|x'\rangle)$$

is an eigenstate of the coordinate operator $x$. What is the corresponding eigenvalue?

**1.31** a. On p. 247, Gottfried (1966) states that

$$[x_i, G(\mathbf{p})] = i\hbar\frac{\partial G}{\partial p_i}, \quad [p_i, F(\mathbf{x})] = -i\hbar\frac{\partial F}{\partial x_i}$$

can be "easily derived" from the fundamental commutation relations for all functions of $F$ and $G$ that can be expressed as power series in their arguments. Verify this statement.

b. Evaluate $[x^2,p^2]$. Compare your result with the classical Poisson bracket $[x^2,p^2]_{classical}$.

**1.32** The translation operator for a finite (spatial) displacement is given by

$$\mathscr{J}(\mathbf{l}) = \exp\left(\frac{-i\mathbf{p}\cdot\mathbf{l}}{\hbar}\right),$$

where $\mathbf{p}$ is the momentum *operator*.

a. Evaluate

$$[x_i, \mathscr{J}(\mathbf{l})].$$

b. Using (a) (or otherwise), demonstrate how the expectation value $\langle\mathbf{x}\rangle$ changes under translation.

**1.33** In the main text we discussed the effect of $\mathscr{J}(d\mathbf{x}')$ on the position and momentum eigenkets and on a more general state ket $|\alpha\rangle$. We can also study the behavior of expectation values $\langle\mathbf{x}\rangle$ and $\langle\mathbf{p}\rangle$ under infinitesimal translation. Using (1.207), (1.227), and $|\alpha\rangle \rightarrow \mathscr{J}(d\mathbf{x}')|\alpha\rangle$ only, prove $\langle\mathbf{x}\rangle \rightarrow \langle\mathbf{x}\rangle + d\mathbf{x}', \langle\mathbf{p}\rangle \rightarrow \langle\mathbf{p}\rangle$ under infinitesimal translation.

**1.34** Starting with a momentum operator $\mathbf{p}$ having eigenstates $|\mathbf{p}'\rangle$, define an infinitesimal *boost* operator $\mathscr{B}(d\mathbf{p}')$ that changes one momentum eigenstate into another, that is

$$\mathscr{B}(d\mathbf{p}')|\mathbf{p}'\rangle = |\mathbf{p}' + d\mathbf{p}'\rangle.$$

Show that the form $\mathscr{B}(d\mathbf{p}') = 1 + i\mathbf{W}\cdot d\mathbf{p}'$, where $\mathbf{W}$ is Hermitian, satisfies the unitary, associative, and inverse properties that are appropriate for $\mathscr{B}(d\mathbf{p}')$. Use dimensional analysis to express $\mathbf{W}$ in terms of the position operator $\mathbf{x}$, and show that the result satisfies the canonical commutation relations $[x_i, p_j] = i\hbar\delta_{ij}$. Derive an expression for the matrix element $\langle\mathbf{p}'|\mathbf{x}|\alpha\rangle$ in terms of a derivative with respect to $\mathbf{p}'$ of $\langle\mathbf{p}'|\alpha\rangle$.

**1.35** a. Verify (1.271a) and (1.271b) for the expectation value of $p$ and $p^2$ from the Gaussian wave packet (1.267).

b. Evaluate the expectation value of $p$ and $p^2$ using the momentum-space wave function (1.274).

**1.36** a. Prove the following:

(i) $\langle p'|x|\alpha\rangle = i\hbar\dfrac{\partial}{\partial p'}\langle p'|\alpha\rangle,$

(ii) $\langle\beta|x|\alpha\rangle = \displaystyle\int dp' \phi_\beta^*(p') i\hbar\frac{\partial}{\partial p'}\phi_\alpha(p'),$

where $\phi_\alpha(p') = \langle p'|\alpha\rangle$ and $\phi_\beta(p') = \langle p'|\beta\rangle$ are momentum-space wave functions.

b. What is the physical significance of

$$\exp\left(\frac{ix\Xi}{\hbar}\right),$$

where $x$ is the position operator and $\Xi$ is some number with the dimension of momentum? Justify your answer.

# 2 Quantum Dynamics

So far we have not discussed how physical systems change with time. This chapter is devoted exclusively to the dynamic development of state kets and/or observables. In other words, we are concerned here with the quantum-mechanical analogue of Newton's (or Lagrange's or Hamilton's) equations of motion.

## 2.1 Time Evolution and the Schrödinger Equation

The first important point we should keep in mind is that time is just a parameter in quantum mechanics, *not* an operator. In particular, time is not an observable in the language of the previous chapter. It is nonsensical to talk about the time operator in the same sense as we talk about the position operator. Ironically, in the historical development of wave mechanics both L. de Broglie and E. Schrödinger were guided by a kind of covariant analogy between energy and time on the one hand and momentum and position (spatial coordinate) on the other. Yet when we now look at quantum mechanics in its finished form, there is no trace of a symmetrical treatment between time and space. The relativistic quantum theory of fields does treat the time and space coordinates on the same footing, but it does so only at the expense of demoting position from the status of being an observable to that of being just a parameter.

### 2.1.1 Time-Evolution Operator

Our basic concern in this section is, How does a state ket change with time? Suppose we have a physical system whose state ket at $t_0$ is represented by $|\alpha\rangle$. At later times, we do not, in general, expect the system to remain in the same state $|\alpha\rangle$. Let us denote the ket corresponding to the state at some later time by

$$|\alpha, t_0; t\rangle \quad (t > t_0), \tag{2.1}$$

where we have written $\alpha, t_0$ to remind ourselves that the system *used to be* in state $|\alpha\rangle$ at some earlier reference time $t_0$. Because time is assumed to be a continuous parameter, we expect

$$\lim_{t \to t_0} |\alpha, t_0; t\rangle = |\alpha\rangle \tag{2.2}$$

and we may as well use a shorthand notation,

$$|\alpha, t_0; t_0\rangle = |\alpha, t_0\rangle, \tag{2.3}$$

for this. Our basic task is to study the time evolution of a state ket:

$$|\alpha, t_0\rangle = |\alpha\rangle \xrightarrow{\text{time evolution}} |\alpha, t_0; t\rangle. \tag{2.4}$$

Put in another way, we are interested in asking how the state ket changes under a time displacement $t_0 \to t$.

As in the case of translation, the two kets are related by an operator which we call the **time-evolution operator** $\mathscr{U}(t, t_0)$:

$$|\alpha, t_0; t\rangle = \mathscr{U}(t, t_0)|\alpha, t_0\rangle. \tag{2.5}$$

What are some of the properties we would like to ascribe to the time-evolution operator? The first important property is the unitary requirement for $\mathscr{U}(t, t_0)$ that follows from probability conservation. Suppose that at $t_0$ the state ket is expanded in terms of the eigenkets of some observable $A$:

$$|\alpha, t_0\rangle = \sum_{a'} c_{a'}(t_0)|a'\rangle. \tag{2.6}$$

Likewise, at some later time, we have

$$|\alpha, t_0; t\rangle = \sum_{a'} c_{a'}(t)|a'\rangle. \tag{2.7}$$

In general, we do not expect the modulus of the individual expansion coefficient to remain the same:[1]

$$|c_{a'}(t)| \neq |c_{a'}(t_0)|. \tag{2.8}$$

For instance, consider a spin $\frac{1}{2}$ system with its spin magnetic moment subjected to a uniform magnetic field in the $z$-direction. To be specific, suppose that at $t_0$ the spin is in the positive $x$-direction; that is, the system is prepared in an eigenstate of $S_x$ with eigenvalue $\hbar/2$. As time goes on, the spin precesses in the $xy$-plane, as will be quantitatively demonstrated later in this section. This means that the probability for observing $S_x +$ is no longer unity at $t > t_0$; there is a finite probability for observing $S_x -$ as well. Yet the *sum* of the probabilities for $S_x +$ and $S_x -$ remains unity at all times. Generally, in the notation of (2.6) and (2.7), we must have

$$\sum_{a'} |c_{a'}(t_0)|^2 = \sum_{a'} |c_{a'}(t)|^2 \tag{2.9}$$

despite (2.8) for the individual expansion coefficients. Stated another way, if the state ket is initially normalized to unity, it must remain normalized to unity at all later times:

$$\langle \alpha, t_0 | \alpha, t_0 \rangle = 1 \Rightarrow \langle \alpha, t_0; t | \alpha, t_0; t \rangle = 1. \tag{2.10}$$

[1] We later show, however, that if the Hamiltonian commutes with $A$, then $|c_{a'}(t)|$ is indeed equal to $|c_{a'}(t_0)|$.

As in the translation case, this property is guaranteed if the time-evolution operator is taken to be unitary. For this reason we take unitarity,

$$\mathscr{U}^{\dagger}(t,t_0)\mathscr{U}(t,t_0) = 1, \tag{2.11}$$

to be one of the fundamental properties of the $\mathscr{U}$ operator. It is no coincidence that many authors regard unitarity as being synonymous with probability conservation.

Another feature we require of the $\mathscr{U}$ operator is the composition property:

$$\mathscr{U}(t_2,t_0) = \mathscr{U}(t_2,t_1)\mathscr{U}(t_1,t_0) \quad (t_2 > t_1 > t_0). \tag{2.12}$$

This equation says that if we are interested in obtaining time evolution from $t_0$ to $t_2$, then we can obtain the same result by first considering time evolution from $t_0$ to $t_1$, then from $t_1$ to $t_2$, a reasonable requirement. Note that we read (2.12) from right to left!

It also turns out to be advantageous to consider an infinitesimal time-evolution operator $\mathscr{U}(t_0 + dt, t_0)$:

$$|\alpha, t_0; t_0 + dt\rangle = \mathscr{U}(t_0 + dt, t_0)|\alpha, t_0\rangle. \tag{2.13}$$

Because of continuity [see (2.2)], the infinitesimal time-evolution operator must reduce to the identity operator as $dt$ goes to zero,

$$\lim_{dt \to 0} \mathscr{U}(t_0 + dt, t_0) = 1, \tag{2.14}$$

and as in the translation case, we expect the difference between $\mathscr{U}(t_0 + dt, t_0)$ and 1 to be of first order in $dt$.

We assert that all these requirements are satisfied by

$$\mathscr{U}(t_0 + dt, t_0) = 1 - i\Omega \, dt, \tag{2.15}$$

where $\Omega$ is a Hermitian operator,[2]

$$\Omega^{\dagger} = \Omega. \tag{2.16}$$

With (2.15) the infinitesimal time-displacement operator satisfies the composition property

$$\mathscr{U}(t_0 + dt_1 + dt_2, t_0) = \mathscr{U}(t_0 + dt_1 + dt_2, t_0 + dt_1)\mathscr{U}(t_0 + dt_1, t_0); \tag{2.17}$$

it differs from the identity operator by a term of order $dt$. The unitarity property can also be checked as follows:

$$\mathscr{U}^{\dagger}(t_0 + dt, t_0)\mathscr{U}(t_0 + dt, t_0) = (1 + i\Omega^{\dagger}dt)(1 - i\Omega dt) \simeq 1, \tag{2.18}$$

to the extent that terms of order $(dt)^2$ or higher can be ignored.

The operator $\Omega$ has the dimension of frequency or inverse time. Is there any familiar observable with the dimension of frequency? We recall that in the old quantum theory, angular frequency $\omega$ is postulated to be related to energy by the Planck–Einstein relation

$$E = \hbar\omega. \tag{2.19}$$

---

[2] If the $\Omega$ operator depends on time explicitly, it must be evaluated at $t_0$.

Let us now borrow from classical mechanics the idea that the Hamiltonian is the generator of time evolution (Goldstein et al. (2002), pp. 401–402). It is then natural to relate $\Omega$ to the Hamiltonian operator $H$:

$$\Omega = \frac{H}{\hbar}. \tag{2.20}$$

To sum up, the infinitesimal time-evolution operator is written as

$$\mathscr{U}(t_0 + dt, t_0) = 1 - \frac{iH\,dt}{\hbar}, \tag{2.21}$$

where $H$, the Hamiltonian operator, is assumed to be Hermitian. The reader may ask whether the $\hbar$ introduced here is the same as the $\hbar$ that appears in the expression for the translation operator (1.214). This question can be answered by comparing the quantum-mechanical equation of motion we derive later with the classical equation of motion. It turns out that unless the two $\hbar$ are taken to be the same, we are unable to obtain a relation like

$$\frac{d\mathbf{x}}{dt} = \frac{\mathbf{p}}{m} \tag{2.22}$$

as the classical limit of the corresponding quantum-mechanical relation.

### 2.1.2  The Schrödinger Equation

We are now in a position to derive the fundamental differential equation for the time-evolution operator $\mathscr{U}(t, t_0)$. We exploit the composition property of the time-evolution operator by letting $t_1 \to t$, $t_2 \to t + dt$ in (2.12):

$$\mathscr{U}(t + dt, t_0) = \mathscr{U}(t + dt, t)\mathscr{U}(t, t_0) = \left(1 - \frac{iH\,dt}{\hbar}\right)\mathscr{U}(t, t_0), \tag{2.23}$$

where the time difference $t - t_0$ need not be infinitesimal. We have

$$\mathscr{U}(t + dt, t_0) - \mathscr{U}(t, t_0) = -i\left(\frac{H}{\hbar}\right)dt\,\mathscr{U}(t, t_0), \tag{2.24}$$

which can be written in differential equation form:

$$i\hbar\frac{\partial}{\partial t}\mathscr{U}(t, t_0) = H\mathscr{U}(t, t_0). \tag{2.25}$$

This is the **Schrödinger equation for the time-evolution operator**. Everything that has to do with time development follows from this fundamental equation.

Equation (2.25) immediately leads to the Schrödinger equation for a state ket. Multiplying both sides of (2.25) by $|\alpha, t_0\rangle$ on the right, we obtain

$$i\hbar\frac{\partial}{\partial t}\mathscr{U}(t, t_0)|\alpha, t_0\rangle = H\mathscr{U}(t, t_0)|\alpha, t_0\rangle. \tag{2.26}$$

But $|\alpha, t_0\rangle$ does not depend on $t$, so this is the same as

$$i\hbar\frac{\partial}{\partial t}|\alpha, t_0; t\rangle = H|\alpha, t_0; t\rangle, \tag{2.27}$$

where (2.5) has been used.

If we are given $\mathcal{U}(t,t_0)$ and, in addition, know how $\mathcal{U}(t,t_0)$ acts on the initial state ket $|\alpha,t_0\rangle$, it is not necessary to bother with the Schrödinger equation for the state ket (2.27). All we have to do is apply $\mathcal{U}(t,t_0)$ to $|\alpha,t_0\rangle$; in this manner we can obtain a state ket at any $t$. Our first task is therefore to derive formal solutions to the Schrödinger equation for the time-evolution operator (2.25). There are three cases to be treated separately.

*Case* 1. The Hamiltonian operator is independent of time. By this we mean that even when the parameter $t$ is changed, the $H$ operator remains unchanged. The Hamiltonian for a spin-magnetic moment interacting with a time-independent magnetic field is an example of this. The solution to (2.25) in such a case is given by

$$\mathcal{U}(t,t_0) = \exp\left[\frac{-iH(t-t_0)}{\hbar}\right]. \tag{2.28}$$

To prove this let us expand the exponential as follows:

$$\exp\left[\frac{-iH(t-t_0)}{\hbar}\right] = 1 + \frac{-iH(t-t_0)}{\hbar} + \left[\frac{(-i)^2}{2}\right]\left[\frac{H(t-t_0)}{\hbar}\right]^2 + \cdots. \tag{2.29}$$

Because the time derivative of this expansion is given by

$$\frac{\partial}{\partial t}\exp\left[\frac{-iH(t-t_0)}{\hbar}\right] = \frac{-iH}{\hbar} + \left[\frac{(-i)^2}{2}\right]2\left(\frac{H}{\hbar}\right)^2(t-t_0) + \cdots, \tag{2.30}$$

expression (2.28) obviously satisfies differential equation (2.25). The boundary condition is also satisfied because as $t \to t_0$, (2.28) reduces to the identity operator. An alternative way to obtain (2.28) is to compound successively infinitesimal time-evolution operators just as we did to obtain (1.218) for finite translation:

$$\lim_{N\to\infty}\left[1 - \frac{(iH/\hbar)(t-t_0)}{N}\right]^N = \exp\left[\frac{-iH(t-t_0)}{\hbar}\right]. \tag{2.31}$$

*Case* 2. The Hamiltonian operator $H$ is time dependent but the $H$ at different times commute. As an example, let us consider the spin-magnetic moment subjected to a magnetic field whose strength varies with time but whose direction is always unchanged. The formal solution to (2.25) in this case is

$$\mathcal{U}(t,t_0) = \exp\left[-\left(\frac{i}{\hbar}\right)\int_{t_0}^{t}dt'H(t')\right]. \tag{2.32}$$

This can be proved in a similar way. We simply replace $H(t-t_0)$ in (2.29) and (2.30) by $\int_{t_0}^{t}dt'H(t')$.

*Case* 3. The $H$ at different times do *not* commute. Continuing with the example involving spin-magnetic moment, we suppose, this time, that the magnetic field direction also changes with time: at $t = t_1$ in the $x$-direction, at $t = t_2$ in the $y$-direction, and so forth. Because $S_x$ and $S_y$ do not commute, $H(t_1)$ and $H(t_2)$, which go like $\mathbf{S}\cdot\mathbf{B}$, do not commute either. The formal solution in such a situation is given by

$$\mathcal{U}(t,t_0) = 1 + \sum_{n=1}^{\infty}\left(\frac{-i}{\hbar}\right)^n\int_{t_0}^{t}dt_1\int_{t_0}^{t_1}dt_2\cdots\int_{t_0}^{t_{n-1}}dt_n\,H(t_1)H(t_2)\cdots H(t_n), \tag{2.33}$$

which is sometimes known as the **Dyson series**, after F. J. Dyson, who developed a perturbation expansion of this form in quantum field theory. We do not prove (2.33) now because the proof is very similar to the one presented in Chapter 5 for the time-evolution operator in the interaction picture.

In elementary applications, only case 1 is of practical interest. In the remaining part of this chapter we assume that the $H$ operator is time independent. We will encounter time-dependent Hamiltonians in Chapter 5.

### 2.1.3 Energy Eigenkets

To be able to evaluate the effect of the time-evolution operator (2.28) on a general initial ket $|\alpha\rangle$, we must first know how it acts on the base kets used in expanding $|\alpha\rangle$. This is particularly straightforward if the base kets used are eigenkets of $A$ such that

$$[A,H] = 0; \tag{2.34}$$

then the eigenkets of $A$ are also eigenkets of $H$, called **energy eigenkets**, whose eigenvalues are denoted by $E_{a'}$:

$$H|a'\rangle = E_{a'}|a'\rangle. \tag{2.35}$$

We can now expand the time-evolution operator in terms of $|a'\rangle\langle a'|$. Taking $t_0 = 0$ for simplicity, we obtain

$$\exp\left(\frac{-iHt}{\hbar}\right) = \sum_{a'}\sum_{a''}|a''\rangle\langle a''|\exp\left(\frac{-iHt}{\hbar}\right)|a'\rangle\langle a'|$$

$$= \sum_{a'}|a'\rangle\exp\left(\frac{-iE_{a'}t}{\hbar}\right)\langle a'|. \tag{2.36}$$

The time-evolution operator written in this form enables us to solve any initial-value problem once the expansion of the initial ket in terms of $\{|a'\rangle\}$ is known. As an example, suppose that the initial ket expansion reads

$$|\alpha, t_0 = 0\rangle = \sum_{a'}|a'\rangle\langle a'|\alpha\rangle = \sum_{a'}c_{a'}|a'\rangle. \tag{2.37}$$

We then have

$$|\alpha, t_0 = 0; t\rangle = \exp\left(\frac{-iHt}{\hbar}\right)|\alpha, t_0 = 0\rangle = \sum_{a'}|a'\rangle\langle a'|\alpha\rangle\exp\left(\frac{-iE_{a'}t}{\hbar}\right). \tag{2.38}$$

In other words, the expansion coefficient changes with time as

$$c_{a'}(t=0) \rightarrow c_{a'}(t) = c_{a'}(t=0)\exp\left(\frac{-iE_{a'}t}{\hbar}\right) \tag{2.39}$$

with its modulus unchanged. Notice that the relative phases among various components do vary with time because the oscillation frequencies are different.

A special case of interest is where the initial state happens to be one of $\{|a'\rangle\}$ itself. We have

$$|\alpha, t_0 = 0\rangle = |a'\rangle \tag{2.40}$$

initially, and at a later time

$$|a, t_0 = 0; t\rangle = |a'\rangle \exp\left(\frac{-iE_{a'}t}{\hbar}\right), \tag{2.41}$$

so if the system is initially a simultaneous eigenstate of $A$ and $H$, it remains so at all times. The most that can happen is the phase modulation, $\exp(-iE_{a'}t/\hbar)$. It is in this sense that an observable compatible with $H$ [see (2.34)] is a *constant of the motion*. We will encounter this connection once again in a different form when we discuss the Heisenberg equation of motion.

In the foregoing discussion the basic task in quantum dynamics is reduced to finding an observable that commutes with $H$ and evaluating its eigenvalues. Once that is done, we expand the initial ket in terms of the eigenkets of that observable and just apply the time-evolution operator. This last step merely amounts to changing the phase of each expansion coefficient, as indicated by (2.39).

Even though we worked out the case where there is just one observable $A$ that commutes with $H$, our considerations can easily be generalized when there are several mutually compatible observables all also commuting with $H$:

$$[A, B] = [B, C] = [A, C] = \cdots = 0,$$
$$[A, H] = [B, H] = [C, H] = \cdots = 0. \tag{2.42}$$

Using the collective index notation of Section 1.4 [see (1.130)], we have

$$\exp\left(\frac{-iHt}{\hbar}\right) = \sum_{K'} |K'\rangle \exp\left(\frac{-iE_{K'}t}{\hbar}\right) \langle K'|, \tag{2.43}$$

where $E_{K'}$ is uniquely specified once $a', b', c', \ldots$ are specified. It is therefore of fundamental importance to find *a complete set of mutually compatible observables that also commute with $H$*. Once such a set is found, we express the initial ket as a superposition of the simultaneous eigenkets of $A$, $B$, $C$, $\ldots$ and $H$. The final step is just to apply the time-evolution operator, written as (2.43). In this manner we can solve the most general initial-value problem with a time-independent $H$.

### 2.1.4 Time Dependence of Expectation Values

It is instructive to study how the expectation value of an observable changes as a function of time. Suppose that at $t = 0$ the initial state is one of the eigenstates of an observable $A$ that commutes with $H$, as in (2.40). We now look at the expectation value of some other observable $B$, which need not commute with $A$ nor with $H$. Because at a later time we have

$$|a', t_0 = 0; t\rangle = \mathscr{U}(t, 0)|a'\rangle \tag{2.44}$$

for the state ket, $\langle B \rangle$ is given by

$$\langle B \rangle = (\langle a'|\mathscr{U}^\dagger(t, 0)) \cdot B \cdot (\mathscr{U}(t, 0)|a'\rangle)$$
$$= \langle a'|\exp\left(\frac{iE_{a'}t}{\hbar}\right) B \exp\left(\frac{-iE_{a'}t}{\hbar}\right)|a'\rangle$$
$$= \langle a'|B|a'\rangle, \tag{2.45}$$

which is *independent of t*. So the expectation value of an observable taken with respect to an energy eigenstate does not change with time. For this reason an energy eigenstate is often referred to as a **stationary state**.

The situation is more interesting when the expectation value is taken with respect to a *superposition* of energy eigenstates, or a **nonstationary state**. Suppose that initially we have

$$|\alpha, t_0 = 0\rangle = \sum_{a'} c_{a'} |a'\rangle. \tag{2.46}$$

We easily compute the expectation value of $B$ to be

$$\langle B \rangle = \left[ \sum_{a'} c_{a'}^* \langle a'| \exp\left(\frac{iE_{a'}t}{\hbar}\right) \right] \cdot B \cdot \left[ \sum_{a''} c_{a''} \exp\left(\frac{-iE_{a''}t}{\hbar}\right) |a''\rangle \right]$$

$$= \sum_{a'} \sum_{a''} c_{a'}^* c_{a''} \langle a'|B|a''\rangle \exp\left[ \frac{-i(E_{a''} - E_{a'})t}{\hbar} \right]. \tag{2.47}$$

So this time the expectation value consists of oscillating terms whose angular frequencies are determined by N. Bohr's frequency condition

$$\omega_{a''a'} = \frac{(E_{a''} - E_{a'})}{\hbar}. \tag{2.48}$$

### 2.1.5 Spin Precession

It is appropriate to treat an example here. We consider an extremely simple system which, however, illustrates the basic formalism we have developed.

We start with a Hamiltonian of a spin $\frac{1}{2}$ system with magnetic moment $e\hbar/2m_e c$ subjected to an external magnetic field **B**:

$$H = -\left(\frac{e}{m_e c}\right) \mathbf{S} \cdot \mathbf{B} \tag{2.49}$$

($e < 0$ for the electron). Furthermore, we take **B** to be a static, uniform magnetic field in the $z$-direction. We can then write $H$ as

$$H = -\left(\frac{eB}{m_e c}\right) S_z. \tag{2.50}$$

Because $S_z$ and $H$ differ just by a multiplicative constant, they obviously commute. The $S_z$ eigenstates are also energy eigenstates, and the corresponding energy eigenvalues are

$$E_\pm = \mp \frac{e\hbar B}{2m_e c}, \quad \text{for } S_z \pm. \tag{2.51}$$

It is convenient to define $\omega$ in such a way that the difference in the two energy eigenvalues is $\hbar\omega$:

$$\omega \equiv \frac{|e|B}{m_e c}. \tag{2.52}$$

We can then rewrite the $H$ operator simply as

$$H = \omega S_z. \tag{2.53}$$

All the information on time development is contained in the time-evolution operator

$$\mathscr{U}(t,0) = \exp\left(\frac{-i\omega S_z t}{\hbar}\right). \tag{2.54}$$

We apply this to the initial state. The base kets we must use in expanding the initial ket are obviously the $S_z$ eigenkets, $|+\rangle$ and $|-\rangle$, which are also energy eigenkets. Suppose that at $t = 0$ the system is characterized by

$$|\alpha\rangle = c_+|+\rangle + c_-|-\rangle. \tag{2.55}$$

Upon applying (2.54), we see that the state ket at some later time is

$$|\alpha, t_0 = 0; t\rangle = c_+ \exp\left(\frac{-i\omega t}{2}\right)|+\rangle + c_- \exp\left(\frac{+i\omega t}{2}\right)|-\rangle, \tag{2.56}$$

where we have used

$$H|\pm\rangle = \left(\frac{\pm\hbar\omega}{2}\right)|\pm\rangle. \tag{2.57}$$

Specifically, let us suppose that the initial ket $|\alpha\rangle$ represents the spin-up (or, more precisely, $S_z+$) state $|+\rangle$, which means that

$$c_+ = 1, \quad c_- = 0. \tag{2.58}$$

At a later time, (2.56) tells us that it is still in the spin-up state, which is no surprise because this is a stationary state.

Next, let us suppose that initially the system is in the $S_x+$ state. Comparing (1.110a) with (2.55), we see that

$$c_+ = c_- = \frac{1}{\sqrt{2}}. \tag{2.59}$$

It is straightforward to work out the probabilities for the system to be found in the $S_x\pm$ state at some later time $t$:

$$|\langle S_x; \pm|\alpha, t_0 = 0; t\rangle|^2 = \left|\left[\left(\frac{1}{\sqrt{2}}\right)\langle+| \pm \left(\frac{1}{\sqrt{2}}\right)\langle-|\right] \cdot \left[\left(\frac{1}{\sqrt{2}}\right)\exp\left(\frac{-i\omega t}{2}\right)|+\rangle \right.\right.$$

$$\left.\left. + \left(\frac{1}{\sqrt{2}}\right)\exp\left(\frac{+i\omega t}{2}\right)|-\rangle\right]\right|^2$$

$$= \left|\frac{1}{2}\exp\left(\frac{-i\omega t}{2}\right) \pm \frac{1}{2}\exp\left(\frac{+i\omega t}{2}\right)\right|^2$$

$$= \cos^2\frac{\omega t}{2} \quad \text{for} \quad S_x+, \tag{2.60a}$$

$$= \sin^2\frac{\omega t}{2} \quad \text{for} \quad S_x-. \tag{2.60b}$$

Even though the spin is initially in the positive $x$-direction, the magnetic field in the $z$-direction causes it to rotate; as a result, we obtain a finite probability for finding

$S_x-$ at some later time. The sum of the two probabilities is seen to be unity at all times, in agreement with the unitarity property of the time-evolution operator.

Using (1.99), we can write the expectation value of $S_x$ as

$$\langle S_x \rangle = \left(\frac{\hbar}{2}\right) \cos^2\left(\frac{\omega t}{2}\right) + \left(\frac{-\hbar}{2}\right) \sin^2\left(\frac{\omega t}{2}\right)$$
$$= \left(\frac{\hbar}{2}\right) \cos \omega t, \tag{2.61}$$

so this quantity oscillates with an angular frequency corresponding to the difference of the two energy eigenvalues divided by $\hbar$, in agreement with our general formula (2.47). Similar exercises with $S_y$ and $S_z$ show that

$$\langle S_y \rangle = \left(\frac{\hbar}{2}\right) \sin \omega t \tag{2.62a}$$

and

$$\langle S_z \rangle = 0. \tag{2.62b}$$

Physically this means that the spin precesses in the $xy$-plane. We will comment further on spin precession when we discuss rotation operators in Chapter 3.

Experimentally, spin precession is well established. In fact, it is used as a tool for other investigations of fundamental quantum-mechanical phenomena. For example, the form of the Hamiltonian (2.49) can be derived for pointlike particles, such as electrons or muons, which obey the Dirac equation, for which the gyromagnetic ratio $g = 2$. (See Section 8.2.) However, higher-order corrections from quantum field theory predict a small but precisely calculable deviation from this, and it is a high priority to produce competitively precise measurements of $g - 2$.

Such an experiment has been recently completed. See Bennett et al., *Phys. Rev. D*, **73** (2006) 072003. Muons are injected into a "storage ring" designed so that their spins would precess in lock step with their momentum vector only if $g \equiv 2$. Consequently, observation of their precession measures $g - 2$ directly, facilitating a very precise result. Figure 2.1 shows the experimenters' observation of the muon spin rotation over more than one hundred periods. They determine a value for $g - 2$ to a precision smaller than one part per million, which agrees reasonably well with the theoretical value.

### 2.1.6 Neutrino Oscillations

A lovely example of quantum-mechanical dynamics leading to interference in a two-state system, based on current physics research, is provided by the phenomenon known as *neutrino oscillations*.[3]

Neutrinos are elementary particles with no charge, and very small mass, much smaller than that of an electron. They are known to occur in nature in three distinct "flavors,"

---

[3] The treatment here is the straightforward approach usually covered in the literature, but it has shortcomings. See, for example, Cohen et al., *Phys. Lett. B*, **678** (2009) 191 and Akhmedov (2018) arXiv:1901.05232 [hep-ph].

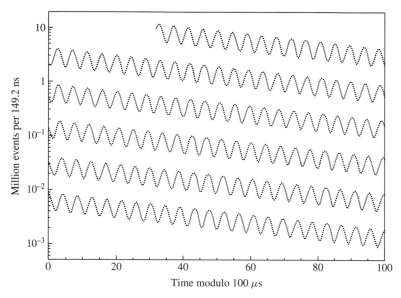

**Fig. 2.1**   Observations of the precession of muon spin by Bennett et al., *Phys. Rev. D*, **73** (2006) 072003. Data points are wrapped around every 100 $\mu$s. The size of the signal decreases with time because the muons decay.

although for this discussion it suffices to only consider two of them. These two flavors are identified by their interactions which may be either with electrons, in which case we write $\nu_e$, or with muons, that is $\nu_\mu$. These are in fact eigenstates of a Hamiltonian which controls those interactions.

On the other hand, it is possible (and, in fact, now known to be true) that neutrinos may have some other interactions, in which case their energy eigenvalues correspond to states that have a well-defined mass. These "mass eigenstates" would have eigenvalues $E_1$ and $E_2$, say, corresponding to masses $m_1$ and $m_2$, and might be denoted as $|\nu_1\rangle$ and $|\nu_2\rangle$. The "flavor eigenstates" are related to these through a simple unitary transformation, specified by some mixing angle $\theta$, as follows:

$$|\nu_e\rangle = \cos\theta|\nu_1\rangle - \sin\theta|\nu_2\rangle \tag{2.63a}$$

$$|\nu_\mu\rangle = \sin\theta|\nu_1\rangle + \cos\theta|\nu_2\rangle. \tag{2.63b}$$

If the mixing angle were zero, then $|\nu_e\rangle$ and $|\nu_\mu\rangle$ would respectively be the same as $|\nu_1\rangle$ and $|\nu_2\rangle$. However, we know of no reason why this should be the case. Indeed, there is no strong theoretical bias for any particular value of $\theta$, and it is a free parameter which, today, can only be determined through experiment.

Neutrino oscillation is the phenomenon by which we can measure the mixing angle. Suppose we prepare, at time $t = 0$, a momentum eigenstate of one flavor of neutrino, say $|\nu_e\rangle$. Then according to (2.63a) the two different mass eigenstate components will evolve with different frequencies, and therefore develop a relative phase difference. If the difference in the masses is small enough, then this phase difference can build up over a macroscopic distance. In fact, by measuring the interference as a function of difference, one can observe oscillations with a period that depends on the difference of masses, and an amplitude that depends on the mixing angle.

It is straightforward (see Problem 2.4 at the end of this chapter) to use (2.63) along with (2.28) and our quantum-mechanical postulates, and find a measurable quantity that exhibits neutrino oscillations. In this case, the Hamiltonian is just that for a free particle, but we need to take some care. Neutrinos are very low mass, so they are highly relativistic for any practical experimental conditions. Therefore, for a fixed momentum $p$, the energy eigenvalue for a neutrino of mass $m$ is given to an extremely good approximation as

$$E = \left[p^2c^2 + m^2c^4\right]^{1/2} \approx pc\left(1 + \frac{m^2c^2}{2p^2}\right). \qquad (2.64)$$

If we then allow our state $|\nu_e\rangle$ to evolve, and then at some later time $t$ ask what is the probability that it still appears as a $|\nu_e\rangle$ (as opposed to a $|\nu_\mu\rangle$), we find

$$P(\nu_e \to \nu_e) = 1 - \sin^2 2\theta \sin^2\left(\Delta m^2 c^4 \frac{L}{4E\hbar c}\right) \qquad (2.65)$$

where $\Delta m^2 \equiv m_1^2 - m_2^2$, $L = ct$ is the flight distance of the neutrino, and $E = pc$ is the nominal neutrino energy.

The oscillations predicted by (2.65) have been dramatically observed by the KamLAND experiment. See Figure 2.2. Neutrinos from a series of nuclear reactors are detected at a distance of $\sim$150km, and the rate is compared to that expected from reactor power and properties. The curve is not a perfect sine wave because the reactors are not all at the same distance from the detector.

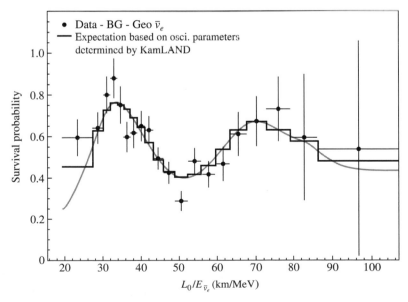

**Fig. 2.2**   Neutrino oscillations as observed by the KamLAND experiment, taken from Abe et al., *Phys. Rev. Lett.*, **100** (2008) 221803. The oscillations as a function of *L/E* demonstrate interference between different mass eigenstates of neutrinos.

## 2.1.7 Correlation Amplitude and the Energy-Time Uncertainty Relation

We conclude this section by asking how state kets at different times are correlated with each other. Suppose the initial state ket at $t = 0$ of a physical system is given by $|\alpha\rangle$. With time it changes into $|\alpha, t_0 = 0; t\rangle$, which we obtain by applying the time-evolution operator. We are concerned with the extent to which the state ket at a later time $t$ is similar to the state ket at $t = 0$; we therefore construct the inner product between the two state kets at different times:

$$C(t) \equiv \langle \alpha | \alpha, t_0 = 0; t \rangle$$
$$= \langle \alpha | \mathscr{U}(t, 0) | \alpha \rangle, \tag{2.66}$$

which is known as the **correlation amplitude**. The modulus of $C(t)$ provides a quantitative measure of the "resemblance" between the state kets at different times.

As an extreme example, consider the very special case where the initial ket $|\alpha\rangle$ is an eigenket of $H$; we then have

$$C(t) = \langle a' | a', t_0 = 0; t \rangle = \exp\left(\frac{-iE_{a'}t}{\hbar}\right), \tag{2.67}$$

so the modulus of the correlation amplitude is unity at all times, which is not surprising for a stationary state. In the more general situation where the initial ket is represented by a superposition of $\{|a'\rangle\}$, as in (2.37), we have

$$C(t) = \left(\sum_{a'} c_{a'}^* \langle a'|\right) \left[\sum_{a''} c_{a''} \exp\left(\frac{-iE_{a''}t}{\hbar}\right) |a''\rangle\right]$$
$$= \sum_{a'} |c_{a'}|^2 \exp\left(\frac{-iE_{a'}t}{\hbar}\right). \tag{2.68}$$

As we sum over many terms with oscillating time dependence of different frequencies, a strong cancellation is possible for moderately large values of $t$. We expect the correlation amplitude that starts with unity at $t = 0$ to decrease in magnitude with time.

To estimate (2.68) in a more concrete manner, let us suppose that the state ket can be regarded as a superposition of so many energy eigenkets with similar energies that we can regard them as exhibiting essentially a quasi-continuous spectrum. It is then legitimate to replace the sum by the integral

$$\sum_{a'} \rightarrow \int dE \rho(E), \quad c_{a'} \rightarrow g(E)\Big|_{E\simeq E_{a'}}, \tag{2.69}$$

where $\rho(E)$ characterizes the density of energy eigenstates. Expression (2.68) now becomes

$$C(t) = \int dE |g(E)|^2 \rho(E) \exp\left(\frac{-iEt}{\hbar}\right), \tag{2.70}$$

subject to the normalization condition

$$\int dE |g(E)|^2 \rho(E) = 1. \tag{2.71}$$

In a realistic physical situation $|g(E)|^2 \rho(E)$ may be peaked around $E = E_0$ with width $\Delta E$. Writing (2.70) as

$$C(t) = \exp\left(\frac{-iE_0 t}{\hbar}\right) \int dE |g(E)|^2 \rho(E) \exp\left[\frac{-i(E - E_0)t}{\hbar}\right], \qquad (2.72)$$

we see that as $t$ becomes large, the integrand oscillates very rapidly unless the energy interval $|E - E_0|$ is small compared with $\hbar/t$. If the interval for which $|E - E_0| \simeq \hbar/t$ holds is much narrower than $\Delta E$, the width of $|g(E)|^2 \rho(E)$, we get essentially no contribution to $C(t)$ because of strong cancellations. The characteristic time at which the modulus of the correlation amplitude starts becoming appreciably different from 1 is given by

$$t \simeq \frac{\hbar}{\Delta E}. \qquad (2.73)$$

Even though this equation is obtained for a superposition state with a quasi-continuous energy spectrum, it also makes sense for a two-level system; in the spin-precession problem considered earlier, the state ket, which is initially $|S_x+\rangle$, starts losing its identity after $\sim 1/\omega = \hbar/(E_+ - E_-)$, as is evident from (2.60).

To summarize, as a result of time evolution the state ket of a physical system ceases to retain its original form after a time interval of order $\hbar/\Delta E$. In the literature this point is often said to illustrate the *energy-time uncertainty relation*

$$\Delta t \Delta E \simeq \hbar. \qquad (2.74)$$

However, it is to be clearly understood that this energy-time uncertainty relation is of a very different nature from the uncertainty relation between two incompatible observables discussed in Section 1.4. In Chapter 5 we will come back to (2.74) in connection with time-dependent perturbation theory.

## 2.2  The Schrödinger Versus the Heisenberg Picture

### 2.2.1  Unitary Operators

In the previous section we introduced the concept of time development by considering the time-evolution operator that affects state kets; that approach to quantum dynamics is known as the **Schrödinger picture**. There is another formulation of quantum dynamics where observables, rather than state kets, vary with time; this second approach is known as the **Heisenberg picture**. Before discussing the differences between the two approaches in detail, we digress to make some general comments on unitary operators.

Unitary operators are used for many different purposes in quantum mechanics. In this book we introduced (Section 1.5) an operator satisfying the unitarity property. In that section we were concerned with the question of how the base kets in one representation are related to those in some other representations. The state kets themselves are assumed not to change as we switch to a different set of base kets even though the numerical values of the expansion coefficients for $|\alpha\rangle$ are, of course, different in different representations.

Subsequently we introduced two unitary operators that actually change the state kets, the translation operator of Section 1.6 and the time-evolution operator of Section 2.1. We have

$$|\alpha\rangle \rightarrow U|\alpha\rangle, \tag{2.75}$$

where $U$ may stand for $\mathcal{T}(d\mathbf{x})$ or $\mathcal{U}(t, t_0)$. Here $U|\alpha\rangle$ is the state ket corresponding to a physical system that actually has undergone translation or time evolution.

It is important to keep in mind that under a unitary transformation that changes the state kets, the inner product of a state bra and a state ket remains unchanged:

$$\langle \beta | \alpha \rangle \rightarrow \langle \beta | U^\dagger U | \alpha \rangle = \langle \beta | \alpha \rangle. \tag{2.76}$$

Using the fact that these transformations affect the state kets but not operators, we can infer how $\langle \beta | X | \alpha \rangle$ must change:

$$\langle \beta | X | \alpha \rangle \rightarrow ((\langle \beta | U^\dagger) \cdot X \cdot (U | \alpha \rangle)) = \langle \beta | U^\dagger X U | \alpha \rangle. \tag{2.77}$$

We now make a very simple mathematical observation that follows from the associative axiom of multiplication.

$$((\langle \beta | U^\dagger) \cdot X \cdot (U | \alpha \rangle)) = \langle \beta | \cdot (U^\dagger X U) \cdot | \alpha \rangle. \tag{2.78}$$

Is there any physics in this observation? This mathematical identity suggests two approaches to unitary transformations.

*Approach* 1:

$$|\alpha\rangle \rightarrow U|\alpha\rangle, \text{ with operators unchanged.} \tag{2.79a}$$

*Approach* 2:

$$X \rightarrow U^\dagger X U, \text{ with state kets unchanged.} \tag{2.79b}$$

In classical physics we do not introduce state kets, yet we talk about translation, time evolution, and the like. This is possible because these operations actually change quantities such as $\mathbf{x}$ and $\mathbf{L}$, which are observables of classical mechanics. We therefore conjecture that a closer connection with classical physics may be established if we follow approach 2.

A simple example may be helpful here. We go back to the infinitesimal translation operator $\mathcal{T}(d\mathbf{x}')$. The formalism presented in Section 1.6 is based on approach 1; $\mathcal{T}(d\mathbf{x}')$ affects the state kets, not the position operator:

$$|\alpha\rangle \rightarrow \left( 1 - \frac{i\mathbf{p} \cdot d\mathbf{x}'}{\hbar} \right) |\alpha\rangle.$$
$$\mathbf{x} \rightarrow \mathbf{x}. \tag{2.80}$$

In contrast, if we follow approach 2, we obtain

$$|\alpha\rangle \rightarrow |\alpha\rangle,$$
$$\mathbf{x} \rightarrow \left( 1 + \frac{i\mathbf{p} \cdot d\mathbf{x}'}{\hbar} \right) \mathbf{x} \left( 1 - \frac{i\mathbf{p} \cdot d\mathbf{x}'}{\hbar} \right)$$
$$= \mathbf{x} + \left( \frac{i}{\hbar} \right) [\mathbf{p} \cdot d\mathbf{x}', \mathbf{x}]$$
$$= \mathbf{x} + d\mathbf{x}'. \tag{2.81}$$

We leave it as an exercise for the reader to show that both approaches lead to the same result for the expectation value of **x**:

$$\langle \mathbf{x} \rangle \rightarrow \langle \mathbf{x} \rangle + \langle d\mathbf{x}' \rangle. \tag{2.82}$$

## 2.2.2 State Kets and Observables in the Schrödinger and the Heisenberg Pictures

We now return to the time-evolution operator $\mathscr{U}(t, t_0)$. In the previous section we examined how state kets evolve with time. This means that we were following approach 1, known as the **Schrödinger picture** when applied to time evolution. Alternatively we may follow approach 2, known as the **Heisenberg picture** when applied to time evolution.

In the Schrödinger picture the operators corresponding to observables like $x$, $p_y$, and $S_z$ are fixed in time, while state kets vary with time, as indicated in the previous section. In contrast, in the Heisenberg picture the operators corresponding to observables vary with time; the state kets are fixed, frozen so to speak, at what they were at $t_0$. It is convenient to set $t_0$ in $\mathscr{U}(t, t_0)$ to zero for simplicity and work with $\mathscr{U}(t)$, which is defined by

$$\mathscr{U}(t, t_0 = 0) \equiv \mathscr{U}(t) = \exp\left(\frac{-iHt}{\hbar}\right). \tag{2.83}$$

Motivated by (2.79b) of approach 2, we define the Heisenberg picture observable by

$$A^{(H)}(t) \equiv \mathscr{U}^{\dagger}(t) A^{(S)} \mathscr{U}(t), \tag{2.84}$$

where the superscripts $H$ and $S$ stand for Heisenberg and Schrödinger, respectively. At $t = 0$, the Heisenberg picture observable and the corresponding Schrödinger picture observable coincide:

$$A^{(H)}(0) = A^{(S)}. \tag{2.85}$$

The state kets also coincide between the two pictures at $t = 0$; at later $t$ the Heisenberg picture state ket is frozen to what it was at $t = 0$:

$$|\alpha, t_0 = 0; t\rangle_H = |\alpha, t_0 = 0\rangle, \tag{2.86}$$

*independent of t.* This is in dramatic contrast with the Schrödinger picture state ket,

$$|\alpha, t_0 = 0; t\rangle_S = \mathscr{U}(t)|\alpha, t_0 = 0\rangle. \tag{2.87}$$

The expectation value $\langle A \rangle$ is obviously the same in both pictures:

$$\begin{aligned}
{}_S\langle \alpha, t_0 = 0; t | A^{(S)} | \alpha, t_0 = 0; t \rangle_S &= \langle \alpha, t_0 = 0 | \mathscr{U}^{\dagger} A^{(S)} \mathscr{U} | \alpha, t_0 = 0 \rangle \\
&= {}_H\langle \alpha, t_0 = 0; t | A^{(H)}(t) | \alpha, t_0 = 0; t \rangle_H.
\end{aligned} \tag{2.88}$$

## 2.2.3  The Heisenberg Equation of Motion

We now derive the fundamental equation of motion in the Heisenberg picture. Assuming that $A^{(S)}$ does not depend explicitly on time, which is the case in most physical situations of interest, we obtain [by differentiating (2.84)]

$$
\begin{aligned}
\frac{dA^{(H)}}{dt} &= \frac{\partial \mathscr{U}^{\dagger}}{\partial t} A^{(S)} \mathscr{U} + \mathscr{U}^{\dagger} A^{(S)} \frac{\partial \mathscr{U}}{\partial t} \\
&= -\frac{1}{i\hbar} \mathscr{U}^{\dagger} H \mathscr{U} \mathscr{U}^{\dagger} A^{(S)} \mathscr{U} + \frac{1}{i\hbar} \mathscr{U}^{\dagger} A^{(S)} \mathscr{U} \mathscr{U}^{\dagger} H \mathscr{U} \\
&= \frac{1}{i\hbar} [A^{(H)}, \mathscr{U}^{\dagger} H \mathscr{U}],
\end{aligned}
\tag{2.89}
$$

where we have used [see (2.25)]

$$
\frac{\partial \mathscr{U}}{\partial t} = \frac{1}{i\hbar} H \mathscr{U},
\tag{2.90a}
$$

$$
\frac{\partial \mathscr{U}^{\dagger}}{\partial t} = -\frac{1}{i\hbar} \mathscr{U}^{\dagger} H.
\tag{2.90b}
$$

Because $H$ was originally introduced in the Schrödinger picture, we may be tempted to define

$$
H^{(H)} = \mathscr{U}^{\dagger} H \mathscr{U}
\tag{2.91}
$$

in accordance with (2.84). But in elementary applications where $\mathscr{U}$ is given by (2.83), $\mathscr{U}$ and $H$ obviously commute; as a result,

$$
\mathscr{U}^{\dagger} H \mathscr{U} = H,
\tag{2.92}
$$

so it is all right to write (2.89) as

$$
\frac{dA^{(H)}}{dt} = \frac{1}{i\hbar} \left[ A^{(H)}, H \right].
\tag{2.93}
$$

This equation is known as the **Heisenberg equation of motion**. Notice that we have derived it using the properties of the time-evolution operator and the defining equation for $A^{(H)}$.

It is instructive to compare (2.93) with the classical equation of motion in Poisson bracket form. In classical physics, for a function $A$ of $q$ and $p$ that does not involve time explicitly, we have (Goldstein et al. (2002), pp. 396–397)

$$
\frac{dA}{dt} = [A, H]_{\text{classical}}.
\tag{2.94}
$$

Again, we see that Dirac's quantization rule (1.6.47) leads to the correct equation in quantum mechanics. Indeed, historically (2.93) was first written by P. A. M. Dirac, who, with his characteristic modesty, called it the Heisenberg equation of motion. It is worth noting, however, that (2.93) makes sense whether or not $A^{(H)}$ has a classical analogue. For example, the spin operator in the Heisenberg picture satisfies

$$\frac{dS_i^{(H)}}{dt} = \frac{1}{i\hbar} \left[ S_i^{(H)}, H \right], \tag{2.95}$$

which can be used to discuss spin precession, but this equation has no classical counterpart because $S_z$ *cannot* be written as a function of $q$ and $p$. Rather than insisting on Dirac's rule, (1.229), we may argue that for quantities possessing classical counterparts, the correct classical equation can be obtained from the corresponding quantum-mechanical equation via the ansatz,

$$\frac{[\,,\,]}{i\hbar} \to [\,,\,]_{\text{classical}}. \tag{2.96}$$

Classical mechanics can be derived from quantum mechanics, but the opposite is not true.[4]

### 2.2.4  Free Particles: Ehrenfest's Theorem

Whether we work in the Schrödinger picture or in the Heisenberg picture, to be able to use the equations of motion we must first learn how to construct the appropriate Hamiltonian operator. For a physical system with classical analogues, we assume the Hamiltonian to be of the same form as in classical physics; we merely replace the classical $x_i$ and $p_i$ by the corresponding operators in quantum mechanics. With this assumption we can reproduce the correct classical equations in the classical limit. Whenever an ambiguity arises because of noncommuting observables, we attempt to resolve it by requiring $H$ to be Hermitian; for instance, we write the quantum-mechanical analogue of the classical product $xp$ as $\frac{1}{2}(xp + px)$. When the physical system in question has no classical analogues, we can only guess the structure of the Hamiltonian operator. We try various forms until we get the Hamiltonian that leads to results agreeing with empirical observation.

In practical applications it is often necessary to evaluate the commutator of $x_i$ (or $p_i$) with functions of $x_j$ and $p_j$. To this end the following formulas are found to be useful:

$$[x_i, F(\mathbf{p})] = i\hbar \frac{\partial F}{\partial p_i} \tag{2.97a}$$

and

$$[p_i, G(\mathbf{x})] = -i\hbar \frac{\partial G}{\partial x_i}, \tag{2.97b}$$

where $F$ and $G$ are functions that can be expanded in powers of $p_j$ and $x_j$, respectively. We can easily prove both formulas by repeatedly applying (1.232e).

We are now in a position to apply the Heisenberg equation of motion to a free particle of mass $m$. The Hamiltonian is taken to be of the same form as in classical mechanics:

$$H = \frac{\mathbf{p}^2}{2m} = \frac{\left( p_x^2 + p_y^2 + p_z^2 \right)}{2m}. \tag{2.98}$$

---

[4] In this book we follow the order: the Schrödinger picture → the Heisenberg picture → classical. For an enlightening treatment of the same subject in opposite order, classical → the Heisenberg picture → the Schrödinger picture, see Finkelstein (1973), pp. 68–70 and 109.

We look at the observables $p_i$ and $x_i$, which are understood to be the momentum and the position operator in the Heisenberg picture even though we omit the superscript $(H)$. Because $p_i$ commutes with any function of $p_j$, we have

$$\frac{dp_i}{dt} = \frac{1}{i\hbar}[p_i, H] = 0. \tag{2.99}$$

Thus for a free particle, the momentum operator is a constant of the motion, which means that $p_i(t)$ is the same as $p_i(0)$ at all times. Quite generally, it is evident from the Heisenberg equation of motion (2.93) that whenever $A^{(H)}$ commutes with the Hamiltonian, $A^{(H)}$ is a constant of the motion. Next,

$$\frac{dx_i}{dt} = \frac{1}{i\hbar}[x_i, H] = \frac{1}{i\hbar}\frac{1}{2m}i\hbar\frac{\partial}{\partial p_i}\left(\sum_{j=1}^{3} p_j^2\right)$$

$$= \frac{p_i}{m} = \frac{p_i(0)}{m}, \tag{2.100}$$

where we have taken advantage of (2.97a), so we have the solution

$$x_i(t) = x_i(0) + \left(\frac{p_i(0)}{m}\right)t, \tag{2.101}$$

which is reminiscent of the classical trajectory equation for a uniform rectilinear motion. It is important to note that even though we have

$$[x_i(0), x_j(0)] = 0 \tag{2.102}$$

at equal times, the commutator of the $x_i$ at *different* times does *not* vanish; specifically,

$$[x_i(t), x_i(0)] = \left[\frac{p_i(0)t}{m}, x_i(0)\right] = \frac{-i\hbar t}{m}. \tag{2.103}$$

Applying the uncertainty relation (1.146) to this commutator, we obtain

$$\langle(\Delta x_i)^2\rangle_t\langle(\Delta x_i)^2\rangle_{t=0} \geq \frac{\hbar^2 t^2}{4m^2}. \tag{2.104}$$

Among other things, this relation implies that even if the particle is well localized at $t = 0$, its position becomes more and more uncertain with time, a conclusion which can also be obtained by studying the time-evolution behavior of free-particle wave packets in wave mechanics.

We now add a potential $V(\mathbf{x})$ to our earlier free-particle Hamiltonian:

$$H = \frac{\mathbf{p}^2}{2m} + V(\mathbf{x}). \tag{2.105}$$

Here $V(\mathbf{x})$ is to be understood as a function of the $x$-, $y$-, and $z$-*operators*. Using (2.97b) this time, we obtain

$$\frac{dp_i}{dt} = \frac{1}{i\hbar}[p_i, V(\mathbf{x})] = -\frac{\partial}{\partial x_i}V(\mathbf{x}). \tag{2.106}$$

On the other hand, we see that

$$\frac{dx_i}{dt} = \frac{p_i}{m} \tag{2.107}$$

still holds because $x_i$ commutes with the newly added term $V(\mathbf{x})$. We can use the Heisenberg equation of motion once again to deduce

$$\frac{d^2 x_i}{dt^2} = \frac{1}{i\hbar}\left[\frac{dx_i}{dt}, H\right] = \frac{1}{i\hbar}\left[\frac{p_i}{m}, H\right]$$
$$= \frac{1}{m}\frac{dp_i}{dt}. \tag{2.108}$$

Combining this with (2.32), we finally obtain in vectorial form

$$m\frac{d^2\mathbf{x}}{dt^2} = -\nabla V(\mathbf{x}). \tag{2.109}$$

This is the quantum-mechanical analogue of Newton's second law. By taking the expectation values of both sides with respect to a Heisenberg state ket that does *not* move with time, we obtain

$$m\frac{d^2}{dt^2}\langle\mathbf{x}\rangle = \frac{d\langle\mathbf{p}\rangle}{dt} = -\langle\nabla V(\mathbf{x})\rangle. \tag{2.110}$$

This is known as the **Ehrenfest theorem** after P. Ehrenfest, who derived it in 1927 using the formalism of wave mechanics. When written in this expectation form, its validity is independent of whether we are using the Heisenberg or the Schrödinger picture; after all, the expectation values are the same in the two pictures. In contrast, the operator form (2.109) is meaningful only if we understand $\mathbf{x}$ and $\mathbf{p}$ to be Heisenberg picture operators.

We note that in (2.110) the $\hbar$ have completely disappeared. It is therefore not surprising that the center of a wave packet moves like a *classical* particle subjected to $V(\mathbf{x})$.

## 2.2.5  Base Kets and Transition Amplitudes

So far we have avoided asking how the base kets evolve in time. A common misconception is that as time goes on, all kets move in the Schrödinger picture and are stationary in the Heisenberg picture. This is *not* the case, as we will make clear shortly. The important point is to distinguish the behavior of state kets from that of base kets.

We started our discussion of ket spaces in Section 1.2 by remarking that the eigenkets of observables are to be used as base kets. What happens to the defining eigenvalue equation

$$A|a'\rangle = a'|a'\rangle \tag{2.111}$$

with time? In the Schrödinger picture, $A$ does not change, so the base kets, obtained as the solutions to this eigenvalue equation at $t = 0$, for instance, must remain unchanged. Unlike state kets, the base kets do *not* change in the Schrödinger picture.

The whole situation is very different in the Heisenberg picture, where the eigenvalue equation we must study is for the time-dependent operator

$$A^{(H)}(t) = \mathscr{U}^\dagger A(0)\mathscr{U}. \tag{2.112}$$

From (2.111) evaluated at $t = 0$, when the two pictures coincide, we deduce

$$\mathscr{U}^\dagger A(0)\mathscr{U}\mathscr{U}^\dagger|a'\rangle = a'\mathscr{U}^\dagger|a'\rangle, \tag{2.113}$$

which implies an eigenvalue equation for $A^{(H)}$:

$$A^{(H)}(\mathscr{U}^\dagger|a'\rangle) = a'(\mathscr{U}^\dagger|a'\rangle). \tag{2.114}$$

If we continue to maintain the view that the eigenkets of observables form the base kets, then $\{\mathscr{U}^\dagger|a'\rangle\}$ must be used as the base kets in the Heisenberg picture. As time goes on, the Heisenberg picture base kets, denoted by $|a',t\rangle_H$, move as follows:

$$|a',t\rangle_H = \mathscr{U}^\dagger|a'\rangle. \tag{2.115}$$

Because of the appearance of $\mathscr{U}^\dagger$ rather than $\mathscr{U}$ in (2.115), the Heisenberg picture base kets are seen to rotate oppositely when compared with the Schrödinger picture state kets; specifically, $|a',t\rangle_H$ satisfies the "wrong-sign Schrödinger equation"

$$i\hbar\frac{\partial}{\partial t}|a',t\rangle_H = -H|a',t\rangle_H. \tag{2.116}$$

As for the eigenvalues themselves, we see from (2.114) that they are unchanged with time. This is consistent with the theorem on unitary equivalent observables discussed in Section 1.5. Notice also the following expansion for $A^{(H)}(t)$ in terms of the base kets and bras of the Heisenberg picture:

$$\begin{aligned}
A^{(H)}(t) &= \sum_{a'}|a',t\rangle_H a'\,{}_H\langle a',t| \\
&= \sum_{a'}\mathscr{U}^\dagger|a'\rangle a'\langle a'|\mathscr{U} \\
&= \mathscr{U}^\dagger A^{(S)}\mathscr{U},
\end{aligned} \tag{2.117}$$

which shows that everything is quite consistent provided that the Heisenberg base kets change as in (2.115).

We see that the expansion coefficients of a state ket in terms of base kets are the same in both pictures:

$$c_{a'}(t) = \underbrace{\langle a'|}_{\text{base bra}} \cdot \underbrace{(\mathscr{U}|\alpha,t_0=0\rangle)}_{\text{state ket}} \quad \text{(the Schrödinger picture)} \tag{2.118a}$$

$$c_{a'}(t) = \underbrace{(\langle a'|\mathscr{U})}_{\text{base bra}} \cdot \underbrace{|\alpha,t_0=0\rangle}_{\text{state ket}} \quad \text{(the Heisenberg picture).} \tag{2.118b}$$

Pictorially, we may say that the cosine of the angle between the state ket and the base ket is the same whether we rotate the state ket counterclockwise or the base ket clockwise. These considerations apply equally well to base kets that exhibit a continuous spectrum; in particular, the wave function $\langle \mathbf{x}'|\alpha\rangle$ can be regarded either as (1) the inner product of the stationary position eigenbra with the moving state ket (the Schrödinger picture) or as (2) the inner product of the moving position eigenbra with the stationary state ket (the Heisenberg picture). We will discuss the time dependence of the wave function in Section 2.4, where we will derive the celebrated wave equation of Schrödinger.

To illustrate further the equivalence between the two pictures, we study transition amplitudes, which will play a fundamental role in Section 2.6. Suppose there is a physical system prepared at $t = 0$ to be in an eigenstate of observable $A$ with eigenvalue $a'$. At some later time $t$ we may ask: What is the probability amplitude, known as the **transition**

| **Table 2.1** The Schrödinger Picture Versus the Heisenberg Picture | | |
|---|---|---|
|  | Schrödinger picture | Heisenberg picture |
| State ket | Moving: (2.5), (2.27) | Stationary |
| Observable | Stationary | Moving: (2.84), (2.93) |
| Base ket | Stationary | Moving oppositely: (2.115), (2.116) |

**amplitude**, for the system to be found in an eigenstate of observable $B$ with eigenvalue $b'$? Here $A$ and $B$ can be the same or different. In the Schrödinger picture the state ket at $t$ is given by $\mathscr{U}|a'\rangle$, while the base kets $|a'\rangle$ and $|b'\rangle$ do not vary with time; so we have

$$\underbrace{\langle b'|}_{\text{base bra}} \cdot \underbrace{(\mathscr{U}|a'\rangle)}_{\text{state ket}} \tag{2.119}$$

for this transition amplitude. In contrast, in the Heisenberg picture the state ket is stationary, that is, it remains as $|a'\rangle$ at all times, but the base kets evolve oppositely. So the transition amplitude is

$$\underbrace{(\langle b'|\mathscr{U})}_{\text{base bra}} \cdot \underbrace{|a'\rangle}_{\text{state ket}} . \tag{2.120}$$

Obviously (2.119) and (2.120) are the same. They can both be written as

$$\langle b'|\mathscr{U}(t,0)|a'\rangle. \tag{2.121}$$

In some loose sense this is the transition amplitude for "going" from state $|a'\rangle$ to state $|b'\rangle$.

To conclude this section let us summarize the differences between the Schrödinger picture and the Heisenberg picture; see Table 2.1.

## 2.3 Simple Harmonic Oscillator

The simple harmonic oscillator is one of the most important problems in quantum mechanics. It not only illustrates many of the basic concepts and methods of quantum mechanics, it also has much practical value. Essentially any potential well can be approximated by a simple harmonic oscillator, so it describes phenomena from molecular vibrations to nuclear structure. Furthermore, since the Hamiltonian is basically the sum of squares of two canonically conjugate variables, it is also an important starting point for much of quantum field theory.

### 2.3.1 Energy Eigenkets and Energy Eigenvalues

We begin our discussion with Dirac's elegant operator method, which is based on the earlier work of M. Born and N. Wiener, to obtain the energy eigenkets and energy eigenvalues of the simple harmonic oscillator. The basic Hamiltonian is

$$H = \frac{p^2}{2m} + \frac{m\omega^2 x^2}{2}, \tag{2.122}$$

where $\omega$ is the angular frequency of the classical oscillator related to the spring constant $k$ in Hooke's law via $\omega = \sqrt{k/m}$. The operators $x$ and $p$ are, of course, Hermitian. It is convenient to define two non-Hermitian operators,

$$a = \sqrt{\frac{m\omega}{2\hbar}} \left( x + \frac{ip}{m\omega} \right), \quad a^\dagger = \sqrt{\frac{m\omega}{2\hbar}} \left( x - \frac{ip}{m\omega} \right), \tag{2.123}$$

known as the **annihilation operator** and the **creation operator**, respectively, for reasons that will become evident shortly. Using the canonical commutation relations, we readily obtain

$$[a, a^\dagger] = \left( \frac{1}{2\hbar} \right) (-i[x,p] + i[p,x]) = 1. \tag{2.124}$$

We also define the number operator

$$N = a^\dagger a, \tag{2.125}$$

which is obviously Hermitian. It is straightforward to show that

$$a^\dagger a = \left( \frac{m\omega}{2\hbar} \right) \left( x^2 + \frac{p^2}{m^2\omega^2} \right) + \left( \frac{i}{2\hbar} \right) [x,p]$$

$$= \frac{H}{\hbar\omega} - \frac{1}{2}, \tag{2.126}$$

so we have an important relation between the number operator and the Hamiltonian operator:

$$H = \hbar\omega \left( N + \tfrac{1}{2} \right). \tag{2.127}$$

Because $H$ is just a linear function of $N$, $N$ can be diagonalized simultaneously with $H$. We denote an energy eigenket of $N$ by its eigenvalue $n$, so

$$N|n\rangle = n|n\rangle. \tag{2.128}$$

We will later show that $n$ must be a nonnegative integer. Because of (2.127) we also have

$$H|n\rangle = (n + \tfrac{1}{2})\hbar\omega|n\rangle, \tag{2.129}$$

which means that the energy eigenvalues are given by

$$E_n = (n + \tfrac{1}{2})\hbar\omega. \tag{2.130}$$

To appreciate the physical significance of $a$, $a^\dagger$, and $N$, let us first note that

$$[N, a] = [a^\dagger a, a] = a^\dagger [a, a] + [a^\dagger, a] a = -a, \tag{2.131}$$

where we have used (2.124). Likewise, we can derive

$$[N, a^\dagger] = a^\dagger. \tag{2.132}$$

As a result, we have

$$Na^\dagger|n\rangle = ([N, a^\dagger] + a^\dagger N)|n\rangle = (n+1)a^\dagger|n\rangle \tag{2.133a}$$

and

$$Na|n\rangle = ([N, a] + aN)|n\rangle = (n-1)a|n\rangle. \tag{2.133b}$$

These relations imply that $a^\dagger|n\rangle\,(a|n\rangle)$ is also an eigenket of $N$ with eigenvalue increased (decreased) by one. Because the increase (decrease) of $n$ by one amounts to the creation (annihilation) of one quantum unit of energy $\hbar\omega$, the term *creation operator (annihilation operator)* for $a^\dagger$ ($a$) is deemed appropriate.

Equation (2.133b) implies that $a|n\rangle$ and $|n-1\rangle$ are the same up to a multiplicative constant. We write

$$a|n\rangle = c|n-1\rangle, \tag{2.134}$$

where $c$ is a numerical constant to be determined from the requirement that both $|n\rangle$ and $|n-1\rangle$ be normalized. First, note that

$$\langle n|a^\dagger a|n\rangle = |c|^2. \tag{2.135}$$

We can evaluate the left-hand side of (2.135) by noting that $a^\dagger a$ is just the number operator, so

$$n = |c|^2. \tag{2.136}$$

Taking $c$ to be real and positive by convention, we finally obtain

$$a|n\rangle = \sqrt{n}|n-1\rangle. \tag{2.137}$$

Similarly, it is easy to show that

$$a^\dagger|n\rangle = \sqrt{n+1}|n+1\rangle. \tag{2.138}$$

Suppose that we keep on applying the annihilation operator $a$ to both sides of (2.137):

$$a^2|n\rangle = \sqrt{n(n-1)}|n-2\rangle,$$
$$a^3|n\rangle = \sqrt{n(n-1)(n-2)}|n-3\rangle, \tag{2.139}$$
$$\vdots$$

We can obtain numerical operator eigenkets with smaller and smaller $n$ until the sequence terminates, which is bound to happen whenever we start with a positive integer $n$. One may argue that if we start with a noninteger $n$, the sequence will not terminate, leading to eigenkets with a negative value of $n$. But we also have the positivity requirement for the norm of $a|n\rangle$:

$$n = \langle n|N|n\rangle = ((\langle n|a^\dagger) \cdot (a|n\rangle)) \geq 0, \tag{2.140}$$

which implies that $n$ can never be negative! So we conclude that the sequence must terminate with $n = 0$ and that the allowed values of $n$ are nonnegative integers.

Because the smallest possible value of $n$ is zero, the ground state of the harmonic oscillator has

$$E_0 = \frac{1}{2}\hbar\omega. \tag{2.141}$$

We can now successively apply the creation operator $a^\dagger$ to the ground state $|0\rangle$. Using (2.138), we obtain

$$|1\rangle = a^\dagger|0\rangle,$$

$$|2\rangle = \left(\frac{a^\dagger}{\sqrt{2}}\right)|1\rangle = \left[\frac{(a^\dagger)^2}{\sqrt{2}}\right]|0\rangle,$$

$$|3\rangle = \left(\frac{a^\dagger}{\sqrt{3}}\right)|2\rangle = \left[\frac{(a^\dagger)^3}{\sqrt{3!}}\right]|0\rangle, \tag{2.142}$$

$$\vdots$$

$$|n\rangle = \left[\frac{(a^\dagger)^n}{\sqrt{n!}}\right]|0\rangle.$$

In this way we have succeeded in constructing simultaneous eigenkets of $N$ and $H$ with energy eigenvalues

$$E_n = \left(n + \tfrac{1}{2}\right)\hbar\omega \quad (n = 0, 1, 2, 3, \ldots). \tag{2.143}$$

From (2.137), (2.138), and the orthonormality requirement for $\{|n\rangle\}$, we obtain the matrix elements

$$\langle n'|a|n\rangle = \sqrt{n}\,\delta_{n',n-1}, \quad \langle n'|a^\dagger|n\rangle = \sqrt{n+1}\,\delta_{n',n+1}. \tag{2.144}$$

Using these together with

$$x = \sqrt{\frac{\hbar}{2m\omega}}(a + a^\dagger), \quad p = i\sqrt{\frac{m\hbar\omega}{2}}(-a + a^\dagger), \tag{2.145}$$

we derive the matrix elements of the $x$ and $p$ operators:

$$\langle n'|x|n\rangle = \sqrt{\frac{\hbar}{2m\omega}}(\sqrt{n}\,\delta_{n',n-1} + \sqrt{n+1}\,\delta_{n',n+1}), \tag{2.146a}$$

$$\langle n'|p|n\rangle = i\sqrt{\frac{m\hbar\omega}{2}}(-\sqrt{n}\,\delta_{n',n-1} + \sqrt{n+1}\,\delta_{n',n+1}). \tag{2.146b}$$

Notice that neither $x$ nor $p$ is diagonal in the $N$-representation we are using. This is not surprising because $x$ and $p$, like $a$ and $a^\dagger$, do not commute with $N$.

The operator method can also be used to obtain the energy eigenfunctions in position space. Let us start with the ground state defined by

$$a|0\rangle = 0, \tag{2.147}$$

which, in the $x$-representation, reads

$$\langle x'|a|0\rangle = \sqrt{\frac{m\omega}{2\hbar}}\langle x'|\left(x + \frac{ip}{m\omega}\right)|0\rangle = 0. \tag{2.148}$$

Recalling (1.249), we can regard this as a differential equation for the ground-state wave function $\langle x'|0\rangle$:

$$\left(x' + x_0^2 \frac{d}{dx'}\right)\langle x'|0\rangle = 0, \tag{2.149}$$

where we have introduced

$$x_0 \equiv \sqrt{\frac{\hbar}{m\omega}}, \tag{2.150}$$

which sets the length scale of the oscillator. We see that the normalized solution to (2.149) is

$$\langle x'|0\rangle = \left(\frac{1}{\pi^{1/4}\sqrt{x_0}}\right)\exp\left[-\frac{1}{2}\left(\frac{x'}{x_0}\right)^2\right]. \tag{2.151}$$

We can also obtain the energy eigenfunctions for excited states by evaluating

$$\langle x'|1\rangle = \langle x'|a^\dagger|0\rangle = \left(\frac{1}{\sqrt{2}x_0}\right)\left(x' - x_0^2\frac{d}{dx'}\right)\langle x'|0\rangle,$$

$$\langle x'|2\rangle = \left(\frac{1}{\sqrt{2}}\right)\langle x'|(a^\dagger)^2|0\rangle = \left(\frac{1}{\sqrt{2!}}\right)\left(\frac{1}{\sqrt{2}x_0}\right)^2\left(x' - x_0^2\frac{d}{dx'}\right)^2\langle x'|0\rangle,\ldots \tag{2.152}$$

In general, we obtain

$$\langle x'|n\rangle = \left(\frac{1}{\pi^{1/4}\sqrt{2^n n!}}\right)\left(\frac{1}{x_0^{n+1/2}}\right)\left(x' - x_0^2\frac{d}{dx'}\right)^n\exp\left[-\frac{1}{2}\left(\frac{x'}{x_0}\right)^2\right]. \tag{2.153}$$

It is instructive to look at the expectation values of $x^2$ and $p^2$ for the ground state. First, note that

$$x^2 = \left(\frac{\hbar}{2m\omega}\right)(a^2 + a^{\dagger 2} + a^\dagger a + aa^\dagger). \tag{2.154}$$

When we take the expectation value of $x^2$, only the last term in (2.154) yields a nonvanishing contribution:

$$\langle x^2\rangle = \frac{\hbar}{2m\omega} = \frac{x_0^2}{2}. \tag{2.155}$$

Likewise,

$$\langle p^2\rangle = \frac{\hbar m\omega}{2}. \tag{2.156}$$

It follows that the expectation values of the kinetic and the potential energies are, respectively,

$$\left\langle\frac{p^2}{2m}\right\rangle = \frac{\hbar\omega}{4} = \frac{\langle H\rangle}{2} \quad\text{and}\quad \left\langle\frac{m\omega^2 x^2}{2}\right\rangle = \frac{\hbar\omega}{4} = \frac{\langle H\rangle}{2}, \tag{2.157}$$

as expected from the virial theorem. From (2.146a) and (2.146b), it follows that

$$\langle x\rangle = \langle p\rangle = 0, \tag{2.158}$$

which also holds for the excited states. We therefore have

$$\langle(\Delta x)^2\rangle = \langle x^2\rangle = \frac{\hbar}{2m\omega} \quad \text{and} \quad \langle(\Delta p)^2\rangle = \langle p^2\rangle = \frac{\hbar m\omega}{2}, \tag{2.159}$$

and we see that the uncertainty relation is satisfied in the minimum uncertainty product form:

$$\langle(\Delta x)^2\rangle\langle(\Delta p)^2\rangle = \frac{\hbar^2}{4}. \tag{2.160}$$

This is not surprising because the ground-state wave function has a Gaussian shape. In contrast, the uncertainty products for the excited states are larger:

$$\langle(\Delta x)^2\rangle\langle(\Delta p)^2\rangle = \left(n+\frac{1}{2}\right)^2\hbar^2, \tag{2.161}$$

as the reader may easily verify.

## 2.3.2 Time Development of the Oscillator

So far we have not discussed the time evolution of oscillator state kets nor of observables like $x$ and $p$. Everything we have done is supposed to hold at some instant of time, say at $t = 0$; the operators $x$, $p$, $a$, and $a^\dagger$ are to be regarded either as Schrödinger picture operators (at all $t$) or as Heisenberg picture operators at $t = 0$. In the remaining part of this section, we work exclusively in the Heisenberg picture, which means that $x$, $p$, $a$, and $a^\dagger$ are all time dependent even though we do not explicitly write $x^{(H)}(t)$, and so forth.

The Heisenberg equations of motion for $p$ and $x$ are, from (2.106) and (2.107),

$$\frac{dp}{dt} = -m\omega^2 x \tag{2.162a}$$

and

$$\frac{dx}{dt} = \frac{p}{m}. \tag{2.162b}$$

This pair of coupled differential equations is equivalent to two uncoupled differential equations for $a$ and $a^\dagger$, namely,

$$\frac{da}{dt} = \sqrt{\frac{m\omega}{2\hbar}}\left(\frac{p}{m} - i\omega x\right) = -i\,\omega\,a \tag{2.163a}$$

and

$$\frac{da^\dagger}{dt} = i\,\omega\,a^\dagger, \tag{2.163b}$$

whose solutions are

$$a(t) = a(0)\exp(-i\omega t) \quad \text{and} \quad a^\dagger(t) = a^\dagger(0)\exp(i\omega t). \tag{2.164}$$

Incidentally, these relations explicitly show that $N$ and $H$ are *time-independent* operators even in the Heisenberg picture, as they must be. In terms of $x$ and $p$, we can rewrite (2.164) as

$$x(t) + \frac{ip(t)}{m\omega} = x(0)\exp(-i\omega t) + i\left[\frac{p(0)}{m\omega}\right]\exp(-i\omega t),$$

$$x(t) - \frac{ip(t)}{m\omega} = x(0)\exp(i\omega t) - i\left[\frac{p(0)}{m\omega}\right]\exp(i\omega t). \tag{2.165}$$

Equating the Hermitian and anti-Hermitian parts of both sides separately, we deduce

$$x(t) = x(0)\cos\omega t + \left[\frac{p(0)}{m\omega}\right]\sin\omega t \tag{2.166a}$$

and

$$p(t) = -m\omega x(0)\sin\omega t + p(0)\cos\omega t. \tag{2.166b}$$

These look the same as the classical equations of motion. We see that the $x$ and $p$ operators "oscillate" just like their classical analogues.

For pedagogical reasons we now present an alternative derivation of (2.166a). Instead of solving the Heisenberg equation of motion, we attempt to evaluate

$$x(t) = \exp\left(\frac{iHt}{\hbar}\right)x(0)\exp\left(\frac{-iHt}{\hbar}\right). \tag{2.167}$$

To this end we record a very useful formula:

$$\exp(iG\lambda)A\exp(-iG\lambda) = A + i\lambda\,[G,A] + \left(\frac{i^2\lambda^2}{2!}\right)[G,[G,A]] +$$

$$\cdots + \left(\frac{i^n\lambda^n}{n!}\right)[G,[G,[G,\ldots[G,A]]]\ldots] + \cdots, \tag{2.168}$$

where $G$ is a Hermitian operator and $\lambda$ is a real parameter. We leave the proof of this formula, known as the **Baker–Hausdorff lemma** as an exercise. Applying this formula to (2.167), we obtain

$$\exp\left(\frac{iHt}{\hbar}\right)x(0)\exp\left(\frac{-iHt}{\hbar}\right)$$

$$= x(0) + \left(\frac{it}{\hbar}\right)[H,x(0)] + \left(\frac{i^2t^2}{2!\hbar^2}\right)[H,[H,x(0)]] + \cdots. \tag{2.169}$$

Each term on the right-hand side can be reduced to either $x$ or $p$ by repeatedly using

$$[H,x(0)] = \frac{-i\hbar p(0)}{m} \tag{2.170a}$$

and

$$[H,p(0)] = i\hbar m\omega^2 x(0). \tag{2.170b}$$

Thus

$$\exp\left(\frac{iHt}{\hbar}\right)x(0)\exp\left(\frac{-iHt}{\hbar}\right) = x(0) + \left[\frac{p(0)}{m}\right]t - \left(\frac{1}{2!}\right)t^2\omega^2 x(0)$$

$$- \left(\frac{1}{3!}\right)\frac{t^3\omega^2 p(0)}{m} + \cdots$$

$$= x(0)\cos\omega t + \left[\frac{p(0)}{m\omega}\right]\sin\omega t, \tag{2.171}$$

in agreement with (2.166a).

From (2.166a) and (2.166b), one may be tempted to conclude that $\langle x \rangle$ and $\langle p \rangle$ always oscillate with angular frequency $\omega$. However, this inference is not correct. Take any energy eigenstate characterized by a definite value of $n$; the expectation value $\langle n|x(t)|n \rangle$ vanishes because the operators $x(0)$ and $p(0)$ change $n$ by $\pm 1$ and $|n\rangle$ and $|n \pm 1\rangle$ are orthogonal. This point is also obvious from our earlier conclusion (see Section 2.1) that the expectation value of an observable taken with respect to a stationary state does not vary with time. To observe oscillations reminiscent of the classical oscillator, we must look at a *superposition* of energy eigenstates such as

$$|\alpha\rangle = c_0|0\rangle + c_1|1\rangle. \tag{2.172}$$

The expectation value of $x(t)$ taken with respect to (2.172) does oscillate, as the reader may readily verify.

We have seen that an energy eigenstate does not behave like the classical oscillator – in the sense of oscillating expectation values for $x$ and $p$ – no matter how large $n$ may be. We may logically ask: How can we construct a superposition of energy eigenstates that most closely imitates the classical oscillator? In wave function language, we want a wave packet that bounces back and forth without spreading in shape. It turns out that a *coherent state* defined by the eigenvalue equation for the non-Hermitian annihilation operator $a$,

$$a|\lambda\rangle = \lambda|\lambda\rangle, \tag{2.173}$$

with, in general, a complex eigenvalue $\lambda$ does the desired job. The coherent state has many other remarkable properties.

1. When expressed as a superposition of energy (or $N$) eigenstates,

$$|\lambda\rangle = \sum_{n=0}^{\infty} f(n)|n\rangle, \tag{2.174}$$

the distribution of $|f(n)|^2$ with respect to $n$ is of the Poisson type about some mean value $\bar{n}$:

$$|f(n)|^2 = \left(\frac{\bar{n}^n}{n!}\right)\exp(-\bar{n}). \tag{2.175}$$

2. It can be obtained by translating the oscillator ground state by some finite distance.
3. It satisfies the minimum uncertainty product relation at all times.

A systematic study of coherent states, pioneered by R. Glauber, is very rewarding; the reader is urged to work out Exercise 2.21 on this subject at the end of this chapter.[5]

## 2.4 Schrödinger's Wave Equation

### 2.4.1 Time-Dependent Wave Equation

We now turn to the Schrödinger picture and examine the time evolution of $|\alpha, t_0; t\rangle$ in the $x$-representation. In other words, our task is to study the behavior of the wave function

$$\psi(\mathbf{x}', t) = \langle \mathbf{x}' | \alpha, t_0; t \rangle \tag{2.176}$$

as a function of time, where $|\alpha, t_0; t\rangle$ is a state ket in the Schrödinger picture at time $t$, and $\langle \mathbf{x}' |$ is a time-independent position eigenbra with eigenvalue $\mathbf{x}'$. The Hamiltonian operator is taken to be

$$H = \frac{\mathbf{p}^2}{2m} + V(\mathbf{x}). \tag{2.177}$$

The potential $V(\mathbf{x})$ is a Hermitian operator; it is also local in the sense that in the $\mathbf{x}$-representation we have

$$\langle \mathbf{x}'' | V(\mathbf{x}) | \mathbf{x}' \rangle = V(\mathbf{x}') \delta^3(\mathbf{x}' - \mathbf{x}''), \tag{2.178}$$

where $V(\mathbf{x}')$ is a real function of $\mathbf{x}'$. Later in this book we will consider more complicated Hamiltonians: a time-dependent potential $V(\mathbf{x}, t)$; a nonlocal but separable potential where the right-hand side of (2.178) is replaced by $v_1(\mathbf{x}'')v_2(\mathbf{x}')$; a momentum-dependent interaction of the form $\mathbf{p} \cdot \mathbf{A} + \mathbf{A} \cdot \mathbf{p}$, where $\mathbf{A}$ is the vector potential in electrodynamics, and so on.

We now derive Schrödinger's time-dependent wave equation. We first write the Schrödinger equation for a state ket (2.27) in the $\mathbf{x}$-representation:

$$i\hbar \frac{\partial}{\partial t} \langle \mathbf{x}' | \alpha, t_0; t \rangle = \langle \mathbf{x}' | H | \alpha, t_0; t \rangle, \tag{2.179}$$

where we have used the fact that the position eigenbras in the Schrödinger picture do not change with time. Using (1.252), we can write the kinetic-energy contribution to the right-hand side of (2.179) as

$$\left\langle \mathbf{x}' \left| \frac{\mathbf{p}^2}{2m} \right| \alpha, t_0; t \right\rangle = -\left( \frac{\hbar^2}{2m} \right) \nabla'^2 \langle \mathbf{x}' | \alpha, t_0; t \rangle. \tag{2.180}$$

As for $V(\mathbf{x})$, we simply use

$$\langle \mathbf{x}' | V(\mathbf{x}) = \langle \mathbf{x}' | V(\mathbf{x}'), \tag{2.181}$$

---

[5] For applications to laser physics, see Sargent, et al. (1974) and Loudon (2000). See also the discussion on squeezed light at the end of Section 7.8 of this book.

where $V(\mathbf{x}')$ is no longer an operator. Combining everything, we deduce

$$i\hbar \frac{\partial}{\partial t} \langle \mathbf{x}' | \alpha, t_0; t \rangle = -\left(\frac{\hbar^2}{2m}\right) \nabla'^2 \langle \mathbf{x}' | \alpha, t_0; t \rangle + V(\mathbf{x}') \langle \mathbf{x}' | \alpha, t_0; t \rangle, \qquad (2.182)$$

which we recognize to be the celebrated time-dependent wave equation of E. Schrödinger, usually written as

$$i\hbar \frac{\partial}{\partial t} \psi(\mathbf{x}', t) = -\left(\frac{\hbar^2}{2m}\right) \nabla'^2 \psi(\mathbf{x}', t) + V(\mathbf{x}') \psi(\mathbf{x}', t). \qquad (2.183)$$

The quantum mechanics based on wave equation (2.183) is known as **wave mechanics**. This equation is, in fact, the starting point of many textbooks on quantum mechanics. In our formalism, however, this is just the Schrödinger equation for a state ket written explicitly in the $x$-basis when the Hamiltonian operator is taken to be (2.177).

## 2.4.2  The Time-Independent Wave Equation

We now derive the partial differential equation satisfied by energy eigenfunctions. We showed in Section 2.1 that the time dependence of a stationary state is given by $\exp(-iE_{a'}t/\hbar)$. This enables us to write its wave function as

$$\langle \mathbf{x}' | a', t_0; t \rangle = \langle \mathbf{x}' | a' \rangle \exp\left(\frac{-iE_{a'}t}{\hbar}\right), \qquad (2.184)$$

where it is understood that initially the system is prepared in a simultaneous eigenstate of $A$ and $H$ with eigenvalues $a'$ and $E_{a'}$, respectively. Let us now substitute (2.184) into the time-dependent Schrödinger equation (2.182). We are then led to

$$-\left(\frac{\hbar^2}{2m}\right) \nabla'^2 \langle \mathbf{x}' | a' \rangle + V(\mathbf{x}') \langle \mathbf{x}' | a' \rangle = E_{a'} \langle \mathbf{x}' | a' \rangle. \qquad (2.185)$$

This partial differential equation is satisfied by the energy eigenfunction $\langle \mathbf{x}' | a' \rangle$ with energy eigenvalue $E_{a'}$. Actually, in wave mechanics where the Hamiltonian operator is given as a function of $\mathbf{x}$ and $\mathbf{p}$, as in (2.177), it is not necessary to refer explicitly to observable $A$ that commutes with $H$ because we can always choose $A$ to be that function of the observables $\mathbf{x}$ and $\mathbf{p}$ which coincides with $H$ itself. We may therefore omit reference to $a'$ and simply write (2.185) as the partial differential equation to be satisfied by the energy eigenfunction $u_E(\mathbf{x}')$:

$$-\left(\frac{\hbar^2}{2m}\right) \nabla'^2 u_E(\mathbf{x}') + V(\mathbf{x}') u_E(\mathbf{x}') = E u_E(\mathbf{x}'). \qquad (2.186)$$

This is the **time-independent wave equation** of E. Schrödinger, announced in the first of four monumental papers, all written in the first half of 1926, that laid the foundations of wave mechanics. In the same paper, Schrödinger immediately applied (2.186) to derive the energy spectrum of the hydrogen atom.

To solve (2.186) some boundary condition has to be imposed. Suppose we seek a solution to (2.186) with

$$E < \lim_{|\mathbf{x}'| \to \infty} V(\mathbf{x}'), \qquad (2.187)$$

where the inequality relation is to hold for $|\mathbf{x}'| \to \infty$ in any direction. The appropriate boundary condition to be used in this case is

$$u_E(\mathbf{x}') \to 0 \quad \text{as} \quad |\mathbf{x}'| \to \infty. \qquad (2.188)$$

Physically this means that the particle is bound or confined within a finite region of space. We know from the theory of partial differential equations that (2.186) subject to boundary condition (2.188) allows nontrivial solutions only for a discrete set of values of $E$. It is in this sense that the time-independent Schrödinger equation (2.186) yields the *quantization of energy levels*.[6] Once the partial differential equation (2.186) is written, the problem of finding the energy levels of microscopic physical systems is as straightforward as that of finding the characteristic frequencies of vibrating strings or membranes. In both cases we solve boundary-value problems in mathematical physics.

A short digression on the history of quantum mechanics is in order here. The fact that exactly soluble eigenvalue problems in the theory of partial differential equations can also be treated using matrix methods was already known to mathematicians in the first quarter of the twentieth century. Furthermore, theoretical physicists like M. Born frequently consulted great mathematicians of the day – D. Hilbert and H. Weyl, in particular. Yet when matrix mechanics was born in the summer of 1925, it did not immediately occur to the theoretical physicists or to the mathematicians to reformulate it using the language of partial differential equations. Six months after Heisenberg's pioneering paper, wave mechanics was proposed by Schrödinger. However, a close inspection of his papers shows that he was not at all influenced by the earlier works of Heisenberg, Born, and Jordan. Instead, the train of reasoning that led Schrödinger to formulate wave mechanics has its roots in W. R. Hamilton's analogy between optics and mechanics, on which we will comment later, and the particle-wave hypothesis of L. de Broglie. Once wave mechanics was formulated, many people, including Schrödinger himself, showed the equivalence between wave mechanics and matrix mechanics.

It is assumed that the reader of this book has some experience in solving the time-dependent and time-independent wave equations. He or she should be familiar with the time evolution of a Gaussian wave packet in a force-free region; should be able to solve one-dimensional transmission-reflection problems involving a rectangular potential barrier, and the like; should have seen derived some simple solutions of the time-independent wave equation – a particle in a box, a particle in a square well, the simple harmonic oscillator, the hydrogen atom, and so on – and should also be familiar with some general properties of the energy eigenfunctions and energy eigenvalues, such as (1) the fact that the energy levels exhibit a discrete or continuous spectrum depending on whether or not (2.187) is

---

[6] Schrödinger's paper that announced (2.186) is appropriately entitled *Quantisierung als Eigenwertproblem (Quantization as an Eigenvalue Problem)*.

satisfied and (2) the property that the energy eigenfunction in one dimension is *sinusoidal* or *damped* depending on whether $E - V(\mathbf{x}')$ is positive or negative.

In this book, we do not thoroughly cover these more elementary topics and solutions. Some of these are pursued, for example the harmonic oscillator and hydrogen atom, but at a mathematical level somewhat higher than what is usually seen in undergraduate courses. In any case, a brief summary of elementary solutions to Schrödinger's equations is presented in Appendix B.

## 2.4.3  Interpretations of the Wave Function

We now turn to discussions of the physical interpretations of the wave function. In Section 1.7 we commented on the probabilistic interpretation of $|\psi|^2$ that follows from the fact that $\langle \mathbf{x}'|\alpha, t_0; t \rangle$ is to be regarded as an expansion coefficient of $|\alpha, t_0; t \rangle$ in terms of the position eigenkets $\{|\mathbf{x}'\rangle\}$. The quantity $\rho(\mathbf{x}', t)$ defined by

$$\rho(\mathbf{x}', t) = |\psi(\mathbf{x}', t)|^2 = |\langle \mathbf{x}'|\alpha, t_0; t \rangle|^2 \tag{2.189}$$

is therefore regarded as the **probability density** in wave mechanics. Specifically, when we use a detector that ascertains the presence of the particle within a small volume element $d^3x'$ around $\mathbf{x}'$, the probability of recording a positive result at time $t$ is given by $\rho(\mathbf{x}', t)d^3x'$.

In the remainder of this section we use $\mathbf{x}$ for $\mathbf{x}'$ because the position operator will not appear. Using Schrödinger's time-dependent wave equation, it is straightforward to derive the continuity equation

$$\frac{\partial \rho}{\partial t} + \nabla \cdot \mathbf{j} = 0, \tag{2.190}$$

where $\rho(\mathbf{x}, t)$ stands for $|\psi|^2$ as before, and $\mathbf{j}(\mathbf{x}, t)$, known as the **probability flux**, is given by

$$\mathbf{j}(\mathbf{x}, t) = -\left(\frac{i\hbar}{2m}\right)[\psi^*\nabla\psi - (\nabla\psi^*)\psi]$$

$$= \left(\frac{\hbar}{m}\right)\mathrm{Im}(\psi^*\nabla\psi). \tag{2.191}$$

The reality of the potential $V$ (or the Hermiticity of the $V$ operator) has played a crucial role in our obtaining this result. Conversely, a complex potential can phenomenologically account for the disappearance of a particle; such a potential is often used for nuclear reactions where incident particles get absorbed by nuclei.

We may intuitively expect that the probability flux $\mathbf{j}$ is related to momentum. This is indeed the case for $\mathbf{j}$ *integrated over all space*. From (2.191) we obtain

$$\int d^3x\, \mathbf{j}(\mathbf{x}, t) = \frac{\langle \mathbf{p} \rangle_t}{m}, \tag{2.192}$$

where $\langle \mathbf{p} \rangle_t$ is the expectation value of the momentum operator at time $t$.

Equation (2.190) is reminiscent of the continuity equation in fluid dynamics that characterizes a hydrodynamic flow of a fluid in a source-free, sink-free region. Indeed, historically Schrödinger was first led to interpret $|\psi|^2$ as the actual matter density, or $e|\psi|^2$ as the actual electric charge density. If we adopt such a view, we are led to face some bizarre consequences.

A typical argument for a position measurement might go as follows. An atomic electron is to be regarded as a continuous distribution of matter filling up a finite region of space around the nucleus; yet, when a measurement is made to make sure that the electron is at some particular point, this continuous distribution of matter suddenly shrinks to a pointlike particle with no spatial extension. The more satisfactory *statistical interpretation* of $|\psi|^2$ as the probability density was first given by M. Born.

To understand the physical significance of the wave function, let us write it as

$$\psi(\mathbf{x},t) = \sqrt{\rho(\mathbf{x},t)}\exp\left[\frac{iS(\mathbf{x},t)}{\hbar}\right], \tag{2.193}$$

with $S$ real and $\rho > 0$, which can always be done for any complex function of $\mathbf{x}$ and $t$. The meaning of $\rho$ has already been given. What is the physical interpretation of $S$? Noting

$$\psi^*\nabla\psi = \sqrt{\rho}\nabla(\sqrt{\rho}) + \left(\frac{i}{\hbar}\right)\rho\nabla S, \tag{2.194}$$

we can write the probability flux as [see (2.191)]

$$\mathbf{j} = \frac{\rho\nabla S}{m}. \tag{2.195}$$

We now see that there is more to the wave function than the fact that $|\psi|^2$ is the probability density; the gradient of the phase $S$ contains a vital piece of information. From (2.195) we see that the *spatial variation of the phase* of the wave function characterizes the probability flux; the stronger the phase variation, the more intense the flux. The direction of $\mathbf{j}$ at some point $\mathbf{x}$ is seen to be normal to the surface of a constant phase that goes through that point. In the particularly simple example of a plane wave (a momentum eigenfunction)

$$\psi(\mathbf{x},t) \propto \exp\left(\frac{i\mathbf{p}\cdot\mathbf{x}}{\hbar} - \frac{iEt}{\hbar}\right), \tag{2.196}$$

where $\mathbf{p}$ stands for the eigenvalue of the momentum operator. All this is evident because

$$\nabla S = \mathbf{p}. \tag{2.197}$$

More generally, it is tempting to regard $\nabla S/m$ as some kind of "velocity,"

$$\text{``}\mathbf{v}\text{''} = \frac{\nabla S}{m}, \tag{2.198}$$

and to write the continuity equation (2.190) as

$$\frac{\partial\rho}{\partial t} + \nabla\cdot(\rho\,\text{``}\mathbf{v}\text{''}) = 0, \tag{2.199}$$

just as in fluid dynamics. However, we would like to caution the reader against a too literal interpretation of $\mathbf{j}$ as $\rho$ times the velocity defined at every point in space, because a simultaneous precision measurement of position and velocity would necessarily violate the uncertainty principle.

## 2.4.4 The Classical Limit

We now discuss the classical limit of wave mechanics. First, we substitute $\psi$ written in form (2.193) into both sides of the time-dependent wave equation. Straightforward differentiations lead to

$$-\left(\frac{\hbar^2}{2m}\right)\left[\nabla^2\sqrt{\rho}+\left(\frac{2i}{\hbar}\right)(\nabla\sqrt{\rho})\cdot(\nabla S)-\left(\frac{1}{\hbar^2}\right)\sqrt{\rho}|\nabla S|^2+\left(\frac{i}{\hbar}\right)\sqrt{\rho}\,\nabla^2 S\right]+\sqrt{\rho}V$$
$$=i\hbar\left[\frac{\partial\sqrt{\rho}}{\partial t}+\left(\frac{i}{\hbar}\right)\sqrt{\rho}\frac{\partial S}{\partial t}\right]. \tag{2.200}$$

So far everything has been exact. Let us suppose now that $\hbar$ can, in some sense, be regarded as a small quantity. The precise physical meaning of this approximation, to which we will come back later, is not evident now, but let us assume

$$\hbar|\nabla^2 S|\ll|\nabla S|^2, \tag{2.201}$$

and so forth. We can then collect terms in (2.200) that do not explicitly contain $\hbar$ to obtain a nonlinear partial differential equation for $S$:

$$\frac{1}{2m}|\nabla S(\mathbf{x},t)|^2+V(\mathbf{x})+\frac{\partial S(\mathbf{x},t)}{\partial t}=0. \tag{2.202}$$

We recognize this to be the **Hamilton–Jacobi equation** in classical mechanics, first written in 1836, where $S(\mathbf{x},t)$ stands for Hamilton's principal function. So, not surprisingly, in the $\hbar\to 0$ limit, classical mechanics is contained in Schrödinger's wave mechanics. We have a semiclassical interpretation of the phase of the wave function: $\hbar$ times the phase is equal to Hamilton's principal function provided that $\hbar$ can be regarded as a small quantity.

Let us now look at a stationary state with time dependence $\exp(-iEt/\hbar)$. This time dependence is anticipated from the fact that for a classical system with a constant Hamiltonian, Hamilton's principal function $S$ is separable:

$$S(x,t)=W(x)-Et, \tag{2.203}$$

where $W(x)$ is called **Hamilton's characteristic function** (Goldstein et al. (2002), pp. 440–444). As time goes on, a surface of a constant $S$ advances in much the same way as a surface of a constant phase in wave optics, a "wave front," advances. The momentum in the classical Hamilton–Jacobi theory is given by

$$\mathbf{P}_{\text{class}}=\nabla S=\nabla W, \tag{2.204}$$

which is consistent with our earlier identification of $\nabla S/m$ with some kind of velocity. In classical mechanics the velocity vector is tangential to the particle trajectory, and as a result we can trace the trajectory by following continuously the direction of the velocity vector. The particle trajectory is like a ray in geometric optics because the $\nabla S$ that traces the trajectory is normal to the wave front defined by a constant $S$. In this sense geometrical optics is to wave optics what classical mechanics is to wave mechanics.

One might wonder, in hindsight, why this optical-mechanical analogy was not fully exploited in the nineteenth century. The reason is that there was no motivation for regarding

Hamilton's principal function as the phase of some traveling wave; the wave nature of a material particle did not become apparent until the 1920s. Besides, the basic unit of action $\hbar$, which must enter into (2.193) for dimensional reasons, was missing in the physics of the nineteenth century.

## 2.5  Elementary Solutions to Schrödinger's Wave Equation

It is both instructive and useful to look at some relatively elementary solutions to (2.186) for particular choices of the potential energy function $V(\mathbf{x})$. In this section we choose some particular examples to illustrate contemporary physics and/or which will be useful in later chapters of this textbook.

### 2.5.1  Free Particle in Three Dimensions

The case $V(\mathbf{x}) = 0$ has fundamental significance. We will consider the solution to Schrödinger's equation here in three dimensions using Cartesian coordinates. The solution in spherical coordinates will be left until our treatment of angular momentum is presented in the next chapter. Equation (2.186) becomes

$$\nabla^2 u_E(\mathbf{x}) = -\frac{2mE}{\hbar^2} u_E(\mathbf{x}). \tag{2.205}$$

Define a vector $\mathbf{k}$ where

$$\mathbf{k}^2 = k_x^2 + k_y^2 + k_z^2 \equiv \frac{2mE}{\hbar^2} = \frac{\mathbf{p}^2}{\hbar^2} \tag{2.206}$$

that is, $\mathbf{p} = \hbar\mathbf{k}$. Differential equation (2.205) is easily solved using the technique known as "separation of variables." Writing

$$u_E(\mathbf{x}) = u_x(x)u_y(y)u_z(z) \tag{2.207}$$

we arrive at

$$\left[\frac{1}{u_x}\frac{d^2u_x}{dx^2} + k_x^2\right] + \left[\frac{1}{u_y}\frac{d^2u_y}{dy^2} + k_y^2\right] + \left[\frac{1}{u_z}\frac{d^2u_z}{dz^2} + k_z^2\right] = 0. \tag{2.208}$$

This leads to individual plane wave solutions $u_w(w) = c_w e^{ik_w w}$ for $w = x, y, z$. Note that one gets the same energy $E$ for values $\pm k_w$.

Collecting these solutions, and combining the normalization constants, we obtain

$$u_E(\mathbf{x}) = c_x c_y c_z e^{ik_x x + ik_y y + ik_z z} = C e^{i\mathbf{k}\cdot\mathbf{x}}. \tag{2.209}$$

The normalization constant $C$ presents the usual difficulties, which are generally handled by using a $\delta$-function normalization condition. It is convenient in many case, however, to use a "big box" normalization, where all space is contained within a cube of side length $L$. We impose periodic boundary conditions on the box, and thereby obtain a finite normalization constant $C$. For any real calculation, we simply let the size $L \to \infty$ at the end of the calculation.

Imposing the condition $u_x(x+L) = u_x(x)$ we have $k_x L = 2\pi n_x$ where $n_x$ is an integer. That is

$$k_x = \frac{2\pi}{L} n_x \qquad k_y = \frac{2\pi}{L} n_y \qquad k_z = \frac{2\pi}{L} n_z \qquad (2.210)$$

and the normalization criterion becomes

$$1 = \int_0^L dx \int_0^L dy \int_0^L dz\, u_E^*(\mathbf{x}) u_E(\mathbf{x}) = L^3 |C|^2 \qquad (2.211)$$

in which case $C = 1/L^{3/2}$ and

$$u_E(\mathbf{x}) = \frac{1}{L^{3/2}} e^{i\mathbf{k}\cdot\mathbf{x}}. \qquad (2.212)$$

The energy eigenvalue is

$$E = \frac{\mathbf{p}^2}{2m} = \frac{\hbar^2 \mathbf{k}^2}{2m} = \frac{\hbar^2}{2m} \left(\frac{2\pi}{L}\right)^2 \left(n_x^2 + n_y^2 + n_z^2\right). \qquad (2.213)$$

The sixfold degeneracy we mentioned earlier corresponds to the six combinations of $(\pm n_x, \pm n_y, \pm n_z)$, but the degeneracy can actually be much larger since, in some cases, there are various combinations of $n_x$, $n_y$, and $n_z$ which can give the same $E$. In fact, in the (realistic) limit where $L$ is very large, there can be a large number of states $N$ which have an energy between $E$ and $E + dE$. This "density of states" $dN/dE$ is an important quantity for calculations of processes which include free particles. See, for example, the discussion of the photoelectric effect in Section 5.8.

To calculate the density of states, imagine a spherical shell in $\mathbf{k}$ space with radius $|\mathbf{k}| = 2\pi|\mathbf{n}|/L$ and thickness $d|\mathbf{k}| = 2\pi d|\mathbf{n}|/L$. All states within this shell have energy $E = \hbar^2 \mathbf{k}^2/2m$. The number of states $dN$ within this shell is $4\pi \mathbf{n}^2 d|\mathbf{n}|$. Therefore

$$\frac{dN}{dE} = \frac{4\pi \mathbf{n}^2 d|\mathbf{n}|}{\hbar^2 |\mathbf{k}| d|\mathbf{k}|/m} = \frac{4\pi}{\hbar^2} m \left(\frac{L}{2\pi}\right)^2 |\mathbf{k}| \frac{L}{2\pi}$$

$$= \frac{m^{3/2} E^{1/2} L^3}{\sqrt{2}\pi^2 \hbar^3}. \qquad (2.214)$$

In a typical "real" calculation, the density of states will be multiplied by some probability that involves $u_E^*(\mathbf{x}) u_E(\mathbf{x})$. In this case, the factors of $L^3$ will cancel explicitly, so the limit $L \to \infty$ is trivial. This "big box" normalization also yields the correct answer for the probability flux. Rewriting (2.196) with this normalization, we have

$$\psi(\mathbf{x}, t) = \frac{1}{L^{3/2}} \exp\left(\frac{i\mathbf{p}\cdot\mathbf{x}}{\hbar} - \frac{iEt}{\hbar}\right) \qquad (2.215)$$

in which case we find

$$\mathbf{j}(\mathbf{x}, t) = \frac{\hbar}{m} \mathrm{Im}(\psi^* \nabla \psi) = \frac{\hbar \mathbf{k}}{m} \frac{1}{L^3} = \mathbf{v}\rho \qquad (2.216)$$

where $\rho = 1/L^3$ is indeed the probability density.

## 2.5.2  The Simple Harmonic Oscillator

We saw an elegant solution for the case $V(x) = m\omega^2 x^2/2$ in Section 2.3, which yielded the energy eigenvalues, eigenstates, and wave functions. Here, we demonstrate a different approach which solves the differential equation

$$-\frac{\hbar^2}{2m}\frac{d^2}{dx^2}u_E(x) + \frac{1}{2}m\omega^2 x^2 u_E(x) = Eu_E(x). \tag{2.217}$$

Our approach will introduce the concept of *generating functions*, a generally useful technique which arises in many treatments of differential eigenvalue problems.

First, transform (2.217) using the dimensionless position $y \equiv x/x_0$ where $x_0 \equiv \sqrt{\hbar/m\omega}$. Also introduce a dimensionless energy variable $\varepsilon \equiv 2E/\hbar\omega$. The differential equation we need to solve becomes therefore

$$\frac{d^2}{dy^2}u(y) + (\varepsilon - y^2)u(y) = 0. \tag{2.218}$$

For $y \to \pm\infty$, the solution must tend to zero, otherwise the wave function will not be normalizable and hence unphysical. The differential equation $w''(y) - y^2 w(y) = 0$ has solutions $w(y) \propto \exp(\pm y^2/2)$, so we would have to choose the minus sign. We then "remove" the asymptotic behavior of the wave function by writing

$$u(y) = h(y)e^{-y^2/2} \tag{2.219}$$

where the function $h(y)$ satisfies the differential equation

$$\frac{d^2 h}{dy^2} - 2y\frac{dh}{dy} + (\varepsilon - 1)h(y) = 0. \tag{2.220}$$

To this point, we have followed the traditional solution of the simple harmonic oscillator as found in many textbooks. Typically, one would now look for a series solution for $h(y)$ and discover that a normalizable solution is only possible if the series terminates. (In fact, we use this approach for the three-dimensional isotropic harmonic oscillator in this book. See Section 3.7.) One forces this termination by imposing the condition that $\varepsilon - 1$ be an even, nonnegative integer $2n$, $n = 0, 1, 2, \ldots$. The solutions are then written using the resulting polynomials $h_n(y)$. Of course, $\varepsilon - 1 = 2n$ is equivalent to $E = (n + \frac{1}{2})\hbar\omega$, the quantization relation (2.143).

Let us take a different approach. Consider the "Hermite polynomials" $H_n(x)$ defined by the "generating function" $g(x,t)$ through

$$g(x,t) \equiv e^{-t^2 + 2tx} \tag{2.221a}$$

$$\equiv \sum_{n=0}^{\infty} H_n(x)\frac{t^n}{n!}. \tag{2.221b}$$

Some properties of the $H_n(x)$ are immediately obvious. For example, $H_0(x) = 1$. Also, since

$$g(0,t) = e^{-t^2} = \sum_{n=0}^{\infty}\frac{(-1)^n}{n!}t^{2n} \tag{2.222}$$

it is clear that $H_n(0) = 0$ if $n$ is odd, since this series only involves even powers of $t$. On the other hand, if we restrict ourselves to even values of $n$, we have

$$g(0,t) = e^{-t^2} = \sum_{n=0}^{\infty} \frac{(-1)^{(n/2)}}{(n/2)!} t^n = \sum_{n=0}^{\infty} \frac{(-1)^{(n/2)}}{(n/2)!} \frac{n!}{n!} t^n \qquad (2.223)$$

and so $H_n(0) = (-1)^{n/2} n! / (n/2)!$. Also, since $g(-x,t)$ reverses the sign only on terms with odd powers of $t$, $H_n(-x) = (-1)^n H_n(x)$.

We can take derivatives of $g(x,t)$ to build the Hermite polynomials using recursion relations between them and their derivatives. The trick is that we can differentiate the analytic form of the generating function (2.221a) or the series form (2.221b) and then compare results. For example, if we take the derivative using (2.221a), then

$$\frac{\partial g}{\partial x} = 2tg(x,t) = \sum_{n=0}^{\infty} 2H_n(x) \frac{t^{n+1}}{n!} = \sum_{n=0}^{\infty} 2(n+1) H_n(x) \frac{t^{n+1}}{(n+1)!} \qquad (2.224)$$

where we insert the series definition of the generating function after taking the derivative. On the other hand, we can take the derivative of (2.221b) directly, in which case

$$\frac{\partial g}{\partial x} = \sum_{n=0}^{\infty} H_n'(x) \frac{t^n}{n!}. \qquad (2.225)$$

Comparing (2.224) and (2.225) shows that

$$H_n'(x) = 2nH_{n-1}(x) \qquad (2.226)$$

This is enough information for us build the Hermite polynomials:

$$H_0(x) = 1$$

so $H_1'(x) = 2$, therefore $H_1(x) = 2x$

so $H_2'(x) = 8x$, therefore $H_2(x) = 4x^2 - 2$

so $H_3'(x) = 24x^2 - 12$, therefore $H_3(x) = 8x^3 - 12x$

$$\cdots$$

So far, this is just a curious mathematical exercise. To see why it is relevant to the simple harmonic oscillator, consider the derivative of the generating function with respect to $t$. If we start with (2.221a) then

$$\frac{\partial g}{\partial t} = -2tg(x,t) + 2xg(x,t)$$

$$= -\sum_{n=0}^{\infty} 2H_n(x) \frac{t^{n+1}}{n!} + \sum_{n=0}^{\infty} 2xH_n(x) \frac{t^n}{n!}$$

$$= -\sum_{n=0}^{\infty} 2nH_{n-1}(x) \frac{t^n}{n!} + \sum_{n=0}^{\infty} 2xH_n(x) \frac{t^n}{n!}. \qquad (2.227)$$

Or, if we differentiate (2.221b) then we have

$$\frac{\partial g}{\partial t} = \sum_{n=0}^{\infty} nH_n(x) \frac{t^{n-1}}{n!} = \sum_{n=0}^{\infty} H_{n+1}(x) \frac{t^n}{n!}. \qquad (2.228)$$

Comparing (2.227) and (2.228) gives us the recursion relation

$$H_{n+1}(x) = 2xH_n(x) - 2nH_{n-1}(x) \tag{2.229}$$

which we combine with (2.226) to find

$$\begin{aligned}
H_n''(x) &= 2n \cdot 2(n-1)H_{n-2}(x) \\
&= 2n\left[2xH_{n-1}(x) - H_n(x)\right] \\
&= 2xH_n'(x) - 2nH_n(x). \tag{2.230}
\end{aligned}$$

In other words, the Hermite polynomials satisfy the differential equation

$$H_n''(x) - 2xH_n'(x) + 2nH_n(x) = 0 \tag{2.231}$$

where $n$ is a nonnegative integer. This, however, is the same as the Schrödinger equation written as (2.220) since $\varepsilon - 1 = 2n$. That is, the wave functions for the simple harmonic oscillator are given by

$$u_n(x) = c_n H_n\left(x\sqrt{\frac{m\omega}{\hbar}}\right)e^{-m\omega x^2/2\hbar} \tag{2.232}$$

up to some normalization constant $c_n$. This constant can be determined from the orthogonality relationship

$$\int_{-\infty}^{\infty} H_n(x)H_m(x)e^{-x^2}\,dx = \pi^{1/2}2^n n!\,\delta_{nm} \tag{2.233}$$

which is easily proved using the generating function. See Problem 2.25 at the end of this chapter.

Generating functions have a usefulness that far outreaches our limited application here. Among other things, many of the orthogonal polynomials which arise from solving the Schrödinger equation for different potentials, can be derived from generating functions. See, for example, Problem 3.30 in Chapter 3. The interested reader is encouraged to pursue this further, probably best from any one of the many excellent texts on mathematical physics.

### 2.5.3 The Linear Potential

Perhaps the first potential energy function, with bound states, to come to mind is the linear potential, namely

$$V(x) = k|x| \tag{2.234}$$

where $k$ is an arbitrary positive constant. Given a total energy $E$, this potential has a classical turning point at a value $x = a$ where $E = ka$. This point will be important for understanding the quantum behavior of a particle of mass $m$ bound by this potential.

The Schrödinger equation becomes

$$-\frac{\hbar^2}{2m}\frac{d^2u_E}{dx^2} + k|x|u_E(x) = Eu_E(x). \tag{2.235}$$

It is easiest to deal with the absolute value by restricting our attention to $x \geq 0$. We can do this because $V(-x) = V(x)$, so there are two types of solutions, namely $u_E(-x) = \pm u_E(x)$. In either case, we need $u_E(x)$ to tend towards zero as $x \to \infty$. If $u_E(-x) = -u_E(x)$, then we need $u_E(0) = 0$. On the other hand, if $u_E(-x) = +u_E(x)$, then we have $u'_E(0) = 0$, since $u_E(\varepsilon) - u_E(-\varepsilon) \equiv 0$, even for $\varepsilon \to 0$. (As we will discuss in Chapter 4, we refer to these solutions as "odd" and "even" parity.)

Once again, we write the differential equation in terms of dimensionless variables, based on appropriate scales for length and energy. In this case, the dimensionless length scale is $x_0 = (\hbar^2/mk)^{1/3}$ and the dimensionless enery scale is $E_0 = kx_0 = (\hbar^2k^2/m)^{1/3}$. Defining $y \equiv x/x_0$ and $\varepsilon \equiv E/E_0$ allows us to rewrite (2.235) as

$$\frac{d^2 u_E}{dy^2} - 2(y - \varepsilon)u_E(y) = 0, \qquad y \geq 0. \tag{2.236}$$

Notice that $y = \varepsilon$ when $x = E/k$, i.e. the classical turning point $x = a$. In fact, defining a translated position variable $z \equiv 2^{1/3}(y - \varepsilon)$, (2.236) becomes

$$\frac{d^2 u_E}{dz^2} - z u_E(z) = 0. \tag{2.237}$$

This is the Airy equation, and the solution is the Airy function $\mathrm{Ai}(z)$, plotted in Figure 2.3. The Airy function has a peculiar behavior, oscillatory for negative values of the argument, and decreasing rapidly towards zero for positive values. Of course, this is exactly the behavior we expect for the wave function, since $z = 0$ is the classical turning point.

Note that the boundary conditions at $x = 0$ translate into zeros for either $\mathrm{Ai}'(z)$ or $\mathrm{Ai}(z)$ where $z = -2^{1/3}\varepsilon$. In other words, the zeros of the Airy function or its derivative determine the quantized energies. One finds that

$$\mathrm{Ai}'(z) = 0 \qquad \text{for } z = -1.019, \; -3.249, \; -4.820, \ldots \qquad \text{(even)} \tag{2.238}$$

$$\mathrm{Ai}(z) = 0 \qquad \text{for } z = -2.338, \; -4.088, \; -5.521, \ldots \qquad \text{(odd)}. \tag{2.239}$$

For example, the ground-state energy is $E = (1.019/2^{1/3})(\hbar^2k^2/m)^{1/3}$.

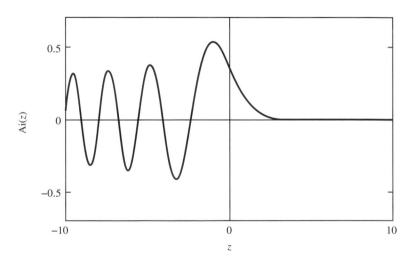

**Fig. 2.3**  The Airy function.

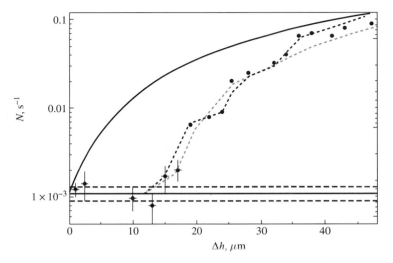

**Fig. 2.4**  Experimental observation of the quantum-mechanical states of a bouncing neutron, from Nesvizhevsky et al., *Phys. Rev. D*, **67** (2003) 102002. The solid curve is a fit to the data based on classical physics. Note that the vertical scale is logarithmic.

The quantum-theoretical treatment of the linear potential may appear to have little to do with the real world. It turns out, however, that a potential of type (2.234) is actually of practical interest in studying the energy spectrum of a quark-antiquark bound system, called **quarkonium**. In this case, the $x$ in (2.234) is replaced by the quark-antiquark separation distance $r$. This constant $k$ is empirically estimated to be in the neighborhood of

$$1 \text{ GeV/fm} \simeq 1.6 \times 10^5 \text{ N}, \tag{2.240}$$

which corresponds to a gravitational force of about 16 tons.

Indeed, another real world example of the linear potential is the "bouncing ball." One interprets (2.234) as the potential energy of a ball of mass $m$ at a height $x$ above the floor, and $k = mg$ where $g$ is the local acceleration due to gravity. Of course, this is the potential energy only for $x \geq 0$ as there is an infinite potential barrier which causes the ball to "bounce." Quantum mechanically, this means that only the odd parity solutions (2.239) are allowed.

The bouncing ball happens to be one of those rare cases where quantum-mechanical effects can be observed macroscopically. The trick is to have a very low mass "ball," which has been achieved with neutrons by a group[7] working at the Institut Laue-Langevin (ILL) in Grenoble, France. For neutrons with $m = 1.68 \times 10^{-27}$ kg, the characteristic length scale is $x_0 = (\hbar^2/m^2 g)^{1/3} = 7.40$ $\mu$m. The "allowed heights" to which a neutron can bounce are $(2.338/2^{1/3})x_0 = 14$ $\mu$m, $(4.088/2^{1/3})x_0 = 24$ $\mu$m, $(5.521/2^{1/3})x_0 = 32$ $\mu$m, and so on. These are small, but measurable with precision mechanical devices and very low energy, aka "ultracold," neutrons. The experimenters' results are shown in Figure 2.4. Plotted is the detected neutron rate as a function of the height of a slit which only allows neutrons

---

[7]  See Nesvizhevsky et al., *Phys. Rev. D*, **67** (2003) 102002, and *Eur. Phys. J.*, **C40** (2005) 479. See also Jenke et al., *Nature Phys.*, **7** (2011) 468.

to pass if they exceed this height. No neutrons are observed unless the height is at least $\approx 14$ $\mu$m, and clear breaks are observed at $\approx 24$ $\mu$m and $\approx 32$ $\mu$m, in excellent agreement with the predictions of quantum mechanics.

## 2.5.4  The WKB (Semiclassical) Approximation

Having solved the problem of a linear potential, it is worthwhile to introduce an important approximation technique known as the WKB solution, after G. Wentzel, A. Kramers, and L. Brillouin.[8] This technique is based on making use of regions where the wavelength is much shorter than the typical distance over which the potential energy varies. Such is *never* the case near classical turning points, but this is where the linear potential solution can be used to join the solutions on either side of them.

Again restricting ourselves to one dimension, we write Schrödinger's wave equation as

$$\frac{d^2 u_E}{dx^2} + \frac{2m}{\hbar^2}(E - V(x))u_E(x) = 0. \tag{2.241}$$

Define the quantities

$$k(x) \equiv \left[\frac{2m}{\hbar^2}(E - V(x))\right]^{1/2} \quad \text{for} \quad E > V(x) \tag{2.242a}$$

$$k(x) \equiv -i\kappa(x) \equiv -i\left[\frac{2m}{\hbar^2}(V(x) - E)\right]^{1/2} \quad \text{for} \quad E < V(x) \tag{2.242b}$$

and so (2.241) becomes

$$\frac{d^2 u_E}{dx^2} + [k(x)]^2 u_E(x) = 0. \tag{2.243}$$

Now, if $V(x)$ were not changing with $x$, then $k(x)$ would be a constant, and $u(x) \propto \exp(\pm ikx)$ would solve (2.243). Consequently, if we assume that $V(x)$ varies only "slowly" with $x$, then we are tempted to try a solution of the form

$$u_E(x) \equiv \exp\left[iW(x)/\hbar\right]. \tag{2.244}$$

(The reason for including the $\hbar$ will become apparent at the end of this section, when we discuss the physical interpretation of the WKB approximation.) In this case, (2.243) becomes

$$i\hbar\frac{d^2 W}{dx^2} - \left(\frac{dW}{dx}\right)^2 + \hbar^2 [k(x)]^2 = 0 \tag{2.245}$$

which is completely equivalent to Schrödinger's equation, although rewritten in what appears to be a nasty form. However, we consider a solution to this equation under the condition that

$$\hbar\left|\frac{d^2 W}{dx^2}\right| \ll \left|\frac{dW}{dx}\right|^2. \tag{2.246}$$

---

[8]  A similar technique was used earlier by H. Jeffreys; this solution is referred to as the JWKB solution in some English books.

This quantifies our notion of a "slowly varying" potential $V(x)$, and we will return soon to the physical significance of this condition.

Forging ahead for now, we use the condition (2.246) with our differential equation (2.245) to write a lowest-order approximation for $W(x)$, namely

$$W_0'(x) = \pm \hbar k(x) \tag{2.247}$$

leading to a first-order approximation for $W(x)$, based on

$$\left(\frac{dW_1}{dx}\right)^2 = \hbar^2 \left[k(x)\right]^2 + i\hbar W_0''(x)$$

$$= \hbar^2 \left[k(x)\right]^2 \pm i\hbar^2 k'(x) \tag{2.248}$$

where the second term in (2.248) is much smaller than the first, so that

$$W(x) \approx W_1(x) = \pm \hbar \int^x dx' \left[k^2(x') \pm ik'(x')\right]^{1/2}$$

$$\approx \pm \hbar \int^x dx' k(x') \left[1 \pm \frac{i}{2} \frac{k'(x')}{k^2(x')}\right]$$

$$= \pm \hbar \int^x dx' k(x') + \frac{i}{2} \hbar \ln \left[k(x)\right]. \tag{2.249}$$

The WKB approximation for the wave function is given by (2.244) and the first-order approximation for (2.249) for $W(x)$, namely

$$u_E(x) \approx \exp\left[iW(x)/\hbar\right] = \frac{1}{\left[k(x)\right]^{1/2}} \exp\left[\pm i \int^x dx' k(x')\right]. \tag{2.250}$$

Note that this specifies a choice of two solutions ($\pm$) either in the region where $E > V(x)$, with $k(x)$ from (2.242a), or in the region where $E < V(x)$, with $k(x)$ from (2.242b). Joining these two solutions across the classical turning point is the next task.

We do not discuss this joining procedure in detail, as it is discussed in many places (Schiff (1968), pp. 268–276, or Merzbacher (1998), Chapter 7, for example). Instead, we content ourselves with presenting the results of such an analysis for a potential well, schematically shown in Figure 2.5, with two turning points, $x_1$ and $x_2$. The wave function must behave like (2.250), with $k(x)$ given by (2.242a) in region II, and by (2.242b) in regions I and III. The solutions in the neighborhood of the turning points, shown as a dashed line in Figure 2.5, are given by Airy functions, since we assume a linear approximation to the potential in these regions. Note that the asymptotic dependences of the Airy function[9] are

$$\text{Ai}(z) \rightarrow \frac{1}{2\sqrt{\pi}} z^{-1/4} \exp\left(-\frac{2}{3} z^{3/2}\right) \qquad z \rightarrow +\infty \tag{2.251a}$$

$$\text{Ai}(z) \rightarrow \frac{1}{\sqrt{\pi}} |z|^{-1/4} \cos\left(\frac{2}{3} |z|^{3/2} - \frac{\pi}{4}\right) \qquad z \rightarrow -\infty. \tag{2.251b}$$

---

[9] There is actually a second Airy function, $\text{Bi}(z)$, which is very similar to $\text{Ai}(z)$ but which is singular at the origin. It is relevant for this discussion, but we are glossing over the details.

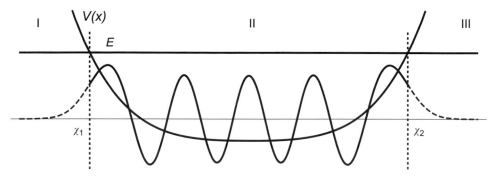

**Fig. 2.5**  Schematic diagram for behavior of the wave function $u_E(x)$ in a potential well $V(x)$ with turning points $x_1$ and $x_2$. Note the similarity with Figure 2.3 near the turning points.

For connecting regions I and II, the correct linear combination of the two solutions (2.250) is determined by choosing the integration constants in such a way that

$$\left\{ \frac{1}{[V(x) - E]^{1/4}} \right\} \exp\left[ -\left(\frac{1}{\hbar}\right) \int_x^{x_1} dx' \sqrt{2m\left[V(x') - E\right]} \right]$$

$$\rightarrow \left\{ \frac{2}{[E - V(x)]^{1/4}} \right\} \cos\left[ \left(\frac{1}{\hbar}\right) \int_{x_1}^{x} dx' \sqrt{2m\left[E - V(x')\right]} - \frac{\pi}{4} \right]. \tag{2.252}$$

Likewise, from region III into region II we have

$$\left\{ \frac{1}{[V(x) - E]^{1/4}} \right\} \exp\left[ -\left(\frac{1}{\hbar}\right) \int_{x_2}^{x} dx' \sqrt{2m\left[V(x') - E\right]} \right]$$

$$\rightarrow \left\{ \frac{2}{[E - V(x)]^{1/4}} \right\} \cos\left[ -\left(\frac{1}{\hbar}\right) \int_x^{x_2} dx' \sqrt{2m\left[E - V(x')\right]} + \frac{\pi}{4} \right]. \tag{2.253}$$

Of course, we must obtain the same form for the wave function in region II, regardless of which turning point is analyzed. This implies that the arguments of the cosine in (2.252) and (2.253) must differ at most by an integer multiple of $\pi$ [not of $2\pi$, because the signs of both sides of (2.253) can be reversed]. In this way we obtain a very interesting consistency condition,

$$\int_{x_1}^{x_2} dx \sqrt{2m\left[E - V(x)\right]} = (n + \tfrac{1}{2})\pi\hbar \quad (n = 0, 1, 2, 3, \ldots). \tag{2.254}$$

Apart from the difference between $n + \tfrac{1}{2}$ and $n$, this equation is simply the quantization condition of the old quantum theory due to A. Sommerfeld and W. Wilson, originally written in 1915 as

$$\oint p\, dq = nh, \tag{2.255}$$

where $h$ is Planck's $h$, not Dirac's $\hbar$, and the integral is evaluated over one whole period of classical motion, from $x_1$ to $x_2$ *and* back.

Equation (2.254) can be used to obtain approximate expressions for the energy levels of a particle confined in a potential well. As an example, we consider the energy spectrum of a ball bouncing up and down over a hard surface, that is the "bouncing neutrons" discussed earlier in this section, namely

$$V = \begin{cases} mgx & \text{for } x > 0 \\ \infty & \text{for } x < 0, \end{cases} \tag{2.256}$$

where $x$ stands for the height of the ball measured from the hard surface. One might be tempted to use (2.254) directly with

$$x_1 = 0, \quad x_2 = \frac{E}{mg}, \tag{2.257}$$

which are the classical turning points of this problem. We note, however, that (2.254) was derived under the assumption that the WKB wave function "leaks into" the $x < x_1$ region, while in our problem the wave function must strictly vanish for $x \le x_1 = 0$. A much more satisfactory approach to this problem is to consider the *odd-parity solutions*, guaranteed to vanish at $x = 0$, of a modified problem defined by

$$V(x) = mg|x| \quad (-\infty < x < \infty) \tag{2.258}$$

whose turning points are

$$x_1 = -\frac{E}{mg}, \quad x_2 = \frac{E}{mg}. \tag{2.259}$$

The energy spectrum of the odd-parity states for this modified problem must clearly be the same as that of the original problem. The quantization condition then becomes

$$\int_{E/mg}^{E/mg} dx \sqrt{2m(E - mg|x|)} = (n_{\text{odd}} + \tfrac{1}{2})\pi\hbar \quad (n_{\text{odd}} = 1, 3, 5, \dots) \tag{2.260}$$

or, equivalently,

$$\int_0^{E/mg} dx \sqrt{2m(E - mgx)} = (n - \tfrac{1}{4})\pi\hbar \quad (n = 1, 2, 3, 4, \dots). \tag{2.261}$$

This integral is elementary, and we obtain

$$E_n = \left\{ \frac{\left[ 3\left(n - \tfrac{1}{4}\right)\pi \right]^{2/3}}{2} \right\} (mg^2\hbar^2)^{1/3} \tag{2.262}$$

for the quantized energy levels of the bouncing ball.

Table 2.2 compares the WKB approximation to the exact solution, using zeros of the Airy function, for the first 10 energy levels. We see that agreement is excellent even for small values of $n$ and essentially exact for $n \simeq 10$.

Before concluding, let us return to the interpretation of the condition (2.246). It is exact in the case $\hbar \to 0$, suggesting a connection between the WKB approximation and the classical limit. In fact, using (2.244) the time-dependent wave function becomes

$$\psi(x, t) \propto u_E(x) \exp(-iEt/\hbar) = \exp\left( iW(x)/\hbar - iEt/\hbar \right). \tag{2.263}$$

| Table 2.2 | The Quantized Energies of a Bouncing Ball in Units of $(mg^2\hbar^2/2)^{1/3}$ | |
|---|---|---|
| $n$ | WKB | Exact |
| 1 | 2.320 | 2.338 |
| 2 | 4.082 | 4.088 |
| 3 | 5.517 | 5.521 |
| 4 | 6.784 | 6.787 |
| 5 | 7.942 | 7.944 |
| 6 | 9.021 | 9.023 |
| 7 | 10.039 | 10.040 |
| 8 | 11.008 | 11.009 |
| 9 | 11.935 | 11.936 |
| 10 | 12.828 | 12.829 |

Comparing this to (2.193) and (2.203) we see that $W(x)$ corresponds directly to Hamilton's characteristic function. Indeed, condition (2.246) is the same as (2.201), the condition for reaching the classical limit. For these reasons, the WKB approximation is frequently referred to as a "semiclassical" approximation.

We also note that condition (2.246) is equivalent to $|k'(x)| \ll |k^2(x)|$. In terms of the de Broglie wavelength divided by $2\pi$, this condition amounts to

$$\lambda = \frac{\hbar}{\sqrt{2m[E - V(x)]}} \ll \frac{2[E - V(x)]}{|dV/dx|}. \tag{2.264}$$

In other words, $\lambda$ must be small compared with the characteristic distance over which the potential varies appreciably. Roughly speaking, the potential must be essentially constant over many wavelengths. Thus we see that the semiclassical picture is reliable *in the short-wavelength limit.*

## 2.6 Propagators and Feynman Path Integrals

### 2.6.1 Propagators in Wave Mechanics

In Section 2.1 we showed how the most general time-evolution problem with a time-independent Hamiltonian can be solved once we expand the initial ket in terms of the eigenkets of an observable that commutes with $H$. Let us translate this statement into the language of wave mechanics. We start with

$$|\alpha, t_0; t\rangle = \exp\left[\frac{-iH(t - t_0)}{\hbar}\right] |\alpha, t_0\rangle$$

$$= \sum_{a'} |a'\rangle\langle a'|\alpha, t_0\rangle \exp\left[\frac{-iE_{a'}(t - t_0)}{\hbar}\right]. \tag{2.265}$$

Multiplying both sides by $\langle \mathbf{x}' |$ on the left, we have

$$\langle \mathbf{x}' | \alpha, t_0; t \rangle = \sum_{a'} \langle \mathbf{x}' | a' \rangle \langle a' | \alpha, t_0 \rangle \exp \left[ \frac{-iE_{a'}(t - t_0)}{\hbar} \right], \tag{2.266}$$

which is of the form

$$\psi(\mathbf{x}', t) = \sum_{a'} c_{a'}(t_0) u_{a'}(\mathbf{x}') \exp \left[ \frac{-iE_{a'}(t - t_0)}{\hbar} \right], \tag{2.267}$$

with

$$u_{a'}(\mathbf{x}') = \langle \mathbf{x}' | a' \rangle \tag{2.268}$$

standing for the eigenfunction of operator $A$ with eigenvalue $a'$. Note also that

$$\langle a' | \alpha, t_0 \rangle = \int d^3x' \langle a' | \mathbf{x}' \rangle \langle \mathbf{x}' | \alpha, t_0 \rangle, \tag{2.269}$$

which we recognize as the usual rule in wave mechanics for getting the expansion coefficients of the initial state:

$$c_{a'}(t_0) = \int d^3x' u_{a'}^*(\mathbf{x}') \psi(\mathbf{x}', t_0). \tag{2.270}$$

All this should be straightforward and familiar. Now (2.266) together with (2.269) can also be visualized as some kind of integral operator acting on the initial wave function to yield the final wave function:

$$\psi(\mathbf{x}'', t) = \int d^3x' K(\mathbf{x}'', t; \mathbf{x}', t_0) \psi(\mathbf{x}', t_0). \tag{2.271}$$

Here the kernel of the integral operator, known as the **propagator** in wave mechanics, is given by

$$K(\mathbf{x}'', t; \mathbf{x}', t_0) = \sum_{a'} \langle \mathbf{x}'' | a' \rangle \langle a' | \mathbf{x}' \rangle \exp \left[ \frac{-iE_{a'}(t - t_0)}{\hbar} \right]. \tag{2.272}$$

In any given problem the propagator depends only on the potential and is independent of the initial wave function. It can be constructed once the energy eigenfunctions and their eigenvalues are given.

Clearly, the time evolution of the wave function is completely predicted if $K(\mathbf{x}'', t; \mathbf{x}', t_0)$ is known and $\psi(\mathbf{x}', t_0)$ is given initially. In this sense Schrödinger's wave mechanics is a *perfectly causal theory*. The time development of a wave function subjected to some potential is as "deterministic" as anything else in classical mechanics *provided that the system is left undisturbed*. The only peculiar feature, if any, is that when a measurement intervenes, the wave function changes abruptly, in an uncontrollable way, into one of the eigenfunctions of the observable being measured.

There are two properties of the propagator worth recording here. First, for $t > t_0$, $K(\mathbf{x}'', t; \mathbf{x}', t_0)$ satisfies Schrödinger's time-dependent wave equation in the variables $\mathbf{x}''$ and $t$, with $\mathbf{x}'$ and $t_0$ fixed. This is evident from (2.272) because $\langle \mathbf{x}'' | a' \rangle \exp[-iE_{a'}(t - t_0)/\hbar]$, being the wave function corresponding to $\mathcal{U}(t, t_0) | a' \rangle$, satisfies the wave equation. Second,

$$\lim_{t \to t_0} K(\mathbf{x}'', t; \mathbf{x}', t_0) = \delta^3(\mathbf{x}'' - \mathbf{x}'), \tag{2.273}$$

which is also obvious; as $t \to t_0$, because of the completeness of $\{|a'\rangle\}$, sum (2.272) just reduces to $\langle \mathbf{x}''|\mathbf{x}'\rangle$.

Because of these two properties, the propagator (2.272), regarded as a function of $\mathbf{x}''$, is simply the wave function at $t$ of a particle which was localized *precisely* at $\mathbf{x}'$ at some earlier time $t_0$. Indeed, this interpretation follows, perhaps more elegantly, from noting that (2.272) can also be written as

$$K(\mathbf{x}'',t;\mathbf{x}',t_0) = \langle \mathbf{x}''|\exp\left[\frac{-iH(t-t_0)}{\hbar}\right]|\mathbf{x}'\rangle, \qquad (2.274)$$

where the time-evolution operator acting on $|\mathbf{x}'\rangle$ is just the state ket at $t$ of a system that was localized precisely at $\mathbf{x}'$ at time $t_0$ ($< t$). If we wish to solve a more general problem where the initial wave function extends over a finite region of space, all we have to do is multiply $\psi(\mathbf{x}',t_0)$ by the propagator $K(\mathbf{x}'',t;\mathbf{x}',t_0)$ and integrate over all space (that is, over $\mathbf{x}'$). In this manner we can add the various contributions from different positions ($\mathbf{x}'$). This situation is analogous to one in electrostatics; if we wish to find the electrostatic potential due to a general charge distribution $\rho(\mathbf{x}')$, we first solve the point-charge problem, multiply the point-charge solution with the charge distribution, and integrate:

$$\phi(\mathbf{x}) = \int d^3 x' \frac{\rho(\mathbf{x}')}{|\mathbf{x}-\mathbf{x}'|}. \qquad (2.275)$$

The reader familiar with the theory of the Green functions must have recognized by this time that the propagator is simply the Green function for the time-dependent wave equation satisfying

$$\left[-\left(\frac{\hbar^2}{2m}\right)\nabla''^2 + V(\mathbf{x}'') - i\hbar\frac{\partial}{\partial t}\right]K(\mathbf{x}'',t;\mathbf{x}',t_0) = -i\hbar\delta^3(\mathbf{x}''-\mathbf{x}')\delta(t-t_0) \qquad (2.276)$$

with the boundary condition

$$K(\mathbf{x}'',t;\mathbf{x}',t_0) = 0, \quad \text{for} \quad t < t_0. \qquad (2.277)$$

The delta function $\delta(t-t_0)$ is needed on the right-hand side of (2.276) because $K$ varies discontinuously at $t = t_0$.

The particular form of the propagator is, of course, dependent on the particular potential to which the particle is subjected. Consider, as an example, a free particle in one dimension. The obvious observable that commutes with $H$ is momentum; $|p'\rangle$ is a simultaneous eigenket of the operators $p$ and $H$:

$$p|p'\rangle = p'|p'\rangle \qquad H|p'\rangle = \left(\frac{p'^2}{2m}\right)|p'\rangle. \qquad (2.278)$$

The momentum eigenfunction is just the transformation function of Section 1.7 [see (1.264)] which is of the plane wave form. Combining everything, we have

$$K(x'',t;x',t_0) = \left(\frac{1}{2\pi\hbar}\right)\int_{-\infty}^{\infty} dp' \exp\left[\frac{ip'(x''-x')}{\hbar} - \frac{ip'^2(t-t_0)}{2m\hbar}\right]. \qquad (2.279)$$

The integral can be evaluated by completing the square in the exponent. Here we simply record the result:

$$K(x'',t;x',t_0) = \sqrt{\frac{m}{2\pi i\hbar(t-t_0)}} \exp\left[\frac{im(x''-x')^2}{2\hbar(t-t_0)}\right]. \qquad (2.280)$$

This expression may be used, for example, to study how a Gaussian wave packet spreads out as a function of time.

For the simple harmonic oscillator, where the wave function of an energy eigenstate is given by

$$u_n(x)\exp\left(\frac{-iE_n t}{\hbar}\right) = \left(\frac{1}{2^{n/2}\sqrt{n!}}\right)\left(\frac{m\omega}{\pi\hbar}\right)^{1/4}\exp\left(\frac{-m\omega x^2}{2\hbar}\right)$$

$$\times H_n\left(\sqrt{\frac{m\omega}{\hbar}}x\right)\exp\left[-i\omega\left(n+\frac{1}{2}\right)t\right], \qquad (2.281)$$

the propagator is given by

$$K(x'',t;x',t_0) = \sqrt{\frac{m\omega}{2\pi i\hbar\sin[\omega(t-t_0)]}} \exp\left[\left\{\frac{im\omega}{2\hbar\sin[\omega(t-t_0)]}\right\}\right.$$

$$\times\left\{(x''^2+x'^2)\cos[\omega(t-t_0)]-2x''x'\right\}\right]. \qquad (2.282)$$

One way to prove this is to use

$$\left(\frac{1}{\sqrt{1-\zeta^2}}\right)\exp\left[\frac{-(\xi^2+\eta^2-2\xi\eta\zeta)}{(1-\zeta^2)}\right]$$

$$= \exp\left[-(\xi^2+\eta^2)\right]\sum_{n=0}\left(\frac{\zeta^n}{2^n n!}\right)H_n(\xi)H_n(\eta), \qquad (2.283)$$

which is found in books on special functions (Morse and Feshbach (1953), p. 786). It can also be obtained using the $a$, $a^\dagger$ operator method (Saxon (1968), pp. 144–145) or, alternatively, the path-integral method to be described later. Notice that (2.282) is a periodic function of $t$ with angular frequency $\omega$, the classical oscillator frequency. This means, among other things, that a particle initially localized precisely at $x'$ will return to its original position with certainty at $2\pi/\omega$ ($4\pi/\omega$, and so forth) later.

Certain space and time integrals derivable from $K(\mathbf{x}'',t;\mathbf{x}',t_0)$ are of considerable interest. Without loss of generality we set $t_0 = 0$ in the following. The first integral we consider is obtained by setting $\mathbf{x}'' = \mathbf{x}'$ and integrating over all space. We have

$$G(t) \equiv \int d^3x' K(\mathbf{x}',t;\mathbf{x}',0)$$

$$= \int d^3x' \sum_{a'} |\langle\mathbf{x}'|a'\rangle|^2 \exp\left(\frac{-iE_{a'}t}{\hbar}\right)$$

$$= \sum_{a'}\exp\left(\frac{-iE_{a'}t}{\hbar}\right). \qquad (2.284)$$

This result is anticipated; recalling (2.274), we observe that setting $\mathbf{x}' = \mathbf{x}''$ and integrating are equivalent to taking the trace of the time-evolution operator in the $\mathbf{x}$-representation. But the trace is independent of representations; it can be evaluated more readily using the

$\{|a'\rangle\}$ basis where the time-evolution operator is diagonal, which immediately leads to the last line of (2.284). Now we see that (2.284) is just the "sum over states," reminiscent of the partition function in statistical mechanics. In fact, if we analytically continue in the $t$ variable and make $t$ purely imaginary, with $\beta$ defined by

$$\beta = \frac{it}{\hbar} \tag{2.285}$$

real and positive, we can identify (2.284) with the partition function itself:

$$Z = \sum_{a'} \exp(-\beta E_{a'}). \tag{2.286}$$

For this reason some of the techniques encountered in studying propagators in quantum mechanics are also useful in statistical mechanics.

Next, let us consider the Laplace–Fourier transform of $G(t)$:

$$\tilde{G}(E) \equiv -i \int_0^\infty dt\, G(t) \exp(iEt/\hbar)/\hbar$$
$$= -i \int_0^\infty dt \sum_{a'} \exp(-iE_{a'}t/\hbar) \exp(iEt/\hbar)/\hbar. \tag{2.287}$$

The integrand here oscillates indefinitely. But we can make the integral meaningful by letting $E$ acquire a small positive imaginary part:

$$E \to E + i\varepsilon. \tag{2.288}$$

We then obtain, in the limit $\varepsilon \to 0$,

$$\tilde{G}(E) = \sum_{a'} \frac{1}{E - E_{a'}}. \tag{2.289}$$

Observe now that the complete energy spectrum is exhibited as simple poles of $\tilde{G}(E)$ in the complex $E$-plane. If we wish to know the energy spectrum of a physical system, it is sufficient to study the analytic properties of $\tilde{G}(E)$.

## 2.6.2 Propagator as a Transition Amplitude

To gain further insight into the physical meaning of the propagator, we wish to relate it to the concept of transition amplitudes introduced in Section 2.2. But first, recall that the wave function which is the inner product of the fixed position bra $\langle \mathbf{x}'|$ with the moving state ket $|\alpha, t_0; t\rangle$ can also be regarded as the inner product of the Heisenberg picture position bra $\langle \mathbf{x}', t|$, which moves "oppositely" with time, with the Heisenberg picture state ket $|\alpha, t_0\rangle$, which is fixed in time. Likewise, the propagator can also be written as

$$K(\mathbf{x}'', t; \mathbf{x}', t_0) = \sum_{a'} \langle \mathbf{x}''|a'\rangle \langle a'|\mathbf{x}'\rangle \exp\left[\frac{-iE_{a'}(t - t_0)}{\hbar}\right]$$
$$= \sum_{a'} \langle \mathbf{x}''| \exp\left(\frac{-iHt}{\hbar}\right) |a'\rangle \langle a'| \exp\left(\frac{iHt_0}{\hbar}\right) |\mathbf{x}'\rangle$$
$$= \langle \mathbf{x}'', t|\mathbf{x}', t_0\rangle, \tag{2.290}$$

where $|\mathbf{x}',t_0\rangle$ and $\langle\mathbf{x}'',t|$ are to be understood as an eigenket and an eigenbra of the position operator in the Heisenberg picture. In Section 2.2 we showed that $\langle b',t|a'\rangle$, in the Heisenberg picture notation, is the probability amplitude for a system originally prepared to be an eigenstate of $A$ with eigenvalue $a'$ at some initial time $t_0 = 0$ to be found at a later time $t$ in an eigenstate of $B$ with eigenvalue $b'$, and we called it the transition amplitude for going from state $|a'\rangle$ to state $|b'\rangle$. Because there is nothing special about the choice of $t_0$, only the time difference $t - t_0$ is relevant, we can identify $\langle\mathbf{x}'',t|\mathbf{x}',t_0\rangle$ as the probability amplitude for the particle prepared at $t_0$ with position eigenvalue $\mathbf{x}'$ to be found at a later time $t$ at $\mathbf{x}''$. Roughly speaking, $\langle\mathbf{x}'',t|\mathbf{x}',t_0\rangle$ is the amplitude for the particle to go from a space-time point $(\mathbf{x}',t_0)$ to another space-time point $(\mathbf{x}'',t)$, so the term *transition amplitude* for this expression is quite appropriate. This interpretation is, of course, in complete accord with the interpretation we gave earlier for $K(\mathbf{x}'',t;\mathbf{x}',t_0)$.

Yet another way to interpret $\langle\mathbf{x}'',t|\mathbf{x}',t_0\rangle$ is as follows. As we emphasized earlier, $|\mathbf{x}',t_0\rangle$ is the position eigenket at $t_0$ with the eigenvalue $\mathbf{x}'$ in the Heisenberg picture. Because at any given time the Heisenberg picture eigenkets of an observable can be chosen as base kets, we can regard $\langle\mathbf{x}'',t|\mathbf{x}',t_0\rangle$ as the transformation function that connects the two sets of base kets at *different* times. So in the Heisenberg picture, time evolution can be viewed as a *unitary transformation*, in the sense of changing bases, that connects one set of base kets formed by $\{|\mathbf{x}',t_0\rangle\}$ to another formed by $\{|\mathbf{x}'',t\rangle\}$. This is reminiscent of classical physics, in which the time development of a classical dynamic variable such as $\mathbf{x}(t)$ is viewed as a canonical (or contact) transformation generated by the classical Hamiltonian (Goldstein et al. (2002), pp. 401–402).

It turns out to be convenient to use a notation that treats the space and time coordinates more symmetrically. To this end we write $\langle\mathbf{x}'',t''|\mathbf{x}',t'\rangle$ in place of $\langle\mathbf{x}'',t|\mathbf{x}',t_0\rangle$. Because at any given time the position kets in the Heisenberg picture form a complete set, it is legitimate to insert the identity operator written as

$$\int d^3x''|\mathbf{x}'',t''\rangle\langle\mathbf{x}'',t''| = 1 \tag{2.291}$$

at any place we desire. For example, consider the time evolution from $t'$ to $t'''$; by dividing the time interval $(t',t''')$ into two parts, $(t',t'')$ and $(t'',t''')$, we have

$$\langle\mathbf{x}''',t'''|\mathbf{x}',t'\rangle = \int d^3x''\langle\mathbf{x}''',t'''|\mathbf{x}'',t''\rangle\langle\mathbf{x}'',t''|\mathbf{x}',t'\rangle$$

$$(t''' > t'' > t'). \tag{2.292}$$

We call this the **composition property** of the transition amplitude.[10] Clearly, we can divide the time interval into as many smaller subintervals as we wish. We have

$$\langle\mathbf{x}'''',t''''|\mathbf{x}',t'\rangle = \int d^3x''' \int d^3x''\langle\mathbf{x}'''',t''''|\mathbf{x}''',t'''\rangle\langle\mathbf{x}''',t'''|\mathbf{x}'',t''\rangle$$

$$\times \langle\mathbf{x}'',t''|\mathbf{x}',t'\rangle \quad (t'''' > t''' > t'' > t'), \tag{2.293}$$

and so on. If we somehow guess the form of $\langle\mathbf{x}'',t''|\mathbf{x}',t'\rangle$ for an *infinitesimal* time interval (between $t'$ and $t'' = t' + dt$), we should be able to obtain the amplitude $\langle\mathbf{x}'',t''|\mathbf{x}',t'\rangle$ for a

---

[10] The analogue of (2.292) in probability theory is known as the Chapman–Kolmogoroff equation, and in diffusion theory, the Smoluchowsky equation.

finite time interval by compounding the appropriate transition amplitudes for infinitesimal time intervals in a manner analogous to (2.293). This kind of reasoning leads to an *independent formulation* of quantum mechanics due to R. P. Feynman, published in 1948, to which we now turn our attention Feynman, *Rev. Mod. Phys.*, **20** (1948) 367.

### 2.6.3  Path Integrals as the Sum over Paths

Without loss of generality we restrict ourselves to one-dimensional problems. Also, we avoid awkward expressions like

$$x^{''''\cdots} \text{ (N times)}$$

by using notation such as $x_N$. With this notation we consider the transition amplitude for a particle going from the initial space-time point $(x_1, t_1)$ to the final space-time point $(x_N, t_N)$. The entire time interval between $t_1$ and $t_N$ is divided into $N-1$ equal parts:

$$t_j - t_{j-1} = \Delta t = \frac{(t_N - t_1)}{(N-1)}. \tag{2.294}$$

Exploiting the composition property, we obtain

$$\langle x_N, t_N | x_1, t_1 \rangle = \int dx_{N-1} \int dx_{N-2} \cdots \int dx_2 \langle x_N, t_N | x_{N-1}, t_{N-1} \rangle$$
$$\times \langle x_{N-1}, t_{N-1} | x_{N-2}, t_{N-2} \rangle \cdots \langle x_2, t_2 | x_1, t_1 \rangle. \tag{2.295}$$

To visualize this pictorially, we consider a space-time plane, as shown in Figure 2.6. The initial and final space-time points are fixed to be $(x_1, t_1)$ and $(x_N, t_N)$, respectively. For each time segment, say between $t_{n-1}$ and $t_n$, we are instructed to consider the transition amplitude to go from $(x_{n-1}, t_{n-1})$ to $(x_n, t_n)$; we then integrate over $x_2, x_3, \ldots, x_{N-1}$. This means that we must *sum over all possible paths* in the space-time plane with the end points fixed.

   Before proceeding further, it is profitable to review here how paths appear in classical mechanics. Suppose we have a particle subjected to a force field derivable from a potential $V(x)$. The *classical* Lagrangian is written as

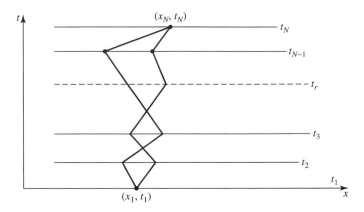

**Fig. 2.6**   Paths in the *xt*-plane.

$$L_{\text{classical}}(x,\dot{x}) = \frac{m\dot{x}^2}{2} - V(x). \tag{2.296}$$

Given this Lagrangian with the end points $(x_1,t_1)$ and $(x_N,t_N)$ specified, we do *not* consider just any path joining $(x_1,t_1)$ and $(x_N,t_N)$ in classical mechanics. On the contrary, there exists a *unique path* that corresponds to the actual motion of the classical particle. For example, given

$$V(x) = mgx, \quad (x_1,t_1) = (h,0), \quad (x_N,t_N) = \left(0, \sqrt{\frac{2h}{g}}\right), \tag{2.297}$$

where $h$ may stand for the height of the Leaning Tower of Pisa, the classical path in the $xt$-plane can *only* be

$$x = h - \frac{gt^2}{2}. \tag{2.298}$$

More generally, according to Hamilton's principle, the unique path is that which minimizes the action, defined as the time integral of the classical Lagrangian:

$$\delta \int_{t_1}^{t_2} dt L_{\text{classical}}(x,\dot{x}) = 0, \tag{2.299}$$

from which Lagrange's equation of motion can be obtained.

### 2.6.4 Feynman's Formulation

The basic difference between classical mechanics and quantum mechanics should now be apparent. In classical mechanics a definite path in the $xt$-plane is associated with the particle's motion; in contrast, in quantum mechanics all possible paths must play roles including those which do not bear any resemblance to the classical path. Yet we must somehow be able to reproduce classical mechanics in a smooth manner in the limit $\hbar \to 0$. How are we to accomplish this?

As a young graduate student at Princeton University, R. P. Feynman tried to attack this problem. In looking for a possible clue, he was said to be intrigued by a mysterious remark in Dirac's book which, in our notation, amounts to the following statement:

$$\exp\left[i \int_{t_1}^{t_2} \frac{dt L_{\text{classical}}(x,\dot{x})}{\hbar}\right] \quad \text{corresponds to} \quad \langle x_2,t_2|x_1,t_1\rangle.$$

Feynman attempted to make sense out of this remark. Is "corresponds to" the same thing as "is equal to" or "is proportional to"? In so doing he was led to formulate a space-time approach to quantum mechanics based on *path integrals*.

In Feynman's formulation the classical action plays a very important role. For compactness, we introduce a new notation:

$$S(n,n-1) \equiv \int_{t_{n-1}}^{t_n} dt L_{\text{classical}}(x,\dot{x}). \tag{2.300}$$

Because $L_{\text{classical}}$ is a function of $x$ and $\dot{x}$, $S(n,n-1)$ is defined only after a definite path is specified along which the integration is to be carried out. So even though the path dependence is not explicit in this notation, it is understood that we are considering a particular path in evaluating the integral. Imagine now that we are following some prescribed path. We concentrate our attention on a small segment along that path, say between $(x_{n-1}, t_{n-1})$ and $(x_n, t_n)$. According to Dirac, we are instructed to associate $\exp[iS(n, n-1)/\hbar]$ with that segment. Going along the definite path we are set to follow, we successively multiply expressions of this type to obtain

$$\prod_{n=2}^{N} \exp\left[\frac{iS(n, n-1)}{\hbar}\right] = \exp\left[\left(\frac{i}{\hbar}\right)\sum_{n=2}^{N} S(n, n-1)\right] = \exp\left[\frac{iS(N, 1)}{\hbar}\right]. \tag{2.301}$$

This does not yet give $\langle x_N, t_N | x_1, t_1 \rangle$; rather, this equation is the contribution to $\langle x_N, t_N | x_1, t_1 \rangle$ arising from the particular path we have considered. We must still integrate over $x_2, x_3, \ldots, x_{N-1}$. At the same time, exploiting the composition property, we let the time interval between $t_{n-1}$ and $t_n$ be infinitesimally small. Thus our candidate expression for $\langle x_N, t_N | x_1, t_1 \rangle$ may be written, in some loose sense, as

$$\langle x_N, t_N | x_1, t_1 \rangle \sim \sum_{\text{all paths}} \exp\left[\frac{iS(N, 1)}{\hbar}\right], \tag{2.302}$$

where the sum is to be taken over an innumerably infinite set of paths!

Before presenting a more precise formulation, let us see whether considerations along this line make sense in the classical limit. As $\hbar \to 0$, the exponential in (2.302) oscillates very violently, so there is a tendency for cancellation among various contributions from neighboring paths. This is because $\exp[iS/\hbar]$ for some definite path and $\exp[iS/\hbar]$ for a slightly different path have very phases because of the smallness of $\hbar$. So most paths do *not* contribute when $\hbar$ is regarded as a small quantity. However, there is an important exception.

Suppose that we consider a path that satisfies

$$\delta S(N, 1) = 0, \tag{2.303}$$

where the change in $S$ is due to a slight deformation of the path with the end points fixed. This is precisely the classical path by virtue of Hamilton's principle. We denote the $S$ that satisfies (2.303) by $S_{\min}$. We now attempt to deform the path a little bit from the classical path. The resulting $S$ is still equal to $S_{\min}$ to first order in deformation. This means that the phase of $\exp[iS/\hbar]$ does not vary very much as we deviate slightly from the classical path even if $\hbar$ is small. As a result, as long as we stay near the classical path, constructive interference between neighboring paths is possible. In the $\hbar \to 0$ limit, the major contributions must then arise from a very narrow strip (or a tube in higher dimensions) containing the classical path, as shown in Figure 2.7. Our (or Feynman's) guess based on Dirac's mysterious remark makes good sense because the classical path is singled out in the $\hbar \to 0$ limit. To formulate Feynman's conjecture more precisely, let us go back to $\langle x_n, t_n | x_{n-1}, t_{n-1} \rangle$, where the time difference $t_n - t_{n-1}$ is assumed to be infinitesimally small. We write

$$\langle x_n, t_n | x_{n-1}, t_{n-1} \rangle = \left[\frac{1}{w(\Delta t)}\right] \exp\left[\frac{iS(n, n-1)}{\hbar}\right], \tag{2.304}$$

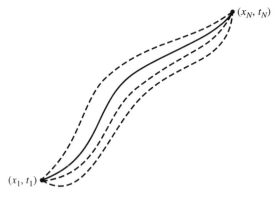

$(x_N, t_N)$

$(x_1, t_1)$

Paths important in the $\hbar \to 0$ limit.

where we evaluate $S(n, n-1)$ in a moment in the $\Delta t \to 0$ limit. Notice that we have inserted a weight factor, $1/w(\Delta t)$, which is assumed to depend only on the time interval $t_n - t_{n-1}$ and not on $V(x)$. That such a factor is needed is clear from dimensional considerations; according to the way we normalized our position eigenkets, $\langle x_n, t_n | x_{n-1}, t_{n-1} \rangle$ must have the dimension of 1/length.

We now look at the exponential in (2.304). Our task is to evaluate the $\Delta t \to 0$ limit of $S(n, n-1)$. Because the time interval is so small, it is legitimate to make a straight-line approximation to the path joining $(x_{n-1}, t_{n-1})$ and $(x_n, t_n)$ as follows:

$$
S(n, n-1) = \int_{t_{n-1}}^{t_n} dt \left[ \frac{m\dot{x}^2}{2} - V(x) \right]
$$
$$
= \Delta t \left\{ \left( \frac{m}{2} \right) \left[ \frac{(x_n - x_{n-1})}{\Delta t} \right]^2 - V\left( \frac{(x_n + x_{n-1})}{2} \right) \right\}. \tag{2.305}
$$

As an example, we consider specifically the free-particle case, $V = 0$. Equation (2.304) now becomes

$$
\langle x_n, t_n | x_{n-1}, t_{n-1} \rangle = \left[ \frac{1}{w(\Delta t)} \right] \exp \left[ \frac{im(x_n - x_{n-1})^2}{2\hbar \Delta t} \right]. \tag{2.306}
$$

We see that the exponent appearing here is completely identical to the one in the expression for the free-particle propagator (2.280). The reader may work out a similar comparison for the simple harmonic oscillator.

We remarked earlier that the weight factor $1/w(\Delta t)$ appearing in (2.304) is assumed to be independent of $V(x)$, so we may as well evaluate it for the free particle. Noting the orthonormality, in the sense of $\delta$-function, of Heisenberg picture position eigenkets at equal times,

$$
\langle x_n, t_n | x_{n-1}, t_{n-1} \rangle |_{t_n = t_{n-1}} = \delta(x_n - x_{n-1}), \tag{2.307}
$$

we obtain

$$
\frac{1}{w(\Delta t)} = \sqrt{\frac{m}{2\pi i \hbar \Delta t}}, \tag{2.308}
$$

where we have used

$$\int_{-\infty}^{\infty} d\xi \exp\left(\frac{im\xi^2}{2\hbar\Delta t}\right) = \sqrt{\frac{2\pi i\hbar\Delta t}{m}} \tag{2.309a}$$

and

$$\lim_{\Delta t \to 0} \sqrt{\frac{m}{2\pi i\hbar\Delta t}} \exp\left(\frac{im\xi^2}{2\hbar\Delta t}\right) = \delta(\xi). \tag{2.309b}$$

This weight factor is, of course, anticipated from the expression for the free-particle propagator (2.280).

To summarize, as $\Delta t \to 0$, we are led to

$$\langle x_n, t_n | x_{n-1}, t_{n-1} \rangle = \sqrt{\frac{m}{2\pi i\hbar\Delta t}} \exp\left[\frac{iS(n, n-1)}{\hbar}\right]. \tag{2.310}$$

The final expression for the transition amplitude with $t_N - t_1$ finite is

$$\langle x_N, t_N | x_1, t_1 \rangle = \lim_{N \to \infty} \left(\frac{m}{2\pi i\hbar\Delta t}\right)^{(N-1)/2}$$
$$\times \int dx_{N-1} \int dx_{N-2} \cdots \int dx_2 \prod_{n=2}^{N} \exp\left[\frac{iS(n, n-1)}{\hbar}\right], \tag{2.311}$$

where the $N \to \infty$ limit is taken with $x_N$ and $t_N$ fixed. It is customary here to define a new kind of multidimensional (in fact, infinite-dimensional) integral operator

$$\int_{x_1}^{x_N} \mathscr{D}[x(t)] \equiv \lim_{N \to \infty} \left(\frac{m}{2\pi i\hbar\Delta t}\right)^{(N-1)/2} \int dx_{N-1} \int dx_{N-2} \cdots \int dx_2 \tag{2.312}$$

and write (2.311) as

$$\langle x_N, t_N | x_1, t_1 \rangle = \int_{x_1}^{x_N} \mathscr{D}[x(t)] \exp\left[i \int_{t_1}^{t_N} dt \frac{L_{\text{classical}}(x, \dot{x})}{\hbar}\right]. \tag{2.313}$$

This expression is known as **Feynman's path integral**. Its meaning as the sum over all possible paths should be apparent from (2.311).

Our steps leading to (2.313) are not meant to be a derivation. Rather, we (or Feynman) have attempted a new formulation of quantum mechanics based on the concept of paths, motivated by Dirac's mysterious remark. The only ideas we borrowed from the conventional form of quantum mechanics are (1) the superposition principle (used in summing the contributions from various alternate paths), (2) the composition property of the transition amplitude, and (3) classical correspondence in the $\hbar \to 0$ limit.

Even though we obtained the same result as the conventional theory for the free-particle case, it is now obvious, from what we have done so far, that Feynman's formulation is completely equivalent to Schrödinger's wave mechanics. We conclude this section by proving that Feynman's expression for $\langle x_N, t_N | x_1, t_1 \rangle$ indeed satisfies Schrödinger's time-dependent wave equation in the variables $x_N, t_N$, just as the propagator defined by (2.272).

We start with

$$
\begin{aligned}
\langle x_N, t_N | x_1, t_1 \rangle &= \int dx_{N-1} \langle x_N, t_N | x_{N-1}, t_{N-1} \rangle \langle x_{N-1}, t_{N-1} | x_1, t_1 \rangle \\
&= \int_{-\infty}^{\infty} dx_{N-1} \sqrt{\frac{m}{2\pi i \hbar \Delta t}} \exp\left[ \left(\frac{im}{2\hbar}\right) \frac{(x_N - x_{N-1})^2}{\Delta t} - \frac{iV\Delta t}{\hbar} \right] \\
&\quad \times \langle x_{N-1}, t_{N-1} | x_1, t_1 \rangle,
\end{aligned}
\tag{2.314}
$$

where we have assumed $t_N - t_{N-1}$ to be infinitesimal. Introducing

$$
\xi = x_N - x_{N-1}
\tag{2.315}
$$

and letting $x_N \to x$ and $t_N \to t + \Delta t$, we obtain

$$
\langle x, t + \Delta t | x_1, t_1 \rangle = \sqrt{\frac{m}{2\pi i \hbar \Delta t}} \int_{-\infty}^{(\infty)} d\xi \exp\left( \frac{im\xi^2}{2\hbar \Delta t} - \frac{iV\Delta t}{\hbar} \right) \langle x - \xi, t | x_1, t_1 \rangle.
\tag{2.316}
$$

As is evident from (2.309b), in the limit $\Delta t \to 0$, the major contribution to this integral comes from the $\xi \simeq 0$ region. It is therefore legitimate to expand $\langle x - \xi, t | x_1, t_1 \rangle$ in powers of $\xi$. We also expand $\langle x, t + \Delta t | x_1, t_1 \rangle$ and $\exp(-iV\Delta t/\hbar)$ in powers of $\Delta t$, so

$$
\begin{aligned}
\langle x, t | x_1, t_1 \rangle + \Delta t \frac{\partial}{\partial t} \langle x, t | x_1, t_1 \rangle &= \sqrt{\frac{m}{2\pi i \hbar \Delta t}} \int_{-\infty}^{\infty} d\xi \exp\left( \frac{im\xi^2}{2\hbar \Delta t} \right) \left( 1 - \frac{iV\Delta t}{\hbar} + \cdots \right) \\
&\quad \times \left[ \langle x, t | x_1, t_1 \rangle + \left( \frac{\xi^2}{2} \right) \frac{\partial^2}{\partial x^2} \langle x, t | x_1, t_1 \rangle + \cdots \right],
\end{aligned}
\tag{2.317}
$$

where we have dropped a term linear in $\xi$ because it vanishes when integrated with respect to $\xi$. The $\langle x, t | x_1, t_1 \rangle$ term on the left-hand side just matches the leading term on the right-hand side because of (2.309a). Collecting terms first order in $\Delta t$, we obtain

$$
\begin{aligned}
\Delta t \frac{\partial}{\partial t} \langle x, t | x_1, t_1 \rangle &= \left( \sqrt{\frac{m}{2\pi i \hbar \Delta t}} \right) (\sqrt{2\pi}) \left( \frac{i\hbar \Delta t}{m} \right)^{3/2} \frac{1}{2} \frac{\partial^2}{\partial x^2} \langle x, t | x_1, t_1 \rangle \\
&\quad - \left( \frac{i}{\hbar} \right) \Delta t V \langle x, t | x_1, t_1 \rangle,
\end{aligned}
\tag{2.318}
$$

where we have used

$$
\int_{-\infty}^{\infty} d\xi\, \xi^2 \exp\left( \frac{im\xi^2}{2\hbar \Delta t} \right) = \sqrt{2\pi} \left( \frac{i\hbar \Delta t}{m} \right)^{3/2},
\tag{2.319}
$$

obtained by differentiating (2.309a) with respect to $\Delta t$. In this manner we see that $\langle x, t | x_1, t_1 \rangle$ satisfies Schrödinger's time-dependent wave equation:

$$
i\hbar \frac{\partial}{\partial t} \langle x, t | x_1, t_1 \rangle = - \left( \frac{\hbar^2}{2m} \right) \frac{\partial^2}{\partial x^2} \langle x, t | x_1, t_1 \rangle + V \langle x, t | x_1, t_1 \rangle.
\tag{2.320}
$$

Thus we can conclude that $\langle x, t | x_1, t_1 \rangle$ constructed according to Feynman's prescription is the same as the propagator in Schrödinger's wave mechanics.

Feynman's space-time approach based on path integrals is not too convenient for attacking practical problems in nonrelativistic quantum mechanics. Even for the simple harmonic oscillator it is rather cumbersome to evaluate explicitly the relevant path integral.[11] However, his approach is extremely gratifying from a conceptual point of view. By imposing a certain set of sensible requirements on a physical theory, we are inevitably led to a formalism equivalent to the usual formulation of quantum mechanics. It makes us wonder whether it is at all possible to construct a sensible alternative theory that is equally successful in accounting for microscopic phenomena.

Methods based on path integrals have been found to be very powerful in other branches of modern physics, such as quantum field theory and statistical mechanics. In this book the path-integral method will appear again when we discuss the Aharonov–Bohm effect.[12]

## 2.7 Potentials and Gauge Transformations

### 2.7.1 Constant Potentials

In classical mechanics it is well known that the zero point of the potential energy is of no physical significance. The time development of dynamic variables such as $\mathbf{x}(t)$ and $\mathbf{L}(t)$ is independent of whether we use $V(\mathbf{x})$ or $V(\mathbf{x}) + V_0$ with $V_0$ constant in both space and time. The force that appears in Newton's second law depends only on the gradient of the potential; an additive constant is clearly irrelevant. What is the analogous situation in quantum mechanics?

We look at the time evolution of a Schrödinger picture state ket subject to some potential. Let $|\alpha, t_0; t\rangle$ be a state ket in the presence of $V(\mathbf{x})$, and $|\widetilde{\alpha, t_0}; t\rangle$, the corresponding state ket appropriate for

$$\tilde{V}(\mathbf{x}) = V(\mathbf{x}) + V_0. \tag{2.321}$$

To be precise, let us agree that the initial conditions are such that both kets coincide with $|\alpha\rangle$ at $t = t_0$. If they represent the same physical situation, this can always be done by a suitable choice of the phase. Recalling that the state ket at $t$ can be obtained by applying the time-evolution operator $\mathscr{U}(t, t_0)$ to the state ket at $t_0$, we obtain

$$|\widetilde{\alpha, t_0}; t\rangle = \exp\left[-i\left(\frac{\mathbf{p}^2}{2m} + V(x) + V_0\right)\frac{(t - t_0)}{\hbar}\right]|\alpha\rangle$$

$$= \exp\left[\frac{-iV_0(t - t_0)}{\hbar}\right]|\alpha, t_0; t\rangle. \tag{2.322}$$

In other words, the ket computed under the influence of $\tilde{V}$ has a time dependence different only by a phase factor $\exp[-iV_0(t - t_0)/\hbar]$. For stationary states, this means that if the time dependence computed with $V(\mathbf{x})$ is $\exp[-iE(t - t_0)/\hbar]$, then the corresponding time

---

[11] The reader is challenged to solve the simple harmonic oscillator problem using the Feynman path-integral method in Problem 2.44 of this chapter.

[12] The reader who is interested in the fundamentals and applications of path integrals may consult Feynman and Hibbs (1965) and also Zee (2010).

dependence computed with $V(\mathbf{x}) + V_0$ is $\exp[-i(E + V_0)(t - t_0)/\hbar]$. In other words, the use of $\tilde{V}$ in place of $V$ just amounts to the following change:

$$E \rightarrow E + V_0, \tag{2.323}$$

which the reader probably guessed immediately. Observable effects such as the time evolution of expectation values of $\langle \mathbf{x} \rangle$ and $\langle \mathbf{S} \rangle$ always depend on energy *differences* [see (2.47)]; the Bohr frequencies that characterize the sinusoidal time dependence of expectation values are the same whether we use $V(\mathbf{x})$ or $V(\mathbf{x}) + V_0$. In general, there can be no difference in the expectation values of observables if every state ket in the world is multiplied by a common factor $\exp[-iV_0(t - t_0)/\hbar]$.

Trivial as it may seem, we see here the first example of a class of transformations known as **gauge transformations**. The change in our convention for the zero point energy of the potential

$$V(\mathbf{x}) \rightarrow V(\mathbf{x}) + V_0 \tag{2.324}$$

must be accompanied by a change in the state ket

$$|\alpha, t_0; t\rangle \rightarrow \exp\left[\frac{-iV_0(t - t_0)}{\hbar}\right] |\alpha, t_0; t\rangle. \tag{2.325}$$

Of course, this change implies the following change in the wave function:

$$\psi(\mathbf{x}', t) \rightarrow \exp\left[\frac{-iV_0(t - t_0)}{\hbar}\right] \psi(\mathbf{x}', t). \tag{2.326}$$

Next we consider $V_0$ that is spatially uniform but dependent on time. We then easily see that the analogue of (2.325) is

$$|\alpha, t_0; t\rangle \rightarrow \exp\left[-i \int_{t_0}^{t} dt' \frac{V_0(t')}{\hbar}\right] |\alpha, t_0; t\rangle. \tag{2.327}$$

Physically, the use of $V(\mathbf{x}) + V_0(t)$ in place of $V(\mathbf{x})$ simply means that we are choosing a new zero point of the energy scale at each instant of time.

Even though the choice of the absolute scale of the potential is arbitrary, *potential differences* are of nontrivial physical significance and, in fact, can be detected in a very striking way. To illustrate this point, let us consider the arrangement shown in Figure 2.8. A beam of charged particles is split into two parts, each of which enters a metallic cage. If we so desire, we can maintain a finite potential difference between the two cages by turning on a switch, as shown. A particle in the beam can be visualized as a wave packet whose dimension is much smaller than the dimension of the cage. Suppose we switch on the potential difference only after the wave packets enter the cages and switch it off before the wave packets leave the cages. The particle in the cage experiences *no force* because inside the cage the potential is spatially uniform; hence no electric field is present. Now let us recombine the two beam components in such a way that they meet in the interference region of Figure 2.8. Because of the existence of the potential, each beam component suffers a phase change, as indicated by (2.327). As a result, there is an observable interference term in the beam intensity in the interference region, namely,

$$\cos(\phi_1 - \phi_2), \quad \sin(\phi_1 - \phi_2), \tag{2.328}$$

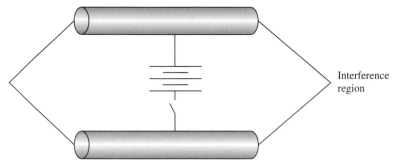

**Fig. 2.8** Quantum-mechanical interference to detect a potential difference.

where

$$\phi_1 - \phi_2 = \left(\frac{1}{\hbar}\right) \int_{t_i}^{t_f} dt [V_2(t) - V_1(t)].$$ (2.329)

So despite the fact that the particle experiences no force, there is an observable effect that depends on whether $V_2(t) - V_1(t)$ has been applied. Notice that this effect is *purely quantum mechanical*; in the limit $\hbar \to 0$, the interesting interference effect is washed out because the oscillation of the cosine becomes infinitely rapid.[13]

## 2.7.2 Gravity in Quantum Mechanics

There is an experiment that exhibits in a striking manner how a gravitational effect appears in quantum mechanics. Before describing it, we first comment on the role of gravity in both classical and quantum mechanics.

Consider the classical equation of motion for a purely falling body:

$$m\ddot{\mathbf{x}} = -m\boldsymbol{\nabla}\Phi_{\text{grav}} = -mg\hat{\mathbf{z}}.$$ (2.330)

The mass term drops out; so in the absence of air resistance, a feather and a stone would behave in the same way – à la Galileo – under the influence of gravity. This is, of course, a direct consequence of the equality of the gravitational and the inertial masses. Because the mass does not appear in the equation of a particle trajectory, gravity in classical mechanics is often said to be a purely geometric theory.

The situation is rather different in quantum mechanics. In the wave-mechanical formulation, the analogue of (2.330) is

$$\left[-\left(\frac{\hbar^2}{2m}\right)\boldsymbol{\nabla}^2 + m\Phi_{\text{grav}}\right]\psi = i\hbar\frac{\partial\psi}{\partial t}.$$ (2.331)

The mass no longer cancels; instead it appears in the combination $\hbar/m$, so in a problem where $\hbar$ appears, $m$ is also expected to appear. We can see this point also using the Feynman path-integral formulation of a falling body based on

---

[13] This gedanken experiment is the Minkowski-rotated form of the Aharonov–Bohm experiment to be discussed later in this section.

$$\langle \mathbf{x}_n, t_n | \mathbf{x}_{n-1}, t_{n-1} \rangle = \sqrt{\frac{m}{2\pi i \hbar \Delta t}} \exp\left[ i \int_{t_{n-1}}^{t_n} dt \frac{(\frac{1}{2}m\dot{\mathbf{x}}^2 - mgz)}{\hbar} \right]$$

$$(t_n - t_{n-1} = \Delta t \to 0). \qquad (2.332)$$

Here again we see that $m$ appears in the combination $m/\hbar$. This is in sharp contrast with Hamilton's classical approach, based on

$$\delta \int_{t_1}^{t_2} dt \left( \frac{m\dot{\mathbf{x}}^2}{2} - mgz \right) = 0, \qquad (2.333)$$

where $m$ can be eliminated in the very beginning.

Starting with the Schrödinger equation (2.331), we may derive the Ehrenfest theorem

$$\frac{d^2}{dt^2} \langle \mathbf{x} \rangle = -g\hat{\mathbf{z}}. \qquad (2.334)$$

However, $\hbar$ does not appear here, nor does $m$. To see a *nontrivial* quantum-mechanical effect of gravity, we must study effects in which $\hbar$ appears explicitly – and consequently where we expect the mass to appear – in contrast with purely gravitational phenomena in classical mechanics.

Until 1975, there had been no direct experiment that established the presence of the $m\Phi_{\text{grav}}$ term in (2.331). To be sure, a free fall of an elementary particle had been observed, but the classical equation of motion, or the Ehrenfest theorem (2.334), where $\hbar$ does not appear, sufficed to account for this. The famous "weight of photon" experiment of V. Pound and collaborators did not test gravity in the quantum domain either because they measured a frequency shift where $\hbar$ does not explicitly appear.

On the microscopic scale, gravitational forces are too weak to be readily observable. To appreciate the difficulty involved in seeing gravity in bound-state problems, let us consider the ground state of an electron and a neutron bound by gravitational forces. This is the gravitational analogue of the hydrogen atom, where an electron and a proton are bound by Coulomb forces. At the same distance, the gravitational force between the electron and the neutron is weaker than the Coulomb force between the electron and the proton by a factor of $\sim 2 \times 10^{39}$. The Bohr radius involved here can be obtained simply:

$$a_0 = \frac{\hbar^2}{e^2 m_e} \to \frac{\hbar^2}{G_N m_e^2 m_n}, \qquad (2.335)$$

where $G_N$ is Newton's gravitational constant. If we substitute numbers in the equation, the Bohr radius of this gravitationally bound system turns out to be $\sim 10^{31}$, or $\sim 10^{13}$ light years, which is larger than the estimated radius of the universe by a few orders of magnitude!

We now discuss a remarkable phenomenon known as **gravity-induced quantum interference**. A nearly monoenergetic beam of particles, in practice, thermal neutrons, is split into two parts and then brought together as shown in Figure 2.9. In actual experiments the neutron beam is split and bent by silicon crystals, but the details of this beautiful art of neutron interferometry do not concern us here. Because the size of the wave packet can be assumed to be much smaller than the macroscopic dimension of the loop formed by the two

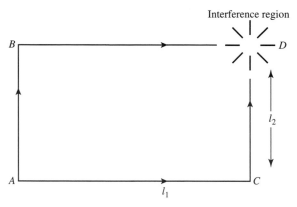

Fig. 2.9
Experiment to detect gravity-induced quantum interference.

alternate paths, we can apply the concept of a classical trajectory. Let us first suppose that path $A \to B \to D$ and path $A \to C \to D$ lie in a horizontal plane. Because the absolute zero of the potential due to gravity is of no significance, we can set $V = 0$ for any phenomenon that takes place in this plane; in other words, it is legitimate to ignore gravity altogether. The situation is very different if the plane formed by the two alternate paths is rotated around segment $AC$ by $\delta$. This time the potential at level $BD$ is higher than that at level $AC$ by $mgl_2 \sin \delta$, which means that the state ket associated with path $BD$ "rotates faster." This leads to a gravity-induced phase difference between the amplitudes for the two wave packets arriving at $D$. Actually there is also a gravity-induced phase change associated with $AB$ and also with $CD$, but the effects cancel as we compare the two alternate paths. The net result is that the wave packet arriving at $D$ via path $ABD$ suffers a phase change

$$\exp\left[\frac{-im_n g l_2 (\sin \delta) T}{\hbar}\right] \tag{2.336}$$

relative to that of the wave packet arriving at $D$ via path $ACD$, where $T$ is the time spent for the wave packet to go from $B$ to $D$ (or from $A$ to $C$) and $m_n$ is the neutron mass. We can control this phase difference by rotating the plane of Figure 2.9; $\delta$ can change from 0 to $\pi/2$, or from 0 to $-\pi/2$. Expressing the time spent $T$, or $l_1/v_{\text{wavepacket}}$, in terms of $\lambdabar$, the de Broglie wavelength of the neutron, we obtain the following expression for the phase difference:

$$\phi_{ABD} - \phi_{ACD} = -\frac{(m_n^2 g l_1 l_2 \lambdabar \sin \delta)}{\hbar^2}. \tag{2.337}$$

In this manner we predict an observable interference effect that depends on angle $\delta$, which is reminiscent of fringes in Michelson-type interferometers in optics.

An alternative, more wave-mechanical way to understand (2.337) follows. Because we are concerned with a time-independent potential, the sum of the kinetic energy and the potential energy is constant:

$$\frac{\mathbf{p}^2}{2m} + mgz = E. \tag{2.338}$$

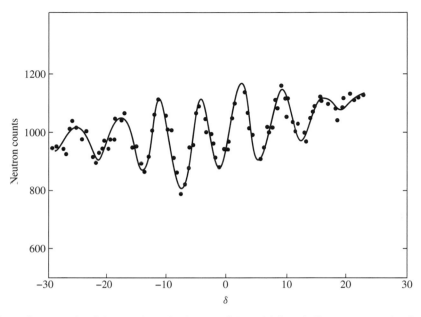

**Fig. 2.10**    Dependence of gravity-induced phase on the angle of rotation $\delta$. From Colella et al. *Phys. Rev. Lett.*, **34** (1975) 1472.

The difference in height between level $BD$ and level $AC$ implies a slight difference in $\mathbf{p}$, or $\lambda$. As a result, there is an accumulation of phase differences due to the $\lambda$ difference. It is left as an exercise to show that this wave-mechanical approach also leads to result (2.337).

What is interesting about expression (2.337) is that its magnitude is neither too small nor too large; it is just right for this interesting effect to be detected with thermal neutrons traveling through paths of "table-top" dimensions. For $\lambda = 1.42\,\text{Å}$ (comparable to interatomic spacing in silicon) and $l_1 l_2 = 10\,\text{cm}^2$, we obtain 55.6 for $m_n^2 g l_1 l_2 \lambda / \hbar^2$. As we rotate the loop plane gradually by $90°$, we predict the intensity in the interference region to exhibit a series of maxima and minima; quantitatively we should see $55.6/2\pi \simeq 9$ oscillations. It is extraordinary that such an effect has indeed been observed experimentally; see Figure 2.10 taken from a 1975 experiment of R. Colella, A. Overhauser, and S. A. Werner. The phase shift due to gravity is seen to be verified to well within 1%.

We emphasize that this effect is purely quantum mechanical because as $\hbar \to 0$, the interference pattern gets washed out. The gravitational potential has been shown to enter into the Schrödinger equation just as expected. This experiment also shows that gravity is not purely geometric at the quantum level because the effect depends on $(m/\hbar)^2$.[14]

---

[14] However, this does not imply that the equivalence principle is unimportant in understanding an effect of this sort. If the gravitational mass ($m_{\text{grav}}$) and inertial mass ($m_{\text{inert}}$) were unequal, $(m/\hbar)^2$ would have to be replaced by $m_{\text{grav}} m_{\text{inert}}/\hbar^2$. The fact that we could correctly predict the interference pattern without making a distinction between $m_{\text{grav}}$ and $m_{\text{inert}}$ shows some support for the equivalence principle at the quantum level.

## 2.7.3　Gauge Transformations in Electromagnetism

Let us now turn to potentials that appear in electromagnetism. We consider an electric and a magnetic field derivable from the time-independent scalar and vector potential, $\phi(\mathbf{x})$ and $\mathbf{A}(\mathbf{x})$:

$$\mathbf{E} = -\nabla\phi, \quad \mathbf{B} = \nabla \times \mathbf{A}. \tag{2.339}$$

The Hamiltonian for a particle of electric charge $e$ ($e < 0$ for the electron) subjected to the electromagnetic field is taken from classical physics to be

$$H = \frac{1}{2m}\left(\mathbf{p} - \frac{e\mathbf{A}}{c}\right)^2 + e\phi. \tag{2.340}$$

In quantum mechanics $\phi$ and $\mathbf{A}$ are understood to be functions of the position *operator* $\mathbf{x}$ of the charged particle. Because $\mathbf{p}$ and $\mathbf{A}$ do not commute, some care is needed in interpreting (2.340). The safest procedure is to write

$$\left(\mathbf{p} - \frac{e\mathbf{A}}{c}\right)^2 \to p^2 - \left(\frac{e}{c}\right)(\mathbf{p}\cdot\mathbf{A} + \mathbf{A}\cdot\mathbf{p}) + \left(\frac{e}{c}\right)^2\mathbf{A}^2. \tag{2.341}$$

In this form the Hamiltonian is obviously Hermitian.

To study the dynamics of a charged particle subjected to $\phi$ and $\mathbf{A}$, let us first proceed in the Heisenberg picture. We can evaluate the time derivative of $\mathbf{x}$ in a straightforward manner as

$$\frac{dx_i}{dt} = \frac{[x_i, H]}{i\hbar} = \frac{(p_i - eA_i/c)}{m}, \tag{2.342}$$

which shows that the operator $\mathbf{p}$, defined in this book to be the generator of translation, is not the same as $m\,d\mathbf{x}/dt$. Quite often $\mathbf{p}$ is called **canonical momentum**, as distinguished from **kinematical** (or mechanical) **momentum**, denoted by $\mathbf{\Pi}$:

$$\mathbf{\Pi} \equiv m\frac{d\mathbf{x}}{dt} = \mathbf{p} - \frac{e\mathbf{A}}{c}. \tag{2.343}$$

Even though we have

$$[p_i, p_j] = 0 \tag{2.344}$$

for canonical momentum, the analogous commutator does not vanish for mechanical momentum. Instead we have

$$[\Pi_i, \Pi_j] = \left(\frac{i\hbar e}{c}\right)\varepsilon_{ijk}B_k, \tag{2.345}$$

as the reader may easily verify. Rewriting the Hamiltonian as

$$H = \frac{\mathbf{\Pi}^2}{2m} + e\phi \tag{2.346}$$

and using the fundamental commutation relation, we can derive the quantum-mechanical version of the **Lorentz force**, namely,

$$m\frac{d^2\mathbf{x}}{dt^2} = \frac{d\mathbf{\Pi}}{dt} = e\left[\mathbf{E} + \frac{1}{2c}\left(\frac{d\mathbf{x}}{dt} \times \mathbf{B} - \mathbf{B} \times \frac{d\mathbf{x}}{dt}\right)\right]. \tag{2.347}$$

This then is Ehrenfest's theorem, written in the Heisenberg picture, for the charged particle in the presence of **E** and **B**.

We now study Schrödinger's wave equation with $\phi$ and **A**. Our first task is to sandwich $H$ between $\langle \mathbf{x}' |$ and $| \alpha, t_0; t \rangle$. The only term with which we have to be careful is

$$\langle \mathbf{x}' | \left[ \mathbf{p} - \frac{e\mathbf{A}(\mathbf{x})}{c} \right]^2 | \alpha, t_0; t \rangle = \left[ -i\hbar \mathbf{\nabla}' - \frac{e\mathbf{A}(\mathbf{x}')}{c} \right] \langle \mathbf{x}' | \left[ \mathbf{p} - \frac{e\mathbf{A}(\mathbf{x})}{c} \right] | \alpha, t_0; t \rangle$$

$$= \left[ -i\hbar \mathbf{\nabla}' - \frac{e\mathbf{A}(\mathbf{x}')}{c} \right] \cdot \left[ -i\hbar \mathbf{\nabla}' - \frac{e\mathbf{A}(\mathbf{x}')}{c} \right] \langle \mathbf{x}' | \alpha, t_0; t \rangle. \tag{2.348}$$

It is important to emphasize that the first $\mathbf{\nabla}'$ in the last line can differentiate *both* $\langle \mathbf{x}' | \alpha, t_0; t \rangle$ and $\mathbf{A}(\mathbf{x}')$. Combining everything, we have

$$\frac{1}{2m} \left[ -i\hbar \mathbf{\nabla}' - \frac{e\mathbf{A}(\mathbf{x}')}{c} \right] \cdot \left[ -i\hbar \mathbf{\nabla}' - \frac{e\mathbf{A}(\mathbf{x}')}{c} \right] \langle \mathbf{x}' | \alpha, t_0; t \rangle$$

$$+ e\phi(\mathbf{x}') \langle \mathbf{x}' | \alpha, t_0; t \rangle = i\hbar \frac{\partial}{\partial t} \langle \mathbf{x}' | \alpha, t_0; t \rangle. \tag{2.349}$$

From this expression we readily obtain the continuity equation

$$\frac{\partial \rho}{\partial t} + \mathbf{\nabla}' \cdot \mathbf{j} = 0, \tag{2.350}$$

where $\rho$ is $|\psi|^2$ as before, with $\langle \mathbf{x}' | \alpha, t_0; t \rangle$ written as $\psi$, but for the probability flux **j** we have

$$\mathbf{j} = \left( \frac{\hbar}{m} \right) \text{Im}(\psi^* \mathbf{\nabla}' \psi) - \left( \frac{e}{mc} \right) \mathbf{A} |\psi|^2, \tag{2.351}$$

which is just what we expect from the substitution

$$\mathbf{\nabla}' \to \mathbf{\nabla}' - \left( \frac{ie}{\hbar c} \right) \mathbf{A}. \tag{2.352}$$

Writing the wave function of $\sqrt{\rho} \exp(iS/\hbar)$ [see (2.193)], we obtain an alternative form for **j**, namely,

$$\mathbf{j} = \left( \frac{\rho}{m} \right) \left( \mathbf{\nabla} S - \frac{e\mathbf{A}}{c} \right), \tag{2.353}$$

which is to be compared with (2.195). We will find this form to be convenient in discussing superconductivity, flux quantization, and so on. We also note that the space integral of **j** is the expectation value of kinematical momentum (not canonical momentum) apart from $1/m$:

$$\int d^3 x' \mathbf{j} = \frac{\langle \mathbf{p} - e\mathbf{A}/c \rangle}{m} = \langle \mathbf{\Pi} \rangle / m. \tag{2.354}$$

We are now in a position to discuss the subject of **gauge transformations** in electromagnetism. First, consider

$$\phi \to \phi + \lambda, \qquad \mathbf{A} \to \mathbf{A}, \tag{2.355}$$

with $\lambda$ constant, that is, independent of $\mathbf{x}$ and $t$. Both $\mathbf{E}$ and $\mathbf{B}$ obviously remain unchanged. This transformation just amounts to a change in the zero point of the energy scale, a possibility treated in the beginning of this section; we just replace $V$ by $e\phi$. We have already discussed the accompanying change needed for the state ket [see (2.325)], so we do not dwell on this transformation any further.

Much more interesting is the transformation

$$\phi \to \phi, \qquad \mathbf{A} \to \mathbf{A} + \nabla \Lambda, \tag{2.356}$$

where $\Lambda$ is a function of $\mathbf{x}$. The static electromagnetic fields $\mathbf{E}$ and $\mathbf{B}$ are unchanged under (2.356). Both (2.355) and (2.356) are special cases of

$$\phi \to \phi - \frac{1}{c}\frac{\partial \Lambda}{\partial t}, \qquad \mathbf{A} \to \mathbf{A} + \nabla \Lambda, \tag{2.357}$$

which leave $\mathbf{E}$ and $\mathbf{B}$, given by

$$\mathbf{E} = -\nabla \phi - \frac{1}{c}\frac{\partial \mathbf{A}}{\partial t}, \qquad \mathbf{B} = \nabla \times \mathbf{A}, \tag{2.358}$$

unchanged, but in the following we do not consider time-dependent fields and potentials. In the remaining part of this section the term *gauge transformation* refers to (2.356).

In classical physics observable effects such as the trajectory of a charged particle are independent of the gauge used, that is, of the particular choice of $\Lambda$ we happen to adopt. Consider a charged particle in a uniform magnetic field in the $z$-direction

$$\mathbf{B} = B\hat{\mathbf{z}}. \tag{2.359}$$

This magnetic field may be derived from

$$A_x = \frac{-By}{2}, \qquad A_y = \frac{Bx}{2}, \qquad A_z = 0 \tag{2.360}$$

or also from

$$A_x = -By, \qquad A_y = 0, \qquad A_z = 0. \tag{2.361}$$

The second form is obtained from the first by

$$\mathbf{A} \to \mathbf{A} - \nabla\left(\frac{Bxy}{2}\right), \tag{2.362}$$

which is indeed of the form of (2.356). Regardless of which $\mathbf{A}$ we may use, the trajectory of the charged particle with a given set of initial conditions is the same; it is just a helix – a uniform circular motion when projected in the $xy$-plane, superposed with a uniform rectilinear motion in the $z$-direction. Yet if we look at $p_x$ and $p_y$, the results are very different. For one thing, $p_x$ is a constant of the motion when (2.361) is used but not when (2.360) is used.

Recall Hamilton's equations of motion:

$$\frac{dp_x}{dt} = -\frac{\partial H}{\partial x}, \qquad \frac{dp_y}{dt} = -\frac{\partial H}{\partial y}, \dots \tag{2.363}$$

In general, the canonical momentum $\mathbf{p}$ is *not* a gauge-invariant quantity; its numerical value depends on the particular gauge used even when we are referring to the same physical situation. In contrast, the *kinematic* momentum $\mathbf{\Pi}$, or $md\mathbf{x}/dt$, that traces the trajectory of the particle *is* a gauge-invariant quantity, as one may explicitly verify. Because $\mathbf{p}$ and $md\mathbf{x}/dt$ are related via (2.343), $\mathbf{p}$ must change to compensate for the change in $\mathbf{A}$ given by (2.362).

We now return to quantum mechanics. We believe that it is reasonable to demand that the expectation values in quantum mechanics behave in a manner similar to the corresponding classical quantities under gauge transformations, so $\langle \mathbf{x} \rangle$ and $\langle \mathbf{\Pi} \rangle$ are *not* to change under gauge transformations, while $\langle \mathbf{p} \rangle$ is expected to change.

Let us denote by $|\alpha\rangle$ the state ket in the presence of $\mathbf{A}$; the state ket for the same physical situation when

$$\tilde{\mathbf{A}} = \mathbf{A} + \boldsymbol{\nabla}\Lambda \tag{2.364}$$

is used in place of $\mathbf{A}$ is denoted by $|\tilde{\alpha}\rangle$. Here $\Lambda$, as well as $\mathbf{A}$, is a function of the position operator $\mathbf{x}$. Our basic requirements are

$$\langle\alpha|\mathbf{x}|\alpha\rangle = \langle\tilde{\alpha}|\mathbf{x}|\tilde{\alpha}\rangle \tag{2.365a}$$

and

$$\langle\alpha|\left(\mathbf{p} - \frac{e\mathbf{A}}{c}\right)|\alpha\rangle = \langle\tilde{\alpha}|\left(\mathbf{p} - \frac{e\tilde{\mathbf{A}}}{c}\right)|\tilde{\alpha}\rangle. \tag{2.365b}$$

In addition, we require, as usual, the norm of the state ket to be preserved:

$$\langle\alpha|\alpha\rangle = \langle\tilde{\alpha}|\tilde{\alpha}\rangle. \tag{2.366}$$

We must construct an operator $\mathscr{G}$ that relates $|\tilde{\alpha}\rangle$ to $|\alpha\rangle$:

$$|\tilde{\alpha}\rangle = \mathscr{G}|\alpha\rangle. \tag{2.367}$$

Invariance properties (2.365a) and (2.365b) are guaranteed if

$$\mathscr{G}^\dagger \mathbf{x} \mathscr{G} = \mathbf{x} \tag{2.368a}$$

and

$$\mathscr{G}^\dagger \left(\mathbf{p} - \frac{e\mathbf{A}}{c} - \frac{e\boldsymbol{\nabla}\Lambda}{c}\right)\mathscr{G} = \mathbf{p} - \frac{e\mathbf{A}}{c}. \tag{2.368b}$$

We assert that

$$\mathscr{G} = \exp\left[\frac{ie\Lambda(\mathbf{x})}{\hbar c}\right] \tag{2.369}$$

will do the job. First, $\mathscr{G}$ is unitary, so (2.366) is all right. Second, (2.368a) is obviously satisfied because $\mathbf{x}$ commutes with any function of $\mathbf{x}$. As for (2.368b), just note that

$$\begin{aligned}
\exp\left(\frac{-ie\Lambda}{\hbar c}\right)\mathbf{p}\exp\left(\frac{ie\Lambda}{\hbar c}\right) &= \exp\left(\frac{-ie\Lambda}{\hbar c}\right)\left[\mathbf{p}, \exp\left(\frac{ie\Lambda}{\hbar c}\right)\right] + \mathbf{p} \\
&= -\exp\left(\frac{-ie\Lambda}{\hbar c}\right)i\hbar\boldsymbol{\nabla}\left[\exp\left(\frac{ie\Lambda}{\hbar c}\right)\right] + \mathbf{p} \\
&= \mathbf{p} + \frac{e\boldsymbol{\nabla}\Lambda}{c},
\end{aligned} \tag{2.370}$$

where we have used (2.97b).

The invariance of quantum mechanics under gauge transformations can also be demonstrated by looking directly at the Schrödinger equation. Let $|\alpha, t_0; t\rangle$ be a solution to the Schrödinger equation in the presence of $\mathbf{A}$:

$$\left[\frac{(\mathbf{p}-e\mathbf{A}/c)^2}{2m}+e\phi\right]|\alpha, t_0; t\rangle = i\hbar\frac{\partial}{\partial t}|\alpha, t_0; t\rangle. \tag{2.371}$$

The corresponding solution in the presence of $\tilde{\mathbf{A}}$ must satisfy

$$\left[\frac{(\mathbf{p}-e\mathbf{A}/c-e\boldsymbol{\nabla}\Lambda/c)^2}{2m}+e\phi\right]|\widetilde{\alpha, t_0}; t\rangle = i\hbar\frac{\partial}{\partial t}|\widetilde{\alpha, t_0}; t\rangle. \tag{2.372}$$

We see that if the new ket is taken to be

$$|\widetilde{\alpha, t_0}; t\rangle = \exp\left(\frac{ie\Lambda}{\hbar c}\right)|\alpha, t_0; t\rangle \tag{2.373}$$

in accordance with (2.369), then the new Schrödinger equation (2.372) will be satisfied; all we have to note is that

$$\exp\left(\frac{-ie\Lambda}{\hbar c}\right)\left(\mathbf{p}-\frac{e\mathbf{A}}{c}-\frac{e\boldsymbol{\nabla}\Lambda}{c}\right)^2\exp\left(\frac{ie\Lambda}{\hbar c}\right) = \left(\mathbf{p}-\frac{e\mathbf{A}}{c}\right)^2, \tag{2.374}$$

which follows from applying (2.370) twice.

Equation (2.373) also implies that the corresponding wave equations are related via

$$\tilde{\psi}(\mathbf{x}',t) = \exp\left[\frac{ie\Lambda(\mathbf{x}')}{\hbar c}\right]\psi(\mathbf{x}',t), \tag{2.375}$$

where $\Lambda(\mathbf{x}')$ is now a real function of the position vector eigenvalue $\mathbf{x}'$. This can, of course, be verified also by directly substituting (2.375) into Schrödinger's wave equation with $\mathbf{A}$ replaced by $\mathbf{A}+\boldsymbol{\nabla}\Lambda$. In terms of $\rho$ and $S$, we see that $\rho$ is unchanged but $S$ is modified as follows:

$$S \to S + \frac{e\Lambda}{c}. \tag{2.376}$$

This is highly satisfactory because we see that the probability flux given by (2.353) is then gauge invariant.

To summarize, when vector potentials in different gauges are used for the same physical situation, the corresponding state kets (or wave functions) must necessarily be different. However, only a simple change is needed; we can go from a gauge specified by $\mathbf{A}$ to another specified by $\mathbf{A}+\boldsymbol{\nabla}\Lambda$ by merely multiplying the old ket (the old wave function) by $\exp[ie\Lambda(\mathbf{x})/\hbar c]$ $(\exp[ie\Lambda(\mathbf{x}')/\hbar c])$. The **canonical momentum**, defined as the generator of translation, is manifestly *gauge dependent* in the sense that its expectation value depends on the particular gauge chosen, while the **kinematic momentum** and the probability flux are *gauge invariant*.

The reader may wonder why invariance under (2.369) is called *gauge invariance*. This word is the translation of the German *Eichinvarianz*, where *Eich* means *gauge*. There is a historical anecdote that goes with the origin of this term.

Consider some function of position at $\mathbf{x}$: $F(\mathbf{x})$. At a neighboring point we obviously have

$$F(\mathbf{x}+d\mathbf{x}) \simeq F(\mathbf{x}) + (\boldsymbol{\nabla}F)\cdot d\mathbf{x}. \tag{2.377}$$

But suppose we apply a scale change as we go from $\mathbf{x}$ to $\mathbf{x} + d\mathbf{x}$ as follows:

$$1|_{\text{at} \mathbf{x}} \rightarrow [1 + \boldsymbol{\Sigma}(\mathbf{x}) \cdot d\mathbf{x}]|_{\text{at} \mathbf{x} + d\mathbf{x}}. \tag{2.378}$$

We must then rescale $F(\mathbf{x})$ as follows:

$$F(\mathbf{x} + d\mathbf{x})|_{\text{rescaled}} \simeq F(\mathbf{x}) + [(\boldsymbol{\nabla} + \boldsymbol{\Sigma})F] \cdot d\mathbf{x} \tag{2.379}$$

instead of (2.377). The combination $\boldsymbol{\nabla} + \boldsymbol{\Sigma}$ is similar to the gauge-invariant combination

$$\boldsymbol{\nabla} - \left(\frac{ie}{\hbar c}\right)\mathbf{A} \tag{2.380}$$

encountered in (2.352) except for the absence of $i$. Historically, H. Weyl unsuccessfully attempted to construct a geometric theory of electromagnetism based on *Eichinvarianz* by identifying the scale function $\boldsymbol{\Sigma}(\mathbf{x})$ in (2.378) and (2.379) with the vector potential $\mathbf{A}$ itself. With the birth of quantum mechanics, V. Fock and F. London realized the importance of the gauge-invariant combination (2.380), and they recalled Weyl's earlier work by comparing $\boldsymbol{\Sigma}$ with $i$ times $\mathbf{A}$. We are stuck with the term *gauge invariance* even though the quantum-mechanical analogue of (2.378)

$$1\left.\right|_{\text{at} \mathbf{x}} \rightarrow \left[1 - \left(\frac{ie}{\hbar c}\right)\mathbf{A} \cdot d\mathbf{x}\right]\Bigg|_{\text{at} \mathbf{x} + d\mathbf{x}} \tag{2.381}$$

would actually correspond to "phase change" rather than to "scale change."

### 2.7.4  The Aharonov–Bohm Effect

The use of vector potential in quantum mechanics has many far-reaching consequences, some of which we are now ready to discuss. We start with a relatively innocuous-looking problem. For filling in the details, plus an extension into a discussion of quantized flux, see Problem 2.37 at the end of this chapter.

Consider a hollow cylindrical shell, as shown in Figure 2.11a. We assume that a particle of charge $e$ can be completely confined to the interior of the shell with rigid walls. The wave function is required to vanish on the inner ($\rho = \rho_a$) and outer ($\rho = \rho_b$) walls as well as at the top and the bottom. It is a straightforward boundary-value problem in mathematical physics to obtain the energy eigenvalues.

Let us now consider a modified arrangement where the cylindrical shell encloses a uniform magnetic field, as shown in Figure 2.11b. Specifically, you may imagine fitting a very long solenoid into the hole in the middle in such a way that no magnetic field leaks into the region $\rho \geq \rho_a$. The boundary conditions for the wave function are taken to be the same as before; the walls are assumed to be just as rigid. Intuitively we may conjecture that the energy spectrum is unchanged because the region with $\mathbf{B} \neq 0$ is completely inaccessible to the charged particle trapped inside the shell. However, quantum mechanics tells us that this conjecture is *not* correct.

Even though the magnetic field vanishes in the interior, the vector potential $\mathbf{A}$ is nonvanishing. We need $\mathbf{A} \neq 0$ for $\rho < \rho_a$, but we cannot set $\mathbf{A} = 0$ for $\rho > \rho_a$ because the discontinuity leads to an unphysical magnetic field. However, we can use a gauge

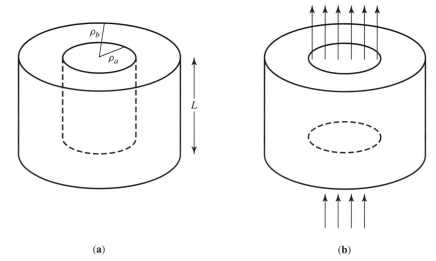

**Fig. 2.11**   Hollow cylindrical shell (a) without a magnetic field, (b) with a uniform magnetic field.

transformation to set $\mathbf{B} = 0$, and then apply (2.375) so that the wave function gains the appropriate phase factor. The correspondingly modified Schrödinger equation is then solved for the energy eigenvalues in the region $\rho_a \leq \rho \leq \rho_b$.

Using Stokes's theorem we can infer that the vector potential needed to produce the magnetic field $\mathbf{B}\,(= B\hat{\mathbf{z}})$ is

$$\mathbf{A} = \left(\frac{B\rho_a^2}{2\rho}\right)\hat{\boldsymbol{\phi}}, \tag{2.382}$$

where $\hat{\boldsymbol{\phi}}$ is the unit vector in the direction of increasing azimuthal angle. In attempting to solve the Schrödinger equation to find the energy eigenvalues for this new problem, we need only to replace the gradient $\nabla$ by $\nabla - (ie/\hbar c)\mathbf{A}$; we can accomplish this in cylindrical coordinates by replacing the partial derivative with respect to $\phi$ as follows:

$$\frac{\partial}{\partial\phi} \to \frac{\partial}{\partial\phi} - \left(\frac{ie}{\hbar c}\right)\frac{B\rho_a^2}{2}; \tag{2.383}$$

recall the expression for gradient in cylindrical coordinates:

$$\nabla = \hat{\boldsymbol{\rho}}\frac{\partial}{\partial\rho} + \hat{\mathbf{z}}\frac{\partial}{\partial z} + \hat{\boldsymbol{\phi}}\frac{1}{\rho}\frac{\partial}{\partial\phi}. \tag{2.384}$$

The replacement (2.383) results in an *observable* change in the energy spectrum, as the reader may verify explicitly. This is quite remarkable because the particle never "touches" the magnetic field; the Lorentz force the particle experiences is identically zero in this problem, yet the energy levels depend on whether or not the magnetic field is finite in the hole region inaccessible to the particle.

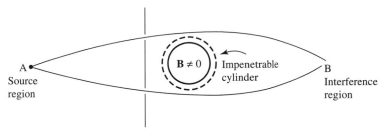

Fig. 2.12   The Aharonov–Bohm effect.

The problem we have just treated is the bound-state version of what is commonly referred to as the *Aharonov–Bohm effect*.[15] We are now in a position to discuss the original form of the Aharonov–Bohm effect itself. Consider a particle of charge $e$ going above or below a very long impenetrable cylinder, as shown in Figure 2.12. Inside the cylinder is a magnetic field parallel to the cylinder axis, taken to be normal to the plane of Figure 2.12. So the particle paths above and below enclose a magnetic flux. Our object is to study how the probability of finding the particle in the interference region B depends on the magnetic flux.

Even though this problem can be attacked by comparing the solutions to the Schrödinger equation in the presence and absence of **B**, for pedagogical reasons we prefer to use the Feynman path-integral method. Let $\mathbf{x}_1$ and $\mathbf{x}_N$ be typical points in source region A and interference region B, respectively. We recall from classical mechanics that the Lagrangian in the presence of the magnetic field can be obtained from that in the absence of the magnetic field, denoted by $L_{\text{classical}}^{(0)}$, as follows:

$$L_{\text{classical}}^{(0)} = \frac{m}{2}\left(\frac{d\mathbf{x}}{dt}\right)^2 \to L_{\text{classical}}^{(0)} + \frac{e}{c}\frac{d\mathbf{x}}{dt}\cdot\mathbf{A}. \tag{2.385}$$

The corresponding change in the action for some definite path segment going from $(\mathbf{x}_{n-1}, t_{n-1})$ to $(\mathbf{x}_n, t_n)$ is then given by

$$S^{(0)}(n, n-1) \to S^{(0)}(n, n-1) + \frac{e}{c}\int_{t_{n-1}}^{t_n} dt\left(\frac{d\mathbf{x}}{dt}\right)\cdot\mathbf{A}. \tag{2.386}$$

But this last integral can be written as

$$\frac{e}{c}\int_{t_{n-1}}^{t_n} dt\left(\frac{d\mathbf{x}}{dt}\right)\cdot\mathbf{A} = \frac{e}{c}\int_{\mathbf{x}_{n-1}}^{\mathbf{x}_n}\mathbf{A}\cdot d\mathbf{s}, \tag{2.387}$$

where $d\mathbf{s}$ is the differential line element along the path segment, so when we consider the entire contribution from $\mathbf{x}_1$ to $\mathbf{x}_N$, we have the following change:

$$\prod \exp\left[\frac{iS^{(0)}(n, n-1)}{\hbar}\right] \to \left\{\prod \exp\left[\frac{iS^{(0)}(n, n-1)}{\hbar}\right]\right\}\exp\left(\frac{ie}{\hbar c}\int_{\mathbf{x}_1}^{\mathbf{x}_N}\mathbf{A}\cdot d\mathbf{s}\right). \tag{2.388}$$

All this is for a particular path, such as going above the cylinder. We must still sum over all possible paths, which may appear to be a formidable task. Fortunately, we know from the

[15] After a 1959 paper by Y. Aharonov and D. Bohm (*Phys. Rev.*, **115** (1959) 485). Essentially the same effect was discussed 10 years earlier by W. Ehrenberg and R. E. Siday.

theory of electromagnetism that the line integral $\int \mathbf{A} \cdot d\mathbf{s}$ is independent of paths, that is, it is dependent only on the end points, as long as the loop formed by a pair of different paths does not enclose a magnetic flux. As a result, the contributions due to $\mathbf{A} \neq 0$ to all paths going above the cylinder are given by a *common* phase factor; similarly, the contributions from all paths going below the cylinder are multiplied by another common phase factor. In the path-integral notation we have, for the entire transition amplitude,

$$\int_{\text{above}} \mathscr{D}[\mathbf{x}(t)] \exp\left[\frac{iS^{(0)}(N,1)}{\hbar}\right] + \int_{\text{below}} \mathscr{D}[\mathbf{x}(t)] \exp\left[\frac{iS^{(0)}(N,1)}{\hbar}\right]$$

$$\rightarrow \int_{\text{above}} \mathscr{D}[\mathbf{x}(t)] \exp\left[\frac{iS^{(0)}(N,1)}{\hbar}\right] \left\{\exp\left[\left(\frac{ie}{\hbar c}\right) \int_{\mathbf{x}_1}^{\mathbf{x}_N} \mathbf{A} \cdot d\mathbf{s}\right]_{\text{above}}\right\}$$

$$+ \int_{\text{below}} \mathscr{D}[\mathbf{x}(t)] \exp\left[\frac{iS^{(0)}(N,1)}{\hbar}\right] \left\{\exp\left[\left(\frac{ie}{\hbar c}\right) \int_{\mathbf{x}_1}^{\mathbf{x}_N} \mathbf{A} \cdot d\mathbf{s}\right]_{\text{below}}\right\}. \qquad (2.389)$$

The probability for finding the particle in the interference region B depends on the modulus squared of the entire transition amplitude and hence on the phase difference between the contribution from the paths going above and below. The phase difference due to the presence of $\mathbf{B}$ is just

$$\left[\left(\frac{e}{\hbar c}\right) \int_{\mathbf{x}_1}^{\mathbf{x}_N} \mathbf{A} \cdot d\mathbf{s}\right]_{\text{above}} - \left[\left(\frac{e}{\hbar c}\right) \int_{\mathbf{x}_1}^{\mathbf{x}_N} \mathbf{A} \cdot d\mathbf{s}\right]_{\text{below}} = \left(\frac{e}{\hbar c}\right) \oint \mathbf{A} \cdot d\mathbf{s}$$

$$= \left(\frac{e}{\hbar c}\right) \Phi_B, \qquad (2.390)$$

where $\Phi_B$ stands for the magnetic flux inside the impenetrable cylinder. This means that as we change the magnetic field strength, there is a sinusoidal component in the probability for observing the particle in region B with a period given by a *fundamental unit of magnetic flux*, namely,

$$\frac{2\pi \hbar c}{|e|} = 4.135 \times 10^{-7} \, \text{Gauss-cm}^2. \qquad (2.391)$$

We emphasize that the interference effect discussed here is purely quantum mechanical. Classically, the motion of a charged particle is determined solely by Newton's second law supplemented by the force law of Lorentz. Here, as in the previous bound-state problem, the particle can never enter the region in which $\mathbf{B}$ is finite; the Lorentz force is identically zero in all regions where the particle wave function is finite. Yet there is a striking interference pattern that depends on the presence or absence of a magnetic field inside the impenetrable cylinder. This point has led some people to conclude that in quantum mechanics it is $\mathbf{A}$ rather than $\mathbf{B}$ that is fundamental. It is to be noted, however, that the observable effects in both examples depend only on $\Phi_B$, which is directly expressible in terms of $\mathbf{B}$. Experiments to verify the Aharonov–Bohm effect have been performed using a thin magnetized iron filament called a *whisker*.[16]

---

[16] One such recent experiment is that of Tonomura et al., *Phys. Rev. Lett.*, **48** (1982) 1443.

## 2.7.5 Magnetic Monopole

We conclude this section with one of the most remarkable predictions of quantum physics, which has yet to be verified experimentally. An astute student of classical electrodynamics may be struck by the fact that there is a strong symmetry between $\mathbf{E}$ and $\mathbf{B}$, yet a magnetic charge, commonly referred to as a **magnetic monopole**, analogous to electric charge is peculiarly absent in Maxwell's equations. The source of a magnetic field observed in nature is either a moving electric charge or a static magnetic dipole, never a static magnetic charge. Instead of

$$\nabla \cdot \mathbf{B} = 4\pi \rho_M \tag{2.392}$$

analogous to

$$\nabla \cdot \mathbf{E} = 4\pi \rho, \tag{2.393}$$

$\nabla \cdot \mathbf{B}$ actually vanishes in the usual way of writing Maxwell's equations. Quantum mechanics does not predict that a magnetic monopole must exist. However, it unambiguously requires that if a magnetic monopole is ever found in nature, the magnitude of magnetic charge must be quantized in terms of $e$, $\hbar$, and $c$, as we now demonstrate.

Suppose there is a point magnetic monopole, situated at the origin, of strength $e_M$ analogous to a point electric charge. The static magnetic field is then given by

$$\mathbf{B} = \left(\frac{e_M}{r^2}\right) \hat{\mathbf{r}}. \tag{2.394}$$

At first sight it may appear that the magnetic field (2.394) can be derived from

$$\mathbf{A} = \left[\frac{e_M(1 - \cos\theta)}{r\sin\theta}\right] \hat{\phi}. \tag{2.395}$$

Recall the expression for curl in spherical coordinates:

$$\nabla \times \mathbf{A} = \hat{\mathbf{r}}\frac{1}{r\sin\theta}\left[\frac{\partial}{\partial\theta}(A_\phi \sin\theta) - \frac{\partial A_\theta}{\partial\phi}\right]$$

$$+ \hat{\theta}\frac{1}{r}\left[\frac{1}{\sin\theta}\frac{\partial A_r}{\partial\phi} - \frac{\partial}{\partial r}(rA_\phi)\right] + \hat{\phi}\frac{1}{r}\left[\frac{\partial}{\partial r}(rA_\theta) - \frac{\partial A_r}{\partial\theta}\right]. \tag{2.396}$$

But vector potential (2.395) has one difficulty – it is singular on the negative $z$-axis $(\theta = \pi)$. In fact, it turns out to be impossible to construct a singularity-free potential valid everywhere for this problem. To see this we first note "Gauss's law"

$$\int_{\text{closed surface}} \mathbf{B} \cdot d\sigma = 4\pi e_M \tag{2.397}$$

for any surface boundary enclosing the origin at which the magnetic monopole is located. On the other hand, if $\mathbf{A}$ were nonsingular, we would have

$$\nabla \cdot (\nabla \times \mathbf{A}) = 0 \tag{2.398}$$

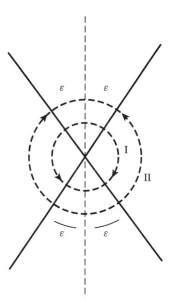

Fig. 2.13 Regions of validity for the potentials $\mathbf{A}^{(\text{I})}$ and $\mathbf{A}^{(\text{II})}$.

everywhere; hence,

$$\int_{\text{closed surface}} \mathbf{B} \cdot d\sigma = \int_{\text{volume inside}} \nabla \cdot (\nabla \times \mathbf{A}) d^3 x = 0, \qquad (2.399)$$

in contradiction with (2.397).

However, one might argue that because the vector potential is just a device for obtaining $\mathbf{B}$, we need not insist on having a single expression for $\mathbf{A}$ valid everywhere. Suppose we construct a pair of potentials,

$$\mathbf{A}^{(\text{I})} = \left[ \frac{e_M(1 - \cos\theta)}{r\sin\theta} \right] \hat{\phi} \qquad (\theta < \pi - \varepsilon) \qquad (2.400\text{a})$$

$$\mathbf{A}^{(\text{II})} = -\left[ \frac{e_M(1 + \cos\theta)}{r\sin\theta} \right] \hat{\phi} \qquad (\theta > \varepsilon), \qquad (2.400\text{b})$$

such that the potential $\mathbf{A}^{(\text{I})}$ can be used everywhere except inside the cone defined by $\theta = \pi - \varepsilon$ around the negative $z$-axis; likewise, the potential $\mathbf{A}^{(\text{II})}$ can be used everywhere except inside the cone $\theta = \varepsilon$ around the positive $z$-axis; see Figure 2.13. Together they lead to the correct expression for $\mathbf{B}$ everywhere.[17]

Consider now what happens in the overlap region $\varepsilon < \theta < \pi - \varepsilon$, where we may use either $\mathbf{A}^{(\text{I})}$ or $\mathbf{A}^{(\text{II})}$. Because the two potentials lead to the same magnetic field, they must be related to each other by a gauge transformation. To find $\Lambda$ appropriate for this problem we first note that

$$\mathbf{A}^{(\text{II})} - \mathbf{A}^{(\text{I})} = -\left( \frac{2e_M}{r\sin\theta} \right) \hat{\phi}. \qquad (2.401)$$

---

[17] An alternative approach to this problem uses $\mathbf{A}^{(\text{I})}$ everywhere, but taking special care of the string of singularities, known as a **Dirac string**, along the negative $z$-axis.

Recalling the expression for gradient in spherical coordinates,

$$\nabla \Lambda = \hat{\mathbf{r}} \frac{\partial \Lambda}{\partial r} + \hat{\theta} \frac{1}{r} \frac{\partial \Lambda}{\partial \theta} + \hat{\phi} \frac{1}{r \sin \theta} \frac{\partial \Lambda}{\partial \phi}, \tag{2.402}$$

we deduce that

$$\Lambda = -2e_M \phi \tag{2.403}$$

will do the job.

Next, we consider the wave function of an electrically charged particle of charge $e$ subjected to magnetic field (2.394). As we emphasized earlier, the particular form of the wave function depends on the particular gauge used. In the overlap region where we may use either $\mathbf{A}^{(I)}$ or $\mathbf{A}^{(II)}$, the corresponding wave functions are, according to (2.375), related to each other by

$$\psi^{(II)} = \exp\left(\frac{-2iee_M \phi}{\hbar c}\right) \psi^{(I)}. \tag{2.404}$$

Wave functions $\psi^{(I)}$ and $\psi^{(II)}$ must each be *single valued* because once we choose a particular gauge, the expansion of the state ket in terms of the position eigenkets must be *unique*. After all, as we have repeatedly emphasized, the wave function is simply an expansion coefficient for the state ket in terms of the position eigenkets.

Let us now examine the behavior of wave function $\psi^{(II)}$ on the equator $\theta = \pi/2$ with some definite radius $r$, which is a constant. When we increase the azimuthal angle $\phi$ along the equator and go around once, say from $\phi = 0$ to $\phi = 2\pi$, $\psi^{(II)}$, as well as $\psi^{(I)}$, must return to its original value because each is single valued. According to (2.404), this is possible only if

$$\frac{2ee_M}{\hbar c} = +N, \qquad N = 0, \pm 1, \pm 2, \ldots. \tag{2.405}$$

So we reach a very far-reaching conclusion: The magnetic charges must be *quantized* in units of

$$\frac{\hbar c}{2|e|} \simeq \left(\frac{137}{2}\right) |e|. \tag{2.406}$$

The smallest magnetic charge possible is $\hbar c/2|e|$, where $e$ is the electronic charge. It is amusing that once a magnetic monopole is assumed to exist, we can use (2.405) backward, so to speak, to explain why the electric charges are quantized, for example, why the proton charge cannot be 0.999972 times $|e|$.[18]

We repeat once again that quantum mechanics does not require magnetic monopoles to exist. However, it unambiguously predicts that a magnetic charge, if it is ever found in nature, must be quantized in units of $\hbar c/2|e|$. The quantization of magnetic charges in quantum mechanics was first shown in 1931 by P. A. M. Dirac. The derivation given here is due to T. T. Wu and C. N. Yang. A different solution, which connects

---

[18] Empirically the equality in magnitude between the electron charge and the proton charge is established to an accuracy of four parts in $10^{19}$.

the Dirac quantization condition to the quantization of angular momentum, is discussed by Lipkin et al., *Ann. Phys.*, **53** (1969) 203. Finally, we will revisit this subject again in Section 5.6 when we discuss Berry's phase in conjunction with the adiabatic approximation.

# Problems

**2.1**  Consider the spin-precession problem discussed in the text. It can also be solved in the Heisenberg picture. Using the Hamiltonian

$$H = -\left(\frac{eB}{mc}\right)S_z = \omega S_z,$$

write the Heisenberg equations of motion for the time-dependent operators $S_x(t)$, $S_y(t)$, and $S_z(t)$. Solve them to obtain $S_{x,y,z}$ as functions of time.

**2.2**  Look again at the Hamiltonian of Chapter 1, Problem 1.13. Suppose the typist made an error and wrote $H$ as

$$H = H_{11}|1\rangle\langle 1| + H_{22}|2\rangle\langle 2| + H_{12}|1\rangle\langle 2|.$$

What principle is now violated? Illustrate your point explicitly by attempting to solve the most general time-dependent problem using an illegal Hamiltonian of this kind. (You may assume $H_{11} = H_{22} = 0$ for simplicity.)

**2.3**  An electron is subject to a uniform, time-independent magnetic field of strength $B$ in the positive $z$-direction. At $t = 0$ the electron is known to be in an eigenstate of $\mathbf{S} \cdot \hat{\mathbf{n}}$ with eigenvalue $\hbar/2$, where $\hat{\mathbf{n}}$ is a unit vector, lying in the $xz$-plane, that makes an angle $\beta$ with the $z$-axis.
  a. Obtain the probability for finding the electron in the $S_x = \hbar/2$ state as a function of time.
  b. Find the expectation value of $S_x$ as a function of time.
  c. For your own peace of mind show that your answers make good sense in the extreme cases (i) $\beta \to 0$ and (ii) $\beta \to \pi/2$.

**2.4**  Derive the neutrino oscillation probability (2.65) and use it, along with the data in Figure 2.2, to estimate the values of $\Delta m^2 c^4$ (in units of $eV^2$) and $\theta$.

**2.5**  Let $x(t)$ be the coordinate operator for a free particle in one dimension in the Heisenberg picture. Evaluate

$$[x(t), x(0)].$$

**2.6**  Consider a particle in one dimension whose Hamiltonian is given by

$$H = \frac{p^2}{2m} + V(x).$$

By calculating $[[H,x],x]$ prove

$$\sum_{a'} |\langle a''|x|a'\rangle|^2 (E_{a'} - E_{a''}) = \frac{\hbar^2}{2m},$$

where $|a'\rangle$ is an energy eigenket with eigenvalue $E_{a'}$.

**2.7** Consider a particle in three dimensions whose Hamiltonian is given by

$$H = \frac{\mathbf{p}^2}{2m} + V(\mathbf{x}).$$

By calculating $[\mathbf{x} \cdot \mathbf{p}, H]$ obtain

$$\frac{d}{dt} \langle \mathbf{x} \cdot \mathbf{p} \rangle = \left\langle \frac{\mathbf{p}^2}{m} \right\rangle - \langle \mathbf{x} \cdot \nabla V \rangle.$$

To identify the preceding relation with the quantum-mechanical analogue of the virial theorem it is essential that the left-hand side vanish. Under what condition would this happen?

**2.8** Consider a free-particle wave packet in one dimension. At $t = 0$ it satisfies the minimum uncertainty relation

$$\langle (\Delta x)^2 \rangle \langle (\Delta p)^2 \rangle = \frac{\hbar^2}{4} \quad (t = 0).$$

In addition, we know

$$\langle x \rangle = \langle p \rangle = 0 \quad (t = 0).$$

*Using the Heisenberg picture*, obtain $\langle (\Delta x)^2 \rangle_t$ as a function of $t (t \geq 0)$ when $\langle (\Delta x)^2 \rangle_{t=0}$ is given. (*Hint:* Take advantage of the property of the minimum-uncertainty wave packet you worked out in Chapter 1, Problem 1.20.)

**2.9** For a wave function $\langle x'|\alpha \rangle = A(x' - a)^2 (x' + a)^2 e^{ikx'}$ for $-a \leq x' \leq a$ and zero otherwise, carry out the following.
a. Find the constant $A$.
b. Find the expectation values $\langle x \rangle$, $\langle p \rangle$, $\langle x^2 \rangle$, and $\langle p^2 \rangle$.
c. Find the expectation values $\langle (\Delta x)^2 \rangle$ and $\langle (\Delta p)^2 \rangle$, and compare their product to that for a Gaussian wave packet.

**2.10** Let $|a'\rangle$ and $|a''\rangle$ be eigenstates of a Hermitian operator $A$ with eigenvalues $a'$ and $a''$, respectively ($a' \neq a''$). The Hamiltonian operator is given by

$$H = |a'\rangle \delta \langle a''| + |a''\rangle \delta \langle a'|,$$

where $\delta$ is just a real number.
a. Clearly, $|a'\rangle$ and $|a''\rangle$ are not eigenstates of the Hamiltonian. Write down the eigenstates of the Hamiltonian. What are their energy eigenvalues?
b. Suppose the system is known to be in state $|a'\rangle$ at $t = 0$. Write down the state vector in the Schrödinger picture for $t > 0$.

c. What is the probability for finding the system in $|a''\rangle$ for $t > 0$ if the system is known to be in state $|a'\rangle$ at $t = 0$?

d. Can you think of a physical situation corresponding to this problem?

**2.11**   A box containing a particle is divided into a right and a left compartment by a thin partition. If the particle is known to be on the right (left) side with certainty, the state is represented by the position eigenket $|R\rangle$ ($|L\rangle$), where we have neglected spatial variations within each half of the box. The most general state vector can then be written as

$$|\alpha\rangle = |R\rangle\langle R|\alpha\rangle + |L\rangle\langle L|\alpha\rangle,$$

where $\langle R|\alpha\rangle$ and $\langle L|\alpha\rangle$ can be regarded as "wave functions." The particle can tunnel through the partition; this tunneling effect is characterized by the Hamiltonian

$$H = \Delta(|L\rangle\langle R| + |R\rangle\langle L|),$$

where $\Delta$ is a real number with the dimension of energy.

a. Find the normalized energy eigenkets. What are the corresponding energy eigenvalues?

b. In the Schrödinger picture the base kets $|R\rangle$ and $|L\rangle$ are fixed, and the state vector moves with time. Suppose the system is represented by $|\alpha\rangle$ as given above at $t = 0$. Find the state vector $|\alpha, t_0 = 0; t\rangle$ for $t > 0$ by applying the appropriate time-evolution operator to $|\alpha\rangle$.

c. Suppose at $t = 0$ the particle is on the right side with certainty. What is the probability for observing the particle on the left side as a function of time?

d. Write down the coupled Schrödinger equations for the wave functions $\langle R|\alpha, t_0 = 0; t\rangle$ and $\langle L|\alpha, t_0 = 0; t\rangle$. Show that the solutions to the coupled Schrödinger equations are just what you expect from (b).

e. Suppose the printer made an error and wrote $H$ as

$$H = \Delta|L\rangle\langle R|.$$

By explicitly solving the most general time-evolution problem with this Hamiltonian, show that probability conservation is violated.

**2.12**   A one-dimensional simple harmonic oscillator with natural frequency $\omega$ is in initial state

$$|\alpha\rangle = \frac{1}{\sqrt{2}}|0\rangle + \frac{e^{i\delta}}{\sqrt{2}}|1\rangle$$

where $\delta$ is a real number.

a. Find the time-dependent wave function $\langle x'|\alpha; t\rangle$ and evaluate the (time-dependent) expectation values $\langle x\rangle$ and $\langle p\rangle$ in the state $|\alpha; t\rangle$, i.e. in the Schrödinger picture.

b. Now calculate $\langle x\rangle$ and $\langle p\rangle$ in the Heisenberg picture and compare the results.

**2.13**   A particle with mass $m$ moves in one dimension and is acted on by a constant force $F$. Find the operators $x(t)$ and $p(t)$ in the Heisenberg picture, and find their expectation values for an arbitrary state $|\alpha\rangle$. Use $\langle x(0)\rangle = x_0$ and $\langle p(0)\rangle = p_0$. The result should be obvious. Comment on how to do this problem in the Schrödinger picture, but do not try to work it through.

**2.14** Consider a particle subject to a one-dimensional simple harmonic oscillator potential. Suppose at $t = 0$ the state vector is given by

$$\exp\left(\frac{-ipa}{\hbar}\right)|0\rangle,$$

where $p$ is the momentum operator, $a$ is some number with dimension of length, and the state $|0\rangle$ is the one for which $\langle x \rangle = 0 = \langle p \rangle$. Using the Heisenberg picture, evaluate the expectation value $\langle x \rangle$ for $t \geq 0$.

**2.15** a. Write down the wave function (in coordinate space) for the state specified in Problem 2.14 at $t = 0$. You may use

$$\langle x'|0\rangle = \pi^{-1/4} x_0^{-1/2} \exp\left[-\frac{1}{2}\left(\frac{x'}{x_0}\right)^2\right], \quad x_0 \equiv \left(\frac{\hbar}{m\omega}\right)^{1/2}.$$

   b. Obtain a simple expression for the probability that the state is found in the ground state at $t = 0$. Does this probability change for $t > 0$?

**2.16** Consider a one-dimensional simple harmonic oscillator.
   a. Using

$$\left.\begin{array}{c} a \\ a^\dagger \end{array}\right\} = \sqrt{\frac{m\omega}{2\hbar}}\left(x \pm \frac{ip}{m\omega}\right), \quad \left.\begin{array}{c} a|n\rangle \\ a^\dagger|n\rangle \end{array}\right\} = \left\{\begin{array}{l} \sqrt{n}|n-1\rangle \\ \sqrt{n+1}|n+1\rangle, \end{array}\right.$$

evaluate $\langle m|x|n\rangle$, $\langle m|p|n\rangle$, $\langle m|\{x,p\}|n\rangle$, $\langle m|x^2|n\rangle$, and $\langle m|p^2|n\rangle$.
   b. Translated from classical physics, the virial theorem states that

$$\left\langle \frac{\mathbf{p}^2}{m} \right\rangle = \langle \mathbf{x} \cdot \nabla V \rangle \quad (3\mathrm{D}) \qquad \text{or} \qquad \left\langle \frac{p^2}{m} \right\rangle = \langle x \frac{dV}{dx} \rangle \quad (1\mathrm{D})$$

Check that the virial theorem holds for the expectation values of the kinetic and the potential energy taken with respect to an energy eigenstate.

**2.17** a. Using

$$\langle x'|p'\rangle = (2\pi\hbar)^{-1/2} e^{ip'x'/\hbar} \qquad \text{(one dimension)}$$

prove

$$\langle p'|x|\alpha\rangle = i\hbar\frac{\partial}{\partial p'}\langle p'|\alpha\rangle.$$

   b. Consider a one-dimensional simple harmonic oscillator. Starting with the Schrödinger equation for the state vector, derive the Schrödinger equation for the *momentum-space* wave function. (Make sure to distinguish the operator $p$ from the eigenvalue $p'$.) Can you guess the energy eigenfunctions in momentum space?

**2.18** Consider a function, known as the **correlation function**, defined by

$$C(t) = \langle x(t)x(0)\rangle,$$

where $x(t)$ is the position operator in the Heisenberg picture. Evaluate the correlation function explicitly for the ground state of a one-dimensional simple harmonic oscillator.

**2.19** Consider again a one-dimensional simple harmonic oscillator. Do the following algebraically, that is, without using wave functions.

a. Construct a linear combination of $|0\rangle$ and $|1\rangle$ such that $\langle x\rangle$ is as large as possible.

b. Suppose the oscillator is in the state constructed in (a) at $t = 0$. What is the state vector for $t > 0$ in the Schrödinger picture? Evaluate the expectation value $\langle x\rangle$ as a function of time for $t > 0$ using (i) the Schrödinger picture and (ii) the Heisenberg picture.

c. Evaluate $\langle(\Delta x)^2\rangle$ as a function of time using either picture.

**2.20** Show for the one-dimensional simple harmonic oscillator

$$\langle 0|e^{ikx}|0\rangle = \exp[-k^2\langle 0|x^2|0\rangle/2],$$

where $x$ is the position *operator*.

**2.21** The problem covers some fundamental concepts in quantum optics. See Glauber, *Phys. Rev.*, **84** (1951) 395 and his Nobel lecture, *Rev. Mod. Phys.*, **78** (2006) 1267; Gottfried (1966), Section 31; Merzbacher (1998), Section 10.7; and Gottfried and Yan (2003), Section 4.2.

A coherent state of a one-dimensional simple harmonic oscillator is defined to be an eigenstate of the (non-Hermitian) annihilation operator $a$:

$$a|\lambda\rangle = \lambda|\lambda\rangle,$$

where $\lambda$ is, in general, a complex number.

a. Prove that

$$|\lambda\rangle = e^{-|\lambda|^2/2}e^{\lambda a^\dagger}|0\rangle$$

is a normalized coherent state.

b. Prove the minimum uncertainty relation for such a state.

c. Write $|\lambda\rangle$ as

$$|\lambda\rangle = \sum_{n=0}^{\infty} f(n)|n\rangle.$$

Show that the distribution of $|f(n)|^2$ with respect to $n$ is in the form of a Poisson distribution, that is $P_n(\mu) = e^{-\mu}\mu^n/n!$ where $\mu$ is the mean of the distribution. Find the most probable (integer) value of $n$, hence of $E$.

d. Show that a coherent state can also be obtained by applying the translation (finite displacement) operator $e^{-ipl/\hbar}$ (where $p$ is the momentum operator, and $l$ is the displacement distance) to the ground state. (See also Gottfried (1966), pp. 262–264; Problem 2.13 in Gottfried and Yan (2003), and Eq. 39 in Glauber's 1951 paper.)

**2.22** Write a computer program to animate the time dependence of an arbitrary linear combination of stationary states for the simple harmonic oscillator in one dimension. The animation should display the time dependence of both the wave function and the probability density distribution for finding the particle, both as a function of position. Check your animation by using pure eigenstates as input, and consider combinations that would approximate classical motion. Also animate the coherent state $|\lambda\rangle$ in Problem 2.21 above, for different mean values of the Poisson distribution.

**2.23** Make the definitions

$$J_\pm \equiv \hbar a_\pm^\dagger a_\mp, \qquad J_z \equiv \frac{\hbar}{2}(a_+^\dagger a_+ - a_-^\dagger a_-), \qquad N \equiv a_+^\dagger a_+ + a_-^\dagger a_-$$

where $a_\pm$ and $a_\pm^\dagger$ are the annihilation and creation operators of two *independent* simple harmonic oscillators satisfying the usual simple harmonic oscillator commutation relations. Also make the definition

$$\mathbf{J}^2 = J_z^2 + \frac{1}{2}(J_+ J_- + J_- J_+).$$

Prove

$$[J_z, J_\pm] = \pm\hbar J_\pm, \qquad [\mathbf{J}^2, J_z] = 0, \qquad \mathbf{J}^2 = \left(\frac{\hbar^2}{2}\right) N\left[\left(\frac{N}{2}\right) + 1\right].$$

**2.24** This exercise has to do with proving conservation laws in classical and quantum physics.

a. Show that if a quantity $Q$ with a density $\rho(\mathbf{x}, t)$ in some region $\mathscr{R}$ can only be changed by a flux density $\mathbf{j}(\mathbf{x}, t)$ through the surface bordering $\mathscr{R}$, then

$$\frac{\partial \rho}{\partial t} + \nabla \cdot \mathbf{j} = 0.$$

b. Prove that Maxwell's equations imply that electric charge is conserved

c. Prove that Schrödinger's equation implies that probability is conserved, i.e. (2.190)

**2.25** Derive the normalization constant $c_n$ in (2.232) by deriving the orthogonality relationship (2.233) using generating functions. Start by working out the integral

$$I = \int_{-\infty}^{\infty} g(x, t) g(x, s) e^{-x^2} dx$$

and then consider the integral again with the generating functions in terms of series with Hermite polynomials.

**2.26** Derive an expression for the action of the position operator on an arbitrary state in the momentum representation. That is, find $\langle p'|x|\alpha\rangle$ in terms of $\langle p'|\alpha\rangle$. Use this to solve the linear potential (2.234) problem in momentum space, and show that the

Fourier transform of your solution is indeed the appropriate Airy function. Note that an alternative form of the Airy function (see http://dlmf.nist.gov/9.5) is

$$Ai(x) = \frac{1}{\pi} \int_0^\infty \cos\left(\frac{1}{3}t^3 + xt\right) dt.$$

You need not be concerned with normalizing the wave function.

**2.27** Consider a particle of mass $m$ subject to a one-dimensional potential of the following form:

$$V = \begin{cases} \dfrac{1}{2}kx^2 & \text{for } x > 0 \\ \infty & \text{for } x < 0. \end{cases}$$

a. What is the ground-state energy?
b. What is the expectation value $\langle x^2 \rangle$ for the ground state?

**2.28** A particle in one dimension is trapped between two rigid walls:

$$V(x) = \begin{cases} 0 & \text{for } 0 < x < L \\ \infty & \text{for } x < 0, \ x > L. \end{cases}$$

At $t = 0$ it is known to be exactly at $x = L/2$ with certainty. What are the *relative* probabilities for the particle to be found in various energy eigenstates? Write down the wave function for $t \geq 0$. (You need not worry about absolute normalization, convergence, and other mathematical subtleties.)

**2.29** Consider a particle in one dimension bound to a fixed center by a $\delta$-function potential of the form

$$V(x) = -v_0 \delta(x)$$

where $v_0$ is real and positive. Find the wave function and the binding energy of the ground state. Are there excited bound states?

**2.30** A particle of mass $m$ in one dimension is bound to a fixed center by an attractive $\delta$-function potential:

$$V(x) = -\lambda \delta(x) \quad (\lambda > 0).$$

At $t = 0$, the potential is suddenly switched off (that is, $V = 0$ for $t > 0$). Find the wave function for $t > 0$. (Be quantitative! But you need not attempt to evaluate an integral that may appear.)

**2.31** A particle in one dimension ($-\infty < x < \infty$) is subjected to a constant force derivable from

$$V = \lambda x \quad (\lambda > 0).$$

a. Is the energy spectrum continuous or discrete? Write down an approximate expression for the energy eigenfunction specified by $E$. Also sketch it crudely.
b. Discuss briefly what changes are needed if $V$ is replaced by

$$V = \lambda |x|.$$

**2.32** A particle of mass $m$ is confined to a one-dimensional square well with finite walls. That is, a potential $V(x) = 0$ for $-a \leq x \leq +a$, and $V(x) = V_0 = \eta(\hbar^2/2ma^2)$ otherwise. You are to find the bound-state energy eigenvalues as $E = \varepsilon V_0$ along with their wave functions.

a. Set the problem up with a wave function $Ae^{\alpha x}$ for $x \leq -a$, $De^{-\alpha x}$ for $x \geq +a$, and $Be^{ikx} + Ce^{-ikx}$ inside the well. Match the boundary conditions at $x = \pm a$ and show that $k$ and $\alpha$ must satisfy $z = \pm z^*$ where $z \equiv e^{iak}(k - i\alpha)$. Proceed to find a purely real or purely imaginary expression for $z$ in terms of $k$ and $\alpha$.

b. Find the wave functions for the two choices of $z$ and show that the purely real (imaginary) choice leads to a wave function that is even (odd) under the exchange $x \to -x$.

c. Find a transcendental equation for each of the two wave functions relating $\eta$ and $\varepsilon$. Show that even a very shallow well ($\eta \to 0$) has at least one solution for the even wave function, but you are not guaranteed any solution for an odd wave function.

d. For $\eta = 10$, find all the energy eigenvalues and plot their normalized wave functions.

**2.33** Consider a particle of mass $m$ moving in one dimension $x$ under the influence of a potential energy function $V(x)$.

a. Show that if $V(-x) = V(x)$, then a solution $u(x)$ to the time-independent Schrödinger equation must have the property $u(-x) = \pm u(x)$. (This is a simple example of parity symmetry, which will be discussed in more detail in Section 4.2.)

b. Consider a potential that is an infinite square well but with a rectangular barrier in the middle. That is $V(x) = V_0 > 0$ for $-b \leq x \leq b$, infinity for $|x| > a$, and zero for $b \leq |x| \leq a$. For $b/a = 1/3$ and $V_0 = 20(\hbar^2/2ma^2)$, find the two energy eigenvalues along with their normalized eigenfunctions. (Plots of the eigenfunctions are shown in Figure 4.3.)

c. For $V_0 = 10(\hbar^2/2ma^2)$ show that there is only one energy eigenstate, and find the eigenvalue.

You will need to write a computer program to calculate the eigenvalues and plot the wave functions. It is easiest to scale the eigenvalues by $\hbar^2/2ma^2$ and find transcendental equations to solve (numerically) for the eigenvalues in the positive and negative parity cases.

**2.34** A particle of mass $m$ moves in one dimension $x$ under a potential energy $V(x)$.

a. For $V(x) = -V_0 b \delta(x)$, $V_0 > 0$, $b > 0$, find the bound-state energy eigenvalue $E$.

b. Generalize this to the "double delta function" potential

$$V(x) = -V_0 \frac{b}{2}\left[\delta\left(x + \frac{a}{2}\right) + \delta\left(x - \frac{a}{2}\right)\right]$$

and find the bound-state energy eigenvalues and plot the corresponding eigenfunctions. Also show that you get the expected results as $a \to 0$.

**2.35** Derive an expression for the density of free-particle states in *two* dimensions, normalized with periodic boundary conditions inside a box of side length $L$.

**2.36** Use the WKB method to find the (approximate) energy eigenvalues for the one-dimensional simple harmonic oscillator potential $V(x) = m\omega^2 x^2/2$.

**2.37** Consider an electron confined to the *interior* of a hollow cylindrical shell whose axis coincides with the z-axis. The wave function is required to vanish on the inner and outer walls, $\rho = \rho_a$ and $\rho_b$, and also at the top and bottom, $z = 0$ and $L$.

    a. Find the energy eigenfunctions. (Do not bother with normalization.) Show that the energy eigenvalues are given by

$$E_{lmn} = \left(\frac{\hbar^2}{2m_e}\right)\left[k_{mn}^2 + \left(\frac{l\pi}{L}\right)^2\right] \quad (l = 1, 2, 3, \ldots, m = 0, 1, 2, \ldots),$$

where $k_{mn}$ is the $n$th root of the transcendental equation

$$J_m(k_{mn}\rho_b)N_m(k_{mn}\rho_a) - N_m(k_{mn}\rho_b)J_m(k_{mn}\rho_a) = 0.$$

    b. Repeat the same problem when there is a uniform magnetic field $\mathbf{B} = B\hat{\mathbf{z}}$ for $0 < \rho < \rho_a$. Note that the energy eigenvalues are influenced by the magnetic field even though the electron never "touches" the magnetic field.

    c. Compare, in particular, the ground state of the $B = 0$ problem with that of the $B \neq 0$ problem. Show that if we require the ground-state energy to be unchanged in the presence of $B$, we obtain "flux quantization"

$$\pi \rho_a^2 B = \frac{2\pi N\hbar c}{e} \quad (N = 0, \pm 1, \pm 2, \ldots).$$

**2.38** Consider a particle moving in one dimension under the influence of a potential $V(x)$. Suppose its wave function can be written as $\exp[iS(x,t)/\hbar]$. Prove that $S(x,t)$ satisfies the classical Hamilton–Jacobi equation to the extent that $\hbar$ can be regarded as small in some sense. Show how one may obtain the correct wave function for a plane wave by starting with the solution of the classical Hamilton–Jacobi equation with $V(x)$ set equal to zero. Why do we get the exact wave function in this particular case?

**2.39** Using spherical coordinates, obtain an expression for $\mathbf{j}$ for the ground and excited states of the hydrogen atom. Show, in particular, that for $m_l \neq 0$ states, there is a circulating flux in the sense that $\mathbf{j}$ is in the direction of increasing or decreasing $\phi$, depending on whether $m_l$ is positive or negative.

**2.40** Derive (2.280) and obtain the three-dimensional generalization of (2.280).

**2.41** A particle of mass $m$ moves along one of two "paths" through space and time connecting the points $(x,t) = (0,0)$ and $(x,t) = (D,T)$. One path is quadratic in time, i.e. $x_1(t) = \frac{1}{2}at^2$ where $a$ is a constant. The second path is linear in time, i.e. $x_2(t) = vt$ where $v$ is a constant. The correct classical path is the quadratic path, that is $x_1(t)$.

    a. Find the acceleration $a$ for the correct classical path. Use freshman physics to find the force $F = ma = -dV/dx$ and then the potential energy function $V(x)$ in terms of $m$, $D$, and $T$. Also find the velocity $v$ for the linear (i.e. incorrect classical) path.

b. Calculate the classical action $S[x(t)] = \int_0^T \left[\frac{1}{2}m\dot{x}^2 - V(x)\right] dt$ for each of the two paths $x_1(t)$ and $x_2(t)$. Confirm that $S_1 \equiv S[x_1(t)] < S_2 \equiv S[x_2(t)]$, and find $\Delta S = S_2 - S_1$.

c. Calculate $\Delta S/\hbar$ for a particle which moves 1 mm in 1 ms for two cases. The particle is a nanoparticle made up of 100 carbon atoms in one case. The other case is an electron. For which of these would you consider the motion "quantum mechanical" and why?

**2.42** Define the partition function as

$$Z = \int d^3x' K(\mathbf{x}', t; \mathbf{x}', 0)|_{\beta=it/\hbar},$$

as in (2.284)–(2.286). Show that the ground-state energy is obtained by taking

$$-\frac{1}{Z}\frac{\partial Z}{\partial \beta} \quad (\beta \to \infty).$$

Illustrate this for a particle in a one-dimensional box.

**2.43** The propagator in momentum space analogous to (2.290) is given by $\langle \mathbf{p}'', t | \mathbf{p}', t_0 \rangle$. Derive an explicit expression for $\langle \mathbf{p}'', t | \mathbf{p}', t_0 \rangle$ for the free-particle case.

**2.44** a. Write down an expression for the classical action for a simple harmonic oscillator for a finite time interval.

b. Construct $\langle x_n, t_n | x_{n-1}, t_{n-1} \rangle$ for a simple harmonic oscillator using Feynman's prescription for $t_n - t_{n-1} = \Delta t$ small. Keeping only terms up to order $(\Delta t)^2$, show that it is in complete agreement with the $t - t_0 \to 0$ limit of the propagator given by (2.290).

**2.45** State the Schwinger action principle (see Finkelstein (1973), p. 155). Obtain the solution for $\langle x_2 t_2 | x_1 t_1 \rangle$ by integrating the Schwinger principle and compare it with the corresponding Feynman expression for $\langle x_2 t_2 | x_1 t_1 \rangle$. Describe the classical limits of these two expressions.

**2.46** Show that the wave-mechanical approach to the gravity-induced problem discussed in Section 2.7 also leads to phase-difference expression (2.337).

**2.47** a. Verify (2.345) and (2.347).

b. Verify continuity equation (2.350) with $\mathbf{j}$ given by (2.351).

**2.48** Consider the Hamiltonian of a spinless particle of charge $e$. In the presence of a static magnetic field, the interaction terms can be generated by

$$\mathbf{P}_{\text{operator}} \to \mathbf{P}_{\text{operator}} - \frac{e\mathbf{A}}{c},$$

where $\mathbf{A}$ is the appropriate vector potential. Suppose, for simplicity, that the magnetic field $\mathbf{B}$ is uniform in the positive $z$-direction. Prove that the above prescription indeed leads to the correct expression for the interaction of the orbital magnetic moment $(e/2mc)\mathbf{L}$ with the magnetic field $\mathbf{B}$. Show that there is also an extra term proportional to $B^2(x^2 + y^2)$, and comment briefly on its physical significance.

**2.49** An electron moves in the presence of a uniform magnetic field in the $z$-direction $(\mathbf{B} = B\hat{z})$.

a. Evaluate

$$[\Pi_x, \Pi_y],$$

where

$$\Pi_x \equiv p_x - \frac{eA_x}{c}, \qquad \Pi_y \equiv p_y - \frac{eA_y}{c}.$$

b. By comparing the Hamiltonian and the commutation relation obtained in (a) with those of the one-dimensional oscillator problem, show how we can immediately write the energy eigenvalues as

$$E_{k,n} = \frac{\hbar^2 k^2}{2m} + \left(\frac{|eB|\hbar}{mc}\right)\left(n + \frac{1}{2}\right),$$

where $\hbar k$ is the continuous eigenvalue of the $p_z$ operator and $n$ is a nonnegative integer including zero.

**2.50** Consider the neutron interferometer.

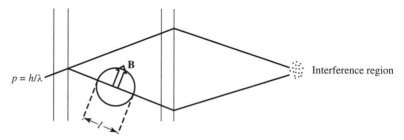

Prove that the difference in the magnetic fields that produce two successive maxima in the counting rates is given by

$$B = \frac{4\pi\hbar c}{|e|g_n \lambdabar l},$$

where $g_n (= -1.91)$ is the neutron magnetic moment in units of $-e\hbar/2m_n c$, and $\lambdabar \equiv \lambda/2\pi$. This problem was in fact analyzed in the paper by Bernstein, *Phys. Rev. Lett.*, **18** (1967) 1102.

# Theory of Angular Momentum

This chapter is concerned with a systematic treatment of angular momentum and related topics. The importance of angular momentum in modern physics can hardly be overemphasized. A thorough understanding of angular momentum is essential in molecular, atomic, and nuclear spectroscopy; angular momentum considerations play an important role in scattering and collision problems as well as in bound-state problems. Furthermore, angular-momentum concepts have important generalizations – isospin in nuclear physics, SU(3), SU(2)⊗U(1) in particle physics, and so forth.

## 3.1 Rotations and Angular Momentum Commutation Relations

### 3.1.1 Finite Versus Infinitesimal Rotations

We recall from elementary physics that rotations about the same axis commute, whereas rotations about different axes do not. For instance, a 30° rotation about the $z$-axis followed by a 60° rotation about the same $z$-axis is obviously equivalent to a 60° rotation followed by a 30° rotation, both about the same axis. However, let us consider a 90° rotation about the $z$-axis, denoted by $R_z(\pi/2)$, followed by a 90° rotation about the $x$-axis, denoted by $R_x(\pi/2)$; compare this with a 90° rotation about the $x$-axis followed by a 90° rotation about the $z$-axis. The net results are different, as we can see from Figure 3.1.

Our first basic task is to work out quantitatively the manner in which rotations about different axes *fail* to commute. To this end, we first recall how to represent rotations in three dimensions by $3 \times 3$ real, orthogonal matrices. Consider a vector $\mathbf{V}$ with components $V_x, V_y$, and $V_z$. When we rotate, the three components become some other set of numbers, $V'_x, V'_y$, and $V'_z$. The old and new components are related via a $3 \times 3$ orthogonal matrix $R$:

$$\begin{pmatrix} V'_x \\ V'_y \\ V'_z \end{pmatrix} = \begin{pmatrix} R \end{pmatrix} \begin{pmatrix} V_x \\ V_y \\ V_z \end{pmatrix}, \tag{3.1a}$$

$$RR^T = R^T R = 1, \tag{3.1b}$$

where the superscript $T$ stands for a transpose of a matrix. It is a property of orthogonal matrices that

$$\sqrt{V_x^2 + V_y^2 + V_z^2} = \sqrt{V_x'^2 + V_y'^2 + V_z'^2} \tag{3.2}$$

is automatically satisfied.

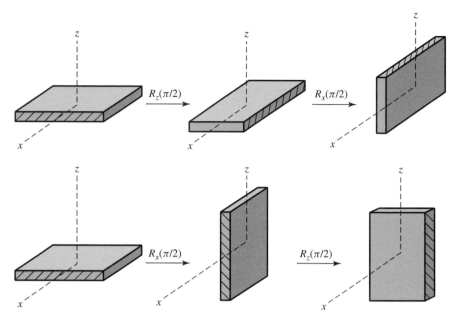

**Fig. 3.1**    Example to illustrate the noncommutativity of finite rotations.

To be definite, we consider a rotation about the $z$-axis by angle $\phi$. The convention we follow throughout this book is that a rotation operation affects a physical system itself, as in Figure 3.1, while the coordinate axes remain *unchanged*. The angle $\phi$ is taken to be positive when the rotation in question is counterclockwise in the $xy$-plane, as viewed from the positive $z$-side. If we associate a right-handed screw with such a rotation, a positive $\phi$ rotation around the $z$-axis means that the screw is advancing in the positive $z$-direction. With this convention, we easily verify that

$$R_z(\phi) = \begin{pmatrix} \cos\phi & -\sin\phi & 0 \\ \sin\phi & \cos\phi & 0 \\ 0 & 0 & 1 \end{pmatrix}. \tag{3.3}$$

Had we adopted a different convention, in which a physical system remained fixed but the coordinate axes rotated, this same matrix with a positive $\phi$ would have represented a *clockwise* rotation of the $x$- and $y$-axes, when viewed from the positive $z$-side. It is obviously important not to mix the two conventions! Some authors distinguish the two approaches by using "active rotations" for physical systems rotated and "passive rotations" for the coordinate axes rotated.

We are particularly interested in an infinitesimal form of $R_z$:

$$R_z(\varepsilon) = \begin{pmatrix} 1 - \dfrac{\varepsilon^2}{2} & -\varepsilon & 0 \\ \varepsilon & 1 - \dfrac{\varepsilon^2}{2} & 0 \\ 0 & 0 & 1 \end{pmatrix}, \tag{3.4}$$

where terms of order $\varepsilon^3$ and higher are ignored. Likewise, we have

$$
R_x(\varepsilon) = \begin{pmatrix} 1 & 0 & 0 \\ 0 & 1 - \dfrac{\varepsilon^2}{2} & -\varepsilon \\ 0 & \varepsilon & 1 - \dfrac{\varepsilon^2}{2} \end{pmatrix}
\tag{3.5a}
$$

and

$$
R_y(\varepsilon) = \begin{pmatrix} 1 - \dfrac{\varepsilon^2}{2} & 0 & \varepsilon \\ 0 & 1 & 0 \\ -\varepsilon & 0 & 1 - \dfrac{\varepsilon^2}{2} \end{pmatrix},
\tag{3.5b}
$$

which may be read from (3.4) by cyclic permutations of $x, y, z$, that is, $x \to y, y \to z, z \to x$. Compare now the effect of a $y$-axis rotation followed by an $x$-axis rotation with that of an $x$-axis rotation followed by a $y$-axis rotation. Elementary matrix manipulations lead to

$$
R_x(\varepsilon)R_y(\varepsilon) = \begin{pmatrix} 1 - \dfrac{\varepsilon^2}{2} & 0 & \varepsilon \\ \varepsilon^2 & 1 - \dfrac{\varepsilon^2}{2} & -\varepsilon \\ -\varepsilon & \varepsilon & 1 - \varepsilon^2 \end{pmatrix}
\tag{3.6a}
$$

and

$$
R_y(\varepsilon)R_x(\varepsilon) = \begin{pmatrix} 1 - \dfrac{\varepsilon^2}{2} & \varepsilon^2 & \varepsilon \\ 0 & 1 & \dfrac{\varepsilon^2}{2} & -\varepsilon \\ -\varepsilon & \varepsilon & 1 - \varepsilon^2 \end{pmatrix}.
\tag{3.6b}
$$

From (3.6a) and (3.6b) we have the first important result: Infinitesimal rotations about different axes do commute if terms of order $\varepsilon^2$ and higher are ignored.[1] The second and even more important result concerns the manner in which rotations about different axes *fail* to commute when terms of order $\varepsilon^2$ are kept:

$$
R_x(\varepsilon)R_y(\varepsilon) - R_y(\varepsilon)R_x(\varepsilon) = \begin{pmatrix} 0 & -\varepsilon^2 & 0 \\ \varepsilon^2 & 0 & 0 \\ 0 & 0 & 0 \end{pmatrix}
$$

$$
= R_z(\varepsilon^2) - 1,
\tag{3.7}
$$

where all terms of order higher than $\varepsilon^2$ have been ignored throughout this derivation. We also have

$$
1 = R_{\text{any}}(0)
\tag{3.8}
$$

---

[1] Actually there is a familiar example of this in elementary mechanics. The angular velocity vector $\boldsymbol{\omega}$ that characterizes an infinitesimal change in rotation angle during an infinitesimal time interval follows the usual rule of vector addition, including commutativity of vector addition. However, we cannot ascribe a vectorial property to a *finite* angular change.

where *any* stands for any rotation axis. Thus the final result can be written as

$$R_x(\varepsilon)R_y(\varepsilon) - R_y(\varepsilon)R_x(\varepsilon) = R_z(\varepsilon^2) - R_{\mathrm{any}}(0). \tag{3.9}$$

This is an example of the commutation relations between rotation operations about different axes, which we will use later in deducing the angular-momentum commutation relations in quantum mechanics.

## 3.1.2 Infinitesimal Rotations in Quantum Mechanics

So far we have not used quantum-mechanical concepts. The matrix $R$ is just a $3 \times 3$ orthogonal matrix acting on a vector $\mathbf{V}$ written in column matrix form. We must now understand how to characterize rotations in quantum mechanics.

Because rotations affect physical systems, the state ket corresponding to a rotated system is expected to look different from the state ket corresponding to the original unrotated system. Given a rotation operation $R$, characterized by a $3 \times 3$ orthogonal matrix $R$, we associate an operator $\mathscr{D}(R)$ in the appropriate ket space such that

$$|\alpha\rangle_R = \mathscr{D}(R)|\alpha\rangle, \tag{3.10}$$

where $|\alpha\rangle_R$ and $|\alpha\rangle$ stand for the kets of the rotated and original system, respectively.[2] Note that the $3 \times 3$ orthogonal matrix $R$ acts on a column matrix made up of the three components of a classical vector, while the operator $\mathscr{D}(R)$ acts on state vectors in ket space. The matrix representation of $\mathscr{D}(R)$, which we will study in great detail in the subsequent sections, depends on the dimensionality $N$ of the particular ket space in question. For $N = 2$, appropriate for describing a spin $\frac{1}{2}$ system with no other degrees of freedom, $\mathscr{D}(R)$ is represented by a $2 \times 2$ matrix; for a spin 1 system, the appropriate representation is a $3 \times 3$ unitary matrix, and so on.

To construct the rotation operator $\mathscr{D}(R)$, it is again fruitful to examine first its properties under an infinitesimal rotation. We can almost guess how we must proceed by analogy. In both translations and time evolution, which we studied in Sections 1.6 and 2.1, respectively, the appropriate infinitesimal operators could be written as

$$U_\varepsilon = 1 - iG\varepsilon \tag{3.11}$$

with a Hermitian operator $G$. Specifically,

$$G \to \frac{p_x}{\hbar}, \quad \varepsilon \to dx' \tag{3.12}$$

for an infinitesimal translation by a displacement $dx'$ in the $x$-direction, and

$$G \to \frac{H}{\hbar}, \quad \varepsilon \to dt \tag{3.13}$$

for an infinitesimal time evolution with time displacement $dt$. We know from classical mechanics that angular momentum is the generator of rotation in much the same way as momentum and Hamiltonian are the generators of translation and time evolution,

---

[2] The symbol $\mathscr{D}$ stems from the German word *Drehung*, meaning *rotation*.

respectively. We therefore *define* the angular-momentum operator $J_k$ in such a way that the operator for an infinitesimal rotation around the $k$th axis by angle $d\phi$ can be obtained by letting

$$G \to \frac{J_k}{\hbar}, \quad \varepsilon \to d\phi \tag{3.14}$$

in (3.11). With $J_k$ taken to be Hermitian, the infinitesimal rotation operator is guaranteed to be unitary and reduces to the identity operator in the limit $d\phi \to 0$. More generally, we have

$$\mathscr{D}(\hat{\mathbf{n}}, d\phi) = 1 - i \left(\frac{\mathbf{J} \cdot \hat{\mathbf{n}}}{\hbar}\right) d\phi \tag{3.15}$$

for a rotation about the direction characterized by a unit vector $\hat{\mathbf{n}}$ by an infinitesimal angle $d\phi$.

We stress that in this book we do not define the angular-momentum operator to be $\mathbf{x} \times \mathbf{p}$. This is important because spin angular momentum, to which our general formalism also applies, has nothing to do with $x_i$ and $p_j$. Put in another way, in classical mechanics one can prove that the angular momentum defined to be $\mathbf{x} \times \mathbf{p}$ is the generator of a rotation; in contrast, in quantum mechanics we *define* $\mathbf{J}$ so that the operator for an infinitesimal rotation takes the form (3.15).

### 3.1.3 Finite Rotations in Quantum Mechanics

A finite rotation can be obtained by compounding successively infinitesimal rotations about the same axis. For instance, if we are interested in a finite rotation about the $z$-axis by angle $\phi$, we consider

$$\mathscr{D}_z(\phi) = \lim_{N \to \infty} \left[1 - i\left(\frac{J_z}{\hbar}\right)\left(\frac{\phi}{N}\right)\right]^N$$

$$= \exp\left(\frac{-iJ_z\phi}{\hbar}\right)$$

$$= 1 - \frac{iJ_z\phi}{\hbar} - \frac{J_z^2\phi^2}{2\hbar^2} + \cdots. \tag{3.16}$$

In order to obtain the angular-momentum commutation relations, we need one more concept. As we remarked earlier, for every rotation $R$ represented by a $3 \times 3$ orthogonal matrix $R$ there exists a rotation operator $\mathscr{D}(R)$ in the appropriate ket space. We further postulate that $\mathscr{D}(R)$ has the same group properties as $R$:

Identity:     $R \cdot 1 = R \Rightarrow \mathscr{D}(R) \cdot 1 = \mathscr{D}(R)$ $\tag{3.17a}$

Closure:     $R_1 R_2 = R_3 \Rightarrow \mathscr{D}(R_1)\mathscr{D}(R_2) = \mathscr{D}(R_3)$ $\tag{3.17b}$

Inverses:     $RR^{-1} = 1 \Rightarrow \mathscr{D}(R)\mathscr{D}^{-1}(R) = 1$
$R^{-1}R = 1 \Rightarrow \mathscr{D}^{-1}(R)\mathscr{D}(R) = 1$ $\tag{3.17c}$

Associativity:        $R_1(R_2R_3) = (R_1R_2)R_3 = R_1R_2R_3$

$$\Rightarrow \mathscr{D}(R_1)[\mathscr{D}(R_2)\mathscr{D}(R_3)]$$

$$= [\mathscr{D}(R_1)\mathscr{D}(R_2)]\mathscr{D}(R_3)$$

$$= \mathscr{D}(R_1)\mathscr{D}(R_2)\mathscr{D}(R_3). \qquad (3.17d)$$

### 3.1.4  Commutation Relations for Angular Momentum

Let us now return to the fundamental commutation relations for rotation operations (3.9) written in terms of the $R$ matrices. Its rotation operator analogue would read

$$\left(1 - \frac{iJ_x\varepsilon}{\hbar} - \frac{J_x^2\varepsilon^2}{2\hbar^2}\right)\left(1 - \frac{iJ_y\varepsilon}{\hbar} - \frac{J_y^2\varepsilon^2}{2\hbar^2}\right)$$

$$- \left(1 - \frac{iJ_y\varepsilon}{\hbar} - \frac{J_y^2\varepsilon^2}{2\hbar^2}\right)\left(1 - \frac{iJ_x\varepsilon}{\hbar} - \frac{J_x^2\varepsilon^2}{2\hbar^2}\right) = 1 - \frac{iJ_z\varepsilon^2}{\hbar} - 1. \qquad (3.18)$$

Terms of order $\varepsilon$ automatically drop out. Equating terms of order $\varepsilon^2$ on both sides of (3.18), we obtain

$$[J_x, J_y] = i\hbar J_z. \qquad (3.19)$$

Repeating this kind of argument with rotations about other axes, we obtain

$$[J_i, J_j] = i\hbar\varepsilon_{ijk}J_k, \qquad (3.20)$$

known as the **fundamental commutation relations of angular momentum**.

In general, when the generators of infinitesimal transformations do not commute, the corresponding group of operations is said to be **non-Abelian**. Because of (3.20), the rotation group in three dimensions is non-Abelian. In contrast, the translation group in three dimensions is Abelian because $p_i$ and $p_j$ commute even with $i \neq j$.

We emphasize that in obtaining the commutation relations (3.20) we have used the following two concepts.

1. $J_k$ is the generator of rotation about the $k$th axis.
2. Rotations about different axes fail to commute.

It is no exaggeration to say that commutation relations (3.20) summarize in a compact manner *all* the basic properties of rotations in three dimensions.

The fundamental commutation relations make it possible to show that the angular-momentum operators themselves transform as expected under rotations. Consider a rotation by a finite angle $\phi$ about the $z$-axis. If the ket of the system before rotation is given by $|\alpha\rangle$, the ket after rotation is given by

$$|\alpha\rangle_R = \mathscr{D}_z(\phi)|\alpha\rangle \qquad (3.21)$$

where $\mathscr{D}_z(\phi)$ is given by (3.16). To see that this operator really rotates the physical system, let us look at its effect on $\langle J_x \rangle$. Under rotation this expectation value changes as follows:

$$\langle J_x \rangle \to {}_R\langle \alpha | J_x | \alpha \rangle_R = \langle \alpha | \mathscr{D}_z^\dagger(\phi) J_x \mathscr{D}_z(\phi) | \alpha \rangle. \tag{3.22}$$

We must therefore compute

$$\exp\left(\frac{iJ_z\phi}{\hbar}\right) J_x \exp\left(\frac{-iJ_z\phi}{\hbar}\right). \tag{3.23}$$

Using (2.168) to evaluate (3.23), we have

$$\exp\left(\frac{iJ_z\phi}{\hbar}\right) J_x \exp\left(\frac{-iJ_z\phi}{\hbar}\right) = J_x + \left(\frac{i\phi}{\hbar}\right) \underbrace{[J_z, J_x]}_{i\hbar J_y}$$

$$+ \left(\frac{1}{2!}\right)\left(\frac{i\phi}{\hbar}\right)^2 \underbrace{[J_z, \underbrace{[J_z, J_x]}_{i\hbar J_y}]}_{\hbar^2 J_x} + \left(\frac{1}{3!}\right)\left(\frac{i\phi}{\hbar}\right)^3 \underbrace{[J_z, \underbrace{[J_z, \underbrace{[J_z, J_x]}_{\hbar^2 J_x}]}_{i\hbar^3 J_y}]} + \cdots$$

$$= J_x \left[1 - \frac{\phi^2}{2!} + \cdots\right] - J_y \left[\phi - \frac{\phi^3}{3!} + \cdots\right]$$

$$= J_x \cos\phi - J_y \sin\phi. \tag{3.24}$$

We will return to transformations of this sort when we discuss symmetry operations in Chapter 4.

## 3.2  Spin $\frac{1}{2}$ Systems and Finite Rotations

### 3.2.1  Rotation Operator for Spin $\frac{1}{2}$

The lowest number, $N$, of dimensions in which the angular-momentum commutation relations (3.20) are realized is $N = 2$. The reader has already checked in Problem 1.10 of Chapter 1 that the operators defined by

$$S_x = \left(\frac{\hbar}{2}\right)\{(|+\rangle\langle-|) + (|-\rangle\langle+|)\},$$

$$S_y = \left(\frac{i\hbar}{2}\right)\{-(|+\rangle\langle-|) + (|-\rangle\langle+|)\}, \tag{3.25}$$

$$S_z = \left(\frac{\hbar}{2}\right)\{(|+\rangle\langle+|) - (|-\rangle\langle-|)\},$$

satisfy commutation relations (3.20) with $J_k$ replaced by $S_k$. It is not a priori obvious that nature takes advantage of the lowest dimensional realization of (3.20), but numerous experiments – from atomic spectroscopy to nuclear magnetic resonance – suffice to convince us that this is in fact the case.

We showed in (3.24) that components of the general angular-momentum operator behave as expected under finite rotations. This is of course also true for spin $\frac{1}{2}$, that is

$$\langle S_x \rangle \rightarrow {}_R\langle \alpha | S_x | \alpha \rangle_R = \langle S_x \rangle \cos\phi - \langle S_y \rangle \sin\phi, \tag{3.26}$$

where the expectation value without subscripts is understood to be taken with respect to the (old) unrotated system. Similarly,

$$\langle S_y \rangle \rightarrow \langle S_y \rangle \cos\phi + \langle S_x \rangle \sin\phi. \tag{3.27}$$

As for the expectation value of $S_z$, there is no change because $S_z$ commutes with $\mathcal{D}_z(\phi)$:

$$\langle S_z \rangle \rightarrow \langle S_z \rangle. \tag{3.28}$$

Relations (3.26), (3.27), and (3.28) are quite reasonable. They show that rotation operator (3.16), when applied to the state ket, does rotate the expectation value of $\mathbf{S}$ around the $z$-axis by angle $\phi$. In other words, the expectation value of the spin operator behaves as though it were a classical vector under rotation:

$$\langle S_k \rangle \rightarrow \sum_l R_{kl} \langle S_l \rangle, \tag{3.29}$$

where $R_{kl}$ are the elements of the $3 \times 3$ orthogonal matrix $R$ that specifies the rotation in question. It should be clear from our derivation that this property is not restricted to the spin operator of spin $\frac{1}{2}$ systems. In general, we have

$$\langle J_k \rangle \rightarrow \sum_l R_{kl} \langle J_l \rangle \tag{3.30}$$

under rotation, where $J_k$ are the generators of rotations satisfying the angular-momentum commutation relations (3.20). Later we will show that relations of this kind can be further generalized to any vector operator.

So far everything has been as expected. But now, be prepared for a surprise! We examine the effect of rotation operator (3.16) on a general ket,

$$|\alpha\rangle = |+\rangle\langle+|\alpha\rangle + |-\rangle\langle-|\alpha\rangle, \tag{3.31}$$

a little more closely. We see that

$$\exp\left(\frac{-iS_z\phi}{\hbar}\right)|\alpha\rangle = e^{-i\phi/2}|+\rangle\langle+|\alpha\rangle + e^{i\phi/2}|-\rangle\langle-|\alpha\rangle. \tag{3.32}$$

The appearance of the half-angle $\phi/2$ here has an extremely interesting consequence.

Let us consider a rotation by $2\pi$. We then have

$$|\alpha\rangle_{R_z(2\pi)} \rightarrow -|\alpha\rangle. \tag{3.33}$$

So the ket for the 360° rotated state differs from the original ket by a minus sign. We would need a 720° ($\phi = 4\pi$) rotation to get back to the same ket with a *plus* sign. Notice that this minus sign disappears for the expectation value of $\mathbf{S}$ because $\mathbf{S}$ is sandwiched by $|\alpha\rangle$ and $\langle\alpha|$, both of which change sign. Will this minus sign ever be observable? We will give the answer to this interesting question after we discuss spin precession once again.

## 3.2.2  Spin Precession Revisited

We now treat the problem of spin precession, already discussed in Section 2.1, from a new point of view. We recall that the basic Hamiltonian of the problem is given by

$$H = - \left( \frac{e}{m_e c} \right) \mathbf{S \cdot B} = \omega S_z, \tag{3.34}$$

where

$$\omega \equiv \frac{|e|B}{m_e c}. \tag{3.35}$$

The time-evolution operator based on this Hamiltonian is given by

$$\mathscr{U}(t,0) = \exp \left( \frac{-iHt}{\hbar} \right) = \exp \left( \frac{-iS_z \omega t}{\hbar} \right). \tag{3.36}$$

Comparing this equation with (3.16), we see that the time-evolution operator here is precisely the same as the rotation operator in (3.16) with $\phi$ set equal to $\omega t$. In this manner we see immediately why this Hamiltonian causes spin precession. Paraphrasing (3.26), (3.27), and (3.28), we obtain

$$\langle S_x \rangle_t = \langle S_x \rangle_{t=0} \cos \omega t - \langle S_y \rangle_{t=0} \sin \omega t, \tag{3.37a}$$

$$\langle S_y \rangle_t = \langle S_y \rangle_{t=0} \cos \omega t + \langle S_x \rangle_{t=0} \sin \omega t, \tag{3.37b}$$

$$\langle S_z \rangle_t = \langle S_z \rangle_{t=0}. \tag{3.37c}$$

After $t = 2\pi/\omega$, the spin returns to its original direction.

This set of equations can be used to discuss the spin precession of a **muon**, an electronlike particle which, however, is 210 times as heavy. The muon magnetic moment can be determined from other experiments – for example, the hyperfine splitting in muonium, a bound state of a positive muon and an electron – to be $e\hbar/2m_\mu c$, just as expected from Dirac's relativistic theory of spin $\frac{1}{2}$ particles. (We will here neglect very small corrections that arise from quantum field theory effects.) Knowing the magnetic moment we can predict the angular frequency of precession. So (3.37) can be and, in fact, has been checked experimentally (see Figure 2.1). In practice, as the external magnetic field causes spin precession, the spin direction is analyzed by taking advantage of the fact that electrons from muon decay tend to be emitted preferentially in the direction opposite to the muon spin.

Let us now look at the time evolution of the state ket itself. Assuming that the initial ($t = 0$) ket is given by (3.31), we obtain after time $t$

$$|\alpha, t_0 = 0; t\rangle = e^{-i\omega t/2} |+\rangle \langle + |\alpha\rangle + e^{+i\omega t/2} |-\rangle \langle - |\alpha\rangle. \tag{3.38}$$

Expression (3.38) acquires a minus sign at $t = 2\pi/\omega$, and we must wait until $t = 4\pi/\omega$ to get back to the original state ket with the same sign. To sum up, the period for the state ket is *twice* as long as the period for spin precession

$$\tau_{\text{precession}} = \frac{2\pi}{\omega}, \tag{3.39a}$$

$$\tau_{\text{state ket}} = \frac{4\pi}{\omega}. \tag{3.39b}$$

### 3.2.3 Neutron Interferometry Experiment to Study $2\pi$ Rotations

We now describe an experiment performed to detect the minus sign in (3.33). Quite clearly, if every state ket in the universe is multiplied by a minus sign, there will be no observable effect. The only way to detect the predicted minus sign is to make a comparison between an unrotated state and a rotated state. As in gravity-induced quantum interference, discussed in Section 2.7, we rely on the art of neutron interferometry to verify this extraordinary prediction of quantum mechanics.

A nearly monoenergetic beam of thermal neutrons is split into two parts, path $A$ and path $B$; see Figure 3.2. Path $A$ always goes through a magnetic-field-free region; in contrast, path $B$ enters a small region where a static magnetic field is present. As a result, the neutron state ket going via path $B$ suffers a phase change $e^{\mp i\omega T/2}$, where $T$ is the time spent in the $\mathbf{B} \neq 0$ region and $\omega$ is the spin-precession frequency

$$\omega = \frac{g_n e B}{m_p c} \quad (g_n \simeq -1.91) \tag{3.40}$$

for the neutron with a magnetic moment of $g_n e\hbar/2m_p c$, as we can see if we compare this with (3.35), which is appropriate for the electron with magnetic moment $e\hbar/2m_e c$. When path $A$ and path $B$ meet again in the interference region of Figure 3.2, the amplitude of the neutron arriving via path $B$ is

$$c_2 = c_2(B = 0)e^{\mp i\omega T/2}, \tag{3.41}$$

while the amplitude of the neutron arriving via path $A$ is $c_1$, independent of $\mathbf{B}$. So the intensity observable in the interference region must exhibit a sinusoidal variation

$$\cos\left(\frac{\mp \omega T}{2} + \delta\right), \tag{3.42}$$

where $\delta$ is the phase difference between $c_1$ and $c_2 \, (B = 0)$. In practice, $T$, the time spent in the $B \neq 0$ region, is fixed but the precession frequency $\omega$ is varied by changing the

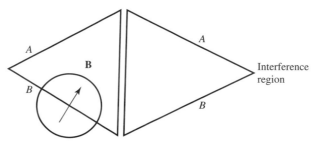

**Fig. 3.2**   Experiment to study the predicted minus sign under a $2\pi$ rotation.

strength of the magnetic field. The intensity in the interference region as a function of $B$ is predicted to have a sinusoidal variation. If we call $\Delta B$ the difference in $B$ needed to produce successive maxima, we can easily show that

$$\Delta B = \frac{4\pi\hbar c}{eg_n \lambda l}, \tag{3.43}$$

where $l$ is the path length.

In deriving this formula we used the fact that a $4\pi$ rotation is needed for the state ket to return to the original ket with the same sign, as required by our formalism. If, on the other hand, our description of spin $\frac{1}{2}$ systems were incorrect and the ket were to return to its original ket with the same sign under a $2\pi$ rotation, the predicted value for $\Delta B$ would be just one-half of (3.43).

Two different groups have conclusively demonstrated experimentally that prediction (3.43) is correct to an accuracy of a fraction of a percent.[3] This is another triumph of quantum mechanics. The nontrivial prediction (3.33) has been experimentally established in a direct manner.

### 3.2.4  Pauli Two-Component Formalism

Manipulations with the state kets of spin $\frac{1}{2}$ systems can be conveniently carried out using the two-component spinor formalism introduced by W. Pauli in 1926. In Section 1.3 we learned how a ket (bra) can be represented by a column (row) matrix; all we have to do is arrange the expansion coefficients in terms of a certain specified set of base kets into a column (row) matrix. In the spin $\frac{1}{2}$ case we have

$$|+\rangle \doteq \begin{pmatrix} 1 \\ 0 \end{pmatrix} \equiv \chi_+ \quad |-\rangle \doteq \begin{pmatrix} 0 \\ 1 \end{pmatrix} \equiv \chi_-$$

$$\langle +| \doteq (1,0) = \chi_+^\dagger \quad \langle -| \doteq (0,1) = \chi_-^\dagger \tag{3.44}$$

for the base kets and bras and

$$|\alpha\rangle = |+\rangle\langle +|\alpha\rangle + |-\rangle\langle -|\alpha\rangle \doteq \begin{pmatrix} \langle +|\alpha\rangle \\ \langle -|\alpha\rangle \end{pmatrix} \tag{3.45a}$$

and

$$\langle \alpha| = \langle \alpha|+\rangle\langle +| + \langle \alpha|-\rangle\langle -| \doteq (\langle \alpha|+\rangle, \langle \alpha|-\rangle) \tag{3.45b}$$

for an arbitrary state ket and the corresponding state bra. Column matrix (3.45a) is referred to as a **two-component spinor** and is written as

$$\chi = \begin{pmatrix} \langle +|\alpha\rangle \\ \langle -|\alpha\rangle \end{pmatrix} \equiv \begin{pmatrix} c_+ \\ c_- \end{pmatrix}$$

$$= c_+\chi_+ + c_-\chi_-, \tag{3.46}$$

---

[3] Rauch et al., *Phys. Lett. A*, **54** (1975) 425; Werner et al., *Phys. Rev. Lett.*, **35** (1975) 1053.

where $c_+$ and $c_-$ are, in general, complex numbers. For $\chi^\dagger$ we have

$$\chi^\dagger = (\langle\alpha|+\rangle, \langle\alpha|-\rangle) = (c_+^*, c_-^*). \tag{3.47}$$

The matrix elements $\langle\pm|S_k|+\rangle$ and $\langle\pm|S_k|-\rangle$, apart from $\hbar/2$, are to be set equal to those of $2 \times 2$ matrices $\sigma_k$, known as the **Pauli matrices**. We identify

$$\langle\pm|S_k|+\rangle \equiv \left(\frac{\hbar}{2}\right)(\sigma_k)_{\pm,+}, \quad \langle\pm|S_k|-\rangle \equiv \left(\frac{\hbar}{2}\right)(\sigma_k)_{\pm,-}. \tag{3.48}$$

We can now write the expectation value $\langle S_k \rangle$ in terms of $\chi$ and $\sigma_k$:

$$\langle S_k \rangle = \langle\alpha|S_k|\alpha\rangle = \sum_{a'=+,-}\sum_{a''=+,-}\langle\alpha|a'\rangle\langle a'|S_k|a''\rangle\langle a''|\alpha\rangle$$
$$= \left(\frac{\hbar}{2}\right)\chi^\dagger \sigma_k \chi, \tag{3.49}$$

where the usual rule of matrix multiplication is used in the last line. Explicitly, we see from (3.25) together with (3.48) that

$$\sigma_1 = \begin{pmatrix} 0 & 1 \\ 1 & 0 \end{pmatrix}, \quad \sigma_2 = \begin{pmatrix} 0 & -i \\ i & 0 \end{pmatrix}, \quad \sigma_3 = \begin{pmatrix} 1 & 0 \\ 0 & -1 \end{pmatrix}, \tag{3.50}$$

where the subscripts 1, 2, and 3 refer to $x$, $y$, and $z$, respectively.

We record some properties of the Pauli matrices. First,

$$\sigma_i^2 = 1 \tag{3.51a}$$

$$\sigma_i\sigma_j + \sigma_j\sigma_i = 0, \quad \text{for} \quad i \neq j, \tag{3.51b}$$

where the right-hand side of (3.51a) is to be understood as the $2 \times 2$ identity matrix. These two relations are, of course, equivalent to the anticommutation relations

$$\{\sigma_i, \sigma_j\} = 2\delta_{ij}. \tag{3.52}$$

We also have the commutation relations

$$[\sigma_i, \sigma_j] = 2i\varepsilon_{ijk}\sigma_k, \tag{3.53}$$

which we see to be the explicit $2 \times 2$ matrix realizations of the angular-momentum commutation relations (3.20). Combining (3.52) and (3.53), we can obtain

$$\sigma_1\sigma_2 = -\sigma_2\sigma_1 = i\sigma_3\ldots. \tag{3.54}$$

Notice also that

$$\sigma_i^\dagger = \sigma_i, \tag{3.55a}$$

$$\det(\sigma_i) = -1, \tag{3.55b}$$

$$\text{Tr}(\sigma_i) = 0. \tag{3.55c}$$

We now consider $\sigma \cdot \mathbf{a}$, where $\mathbf{a}$ is a vector in three dimensions. This is actually to be understood as a $2 \times 2$ matrix. Thus

$$
\sigma \cdot \mathbf{a} \equiv \sum_k a_k \sigma_k
$$
$$
= \begin{pmatrix} +a_3 & a_1 - ia_2 \\ a_1 + ia_2 & -a_3 \end{pmatrix}. \tag{3.56}
$$

There is also a very important identity,

$$
(\sigma \cdot \mathbf{a})(\sigma \cdot \mathbf{b}) = \mathbf{a} \cdot \mathbf{b} + i\sigma \cdot (\mathbf{a} \times \mathbf{b}). \tag{3.57}
$$

To prove this all we need are the anticommutation and commutation relations, (3.52) and (3.53), respectively:

$$
\sum_j \sigma_j a_j \sum_k \sigma_k b_k = \sum_j \sum_k \left( \frac{1}{2} \{\sigma_j, \sigma_k\} + \frac{1}{2} [\sigma_j, \sigma_k] \right) a_j b_k
$$
$$
= \sum_j \sum_k (\delta_{jk} + i\varepsilon_{jkl}\sigma_l) a_j b_k
$$
$$
= \mathbf{a} \cdot \mathbf{b} + i\sigma \cdot (\mathbf{a} \times \mathbf{b}). \tag{3.58}
$$

If the components of $\mathbf{a}$ are real, we have

$$
(\sigma \cdot \mathbf{a})^2 = |\mathbf{a}|^2, \tag{3.59}
$$

where $|\mathbf{a}|$ is the magnitude of the vector $\mathbf{a}$.

## 3.2.5 Rotations in the Two-Component Formalism

Let us now study the $2 \times 2$ matrix representation of the rotation operator $\mathscr{D}(\hat{\mathbf{n}}, \phi)$. We have

$$
\exp\left( \frac{-i\mathbf{S} \cdot \hat{\mathbf{n}}\phi}{\hbar} \right) \doteq \exp\left( \frac{-i\sigma \cdot \hat{\mathbf{n}}\phi}{2} \right). \tag{3.60}
$$

Using

$$
(\sigma \cdot \hat{\mathbf{n}})^n = \begin{cases} 1 & \text{for } n \text{ even} \\ \sigma \cdot \hat{\mathbf{n}} & \text{for } n \text{ odd,} \end{cases} \tag{3.61}
$$

which follows from (3.59), we can write

$$
\exp\left( \frac{-i\sigma \cdot \hat{\mathbf{n}}\phi}{2} \right) = \left[ 1 - \frac{(\sigma \cdot \hat{\mathbf{n}})^2}{2!} \left( \frac{\phi}{2} \right)^2 + \frac{(\sigma \cdot \hat{\mathbf{n}})^4}{4!} \left( \frac{\phi}{2} \right)^4 - \cdots \right]
$$
$$
- i\left[ (\sigma \cdot \hat{\mathbf{n}}) \frac{\phi}{2} - \frac{(\sigma \cdot \hat{\mathbf{n}})^3}{3!} \left( \frac{\phi}{2} \right)^3 + \cdots \right]
$$
$$
= 1 \cos\left( \frac{\phi}{2} \right) - i\sigma \cdot \hat{\mathbf{n}} \sin\left( \frac{\phi}{2} \right). \tag{3.62}
$$

Explicitly, in $2 \times 2$ form we have

$$\exp\left(\frac{-i\boldsymbol{\sigma}\cdot\hat{\mathbf{n}}\phi}{2}\right) = \begin{pmatrix} \cos\left(\dfrac{\phi}{2}\right) - in_z \sin\left(\dfrac{\phi}{2}\right) & (-in_x - n_y)\sin\left(\dfrac{\phi}{2}\right) \\ (-in_x + n_y)\sin\left(\dfrac{\phi}{2}\right) & \cos\left(\dfrac{\phi}{2}\right) + in_z \sin\left(\dfrac{\phi}{2}\right) \end{pmatrix}. \tag{3.63}$$

Just as the operator $\exp(-i\mathbf{S}\cdot\hat{\mathbf{n}}\phi/\hbar)$ acts on a state ket $|\alpha\rangle$, the $2 \times 2$ matrix $\exp(-i\boldsymbol{\sigma}\cdot\hat{\mathbf{n}}\phi/2)$ acts on a two-component spinor $\chi$. Under rotations we change $\chi$ as follows:

$$\chi \to \exp\left(\frac{-i\boldsymbol{\sigma}\cdot\hat{\mathbf{n}}\phi}{2}\right)\chi. \tag{3.64}$$

On the other hand, the $\sigma_k$ themselves are to remain *unchanged* under rotations. So strictly speaking, despite its appearance, $\boldsymbol{\sigma}$ is not to be regarded as a vector; rather, it is $\chi^\dagger \boldsymbol{\sigma} \chi$ which obeys the transformation property of a vector:

$$\chi^\dagger \sigma_k \chi \to \sum_l R_{kl} \chi^\dagger \sigma_l \chi. \tag{3.65}$$

An explicit proof of this may be given using

$$\exp\left(\frac{i\sigma_3 \phi}{2}\right) \sigma_1 \exp\left(\frac{-i\sigma_3 \phi}{2}\right) = \sigma_1 \cos\phi - \sigma_2 \sin\phi \tag{3.66}$$

and so on, which is the $2 \times 2$ matrix analogue of (3.16).

In discussing a $2\pi$ rotation using the ket formalism, we have seen that a spin $\frac{1}{2}$ ket $|\alpha\rangle$ goes into $-|\alpha\rangle$. The $2 \times 2$ analogue of this statement is

$$\exp\left(\frac{-i\boldsymbol{\sigma}\cdot\hat{\mathbf{n}}\phi}{2}\right)\Bigg|_{\phi=2\pi} = -1, \quad \text{for any } \hat{\mathbf{n}}, \tag{3.67}$$

which is evident from (3.62).

As an instructive application of rotation matrix (3.63), let us see how we can construct an eigenspinor of $\boldsymbol{\sigma}\cdot\hat{\mathbf{n}}$ with eigenvalue $+1$, where $\hat{\mathbf{n}}$ is a unit vector in some specified direction. Our object is to construct $\chi$ satisfying

$$\boldsymbol{\sigma}\cdot\hat{\mathbf{n}}\chi = \chi. \tag{3.68}$$

In other words, we look for the two-component column matrix representation of $|\mathbf{S}\cdot\hat{\mathbf{n}}; +\rangle$ defined by

$$\mathbf{S}\cdot\hat{\mathbf{n}}|\mathbf{S}\cdot\hat{\mathbf{n}}; +\rangle = \left(\frac{\hbar}{2}\right)|\mathbf{S}\cdot\hat{\mathbf{n}}; +\rangle. \tag{3.69}$$

Actually this can be solved as a straightforward eigenvalue problem (see Problem 1.11 in Chapter 1), but here we present an alternative method based on rotation matrix (3.63).

Let the polar and the azimuthal angles that characterize $\hat{\mathbf{n}}$ be $\beta$ and $\alpha$, respectively. We start with $\begin{pmatrix} 1 \\ 0 \end{pmatrix}$, the two-component spinor that represents the spin-up state. Given this, we first rotate about the $y$-axis by angle $\beta$; we subsequently rotate by angle $\alpha$ about the $z$-axis. We see that the desired spin state is then obtained; see Figure 3.3. In the Pauli

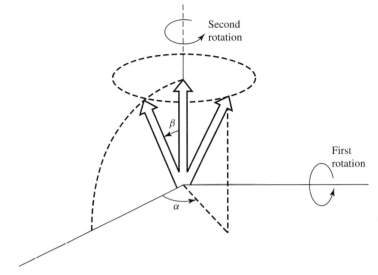

Fig. 3.3    Construction of the $\sigma \cdot \hat{\mathbf{n}}$ eigenspinor.

spinor language this sequence of operations is equivalent to applying $\exp(-i\sigma_2\beta/2)$ to $\begin{pmatrix} 1 \\ 0 \end{pmatrix}$ followed by an application of $\exp(-i\sigma_3\alpha/2)$. The net result is

$$
\begin{aligned}
\chi &= \left[\cos\left(\frac{\alpha}{2}\right) - i\sigma_3\sin\left(\frac{\alpha}{2}\right)\right]\left[\cos\left(\frac{\beta}{2}\right) - i\sigma_2\sin\left(\frac{\beta}{2}\right)\right]\begin{pmatrix} 1 \\ 0 \end{pmatrix} \\
&= \begin{pmatrix} \cos\left(\frac{\alpha}{2}\right) - i\sin\left(\frac{\alpha}{2}\right) & 0 \\ 0 & \cos\left(\frac{\alpha}{2}\right) + i\sin\left(\frac{\alpha}{2}\right) \end{pmatrix}\begin{pmatrix} \cos\left(\frac{\beta}{2}\right) & -\sin\left(\frac{\beta}{2}\right) \\ \sin\left(\frac{\beta}{2}\right) & \cos\left(\frac{\beta}{2}\right) \end{pmatrix}\begin{pmatrix} 1 \\ 0 \end{pmatrix} \\
&= \begin{pmatrix} \cos\left(\frac{\beta}{2}\right)e^{-i\alpha/2} \\ \sin\left(\frac{\beta}{2}\right)e^{i\alpha/2} \end{pmatrix},
\end{aligned}
\tag{3.70}
$$

in complete agreement with Problem 1.11 of Chapter 1 if we realize that a phase common to both the upper and lower components is devoid of physical significance.

## 3.3  SO(3), SU(2), and Euler Rotations

### 3.3.1  Orthogonal Group

We will now study a little more systematically the group properties of the operations with which we have been concerned in the previous two sections.

The most elementary approach to rotations is based on specifying the axis of rotation and the angle of rotation. It is clear that we need three real numbers to characterize a general rotation: the polar and the azimuthal angles of the unit vector $\hat{\mathbf{n}}$ taken in the direction of the rotation axis and the rotation angle $\phi$ itself. Equivalently, the same rotation can be specified by the three Cartesian components of the vector $\hat{\mathbf{n}}\phi$. However, these ways of characterizing rotation are not so convenient from the point of view of studying the group properties of rotations. For one thing, unless $\phi$ is infinitesimal or $\hat{\mathbf{n}}$ is always in the same direction, we cannot add vectors of the form $\hat{\mathbf{n}}\phi$ to characterize a succession of rotations. It is much easier to work with a $3 \times 3$ orthogonal matrix $R$ because the effect of successive rotations can be obtained just by multiplying the appropriate orthogonal matrices.

How many independent parameters are there in a $3 \times 3$ orthogonal matrix? A real $3 \times 3$ matrix has 9 entries, but we have the orthogonality constraint

$$RR^T = 1 \tag{3.71}$$

which corresponds to 6 independent equations because the product $RR^T$, being the same as $R^T R$, is a symmetrical matrix with 6 independent entries. As a result, there are 3 (that is, 9–6) independent numbers in $R$, the same number previously obtained by a more elementary method.

The set of all multiplication operations with orthogonal matrices forms a group. By this we mean that the following four requirements are satisfied.

1. The product of any two orthogonal matrices is another orthogonal matrix, which is satisfied because

$$(R_1 R_2)(R_1 R_2)^T = R_1 R_2 R_2^T R_1^T = 1. \tag{3.72}$$

2. The associative law holds:

$$R_1(R_2 R_3) = (R_1 R_2)R_3. \tag{3.73}$$

3. The identity matrix 1, physically corresponding to no rotation, defined by

$$R1 = 1R = R \tag{3.74}$$

is a member of the class of all orthogonal matrices.

4. The inverse matrix $R^{-1}$, physically corresponding to rotation in the opposite sense, defined by

$$RR^{-1} = R^{-1}R = 1 \tag{3.75}$$

is also a member.

This group has the name SO(3), where S stands for special, O stands for orthogonal, 3 for three dimensions. Note only rotational operations are considered here, hence we have SO(3) rather than O(3) (which can include the inversion operation of Chapter 4 later).

### 3.3.2  Unitary Unimodular Group

In the previous section we learned yet another way to characterize an arbitrary rotation, that is, to look at the $2 \times 2$ matrix (3.63) that acts on the two-component spinor $\chi$. Clearly, (3.63) is unitary. As a result, for the $c_+$ and $c_-$, defined in (3.46),

$$|c_+|^2 + |c_-|^2 = 1 \tag{3.76}$$

is left invariant. Furthermore, matrix (3.63) is unimodular; that is, its determinant is 1, as will be shown explicitly below.

We can write the most general unitary unimodular matrix as

$$U(a,b) = \begin{pmatrix} a & b \\ -b^* & a^* \end{pmatrix}, \tag{3.77}$$

where $a$ and $b$ are *complex* numbers satisfying the unimodular condition

$$|a|^2 + |b|^2 = 1. \tag{3.78}$$

We can easily establish the unitary property of (3.77) as follows:

$$U(a,b)^\dagger U(a,b) = \begin{pmatrix} a^* & -b \\ b^* & a \end{pmatrix} \begin{pmatrix} a & b \\ -b^* & a^* \end{pmatrix} = 1. \tag{3.79}$$

We can readily see that the $2 \times 2$ matrix (3.63) that characterizes a rotation of a spin $\frac{1}{2}$ system can be written as $U(a,b)$. Comparing (3.63) with (3.77), we identify

$$\begin{aligned} \text{Re}(a) &= \cos\left(\frac{\phi}{2}\right), & \text{Im}(a) &= -n_z \sin\left(\frac{\phi}{2}\right), \\ \text{Re}(b) &= -n_y \sin\left(\frac{\phi}{2}\right), & \text{Im}(b) &= -n_x \sin\left(\frac{\phi}{2}\right), \end{aligned} \tag{3.80}$$

from which the unimodular property of (3.78) is immediate. Conversely, it is clear that the most general unitary unimodular matrix of form (3.77) can be interpreted as representing a rotation.

The two complex numbers $a$ and $b$ are known as **Cayley–Klein parameters**. Historically the connection between a unitary unimodular matrix and a rotation was known long before the birth of quantum mechanics. In fact, the Cayley–Klein parameters were used to characterize complicated motions of gyroscopes in rigid-body kinematics.

Without appealing to the interpretations of unitary unimodular matrices in terms of rotations, we can directly check the group properties of multiplication operations with unitary unimodular matrices. Note in particular that

$$U(a_1, b_1) U(a_2, b_2) = U(a_1 a_2 - b_1 b_2^*, a_1 b_2 + a_2^* b_1), \tag{3.81}$$

where the unimodular condition for the product matrix is

$$|a_1 a_2 - b_1 b_2^*|^2 + |a_1 b_2 + a_2^* b_1|^2 = 1. \tag{3.82}$$

For the inverse of $U$ we have

$$U^{-1}(a,b) = U(a^*, -b). \tag{3.83}$$

This group is known as SU(2), where S stands for special, $U$ for unitary, and 2 for dimensionality 2. In contrast, the group defined by multiplication operations with general $2 \times 2$ unitary matrices (not necessarily constrained to be unimodular) is known as U(2). The most general unitary matrix in two dimensions has four independent parameters and can be written as $e^{i\gamma}$ (with $\gamma$ real) times a unitary unimodular matrix:

$$U = e^{i\gamma} \begin{pmatrix} a & b \\ -b^* & a^* \end{pmatrix}, \qquad |a|^2 + |b|^2 = 1, \qquad \gamma^* = \gamma. \tag{3.84}$$

SU(2) is called a **subgroup** of U(2).

Because we can characterize rotations using both the SO(3) language and the SU(2) language, we may be tempted to conclude that the *groups* SO(3) and SU(2) are isomorphic, that is, that there is a one-to-one correspondence between an element of SO(3) and an element of SU(2). This inference is not correct. Consider a rotation by $2\pi$ and another one by $4\pi$. In the SO(3) language, the matrices representing a $2\pi$ rotation and a $4\pi$ rotation are both $3 \times 3$ identity matrices; however, in the SU(2) language the corresponding matrices are $-1$ times the $2 \times 2$ identity matrix and the identity matrix itself, respectively. More generally, $U(a,\ b)$ and $U(-a,-b)$ both correspond to a *single* $3 \times 3$ matrix in the SO(3) language. The correspondence therefore is two-to-one; for a given $R$, the corresponding $U$ is double valued. One can say, however, that the two groups are *locally* isomorphic.

### 3.3.3 Euler Rotations

From classical mechanics the reader may be familiar with the fact that an arbitrary rotation of a rigid body can be accomplished in three steps, known as **Euler rotations**. The Euler rotation language, specified by three Euler angles, provides yet another way to characterize the most general rotation in three dimensions.

The three steps of Euler rotations are as follows. First, rotate the rigid body counter-clockwise (as seen from the positive $z$-side) about the $z$-axis by angle $\alpha$. Imagine now that there is a body $y$-axis embedded, so to speak, in the rigid body such that before the $z$-axis rotation is carried out, the body $y$-axis coincides with the usual $y$-axis, referred to as the **space-fixed $y$-axis**. Obviously, after the rotation about the $z$-axis, the body $y$-axis no longer coincides with the space-fixed $y$-axis; let us call the former the $y'$-axis. To see how all this may appear for a thin disk, refer to Figure 3.4a. We now perform a second rotation, this time about the $y'$-axis by angle $\beta$. As a result, the body $z$-axis no longer points in the space-fixed $z$-axis direction. We call the body-fixed $z$-axis after the second rotation the $z'$-axis; see Figure 3.4b. The third and final rotation is about the $z'$-axis by angle $\gamma$. The body $y$-axis now becomes the $y''$-axis of Figure 3.4c. In terms of $3 \times 3$ orthogonal matrices the product of the three operations can be written as

$$R(\alpha,\beta,\gamma) \equiv R_{z'}(\gamma)R_{y'}(\beta)R_z(\alpha). \tag{3.85}$$

A cautionary remark is in order here. Most textbooks in classical mechanics prefer to perform the second rotation (the middle rotation) about the body $x$-axis rather than about the body $y$-axis (see, for example, Goldstein et al. (2002)). This convention is to be avoided in quantum mechanics for a reason that will become apparent in a moment.

In (3.85) there appear $R_{y'}$ and $R_{z'}$, which are matrices for rotations about body axes. This approach to Euler rotations is rather inconvenient in quantum mechanics because we earlier obtained simple expressions for the space-fixed (unprimed) axis components of the **S** operator, but not for the body-axis components. It is therefore desirable to express the body-axis rotations we considered in terms of space-fixed axis rotations. Fortunately there is a very simple relation, namely,

$$R_{y'}(\beta) = R_z(\alpha)R_y(\beta)R_z^{-1}(\alpha). \tag{3.86}$$

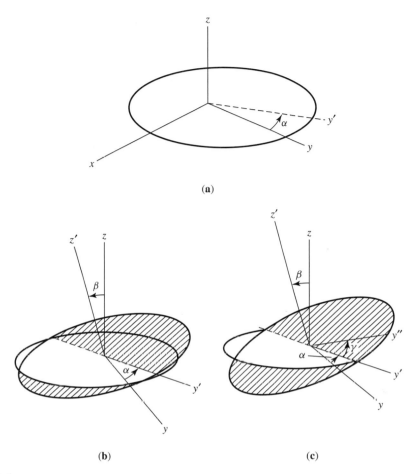

**Fig. 3.4**   Euler rotations.

The meaning of the right-hand side is as follows. First, bring the body $y$-axis of Figure 3.4a (that is, the $y'$-axis) back to the original fixed-space $y$-direction by rotating *clockwise* (as seen from the positive $z$-side) about the $z$-axis by angle $\alpha$ then rotate about the $y$-axis by angle $\beta$. Finally, return the body $y$-axis to the direction of the $y'$-axis by rotating about the fixed-space $z$-axis (*not* about the $z'$-axis!) by angle $\alpha$. Equation (3.86) tells us that the net effect of these rotations is a single rotation about the $y'$-axis by angle $\beta$.

To prove this assertion, let us look more closely at the effect of both sides of (3.86) on the circular disk of Figure 3.4a. Clearly, the orientation of the body $y$-axis is unchanged in both cases, namely, in the $y'$-direction. Furthermore, the orientation of the final body $z$-axis is the same whether we apply $R_{y'}(\beta)$ or $R_z(\alpha)R_y(\beta)R_z^{-1}(\alpha)$. In both cases the final body $z$-axis makes a polar angle $\beta$ with the fixed $z$-axis (the same as the initial $z$-axis), and its azimuthal angle, as measured in the fixed-coordinate system, is just $\alpha$. In other words, the final body $z$-axis is the same as the $z'$-axis of Figure 3.4b. Similarly, we can prove

$$R_{z'}(\gamma) = R_{y'}(\beta)R_z(\gamma)R_{y'}^{-1}(\beta).$$

(3.87)

Using (3.86) and (3.87), we can now rewrite (3.85). We obtain

$$R_{z'}(\gamma)R_{y'}(\beta)R_z(\alpha) = R_{y'}(\beta)R_z(\gamma)R_{y'}^{-1}(\beta)R_{y'}(\beta)R_z(\alpha)$$
$$= R_z(\alpha)R_y(\beta)R_z^{-1}(\alpha)R_z(\gamma)R_z(\alpha)$$
$$= R_z(\alpha)R_y(\beta)R_z(\gamma), \tag{3.88}$$

where in the final step we used the fact that $R_z(\gamma)$ and $R_z(\alpha)$ commute. To summarize,

$$R(\alpha,\beta,\gamma) = R_z(\alpha)R_y(\beta)R_z(\gamma), \tag{3.89}$$

where all three matrices on the right-hand side refer to *fixed*-axis rotations.

Now let us apply this set of operations to spin $\frac{1}{2}$ systems in quantum mechanics. Corresponding to the product of orthogonal matrices in (3.89) there exists a product of rotation operators in the ket space of the spin $\frac{1}{2}$ system under consideration:

$$\mathscr{D}(\alpha,\beta,\gamma) = \mathscr{D}_z(\alpha)\mathscr{D}_y(\beta)\mathscr{D}_z(\gamma). \tag{3.90}$$

The $2 \times 2$ matrix representation of this product is

$$\exp\left(\frac{-i\sigma_3\alpha}{2}\right)\exp\left(\frac{-i\sigma_2\beta}{2}\right)\exp\left(\frac{-i\sigma_3\gamma}{2}\right)$$

$$= \begin{pmatrix} e^{-i\alpha/2} & 0 \\ 0 & e^{i\alpha/2} \end{pmatrix}\begin{pmatrix} \cos(\beta/2) & -\sin(\beta/2) \\ \sin(\beta/2) & \cos(\beta/2) \end{pmatrix}\begin{pmatrix} e^{-i\gamma/2} & 0 \\ 0 & e^{i\gamma/2} \end{pmatrix}$$

$$= \begin{pmatrix} e^{-i(\alpha+\gamma)/2}\cos(\beta/2) & -e^{-i(\alpha-\gamma)/2}\sin(\beta/2) \\ e^{i(\alpha-\gamma)/2}\sin(\beta/2) & e^{i(\alpha+\gamma)/2}\cos(\beta/2) \end{pmatrix}, \tag{3.91}$$

where (3.62) was used. This matrix is clearly of the unitary unimodular form. Conversely, the most general $2\times 2$ unitary unimodular matrix can be written in this Euler angle form.

Notice that the matrix elements of the second (middle) rotation $\exp(-i\sigma_y\phi/2)$ are purely real. This would not have been the case had we chosen to rotate about the *x*-axis rather than the *y*-axis, as done in most textbooks in classical mechanics. In quantum mechanics it pays to stick to our convention because we prefer the matrix elements of the second rotation, which is the only rotation matrix containing off-diagonal elements, to be purely real.[4]

The $2 \times 2$ matrix in (3.91) is called the $j = \frac{1}{2}$ irreducible representation of the rotation operator $\mathscr{D}(\alpha,\beta,\gamma)$. Its matrix elements are denoted by $\mathscr{D}_{m'm}^{(1/2)}(\alpha,\beta,\gamma)$. In terms of the angular-momentum operators we have

$$\mathscr{D}_{m'm}^{(1/2)}(\alpha,\beta,\gamma) = \left\langle j = \frac{1}{2}, m' \left| \exp\left(\frac{-iJ_z\alpha}{\hbar}\right) \times \exp\left(\frac{-iJ_y\beta}{\hbar}\right)\exp\left(\frac{-iJ_z\gamma}{\hbar}\right) \right| j = \frac{1}{2}, m \right\rangle. \tag{3.92}$$

In Section 3.5 we will extensively study higher *j*-analogues of (3.91).

---

[4] This, of course, depends on our convention that the matrix elements of $S_y$ (or, more generally, $J_y$) are taken to be purely imaginary.

# 3.4 Density Operators and Pure Versus Mixed Ensembles

## 3.4.1 Polarized Versus Unpolarized Beams

The formalism of quantum mechanics developed so far makes statistical predictions on an *ensemble*, that is, a collection, of identically prepared physical systems. More precisely, in such an ensemble all members are supposed to be characterized by the same state ket $|\alpha\rangle$. A good example of this is a beam of silver atoms coming out of an SG filtering apparatus. Every atom in the beam has its spin pointing in the same direction, namely, the direction determined by the inhomogeneity of the magnetic field of the filtering apparatus. We have not yet discussed how to describe quantum mechanically an ensemble of physical systems for which some, say 60%, are characterized by $|\alpha\rangle$, and the remaining 40% are characterized by some other ket $|\beta\rangle$.

To illustrate vividly the incompleteness of the formalism developed so far, let us consider silver atoms coming directly out of a hot oven, yet to be subjected to a filtering apparatus of the Stern–Gerlach type. On symmetry grounds we expect that such atoms have *random* spin orientations; in other words, there should be no preferred direction associated with such an ensemble of atoms. According to the formalism developed so far, the most general state ket of a spin $\frac{1}{2}$ system is given by

$$|\alpha\rangle = c_+|+\rangle + c_-|-\rangle. \tag{3.93}$$

Is this equation capable of describing a collection of atoms with random spin orientations? The answer is clearly no; (3.93) characterizes a state ket whose spin is pointing in *some definite direction*, namely, in the direction of $\hat{\mathbf{n}}$, whose polar and azimuthal angles, $\beta$ and $\alpha$, respectively, are obtained by solving

$$\frac{c_+}{c_-} = \frac{\cos(\beta/2)}{e^{i\alpha}\sin(\beta/2)}; \tag{3.94}$$

see (3.70).

To cope with a situation of this kind we introduce the concept of **fractional population**, or probability weight. An ensemble of silver atoms with completely random spin orientation can be viewed as a collection of silver atoms in which 50% of the members of the ensemble are characterized by $|+\rangle$ and the remaining 50% by $|-\rangle$. We specify such an ensemble by assigning

$$w_+ = 0.5, \quad w_- = 0.5, \tag{3.95}$$

where $w_+$ and $w_-$ are the fractional population for spin-up and spin-down, respectively. Because there is no preferred direction for such a beam, it is reasonable to expect that this *same* ensemble can be regarded equally well as a 50:50 mixture of $|S_x; +\rangle$ and $|S_x; -\rangle$. The mathematical formalism needed to accomplish this will appear shortly.

It is very important to note that we are simply introducing two *real* numbers $w_+$ and $w_-$. There is no information on the relative phase between the spin-up and the spin-down ket. Quite often we refer to such a situation as an **incoherent mixture** of spin-up and spin-down states. What we are doing here is to be clearly distinguished from what we did with a coherent linear superposition, for example,

$$\left(\frac{1}{\sqrt{2}}\right)|+\rangle + \left(\frac{1}{\sqrt{2}}\right)|-\rangle, \tag{3.96}$$

where the phase relation between $|+\rangle$ and $|-\rangle$ contains vital information on the spin orientation in the $xy$-plane, in this case in the positive $x$-direction. In general, we should not confuse $w_+$ and $w_-$ with $|c_+|^2$ and $|c_-|^2$. The probability concept associated with $w_+$ and $w_-$ is much closer to that encountered in classical probability theory. The situation encountered in dealing with silver atoms directly from the hot oven may be compared with that of a graduating class in which 50% of the graduating seniors are male, the remaining 50% female. When we pick a student at random, the probability that the particular student is male (or female) is 0.5. Whoever heard of a student referred to as a coherent linear superposition of male and female with a particular phase relation?

The beam of silver atoms coming directly out of the oven is an example of a **completely random ensemble**; the beam is said to be **unpolarized** because there is no preferred direction for spin orientation. In contrast, the beam that has gone through a selective Stern–Gerlach type measurement is an example of a **pure ensemble**; the beam is said to be **polarized** because all members of the ensemble are characterized by a single common ket that describes a state with spin pointing in some definite direction. To appreciate the difference between a completely random ensemble and a pure ensemble, let us consider a rotatable SG apparatus where we can vary the direction of the inhomogeneous **B** just by rotating the apparatus. When a completely unpolarized beam directly out of the oven is subjected to such an apparatus, we *always* obtain two emerging beams of *equal* intensity *no matter what the orientation of the apparatus may be*. In contrast, if a polarized beam is subjected to such an apparatus, the relative intensities of the two emerging beams vary as the apparatus is rotated. For some *particular* orientation the ratio of the intensities actually becomes one to zero. In fact, the formalism we developed in Chapter 1 tells us that the relative intensities are simply $\cos^2(\beta/2)$ and $\sin^2(\beta/2)$, where $\beta$ is the angle between the spin direction of the atoms and the direction of the inhomogeneous magnetic field in the SG apparatus.

A complete random ensemble and a pure ensemble can be regarded as the extremes of what is known as a **mixed ensemble**. In a mixed ensemble a certain fraction, for example, 70%, of the members are characterized by a state ket $|\alpha\rangle$, the remaining 30% by $|\beta\rangle$. In such a case the beam is said to be **partially polarized**. Here $|\alpha\rangle$ and $|\beta\rangle$ need not even be orthogonal; we can, for example, have 70% with spin in the positive $x$-direction and 30% with spin in the negative $z$-direction.[5]

## 3.4.2 Ensemble Averages and Density Operator

We now present the density operator formalism, pioneered by J. von Neumann in 1927, that quantitatively describes physical situations with mixed as well as pure ensembles. Our

---

[5] In the literature what we call pure and mixed ensembles are often referred to as pure and mixed states. In this book, however, we use *state* to mean a physical system described by a definite state ket $|\alpha\rangle$.

general discussion here is not restricted to spin $\frac{1}{2}$ systems, but for illustrative purposes we return repeatedly to spin $\frac{1}{2}$ systems.

A pure ensemble by definition is a collection of physical systems such that every member is characterized by the same ket $|\alpha\rangle$. In contrast, in a mixed ensemble, a fraction of the members with relative population $w_1$ is characterized by $|\alpha^{(1)}\rangle$, some other fraction with relative population $w_2$, by $|\alpha^{(2)}\rangle$, and so on. Roughly speaking, a mixed ensemble can be viewed as a mixture of pure ensembles, just as the name suggests. The fractional populations are constrained to satisfy the normalization condition

$$\sum_i w_i = 1. \tag{3.97}$$

As we mentioned previously, $|\alpha^{(1)}\rangle$ and $|\alpha^{(2)}\rangle$ need not be orthogonal. Furthermore, the number of terms in the $i$ sum of (3.97) need not coincide with the dimensionality $N$ of the ket space; it can easily exceed $N$. For example, for spin $\frac{1}{2}$ systems with $N = 2$, we may consider 40% with spin in the positive $z$-direction, 30% with spin in the positive $x$-direction, and the remaining 30% with spin in the negative $y$-direction.

Suppose we make a measurement on a mixed ensemble of some observable $A$. We may ask what is the average measured value of $A$ when a large number of measurements are carried out. The answer is given by the **ensemble average** of $A$, which is defined by

$$[A] \equiv \sum_i w_i \langle \alpha^{(i)} | A | \alpha^{(i)} \rangle$$
$$= \sum_i \sum_{a'} w_i |\langle a' | \alpha^{(i)} \rangle|^2 a', \tag{3.98}$$

where $|a'\rangle$ is an eigenket of $A$. Recall that $\langle \alpha^{(i)} | A | \alpha^{(i)} \rangle$ is the usual quantum-mechanical expectation value of $A$ taken with respect to state $|\alpha^{(i)}\rangle$. Equation (3.98) tells us that these expectation values must further be weighted by the corresponding fractional populations $w_i$. Notice how probabilistic concepts enter twice; first in $|\langle a' | \alpha^{(i)} \rangle|^2$ for the quantum-mechanical probability for state $|\alpha^{(i)}\rangle$ to be found in an $A$ eigenstate $|a'\rangle$; second, in the probability factor $w_i$ for finding in the ensemble a quantum-mechanical state characterized by $|\alpha^{(i)}\rangle$.[6]

We can now rewrite ensemble average (3.98) using a more general basis, $\{|b'\rangle\}$:

$$[A] = \sum_i w_i \sum_{b'} \sum_{b''} \langle \alpha^{(i)} | b' \rangle \langle b' | A | b'' \rangle \langle b'' | \alpha^{(i)} \rangle$$
$$= \sum_{b'} \sum_{b''} \left( \sum_i w_i \langle b'' | \alpha^{(i)} \rangle \langle \alpha^{(i)} | b' \rangle \right) \langle b' | A | b'' \rangle. \tag{3.99}$$

The number of terms in the sum of the $b'$ ($b''$) is just the dimensionality of the ket space, while the number of terms in the sum of the $i$ depends on how the mixed ensemble is viewed as a mixture of pure ensembles. Notice that in this form the basic property of the ensemble

---

[6] Quite often in the literature the ensemble average is also called the *expectation value*. However, in this book, the term expectation value is reserved for the average measured value when measurements are carried out on a pure ensemble.

which does not depend on the particular observable $A$ is factored out. This motivates us to define the **density operator** $\rho$ as follows:

$$\rho \equiv \sum_i w_i |\alpha^{(i)}\rangle\langle\alpha^{(i)}|. \tag{3.100}$$

The elements of the corresponding **density matrix** have the following form:

$$\langle b''|\rho|b'\rangle = \sum_i w_i \langle b''|\alpha^{(i)}\rangle\langle\alpha^{(i)}|b'\rangle. \tag{3.101}$$

The density operator contains all the physically significant information we can possibly obtain about the ensemble in question. Returning to (3.99), we see that the ensemble average can be written as

$$[A] = \sum_{b'}\sum_{b''}\langle b''|\rho|b'\rangle\langle b'|A|b''\rangle$$
$$= \mathrm{tr}(\rho A). \tag{3.102}$$

Because the trace is independent of representations, $\mathrm{tr}(\rho A)$ can be evaluated using any convenient basis. As a result, (3.102) is an extremely powerful relation.

There are two properties of the density operator worth recording. First, the density operator is Hermitian, as is evident from (3.100). Second, the density operator satisfies the normalization condition

$$\mathrm{tr}(\rho) = \sum_i\sum_{b'} w_i \langle b'|\alpha^{(i)}\rangle\langle\alpha^{(i)}|b'\rangle$$
$$= \sum_i w_i \langle\alpha^{(i)}|\alpha^{(i)}\rangle$$
$$= 1. \tag{3.103}$$

Because of the Hermiticity and the normalization condition, for spin $\frac{1}{2}$ systems with dimensionality 2, the density operator, or the corresponding density matrix, is characterized by three independent real parameters. Four real numbers characterize a $2 \times 2$ Hermitian matrix. However, only three are independent because of the normalization condition. The three numbers needed are $[S_x]$, $[S_y]$, and $[S_z]$; the reader may verify that knowledge of these three ensemble averages is sufficient to reconstruct the density operator. The manner in which a mixed ensemble is formed can be rather involved. We may mix pure ensembles characterized by all kinds of $|\alpha^{(i)}\rangle$ with appropriate $w_i$; yet for spin $\frac{1}{2}$ systems three real numbers completely characterize the ensemble in question. This strongly suggests that a mixed ensemble can be decomposed into pure ensembles in many different ways. A problem to illustrate this point appears at the end of this chapter.

A pure ensemble is specified by $w_i = 1$ for some $|\alpha^{(i)}\rangle$, with $i = n$, for instance, and $w_i = 0$ for all other conceivable state kets, so the corresponding density operator is written as

$$\rho = |\alpha^{(n)}\rangle\langle\alpha^{(n)}| \tag{3.104}$$

with no summation. Clearly, the density operator for a pure ensemble is idempotent, that is,

$$\rho^2 = \rho \tag{3.105}$$

or, equivalently,

$$\rho(\rho - 1) = 0. \tag{3.106}$$

Thus, for a pure ensemble only, we have

$$\mathrm{tr}(\rho^2) = 1 \tag{3.107}$$

in addition to (3.103). The eigenvalues of the density operator for a pure ensemble are zero or one, as can be seen by inserting a complete set of base kets that diagonalize the Hermitian operator $\rho$ between $\rho$ and $(\rho - 1)$ of (3.106). When diagonalized, the density matrix for a pure ensemble must therefore look like

$$\rho \doteq \begin{pmatrix} 0 & & & & & & & & & 0 \\ & 0 & & & & & & & & \\ & & \ddots & & & & & & & \\ & & & 0 & & & & & & \\ & & & & 1 & & & & & \\ & & & & & 0 & & & & \\ & & & & & & 0 & & & \\ & & & & & & & 0 & & \\ & & & & & & & & \ddots & \\ 0 & & & & & & & & & 0 \end{pmatrix} \quad \text{(diagonal form).} \tag{3.108}$$

It can be shown that $\mathrm{tr}(\rho^2)$ is maximal when the ensemble is pure; for a mixed ensemble $\mathrm{tr}(\rho^2)$ is a positive number less than one.

Given a density operator, let us see how we can construct the corresponding density matrix in some specified basis. To this end we first recall that

$$|\alpha\rangle\langle\alpha| = \sum_{b'}\sum_{b''} |b'\rangle\langle b'|\alpha\rangle\langle\alpha|b''\rangle\langle b''|. \tag{3.109}$$

This shows that we can form the square matrix corresponding to $|\alpha^{(i)}\rangle\langle\alpha^{(i)}|$ by combining, in the sense of outer product, the column matrix formed by $\langle b'|\alpha^{(i)}\rangle$ with the row matrix formed by $\langle\alpha^{(i)}|b''\rangle$, which, of course, is equal to $\langle b''|\alpha^{(i)}\rangle^*$. The final step is to sum such square matrices with weighting factors $w_i$, as indicated in (3.100). The final form agrees with (3.101), as expected.

It is instructive to study several examples, all referring to spin $\frac{1}{2}$ systems.

**Example 1**  A completely polarized beam with $S_z+$:

$$\rho = |+\rangle\langle+| \doteq \begin{pmatrix} 1 \\ 0 \end{pmatrix} (1, 0)$$

$$= \begin{pmatrix} 1 & 0 \\ 0 & 0 \end{pmatrix}. \tag{3.110}$$

**Example 2**   A completely polarized beam with $S_x\pm$:

$$\rho = |S_x; \pm\rangle\langle S_x; \pm| = \left(\frac{1}{\sqrt{2}}\right)(|+\rangle \pm |-\rangle)\left(\frac{1}{\sqrt{2}}\right)(\langle +| \pm \langle -|)$$

$$\doteq \begin{pmatrix} \frac{1}{2} & \pm\frac{1}{2} \\ \pm\frac{1}{2} & \frac{1}{2} \end{pmatrix}. \tag{3.111}$$

The ensembles of Examples 1 and 2 are both pure.

**Example 3**   An unpolarized beam. This can be regarded as an incoherent mixture of a spin-up ensemble and a spin-down ensemble with equal weights (50% each):

$$\rho = (\tfrac{1}{2})|+\rangle\langle +| + (\tfrac{1}{2})|-\rangle\langle -|$$

$$\doteq \begin{pmatrix} \frac{1}{2} & 0 \\ 0 & \frac{1}{2} \end{pmatrix}, \tag{3.112}$$

which is just the identity matrix divided by 2. As we remarked earlier, the same ensemble can also be regarded as an incoherent mixture of an $S_x+$ ensemble and an $S_x-$ ensemble with equal weights. It is gratifying that our formalism automatically satisfies the expectation

$$\begin{pmatrix} \frac{1}{2} & 0 \\ 0 & \frac{1}{2} \end{pmatrix} = \frac{1}{2}\begin{pmatrix} \frac{1}{2} & \frac{1}{2} \\ \frac{1}{2} & \frac{1}{2} \end{pmatrix} + \frac{1}{2}\begin{pmatrix} \frac{1}{2} & -\frac{1}{2} \\ -\frac{1}{2} & \frac{1}{2} \end{pmatrix}, \tag{3.113}$$

where we see from Example 2 that the two terms on the right-hand side are the density matrices for pure ensemble with $S_x+$ and $S_x-$. Because $\rho$ in this case is just the identity operator divided by 2 (the dimensionality), we have

$$\mathrm{tr}(\rho S_x) = \mathrm{tr}(\rho S_y) = \mathrm{tr}(\rho S_z) = 0, \tag{3.114}$$

where we used the fact that $S_k$ is traceless. Thus for the ensemble average of **S** we have

$$[\mathbf{S}] = 0. \tag{3.115}$$

This is reasonable because there should be no preferred spin direction in a completely random ensemble of spin $\frac{1}{2}$ systems.

**Example 4**   As an example of a partially polarized beam, let us consider a 75:25 mixture of two pure ensembles, one with $S_z+$ and the other with $S_x+$:

$$w(S_z+) = 0.75, \quad w(S_x+) = 0.25. \tag{3.116}$$

The corresponding $\rho$ can be represented by

$$\rho \doteq \frac{3}{4}\begin{pmatrix} 1 & 0 \\ 0 & 0 \end{pmatrix} + \frac{1}{4}\begin{pmatrix} \frac{1}{2} & \frac{1}{2} \\ \frac{1}{2} & \frac{1}{2} \end{pmatrix}$$

$$= \begin{pmatrix} \frac{7}{8} & \frac{1}{8} \\ \frac{1}{8} & \frac{1}{8} \end{pmatrix}, \tag{3.117}$$

from which follows

$$[S_x] = \frac{\hbar}{8}, \quad [S_y] = 0, \quad [S_z] = \frac{3\hbar}{8}. \tag{3.118}$$

We leave as an exercise for the reader the task of showing that this ensemble can be decomposed in ways other than (3.116).

### 3.4.3  Time Evolution of Ensembles

How does the density operator $\rho$ change as a function of time? Let us suppose that at some time $t_0$ the density operator is given by

$$\rho(t_0) = \sum_i w_i |\alpha^{(i)}\rangle\langle\alpha^{(i)}|. \tag{3.119}$$

If the ensemble is to be left undisturbed, we cannot change the fractional population $w_i$. So the change in $\rho$ is governed solely by the time evolution of state ket $|\alpha^{(i)}\rangle$:

$$|\alpha^{(i)}\rangle \quad \text{at} \quad t_0 \to |\alpha^{(i)}, t_0; t\rangle. \tag{3.120}$$

From the fact that $|\alpha^{(i)}, t_0; t\rangle$ satisfies the Schrödinger equation we obtain

$$i\hbar\frac{\partial\rho}{\partial t} = \sum_i w_i(H|\alpha^{(i)}, t_0; t\rangle\langle\alpha^{(i)}, t_0; t| - |\alpha^{(i)}, t_0; t\rangle\langle\alpha^{(i)}, t_0; t|H)$$

$$= -[\rho, H]. \tag{3.121}$$

This looks like the Heisenberg equation of motion except that the sign is wrong! This is not disturbing because $\rho$ is not a dynamic observable in the Heisenberg picture. On the contrary, $\rho$ is built up of Schrödinger picture state kets and state bras which evolve in time according to the Schrödinger equation.

It is amusing that (3.121) can be regarded as the quantum-mechanical analogue of Liouville's theorem in classical statistical mechanics,

$$\frac{\partial\rho_{\text{classical}}}{\partial t} = -[\rho_{\text{classical}}, H]_{\text{classical}}, \tag{3.122}$$

where $\rho_{\text{classical}}$ stands for the density of representative points in phase space.[7] Thus the name *density operator* for the $\rho$ appearing in (3.121) is indeed appropriate. The classical analogue of (3.102) for the ensemble average of some observable $A$ is given by

$$A_{\text{average}} = \frac{\int \rho_{\text{classical}} A(q, p) d\Gamma_{q,p}}{\int \rho_{\text{classical}} d\Gamma_{q,p}}, \tag{3.123}$$

where $d\Gamma_{q,p}$ stands for a volume element in phase space.

---

[7]  Remember, a pure classical state is one represented by a single moving point in phase space $(q_1, \ldots, q_f, p_1, \ldots, p_f)$ at each instant of time. A classical statistical state, on the other hand, is described by our nonnegative density function $\rho_{\text{classical}}(q_1, \ldots, q_f, p_1, \ldots, p_f, t)$ such that the probability that a system is found in the interval $dq_1, \ldots, dp_f$ at time $t$ is $\rho_{\text{classical}} dq_1, \ldots, dp_f$.

### 3.4.4  Continuum Generalizations

So far we have considered density operators in ket space where the base kets are labeled by the discrete eigenvalues of some observable. The concept of density matrix can be generalized to cases where the base kets used are labeled by continuous eigenvalues. In particular, let us consider the ket space spanned by the position eigenkets $|\mathbf{x}'\rangle$. The analogue of (3.102) is given by

$$[A] = \int d^3x' \int d^3x'' \langle \mathbf{x}''|\rho|\mathbf{x}'\rangle\langle \mathbf{x}'|A|\mathbf{x}''\rangle. \tag{3.124}$$

The density matrix here is actually a function of $\mathbf{x}'$ and $\mathbf{x}''$, namely,

$$\langle \mathbf{x}''|\rho|\mathbf{x}'\rangle = \langle \mathbf{x}''| \left( \sum_i w_i |\alpha^{(i)}\rangle\langle\alpha^{(i)}| \right) |\mathbf{x}'\rangle$$
$$= \sum_i w_i \psi_i(\mathbf{x}'')\psi_i^*(\mathbf{x}'), \tag{3.125}$$

where $\psi_i$ is the wave function corresponding to the state ket $|\alpha^{(i)}\rangle$. Notice that the diagonal element (that is, $\mathbf{x}' = \mathbf{x}''$) of this is just the weighted sum of the probability densities. Once again, the term *density matrix* is indeed appropriate.

In continuum cases, too, it is important to keep in mind that the same mixed ensemble can be decomposed in different ways into pure ensembles. For instance, it is possible to regard a "realistic" beam of particles either as a mixture of plane wave states (monoenergetic free-particle states) or as a mixture of wave packet states.

### 3.4.5  Quantum Statistical Mechanics

We conclude this section with a brief discussion on the connection between the density operator formalism and statistical mechanics. Let us first record some properties of completely random and of pure ensembles. The density matrix of a completely random ensemble looks like

$$\rho \doteq \frac{1}{N} \begin{pmatrix} 1 & & & & & & \\ & 1 & & & & & \\ & & 1 & & & & \\ & & & \ddots & & & \\ & & & & 1 & & \\ & & & & & 1 & \\ & & & & & & 1 \end{pmatrix} \tag{3.126}$$

in any representation [compare Example 3 with (3.112)]. This follows from the fact that all states corresponding to the base kets with respect to which the density matrix is written are equally populated. In contrast, in the basis where $\rho$ is diagonalized, we have (3.108) for the matrix representation of the density operator for a pure ensemble. The two diagonal matrices (3.126) and (3.108), both satisfying the normalization requirement (3.103), cannot look more different. It would be desirable if we could somehow construct a quantity that characterizes this dramatic difference.

Thus we define a quantity called $\sigma$ by

$$\sigma = -\mathrm{tr}(\rho \ln \rho). \tag{3.127}$$

The logarithm of the operator $\rho$ may appear rather formidable, but the meaning of (3.127) is quite unambiguous if we use the basis in which $\rho$ is diagonal:

$$\sigma = -\sum_k \rho_{kk}^{(\mathrm{diag})} \ln \rho_{kk}^{(\mathrm{diag})}. \tag{3.128}$$

Because each element $\rho_{kk}^{(\mathrm{diag})}$ is a real number between 0 and 1, $\sigma$ is necessarily positive semidefinite. For a completely random ensemble (3.126), we have

$$\sigma = -\sum_{k=1}^{N} \frac{1}{N} \ln\left(\frac{1}{N}\right) = \ln N. \tag{3.129}$$

In contrast, for a pure ensemble (3.108) we have

$$\sigma = 0 \tag{3.130}$$

where we have used

$$\rho_{kk}^{(\mathrm{diag})} = 0 \quad \text{or} \quad \ln \rho_{kk}^{(\mathrm{diag})} = 0 \tag{3.131}$$

for each term in (3.128).

We now argue that physically $\sigma$ can be regarded as a quantitative measure of disorder. A pure ensemble is an ensemble with a maximum amount of order because all members are characterized by the same quantum-mechanical state ket; it may be likened to marching soldiers in a well-regimented army. According to (3.130), $\sigma$ vanishes for such an ensemble. On the other extreme, a completely random ensemble, in which all quantum-mechanical states are equally likely, may be likened to drunken soldiers wandering around in random directions. According to (3.129), $\sigma$ is large; indeed, we will show later that $\ln N$ is the maximum possible value for $\sigma$ subject to the normalization condition

$$\sum_k \rho_{kk} = 1. \tag{3.132}$$

In thermodynamics we learn that a quantity called **entropy** measures disorder. It turns out that our $\sigma$ is related to the entropy per constituent member, denoted by $S$, of the ensemble via

$$S = k\sigma, \tag{3.133}$$

where $k$ is a universal constant identifiable with the Boltzmann constant. In fact, (3.133) may be taken as the *definition* of entropy in quantum statistical mechanics.

We now show how the density operator $\rho$ can be obtained for an ensemble in thermal equilibrium. The basic assumption we make is that nature tends to maximize $\sigma$ subject to the constraint that the ensemble average of the Hamiltonian has a certain prescribed value. To justify this assumption would involve us in a delicate discussion of how equilibrium is established as a result of interactions with the environment, which is beyond the scope of this book. In any case, once thermal equilibrium is established, we expect

$$\frac{\partial \rho}{\partial t} = 0. \tag{3.134}$$

Because of (3.121), this means that $\rho$ and $H$ can be simultaneously diagonalized. So the kets used in writing (3.128) may be taken to be energy eigenkets. With this choice $\rho_{kk}$ stands for the fractional population for an energy eigenstate with energy eigenvalue $E_k$.

Let us maximize $\sigma$ by requiring that

$$\delta \sigma = 0. \tag{3.135}$$

However, we must take into account the constraint that the ensemble average of $H$ has a certain prescribed value. In the language of statistical mechanics, $[H]$ is identified with the internal energy per constituent denoted by $U$:

$$[H] = \text{tr}(\rho H) = U. \tag{3.136}$$

In addition, we should not forget the normalization constraint (3.132). So our basic task is to require (3.135) subject to the constraints

$$\delta[H] = \sum_k \delta \rho_{kk} E_k = 0 \tag{3.137a}$$

and

$$\delta(\text{tr}\rho) = \sum_k \delta \rho_{kk} = 0. \tag{3.137b}$$

We can most readily accomplish this by using Lagrange multipliers. We obtain

$$\sum_k \delta \rho_{kk}[(\ln \rho_{kk} + 1) + \beta E_k + \gamma] = 0, \tag{3.138}$$

which for an arbitrary variation is possibly only if

$$\rho_{kk} = \exp(-\beta E_k - \gamma - 1). \tag{3.139}$$

The constant $\gamma$ can be eliminated using the normalization condition (3.132), and our final result is

$$\rho_{kk} = \frac{\exp(-\beta E_k)}{\sum\limits_l^N \exp(-\beta E_l)}, \tag{3.140}$$

which directly gives the fractional population for an energy eigenstate with eigenvalue $E_k$. It is to be understood throughout that the sum is over distinct energy eigenstates; if there is degeneracy we must sum over states with the same energy eigenvalue.

The density matrix element (3.140) is appropriate for what is known in statistical mechanics as a **canonical ensemble**. Had we attempted to maximize $\sigma$ without the internal-energy constraint (3.137a), we would have obtained instead

$$\rho_{kk} = \frac{1}{N} \quad (\text{independent of } k), \tag{3.141}$$

which is the density matrix element appropriate for a completely random ensemble. Comparing (3.140) with (3.141), we infer that a completely random ensemble can be regarded as the $\beta \to 0$ limit (physically the high-temperature limit) of a canonical ensemble.

We recognize the denominator of (3.140) as the partition function

$$Z = \sum_{k}^{N} \exp(-\beta E_k) \tag{3.142}$$

in statistical mechanics. It can also be written as

$$Z = \mathrm{tr}(e^{-\beta H}). \tag{3.143}$$

Knowing $\rho_{kk}$ given in the energy basis, we can write the density operator as

$$\rho = \frac{e^{-\beta H}}{Z}. \tag{3.144}$$

This is the most basic equation from which everything follows. We can immediately evaluate the ensemble average of any observable $A$:

$$\begin{aligned}
[A] &= \frac{\mathrm{tr}(e^{-\beta H} A)}{Z} \\
&= \frac{\left[\sum_{k}^{N} \langle A \rangle_k \exp(-\beta E_k)\right]}{\sum_{k}^{N} \exp(-\beta E_k)}.
\end{aligned} \tag{3.145}$$

In particular, for the internal energy per constituent we obtain

$$\begin{aligned}
U &= \frac{\left[\sum_{k}^{N} E_k \exp(-\beta E_k)\right]}{\sum_{k}^{N} \exp(-\beta E_k)} \\
&= -\frac{\partial}{\partial \beta}(\ln Z),
\end{aligned} \tag{3.146}$$

a formula well known to every student of statistical mechanics.

The parameter $\beta$ is related to the temperature $T$ as follows:

$$\beta = \frac{1}{kT} \tag{3.147}$$

where $k$ is the Boltzmann constant. It is instructive to convince ourselves of this identification by comparing the ensemble average $[H]$ of simple harmonic oscillators with the $kT$ expected for the internal energy in the classical limit, which is left as an exercise. We have already commented that in the high-temperature limit, a canonical ensemble becomes a completely random ensemble in which all energy eigenstates are equally populated. In the opposite low-temperature limit ($\beta \to \infty$), (3.140) tells us that a canonical ensemble becomes a pure ensemble where only the ground state is populated.

As a simple illustrative example, consider a canonical ensemble made up of spin $\frac{1}{2}$ systems, each with a magnetic moment $e\hbar/2m_e c$ subjected to a uniform magnetic field in the $z$-direction. The Hamiltonian relevant to this problem has already been given [see (3.34)]. Because $H$ and $S_z$ commute, the density matrix for this canonical ensemble is diagonal in the $S_z$ basis. Thus

$$\rho \doteq \frac{\begin{pmatrix} e^{-\beta\hbar\omega/2} & 0 \\ 0 & e^{\beta\hbar\omega/2} \end{pmatrix}}{Z}, \tag{3.148}$$

where the partition function is just

$$Z = e^{-\beta\hbar\omega/2} + e^{\beta\hbar\omega/2}. \tag{3.149}$$

From this we compute

$$[S_x] = [S_y] = 0, \quad [S_z] = -\left(\frac{\hbar}{2}\right)\tanh\left(\frac{\beta\hbar\omega}{2}\right). \tag{3.150}$$

The ensemble average of the magnetic moment component is just $e/m_e c$ times $[S_z]$. The paramagnetic susceptibility $\chi$ may be computed from

$$\left(\frac{e}{m_e c}\right)[S_z] = \chi B. \tag{3.151}$$

In this way we arrive at Brillouin's formula for $\chi$:

$$\chi = \left(\frac{|e|\hbar}{2m_e cB}\right)\tanh\left(\frac{\beta\hbar\omega}{2}\right). \tag{3.152}$$

## 3.5  Eigenvalues and Eigenstates of Angular Momentum

Up to now our discussion of angular momentum has been confined exclusively to spin $\frac{1}{2}$ systems with dimensionality $N = 2$. In this and subsequent sections we study more general angular-momentum states. To this end we first work out the eigenvalues and eigenkets of $\mathbf{J}^2$ and $J_z$ and derive the expressions for matrix elements of angular-momentum operators, first obtained in a 1926 paper by M. Born, W. Heisenberg, and P. Jordan.

### 3.5.1  Commutation Relations and the Ladder Operators

Everything we will do follows from the angular-momentum commutation relations (3.20), where we may recall that $J_i$ is defined as the generator of infinitesimal rotation. The first important property we derive from the basic commutation relations is the existence of a new operator $\mathbf{J}^2$, defined by

$$\mathbf{J}^2 \equiv J_x J_x + J_y J_y + J_z J_z, \tag{3.153}$$

that commutes with every one of $J_k$:

$$[\mathbf{J}^2, J_k] = 0 \quad (k = 1, 2, 3). \tag{3.154}$$

To prove this let us look at the $k = 3$ case:

$$\begin{aligned}
[J_x J_x + J_y J_y + J_z J_z, J_z] &= J_x[J_x, J_z] + [J_x, J_z]J_x + J_y[J_y, J_z] + [J_y, J_z]J_y \\
&= J_x(-i\hbar J_y) + (-i\hbar J_y)J_x + J_y(i\hbar J_x) + (i\hbar J_x)J_y \\
&= 0.
\end{aligned} \tag{3.155}$$

The proofs for the cases where $k = 1$ and 2 follow by cyclic permutation ($1 \to 2 \to 3 \to 1$) of the indices. Because $J_x$, $J_y$, and $J_z$ do not commute with each other, we can choose only one of them to be the observable to be diagonalized simultaneously with $\mathbf{J}^2$. By convention we choose $J_z$ for this purpose.

We now look for the simultaneous eigenkets of $\mathbf{J}^2$ and $J_z$. We denote the eigenvalues of $\mathbf{J}^2$ and $J_z$ by $a$ and $b$, respectively:

$$\mathbf{J}^2|a, b\rangle = a|a, b\rangle \tag{3.156a}$$

$$J_z|a, b\rangle = b|a, b\rangle. \tag{3.156b}$$

To determine the allowed values for $a$ and $b$, it is convenient to work with the non-Hermitian operators

$$J_\pm \equiv J_x \pm iJ_y, \tag{3.157}$$

called the **ladder operators**, rather than with $J_x$ and $J_y$. They satisfy the commutation relations

$$[J_+, J_-] - 2\hbar J_z \tag{3.158a}$$

and

$$[J_z, J_\pm] = \pm\hbar J_\pm, \tag{3.158b}$$

which can easily be obtained from (3.20). Note also that

$$[\mathbf{J}^2, J_\pm] = 0, \tag{3.159}$$

which is an obvious consequence of (3.154).

What is the physical meaning of $J_\pm$? To answer this we examine how $J_z$ acts on $J_\pm|a, b\rangle$:

$$\begin{aligned}
J_z(J_\pm|a, b\rangle) &= ([J_z, J_\pm] + J_\pm J_z)|a, b\rangle \\
&= (b \pm \hbar)(J_\pm|a, b\rangle)
\end{aligned} \tag{3.160}$$

where we have used (3.158b). In other words, if we apply $J_+ (J_-)$ to a $J_z$ eigenket, the resulting ket is still a $J_z$ eigenket except that its eigenvalue is now increased (decreased) by one unit of $\hbar$. So now we see why $J_\pm$, which step one step up (down) on the "ladder" of $J_z$ eigenvalues, are known as the ladder operators.

We now digress to recall that the commutation relations in (3.158b) are reminiscent of some commutation relations we encountered in the earlier chapters. In discussing the translation operator $\mathscr{T}(\mathbf{l})$ we had

$$[x_i, \mathscr{T}(\mathbf{l})] = l_i \mathscr{T}(\mathbf{l}); \tag{3.161}$$

also, in discussing the simple harmonic oscillator we had

$$[N, a^\dagger] = a^\dagger, \quad [N, a] = -a. \tag{3.162}$$

We see that both (3.161) and (3.162) have a structure similar to (3.158b). The physical interpretation of the translation operator is that it changes the eigenvalue of the position operator $\mathbf{x}$ by $\mathbf{l}$ in much the same way as the ladder operator $J_+$ changes the eigenvalue of $J_z$ by one unit of $\hbar$. Likewise, the oscillator creation operator $a^\dagger$ increases the eigenvalue of the number operator $N$ by unity.

Even though $J_\pm$ changes the eigenvalue of $J_z$ by one unit of $\hbar$, it does not change the eigenvalue of $\mathbf{J}^2$:

$$\mathbf{J}^2(J_\pm|a,b\rangle) = J_\pm\mathbf{J}^2|a,b\rangle$$
$$= a(J_\pm|a,b\rangle), \tag{3.163}$$

where we have used (3.159). To summarize, $J_\pm|a,b\rangle$ are simultaneous eigenkets of $\mathbf{J}^2$ and $J_z$ with eigenvalues $a$ and $b \pm \hbar$. We may write

$$J_\pm|a,b\rangle = c_\pm|a,b\pm\hbar\rangle, \tag{3.164}$$

where the proportionality constant $c_\pm$ will be determined later from the normalization requirement of the angular-momentum eigenkets.

## 3.5.2  Eigenvalues of $J^2$ and $J_z$

We now have the machinery needed to construct angular-momentum eigenkets and to study their eigenvalue spectrum. Suppose we apply $J_+$ successively, say $n$ times, to a simultaneous eigenket of $\mathbf{J}^2$ and $J_z$. We then obtain another eigenket of $\mathbf{J}^2$ and $J_z$ with the $J_z$ eigenvalue increased by $n\hbar$, while its $\mathbf{J}^2$ eigenvalue is unchanged. However, this process cannot go on indefinitely. It turns out that there exists an upper limit to $b$ (the $J_z$ eigenvalue) for a given $a$ (the $\mathbf{J}^2$ eigenvalue):

$$a \geq b^2. \tag{3.165}$$

To prove this assertion we first note that

$$\mathbf{J}^2 - J_z^2 = \tfrac{1}{2}(J_+J_- + J_-J_+)$$
$$= \tfrac{1}{2}(J_+J_+^\dagger + J_+^\dagger J_+). \tag{3.166}$$

Now $J_+J_+^\dagger$ and $J_+^\dagger J_+$ must have nonnegative expectation values because

$$J_+^\dagger|a,b\rangle \overset{\text{DC}}{\leftrightarrow} \langle a,b|J_+, \quad J_+|a,b\rangle \overset{\text{DC}}{\leftrightarrow} \langle a,b|J_+^\dagger; \tag{3.167}$$

thus

$$\langle a,b|(\mathbf{J}^2 - J_z^2)|a,b\rangle \geq 0, \tag{3.168}$$

which, in turn, implies (3.165). It therefore follows that there must be a $b_{max}$ such that

$$J_+|a,b_{max}\rangle = 0. \tag{3.169}$$

Stated another way, the eigenvalue of $b$ cannot be increased beyond $b_{max}$. Now (3.169) also implies

$$J_-J_+|a,b_{max}\rangle = 0. \tag{3.170}$$

But

$$J_-J_+ = J_x^2 + J_y^2 - i(J_yJ_x - J_xJ_y)$$
$$= \mathbf{J}^2 - J_z^2 - \hbar J_z. \tag{3.171}$$

So

$$(\mathbf{J}^2 - J_z^2 - \hbar J_z)|a,b_{max}\rangle = 0. \tag{3.172}$$

Because $|a,b_{max}\rangle$ itself is not a null ket, this relationship is possible only if

$$a \quad b_{max}^2 \quad b_{max}\hbar = 0 \tag{3.173}$$

or

$$a = b_{max}(b_{max} + \hbar). \tag{3.174}$$

In a similar manner we argue from (3.165) that there must also exist a $b_{min}$ such that

$$J_-|a,b_{min}\rangle = 0. \tag{3.175}$$

By writing $J_+J_-$ as

$$J_+J_- = \mathbf{J}^2 - J_z^2 + \hbar J_z \tag{3.176}$$

in analogy with (3.171), we conclude that

$$a = b_{min}(b_{min} - \hbar). \tag{3.177}$$

By comparing (3.174) with (3.177) we infer that

$$b_{max} = -b_{min}, \tag{3.178}$$

with $b_{max}$ positive, and that the allowed values of $b$ lie within

$$-b_{max} \leq b \leq b_{max}. \tag{3.179}$$

Clearly, we must be able to reach $|a,b_{max}\rangle$ by applying $J_+$ successively to $|a,b_{min}\rangle$ a finite number of times. We must therefore have

$$b_{max} = b_{min} + n\hbar, \tag{3.180}$$

where $n$ is some integer. As a result, we get

$$b_{max} = \frac{n\hbar}{2}. \tag{3.181}$$

It is more conventional to work with $j$, defined to be $b_{max}/\hbar$, instead of with $b_{max}$ so that

$$j = \frac{n}{2}. \tag{3.182}$$

The maximum value of the $J_z$ eigenvalue is $j\hbar$, where $j$ is either an integer or a half-integer. Equation (3.174) implies that the eigenvalue of $\mathbf{J}^2$ is given by

$$a = \hbar^2 j(j+1). \tag{3.183}$$

Let us also define $m$ so that

$$b \equiv m\hbar. \tag{3.184}$$

If $j$ is an integer, all $m$ values are integers; if $j$ is a half-integer, all $m$ values are half-integers. The allowed $m$-values for a given $j$ are

$$m = \underbrace{-j, -j+1, \ldots, j-1, j}_{2j+1 \text{ states}}. \tag{3.185}$$

Instead of $|a, b\rangle$ it is more convenient to denote a simultaneous eigenket of $\mathbf{J}^2$ and $J_z$ by $|j, m\rangle$. The basic eigenvalue equations now read

$$\mathbf{J}^2|j, m\rangle = j(j+1)\hbar^2|j, m\rangle \tag{3.186a}$$

and

$$J_z|j, m\rangle = m\hbar|j, m\rangle, \tag{3.186b}$$

with $j$ either an integer or a half-integer and $m$ given by (3.185). It is very important to recall here that we have used only the commutation relations (3.20) to obtain these results. The quantization of angular momentum, manifested in (3.186), is a direct consequence of the angular-momentum commutation relations, which, in turn, follow from the properties of rotations together with the definition of $J_k$ as the generator of rotation.

### 3.5.3 Matrix Elements of Angular-Momentum Operators

Let us work out the matrix elements of the various angular-momentum operators. Assuming $|j, m\rangle$ to be normalized, we obviously have from (3.186)

$$\langle j', m'|\mathbf{J}^2|j, m\rangle = j(j+1)\hbar^2 \delta_{j'j}\delta_{m'm} \tag{3.187a}$$

and

$$\langle j', m'|J_z|j, m\rangle = m\hbar\delta_{j'j}\delta_{m'm}. \tag{3.187b}$$

To obtain the matrix elements of $J_\pm$, we first consider

$$\langle j, m|J_+^\dagger J_+|j, m\rangle = \langle j, m|(\mathbf{J}^2 - J_z^2 - \hbar J_z)|j, m\rangle$$
$$= \hbar^2[j(j+1) - m^2 - m]. \tag{3.188}$$

Now $J_+|j,m\rangle$ must be the same as $|j,m+1\rangle$ (normalized) up to a multiplicative constant [see (3.164)]. Thus

$$J_+|j,m\rangle = c_{jm}^+|j,m+1\rangle. \tag{3.189}$$

Comparison with (3.188) leads to

$$\begin{aligned}|c_{jm}^+|^2 &= \hbar^2[j(j+1)-m(m+1)]\\ &= \hbar^2(j-m)(j+m+1).\end{aligned} \tag{3.190}$$

So we have determined $c_{jm}^\downarrow$ up to an arbitrary phase factor. It is customary to choose $c_{jm}^+$ to be real and positive by convention. So

$$J_+|j,m\rangle = \sqrt{(j-m)(j+m+1)}\,\hbar|j,m+1\rangle. \tag{3.191}$$

Similarly, we can derive

$$J_-|j,m\rangle = \sqrt{(j+m)(j-m+1)}\,\hbar|j,m-1\rangle. \tag{3.192}$$

Finally, we determine the matrix elements of $J_\pm$ to be

$$\langle j',m'|J_\pm|j,m\rangle = \sqrt{(j\mp m)(j\pm m+1)}\,\hbar\delta_{j'j}\delta_{m',m+1}. \tag{3.193}$$

## 3.5.4 Representations of the Rotation Operator

Having obtained the matrix elements of $J_z$ and $J_\pm$, we are now in a position to study the matrix elements of the rotation operator $\mathcal{D}(R)$. If a rotation $R$ is specified by $\hat{\mathbf{n}}$ and $\phi$, we can define its matrix elements by

$$\mathcal{D}_{m'm}^{(j)}(R) = \langle j,m'|\exp\left(\frac{-i\mathbf{J}\cdot\hat{\mathbf{n}}\phi}{\hbar}\right)|j,m\rangle. \tag{3.194}$$

These matrix elements are sometimes called **Wigner functions** after E. P. Wigner, who made pioneering contributions to the group-theoretical properties of rotations in quantum mechanics. Notice here that the same $j$-value appears in the ket and bra of (3.194); we need not consider matrix elements of $\mathcal{D}(R)$ between states with different $j$-values because they all vanish trivially. This is because $\mathcal{D}(R)|j,m\rangle$ is still an eigenket of $\mathbf{J}^2$ with the same eigenvalue $j(j+1)\hbar^2$:

$$\begin{aligned}\mathbf{J}^2\mathcal{D}(R)|j,m\rangle &= \mathcal{D}(R)\mathbf{J}^2|j,m\rangle\\ &= j(j+1)\hbar^2[\mathcal{D}(R)|j,m\rangle],\end{aligned} \tag{3.195}$$

which follows directly from the fact that $\mathbf{J}^2$ commutes with $J_k$ (hence with any function of $J_k$). Simply stated, rotations cannot change the $j$-value, which is an eminently sensible result.

Often in the literature the $(2j+1)\times(2j+1)$ matrix formed by $\mathcal{D}_{m'm}^{(j)}(R)$ is referred to as the $(2j+1)$-*dimensional irreducible representation* of the rotation operator $\mathcal{D}(R)$. This means that the matrix which corresponds to an arbitrary rotation operator in ket space *not* necessarily characterized by a single $j$-value can, with a suitable choice of basis, be brought to block-diagonal form:

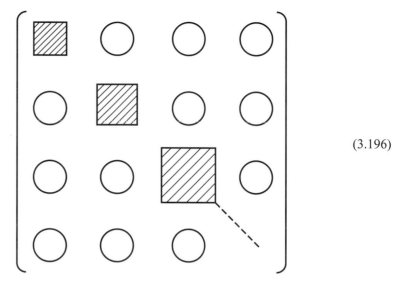

$$(3.196)$$

where each shaded square is a $(2j+1) \times (2j+1)$ square matrix formed by $\mathscr{D}^{(j)}_{m'm}$ with some definite value of $j$. Furthermore, each square matrix itself cannot be broken into smaller blocks

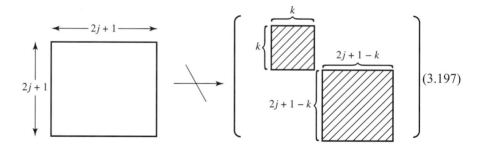

$$(3.197)$$

with any choice of basis.

The rotation matrices characterized by definite $j$ form a group. First, the identity is a member because the rotation matrix corresponding to no rotation ($\phi = 0$) is the $(2j+1) \times (2j+1)$ identity matrix. Second, the inverse is also a member; we simply reverse the rotation angle $\phi \to -\phi$ without changing the rotation axis $\hat{\mathbf{n}}$. Third, the product of any two members is also a member; explicitly we have

$$\sum_{m'} \mathscr{D}^{(j)}_{m''m'}(R_1)\mathscr{D}^{(j)}_{m'm}(R_2) = \mathscr{D}^{(j)}_{m''m}(R_1 R_2),$$

$$(3.198)$$

where the product $R_1 R_2$ represents a single rotation. We also note that the rotation matrix is unitary because the corresponding rotation operator is unitary; explicitly we have

$$\mathscr{D}_{m'm}(R^{-1}) = \mathscr{D}^*_{mm'}(R).$$

$$(3.199)$$

To appreciate the physical significance of the rotation matrix let us start with a state represented by $|j,m\rangle$. We now rotate it:

$$|j,m\rangle \rightarrow \mathscr{D}(R)|j,m\rangle. \tag{3.200}$$

Even though this rotation operation does not change $j$, we generally obtain states with $m$-values other than the original $m$. To find the amplitude for being found in $|j,m'\rangle$, we simply expand the rotated state as follows:

$$\mathscr{D}(R)|j,m\rangle = \sum_{m'} |j,m'\rangle\langle j,m'|\mathscr{D}(R)|j,m\rangle$$

$$= \sum_{m'} |j,m'\rangle \mathscr{D}^{(j)}_{m'm}(R), \tag{3.201}$$

where, in using the completeness relation, we took advantage of the fact that $\mathscr{D}(R)$ connects only states with the same $j$. So the matrix element $\mathscr{D}^{(j)}_{m'm}(R)$ is simply the amplitude for the rotated state to be found in $|j,m'\rangle$ when the original unrotated state is given by $|j,m\rangle$.

In Section 3.3 we saw how Euler angles may be used to characterize the most general rotation. We now consider the matrix realization of (3.90) for an arbitrary $j$ (not necessarily $\frac{1}{2}$):

$$\mathscr{D}^{(j)}_{m'm}(\alpha,\beta,\gamma) = \langle j,m'|\exp\left(\frac{-iJ_z\alpha}{\hbar}\right)\exp\left(\frac{-iJ_y\beta}{\hbar}\right)\exp\left(\frac{-iJ_z\gamma}{\hbar}\right)|j,m\rangle$$

$$= e^{-i(m'\alpha+m\gamma)}\langle j,m'|\exp\left(\frac{-iJ_y\beta}{\hbar}\right)|j,m\rangle. \tag{3.202}$$

Notice that the only nontrivial part is the middle rotation about the $y$-axis, which mixes different $m$-values. It is convenient to define a new matrix $d^{(j)}(\beta)$ as

$$d^{(j)}_{m'm}(\beta) \equiv \langle j,m'|\exp\left(\frac{-iJ_y\beta}{\hbar}\right)|j,m\rangle. \tag{3.203}$$

Finally, let us turn to some examples. The $j=\frac{1}{2}$ case has already been worked out in Section 3.3. See the middle matrix of (3.91),

$$d^{1/2} = \begin{pmatrix} \cos\left(\dfrac{\beta}{2}\right) & -\sin\left(\dfrac{\beta}{2}\right) \\ \sin\left(\dfrac{\beta}{2}\right) & \cos\left(\dfrac{\beta}{2}\right) \end{pmatrix}. \tag{3.204}$$

The next simplest case is $j=1$, which we consider in some detail. Clearly, we must first obtain the $3 \times 3$ matrix representation of $J_y$. Because

$$J_y = \frac{(J_+ - J_-)}{2i} \tag{3.205}$$

from the defining equation (3.157) for $J_\pm$, we can use (3.193) to obtain

$$m=1, \quad m=0, \quad m=-1,$$

$$J_y^{(j=1)} = \left(\frac{\hbar}{2}\right) \begin{pmatrix} 0 & -\sqrt{2}i & 0 \\ \sqrt{2}i & 0 & -\sqrt{2}i \\ 0 & \sqrt{2}i & 0 \end{pmatrix} \begin{array}{l} m' = 1 \\ m' = 0 \\ m' = -1. \end{array} \tag{3.206}$$

Our next task is to work out the Taylor expansion of $\exp(-iJ_y\beta/\hbar)$. Unlike the case $j = \frac{1}{2}$, $[J_y^{(j=1)}]^2$ is *independent* of 1 and $J_y^{(j=1)}$. However, it is easy to work out

$$\left(\frac{J_y^{(j=1)}}{\hbar}\right)^3 = \frac{J_y^{(j=1)}}{\hbar}. \tag{3.207}$$

Consequently, for $j = 1$ *only*, it is legitimate to replace

$$\exp\left(\frac{-iJ_y\beta}{\hbar}\right) \to 1 - \left(\frac{J_y}{\hbar}\right)^2 (1 - \cos\beta) - i\left(\frac{J_y}{\hbar}\right)\sin\beta, \tag{3.208}$$

as the reader may verify in detail. Explicitly we have

$$d^{(1)}(\beta) = \begin{pmatrix} \left(\frac{1}{2}\right)(1+\cos\beta) & -\left(\frac{1}{\sqrt{2}}\right)\sin\beta & \left(\frac{1}{2}\right)(1-\cos\beta) \\ \left(\frac{1}{\sqrt{2}}\right)\sin\beta & \cos\beta & -\left(\frac{1}{\sqrt{2}}\right)\sin\beta \\ \left(\frac{1}{2}\right)(1-\cos\beta) & \left(\frac{1}{\sqrt{2}}\right)\sin\beta & \left(\frac{1}{2}\right)(1+\cos\beta) \end{pmatrix}. \tag{3.209}$$

Clearly, this method becomes time consuming for large $j$. Other, much easier methods are possible, but we will not pursue them in this book.

## 3.6 Orbital Angular Momentum

We introduced the concept of angular momentum by defining it to be the generator of an infinitesimal rotation. There is another way to approach the subject of angular momentum when spin-angular momentum is zero or can be ignored. The angular momentum **J** for a single particle is then the same as orbital angular momentum, which is defined as

$$\mathbf{L} = \mathbf{x} \times \mathbf{p}. \tag{3.210}$$

In this section we explore the connection between the two approaches.

### 3.6.1 Orbital Angular Momentum as Rotation Generator

We first note that the orbital angular-momentum operator defined as (3.210) satisfies the angular-momentum commutation relations

$$[L_i, L_j] = i\varepsilon_{ijk}\hbar L_k \tag{3.211}$$

by virtue of the commutation relations among the components of **x** and **p**. This can easily be proved as follows:

$$[L_x, L_y] = [yp_z - zp_y, zp_x - xp_z]$$
$$= [yp_z, zp_x] + [zp_y, xp_z]$$
$$= yp_x[p_z, z] + p_y x[z, p_z]$$
$$= i\hbar(xp_y - yp_x)$$
$$= i\hbar L_z$$

$$\vdots$$

$$(3.212)$$

Next we let

$$1 - i\left(\frac{\delta\phi}{\hbar}\right)L_z = 1 - i\left(\frac{\delta\phi}{\hbar}\right)(xp_y - yp_x) \qquad (3.213)$$

act on an arbitrary position eigenket $|x', y', z'\rangle$ to examine whether it can be interpreted as the infinitesimal rotation operator about the $z$-axis by angle $\delta\phi$. Using the fact that momentum is the generator of translation, we obtain [see (1.214)]

$$\left[1 - i\left(\frac{\delta\phi}{\hbar}\right)L_z\right]|x', y', z'\rangle = \left[1 - i\left(\frac{p_y}{\hbar}\right)(\delta\phi x') + i\left(\frac{p_x}{\hbar}\right)(\delta\phi y')\right]|x', y', z'\rangle$$
$$= |x' - y'\delta\phi, y' + x'\delta\phi, z'\rangle. \qquad (3.214)$$

This is precisely what we expect if $L_z$ generates an infinitesimal rotation about the $z$-axis. So we have demonstrated that if **p** generates translation, then **L** generates rotation.

Suppose the wave function for an arbitrary physical state of a spinless particle is given by $\langle x', y', z'|\alpha\rangle$. After an infinitesimal rotation about the $z$-axis is performed, the wave function for the rotated state is

$$\langle x', y', z'|\left[1 - i\left(\frac{\delta\phi}{\hbar}\right)L_z\right]|\alpha\rangle = \langle x' + y'\delta\phi, y' - x'\delta\phi, z'|\alpha\rangle. \qquad (3.215)$$

It is actually more transparent to change the coordinate basis

$$\langle x', y', z'|\alpha\rangle \rightarrow \langle r, \theta, \phi|\alpha\rangle. \qquad (3.216)$$

For the rotated state we have, according to (3.215),

$$\langle r, \theta, \phi|\left[1 - i\left(\frac{\delta\phi}{\hbar}\right)L_z\right]|\alpha\rangle = \langle r, \theta, \phi - \delta\phi|\alpha\rangle$$
$$= \langle r, \theta, \phi|\alpha\rangle - \delta\phi\frac{\partial}{\partial\phi}\langle r, \theta, \phi|\alpha\rangle. \qquad (3.217)$$

Because $\langle r, \theta, \phi|$ is an arbitrary position eigenket, we can identify

$$\langle \mathbf{x}'|L_z|\alpha\rangle = -i\hbar\frac{\partial}{\partial\phi}\langle \mathbf{x}'|\alpha\rangle, \qquad (3.218)$$

which is a well-known result from wave mechanics. Even though this relation can also be obtained just as easily using the position representation of the momentum operator, the derivation given here emphasizes the role of $L_z$ as the generator of rotation.

We next consider a rotation about the $x$-axis by angle $\delta\phi_x$. In analogy with (3.215) we have

$$\langle x', y', z' | \left[ 1 - i \left( \frac{\delta\phi_x}{\hbar} \right) L_x \right] |\alpha\rangle = \langle x', y' + z'\delta\phi_x, z' - y'\delta\phi_x | \alpha\rangle. \tag{3.219}$$

By expressing $x'$, $y'$, and $z'$ in spherical coordinates, we can show that

$$\langle \mathbf{x}' | L_x | \alpha\rangle = -i\hbar \left( -\sin\phi \frac{\partial}{\partial\theta} - \cot\theta \cos\phi \frac{\partial}{\partial\phi} \right) \langle \mathbf{x}' | \alpha\rangle. \tag{3.220}$$

Likewise,

$$\langle \mathbf{x}' | L_y | \alpha\rangle = -i\hbar \left( \cos\phi \frac{\partial}{\partial\theta} - \cot\theta \sin\phi \frac{\partial}{\partial\phi} \right) \langle \mathbf{x}' | \alpha\rangle. \tag{3.221}$$

Using (3.220) and (3.221), for the ladder operator $L_\pm$ defined as in (3.157), we have

$$\langle \mathbf{x}' | L_\pm | \alpha\rangle = -i\hbar e^{\pm i\phi} \left( \pm i \frac{\partial}{\partial\theta} - \cot\theta \frac{\partial}{\partial\phi} \right) \langle \mathbf{x}' | \alpha\rangle. \tag{3.222}$$

Finally, it is possible to write $\langle \mathbf{x}' | \mathbf{L}^2 | \alpha\rangle$ using

$$\mathbf{L}^2 = L_z^2 + \left( \frac{1}{2} \right) (L_+ L_- + L_- L_+), \tag{3.223}$$

(3.218), and (3.222), as follows:

$$\langle \mathbf{x}' | \mathbf{L}^2 | \alpha\rangle = -\hbar^2 \left[ \frac{1}{\sin^2\theta} \frac{\partial^2}{\partial\phi^2} + \frac{1}{\sin\theta} \frac{\partial}{\partial\theta} \left( \sin\theta \frac{\partial}{\partial\theta} \right) \right] \langle \mathbf{x}' | \alpha\rangle. \tag{3.224}$$

Apart from $1/r^2$, we recognize the differential operator that appears here to be just the angular part of the Laplacian in spherical coordinates.

It is instructive to establish this connection between the $\mathbf{L}^2$ operator and the angular part of the Laplacian in another way by looking directly at the kinetic-energy operator. We first record an important operator identity,

$$\mathbf{L}^2 = \mathbf{x}^2 \mathbf{p}^2 - (\mathbf{x} \cdot \mathbf{p})^2 + i\hbar \mathbf{x} \cdot \mathbf{p}, \tag{3.225}$$

where $\mathbf{x}^2$ is understood to be the operator $\mathbf{x} \cdot \mathbf{x}$, just as $\mathbf{p}^2$ stands for the operator $\mathbf{p} \cdot \mathbf{p}$. The proof of this is straightforward:

$$\begin{aligned}
\mathbf{L}^2 &= \sum_{ijlmk} \varepsilon_{ijk} x_i p_j \varepsilon_{lmk} x_l p_m \\
&= \sum_{ijlm} (\delta_{il}\delta_{jm} - \delta_{im}\delta_{jl}) x_i p_j x_l p_m \\
&= \sum_{ijlm} [\delta_{il}\delta_{jm} x_i (x_l p_j - i\hbar\delta_{jl}) p_m - \delta_{im}\delta_{jl} x_i p_j (p_m x_l + i\hbar\delta_{lm})] \\
&= \mathbf{x}^2 \mathbf{p}^2 - i\hbar\mathbf{x} \cdot \mathbf{p} - \sum_{ijlm} \delta_{im}\delta_{jl} [x_i p_m (x_l p_j - i\hbar\delta_{jl}) + i\hbar\delta_{lm} x_i p_j] \\
&= \mathbf{x}^2 \mathbf{p}^2 - (\mathbf{x} \cdot \mathbf{p})^2 + i\hbar\mathbf{x} \cdot \mathbf{p}.
\end{aligned} \tag{3.226}$$

Before taking the preceding expression between $\langle \mathbf{x}' |$ and $| \alpha \rangle$, first note that

$$\langle \mathbf{x}' | \mathbf{x} \cdot \mathbf{p} | \alpha \rangle = \mathbf{x}' \cdot (-i\hbar \nabla' \langle \mathbf{x}' | \alpha \rangle)$$

$$= -i\hbar r \frac{\partial}{\partial r} \langle \mathbf{x}' | \alpha \rangle. \tag{3.227}$$

Likewise,

$$\langle \mathbf{x}' | (\mathbf{x} \cdot \mathbf{p})^2 | \alpha \rangle = -\hbar^2 r \frac{\partial}{\partial r} \left( r \frac{\partial}{\partial r} \langle \mathbf{x}' | \alpha \rangle \right)$$

$$= -\hbar^2 \left( r^2 \frac{\partial^2}{\partial r^2} \langle \mathbf{x}' | \alpha \rangle + r \frac{\partial}{\partial r} \langle \mathbf{x}' | \alpha \rangle \right). \tag{3.228}$$

Thus

$$\langle \mathbf{x}' | \mathbf{L}^2 | \alpha \rangle = r^2 \langle \mathbf{x}' | \mathbf{p}^2 | \alpha \rangle + \hbar^2 \left( r^2 \frac{\partial^2}{\partial r^2} \langle \mathbf{x}' | \alpha \rangle + 2r \frac{\partial}{\partial r} \langle \mathbf{x}' | \alpha \rangle \right). \tag{3.229}$$

In terms of the kinetic energy $\mathbf{p}^2 / 2m$, we have

$$\frac{1}{2m} \langle \mathbf{x}' | \mathbf{p}^2 | \alpha \rangle = -\left( \frac{\hbar^2}{2m} \right) \nabla'^2 \langle \mathbf{x}' | \alpha \rangle$$

$$= -\left( \frac{\hbar^2}{2m} \right) \left( \frac{\partial^2}{\partial r^2} \langle \mathbf{x}' | \alpha \rangle + \frac{2}{r} \frac{\partial}{\partial r} \langle \mathbf{x}' | \alpha \rangle - \frac{1}{\hbar^2 r^2} \langle \mathbf{x}' | \mathbf{L}^2 | \alpha \rangle \right). \tag{3.230}$$

The first two terms in the last line are just the radial part of the Laplacian acting on $\langle \mathbf{x}' | \alpha \rangle$. The last term must then be the angular part of the Laplacian acting on $\langle \mathbf{x}' | \alpha \rangle$, in complete agreement with (3.224).

## 3.6.2  Spherical Harmonics

Consider a spinless particle subjected to a spherical symmetrical potential. The wave equation is known to be separable in spherical coordinates and the energy eigenfunctions can be written as

$$\langle \mathbf{x}' | n, l, m \rangle = R_{nl}(r) Y_l^m(\theta, \phi), \tag{3.231}$$

where the position vector $\mathbf{x}'$ is specified by the spherical coordinates $r$, $\theta$, and $\phi$, and $n$ stands for some quantum number other than $l$ and $m$, for example, the radial quantum number for bound-state problems or the energy for a free-particle spherical wave. As will be made clearer in Section 3.11, this form can be regarded as a direct consequence of the rotational invariance of the problem. When the Hamiltonian is spherically symmetrical, $H$ commutes with $L_z$ and $\mathbf{L}^2$, and the energy eigenkets are expected to be eigenkets of $\mathbf{L}^2$ and $L_z$ also. Because $L_k$ with $k = 1, 2, 3$ satisfy the angular-momentum commutation relations, the eigenvalues of $\mathbf{L}^2$ and $L_z$ are expected to be $l(l+1)\hbar^2$, and $m\hbar = [-l\hbar, (-l+1)\hbar, \dots, (l-1)\hbar, l\hbar]$.

Because the angular dependence is common to all problems with spherical symmetry, we can isolate it and consider

$$\langle \hat{\mathbf{n}} | l, m \rangle = Y_l^m(\theta, \phi) = Y_l^m(\hat{\mathbf{n}}), \tag{3.232}$$

where we have defined a **direction eigenket** $|\hat{\mathbf{n}}\rangle$. From this point of view, $Y_l^m(\theta, \phi)$ is the amplitude for a state characterized by $l$, $m$ to be found in the direction $\hat{\mathbf{n}}$ specified by $\theta$ and $\phi$.

Suppose we have relations involving orbital angular-momentum eigenkets. We can immediately write the corresponding relations involving the spherical harmonics. For example, take the eigenvalue equation

$$L_z|l, m\rangle = m\hbar|l, m\rangle. \tag{3.233}$$

Multiplying $\langle\hat{\mathbf{n}}|$ on the left and using (3.218), we obtain

$$-i\hbar\frac{\partial}{\partial\phi}\langle\hat{\mathbf{n}}|l, m\rangle = m\hbar\langle\hat{\mathbf{n}}|l, m\rangle. \tag{3.234}$$

We recognize this equation to be

$$-i\hbar\frac{\partial}{\partial\phi}Y_l^m(\theta, \phi) = m\hbar Y_l^m(\theta, \phi), \tag{3.235}$$

which implies that the $\phi$-dependence $Y_l^m(\theta, \phi)$ must behave like $e^{im\phi}$. Likewise, corresponding to

$$\mathbf{L}^2|l, m\rangle = l(l+1)\hbar^2|l, m\rangle, \tag{3.236}$$

we have [see (3.224)]

$$\left[\frac{1}{\sin\theta}\frac{\partial}{\partial\theta}\left(\sin\theta\frac{\partial}{\partial\theta}\right) + \frac{1}{\sin^2\theta}\frac{\partial^2}{\partial\phi^2} + l(l+1)\right]Y_l^m = 0, \tag{3.237}$$

which is simply the partial differential equation satisfied by $Y_l^m$ itself. The orthogonality relation

$$\langle l', m'|l, m\rangle = \delta_{ll'}\delta_{mm'} \tag{3.238}$$

leads to

$$\int_0^{2\pi}d\phi\int_{-1}^1 d(\cos\theta)\,Y_{l'}^{m'*}(\theta, \phi)Y_l^m(\theta, \phi) = \delta_{ll'}\delta_{mm'}, \tag{3.239}$$

where we have used the completeness relation for the direction eigenkets,

$$\int d\Omega_{\hat{\mathbf{n}}}|\hat{\mathbf{n}}\rangle\langle\hat{\mathbf{n}}| = 1. \tag{3.240}$$

To obtain the $Y_l^m$ themselves, we may start with the $m = l$ case. We have

$$L_+|l, l\rangle = 0, \tag{3.241}$$

which, because of (3.222), leads to

$$-i\hbar e^{i\phi}\left[i\frac{\partial}{\partial\theta} - \cot\theta\frac{\partial}{\partial\phi}\right]\langle\hat{\mathbf{n}}|l, l\rangle = 0. \tag{3.242}$$

Remembering that the $\phi$-dependence must behave like $e^{il\phi}$, we can easily show that this partial differential equation is satisfied by

$$\langle\hat{\mathbf{n}}|l, l\rangle = Y_l^l(\theta, \phi) = c_l e^{il\phi}\sin^l\theta, \tag{3.243}$$

where $c_l$ is the normalization constant determined from (3.239) to be[8]

$$c_l = \left[\frac{(-1)^l}{2^l l!}\right] \sqrt{\frac{[(2l+1)(2l)!]}{4\pi}}. \qquad (3.244)$$

Starting with (3.243) we can use

$$\langle \hat{\mathbf{n}}|l,m-1\rangle = \frac{\langle \hat{\mathbf{n}}|L_-|l,m\rangle}{\sqrt{(l+m)(l-m+1)}\hbar}$$

$$= \frac{1}{\sqrt{(l+m)(l-m+1)}} e^{-i\phi}\left(-\frac{\partial}{\partial\theta} + i\cot\theta\frac{\partial}{\partial\phi}\right)\langle \hat{\mathbf{n}}|l,m\rangle \qquad (3.245)$$

successively to obtain all $Y_l^m$ with $l$ fixed. Because this is done in many textbooks on elementary quantum mechanics, for example Townsend (2000), we will not work out the details here. The result for $m \geq 0$ is

$$Y_l^m(\theta,\phi) = \frac{(-1)^l}{2^l l!}\sqrt{\frac{(2l+1)}{4\pi}\frac{(l+m)!}{(l-m)!}}e^{im\phi}\frac{1}{\sin^m\theta}\frac{d^{l-m}}{d(\cos\theta)^{l-m}}(\sin\theta)^{2l}, \qquad (3.246)$$

and we define $Y_l^{-m}$ by

$$Y_l^{-m}(\theta,\phi) = (-1)^m[Y_l^m(\theta,\phi)]^*. \qquad (3.247)$$

Regardless of whether $m$ is positive or negative, the $\theta$-dependent part of $Y_l^m(\theta,\phi)$ is $|\sin\theta|^{|m|}$ times a polynomial in $\cos\theta$ with a highest power of $l-|m|$. For $m=0$, we obtain

$$Y_l^0(\theta,\phi) = \sqrt{\frac{2l+1}{4\pi}}P_l(\cos\theta). \qquad (3.248)$$

From the point of view of the angular-momentum commutation relations alone, it might not appear obvious why $l$ cannot be a half-integer. It turns out that several arguments can be advanced against half-integer $l$-values. First, for half-integer $l$, and hence for half-integer $m$, the wave function would acquire a minus sign,

$$e^{im(2\pi)} = -1, \qquad (3.249)$$

under a $2\pi$ rotation. As a result, the wave function would not be single valued; we pointed out in Section 2.4 that the wave function must be single valued because of the requirement that the expansion of a state ket in terms of position eigenkets be unique. We can prove that if $\mathbf{L}$, defined to be $\mathbf{x}\times\mathbf{p}$, is to be identified as the generator of rotation, then the wave function must acquire a plus sign under a $2\pi$ rotation. This follows from the fact that the wave function for a $2\pi$-rotated state is the original wave function itself with no sign change:

$$\langle \mathbf{x}'|\exp\left(\frac{-iL_z 2\pi}{\hbar}\right)|\alpha\rangle = \langle x'\cos 2\pi + y'\sin 2\pi, y'\cos 2\pi - x'\sin 2\pi, z'|\alpha\rangle$$

$$= \langle \mathbf{x}'|\alpha\rangle, \qquad (3.250)$$

---

[8] Normalization condition (3.239), of course, does not determine the phase of $c_l$. The factor $(-1)^l$ is inserted so that when we use the $L_-$ operator successively to reach the state $m=0$, we obtain $Y_l^0$ with the same sign as the Legendre polynomial $P_l(\cos\theta)$ whose phase is fixed by $P_l(1)=1$ [see (3.248)].

where we have used the finite-angle version of (3.215). Next, let us suppose $Y_l^m(\theta, \phi)$ with a half-integer $l$ were possible. To be specific, we choose the simplest case, $l = m = \frac{1}{2}$. According to (3.243) we would have

$$Y_{1/2}^{1/2}(\theta, \phi) = c_{1/2} e^{i\phi/2} \sqrt{\sin \theta}. \tag{3.251}$$

From the property of $L_-$ [see (3.245)] we would then obtain

$$Y_{1/2}^{-1/2}(\theta, \phi) = e^{-i\phi} \left( -\frac{\partial}{\partial \theta} + i \cot \theta \frac{\partial}{\partial \phi} \right) \left( c_{1/2} e^{i\phi/2} \sqrt{\sin \theta} \right)$$

$$= -c_{1/2} e^{-i\phi/2} \cot \theta \sqrt{\sin \theta}. \tag{3.252}$$

This expression is not permissible because it is singular at $\theta = 0, \pi$. What is worse, from the partial differential equation

$$\left\langle \hat{\mathbf{n}} \middle| L_- \middle| \frac{1}{2}, -\frac{1}{2} \right\rangle = -i\hbar e^{-i\phi} \left( -i \frac{\partial}{\partial \theta} - \cot \theta \frac{\partial}{\partial \phi} \right) \left\langle \hat{\mathbf{n}} \middle| \frac{1}{2}, -\frac{1}{2} \right\rangle$$

$$= 0 \tag{3.253}$$

we directly obtain

$$Y_{1/2}^{-1/2} = c'_{1/2} e^{-i\phi/2} \sqrt{\sin \theta}, \tag{3.254}$$

in sharp contradiction with (3.252). Finally, we know from the Sturm–Liouville theory of differential equations that the solutions of (3.237) with $l$ integer form a complete set. An arbitrary function of $\theta$ and $\phi$ can be expanded in terms of $Y_l^m$ with integer $l$ and $m$ only. For all these reasons it is futile to contemplate orbital angular momentum with half-integer $l$-values.

### 3.6.3 Spherical Harmonics as Rotation Matrices

We conclude this section on orbital angular momentum by discussing the spherical harmonics from the point of view of the rotation matrices introduced in the last section. We can readily establish the desired connection between the two approaches by constructing the most general direction eigenket $|\hat{\mathbf{n}}\rangle$ by applying appropriate rotation operators to $|\hat{\mathbf{z}}\rangle$, the direction eigenket in the positive $z$-direction. We wish to find $\mathscr{D}(R)$ such that

$$|\hat{\mathbf{n}}\rangle = \mathscr{D}(R)|\hat{\mathbf{z}}\rangle. \tag{3.255}$$

We can rely on the technique used in constructing the eigenspinor of $\sigma \cdot \hat{\mathbf{n}}$ in Section 3.2. We first rotate about the $y$-axis by angle $\theta$, then around the $z$-axis by angle $\phi$; see Figure 3.3 with $\beta \to \theta$, $\alpha \to \phi$. In the notation of Euler angles we have

$$\mathscr{D}(R) = \mathscr{D}(\alpha = \phi, \beta = \theta, \gamma = 0). \tag{3.256}$$

Writing (3.255) as

$$|\hat{\mathbf{n}}\rangle = \sum_l \sum_m \mathscr{D}(R)|l, m\rangle \langle l, m|\hat{\mathbf{z}}\rangle, \tag{3.257}$$

we see that $|\hat{\mathbf{n}}\rangle$, when expanded in terms of $|l,m\rangle$, contains all possible $l$-values. However, when this equation is multiplied by $\langle l,m'|$ on the left, only one term in the $l$-sum contributes, namely,

$$\langle l,m'|\hat{\mathbf{n}}\rangle = \sum_m \mathscr{D}^{(l)}_{m'm}(\alpha = \phi, \beta = \theta, \gamma = 0)\langle l,m|\hat{\mathbf{z}}\rangle. \tag{3.258}$$

Now $\langle l,m|\hat{\mathbf{z}}\rangle$ is just a number; in fact, it is precisely $Y_l^{m^*}(\theta,\phi)$ evaluated at $\theta = 0$ with $\phi$ undetermined. At $\theta = 0$, $Y_l^m$ is known to vanish for $m \neq 0$, which can also be seen directly from the fact that $|\hat{\mathbf{z}}\rangle$ is an eigenket of $L_z$ (which equals $xp_y - yp_x$) with eigenvalue zero. So we can write

$$\langle l,m|\hat{\mathbf{z}}\rangle = Y_l^{m^*}(\theta = 0, \phi\text{ undetermined})\delta_{m0}$$

$$= \left.\sqrt{\frac{(2l+1)}{4\pi}}P_l(\cos\theta)\right|_{\cos\theta - 1}\delta_{m0}$$

$$= \sqrt{\frac{(2l+1)}{4\pi}}\delta_{m0}. \tag{3.259}$$

Returning to (3.258), we have

$$Y_l^{m'^*}(\theta,\phi) = \sqrt{\frac{(2l+1)}{4\pi}}\mathscr{D}^{(l)}_{m'0}(\alpha = \phi, \beta = \theta, \gamma = 0) \tag{3.260}$$

or

$$\mathscr{D}^{(l)}_{m0}(\alpha,\beta,\gamma = 0) = \left.\sqrt{\frac{4\pi}{(2l+1)}}Y_l^{m^*}(\theta,\phi)\right|_{\theta=\beta,\phi=\alpha}. \tag{3.261}$$

Notice the $m = 0$ case, which is of particular importance:

$$\left.d^{(l)}_{00}(\beta)\right|_{\beta-\theta} = P_l(\cos 0). \tag{3.262}$$

## 3.7 Schrödinger's Equation for Central Potentials

Problems described by Hamiltonians of the form

$$H = \frac{\mathbf{p}^2}{2m} + V(r) \qquad r^2 = \mathbf{x}^2 \tag{3.263}$$

form the basis for very many situations in the physical world. The fundamental importance of this Hamiltonian lies in the fact that it is spherically symmetric. Classically, we expect orbital angular momentum to be conserved in such a system. This is also true quantum mechanically, since it is easy to show that

$$[\mathbf{L},\mathbf{p}^2] = [\mathbf{L},\mathbf{x}^2] = 0 \tag{3.264}$$

and therefore

$$[\mathbf{L},H] = [\mathbf{L}^2,H] = 0 \tag{3.265}$$

if $H$ is given by (3.263). We refer to such problems as "central potential" or "central force" problems. Even if the Hamiltonian is not strictly of this form, it is often the case that this is a good starting point when we consider approximation schemes that build on "small" corrections to central potential problems.

In this section we will discuss some general properties of bound-state eigenfunctions generated by (3.263), and a few representative central potential problems. (Scattering solutions will be discussed in Chapter 6.) For more detail, the reader is referred to any number of excellent texts which explore such problems in greater depth.

## 3.7.1 The Radial Equation

Equation (3.265) makes it clear that we should search for energy eigenstates $|\alpha\rangle = |Elm\rangle$ where

$$H|Elm\rangle = E|Elm\rangle \tag{3.266}$$

$$\mathbf{L}^2|Elm\rangle = l(l+1)\hbar^2|Elm\rangle \tag{3.267}$$

$$L_z|Elm\rangle = m\hbar|Elm\rangle. \tag{3.268}$$

It is easiest to work in the coordinate representation, and solve the appropriate differential equation for eigenfunctions in terms of a radial function $R_{El}(r)$ and spherical harmonics, as shown in (3.231). Combining (3.263), (3.266), and (3.267) with (3.230) and (3.231) we arrive at the **radial equation**[9]

$$\left[ -\frac{\hbar^2}{2mr^2}\frac{d}{dr}\left(r^2\frac{d}{dr}\right) + \frac{l(l+1)\hbar^2}{2mr^2} + V(r) \right]R_{El}(r) = ER_{El}(r). \tag{3.269}$$

Depending on the specific form of $V(r)$, we may work with this equation or some variant of it, to identify the radial part $R_{El}(r)$ of the eigenfunction and/or the energy eigenvalues $E$.

In fact, we can immediately gain some insight to the effects of angular momentum on the eigenfunctions by making the substitution

$$R_{El}(r) = \frac{u_{El}(r)}{r} \tag{3.270}$$

which reduces (3.269) to

$$-\frac{\hbar^2}{2m}\frac{d^2u_{El}}{dr^2} + \left[ \frac{l(l+1)\hbar^2}{2mr^2} + V(r) \right]u_{El}(r) = Eu_{El}(r). \tag{3.271}$$

Coupled with the fact that the spherical harmonics are separately normalized, so that the overall normalization condition becomes

$$1 = \int r^2 dr\, R^*_{El}(r)R_{El}(r) = \int dr\, u^*_{El}(r)u_{El}(r), \tag{3.272}$$

we see that $u_{El}(r)$ can be interpreted as a wave function in one dimension for a particle moving in an "effective potential"

---

[9]  We apologize for using $m$ for both "mass" and the quantum number for angular momentum. However, in this section, it should be clear from the context which is which.

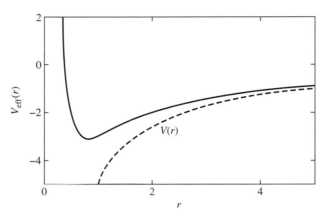

The "effective potential" which governs the behavior of the "radial wave function" $u_{El}(r)$. If the potential energy $V(r)$ (shown as a dashed line) is not too singular at the origin, then there is an angular momentum barrier for all states with $l \neq 0$, which makes it very improbable for a particle to located near the origin.

$$V_{\text{eff}}(r) = V(r) + \frac{l(l+1)\hbar^2}{2mr^2}. \tag{3.273}$$

Equation (3.273) demonstrates the existence of an "angular momentum barrier" if $l \neq 0$, as shown in Figure 3.5. Quantum mechanically, this means that the amplitude (and therefore the probability) is small for locating the particle near the origin, except for $s$-states. As we will see later on, this fact has important physical consequences in atoms, for example.

We can be more quantitative about this interpretation. Let us assume that the potential energy function $V(r)$ is not so singular so that $\lim_{r\to 0} r^2 V(r) = 0$. Then, for small values of $r$, (3.271) becomes

$$\frac{d^2 u_{El}}{dr^2} = \frac{l(l+1)}{r^2} u_{El}(r) \qquad (r \to 0) \tag{3.274}$$

which has the general solution

$$u_{El}(r) = Ar^{l+1} + \frac{B}{r^l}. \tag{3.275}$$

It is tempting to set $B = 0$ out of hand, because $1/r^l$ produces severe singularities as $r \to 0$, especially for large $l$. However, there are better reasons for setting $B = 0$, that are rooted in the foundations of quantum mechanics. For example, under certain conditions, one can show that the interference between the two terms leads to nonconservation of probability. (See Problem 3.27 at the end of this chapter.) Other rather sophisticated explanations are given in some textbooks, but we give a relatively simple approach here.

If $B \neq 0$ in (3.275), then the radial wave function $R_{El}(r) \to 1/r^{l+1}$ as $r \to 0$, and this is not normalizable for $l \geq 1$. The wave function can be normalized for $l = 0$, but in this case the Schrödinger equation involves $\nabla^2(1/r) = -4\pi\delta(\mathbf{x})$ and solutions cannot be found (unless the potential is a $\delta$-function).

Consequently, we are left with

$$R_{El}(r) \to r^l \qquad \text{as} \quad r \to 0. \tag{3.276}$$

This relation has profound consequences. Firstly, it embodies the "angular-momentum barrier" shown in Figure 3.5, since the wave function goes to zero except for s-states. More practically, it means that the probability of finding, say, an electron in an atom in the region of the nucleus, goes like $(R/a_0)^{2l}$ where $R \ll a_0$ is the size of the nucleus and $a_0$ is the Bohr radius. These concepts will become explicit when we come to the study of atomic structure.

When considering bound states of potential energy functions $V(r)$ which tend to zero at large $r$, there is another form of the radial equation which we can consider. For $r \to \infty$, (3.271) becomes

$$\frac{d^2 u_E}{dr^2} = \kappa^2 u \qquad \kappa^2 \equiv -2mE/\hbar^2 > 0 \qquad r \to \infty \qquad (3.277)$$

since $E < 0$ for bound states. The solution to this equation is simply $u_E(r) \propto e^{-\kappa r}$. Also, it makes it clear that the dimensionless variable $\rho \equiv \kappa r$ would be useful to recast the radial equation. Consequently, we remove both the short and long distance behavior of the wave function and write[10]

$$u_{El}(\rho) = \rho^{l+1} e^{-\rho} w(\rho) \qquad (3.278)$$

where the function $w(\rho)$ is "well behaved" and satisfies

$$\frac{d^2 w}{d\rho^2} + 2\left(\frac{l+1}{\rho} - 1\right)\frac{dw}{d\rho} + \left[\frac{V}{E} - \frac{2(l+1)}{\rho}\right]w = 0. \qquad (3.279)$$

(The manipulations which lead to this equation are left to the reader.) One then attacks the solution $w(\rho)$ of (3.279) for the particular function $V(r = \rho/\kappa)$.

### 3.7.2 The Free Particle and Infinite Spherical Well

In Section 2.5.1 we saw the solution to the free-particle problem in three dimensions, using Cartesian coordinates. We can of course approach the same problem by exploiting spherical symmetry and angular momentum. Starting from (3.269) we write

$$E \equiv \frac{\hbar^2 k^2}{2m} \qquad \text{and} \qquad \rho \equiv kr \qquad (3.280)$$

and arrive at the modified radial equation

$$\frac{d^2 R}{d\rho^2} + \frac{2}{\rho}\frac{dR}{d\rho} + \left[1 - \frac{l(l+1)}{\rho^2}\right]R = 0. \qquad (3.281)$$

This is a well-known differential equation whose solutions are called *spherical Bessel functions* $j_l(\rho)$ and $n_l(\rho)$, where

---

[10] We are being a bit careless with notation when switching arguments from $r$ to $\rho$, as is typical in these kinds of treatments. For example, going from (3.277) to (3.278) and ignoring subscripts, we should define the dimensionless function $\tilde{u}(\rho) \equiv \kappa^{-1/2} u(\rho/\kappa)$ in which case $\int_0^\infty \tilde{u}^2(\rho)d\rho = 1$. We can otherwise ignore the factor $\kappa^{-1/2}$ which gets absorbed into the normalization, and then drop the "tilde," that is $\tilde{u}(\rho) \to u(\rho)$.

$$j_l(\rho) = (-\rho)^l \left[\frac{1}{\rho}\frac{d}{d\rho}\right]^l \left(\frac{\sin\rho}{\rho}\right) \tag{3.282a}$$

$$n_l(\rho) = -(-\rho)^l \left[\frac{1}{\rho}\frac{d}{d\rho}\right]^l \left(\frac{\cos\rho}{\rho}\right). \tag{3.282b}$$

It is easy to show that as $\rho \to 0$, $j_l(\rho) \to \rho^l$ and $n_l(\rho) \to \rho^{-l-1}$. Hence, $j_l(\rho)$ corresponds to (3.276) and these are the only solutions we consider here.[11] It is also useful to point out that the spherical Bessel functions are defined over the entire complex plane, and it can be shown that

$$j_l(z) = \frac{1}{2i^l} \int_{-1}^{1} ds \, e^{izs} P_l(s). \tag{3.283}$$

The first few spherical Bessel functions are

$$j_0(\rho) = \frac{\sin\rho}{\rho} \tag{3.284}$$

$$j_1(\rho) = \frac{\sin\rho}{\rho^2} - \frac{\cos\rho}{\rho} \tag{3.285}$$

$$j_2(\rho) = \left[\frac{3}{\rho^3} - \frac{1}{\rho}\right]\sin\rho - \frac{3\cos\rho}{\rho^2}. \tag{3.286}$$

This result can be immediately applied to the case of a particle confined to an infinite spherical well, that is, a potential energy function $V(r) = 0$ within $r < a$, but with the wave function constrained to be zero at $r = a$. For any given value of $l$, this leads to the "quantization condition" $j_l(ka) = 0$, that is, $ka$ equals the set of zeros of the spherical Bessel function. For $l = 0$ these are obviously $ka = \pi, 2\pi, 3\pi, \ldots$. For other values of $l$, computer programs are readily available which can compute the zeros. We find that

$$E_{l=0} = \frac{\hbar^2}{2ma^2}\left[\pi^2, (2\pi)^2, (3\pi)^2, \ldots\right] \tag{3.287}$$

$$E_{l=1} = \frac{\hbar^2}{2ma^2}\left[4.49^2, 7.73^2, 10.90^2, \ldots\right] \tag{3.288}$$

$$E_{l=2} = \frac{\hbar^2}{2ma^2}\left[5.76^2, 9.10^2, 12.32^2, \ldots\right]. \tag{3.289}$$

It should be noted that this series of energy levels shows no degeneracies in $l$. Indeed, such degenerate energy levels are impossible, except for any accidental equality between zeros of spherical Bessel functions of different orders.

### 3.7.3 The Isotropic Harmonic Oscillator

Energy eigenvalues for the Hamiltonian

$$H = \frac{\mathbf{p}^2}{2m} + \frac{1}{2}m\omega^2 r^2 \tag{3.290}$$

---

[11] In a treatment of "hard sphere scattering" problems, the origin is explicitly excluded, and the solutions $n_l(\rho)$ are also kept. The relative phase between the two solutions for a given $l$ is called the "phase shift."

are straightforward to determine. Introducing dimensionless energy $\lambda$ and radial coordinate $\rho$ through

$$E = \frac{1}{2}\hbar\omega\lambda \quad \text{and} \quad r = \left[\frac{\hbar}{m\omega}\right]^{1/2}\rho \tag{3.291}$$

we transform (3.271) into

$$\frac{d^2u}{d\rho^2} - \frac{l(l+1)}{\rho^2}u(\rho) + (\lambda - \rho^2)u(\rho) = 0. \tag{3.292}$$

It is again worthwhile to explicitly remove the behavior for large (and small) $\rho$, although we cannot use (3.278) because $V(r)$ does not tend to zero for large $r$. Instead, we write

$$u(\rho) = \rho^{l+1}e^{-\rho^2/2}f(\rho). \tag{3.293}$$

This yields the following differential equation for the function $f(\rho)$:

$$\rho\frac{d^2f}{d\rho^2} + 2[(l+1) - \rho^2]\frac{df}{d\rho} + [\lambda - (2l+3)]\rho f(\rho) = 0. \tag{3.294}$$

We solve (3.294) by writing $f(\rho)$ as an infinite series, namely

$$f(\rho) = \sum_{n=0}^{\infty} a_n\rho^n. \tag{3.295}$$

We insert this into the differential equation, and set each term to zero by powers of $\rho$. The only surviving term in $\rho^0$ is $2(l+1)a_1$, so

$$a_1 = 0. \tag{3.296}$$

The terms proportional to $\rho^1$ allow us to relate $a_2$ to $a_0$, which in turn can be set through the normalization condition. Continuing, (3.294) becomes

$$\sum_{n=2}^{\infty} \left\{(n+2)(n+1)a_{n+2} + 2(l+1)(n+2)a_{n+2} - 2na_n + [\lambda - (2l+3)a_n]\right\}\rho^{n+1} = 0 \tag{3.297}$$

which leads, finally, to the recursion relation

$$a_{n+2} = \frac{2n+2l+3-\lambda}{(n+2)(n+2l+3)}a_n. \tag{3.298}$$

Immediately we see that $f(\rho)$ involves only even powers of $\rho$, since (3.296) and (3.298) imply that $a_n = 0$ for odd $n$. Also, as $n \to \infty$, we have

$$\frac{a_{n+2}}{a_n} \to \frac{2}{n} = \frac{1}{q} \tag{3.299}$$

where $q = n/2$ includes both odd and even integers. Therefore, for large values of $\rho$, (3.295) becomes

$$f(\rho) \to \text{constant} \times \sum_q \frac{1}{q!}(\rho^2)^q \propto e^{\rho^2}. \tag{3.300}$$

In other words, $u(\rho)$ from (3.293) would grow exponentially for large $\rho$ (and therefore be unable to meet the normalization condition) unless the series terminates. Therefore

$$2n + 2l + 3 - \lambda = 0 \qquad (3.301)$$

for some even value of $n = 2q$, and the energy eigenvalues are

$$E_{ql} = \left(2q + l + \frac{3}{2}\right)\hbar\omega \equiv \left(N + \frac{3}{2}\right)\hbar\omega \qquad (3.302)$$

for $q = 0, 1, 2, \ldots$ and $l = 0, 1, 2, \ldots$, and $N \equiv 2q + l$. One frequently refers to $N$ as the "principal" quantum number. It can be shown that $q$ counts the number of nodes in the radial function.

Quite unlike the square well, the three-dimensional isotropic harmonic oscillator has degenerate energy eigenvalues in the $l$ quantum number. There are three states (all $l = 1$) for $N = 1$. For $N = 2$ there are five states with $l = 2$, plus one state with $q = 1$ and $l = 0$, giving a total of six. Notice that for even (odd) values of $N$, only even (odd) values of $l$ are allowed. Therefore, the parity of the wave function is even or odd with the value of $N$.

These wave functions are popular basis states for calculations of various natural phenomena, when the potential energy function is a "well" of some finite size. One of the greatest successes of such an approach is the nuclear shell model, where individual protons and neutrons are pictured to move independently in a potential energy function generated by the cumulative effect of all nucleons in the nucleus. Figure 3.6 compares the energy levels observed in nuclei with those obtained for the isotropic harmonic oscillator and for the infinite spherical well.

It is natural to label the eigenstates of the Hamiltonian (3.290) as $|qlm\rangle$ or $|Nlm\rangle$. However, this Hamiltonian may also be written as

$$H = H_x + H_y + H_z \qquad (3.303)$$

where $H_i = a_i^\dagger a_i + \frac{1}{2}$ is an independent one-dimensional harmonic oscillator in direction $i = x, y, z$. In this way, we would label the eigenstates $|n_x, n_y, n_z\rangle$ and the energy eigenvalues are

$$E = \left(n_x + \frac{1}{2} + n_x + \frac{1}{2} + n_x + \frac{1}{2}\right)\hbar\omega$$
$$= \left(N + \frac{3}{2}\right)\hbar\omega \qquad (3.304)$$

where, now, $N = n_x + n_y + n_z$. It is simple to show numerically for the first few energy levels, that the degeneracy is the same regardless of which basis is used. It is an interesting exercise to show this in general, and also to derive the unitary transformation matrix $\langle n_x, n_y, n_z | qlm \rangle$ which changes from one basis to the other. (See Problem 3.29 at the end of this chapter.)

### 3.7.4 The Coulomb Potential

Perhaps the most important potential energy function in physics is

$$V(\mathbf{x}) = -\frac{Ze^2}{r} \qquad (3.305)$$

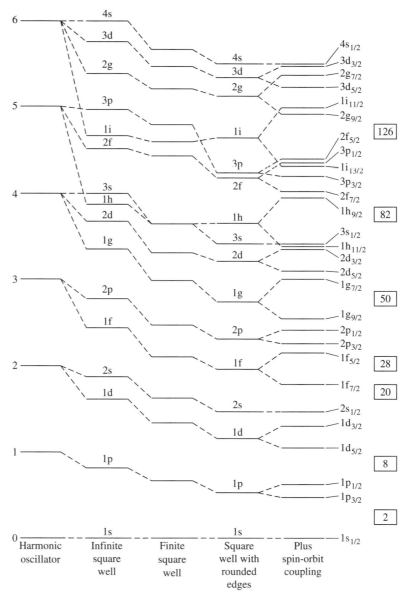

**Fig. 3.6** Energy levels in the nuclear shell model, adapted from Haxel et al., *Z. Phys.*, **128** (1950) 295. Energy levels of the three-dimensional isotropic harmonic oscillator are on the left, followed by the infinite spherical well. Modifications of the infinite square well, for finite walls and then for "rounded corners" follow. The rightmost plot of energy levels shows those obtained by including the interaction between the nucleon spin and the orbital angular momentum. The final column indicates the total angular-momentum quantum number.

where the constant $Ze^2$ is obviously chosen so that (3.305) represents the potential for a one-electron atom with atomic number $Z$. In addition to Coulomb forces, and classical gravity, it is widely used in models applied to very many physical systems.[12] We consider here the radial equation based on such a function and the resulting energy eigenvalues.

The $1/r$ potential satisfies all the requirements that led us to (3.279). We therefore search for solutions of the form (3.278) by determining the function $w(\rho)$. Making the definition

$$\rho_0 = \left[\frac{2m}{-E}\right]^{1/2} \frac{Ze^2}{\hbar} = \left[\frac{2mc^2}{-E}\right]^{1/2} Z\alpha \qquad (3.306)$$

where $\alpha \equiv e^2/\hbar c \approx 1/137$ is the fine-structure constant, (3.279) becomes

$$\rho\frac{d^2w}{d\rho^2} + 2(l+1-\rho)\frac{dw}{d\rho} + [\rho_0 - 2(l+1)]w(\rho) = 0. \qquad (3.307)$$

We could of course proceed to solve (3.307) using a series approach and derive a recursion relation for the coefficients, just as we did with (3.294). However, it turns out that the solution is in fact already well known.

Equation (3.307) can be written as *Kummer's equation*

$$x\frac{d^2F}{dx^2} + (c-x)\frac{dF}{dx} - aF = 0 \qquad (3.308)$$

where

$$\begin{aligned} x &= 2\rho \\ c &= 2(l+1) \\ 2a &= 2(l+1) - \rho_0. \end{aligned} \qquad (3.309)$$

The solution to (3.308) is called the *confluent hypergeometric function*, written as the series

$$F(a; c; x) = 1 + \frac{a}{c}\frac{x}{1!} + \frac{a(a+1)}{c(c+1)}\frac{x^2}{2!} + \cdots \qquad (3.310)$$

and so

$$w(\rho) = F\left(l+1 - \frac{\rho_0}{2} \; ; \; 2(l+1) \; ; \; 2\rho\right). \qquad (3.311)$$

Note that for large $\rho$, we have

$$\begin{aligned} w(\rho) &\approx \sum_{\text{Large } N} \frac{a(a+1)\cdots}{c(c+1)\cdots} \frac{(2\rho)^N}{N!} \\ &\approx \sum_{\text{Large } N} \frac{(N/2)^N}{N^N} \frac{(2\rho)^N}{N!} \approx \sum_{\text{Large } N} \frac{(\rho)^N}{N!} \approx e^\rho. \end{aligned}$$

---

[12] Indeed, $1/r$ potential energy functions result from any quantum field theory in three spatial dimensions with massless intermediate exchange particles. See Chapter I.6 in Zee (2010).

Therefore, once again, (3.278) gives a radial wave function which would grow without bound unless the series (3.310) terminates. So, for some integer $N$ we must have $a + N = 0$ which leads to

$$\rho_0 = 2(N + l + 1)$$

where     $N = 0, 1, 2 \ldots,$     and     $l = 0, 1, 2, \ldots.$     (3.312)

It is customary (and, as we shall soon see, instructive) to define the *principal quantum number n* as

$$n \equiv N + l + 1 = 1, 2, 3, \ldots$$

where     $l = 0, 1, \ldots, n - 1.$     (3.313)

We point out that it is possible to solve the radial equation for the Coulomb problem using the generating function techniques described in Section 2.5. See Problem 3.30 at the end of this chapter.

Energy eigenvalues arise by combining (3.306) and (3.312) in terms of the principal quantum number, that is

$$\rho_0 = \left[ \frac{2mc^2}{-E} \right]^{1/2} Z\alpha = 2n$$     (3.314)

which leads to

$$E = -\frac{1}{2} mc^2 \frac{Z^2 \alpha^2}{n^2} = -13.6 \text{ eV} \frac{Z^2}{n^2}$$     (3.315)

where the numerical result is for a one-electron atom, i.e. $mc^2 = 511$ keV. Equation (3.315) is of course the familiar Balmer formula.

It is time to make various points. First, there is a stark disagreement between the energy level properties predicted by modern quantum mechanics, and those of the old Bohr model of the atom. The Bohr model had a one-to-one correspondence between angular-momentum eigenvalues $l$ and principal quantum number $n$. In fact the ground state corresponded to $n = l = 1$. We see instead that only $l = 0$ is allowed for $n = 1$, and that different values of $l$ are allowed for higher energy levels.

Second, a natural length scale $a_0$ has emerged. Since $\rho = \kappa r$ where $\kappa = \sqrt{-2mE/\hbar^2}$ (see (3.277) we have

$$\frac{1}{\kappa} = \frac{\hbar}{mc\alpha} \frac{n}{Z} \equiv a_0 \frac{n}{Z}$$     (3.316)

where

$$a_0 = \frac{\hbar}{mc\alpha} = \frac{\hbar^2}{me^2}$$     (3.317)

is called the *Bohr radius*. For an electron, $a_0 = 0.53 \times 10^{-8}$ cm $= 0.53$ Å. This is indeed the typical size of an atom.

Finally, the energy eigenvalues (3.315) demonstrate an interesting degeneracy. The eigenvalues depend only on $n$, and not on $l$ or $m$. The level of degeneracy for a state $|nlm\rangle$ is therefore given by

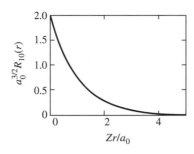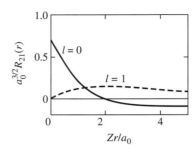

**Fig. 3.7** Radial wave functions for the Coulomb potential and principal quantum numbers $n = 1$ (left) and $n = 2$ (right).

$$\text{Degeneracy} = \sum_{l=0}^{n-1}(2l+1) = n^2. \tag{3.318}$$

This degeneracy is in fact not accidental, but reflects a subtle symmetry of the Coulomb potential. We will return to this question in Chapter 4.

We can now write down the hydrogen atom wave functions explicitly. Going back to (3.231) and putting in the appropriate normalization (see Problem 3.30), we have

$$\psi_{nlm}(\mathbf{x}) = \langle \mathbf{x}|nlm\rangle = R_{nl}(r)Y_l^m(\theta,\phi) \tag{3.319}$$

where

$$R_{nl}(r) = \frac{1}{(2l+1)!}\left(\frac{2Zr}{na_0}\right)^l e^{-Zr/na_0}\left[\left(\frac{2Z}{na_0}\right)^3\frac{(n+l)!}{2n(n-l-1)!}\right]^{1/2}$$
$$\times F(-n+l+1; 2l+2; 2Zr/na_0). \tag{3.320}$$

Figure 3.7 plots these radial wave functions for $n = 1$ and $n = 2$. As we have discussed, only the $l = 0$ wave functions are nonzero at the origin. Also note that there are $n-1$ nodes in the wave function for $l = 0$, and no nodes for the wave function with $l = n-1$.

# 3.8  Addition of Angular Momenta

Angular-momentum addition has important applications in all areas of modern physics – from atomic spectroscopy to nuclear and particle collisions. Furthermore, a study of angular-momentum addition provides an excellent opportunity to illustrate the concept of change of basis, which we discussed extensively in Chapter 1.

### 3.8.1  Simple Examples of Angular-Momentum Addition

Before studying a formal theory of angular-momentum addition, it is worth looking at two simple examples with which the reader may be familiar: (1) how to add orbital angular momentum and spin-angular momentum and (2) how to add the spin-angular momenta of two spin $\frac{1}{2}$ particles.

Previously we studied both spin $\frac{1}{2}$ systems with all quantum-mechanical degrees of freedom other than spin, such as position and momentum, ignored and quantum-mechanical particles with the space degrees of freedom (such as position or momentum) taken into account but the internal degrees of freedom (such as spin) ignored. A realistic description of a particle with spin must of course take into account both the space degree of freedom and the internal degrees of freedom. The base ket for a spin $\frac{1}{2}$ particle may be visualized to be in the direct-product space of the infinite-dimensional ket space spanned by the position eigenkets $\{|\mathbf{x}'\rangle\}$ and the two-dimensional spin space spanned by $|+\rangle$ and $|-\rangle$. Explicitly, we have for the base ket

$$|\mathbf{x}', \pm\rangle = |\mathbf{x}'\rangle \otimes |\pm\rangle, \tag{3.321}$$

where any operator in the space spanned by $\{|\mathbf{x}'\rangle\}$ commutes with any operator in the two-dimensional space spanned by $|\pm\rangle$.

The rotation operator still takes the form $\exp(-i\mathbf{J}\cdot\hat{\mathbf{n}}\phi/\hbar)$ but $\mathbf{J}$, the generator of rotations, is now made up of two parts, namely,

$$\mathbf{J} = \mathbf{L} + \mathbf{S}. \tag{3.322}$$

It is actually more obvious to write (3.322) as

$$\mathbf{J} = \mathbf{L} \otimes 1 + 1 \otimes \mathbf{S}, \tag{3.323}$$

where the 1 in $\mathbf{L} \otimes 1$ stands for the identity operator in the spin space, while the 1 in $1 \otimes \mathbf{S}$ stands for the identity operator in the infinite-dimensional ket space spanned by the position eigenkets. Because $\mathbf{L}$ and $\mathbf{S}$ commute, we can write

$$\mathscr{D}(R) = \mathscr{D}^{(\mathrm{orb})}(R) \otimes \mathscr{D}^{(\mathrm{spin})}(R) = \exp\left(\frac{-i\mathbf{L}\cdot\hat{\mathbf{n}}\phi}{\hbar}\right) \otimes \exp\left(\frac{-i\mathbf{S}\cdot\hat{\mathbf{n}}\phi}{\hbar}\right). \tag{3.324}$$

The wave function for a particle with spin is written as

$$\langle \mathbf{x}', \pm|\alpha\rangle = \psi_\pm(\mathbf{x}'). \tag{3.325}$$

The two components $\psi_\pm$ are often arranged in column matrix form as follows:

$$\begin{pmatrix} \psi_+(\mathbf{x}') \\ \psi_-(\mathbf{x}') \end{pmatrix}, \tag{3.326}$$

where $|\psi_\pm(\mathbf{x}')|^2$ stands for the probability density for the particle to be found at $\mathbf{x}'$ with spin up and spin down, respectively. Instead of $|\mathbf{x}'\rangle$ as the base kets for the space part, we may use $|n, l, m\rangle$, which are eigenkets of $\mathbf{L}^2$ and $L_z$ with eigenvalues $\hbar^2 l(l+1)$ and $m_l\hbar$, respectively. For the spin part, $|\pm\rangle$ are eigenkets of $\mathbf{S}^2$ and $S_z$ with eigenvalues $3\hbar^2/4$ and $\pm\hbar/2$, respectively. However, as we will show later, we can also use base kets which are eigenkets of $\mathbf{J}^2$, $J_z$, $\mathbf{L}^2$, and $\mathbf{S}^2$. In other words, we can expand a state ket of a particle with spin in terms of simultaneous eigenkets of $\mathbf{L}^2$, $\mathbf{S}^2$, $L_z$, and $S_z$ *or* in terms of simultaneous eigenkets of $\mathbf{J}^2$, $J_z$, $\mathbf{L}^2$, and $\mathbf{S}^2$. We will study in detail how the two descriptions are related.

As a second example, we study two spin $\frac{1}{2}$ particles, say two electrons, with the orbital degree of freedom suppressed. The total spin operator is usually written as

$$\mathbf{S} = \mathbf{S}_1 + \mathbf{S}_2, \tag{3.327}$$

but again it is to be understood as

$$\mathbf{S}_1 \otimes 1 + 1 \otimes \mathbf{S}_2, \tag{3.328}$$

where the 1 in the first (second) term stands for the identity operator in the spin space of electron 2 (1). We, of course, have

$$[S_{1x}, S_{2y}] = 0 \tag{3.329}$$

and so forth. Within the space of electron 1(2) we have the usual commutation relations

$$[S_{1x}, S_{1y}] = i\hbar S_{1z}, [S_{2x}, S_{2y}] = i\hbar S_{2z}, \ldots. \tag{3.330}$$

As a direct consequence of (3.329) and (3.330), we have

$$[S_x, S_y] = i\hbar S_z \tag{3.331}$$

and so on for the *total* spin operator.

The eigenvalues of the various spin operators are denoted as follows:

$$\begin{aligned}
\mathbf{S}^2 &= (\mathbf{S}_1 + \mathbf{S}_2)^2 : s(s+1)\hbar^2 \\
S_z &= S_{1z} + S_{2z} \quad : m\hbar \\
S_{1z} &\qquad\qquad\quad : m_1\hbar \\
S_{2z} &\qquad\qquad\quad : m_2\hbar.
\end{aligned} \tag{3.332}$$

Again, we can expand the ket corresponding to an arbitrary spin state of two electrons in terms of either the eigenkets of $\mathbf{S}^2$ and $S_z$ *or* the eigenkets of $S_{1z}$ and $S_{2z}$. The two possibilities are as follows.

1. The $\{m_1, m_2\}$ representation based on the eigenkets of $S_{1z}$ and $S_{2z}$:

$$|++\rangle, |+-\rangle, |-+\rangle, \quad \text{and} \quad |--\rangle, \tag{3.333}$$

where $|+-\rangle$ stands for $m_1 = \frac{1}{2}, m_2 = -\frac{1}{2}$, and so forth.

2. The $\{s, m\}$ representation (or the triplet-singlet representation) based on the eigenkets of $\mathbf{S}^2$ and $S_z$:

$$|s = 1, m = \pm 1, 0\rangle, |s = 0, m = 0\rangle, \tag{3.334}$$

where $s = 1$ ($s = 0$) is referred to as the spin triplet (spin singlet).

Notice that in each set there are four base kets. The relationship between the two sets of base kets is as follows:

$$|s = 1, m = 1\rangle = |++\rangle, \tag{3.335a}$$

$$|s = 1, m = 0\rangle = \left(\frac{1}{\sqrt{2}}\right)(|+-\rangle + |-+\rangle), \tag{3.335b}$$

$$|s = 1, m = -1\rangle = |--\rangle, \tag{3.335c}$$

$$|s = 0, m = 0\rangle = \left(\frac{1}{\sqrt{2}}\right)(|+-\rangle - |-+\rangle). \tag{3.335d}$$

The right-hand side of (3.335a) tells us that we have both electrons with spin up; this situation can correspond only to $s = 1$, $m = 1$. We can obtain (3.335b) from (3.335a) by applying the ladder operator

$$S_- \equiv S_{1-} + S_{2-}$$
$$= (S_{1x} - iS_{1y}) + (S_{2x} - iS_{2y}) \qquad (3.336)$$

to both sides of (3.335a). In doing so we must remember that an electron 1 operator like $S_{1-}$ affects just the first entry of $|++\rangle$, and so on. We can write

$$S_-|s = 1, m = 1\rangle = (S_{1-} + S_{2-})|++\rangle \qquad (3.337)$$

as

$$\sqrt{(1+1)(1-1+1)}|s = 1, m = 0\rangle = \sqrt{\left(\tfrac{1}{2} + \tfrac{1}{2}\right)\left(\tfrac{1}{2} - \tfrac{1}{2} + 1\right)} \times |-+\rangle$$
$$+ \sqrt{\left(\tfrac{1}{2} + \tfrac{1}{2}\right)\left(\tfrac{1}{2} - \tfrac{1}{2} + 1\right)}|+-\rangle, \qquad (3.338)$$

which immediately leads to (3.335b). Likewise, we can obtain $|s = 1, m = -1\rangle$ by applying (3.336) once again to (3.335b). Finally, we can obtain (3.335d) by requiring it to be orthogonal to the other three kets, in particular to (3.335b).

The coefficients that appear on the right-hand side of (3.335) are the simplest example of **Clebsch–Gordan coefficients** to be discussed further at a later time. They are simply the elements of the transformation matrix that connects the $\{m_1, m_2\}$ basis to the $\{s, m\}$ basis. It is instructive to derive these coefficients in another way. Suppose we write the $4 \times 4$ matrix corresponding to

$$\mathbf{S}^2 = \mathbf{S}_1^2 + \mathbf{S}_2^2 + 2\mathbf{S}_1 \cdot \mathbf{S}_2$$
$$= \mathbf{S}_1^2 + \mathbf{S}_2^2 + 2S_{1z}S_{2z} + S_{1+}S_{2-} + S_{1-}S_{2+} \qquad (3.339)$$

using the $(m_1, m_2)$ basis. The square matrix is obviously not diagonal because an operator like $S_{1+}$ connects $|-+\rangle$ with $|++\rangle$. The unitary matrix that diagonalizes this matrix carries the $|m_1, m_2\rangle$ base kets into the $|s, m\rangle$ base kets. The elements of this unitary matrix are precisely the Clebsch–Gordan coefficients for this problem. The reader is encouraged to work out all this in detail. See Problem 3.33 at the end of this chapter.

### 3.8.2  Formal Theory of Angular-Momentum Addition

Having gained some physical insight by considering simple examples, we are now in a position to study more systematically the formal theory of angular-momentum addition. Consider two angular-momentum operators $\mathbf{J}_1$ and $\mathbf{J}_2$ in different subspaces. The components of $\mathbf{J}_1(\mathbf{J}_2)$ satisfy the usual angular-momentum commutation relations:

$$[J_{1i}, J_{1j}] = i\hbar\varepsilon_{ijk}J_{1k} \qquad (3.340a)$$

and

$$[J_{2i}, J_{2j}] = i\hbar\varepsilon_{ijk}J_{2k}. \qquad (3.340b)$$

However, we have

$$[J_{1k}, J_{2l}] = 0 \qquad (3.341)$$

between any pair of operators from different subspaces.

The infinitesimal rotation operator that affects both subspace 1 and subspace 2 is written as

$$\left(1 - \frac{i\mathbf{J}_1 \cdot \hat{\mathbf{n}}\delta\phi}{\hbar}\right) \otimes \left(1 - \frac{i\mathbf{J}_2 \cdot \hat{\mathbf{n}}\delta\phi}{\hbar}\right) = 1 - \frac{i(\mathbf{J}_1 \otimes 1 + 1 \otimes \mathbf{J}_2) \cdot \hat{\mathbf{n}}\delta\phi}{\hbar}. \qquad (3.342)$$

We define the total angular momentum by

$$\mathbf{J} \equiv \mathbf{J}_1 \otimes 1 + 1 \otimes \mathbf{J}_2, \qquad (3.343)$$

which is more commonly written as

$$\mathbf{J} = \mathbf{J}_1 + \mathbf{J}_2. \qquad (3.344)$$

The finite-angle version of (3.342) is

$$\mathcal{D}_1(R) \otimes \mathcal{D}_2(R) = \exp\left(\frac{-i\mathbf{J}_1 \cdot \hat{\mathbf{n}}\phi}{\hbar}\right) \otimes \exp\left(\frac{-i\mathbf{J}_2 \cdot \hat{\mathbf{n}}\phi}{\hbar}\right). \qquad (3.345)$$

Notice the appearance of the same axis of rotation and the same angle of rotation.

It is very important to note that the total $\mathbf{J}$ satisfies the angular-momentum commutation relations

$$[J_i, J_j] = i\hbar\varepsilon_{ijk}J_k \qquad (3.346)$$

as a direct consequence of (3.340) and (3.341). In other words, $\mathbf{J}$ is an angular momentum in the sense of Section 3.1. Physically this is reasonable because $\mathbf{J}$ is the generator for the *entire* system. Everything we learned in Section 3.5, for example, the eigenvalue spectrum of $\mathbf{J}^2$ and $J_z$ and the matrix elements of the ladder operators, also holds for the total $\mathbf{J}$.

As for the choice of base kets we have two options.

*Option A*: Simultaneous eigenkets of $\mathbf{J}_1^2$, $\mathbf{J}_2^2$, $J_{1z}$, and $J_{2z}$, denoted by $|j_1 j_2; m_1 m_2\rangle$. Obviously the four operators commute with each other. The defining equations are

$$\mathbf{J}_1^2|j_1 j_2; m_1 m_2\rangle = j_1(j_1 + 1)\hbar^2|j_1 j_2; m_1 m_2\rangle, \qquad (3.347a)$$

$$J_{1z}|j_1 j_2; m_1 m_2\rangle = m_1\hbar|j_1 j_2; m_1 m_2\rangle, \qquad (3.347b)$$

$$\mathbf{J}_2^2|j_1 j_2; m_1 m_2\rangle = j_2(j_2 + 1)\hbar^2|j_1 j_2; m_1 m_2\rangle, \qquad (3.347c)$$

$$J_{2z}|j_1 j_2; m_1 m_2\rangle = m_2\hbar|j_1 j_2; m_1 m_2\rangle. \qquad (3.347d)$$

*Option B*: Simultaneous eigenkets of $\mathbf{J}^2$, $\mathbf{J}_1^2$, $\mathbf{J}_2^2$, and $J_z$. First, note that this set of operators mutually commute. In particular, we have

$$[\mathbf{J}^2, \mathbf{J}_1^2] = 0, \qquad (3.348)$$

which can readily be seen by writing $\mathbf{J}^2$ as

$$\mathbf{J}^2 = \mathbf{J}_1^2 + \mathbf{J}_2^2 + 2J_{1z}J_{2z} + J_{1+}J_{2-} + J_{1-}J_{2+}. \qquad (3.349)$$

We use $|j_1,j_2;jm\rangle$ to denote the base kets of option B:

$$\mathbf{J}_1^2|j_1j_2;jm\rangle = j_1(j_1+1)\hbar^2|j_1j_2;jm\rangle, \tag{3.350a}$$

$$\mathbf{J}_2^2|j_1j_2;jm\rangle = j_2(j_2+1)\hbar^2|j_1j_2;jm\rangle, \tag{3.350b}$$

$$\mathbf{J}^2|j_1j_2;jm\rangle = j(j+1)\hbar^2|j_1j_2;jm\rangle, \tag{3.350c}$$

$$J_z|j_1j_2;jm\rangle = m\hbar|j_1j_2;jm\rangle. \tag{3.350d}$$

Quite often $j_1$, $j_2$ are understood, and the base kets are written simply as $|j,m\rangle$.

It is very important to note that even though

$$[\mathbf{J}^2, J_z] = 0, \tag{3.351}$$

we have

$$[\mathbf{J}^2, J_{1z}] \neq 0, \qquad [\mathbf{J}^2, J_{2z}] \neq 0, \tag{3.352}$$

as the reader may easily verify using (3.349). This means that we cannot add $\mathbf{J}^2$ to the set of operators of option A. Likewise, we cannot add $J_{1z}$ and/or $J_{2z}$ to the set of operators of option B. We have two possible sets of base kets corresponding to the two maximal sets of mutually compatible observables we have constructed.

Let us consider the unitary transformation in the sense of Section 1.5 that connects the two bases:

$$|j_1j_2;jm\rangle = \sum_{m_1}\sum_{m_2}|j_1j_2;m_1m_2\rangle\langle j_1j_2;m_1m_2|j_1j_2;jm\rangle, \tag{3.353}$$

where we have used

$$\sum_{m_1}\sum_{m_2}|j_1j_2;m_1m_2\rangle\langle j_1j_2;m_1m_2| = 1 \tag{3.354}$$

and where the right-hand side is the identity operator in the ket space of given $j_1$ and $j_2$. The elements of this transformation matrix $\langle j_1j_2;m_1m_2|j_1j_2;jm\rangle$ are Clebsch–Gordan coefficients.

There are many important properties of Clebsch–Gordan coefficients that we are now ready to study. First, the coefficients vanish unless

$$m = m_1 + m_2. \tag{3.355}$$

To prove this, first note that

$$(J_z - J_{1z} - J_{2z})|j_1j_2;jm\rangle = 0. \tag{3.356}$$

Multiplying $\langle j_1j_2;m_1m_2|$ on the left, we obtain

$$(m - m_1 - m_2)\langle j_1j_2;m_1m_2|j_1j_2;jm\rangle = 0, \tag{3.357}$$

which proves our assertion. Admire the power of the Dirac notation! It really pays to write the Clebsch–Gordan coefficients in Dirac's bracket form, as we have done.

Second, the coefficients vanish unless

$$|j_1 - j_2| \leq j \leq j_1 + j_2. \tag{3.358}$$

This property may appear obvious from the vector model of angular-momentum addition, where we visualize $\mathbf{J}$ to be the vectorial sum of $\mathbf{J}_1$ and $\mathbf{J}_2$. However, it is worth checking this point by showing that if (3.358) holds, then the dimensionality of the space spanned by $\{|j_1 j_2; m_1 m_2\rangle\}$ is the same as that of the space spanned by $\{|j_1 j_2; jm\rangle\}$. For the $(m_1, m_2)$ way of counting we obtain

$$N = (2j_1 + 1)(2j_2 + 1) \tag{3.359}$$

because for given $j_1$ there are $2j_1 + 1$ possible values of $m_1$; a similar statement is true for the other angular momentum $j_2$. As for the $(j, m)$ way of counting, we note that for each $j$, there are $2j + 1$ states, and according to (3.358), $j$ itself runs from $j_1 - j_2$ to $j_1 + j_2$, where we have assumed, without loss of generality, that $j_1 \geq j_2$. We therefore obtain

$$
\begin{aligned}
N &= \sum_{j=j_1-j_2}^{j_1+j_2} (2j+1) \\
&= \tfrac{1}{2}[\{2(j_1 - j_2) + 1\} + \{2(j_1 + j_2) + 1\}](2j_2 + 1) \\
&= (2j_1 + 1)(2j_2 + 1). \tag{3.360}
\end{aligned}
$$

Because both ways of counting give the same $N$-value, we see that (3.358) is quite consistent.[13]

The Clebsch–Gordan coefficients form a unitary matrix. Furthermore, the matrix elements are taken to be real by convention. An immediate consequence of this is that the inverse coefficient $\langle j_1 j_2; jm | j_1 j_2; m_1 m_2 \rangle$ is the same as $\langle j_1 j_2; m_1 m_2 | j_1 j_2; jm \rangle$ itself. A real unitary matrix is orthogonal, so we have the orthogonality condition

$$\sum_j \sum_m \langle j_1 j_2; m_1 m_2 | j_1 j_2; jm \rangle \langle j_1 j_2; m_1' m_2' | j_1 j_2; jm \rangle = \delta_{m_1 m_1'} \delta_{m_2 m_2'}, \tag{3.361}$$

which is obvious from the orthonormality of $\{|j_1 j_2; m_1 m_2\rangle\}$ together with the reality of the Clebsch–Gordan coefficients. Likewise, we also have

$$\sum_{m_1} \sum_{m_2} \langle j_1 j_2; m_1 m_2 | j_1 j_2; jm \rangle \langle j_1 j_2; m_1 m_2 | j_1 j_2; j'm' \rangle = \delta_{jj'} \delta_{mm'}. \tag{3.362}$$

As a special case of this we may set $j' = j, m' = m = m_1 + m_2$. We then obtain

$$\sum_{m_1, m_2 = m - m_1} |\langle j_1 j_2; m_1 m_2 | j_1 j_2; jm \rangle|^2 = 1, \tag{3.363}$$

which is just the normalization condition for $|j_1 j_2; jm\rangle$.

Some authors use somewhat different notations for the Clebsch–Gordan coefficients. Instead of $\langle j_1 j_2; m_1 m_2 | j_1 j_2; jm \rangle$ we sometimes see $\langle j_1 m_1 j_2 m_2 | j_1 j_2 jm \rangle$, $C(j_1 j_2 j; m_1 m_2 m)$,

---

[13] A complete proof of (3.358) is given in Gottfried (1966), p. 215, and also in Appendix D of this book.

$C_{j_1 j_2}(jm; m_1 m_2)$, and so on. They can also be written in terms of **Wigner's 3-j symbol**, which is occasionally found in the literature:

$$\langle j_1 j_2; m_1 m_2 | j_1 j_2; jm \rangle = (-1)^{j_1 - j_2 + m} \sqrt{2j+1} \begin{pmatrix} j_1 & j_2 & j \\ m_1 & m_2 & -m \end{pmatrix}. \tag{3.364}$$

### 3.8.3 Recursion Relations for the Clebsch–Gordan Coefficients[14]

With $j_1, j_2$, and $j$ fixed, the coefficients with different $m_1$ and $m_2$ are related to each other by **recursion relations**. We start with

$$J_\pm |j_1 j_2; jm \rangle = (J_{1\pm} + J_{2\pm}) \sum_{m_1} \sum_{m_2} |j_1 j_2; m_1 m_2 \rangle \langle j_1 j_2; m_1 m_2 | j_1 j_2; jm \rangle. \tag{3.365}$$

Using (3.191) and (3.192) we obtain (with $m_1 \to m_1', m_2 \to m_2'$)

$$\sqrt{(j \mp m)(j \pm m + 1)} |j_1 j_2; j, m \pm 1\rangle$$
$$= \sum_{m_1'} \sum_{m_2'} \left( \sqrt{(j_1 \mp m_1')(j_1 \pm m_1' + 1)} \; |j_1 j_2; m_1' \pm 1, m_2'\rangle \right.$$
$$\left. + \sqrt{(j_2 \mp m_2')(j_2 \pm m_2' + 1)} |j_1 j_2; m_1', m_2' \pm 1\rangle \right)$$
$$\times \langle j_1 j_2; m_1' m_2' | j_1 j_2; jm \rangle. \tag{3.366}$$

Our next step is to multiply by $\langle j_1 j_2; m_1 m_2 |$ on the left and use orthonormality, which means that nonvanishing contributions from the right-hand side are possible only with

$$m_1 = m_1' \pm 1, \qquad m_2 = m_2' \tag{3.367}$$

for the first term and

$$m_1 = m_1', \qquad m_2 = m_2' \pm 1 \tag{3.368}$$

for the second term. In this manner we obtain the desired recursion relations:

$$\sqrt{(j \mp m)(j \pm m + 1)} \langle j_1 j_2; m_1 m_2 | j_1 j_2; j, m \pm 1\rangle$$
$$= \sqrt{(j_1 \mp m_1 + 1)(j_1 \pm m_1)} \langle j_1 j_2; m_1 \mp 1, m_2 | j_1 j_2; jm \rangle$$
$$+ \sqrt{(j_2 \mp m_2 + 1)(j_2 \pm m_2)} \langle j_1 j_2; m_1, m_2 \mp 1 | j_1 j_2; jm \rangle. \tag{3.369}$$

It is important to note that because the $J_\pm$ operators have shifted the $m$-values, the nonvanishing condition (3.355) for the Clebsch–Gordan coefficients has now become [when applied to (3.369)]

$$m_1 + m_2 = m \pm 1. \tag{3.370}$$

We can appreciate the significance of the recursion relations by looking at (3.369) in an $m_1 m_2$-plane. The $J_+$ recursion relation (upper sign) tells us that the coefficient at $(m_1, m_2)$

---

[14] More detailed discussion of Clebsch–Gordan and Racah coefficients, recoupling, and the like is given in Edmonds (1974), for instance.

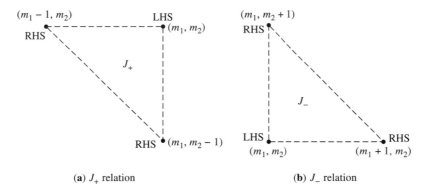

**(a)** $J_+$ relation                    **(b)** $J_-$ relation

**Fig. 3.8**   $m_1m_2$-plane showing the Clebsch–Gordan coefficients related by the recursion relations (3.369).

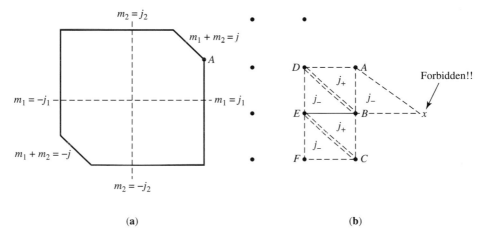

**(a)**                                          **(b)**

**Fig. 3.9**   Use of the recursion relations to obtain the Clebsch–Gordan coefficients.

is related to the coefficients at $(m_1 - 1, m_2)$ and $(m_1, m_2 - 1)$, as shown in Figure 3.8a. Likewise, the $J_-$ recursion relation (lower sign) relates the three coefficients whose $m_1, m_2$ values are given in Figure 3.8b.

Recursion relations (3.369), together with normalization condition (3.363), almost uniquely determine all Clebsch–Gordan coefficients. (We say "almost uniquely" because certain sign conventions have yet to be specified.) Our strategy is as follows. We go back to the $m_1m_2$-plane, again for fixed $j_1, j_2$, and $j$, and plot the boundary of the allowed region determined by

$$|m_1| \leq j_1, \qquad |m_2| \leq j_2, \qquad -j \leq m_1 + m_2 \leq j; \qquad (3.371)$$

see Figure 3.9a. We may start with the upper right-hand corner, denoted by $A$. Because we work near $A$ at the start, a more detailed "map" is in order; see Figure 3.6b. We apply the $J_-$ recursion relation (3.369) (lower sign), with $(m_1, m_2 + 1)$ corresponding to $A$. Observe now that the recursion relation connects $A$ with only $B$ because the site corresponding to $(m_1 + 1, m_2)$ is forbidden by $m_1 \leq j_1$. As a result, we can obtain the Clebsch–Gordan coefficient of $B$ in terms of the coefficient of $A$. Next, we form a $J_+$ triangle made up of $A$,

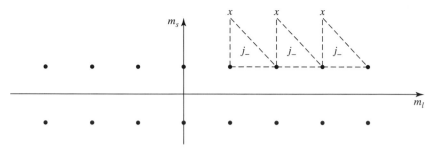

**Fig. 3.10**   Recursion relations used to obtain the Clebsch–Gordan coefficients for $j_i = l$ and $j_2 = s = \frac{1}{2}$.

$B$, and $D$. This enables us to obtain the coefficient of $D$ once the coefficient of $A$ is specified. We can continue in this fashion: Knowing $B$ and $D$, we can get to $E$; knowing $B$ and $E$ we can get to $C$, and so on. With enough patience we can obtain the Clebsch-Gordan coefficient of every site in terms of the coefficient of starting site, $A$. For overall normalization we use (3.363). The final overall sign is fixed by convention. (See the following example.)

As an important practical example we consider the problem of adding the orbital and spin-angular momenta of a single spin $\frac{1}{2}$ particle. We have

$$
\begin{aligned}
j_1 &= l \ (\text{integer}), & m_1 &= m_l, \\
j_2 &= s = \tfrac{1}{2}, & m_2 &= m_s = \pm\tfrac{1}{2}.
\end{aligned}
\tag{3.372}
$$

The allowed values of $j$ are given by

$$
j = l \pm \tfrac{1}{2}, \quad l > 0; \qquad j = \tfrac{1}{2}, \quad l = 0.
\tag{3.373}
$$

So for each $l$ there are two possible $j$-values; for example, for $l = 1$ ($p$ state) we get, in spectroscopic notation, $p_{3/2}$ and $p_{1/2}$, where the subscript refers to $j$. The $m_1 m_2$-plane, or better the $m_l m_s$-plane, of this problem is particularly simple. The allowed sites form only two rows: the upper row for $m_s = \frac{1}{2}$ and the lower row for $m_s = -\frac{1}{2}$ (see Figure 3.10). Specifically, we work out the case $j = l + \frac{1}{2}$. Because $m_s$ cannot exceed $\frac{1}{2}$, we can use the $J_-$ recursion in such a way that we always stay in the upper row ($m_2 = m_s = \frac{1}{2}$), while the $m_l$-value changes by one unit each time we consider a new $J_-$ triangle. Suppressing $j_1 = l, j_2 = \frac{1}{2}$, in writing the Clebsch–Gordan coefficient, we obtain from (3.369) (lower sign)

$$
\sqrt{\left(l + \tfrac{1}{2} + m + 1\right)\left(l + \tfrac{1}{2} - m\right)} \left\langle m - \tfrac{1}{2}, \tfrac{1}{2} \middle| l + \tfrac{1}{2}, m \right\rangle
$$
$$
= \sqrt{\left(l + m + \tfrac{1}{2}\right)\left(l - m + \tfrac{1}{2}\right)} \left\langle m + \tfrac{1}{2}, \tfrac{1}{2} \middle| l + \tfrac{1}{2}, m + 1 \right\rangle,
\tag{3.374}
$$

where we have used

$$
m_1 = m_l = m - \tfrac{1}{2}, \qquad m_2 = m_s = \tfrac{1}{2}.
\tag{3.375}
$$

In this way we can move horizontally by one unit:

$$
\left\langle m - \frac{1}{2}, \frac{1}{2} \middle| l + \frac{1}{2}, m \right\rangle = \sqrt{\frac{l + m + \frac{1}{2}}{l + m + \frac{3}{2}}} \left\langle m + \frac{1}{2}, \frac{1}{2} \middle| l + \frac{1}{2}, m + 1 \right\rangle.
\tag{3.376}
$$

We can in turn express $\langle m + \frac{1}{2}, \frac{1}{2} | l + \frac{1}{2}, m+1 \rangle$ in terms of $\langle m + \frac{3}{2}, \frac{1}{2} | l + \frac{1}{2}, m+2 \rangle$, and so forth. Clearly, this procedure can be continued until $m_l$ reaches $l$, the maximum possible value:

$$
\begin{aligned}
\left\langle m - \frac{1}{2}, \frac{1}{2} \middle| l + \frac{1}{2}, m \right\rangle &= \sqrt{\frac{l+m+\frac{1}{2}}{l+m+\frac{3}{2}}} \sqrt{\frac{l+m+\frac{3}{2}}{l+m+\frac{5}{2}}} \left\langle m + \frac{3}{2}, \frac{1}{2} \middle| l + \frac{1}{2}, m+2 \right\rangle \\
&= \sqrt{\frac{l+m+\frac{1}{2}}{l+m+\frac{3}{2}}} \sqrt{\frac{l+m+\frac{3}{2}}{l+m+\frac{5}{2}}} \sqrt{\frac{l+m+\frac{5}{2}}{l+m+\frac{7}{2}}} \\
&\quad \times \left\langle m + \frac{5}{2}, \frac{1}{2} \middle| l + \frac{1}{2}, m+3 \right\rangle \\
&\quad \vdots \\
&= \sqrt{\frac{l+m+\frac{1}{2}}{2l+1}} \left\langle l, \frac{1}{2} \middle| l + \frac{1}{2}, l + \frac{1}{2} \right\rangle.
\end{aligned}
\tag{3.377}
$$

Consider the angular-momentum configuration in which $m_l$ and $m_s$ are both maximal, that is, $l$ and $\frac{1}{2}$, respectively. The total $m = m_l + m_s$ is $l + \frac{1}{2}$, which is possible only for $j = l + \frac{1}{2}$ and not for $j = l - \frac{1}{2}$. So $|m_l = l, m_s = \frac{1}{2}\rangle$ must be equal to $|j = l + \frac{1}{2}, m = l + \frac{1}{2}\rangle$, up to a phase factor. We take this phase factor to be real and positive by convention. With this choice we have

$$
\left\langle l, \frac{1}{2} \middle| l + \frac{1}{2}, l + \frac{1}{2} \right\rangle = 1.
\tag{3.378}
$$

Returning to (3.377), we finally obtain

$$
\left\langle m - \frac{1}{2}, \frac{1}{2} \middle| l + \frac{1}{2}, m \right\rangle = \sqrt{\frac{l+m+\frac{1}{2}}{2l+1}}.
\tag{3.379}
$$

But this is only about one-fourth of the story. We must still determine the value of the question marks that appear in the following:

$$
\begin{aligned}
\left| j = l + \frac{1}{2}, m \right\rangle &= \sqrt{\frac{l+m+\frac{1}{2}}{2l+1}} \left| m_l = m - \frac{1}{2}, m_s = \frac{1}{2} \right\rangle \\
&\quad + ? \left| m_l = m + \frac{1}{2}, m_s = -\frac{1}{2} \right\rangle, \\
\left| j = l - \frac{1}{2}, m \right\rangle &= ? \left| m_l = m - \frac{1}{2}, m_s = \frac{1}{2} \right\rangle + ? \left| m_l = m + \frac{1}{2}, m_s = -\frac{1}{2} \right\rangle.
\end{aligned}
\tag{3.380}
$$

We note that the transformation matrix with fixed $m$ from the $(m_l, m_s)$ basis to the $(j, m)$ basis is, because of orthogonality, expected to have the form

$$
\begin{pmatrix} \cos\alpha & \sin\alpha \\ -\sin\alpha & \cos\alpha \end{pmatrix}.
\tag{3.381}
$$

Comparison with (3.380) shows that $\cos\alpha$ is (3.379) itself; so we can readily determine $\sin\alpha$ up to a sign ambiguity:

$$\sin^2\alpha = 1 - \frac{\left(l+m+\frac{1}{2}\right)}{(2l+1)} = \frac{\left(l-m+\frac{1}{2}\right)}{(2l+1)}. \tag{3.382}$$

We claim that $\langle m_l = m+\frac{1}{2}, m_s = -\frac{1}{2}|j = l+\frac{1}{2}, m\rangle$ must be positive because all $j = l+\frac{1}{2}$ states are reachable by applying the $J_-$ operator successively to $|j = l+\frac{1}{2}, m = l+\frac{1}{2}\rangle$, and the matrix elements of $J_-$ are always positive by convention. So the $2 \times 2$ transformation matrix (3.381) can be only

$$\begin{pmatrix} \sqrt{\dfrac{l+m+\frac{1}{2}}{2l+1}} & \sqrt{\dfrac{l-m+\frac{1}{2}}{2l+1}} \\[3mm] -\sqrt{\dfrac{l-m+\frac{1}{2}}{2l+1}} & \sqrt{\dfrac{l+m+\frac{1}{2}}{2l+1}} \end{pmatrix}. \tag{3.383}$$

We define **spin-angular functions** in two-component form as follows:

$$\begin{aligned} \mathscr{Y}_l^{j=l\pm 1/2, m} &= \pm\sqrt{\frac{l\pm m+\frac{1}{2}}{2l+1}}\, Y_l^{m-1/2}(\theta,\phi)\chi_+ \\[2mm] &+ \sqrt{\frac{l\mp m+\frac{1}{2}}{2l+1}}\, Y_l^{m+1/2}(\theta,\phi)\chi_- \\[2mm] &= \frac{1}{\sqrt{2l+1}}\begin{pmatrix} \pm\sqrt{l\pm m+\frac{1}{2}}\, Y_l^{m-1/2}(\theta,\phi) \\[2mm] \sqrt{l\mp m+\frac{1}{2}}\, Y_l^{m+1/2}(\theta,\phi) \end{pmatrix}. \end{aligned} \tag{3.384}$$

They are, by construction, simultaneous eigenfunctions of $\mathbf{L}^2$, $\mathbf{S}^2$, $\mathbf{J}^2$, and $J_z$. They are also eigenfunctions of $\mathbf{L}\cdot\mathbf{S}$ but $\mathbf{L}\cdot\mathbf{S}$, being just

$$\mathbf{L}\cdot\mathbf{S} = \left(\frac{1}{2}\right)\left(\mathbf{J}^2 - \mathbf{L}^2 - \mathbf{S}^2\right), \tag{3.385}$$

is not independent. Indeed, its eigenvalue can easily be computed as follows:

$$\left(\frac{\hbar^2}{2}\right)\left[j(j+1) - l(l+1) - \frac{3}{4}\right] = \begin{cases} \dfrac{l\hbar^2}{2} & \text{for } j = l+\frac{1}{2}, \\[3mm] -\dfrac{(l+1)\hbar^2}{2} & \text{for } j = l-\frac{1}{2}. \end{cases} \tag{3.386}$$

### 3.8.4 Clebsch–Gordan Coefficients and Rotation Matrices

Angular-momentum addition may be discussed from the point of view of rotation matrices. Consider the rotation operator $\mathscr{D}^{(j_1)}(R)$ in the ket space spanned by the angular-momentum eigenkets with eigenvalue $j_1$. Likewise, consider $\mathscr{D}^{(j_2)}(R)$. The product $\mathscr{D}^{(j_1)} \otimes \mathscr{D}^{(j_2)}$ is reducible in the sense that after suitable choice of base kets, its matrix representation can take the following form:

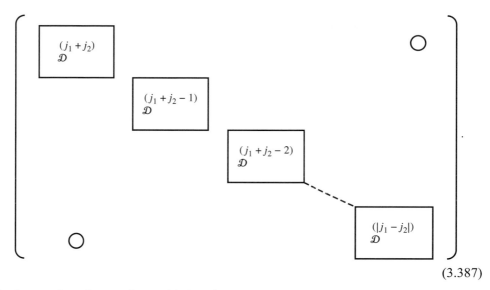

$$(3.387)$$

In the notation of group theory this is written as

$$\mathscr{D}^{(j_1)} \otimes \mathscr{D}^{(j_2)} = \mathscr{D}^{(j_1+j_2)} \oplus \mathscr{D}^{(j_1+j_2-1)} \oplus \cdots \oplus \mathscr{D}^{(|j_1-j_2|)}. \tag{3.388}$$

In terms of the elements of rotation matrices, we have an important expansion known as the **Clebsch–Gordan series**:

$$\mathscr{D}^{(j_1)}_{m_1 m_1'}(R)\mathscr{D}^{(j_2)}_{m_2 m_2'}(R) = \sum_j \sum_m \sum_{m'} \langle j_1 j_2; m_1 m_2 | j_1 j_2; jm \rangle$$
$$\times \langle j_1 j_2; m_1' m_2' | j_1 j_2; jm' \rangle \mathscr{D}^{(j)}_{mm'}(R), \tag{3.389}$$

where the $j$-sum runs from $|j_1 - j_2|$ to $j_1 + j_2$ . The proof of this equation follows. First, note that the left-hand side of (3.389) is the same as

$$\langle j_1 j_2; m_1 m_2 | \mathscr{D}(R) | j_1 j_2; m_1' m_2' \rangle = \langle j_1 m_1 | \mathscr{D}(R) | j_1 m_1' \rangle \langle j_2 m_2 | \mathscr{D}(R) | j_2 m_2' \rangle$$
$$= \mathscr{D}^{(j_1)}_{m_1 m_1'}(R)\mathscr{D}^{(j_2)}_{m_2 m_2'}(R). \tag{3.390}$$

But the same matrix element is also computable by inserting a complete set of states in the $(j, m)$ basis. Thus

$$\langle j_1 j_2; m_1 m_2 | \mathscr{D}(R) | j_1 j_2; m_1' m_2' \rangle$$
$$= \sum_j \sum_m \sum_{j'} \sum_{m'} \langle j_1 j_2; m_1 m_2 | j_1 j_2; jm \rangle \langle j_1 j_2; jm | \mathscr{D}(R) | j_1 j_2; j'm' \rangle$$
$$\times \langle j_1 j_2; j'm' | j_1 j_2; m_1' m_2' \rangle$$
$$= \sum_j \sum_m \sum_{j'} \sum_{m'} \langle j_1 j_2; m_1 m_2 | j_1 j_2; jm \rangle \mathscr{D}^{(j)}_{mm'}(R)\delta_{jj'}$$
$$\times \langle j_1 j_2; m_1' m_2' | j_1 j_2; j'm' \rangle, \tag{3.391}$$

which is just the right-hand side of (3.389).

As an interesting application of (3.389), we derive an important formula for an integral involving three spherical harmonics. First, recall the connection between $\mathscr{D}^{(l)}_{m0}$ and $Y_l^{m*}$

given by (3.261). Letting $j_1 \to l_1, j_2 \to l_2, m'_1 \to 0, m'_2 \to 0$ (hence $m' \to 0$) in (3.389), we obtain, after complex conjugation,

$$Y_{l_1}^{m_1}(\theta,\phi) Y_{l_2}^{m_2}(\theta,\phi) = \frac{\sqrt{(2l_1+1)(2l_2+1)}}{4\pi} \sum_{l'} \sum_{m'} \langle l_1 l_2; m_1 m_2 | l_1 l_2; l'm' \rangle$$

$$\times \langle l_1 l_2; 00 | l_1 l_2; l'0 \rangle \sqrt{\frac{4\pi}{2l'+1}} Y_{l'}^{m'}(\theta,\phi). \tag{3.392}$$

We multiply both sides by $Y_l^{m*}(\theta,\phi)$ and integrate over solid angles. The summations drop out because of the orthogonality of spherical harmonics, and we are left with

$$\int d\Omega \, Y_l^{m*}(\theta,\phi) Y_{l_1}^{m_1}(\theta,\phi) Y_{l_2}^{m_2}(\theta,\phi)$$

$$= \sqrt{\frac{(2l_1+1)(2l_2+1)}{4\pi(2l+1)}} \langle l_1 l_2; 00 | l_1 l_2; l0 \rangle \langle l_1 l_2; m_1 m_2 | l_1 l_2; lm \rangle. \tag{3.393}$$

The square root factor times the first Clebsch–Gordan coefficient is independent of orientations, that is, of $m_1$ and $m_2$. The second Clebsch–Gordan coefficient is the one appropriate for adding $l_1$ and $l_2$ to obtain total $l$. Equation (3.393) turns out to be a special case of the Wigner–Eckart theorem to be derived in Section 3.11. This formula is extremely useful in evaluating multipole matrix elements in atomic and nuclear spectroscopy.

## 3.9 Schwinger's Oscillator Model of Angular Momentum

### 3.9.1 Angular Momentum and Uncoupled Oscillators

There exists a very interesting connection between the algebra of angular momentum and the algebra of two independent (that is, uncoupled) oscillators, which was worked out in J. Schwinger's notes. See Biedenharn and Van Dam (1965), p. 229. Let us consider two simple harmonic oscillators, which we call the *plus type* and the *minus type*. We have the annihilation and creation operators, denoted by $a_+$ and $a_+^\dagger$ for the plus-type oscillator; likewise, we have $a_-$ and $a_-^\dagger$ for the minus-type oscillator. We also define the number operators $N_+$ and $N_-$ as follows:

$$N_+ \equiv a_+^\dagger a_+, \quad N_- \equiv a_-^\dagger a_-. \tag{3.394}$$

We assume that the usual commutation relations among $a$, $a^\dagger$, and $N$ hold for oscillators of the same type (see Section 2.3):

$$[a_+, a_+^\dagger] = 1, \qquad [a_-, a_-^\dagger] = 1, \tag{3.395a}$$

$$[N_+, a_+] = -a_+, \qquad [N_-, a_-] = -a_-, \tag{3.395b}$$

$$[N_+, a_+^\dagger] = a_+^\dagger, \qquad [N_-, a_-^\dagger] = a_-^\dagger. \tag{3.395c}$$

However, we assume that any pair of operators between different oscillators commute:

$$[a_+, a_-^\dagger] = [a_-, a_+^\dagger] = 0 \tag{3.396}$$

and so forth. So it is in this sense that we say the two oscillators are uncoupled.

Because $N_+$ and $N_-$ commute by virtue of (3.396), we can build up simultaneous eigenkets of $N_+$ and $N_-$ with eigenvalues $n_+$ and $n_-$, respectively. So we have the following eigenvalue equations for $N_\pm$:

$$N_+|n_+, n_-\rangle = n_+|n_+, n_-\rangle, \qquad N_-|n_+, n_-\rangle = n_-|n_+, n_-\rangle. \tag{3.397}$$

In complete analogy with (2.137) and (2.138), the creation and annihilation operators, $a_\pm^\dagger$ and $a_\pm$, act on $|n_+, n_-\rangle$ as follows:

$$a_+^\dagger|n_+, n_-\rangle = \sqrt{n_+ + 1}|n_+ + 1, n_-\rangle, \quad a_-^\dagger|n_+, n_-\rangle = \sqrt{n_- + 1}|n_+, n_- + 1\rangle, \tag{3.398a}$$

$$a_+|n_+, n_-\rangle = \sqrt{n_+}|n_+ - 1, n_-\rangle, \quad a_-|n_+, n_-\rangle = \sqrt{n_-}|n_+, n_- - 1\rangle. \tag{3.398b}$$

We can obtain the most general eigenkets of $N_+$ and $N_-$ by applying $a_+^\dagger$ and $a_-^\dagger$ successively to the **vacuum ket** defined by

$$a_+|0,0\rangle = 0, \quad a_-|0,0\rangle = 0. \tag{3.399}$$

In this way we obtain

$$|n_+, n_-\rangle = \frac{(a_+^\dagger)^{n_+}(a_-^\dagger)^{n_-}}{\sqrt{n_+!}\sqrt{n_-!}}|0,0\rangle. \tag{3.400}$$

Next, we *define*

$$J_+ \equiv \hbar a_+^\dagger a_-, \qquad J_- \equiv \hbar a_-^\dagger a_+, \tag{3.401a}$$

and

$$J_z \equiv \left(\frac{\hbar}{2}\right)\left(a_+^\dagger a_+ - a_-^\dagger a_-\right) = \left(\frac{\hbar}{2}\right)(N_+ - N_-). \tag{3.401b}$$

We can readily prove that these operators satisfy the angular-momentum commutation relations of the usual form

$$[J_z, J_\pm] = \pm\hbar J_\pm, \tag{3.402a}$$

$$[J_+, J_-] = 2\hbar J_z. \tag{3.402b}$$

For example, we prove (3.402) as follows:

$$\hbar^2[a_+^\dagger a_-, a_-^\dagger a_+] = \hbar^2 a_+^\dagger a_- a_-^\dagger a_+ - \hbar^2 a_-^\dagger a_+ a_+^\dagger a_-$$
$$= \hbar^2 a_+^\dagger (a_-^\dagger a_- + 1)a_+ - \hbar^2 a_-^\dagger (a_+^\dagger a_+ + 1)a_-$$
$$= \hbar^2 (a_+^\dagger a_+ - a_-^\dagger a_-) = 2\hbar J_z. \tag{3.403}$$

Defining the *total N* to be

$$N \equiv N_+ + N_- = a_+^\dagger a_+ + a_-^\dagger a_-, \tag{3.404}$$

we can also prove

$$\mathbf{J}^2 \equiv J_z^2 + \left(\frac{1}{2}\right)(J_+J_- + J_-J_+)$$

$$= \left(\frac{\hbar^2}{2}\right) N\left(\frac{N}{2}+1\right), \tag{3.405}$$

which is left as an exercise.

What are the physical interpretations of all this? We associate spin up ($m = \frac{1}{2}$) with one quantum unit of the plus-type oscillator and spin down ($m = -\frac{1}{2}$) with one quantum unit of the minus-type oscillator. If you like, you may imagine one spin $\frac{1}{2}$ "particle" with spin up (down) with each quantum unit of the plus- (minus-) type oscillator. The eigenvalues $n_+$ and $n_-$ are just the number of spins up and spins down, respectively. The meaning of $J_+$ is that it destroys one unit of spin down with the $z$-component of spin-angular momentum $-\hbar/2$ and creates one unit of spin up with the $z$-component of spin-angular momentum $+\hbar/2$; the $z$-component of angular momentum is therefore increased by $\hbar$. Likewise $J_-$ destroys one unit of spin up and creates one unit of spin down; the $z$-component of angular momentum is therefore decreased by $\hbar$. As for the $J_z$ operator, it simply counts $\hbar/2$ times the difference of $n_+$ and $n_-$, just the $z$-component of the total angular momentum. With (3.398) at our disposal we can easily examine how $J_\pm$ and $J_z$ act on $|n_+, n_-\rangle$ as follows:

$$J_+|n_+, n_-\rangle = \hbar a_+^\dagger a_- |n_+, n_-\rangle = \sqrt{n_-(n_+ + 1)}\hbar|n_+ + 1, n_- - 1\rangle, \tag{3.406a}$$

$$J_-|n_+, n_-\rangle = \hbar a_-^\dagger a_+ |n_+, n_-\rangle = \sqrt{n_+(n_- + 1)}\hbar|n_+ - 1, n_- + 1\rangle, \tag{3.406b}$$

$$J_z|n_+, n_-\rangle = \left(\frac{\hbar}{2}\right)(N_+ - N_-)|n_+, n_-\rangle = \left(\frac{1}{2}\right)(n_+ - n_-)\hbar|n_+, n_-\rangle. \tag{3.406c}$$

Notice that in all these operations, the sum $n_+ + n_-$, which corresponds to the total number of spin $\frac{1}{2}$ particles, remains unchanged.

Observe now that (3.406a), (3.406b), and (3.406c) reduce to the familiar expressions for the $J_\pm$ and $J_z$ operators we derived in Section 3.5, provided we substitute

$$n_+ \to j + m, \quad n_- \to j - m. \tag{3.407}$$

The square root factors in (3.406a) and (3.406b) change to

$$\sqrt{n_-(n_+ + 1)} \to \sqrt{(j-m)(j+m+1)},$$
$$\sqrt{n_+(n_- + 1)} \to \sqrt{(j+m)(j-m+1)}, \tag{3.408}$$

which are exactly the square root factors appearing in (3.191) and (3.193).

Notice also that the eigenvalue of the $\mathbf{J}^2$ operator defined by (3.405) changes as follows:

$$\left(\frac{\hbar^2}{2}\right)(n_+ + n_-)\left[\frac{(n_+ + n_-)}{2} + 1\right] \to \hbar^2 j(j+1). \tag{3.409}$$

All this may not be too surprising because we have already proved that $J_\pm$ and $\mathbf{J}^2$ operators we constructed out of the oscillator operators satisfy the usual angular-momentum commutation relations. But it is instructive to see in an explicit manner the connection between the oscillator matrix elements and angular-momentum matrix elements. In any case, it is now natural to use

$$j \equiv \frac{(n_+ + n_-)}{2}, \quad m \equiv \frac{(n_+ - n_-)}{2} \tag{3.410}$$

in place of $n_+$ and $n_-$ to characterize simultaneous eigenkets of $\mathbf{J}^2$ and $J_z$. According to (3.406a) the action of $J_+$ changes $n_+$ into $n_+ + 1$, $n_-$ into $n_- - 1$, which means that $j$ is unchanged and $m$ goes into $m + 1$. Likewise, we see that the $J_-$ operator that changes $n_+$ into $n_+ - 1$ and $n_-$ into $n_+ - 1$ lowers $m$ by one unit without changing $j$. We can now write as (3.400) for the most general $N_+$, $N_-$ eigenket

$$|j, m\rangle = \frac{(a_+^\dagger)^{j+m} (a_-^\dagger)^{j-m}}{\sqrt{(j+m)!\,(j-m)!}} |0\rangle, \tag{3.411}$$

where we have used $|0\rangle$ for the vacuum ket, earlier denoted by $|0, 0\rangle$.

A special case of (3.411) is of interest. Let us set $m = j$, which physically means that the eigenvalue of $J_z$ is as large as possible for a given $j$. We have

$$|j, j\rangle = \frac{(a_+^\dagger)^{2j}}{\sqrt{(2j)!}} |0\rangle. \tag{3.412}$$

We can imagine this state to be built up of $2j$ spin $\frac{1}{2}$ particles with their spins all pointing in the positive $z$-direction.

In general, we note that a complicated object of high $j$ can be visualized as being made up of primitive spin $\frac{1}{2}$ particles, $j + m$ of them with spin up and the remaining $j - m$ of them with spin down. This picture is extremely convenient even though we obviously cannot always regard an object of angular momentum $j$ literally as a composite system of spin $\frac{1}{2}$ particles. All we are saying is that, *as far as the transformation properties under rotations are concerned*, we can visualize any object of angular momentum $j$ as a composite system of $2j$ spin $\frac{1}{2}$ particles formed in the manner indicated by (3.411).

From the point of view of angular-momentum addition developed in the previous section, we can add the spins of $2j$ spin $\frac{1}{2}$ particles to obtain states with angular momentum $j, j - 1, j - 2, \ldots$. As a simple example, we can add the spin-angular momenta of two spin $\frac{1}{2}$ particles to obtain a total angular momentum of zero as well as one. In Schwinger's oscillator scheme, however, we obtain only states with angular momentum $j$ when we start with $2j$ spin $\frac{1}{2}$ particles. In the language of permutation symmetry to be developed in Chapter 7, only totally symmetrical states are constructed by this method. The primitive spin $\frac{1}{2}$ particles appearing here are actually *bosons*! This method is quite adequate if our purpose is to examine the properties under rotations of states characterized by $j$ and $m$ without asking how such states are built up initially.

The reader who is familiar with isospin in nuclear and particle physics may note that what we are doing here provides a new insight into the isospin (or isotopic spin) formalism. The operator $J_+$ that destroys one unit of the minus type and creates one unit of the plus type is completely analogous to the isospin ladder operator $T_+$ (sometimes denoted by $I_+$) that annihilates a neutron (isospin down) and creates a proton (isospin up), thus raising the

$z$-component of isospin by one unit. In contrast, $J_z$ is analogous to $T_z$, which simply counts the difference between the number of protons and neutrons in nuclei.

## 3.9.2  Explicit Formula for Rotation Matrices

Schwinger's scheme can be used to derive, in a very simple way, a closed formula for rotation matrices, first obtained by E. P. Wigner using a similar (but not identical) method. We apply the rotation operator $\mathscr{D}(R)$ to $|j,m\rangle$, written as (3.411). In the Euler angle notation the only nontrivial rotation is the second one about the $y$-axis, so we direct our attention to

$$\mathscr{D}(R) = \mathscr{D}(\alpha,\beta,\gamma)|_{\alpha=\gamma=0} = \exp\left(\frac{-iJ_y\beta}{\hbar}\right). \tag{3.413}$$

We have

$$\mathscr{D}(R)|j,m\rangle = \frac{[\mathscr{D}(R)a_+^\dagger \mathscr{D}^{-1}(R)]^{j+m}[\mathscr{D}(R)a_-^\dagger \mathscr{D}^{-1}(R)]^{j-m}}{\sqrt{(j+m)!\,(j-m)!}} \mathscr{D}(R)|0\rangle. \tag{3.414}$$

Now, $\mathscr{D}(R)$ acting on $|0\rangle$ just reproduces $|0\rangle$ because, by virtue of (3.399), only the leading term, 1, in the expansion of exponential (3.413) contributes. So

$$\mathscr{D}(R)a_\pm^\dagger \mathscr{D}^{-1}(R) = \exp\left(\frac{-iJ_y\beta}{\hbar}\right) a_\pm^\dagger \exp\left(\frac{iJ_y\beta}{\hbar}\right). \tag{3.415}$$

Thus we may use formula (2.168). Letting

$$G \to \frac{-J_y}{\hbar}, \quad \lambda \to \beta \tag{3.416}$$

in (2.168), we realize that we must look at various commutators, namely,

$$\left[\frac{-J_y}{\hbar}, a_+^\dagger\right] = \left(\frac{1}{2i}\right)[a_-^\dagger a_+, a_+^\dagger] = \left(\frac{1}{2i}\right) a_-^\dagger,$$

$$\left[\frac{-J_y}{\hbar}, \left[\frac{-J_y}{\hbar}, a_+^\dagger\right]\right] = \left[\frac{-J_y}{\hbar}, \frac{a_-^\dagger}{2i}\right] = \left(\frac{1}{4}\right) a_+^\dagger \tag{3.417}$$

and so forth. Clearly, we always obtain either $a_+^\dagger$ or $a_-^\dagger$. Collecting terms, we get

$$\mathscr{D}(R)a_+^\dagger \mathscr{D}^{-1}(R) = a_+^\dagger \cos\left(\frac{\beta}{2}\right) + a_-^\dagger \sin\left(\frac{\beta}{2}\right). \tag{3.418}$$

Likewise,

$$\mathscr{D}(R)a_-^\dagger \mathscr{D}^{-1}(R) = a_-^\dagger \cos\left(\frac{\beta}{2}\right) - a_+^\dagger \sin\left(\frac{\beta}{2}\right). \tag{3.419}$$

Actually this result is not surprising. After all, the basic spin-up state is supposed to transform as

$$a_+^\dagger|0\rangle \to \cos\left(\frac{\beta}{2}\right) a_+^\dagger|0\rangle + \sin\left(\frac{\beta}{2}\right) a_-^\dagger|0\rangle \tag{3.420}$$

under a rotation about the $y$-axis. Substituting (3.418) and (3.419) into (3.414) and recalling the binomial theorem

$$(x+y)^N = \sum_k \frac{N!\, x^{N-k} y^k}{(N-k)!\, k!},\tag{3.421}$$

we obtain

$$\mathscr{D}(\alpha=0,\beta,\gamma=0|j,m) = \sum_k \sum_l \frac{(j+m)!\,(j-m)!}{(j+m-k)!\,k!\,(j-m-l)!\,l!}$$
$$\times \frac{[a_+^\dagger \cos(\beta/2)]^{j+m-k}[a_-^\dagger \sin(\beta/2)]^k}{\sqrt{(j+m)!\,(j-m)!}}$$
$$\times [-a_+^\dagger \sin(\beta/2)]^{j-m-l}[a_-^\dagger \cos(\beta/2)]^l|0\rangle.\tag{3.422}$$

We may compare (3.422) with

$$\mathscr{D}(\alpha=0,\beta,\gamma=0)|j,m\rangle = \sum_{m'} |j,m'\rangle d_{m'm}^{(j)}(\beta)$$

$$= \sum_{m'} d_{m'm}^{(j)}(\beta) \frac{(a_+^\dagger)^{j+m'}(a_-^\dagger)^{j-m'}}{\sqrt{(j+m')!\,(j-m')!}}|0\rangle.\tag{3.423}$$

We can obtain an explicit form for $d_{m'm}^{(j)}(\beta)$ by equating the coefficients of powers of $a_+^\dagger$ in (3.422) and (3.423). Specifically, we want to compare $a_+^\dagger$ raised to $j+m'$ in (3.423) with $a_+^\dagger$ raised to $2j-k-l$, so we identify

$$l = j-k-m'.\tag{3.424}$$

We are seeking $d_{m'm}(\beta)$ with $m'$ fixed. The $k$-sum and the $l$-sum in (3.422) are not independent of each other; we eliminate $l$ in favor of $k$ by taking advantage of (3.424). As for the powers of $a_-^\dagger$, we note that $a_-^\dagger$ raised to $j-m'$ in (3.423) automatically matches with $a_-^\dagger$ raised to $k+l$ in (3.422) when (3.424) is imposed. The last step is to identify the exponents of $\cos(\beta/2)$, $\sin(\beta/2)$, and $(-1)$, which are, respectively,

$$j+m-k+l = 2j-2k+m-m',\tag{3.425a}$$
$$k+j-m-l = 2k-m+m',\tag{3.425b}$$
$$j-m-l = k-m+m',\tag{3.425c}$$

where we have used (3.424) to eliminate $l$. In this way we obtain **Wigner's formula** for $d_{m'm}^{(j)}(\beta)$:

$$d_{m'm}^{(j)}(\beta) = \sum_k (-1)^{k-m+m'} \frac{\sqrt{(j+m)!\,(j-m)!\,(j+m')!\,(j-m')!}}{(j+m-k)!\,k!\,(j-k-m')!\,(k-m+m')!}$$
$$\times \left(\cos\frac{\beta}{2}\right)^{2j-2k+m-m'} \left(\sin\frac{\beta}{2}\right)^{2k-m+m'},\tag{3.426}$$

where we take the sum over $k$ whenever none of the arguments of factorials in the denominator is negative.

# 3.10 Spin Correlation Measurements and Bell's Inequality

### 3.10.1 Correlations in Spin-Singlet States

The simplest example of angular-momentum addition we encountered in Section 3.8 was concerned with a composite system made up of spin $\frac{1}{2}$ particles. In this section we use such a system to illustrate one of the most astonishing consequences of quantum mechanics.

Consider a two-electron system in a spin-singlet state, that is, with a total spin of zero. We have already seen that the state ket can be written as [see (3.335d)]

$$|\text{spin singlet}\rangle = \left(\frac{1}{\sqrt{2}}\right)(|\hat{\mathbf{z}}+;\hat{\mathbf{z}}-\rangle - |\hat{\mathbf{z}}-;\hat{\mathbf{z}}+\rangle), \tag{3.427}$$

where we have explicitly indicated the quantization direction. Recall that $|\hat{\mathbf{z}}+;\hat{\mathbf{z}}-\rangle$ means that electron 1 is in the spin-up state and electron 2 is in the spin-down state. The same is true for $|\hat{\mathbf{z}}-;\hat{\mathbf{z}}+\rangle$.

Suppose we make a measurement on the spin component of one of the electrons. Clearly, there is a 50% chance of getting either up or down because the composite system may be in $|\hat{\mathbf{z}}+;\hat{\mathbf{z}}-\rangle$ or $|\hat{\mathbf{z}}-;\hat{\mathbf{z}}+\rangle$ with equal probabilities. But if one of the components is shown to be in the spin-up state, the other is necessarily in the spin-down state, and vice versa. When the spin component of electron 1 is shown to be up, the measurement apparatus has selected the first term, $|\hat{\mathbf{z}}+;\hat{\mathbf{z}}-\rangle$ of (3.427); a subsequent measurement of the spin component of electron 2 must ascertain that the state ket of the composite system is given by $|\hat{\mathbf{z}}+;\hat{\mathbf{z}}-\rangle$.

It is remarkable that this kind of correlation can persist even if the two particles are well separated and have ceased to interact provided that as they fly apart, there is no change in their spin states. This is certainly the case for a $J=0$ system disintegrating spontaneously into two spin $\frac{1}{2}$ particles with no relative orbital angular momentum, because angular-momentum conservation must hold in the disintegration process. An example of this would be a rare decay of the $\eta$ meson (mass 549 MeV/$c^2$) into a muon pair

$$\eta \to \mu^+ + \mu^- \tag{3.428}$$

which, unfortunately, has a branching ratio of only approximately $6 \times 10^{-6}$. More realistically, in proton-proton scattering at low kinetic energies, the Pauli principle to be discussed in Chapter 7 forces the interacting protons to be in $^1S_0$ (orbital angular momentum 0, spin-singlet state), and the spin states of the scattered protons must be correlated in the manner indicated by (3.427) even after they become separated by a *macroscopic distance*.

To be more pictorial we consider a system of two spin $\frac{1}{2}$ particles moving in opposite directions, as in Figure 3.11. Observer A specializes in measuring $S_z$ of particle 1 (flying

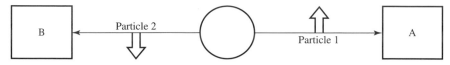

**Fig. 3.11**    Spin correlation in a spin-singlet state.

to the right), while observer B specializes in measuring $S_z$ of particle 2 (flying to the left). To be specific, let us assume that observer A finds $S_z$ to be positive for particle 1. Then he or she can predict, even before B performs any measurement, the outcome of B's measurement with certainty: B must find $S_z$ to be negative for particle 2. On the other hand, if A makes no measurement, B has a 50% chance of getting $S_z+$ or $S_z-$.

This by itself might not be so peculiar. One may say, "It is just like an urn known to contain one black ball and one white ball. When we blindly pick one of them, there is a 50% chance of getting black or white. But if the first ball we pick is black, then we can predict with certainty that the second ball will be white."

It turns out that this analogy is too simple. The actual quantum-mechanical situation is far more sophisticated than that! This is because observers may choose to measure $S_x$ in place of $S_z$. The *same* pair of "quantum-mechanical balls" can be analyzed either in terms of black and white *or* in terms of blue and red!

Recall now that for a single spin $\frac{1}{2}$ system the $S_x$ eigenkets and $S_z$ eigenkets are related as follows:

$$|\hat{x}\pm\rangle = \left(\frac{1}{\sqrt{2}}\right)(|\hat{z}+\rangle \pm |\hat{z}-\rangle), \quad |\hat{z}\pm\rangle = \left(\frac{1}{\sqrt{2}}\right)(|\hat{x}+\rangle \pm |\hat{x}-\rangle). \tag{3.429}$$

Returning now to our composite system, we can rewrite spin-singlet ket (3.427) by choosing the $x$-direction as the axis of quantization:

$$|\text{spin singlet}\rangle = \left(\frac{1}{\sqrt{2}}\right)(|\hat{x}-;\hat{x}+\rangle - |\hat{x}+;\hat{x}-\rangle). \tag{3.430}$$

Apart from the overall sign, which in any case is a matter of convention, we could have guessed this form directly from (3.427) because spin-singlet states have no preferred direction in space. Let us now suppose that observer A can choose to measure $S_z$ or $S_x$ of particle 1 by changing the orientation of his or her spin analyzer, while observer B always specializes in measuring $S_x$ of particle 2. If A determines $S_z$ of particle 1 to be positive, B clearly has a 50% chance for getting $S_x+$ or $S_x-$; even though $S_z$ of particle 2 is known to be negative with certainty, its $S_x$ is completely undetermined. On the other hand, let us suppose that A also chooses to measure $S_x$; if observer A determines $S_x$ of particle 1 to be positive, then without fail, observer B will measure $S_x$ of particle 2 to be negative. Finally, if A chooses to make no measurement, B, of course, will have a 50% chance of getting $S_x+$ or $S_x-$. To sum up, we have the following.

1. If A measures $S_z$ and B measures $S_x$, there is a completely random correlation between the two measurements.
2. If A measures $S_x$ and B measures $S_x$, there is a 100% (opposite sign) correlation between the two measurements.
3. If A makes no measurement, B's measurements show random results.

Table 3.1 shows all possible results of such measurements when B and A are allowed to choose to measure $S_x$ or $S_z$. These considerations show that the outcome of B's measurement appears to depend on what kind of measurement A decides to perform: an $S_x$ measurement, an $S_z$ measurement, or no measurement. Notice again that A and B can be

| Table 3.1 Spin-Correlation Measurements | | | |
|---|---|---|---|
| Spin component measured by A | A's result | Spin component measured by B | B's result |
| z | + | z | − |
| z | − | x | + |
| x | − | z | − |
| x | − | z | + |
| z | + | x | − |
| x | + | x | − |
| z | + | x | + |
| x | − | x | + |
| z | − | z | + |
| z | − | x | − |
| x | + | z | + |
| x | + | z | − |

miles apart with no possibility of communications or mutual interactions. Observer A can decide how to orient his or her spin-analyzer apparatus long after the two particles have separated. It is as though particle 2 "knows" which spin component of particle 1 is being measured.

The orthodox quantum-mechanical interpretation of this situation is as follows. A measurement is a selection (or filtration) process. When $S_z$ of particle 1 is measured to be positive, then component $|\hat{\mathbf{z}}+; \hat{\mathbf{z}}-\rangle$ is selected. A subsequent measurement of the other particle's $S_z$ merely ascertains that the system is still in $|\hat{\mathbf{z}}+; \hat{\mathbf{z}}-\rangle$. We must accept that a measurement on what appears to be a part of the system is to be regarded as a measurement on the whole system.

### 3.10.2 Einstein's Locality Principle and Bell's Inequality

Many physicists have felt uncomfortable with the preceding orthodox interpretation of spin-correlation measurements. Their feelings are typified in the following frequently quoted remarks by A. Einstein, which we call **Einstein's locality principle**: "But on one supposition we should, in my opinion, absolutely hold fast: The real factual situation of the system $S_2$ is independent of what is done with the system $S_1$, which is spatially separated from the former." Because this problem was first discussed in a 1935 paper of A. Einstein, B. Podolsky, and N. Rosen, it is sometimes known as the Einstein–Podolsky–Rosen paradox.[15]

Some have argued that the difficulties encountered here are inherent in the probabilistic interpretations of quantum mechanics and that the dynamic behavior at the microscopic level appears probabilistic only because some yet unknown parameters, so-called hidden

---

[15] To be historically accurate, the original Einstein–Podolsky–Rosen paper dealt with measurements of $x$ and $p$. The use of composite spin $\frac{1}{2}$ systems to illustrate the Einstein–Podolsky–Rosen paradox started with D. Bohm.

variables, have not been specified. It is not our purpose here to discuss various alternatives to quantum mechanics based on hidden-variable or other considerations. Rather, let us ask: Do such theories make predictions different from those of quantum mechanics? Until 1964, it could be thought that the alternative theories could be concocted in such a way that they would give no predictions, other than the usual quantum-mechanical predictions, that could be verified experimentally. The whole debate would have belonged to the realm of metaphysics rather than physics. It was then pointed out by J. S. Bell that the alternative theories based on Einstein's locality principle actually predict a *testable inequality relation* among the observables of spin-correlation experiments that *disagrees* with the predictions of quantum mechanics.

We derive Bell's inequality within the framework of a simple model, conceived by E. P. Wigner, that incorporates the essential features of the various alternative theories. Proponents of this model agree that it is impossible to determine $S_x$ and $S_z$ simultaneously. However, when we have a large number of spin $\frac{1}{2}$ particles, we assign a certain fraction of them to have the following property.

If $S_z$ is measured, we obtain a plus sign with certainty.

If $S_x$ is measured, we obtain a minus sign with certainty.

A particle satisfying this property is said to belong to type $(\hat{\mathbf{z}}+, \hat{\mathbf{x}}-)$. Notice that we are not asserting that we can simultaneously measure $S_z$ and $S_x$ to be $+$ and $-$, respectively. When we measure $S_z$, we do not measure $S_x$, and vice versa. We are assigning definite values of spin components *in more than one direction* with the understanding that only one or the other of the components can actually be measured. Even though this approach is fundamentally different from that of quantum mechanics, the quantum-mechanical predictions for $S_z$ and $S_x$ measurements performed on the spin-up ($S_z+$) state are reproduced provided there are as many particles belonging to type $(\hat{\mathbf{z}}+, \hat{\mathbf{x}}+)$ as to type $(\hat{\mathbf{z}}+, \hat{\mathbf{x}}-)$.

Let us now examine how this model can account for the results of spin-correlation measurements made on composite spin-singlet systems. Clearly, for a particular pair, there must be a perfect matching between particle 1 and particle 2 to ensure zero total angular momentum: If particle 1 is of type $(\hat{\mathbf{z}}+, \hat{\mathbf{x}}-)$, then particle 2 must belong to type $(\hat{\mathbf{z}}-, \hat{\mathbf{x}}+)$, and so forth. The results of correlation measurements, such as in Table 3.1, can be reproduced if particle 1 and particle 2 are matched as follows:

$$\begin{aligned} & \text{particle 1} \quad \text{particle 2} \\ & (\hat{\mathbf{z}}+, \hat{\mathbf{x}}-) \leftrightarrow (\hat{\mathbf{z}}-, \hat{\mathbf{x}}+), \end{aligned} \tag{3.431a}$$

$$(\hat{\mathbf{z}}+, \hat{\mathbf{x}}+) \leftrightarrow (\hat{\mathbf{z}}-, \hat{\mathbf{x}}-), \tag{3.431b}$$

$$(\hat{\mathbf{z}}-, \hat{\mathbf{x}}+) \leftrightarrow (\hat{\mathbf{z}}+, \hat{\mathbf{x}}-), \tag{3.431c}$$

$$(\hat{\mathbf{z}}-, \hat{\mathbf{x}}-) \leftrightarrow (\hat{\mathbf{z}}+, \hat{\mathbf{x}}+), \tag{3.431d}$$

with equal populations, that is, 25% each. A very important assumption is implied here. Suppose a particular pair belongs to type (3.431a) and observer A decides to measure $S_z$ of

| Table 3.2 Spin-Component Matching in the Alternative Theories | | |
|---|---|---|
| Population | Particle 1 | Particle 2 |
| $N_1$ | $(\hat{\mathbf{a}}+,\hat{\mathbf{b}}+,\hat{\mathbf{c}}+)$ | $(\hat{\mathbf{a}}-,\hat{\mathbf{b}}-,\hat{\mathbf{c}}-)$ |
| $N_2$ | $(\hat{\mathbf{a}}+,\hat{\mathbf{b}}+,\hat{\mathbf{c}}-)$ | $(\hat{\mathbf{a}}-,\hat{\mathbf{b}}-,\hat{\mathbf{c}}+)$ |
| $N_3$ | $(\hat{\mathbf{a}}+,\hat{\mathbf{b}}-,\hat{\mathbf{c}}+)$ | $(\hat{\mathbf{a}}-,\hat{\mathbf{b}}+,\hat{\mathbf{c}}-)$ |
| $N_4$ | $(\hat{\mathbf{a}}+,\hat{\mathbf{b}}-,\hat{\mathbf{c}}-)$ | $(\hat{\mathbf{a}}-,\hat{\mathbf{b}}+,\hat{\mathbf{c}}+)$ |
| $N_5$ | $(\hat{\mathbf{a}}-,\hat{\mathbf{b}}+,\hat{\mathbf{c}}+)$ | $(\hat{\mathbf{a}}+,\hat{\mathbf{b}}-,\hat{\mathbf{c}}-)$ |
| $N_6$ | $(\hat{\mathbf{a}}-,\hat{\mathbf{b}}+,\hat{\mathbf{c}}-)$ | $(\hat{\mathbf{a}}+,\hat{\mathbf{b}}-,\hat{\mathbf{c}}+)$ |
| $N_7$ | $(\hat{\mathbf{a}}-,\hat{\mathbf{b}}\ ,\hat{\mathbf{c}}\,|\,)$ | $(\hat{\mathbf{a}}+,\hat{\mathbf{b}}+,\hat{\mathbf{c}}-)$ |
| $N_8$ | $(\hat{\mathbf{a}}-,\hat{\mathbf{b}}-,\hat{\mathbf{c}}-)$ | $(\hat{\mathbf{a}}+,\hat{\mathbf{b}}+,\hat{\mathbf{c}}+)$ |

particle 1; then he or she necessarily obtains a plus sign regardless of whether B decides to measure $S_z$ or $S_x$. It is in this sense that Einstein's locality principle is incorporated in this model: A's result is predetermined independently of B's choice as to what to measure.

In the examples considered so far, this model has been successful in reproducing the predictions of quantum mechanics. We now consider more complicated situations where the model leads to predictions different from the usual quantum-mechanical predictions. This time we start with three unit vectors $\hat{\mathbf{a}}$, $\hat{\mathbf{b}}$, and $\hat{\mathbf{c}}$, which are, in general, not mutually orthogonal. We imagine that one of the particles belongs to some definite type, say $(\hat{\mathbf{a}}-,\hat{\mathbf{b}}+,\hat{\mathbf{c}}+)$, which means that if $\mathbf{S}\cdot\hat{\mathbf{a}}$ is measured, we obtain a minus sign with certainty; if $\mathbf{S}\cdot\hat{\mathbf{b}}$ is measured, we obtain a plus sign with certainty; if $\mathbf{S}\cdot\hat{\mathbf{c}}$ is measured, we obtain a plus sign with certainty. Again there must be a perfect matching in the sense that the other particle necessarily belongs to type $(\hat{\mathbf{a}}+,\hat{\mathbf{b}}-,\hat{\mathbf{c}}-)$ to ensure zero total angular momentum. In any given event, the particle pair in question must be a member of one of the eight types shown in Table 3.2. These eight possibilities are mutually exclusive and disjoint. The population of each type is indicated in the first column.

Let us suppose that observer $A$ finds $\mathbf{S}_1\cdot\hat{\mathbf{a}}$ to be plus and observer $B$ finds $\mathbf{S}_2\cdot\hat{\mathbf{b}}$ to be plus also. It is clear from Table 3.2 that the pair belong to either type 3 or type 4, so the number of particle pairs for which this situation is realized is $N_3 + N_4$. Because $N_i$ is positive semidefinite, we must have inequality relations like

$$N_3 + N_4 \leq (N_2 + N_4) + (N_3 + N_7). \tag{3.432}$$

Let $P(\hat{\mathbf{a}}+;\hat{\mathbf{b}}+)$ be the probability that, in a random selection, observer A measures $\mathbf{S}_1\cdot\hat{\mathbf{a}}$ to be $+$ and observer B measures $\mathbf{S}_2\cdot\hat{\mathbf{b}}$ to be $+$, and so on.

Clearly, we have

$$P(\hat{\mathbf{a}}+;\hat{\mathbf{b}}+) = \frac{(N_3 + N_4)}{\Sigma_i^8 N_i}. \tag{3.433}$$

In a similar manner, we obtain

$$P(\hat{\mathbf{a}}+;\hat{\mathbf{c}}+) = \frac{(N_2 + N_4)}{\Sigma_i^8 N_i} \quad \text{and} \quad P(\hat{\mathbf{c}}+;\hat{\mathbf{b}}+) = \frac{(N_3 + N_7)}{\Sigma_i^8 N_i}. \tag{3.434}$$

The positivity condition (3.432) now becomes

$$P(\hat{\mathbf{a}}+;\hat{\mathbf{b}}+) \leq P(\hat{\mathbf{a}}+;\hat{\mathbf{c}}+) + P(\hat{\mathbf{c}}+;\hat{\mathbf{b}}+). \tag{3.435}$$

This is **Bell's inequality**, which follows from Einstein's locality principle.

### 3.10.3  Quantum Mechanics and Bell's Inequality

We now return to the world of quantum mechanics. In quantum mechanics we do not talk about a certain fraction of particle pairs, say $N_3/\sum_i^8 N_i$, belonging to type 3. Instead, we characterize all spin-singlet systems by the same ket (3.427); in the language of Section 3.4 we are concerned here with a pure ensemble. Using this ket and the rules of quantum mechanics we have developed, we can unambiguously calculate each of the three terms in inequality (3.435).

We first evaluate $P(\hat{\mathbf{a}}+;\hat{\mathbf{b}}+)$. Suppose observer A finds $\mathbf{S}_1 \cdot \hat{\mathbf{a}}$ to be positive; because of the 100% (opposite sign) correlation we discussed earlier, B's measurement of $\mathbf{S}_2 \cdot \hat{\mathbf{a}}$ will yield a minus sign with certainty. But to calculate $P(\hat{\mathbf{a}}+;\hat{\mathbf{b}}+)$ we must consider a new quantization axis $\hat{\mathbf{b}}$ that makes an angle $\theta_{ab}$ with $\hat{\mathbf{a}}$; see Figure 3.12. According to the formalism of Section 3.2, the probability that the $\mathbf{S}_2 \cdot \hat{\mathbf{b}}$ measurement yields $+$ when particle 2 is known to be in an eigenket of $\mathbf{S}_2 \cdot \hat{\mathbf{a}}$ with negative eigenvalue is given by

$$\cos^2\left[\frac{(\pi - \theta_{ab})}{2}\right] = \sin^2\left(\frac{\theta_{ab}}{2}\right). \tag{3.436}$$

As a result, we obtain

$$P(\hat{\mathbf{a}}+;\hat{\mathbf{b}}+) = \left(\frac{1}{2}\right)\sin^2\left(\frac{\theta_{ab}}{2}\right), \tag{3.437}$$

where the factor $\frac{1}{2}$ arises from the probability of initially obtaining $\mathbf{S}_1 \cdot \hat{\mathbf{a}}$ with $+$. Using (3.437) and its generalization to the other two terms of (3.435), we can write Bell's inequality as

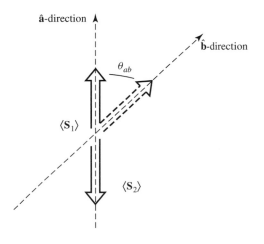

**Fig. 3.12**  Evaluation of $P(\hat{\mathbf{a}}+;\hat{\mathbf{b}}+)$.

$$\sin^2\left(\frac{\theta_{ab}}{2}\right) \leq \sin^2\left(\frac{\theta_{ac}}{2}\right) + \sin^2\left(\frac{\theta_{cb}}{2}\right). \tag{3.438}$$

We now show that inequality (3.438) is not always possible from a geometric point of view. For simplicity let us choose $\hat{\mathbf{a}}$, $\hat{\mathbf{b}}$, and $\hat{\mathbf{c}}$ to lie in a plane, and let $\hat{\mathbf{c}}$ bisect the two directions defined by $\hat{\mathbf{a}}$ and $\hat{\mathbf{b}}$:

$$\theta_{ab} = 2\theta, \quad \theta_{ac} = \theta_{cb} = \theta. \tag{3.439}$$

Inequality (3.438) is then violated for

$$0 < \theta < \frac{\pi}{2}. \tag{3.440}$$

For example, take $\theta = \pi/4$; we then obtain

$$0.500 \leq 0.292 \ ?? \tag{3.441}$$

So the quantum-mechanical predictions are not compatible with Bell's inequality. There is a real observable, in the sense of being experimentally verifiable, difference between quantum mechanics and the alternative theories satisfying Einstein's locality principle.

Several experiments have been performed to test Bell's inequality. For a review, see "Bell's inequality test: more ideal than ever" by Aspect, *Nature*, **398** (1999) 189. In one of the experiments spin correlations between the final protons in low-energy proton-proton scattering were measured. In all other experiments photon polarization correlations between a pair of photons in a cascade transition of an excited atom (Ca, Hg,...),

$$(j = 0) \xrightarrow{\gamma} (j = 1) \xrightarrow{\gamma} (j = 0), \tag{3.442}$$

or in the decay of a positronium (an $e^+ e^-$ bound state in $^1S_0$) were measured; studying photon polarization correlations should be just as good in view of the analogy developed in Section 1.1

$$S_z+ \rightarrow \hat{\varepsilon} \quad \text{in } x\text{-direction,} \tag{3.443a}$$

$$S_z- \rightarrow \hat{\varepsilon} \quad \text{in } y\text{-direction,} \tag{3.443b}$$

$$S_x+ \rightarrow \hat{\varepsilon} \quad \text{in } 45° \text{ diagonal direction,} \tag{3.443c}$$

$$S_x- \rightarrow \hat{\varepsilon} \quad \text{in } 135° \text{ diagonal direction.} \tag{3.443d}$$

The results of all recent precision experiments have conclusively established that Bell's inequality was violated, in one case by more than nine standard deviations. Furthermore, in all these experiments the inequality relation was violated in such a way that the quantum-mechanical predictions were fulfilled within error limits. In this controversy, quantum mechanics has triumphed with flying colors.

The fact that the quantum-mechanical predictions have been verified does not mean that the whole subject is now a triviality. Despite the experimental verdict we may still feel psychologically uncomfortable about many aspects of measurements of this kind. Consider in particular the following point: Right after observer A performs a measurement on particle 1, how does particle 2 – which may, in principle, be many light years away from particle 1 – get to "know" how to orient its spin so that the remarkable correlations apparent in Table 3.1 are realized? In one of the experiments to test Bell's inequality (performed by A. Aspect and collaborators) the analyzer settings were changed so rapidly that A's decision as to what to measure could not be made until it was too late for any kind of influence, traveling slower than light, to reach B.

We conclude this section by showing that despite these peculiarities, we cannot use spin-correlation measurements to transmit any useful information between two macroscopically separated points. In particular, superluminal (faster than light) communications are impossible.

Suppose A and B both agree in advance to measure $S_z$; then, without asking A, B knows precisely what A is getting. But this does not mean that A and B are communicating; B just observes a random sequence of positive and negative signs. There is obviously no useful information contained in it. B verifies the remarkable correlations predicted by quantum mechanics only after he or she gets together with A and compares the notes (or computer sheets).

It might be thought that A and B can communicate if one of them suddenly changes the orientation of his or her analyzing apparatus. Let us suppose that A agrees initially to measure $S_z$, and $B$, $S_x$. The results of A's measurements are completely uncorrelated with the results of B's measurements, so there is no information transferred. But then, suppose A suddenly breaks his or her promise and without telling B starts measuring $S_x$. There are now complete correlations between A's results and B's results. However, B has no way of inferring that A has changed the orientation of his or her analyzer. B continues to see just a random sequence of $+$ and $-$ by looking at his or her own notebook *only*. So again there is no information transferred.

## 3.11 Tensor Operators

### 3.11.1 Vector Operator

We have been using notations such as **x**, **p**, **S**, and **L**, but as yet we have not systematically discussed their rotational properties. They are vector operators, but what are their properties under rotations? In this section we give a precise quantum-mechanical definition of vector operators based on their commutation relations with the angular-momentum operator. We then generalize to tensor operators with more complicated transformation properties and derive an important theorem on the matrix elements of vector and tensor operators.

We all know that a **vector** in classical physics is a quantity with three components that transforms by definition like $V_i \rightarrow \Sigma_j R_{ij} V_j$ under a rotation. It is reasonable to demand that the expectation value of a vector operator $V$ in quantum mechanics be transformed like a classical vector under rotation. Specifically, as the state ket is changed under rotation according to

$$|\alpha\rangle \rightarrow \mathscr{D}(R)|\alpha\rangle, \tag{3.444}$$

the expectation value of **V** is assumed to change as follows:

$$\langle \alpha | V_i | \alpha \rangle \to \langle \alpha | \mathscr{D}^\dagger(R) V_i \mathscr{D}(R) | \alpha \rangle = \sum_j R_{ij} \langle \alpha | V_j | \alpha \rangle. \tag{3.445}$$

This must be true for an arbitrary ket $|\alpha\rangle$. Therefore,

$$\mathscr{D}^\dagger(R) V_i \mathscr{D}(R) = \sum_j R_{ij} V_j \tag{3.446}$$

must hold as an **operator equation**, where $R_{ij}$ is the $3 \times 3$ matrix that corresponds to rotation $R$.

Let us now consider a specific case, an infinitesimal rotation. When the rotation is infinitesimal, we have

$$\mathscr{D}(R) = 1 - \frac{i\varepsilon \mathbf{J} \cdot \hat{\mathbf{n}}}{\hbar}. \tag{3.447}$$

We can now write (3.446) as

$$V_i + \frac{\varepsilon}{i\hbar} [V_i, \mathbf{J} \cdot \hat{\mathbf{n}}] = \sum_j R_{ij}(\hat{\mathbf{n}}; \varepsilon) V_j. \tag{3.448}$$

In particular, for $\hat{\mathbf{n}}$ along the $z$-axis, we have

$$R(\hat{\mathbf{z}}; \varepsilon) = \begin{pmatrix} 1 & -\varepsilon & 0 \\ \varepsilon & 1 & 0 \\ 0 & 0 & 1 \end{pmatrix}, \tag{3.449}$$

so

$$i = 1, \quad V_x + \frac{\varepsilon}{i\hbar}[V_x, J_z] = V_x - \varepsilon V_y \tag{3.450a}$$

$$i = 2, \quad V_y + \frac{\varepsilon}{i\hbar}[V_y, J_z] = \varepsilon V_x + V_y \tag{3.450b}$$

$$i = 3, \quad V_z + \frac{\varepsilon}{i\hbar}[V_z, J_z] = V_z. \tag{3.450c}$$

This means that **V** must satisfy the commutation relations

$$[V_i, J_j] = i\varepsilon_{ijk}\hbar V_k. \tag{3.451}$$

Clearly, the behavior of **V** under a *finite* rotation is completely determined by the preceding commutation relations; we just apply the by now familiar formula (2.168) to

$$\exp\left(\frac{iJ_j\phi}{\hbar}\right) V_i \exp\left(\frac{-iJ_j\phi}{\hbar}\right). \tag{3.452}$$

We simply need to calculate

$$[J_j, [J_j, [\cdots [J_j, V_i]\cdots]]]. \tag{3.453}$$

Multiple commutators keep on giving back to us $V_i$ or $V_k$ ($k \neq i,j$) as in spin case (3.24).

We can use (3.451) as the *defining* property of a vector operator. Notice that the angular-momentum commutation relations are a special case of (3.451) in which we let $V_i \to J_i, V_k \to J_k$. Other special cases are $[y, L_z] = i\hbar x, [x, L_z] = -i\hbar y, [p_x, L_z] = -i\hbar p_y, [p_y, L_z] = i\hbar p_x$; these can be proved explicitly.

## 3.11.2 Cartesian Tensors Versus Irreducible Tensors

In classical physics it is customary to define a tensor $T_{ijk\cdots}$ by generalizing $V_i \rightarrow \Sigma_j R_{ij} V_j$ as follows:

$$T_{ijk\cdots} \rightarrow \sum_{i'}\sum_{j'}\sum_{k'} \cdots R_{ii'} R_{jj'} \cdots T_{i'j'k'\cdots} \tag{3.454}$$

under a rotation specified by the $3 \times 3$ orthogonal matrix $R$. The number of indices is called the **rank** of a tensor. Such a tensor is known as a **Cartesian tensor**.

The simplest example of a Cartesian tensor of rank 2 is a **dyadic** formed out of two vectors **U** and **V**. One simply takes a Cartesian component of **U** and a Cartesian component of **V** and puts them together:

$$T_{ij} \equiv U_i V_j. \tag{3.455}$$

Notice that we have nine components altogether. They obviously transform like (3.454) under rotation.

The trouble with a Cartesian tensor like (3.455) is that it is reducible, that is, it can be decomposed into objects that transform differently under rotations. Specifically, for the dyadic in (3.455) we have

$$U_i V_j = \frac{\mathbf{U} \cdot \mathbf{V}}{3} \delta_{ij} + \frac{(U_i V_j - U_j V_i)}{2} + \left( \frac{U_i V_j + U_j V_i}{2} - \frac{\mathbf{U} \cdot \mathbf{V}}{3} \delta_{ij} \right). \tag{3.456}$$

The first term on the right-hand side, $\mathbf{U} \cdot \mathbf{V}$, is a scalar product invariant under rotation. The second is an antisymmetric tensor which can be written as a vector product $\varepsilon_{ijk} (\mathbf{U} \times \mathbf{V})_k$. There are altogether 3 independent components. The last is a $3 \times 3$ symmetric traceless tensor with 5 ($= 6 - 1$, where 1 comes from the traceless condition) independent components. The number of independent components checks:

$$3 \times 3 = 1 + 3 + 5. \tag{3.457}$$

We note that the numbers appearing on the right-hand side of (3.457) are precisely the multiplicities of objects with angular momentum $l = 0$, $l = 1$, and $l = 2$, respectively. This suggests that the dyadic has been decomposed into tensors that can transform like spherical harmonics with $l = 0$, 1, and 2. In fact, (3.456) is the simplest nontrivial example to illustrate the reduction of a Cartesian tensor into irreducible **spherical tensors**.

Before presenting the precise definition of a spherical tensor, we first give an example of a spherical tensor of rank $k$. Suppose we take a spherical harmonic $Y_l^m(\theta, \phi)$. We have already seen that it can be written as $Y_l^m(\hat{\mathbf{n}})$, where the orientation of $\hat{\mathbf{n}}$ is characterized by $\theta$ and $\phi$. We now replace $\hat{\mathbf{n}}$ by some vector **V**. The result is that we have a spherical tensor of rank $k$ (in place of $l$) with magnetic quantum number $q$ (in place of $m$), namely

$$T_q^{(k)} = Y_{l=k}^{m=q}(\mathbf{V}). \tag{3.458}$$

Specifically, in the case $k = 1$ we take spherical harmonics with $l = 1$ and replace $(z/r) = (\hat{\mathbf{n}})_z$ by $V_z$, and so on.

$$Y_1^0 = \sqrt{\frac{3}{4\pi}} \cos\theta = \sqrt{\frac{3}{4\pi}} \frac{z}{r} \rightarrow T_0^{(1)} = \sqrt{\frac{3}{4\pi}} V_z,$$

$$Y_1^{\pm 1} = \mp \sqrt{\frac{3}{4\pi}} \frac{x \pm iy}{\sqrt{2}r} \rightarrow T_{\pm 1}^{(1)} = \sqrt{\frac{3}{4\pi}} \left( \mp \frac{V_x \pm iV_y}{\sqrt{2}} \right).$$

(3.459)

Obviously this can be generalized for higher $k$, for example,

$$Y_2^{\pm 2} = \sqrt{\frac{15}{32\pi}} \frac{(x \pm iy)^2}{r^2} \rightarrow T_{\pm 2}^{(2)} = \sqrt{\frac{15}{32\pi}} (V_x \pm iV_y)^2$$

(3.460)

$T_q^{(k)}$ are irreducible, just as $Y_l^{(m)}$ are. For this reason, working with spherical tensors is more satisfactory than working with Cartesian tensors.

To see the transformation of spherical tensors constructed in this manner, let us first review how $Y_l^m$ transform under rotations. First, we have for the direction eigenket;

$$|\hat{\mathbf{n}}\rangle \rightarrow \mathscr{D}(R)|\hat{\mathbf{n}}\rangle \equiv |\hat{\mathbf{n}}'\rangle,$$

(3.461)

which defines the rotated eigenket $|\hat{\mathbf{n}}'\rangle$. We wish to examine how $Y_l^m(\hat{\mathbf{n}}') = \langle \hat{\mathbf{n}}'|l,m\rangle$ would look in terms of $Y_l^m(\hat{\mathbf{n}})$. We can easily see this by starting with

$$\mathscr{D}(R^{-1})|l,m\rangle = \sum_{m'} |l,m'\rangle \mathscr{D}_{m'm}^{(l)}(R^{-1})$$

(3.462)

and contracting with $\langle \hat{\mathbf{n}}|$ on the left, using (3.461):

$$Y_l^m(\hat{\mathbf{n}}') = \sum_{m'} Y_l^{m'}(\hat{\mathbf{n}}) \mathscr{D}_{m'm}^{(l)}(R^{-1}).$$

(3.463)

If there is an operator that acts like $Y_l^m(\mathbf{V})$, it is then reasonable to expect

$$\mathscr{D}^\dagger(R) Y_l^m(\mathbf{V}) \mathscr{D}(R) = \sum_{m'} Y_l^{m'}(\mathbf{V}) \mathscr{D}_{mm'}^{(l)*}(R),$$

(3.464)

where we have used the unitarity of the rotation operator to rewrite $\mathscr{D}_{m'm}^{(l)}(R^{-1})$.

All this work is just to motivate the definition of a spherical tensor. We now consider spherical tensors in quantum mechanics. Motivated by (3.464) we define a spherical tensor operator of rank $k$ with $(2k+1)$ components as

$$\mathscr{D}^\dagger(R) T_q^{(k)} \mathscr{D}(R) = \sum_{q'=-k}^{k} \mathscr{D}_{qq'}^{(k)*}(R) T_{q'}^{(k)}$$

(3.465a)

or, equivalently,

$$\mathscr{D}(R) T_q^{(k)} \mathscr{D}^\dagger(R) = \sum_{q'=-k}^{k} \mathscr{D}_{q'q}^{(k)}(R) T_{q'}^{(k)}.$$

(3.465b)

This definition holds regardless of whether $T_q^{(k)}$ can be written as $Y_{l=k}^{m=q}(\mathbf{V})$; for example, $(U_x + iU_y)(V_x + iV_y)$ is the $q = +2$ component of a spherical tensor of rank 2 even though, unlike $(V_x + iV_y)^2$, it cannot be written as $Y_k^q(\mathbf{V})$.

A more convenient definition of a spherical tensor is obtained by considering the infinitesimal form of (3.465b), namely,

$$\left(1 - \frac{i\mathbf{J} \cdot \hat{\mathbf{n}}\varepsilon}{\hbar}\right) T_q^{(k)} \left(1 + \frac{i\mathbf{J} \cdot \hat{\mathbf{n}}\varepsilon}{\hbar}\right) = \sum_{q'=-k}^{k} T_{q'}^{(k)} \langle kq' | \left(1 - \frac{i\mathbf{J} \cdot \hat{\mathbf{n}}\varepsilon}{\hbar}\right) | kq \rangle \qquad (3.466)$$

or

$$[\mathbf{J} \cdot \hat{\mathbf{n}}, T_q^{(k)}] = \sum_{q'} T_{q'}^{(k)} \langle kq' | \mathbf{J} \cdot \hat{\mathbf{n}} | kq \rangle. \qquad (3.467)$$

By taking $\hat{\mathbf{n}}$ in the $\hat{\mathbf{z}}$- and in the $(\hat{\mathbf{x}} \pm i\hat{\mathbf{y}})$ directions and using the nonvanishing matrix elements of $J_z$ and $J_{\pm}$ [see (3.187b) and (3.193)], we obtain

$$[J_z, T_q^{(k)}] = \hbar q T_q^{(k)} \qquad (3.468a)$$

and

$$[J_{\pm}, T_q^{(k)}] = \hbar \sqrt{(k \mp q)(k \pm q + 1)} T_{q \pm 1}^{(k)}. \qquad (3.468b)$$

These commutation relations can be considered as a definition of spherical tensors in place of (3.465).

### 3.11.3  Product of Tensors

We have seen how to form a scalar, vector (or antisymmetric tensor), and a traceless symmetric tensor out of two vectors using the Cartesian tensor language. Of course, spherical tensor language can also be used (Baym (1969), Chapter 17), for example,

$$
\begin{aligned}
T_0^{(0)} &= \frac{-\mathbf{U} \cdot \mathbf{V}}{3} = \frac{(U_{+1}V_{-1} + U_{-1}V_{+1} - U_0 V_0)}{3}, \\
T_q^{(1)} &= \frac{(\mathbf{U} \times \mathbf{V})_q}{i\sqrt{2}}, \\
T_{\pm 2}^{(2)} &= U_{\pm 1} V_{\pm 1}, \qquad\qquad\qquad\qquad\qquad\qquad (3.469) \\
T_{\pm 1}^{(2)} &= \frac{U_{\pm 1} V_0 + U_0 V_{\pm 1}}{\sqrt{2}}, \\
T_0^{(2)} &= \frac{U_{+1} V_{-1} + 2U_0 V_0 + U_{-1} V_{+1}}{\sqrt{6}},
\end{aligned}
$$

where $U_q(V_q)$ is the $q$th component of a spherical tensor of rank 1, corresponding to vector $\mathbf{U}(\mathbf{V})$. The preceding transformation properties can be checked by comparing with $Y_l^m$ and remembering that $U_{+1} = -(U_x + iU_y)/\sqrt{2}, U_{-1} = (U_x - iU_y)/\sqrt{2}, U_0 = U_z$. A similar check can be made for $V_{\pm 1,0}$. For instance,

$$Y_2^0 = \sqrt{\frac{5}{16\pi}} \frac{3z^2 - r^2}{r^2},$$

where $3z^2 - r^2$ can be written as

$$2z^2 + 2\left[-\frac{(x + iy)}{\sqrt{2}} \frac{(x - iy)}{\sqrt{2}}\right];$$

hence, $Y_2^0$ is just a special case of $T_0^{(2)}$ for $\mathbf{U} = \mathbf{V} = \mathbf{r}$.

A more systematic way of forming tensor products goes as follows. We start by stating a theorem.

**Theorem 5** *Let* $X_{q_1}^{(k_1)}$ *and* $Z_{q_2}^{(k_2)}$ *be irreducible spherical tensors of rank* $k_1$ *and* $k_2$, *respectively. Then*

$$T_q^{(k)} = \sum_{q_1} \sum_{q_2} \langle k_1 k_2; q_1 q_2 | k_1 k_2; kq \rangle X_{q_1}^{(k_1)} Z_{q_2}^{(k_2)} \tag{3.470}$$

*is a spherical (irreducible) tensor of rank k.*

**Proof** We must show that under rotation $T_q^{(k)}$ must transform according to (3.465)

$$\mathscr{D}^\dagger(R) T_q^{(k)} \mathscr{D}(R) = \sum_{q_1} \sum_{q_2} \langle k_1 k_2; q_1 q_2 | k_1 k_2; kq \rangle$$
$$\times \mathscr{D}^\dagger(R) X_{q_1}^{(k_1)} \mathscr{D}(R) \mathscr{D}^\dagger(R) Z_{q_2}^{(k_2)} \mathscr{D}(R)$$
$$= \sum_{q_1} \sum_{q_2} \sum_{q_1'} \sum_{q_2'} \langle k_1 k_2; q_1 q_2 | k_1 k_2; kq \rangle$$
$$\times X_{q_1'}^{(k_1)} \mathscr{D}_{q_1' q_1}^{(k_1)}(R^{-1}) Z_{q_2'}^{(k_2)} \mathscr{D}_{q_2' q_2}^{(k_2)}(R^{-1})$$
$$= \sum_{k''} \sum_{q_1} \sum_{q_2} \sum_{q_1'} \sum_{q_2'} \sum_{q''} \sum_{q'} \langle k_1 k_2; q_1 q_2 | k_1 k_2; kq \rangle$$
$$\times \langle k_1 k_2; q_1' q_2' | k_1 k_2; k'' q' \rangle$$
$$\times \langle k_1 k_2; q_1 q_2 | k_1 k_2; k'' q'' \rangle \mathscr{D}_{q' q''}^{(k'')}(R^{-1}) X_{q_1'}^{(k_1)} Z_{q_2'}^{(k_2)},$$

where we have used the Clebsch–Gordan series formula (3.389). The preceding expression becomes

$$= \sum_{k''} \sum_{q_1'} \sum_{q_2'} \sum_{q''} \sum_{q'} \delta_{kk''} \delta_{qq''} \langle k_1 k_2; q_1' q_2' | k_1 k_2; k'' q' \rangle \mathscr{D}_{q' q''}^{(k'')}(R^{-1}) X_{q_1'}^{(k_1)} Z_{q_2'}^{(k_2)},$$

where we have used the orthogonality of Clebsch–Gordan coefficients (3.362). Finally, this expression reduces to

$$= \sum_{q'} \left( \sum_{q_1'} \sum_{q_2'} \langle k_1 k_2; q_1' q_2' | k_1 k_2; kq' \rangle X_{q_1'}^{(k_1)} Z_{q_2'}^{(k_2)} \right) \mathscr{D}_{q' q}^{(k)}(R^{-1})$$
$$= \sum_{q'} T_{q'}^{(k)} \mathscr{D}_{q' q}^{(k)}(R^{-1}) = \sum_{q'} \mathscr{D}_{qq'}^{(k)*}(R) T_{q'}^{(k)}.$$

The foregoing shows how we can construct tensor operators of higher or lower ranks by multiplying two tensor operators. Furthermore, the manner in which we construct tensor products out of two tensors is completely analogous to the manner in which we construct an angular-momentum eigenstate by adding two angular momenta; exactly the same Clebsch–Gordan coefficients appear if we let $k_{1,2} \to j_{1,2}$ and $q_{1,2} \to m_{1,2}$.

### 3.11.4  Matrix Elements of Tensor Operators; the Wigner–Eckart Theorem

In considering the interactions of an electromagnetic field with atoms and nuclei, it is often necessary to evaluate matrix elements of tensor operators with respect to angular-momentum eigenstates. Examples of this will be given in Chapter 5. In general, it is a formidable dynamic task to calculate such matrix elements. However, there are certain properties of these matrix elements that follow purely from kinematic or geometric considerations, which we now discuss.

First, there is a very simple $m$-selection rule:

$$\langle \alpha', j'm'|T_q^{(k)}|\alpha, jm\rangle = 0, \quad \text{unless} \quad m' = q + m. \tag{3.471}$$

*Proof.* Using (3.468a), we have

$$\langle \alpha', j'm'|\left(\left[J_z, T_q^{(k)}\right] - \hbar q T_q^{(k)}\right)|\alpha, jm\rangle = [(m' - m)\hbar - \hbar q]$$
$$\times \langle \alpha', j'm'|T_q^{(k)}|\alpha, jm\rangle = 0;$$

hence,

$$\langle \alpha', j'm'|T_q^{(k)}|\alpha, jm\rangle = 0, \quad \text{unless} \quad m' = q + m.$$

Another way to see this is to note the transformation property of $T_q^{(k)}|\alpha, jm\rangle$ under rotation, namely,

$$\mathcal{D} T_q^{(k)}|\alpha, jm\rangle = \mathcal{D} T_q^{(k)} \mathcal{D}^\dagger \mathcal{D}|\alpha, jm\rangle. \tag{3.472}$$

If we now let $\mathcal{D}$ stand for a rotation operator around the $z$-axis, we get [see (3.465b) and (3.16)]

$$\mathcal{D}(\hat{\mathbf{z}}, \phi) T_q^{(k)}|\alpha, jm\rangle = e^{-iq\phi} e^{-im\phi} T_q^{(k)}|\alpha, jm\rangle, \tag{3.473}$$

which is orthogonal to $|\alpha', j'm'\rangle$ unless $q + m = m'$.

We are going to prove one of the most important theorems in quantum mechanics, the **Wigner–Eckart theorem**.

**The Wigner–Eckart Theorem**    *The matrix elements of tensor operators with respect to angular-momentum eigenstates satisfy*

$$\langle \alpha', j'm'|T_q^{(k)}|\alpha, jm\rangle = \langle jk; mq|jk; j'm'\rangle \frac{\langle \alpha'j'||T^{(k)}||\alpha j\rangle}{\sqrt{2j'+1}}, \tag{3.474}$$

*where the* **double-bar matrix element** *is independent of $m$ and $m'$, and $q$.*

Before we present a proof of this theorem, let us look at its significance. First, we see that the matrix element is written as the product of two factors. The first factor is a Clebsch–Gordan coefficient for adding $j$ and $k$ to get $j'$. It depends only on the geometry, that is, the way the system is oriented with respect to the $z$-axis. There is no reference whatsoever to the particular nature of the tensor operator. The second factor does depend on the dynamics, for instance, $\alpha$ may stand for the radial quantum number and its evaluation may evolve, for example, evaluation of radial integrals. On the other hand, it is completely independent of the magnetic quantum numbers $m$, $m'$, and $q$, which specify the orientation of the physical system. To evaluate $\langle \alpha', j'm'|T_q^{(k)}|\alpha, jm\rangle$ with various combinations of $m$, $m'$, and $q'$ it is sufficient to know just one of them; all others can be related geometrically because they are proportional to Clebsch–Gordan coefficients, which are known. The common proportionality factor is $\langle \alpha'j'||T^{(k)}||\alpha j \rangle$, which makes no reference whatsoever to the geometric features.

The selection rules for the tensor operator matrix element can be immediately read off from the selection rules for adding angular momentum. Indeed, from the requirement that the Clebsch–Gordan coefficient be nonvanishing, we immediately obtain the $m$-selection rule (3.471) derived before and also the triangular relation

$$|j - k| \leq j' \leq j + k. \tag{3.475}$$

Now we prove the theorem.

**Proof**  Using (3.468b) we have

$$\langle \alpha',j'm'|[J_{\pm}, T_q^{(k)}]|\alpha,jm \rangle = \hbar \sqrt{(k \mp q)(k \pm q + 1)} \langle \alpha',j'm'|T_{q\pm1}^{(k)}|\alpha,jm \rangle, \tag{3.476}$$

or using (3.191) and (3.192) we have

$$\sqrt{(j' \pm m')(j' \mp m' + 1)} \langle \alpha',j',m' \mp 1|T_q^{(k)}|\alpha,jm \rangle$$
$$= \sqrt{(j \mp m)(j \pm m + 1)} \langle \alpha',j'm'|T_q^{(k)}|\alpha,j,m \pm 1 \rangle$$
$$+ \sqrt{(k \mp q)(k \pm q + 1)} \langle \alpha',j'm'|T_{q\pm1}^{(k)}|\alpha,jm \rangle. \tag{3.477}$$

Compare this with the recursion relation for the Clebsch–Gordan coefficient (3.369). Note the striking similarity if we substitute $j' \to j$, $m' \to m$, $j \to j_1$, $m \to m_1$, $k \to j_2$, and $q \to m_2$. Both recursion relations are of the form $\sum_j a_{ij} x_j = 0$, that is, first-order linear homogeneous equations with the same coefficients $a_{ij}$. Whenever we have

$$\sum_j a_{ij} x_j = 0, \quad \sum_j a_{ij} y_j = 0, \tag{3.478}$$

we cannot solve for the $x_j$ (or $y_j$) individually but we can solve for the ratios; so

$$\frac{x_j}{x_k} = \frac{y_j}{y_k} \quad \text{or} \quad x_j = c y_j, \tag{3.479}$$

where $c$ is a universal proportionality factor. Noting that $\langle j_1 j_2; m_1, m_2 \pm 1|j_1 j_2; jm \rangle$ in the Clebsch–Gordan recursion relation (3.369) corresponds to $\langle \alpha',j'm'|T_{q\pm1}^{(k)}|\alpha,jm \rangle$, we see that

$$\langle \alpha',j'm'|T_{q\pm1}^{(k)}|\alpha,jm \rangle = (\text{universal proportionality constant independent of}$$
$$m, q, \text{and} \, m') \langle jk; mq \pm 1|jk; j'm' \rangle, \tag{3.480}$$

which proves the theorem.  □

There are several conventions for the "universal proportionality constant." Our choice (3.474) is consistent with Gottfried and Yan (2003). Edmonds (1974) uses a different, popular convention, and also includes a guide to other choices.

Let us now look at two simple examples of the Wigner–Eckart theorem.

**Example 6**   Tensor of rank 0, that is, scalar $T_0^{(0)} = S$. The matrix element of a scalar operator satisfies

$$\langle \alpha', j'm'|S|\alpha, jm \rangle = \delta_{jj'}\delta_{mm'}\frac{\langle \alpha'j'||S||\alpha j \rangle}{\sqrt{2j'+1}} \tag{3.481}$$

because $S$ acting on $|\alpha, jm \rangle$ is like adding an angular momentum of zero. Thus the scalar operator cannot change $j, m$ values.

**Example 7**   Vector operator which in the spherical tensor language is a rank 1 tensor. The spherical component of $\mathbf{V}$ can be written as $V_{q=\pm 1,0}$, so we have the selection rule

$$\Delta m \equiv m' - m = \pm 1, 0, \quad \Delta j \equiv j' - j = \begin{cases} \pm 1 \\ 0. \end{cases} \tag{3.482}$$

In addition, the $0 \to 0$ transition is forbidden. This selection rule is of fundamental importance in the theory of radiation; it is the dipole selection rule obtained in the long-wavelength limit of emitted photons.

For $j = j'$ the Wigner–Eckart theorem, when applied to the vector operator, takes a particularly simple form, often known as the **projection theorem** for obvious reasons.

**The Projection Theorem**   This theorem states that

$$\langle \alpha', jm'|V_q|\alpha, jm \rangle = \frac{\langle \alpha', jm|\mathbf{J}\cdot\mathbf{V}|\alpha, jm \rangle}{\hbar^2 j(j+1)}\langle jm'|J_q|jm \rangle, \tag{3.483}$$

where analogous to our discussion after (3.469) we choose

$$J_{\pm 1} = +\frac{1}{\sqrt{2}}(J_x \pm iJ_y) = \mp\frac{1}{\sqrt{2}}J_{\pm}, \quad J_0 = J_z. \tag{3.484}$$

**Proof**   Noting (3.469) we have

$$\langle \alpha', jm|\mathbf{J}\cdot\mathbf{V}|\alpha, jm \rangle = \langle \alpha', jm|(J_0V_0 - J_{+1}V_{-1} - J_{-1}V_{+1})|\alpha, jm \rangle$$

$$= m\hbar\langle \alpha', jm|V_0|\alpha, jm \rangle + \frac{\hbar}{\sqrt{2}}\sqrt{(j+m)(j-m+1)}$$

$$\times \langle \alpha', jm-1|V_{-1}|\alpha, jm \rangle$$

$$-\frac{\hbar}{\sqrt{2}}\sqrt{(j-m)(j+m+1)}\langle \alpha', jm+1|V_{+1}|\alpha, jm \rangle$$

$$= c_{jm}\langle \alpha'j||\mathbf{V}||\alpha j \rangle \tag{3.485}$$

by the Wigner–Eckart theorem (3.474), where $c_{jm}$ is independent of $\alpha$, $\alpha'$, and $\mathbf{V}$, and the matrix elements of $V_{0,\pm 1}$ are all proportional to the double-bar matrix element (sometimes also called the **reduced matrix element**). Furthermore, $c_{jm}$ is independent of $m$ because $\mathbf{J}\cdot\mathbf{V}$ is a scalar operator, so we may as well write it as $c_j$. Because $c_j$ does not depend on $\mathbf{V}$, (3.485) holds even if we let $\mathbf{V} \to \mathbf{J}$ and $\alpha' \to \alpha$, that is,

$$\langle \alpha, jm|\mathbf{J}^2|\alpha, jm \rangle = c_j\langle \alpha j||\mathbf{J}||\alpha j \rangle. \tag{3.486}$$

Returning to the Wigner–Eckart theorem applied to $V_q$ and $J_q$, we have

$$\frac{\langle \alpha',jm'|V_q|\alpha,jm\rangle}{\langle \alpha,jm'|J_q|\alpha,jm\rangle} = \frac{\langle \alpha'j||\mathbf{V}||\alpha j\rangle}{\langle \alpha j||\mathbf{J}||\alpha j\rangle}. \tag{3.487}$$

But the right-hand side of (3.487) is the same as $\langle \alpha',jm|\mathbf{J}\cdot\mathbf{V}|\alpha,jm\rangle/\langle \alpha,jm|\mathbf{J}^2|\alpha,jm\rangle$ by (3.485) and (3.486). Moreover, the left-hand side of (3.486) is just $j(j+1)\hbar^2$. So

$$\langle \alpha',jm'|V_q|\alpha,jm\rangle = \frac{\langle \alpha',jm|\mathbf{J}\cdot\mathbf{V}|\alpha,jm\rangle}{\hbar^2 j(j+1)} \langle jm'|J_q|jm\rangle, \tag{3.488}$$

which proves the projection theorem.     □

We will give applications of the theorem in subsequent sections.

# Problems

**3.1**   Use the specific form of $S_x$ given by (3.25) to evaluate (3.23) and show that $S_x$ rotates as expected through an angle $\phi$ about the $z$-axis.

**3.2**   Find the eigenvalues and eigenvectors of $\sigma_y = \begin{pmatrix} 0 & -i \\ i & 0 \end{pmatrix}$. Suppose an electron is in the spin state $\begin{pmatrix} \alpha \\ \beta \end{pmatrix}$. If $s_y$ is measured, what is the probability of the result $\hbar/2$?

**3.3**   Find, by explicit construction using Pauli matrices, the eigenvalues for the Hamiltonian

$$H = -\frac{2\mu}{\hbar}\mathbf{S}\cdot\mathbf{B}$$

for a spin $\frac{1}{2}$ particle in the presence of a magnetic field $\mathbf{B} = B_x\hat{\mathbf{x}} + B_y\hat{\mathbf{y}} + B_z\hat{\mathbf{z}}$.

**3.4**   Consider the $2\times 2$ matrix defined by

$$U = \frac{a_0 + i\boldsymbol{\sigma}\cdot\mathbf{a}}{a_0 - i\boldsymbol{\sigma}\cdot\mathbf{a}},$$

where $a_0$ is a real number and $\mathbf{a}$ is a three-dimensional vector with real components.
a. Prove that $U$ is unitary and unimodular.
b. In general, a $2\times 2$ unitary unimodular matrix represents a rotation in three dimensions. Find the axis and angle of rotation appropriate for $U$ in terms of $a_0, a_1, a_2$, and $a_3$.

**3.5**   The spin-dependent Hamiltonian of an electron-positron system in the presence of a uniform magnetic field in the $z$-direction can be written as

$$H = A\mathbf{S}^{(e^-)}\cdot\mathbf{S}^{(e^+)} + \left(\frac{eB}{mc}\right)\left(S_z^{(e^-)} - S_z^{(e^+)}\right).$$

Suppose the spin function of the system is given by $\chi_+^{(e^-)}\chi_-^{(e^+)}$.

a. Is this an eigenfunction of $H$ in the limit $A \to 0$, $eB/mc \neq 0$? If it is, what is the energy eigenvalue? If it is not, what is the expectation value of $H$?

b. Same problem when $eB/mc \to 0$, $A \neq 0$.

**3.6**  Consider a *spin* 1 particle. Evaluate the matrix elements of

$$S_z(S_z + \hbar)(S_z - \hbar) \quad \text{and} \quad S_x(S_x + \hbar)(S_x - \hbar).$$

**3.7**  Let the Hamiltonian of a rigid body be

$$H = \frac{1}{2}\left(\frac{K_1^2}{I_1} + \frac{K_2^2}{I_2} + \frac{K_3^2}{I_3}\right),$$

where $\mathbf{K}$ is the angular momentum in the body frame. From this expression obtain the Heisenberg equation of motion for $\mathbf{K}$ and then find Euler's equation of motion in the correspondence limit.

**3.8**  Let $U = e^{iG_3\alpha}e^{iG_2\beta}e^{iG_3\gamma}$, where $(\alpha, \beta, \gamma)$ are the Eulerian angles. In order that $U$ represent a rotation $(\alpha, \beta, \gamma)$, what are the commutation rules satisfied by the $G_k$? Relate $\mathbf{G}$ to the angular-momentum operators.

**3.9**  What is the meaning of the following equation:

$$U^{-1}A_k U = \sum_l R_{kl}A_l,$$

where the three components of $\mathbf{A}$ are matrices? From this equation show that matrix elements $\langle m|A_k|n\rangle$ transform like vectors.

**3.10**  Consider a sequence of Euler rotations represented by

$$\mathscr{D}^{(1/2)}(\alpha, \beta, \gamma) = \exp\left(\frac{-i\sigma_3\alpha}{2}\right)\exp\left(\frac{-i\sigma_2\beta}{2}\right)\exp\left(\frac{-i\sigma_3\gamma}{2}\right)$$

$$= \begin{pmatrix} e^{-i(\alpha+\gamma)/2}\cos\dfrac{\beta}{2} & -e^{-i(\alpha-\gamma)/2}\sin\dfrac{\beta}{2} \\ e^{i(\alpha-\gamma)/2}\sin\dfrac{\beta}{2} & e^{i(\alpha+\gamma)/2}\cos\dfrac{\beta}{2} \end{pmatrix}.$$

Because of the group properties of rotations, we expect that this sequence of operations is equivalent to a *single* rotation about some axis by an angle $\theta$. Find $\theta$.

**3.11**  Use the triangle inequality (1.147) and the definition (3.100) of the density operator $\rho$ to prove that $0 \leq \text{Tr}(\rho^2) \leq 1$.

**3.12**  A large collection of spin $\frac{1}{2}$ particles is in a mixture of the two states $|S_z; +\rangle$ and $|S_y; -\rangle$. The fraction of particles in the state $|S_z; +\rangle$ is $a$. Find the ensemble averages $[S_x]$, $[S_y]$, and $[S_z]$ in terms of $a$. Confirm that your expression gives the answers you expect for $a = 0$ and $a = 1$.

**3.13**  a. Consider a pure ensemble of identically prepared spin $\frac{1}{2}$ systems. Suppose the expectation values $\langle S_x \rangle$ and $\langle S_z \rangle$ and the sign of $\langle S_y \rangle$ are known. Show how we may determine the state vector. Why is it unnecessary to know the magnitude of $\langle S_y \rangle$?

b. Consider a mixed ensemble of spin $\frac{1}{2}$ systems. Suppose the ensemble averages $[S_x], [S_y]$, and $[S_z]$ are all known. Show how we may construct the $2 \times 2$ density matrix that characterizes the ensemble.

**3.14** Consider a one-dimensional simple harmonic oscillator with frequency $\omega$ and eigenstates $|0\rangle, |1\rangle, |2\rangle,\dots$. A mixed ensemble is formed with equal parts of each of the three states

$$|\alpha\rangle \equiv \frac{1}{\sqrt{2}}[|0\rangle + |1\rangle], \qquad |\beta\rangle \equiv \frac{1}{\sqrt{2}}[|1\rangle + |2\rangle], \qquad \text{and} \qquad |2\rangle.$$

Find the density operator $\rho$ and calculate the ensemble average of the energy.

**3.15** a. Prove that the time evolution of the density operator $\rho$ (in the Schrödinger picture) is given by

$$\rho(t) = \mathscr{U}(t, t_0)\rho(t_0)\mathscr{U}^\dagger(t, t_0).$$

b. Suppose we have a pure ensemble at $t = 0$. Prove that it cannot evolve into a mixed ensemble as long as the time evolution is governed by the Schrödinger equation.

**3.16** Consider an ensemble of spin 1 systems. The density matrix is now a $3 \times 3$ matrix. How many independent (real) parameters are needed to characterize the density matrix? What must we know in addition to $[S_x], [S_y]$, and $[S_z]$ to characterize the ensemble completely?

**3.17** An angular-momentum eigenstate $|j, m = m_{\max} = j\rangle$ is rotated by an infinitesimal angle $\varepsilon$ about the $y$-axis. Without using the explicit form of the $d_{m'm}^{(j)}$ function, obtain an expression for the probability for the new rotated state to be found in the original state up to terms of order $\varepsilon^2$.

**3.18** Show that the $3 \times 3$ matrices $G_i(i = 1, 2, 3)$ whose elements are given by

$$(G_i)_{jk} = -i\hbar \varepsilon_{ijk},$$

where $j$ and $k$ are the row and column indices, satisfy the angular-momentum commutation relations. What is the physical (or geometric) significance of the transformation matrix that connects $G_i$ to the more usual $3 \times 3$ representations of the angular-momentum operator $J_i$ with $J_3$ taken to be diagonal? Relate your result to

$$\mathbf{V} \to \mathbf{V} + \hat{\mathbf{n}}\delta\phi \times \mathbf{V}$$

under infinitesimal rotations. (*Note*: This problem may be helpful in understanding the photon spin.)

**3.19** a. Using the fact that $J_x, J_y, J_z$, and $J_\pm \equiv J_x \pm iJ_y$ satisfy the usual angular-momentum commutation relations, prove that

$$\mathbf{J}^2 = J_z^2 + J_+J_- - \hbar J_z.$$

b. Using this result, or otherwise, derive the coefficient $c_-$ that appears in

$$J_-|jm\rangle = c_-|j,m-1\rangle.$$

**3.20** Construct the matrix representations of the operators $J_x$ and $J_y$ for a spin 1 system, in the $J_z$ basis, spanned by the kets $|+\rangle \equiv |1,1\rangle$, $|0\rangle \equiv |1,0\rangle$, and $|-\rangle \equiv |1,-1\rangle$. Use these matrices to find the three analogous eigenstates for each of the two operators $J_x$ and $J_y$ in terms of $|+\rangle$, $|0\rangle$, and $|-\rangle$.

**3.21** Show that the orbital angular-momentum operator $\mathbf{L}$ commutes both with the operators $\mathbf{p}^2$ and $\mathbf{x}^2$, that is prove (3.264).

**3.22** For the orbital angular-momentum operator $\mathbf{L} = \mathbf{x} \times \mathbf{p}$, derive (3.218), that is

$$\langle \mathbf{x}'|L_z|\alpha\rangle = -i\hbar\frac{\partial}{\partial\phi}\langle\mathbf{x}'|\alpha\rangle$$

by using the standard spherical coordinate transformation

$$x' = r\cos\phi\sin\theta, \qquad y' = r\sin\phi\sin\theta, \qquad z' = r\cos\theta.$$

**3.23** The wave function of a particle subjected to a spherically symmetrical potential $V(r)$ is given by

$$\psi(\mathbf{x}) = (x+y+3z)f(r).$$

a. Is $\psi$ an eigenfunction of $\mathbf{L}^2$? If so, what is the $l$-value? If not, what are the possible values of $l$ we may obtain when $\mathbf{L}^2$ is measured?
b. What are the probabilities for the particle to be found in various $m_l$ states?
c. Suppose it is known somehow that $\psi(\mathbf{x})$ is an energy eigenfunction with eigenvalue $E$. Indicate how we may find $V(r)$.

**3.24** A particle in a spherically symmetrical potential is known to be in an eigenstate of $\mathbf{L}^2$ and $L_z$ with eigenvalues $\hbar^2 l(l+1)$ and $m\hbar$, respectively. Prove that the expectation values between $|lm\rangle$ states satisfy

$$\langle L_x\rangle = \langle L_y\rangle = 0, \quad \langle L_x^2\rangle = \langle L_y^2\rangle = \frac{[l(l+1)\hbar^2 - m^2\hbar^2]}{2}.$$

Interpret this result semiclassically.

**3.25** Suppose a half-integer $l$-value, say $\frac{1}{2}$, were allowed for orbital angular momentum. From

$$L_+ Y_{1/2}^{1/2}(\theta,\phi) = 0,$$

we may deduce, as usual,

$$Y_{1/2}^{1/2}(\theta,\phi) \propto e^{i\phi/2}\sqrt{\sin\theta}.$$

Now try to construct $Y_{1/2}^{-1/2}(\theta,\phi)$ (a) by applying $L_-$ to $Y_{1/2}^{1/2}(\theta,\phi)$ and (b) using $L_- Y_{1/2}^{-1/2}(\theta,\phi) = 0$. Show that the two procedures lead to contradictory results. (This gives an argument against half-integer $l$-values for orbital angular momentum.)

**3.26** Consider an orbital angular-momentum eigenstate $|l = 2, m = 0\rangle$. Suppose this state is rotated by an angle $\beta$ about the $y$-axis. Find the probability for the new state to be found in $m = 0, \pm 1$, and $\pm 2$. (The spherical harmonics for $l = 0, 1$, and 2 given in Section B.5 in Appendix B may be useful.)

**3.27** Show that by keeping both terms in (3.275), the origin becomes a source of probability, provided that there is a relative phase between the constants $A$ and $B$.

**3.28** Consider the energy eigenvalues for a spherically symmetric "box" of radius $a$.
     a. For the box with infinite walls, check the eigenvalues for the $l = 0$, $l = 1$, and $l = 2$ states, given in (3.287), (3.288), and (3.289).
     b. Find the lowest energy eigenvalues with $l = 0$ for a finite spherical box with potential wall height $V_0 = \hbar^2 \beta^2 / 2ma^2$ where $\beta = 4$, 10, 25, and 100, and show that your numerical results approach the appropriate value given in (a).

**3.29** The goal of this problem is to determine degenerate eigenstates of the three-dimensional isotropic harmonic oscillator written as eigenstates of $\mathbf{L}^2$ and $L_z$, in terms of the Cartesian eigenstates $|n_x n_y n_z\rangle$.
     a. Show that the angular-momentum operators are given by

$$L_i = i\hbar \varepsilon_{ijk} a_j a_k^\dagger$$
$$\mathbf{L}^2 = \hbar^2 \left[ N(N+1) - a_k^\dagger a_k^\dagger a_j a_j \right]$$

where summation is implied over repeated indices, $\varepsilon_{ijk}$ is the totally antisymmetric symbol, and $N \equiv a_j^\dagger a_j$ counts the total number of quanta.
     b. Use these relations to express the states $|qlm\rangle = |01m\rangle$, $m = 0, \pm 1$, in terms of the three eigenstates $|n_x n_y n_z\rangle$ that are degenerate in energy. Write down the representation of your answer in coordinate space, and check that the angular and radial dependences are correct.
     c. Repeat for $|qlm\rangle = |100\rangle$.
     d. Repeat for $|qlm\rangle = |02m\rangle$, with $m = 0$, 1, and 2.

**3.30** By considering the associated Laguerre polynomials $L_p^q(x)$, which are also solutions to Kummer's equation (3.308), we can derive the normalization constant in (3.320).
     a. First consider the Laguerre polynomials $L_p(x)$, which are defined according to a generating function as

$$g(x,t) = \frac{e^{-xt/(1-t)}}{1-t} = \sum_{p=0}^{\infty} L_p(x) \frac{t^p}{p!}$$

where $0 < t < 1$. Prove that $L_n(0) = n!$ and $L_0(x) = 1$. Differentiate $g(x,t)$ with respect to $x$ to show that

$$L_p'(x) - pL_{p-1}'(x) = -pL_{p-1}(x).$$

Then differentiate $g(x,t)$ with respect to $t$ to show that

$$L_{p+1}(x) - (2p + 1 - x)L_p(x) + p^2 L_{p-1}(x) = 0.$$

Combine these equations to derive the differential equation for the $L_p(x)$, namely

$$xL_p''(x) + (1-x)L_p'(x) + pL_p(x) = 0.$$

b. The associated laguerre Polynomials $L_p^q(x)$ are defined from the $L_p(x)$ as

$$L_p^q(x) = (-1)^q \frac{d^q}{dx^q} [L_{p+q}(x)].$$

Use this to show that the $L_p^q(x)$ satisfy the differential equation

$$xL_p^{q''}(x) + (q+1-x)L_p^{q'}(x) + pL_p^q(x) = 0$$

which is the same as (3.308) where $q = c - 1$ and $p = -a$. Also show that

$$h(x,t) = \frac{(-1)^q}{(1-t)^{q+1}} e^{-xt/(1-t)} = \sum_{p=0}^{\infty} L_p^q(x) \frac{t^p}{(p+q)!}$$

which gives $L_p^q(0) = (-1)^q [(p+q)!]^2/p! \, q!$ through a generating function $h(x,t)$.

c. Now find the normalization constant in (3.320). Start with the relationship

$$\int_0^{\infty} x^{q+1} e^{-x} h(x,t) h(x,s) \, dx = \sum_{p=0}^{\infty} \sum_{p'=0}^{\infty} \frac{t^p}{(p+q)!} \frac{s^{p'}}{(p'+q)!} I_{pp'}^q$$

where $I_{pp'}^q \equiv \int_0^{\infty} x^{q+1} e^{-x} L_p^q(x) L_{p'}^q(x) \, dx$, and determine $I_{pp}^q$ by isolating the terms with the same powers of $s$ and $t$. Make use of the generalized binomial expansion

$$\frac{1}{(1-x)^n} = \sum_{p=0}^{\infty} \binom{n+p-1}{p} x^p = \sum_{p=0}^{\infty} \frac{(n+p-1)!}{p!\,(n-1)!} x^p.$$

**3.31** Consider the Coulomb potential $V(\mathbf{x}) = -Ze^2/r$ and define the (quantum-mechanical operator analogue of the) Runge–Lenz vector

$$\mathbf{M} = \frac{1}{2m}(\mathbf{p} \times \mathbf{L} - \mathbf{L} \times \mathbf{p}) - \frac{Ze^2}{r}\mathbf{x}.$$

Prove that $\mathbf{M}$ is Hermitian and that it commutes with the Hamiltonian. We will return to $\mathbf{M}$ when we go through Pauli's algebraic solution for this Hamiltonian in Section 4.1.4.

**3.32** What is the physical significance of the operators

$$K_+ \equiv a_+^\dagger a_-^\dagger \quad \text{and} \quad K_- \equiv a_+ a_-$$

in Schwinger's scheme for angular momentum? Give the nonvanishing matrix elements of $K_+$.

**3.33** Carry through the argument outlined on p. 208 for adding two spin $\frac{1}{2}$ particles by diagonalizing the $4 \times 4$ matrix corresponding to the operator $\mathbf{S}^2$ given in (3.339). That is, construct the matrix representation of $\mathbf{S}^2$ in the $|\pm \pm\rangle$ basis, and find the eigenvalues and eigenvectors. Your result should agree with (3.335).

**3.34** Find all nine states $|j,m\rangle$ for $j = 2$, 1, and 0 formed by adding $j_1 = 1$ and $j_2 = 1$. Use a simplified notation, where $|j,m\rangle$ is explicit and $\pm,0$ stand for $m_{1,2} = \pm1,0$, respectively, for example

$$|1,1\rangle = \frac{1}{\sqrt{2}}|+0\rangle - \frac{1}{\sqrt{2}}|0+\rangle.$$

You may also want to make use of the ladder operators $J_\pm$, or recursion relations, as well as orthonormality. Check your answers by finding a table of Clebsch–Gordan coefficients for comparison; see Appendix E.

**3.35** A spin $\frac{1}{2}$ particle is in an orbital angular momentum $l = 1$ state.

a. Starting with the total angular momentum state $|j,m\rangle = |j,j\rangle$ for $j = 3/2$, use the operator $\mathbf{J} = \mathbf{L} + \mathbf{S}$ to construct all the states $|j,m_j\rangle$ in terms of the eigenstates $|l,m_l\rangle$ for $\mathbf{L}^2$ and $L_z$, and the eigenstates $|1/2,\pm1/2\rangle$ for $\mathbf{S}^2$ and $S_z$. Check your answers against any available table of Clebsch–Gordan coefficients; see Appendix E.

b. If the particle is in a total angular-momentum eigenstate with $z$-component $+\hbar/2$, calculate the probability of finding the $z$-component of the spin of the particle to have the value $m_s = +1/2$.

**3.36** The "spin-angular functions" (aka "spinor spherical harmonics") are defined as

$$\mathscr{Y}_l^{j=l\pm1/2,m} = \frac{1}{\sqrt{2l+1}} \left[ \begin{array}{c} \pm\sqrt{l\pm m+\frac{1}{2}}Y_l^{m-1/2}(\theta,\phi) \\ \sqrt{l\mp m+\frac{1}{2}}Y_l^{m+1/2}(\theta,\phi) \end{array} \right].$$

See (3.384). These were constructed to be properly normalized eigenfunctions of $\mathbf{L}^2$, $\mathbf{S}^2$, $\mathbf{J}^2$, and $J_z$, where $\mathbf{J} \equiv \mathbf{L} + \mathbf{S}$. Use explicit calculations to prove the following:

a. The $\mathscr{Y}_l^{j=l\pm1/2,m}$ are normalized, that is

$$\int_0^{2\pi} d\phi \int_0^{\pi} \sin\theta d\theta \left(\mathscr{Y}_l^{j=l\pm1/2,m}\right)^\dagger \mathscr{Y}_l^{j=l\pm1/2,m} = 1.$$

b. The $\mathscr{Y}_l^{j=l\pm1/2,m}$ have the correct eigenvalues for $\mathbf{J}^2 = \mathbf{L}^2 + \mathbf{S}^2 + 2L_zS_z + L_+S_- + L_-S_+$, that is see (3.349), and $J_z$.

**3.37** a. Evaluate

$$\sum_{m=-j}^{j} |d_{mm'}^{(j)}(\beta)|^2 m$$

for *any* $j$ (integer or half-integer); then check your answer for $j = \frac{1}{2}$.

b. Prove, for any $j$,

$$\sum_{m=-j}^{j} m^2 |d_{m'm}^{(j)}(\beta)|^2 = \frac{1}{2}j(j+1)\sin^2\beta + m'^2\frac{1}{2}(3\cos^2\beta - 1).$$

[*Hint*: This can be proved in many ways. You may, for instance, examine the rotational properties of $J_z^2$ using the spherical (irreducible) tensor language.]

**3.38**  a. Consider a system with $j = 1$. Explicitly write

$$\langle j = 1, m' | J_y | j = 1, m \rangle$$

in $3 \times 3$ matrix form.

b. Show that for $j = 1$ only, it is legitimate to replace $e^{-iJ_y\beta/\hbar}$ by

$$1 - i\left(\frac{J_y}{\hbar}\right)\sin\beta - \left(\frac{J_y}{\hbar}\right)^2 (1 - \cos\beta).$$

c. Using (b), prove

$$d^{(j=1)}(\beta) = \begin{pmatrix} \left(\frac{1}{2}\right)(1+\cos\beta) & \left(\frac{1}{\sqrt{2}}\right)\sin\beta & \left(\frac{1}{2}\right)(1-\cos\beta) \\ \left(\frac{1}{\sqrt{2}}\right)\sin\beta & \cos\beta & -\left(\frac{1}{\sqrt{2}}\right)\sin\beta \\ \left(\frac{1}{2}\right)(1-\cos\beta) & \left(\frac{1}{\sqrt{2}}\right)\sin\beta & \left(\frac{1}{2}\right)(1+\cos\beta) \end{pmatrix}.$$

**3.39**  Express the matrix element $\langle \alpha_2\beta_2\gamma_2 | J_3^2 | \alpha_1\beta_1\gamma_1 \rangle$ in terms of a series in

$$\mathscr{D}_{mn}^j(\alpha\beta\gamma) = \langle \alpha\beta\gamma | jmn \rangle.$$

**3.40**  Consider a system made up of two spin $\frac{1}{2}$ particles. Observer A specializes in measuring the spin components of one of the particles ($s_{1z}, s_{1x}$ and so on), while observer B measures the spin components of the other particle. Suppose the system is known to be in a spin-singlet state, that is, $S_{\text{total}} = 0$.

a. What is the probability for observer A to obtain $s_{1z} = \hbar/2$ when observer B makes no measurement? Same problem for $s_{1x} = \hbar/2$.

b. Observer B determines the spin of particle 2 to be in the $s_{2z} = \hbar/2$ state with certainty. What can we then conclude about the outcome of observer A's measurement if (i) A measures $s_{1z}$ and (ii) A measures $s_{1x}$? Justify your answer.

**3.41**  Using the defining property (3.451) for a vector operator, prove that the momentum operator **p** is a vector, based on its commutation relations with angular momentum **L**.

**3.42**  Consider a spherical tensor of rank 1 (that is, a vector)

$$V_{\pm 1}^{(1)} = \mp \frac{V_x \pm iV_y}{\sqrt{2}}, \quad V_0^{(1)} = V_z.$$

Using the expression for $d^{(j=1)}$ given in Problem 3.38, evaluate

$$\sum_{q'} d_{qq'}^{(1)}(\beta) V_{q'}^{(1)}$$

and show that your results are just what you expect from the transformation properties of $V_{x,y,z}$ under rotations about the $y$-axis.

**3.43**  a. Construct a spherical tensor of rank 1 out of two different vectors $\mathbf{U} = (U_x, U_y, U_z)$ and $\mathbf{V} = (V_x, V_y, V_z)$. Explicitly write $T_{\pm 1,0}^{(1)}$ in terms of $U_{x,y,z}$ and $V_{x,y,z}$.

b. Construct a spherical tensor of rank 2 out of two different vectors $\mathbf{U}$ and $\mathbf{V}$. Write down explicitly $T_{\pm 2, \pm 1, 0}^{(2)}$ in terms of $U_{x,y,z}$ and $V_{x,y,z}$.

**3.44** Consider a spinless particle bound to a fixed center by a central force potential.
   a. Relate, as much as possible, the matrix elements

$$\langle n',l',m'|\mp \frac{1}{\sqrt{2}}(x\pm iy)|n,l,m\rangle \quad \text{and} \quad \langle n',l',m'|z|n,l,m\rangle$$

   using *only* the Wigner–Eckart theorem. Make sure to state under what conditions the matrix elements are nonvanishing.
   b. Do the same problem using wave functions $\psi(\mathbf{x})=R_{nl}(r)Y_l^m(\theta,\phi)$.

**3.45** a. Write $xy$, $xz$, and $(x^2-y^2)$ as components of the components of a spherical (irreducible) tensor of rank 2.
   b. The expectation value

$$Q \equiv e\langle \alpha,j,m=j|(3z^2-r^2)|\alpha,j,m=j\rangle$$

   is known as the *quadrupole moment*. Evaluate

$$e\langle \alpha,j,m'|(x^2-y^2)|\alpha,j,m=j\rangle,$$

   (where $m'=j, j-1, j-2,\ldots$) in terms of $Q$ and appropriate Clebsch–Gordan coefficients.

**3.46** A spin $\frac{3}{2}$ nucleus situated at the origin is subjected to an external inhomogeneous electric field. The basic electric quadrupole interaction may by taken to be

$$H_{\text{int}} = \frac{eQ}{2s(s-1)\hbar^2}\left[\left(\frac{\partial^2\phi}{\partial x^2}\right)_0 S_x^2 + \left(\frac{\partial^2\phi}{\partial y^2}\right)_0 S_y^2 + \left(\frac{\partial^2\phi}{\partial z^2}\right)_0 S_z^2\right],$$

where $\phi$ is the electrostatic potential satisfying Laplace's equation and the coordinate axes are so chosen that

$$\left(\frac{\partial^2\phi}{\partial x\partial y}\right)_0 = \left(\frac{\partial^2\phi}{\partial y\partial z}\right)_0 = \left(\frac{\partial^2\phi}{\partial x\partial z}\right)_0 = 0.$$

Show that the interaction energy can be written as

$$A(3S_z^2-\mathbf{S}^2)+B(S_+^2+S_-^2),$$

and express $A$ and $B$ in terms of $(\partial^2\phi/\partial x^2)_0$ and so on. Determine the energy eigenkets (in terms of $|m\rangle$, where $m=\pm\frac{3}{2},\pm\frac{1}{2}$) and the corresponding energy eigenvalues. Is there any degeneracy?

# 4 | Symmetry in Quantum Mechanics

Having studied the theory of rotation in detail, we are in a position to discuss, in more general terms, the connection between symmetries, degeneracies, and conservation laws. We have deliberately postponed this very important topic until now so that we can discuss it using the rotation symmetry of Chapter 3 as an example.

## 4.1 Symmetries, Conservation Laws, and Degeneracies

### 4.1.1 Symmetries in Classical Physics

We begin with an elementary review of the concepts of symmetry and conservation law in classical physics. In the Lagrangian formulation of quantum mechanics, we start with the Lagrangian $L$, which is a function of a generalized coordinate $q_i$ and the corresponding generalized velocity $\dot{q}_i$. If $L$ is unchanged under displacement,

$$q_i \to q_i + \delta q_i, \tag{4.1}$$

then we must have

$$\frac{\partial L}{\partial q_i} = 0. \tag{4.2}$$

It then follows, by virtue of the Lagrange equation, $d/dt(\partial L/\partial \dot{q}_i) - \partial L/\partial q_i = 0$, that

$$\frac{dp_i}{dt} = 0, \tag{4.3}$$

where the canonical momentum is defined as

$$p_i = \frac{\partial L}{\partial \dot{q}_i}. \tag{4.4}$$

So if $L$ is unchanged under displacement (4.1), then we have a conserved quantity, the canonical momentum conjugate to $q_i$.

Likewise, in the Hamiltonian formulation based on $H$ regarded as a function of $q_i$ and $p_i$, we have

$$\frac{dp_i}{dt} = 0 \tag{4.5}$$

whenever

$$\frac{\partial H}{\partial q_i} = 0. \tag{4.6}$$

So if the Hamiltonian does not explicitly depend on $q_i$, which is another way of saying $H$ has a symmetry under $q_i \rightarrow q_i + \delta q_i$, we have a conserved quantity.

## 4.1.2  Symmetry in Quantum Mechanics

In quantum mechanics we have learned to associate a **unitary operator**, say $\mathscr{S}$, with an operation like translation or rotation. It has become customary to call $\mathscr{S}$ a **symmetry operator** regardless of whether the physical system itself possesses the symmetry corresponding to $\mathscr{S}$. Further, we have learned that for symmetry operations that differ infinitesimally from the identity transformation, we can write

$$\mathscr{S} = 1 - \frac{i\varepsilon}{\hbar}G, \tag{4.7}$$

where $G$ is the Hermitian generator of the symmetry operator in question. Let us now suppose that $H$ is invariant under $\mathscr{S}$. We then have

$$\mathscr{S}^\dagger H \mathscr{S} = H. \tag{4.8}$$

But this is equivalent to

$$[G, H] = 0. \tag{4.9}$$

By virtue of the Heisenberg equation of motion, we have

$$\frac{dG}{dt} = 0; \tag{4.10}$$

hence, $G$ is a constant of the motion. For instance, if $H$ is invariant under translation, then momentum is a constant of the motion; if $H$ is invariant under rotation, then angular momentum is a constant of the motion.

It is instructive to look at the connection between (4.9) and conservation of $G$ from the point of view of an eigenket of $G$ when $G$ commutes with $H$. Suppose at $t_0$ the system is in an eigenstate of $G$. Then the ket at a later time obtained by applying the time-evolution operator

$$|g', t_0; t\rangle = U(t, t_0)|g'\rangle \tag{4.11}$$

is also an eigenket of $G$ with the same eigenvalue $g'$. In other words, once a ket is a $G$ eigenket, it is always a $G$ eigenket with the same eigenvalue. The proof of this is extremely simple once we realize that (4.9) and (4.10) also imply that $G$ commutes with the time-evolution operator, namely,

$$G[U(t, t_0)|g'\rangle] = U(t, t_0)G|g'\rangle = g'[U(t, t_0)|g'\rangle]. \tag{4.12}$$

## 4.1.3 Degeneracies

Let us now turn to the concept of degeneracies. Even though degeneracies may be discussed at the level of classical mechanics – for instance in discussing closed (non-precessing) orbits in the Kepler problem (Goldstein et al. (2002)) – this concept plays a far more important role in quantum mechanics. Let us suppose that

$$[H, \mathscr{S}] = 0 \tag{4.13}$$

for some symmetry operator and $|n\rangle$ is an energy eigenket with eigenvalue $E_n$. Then $\mathscr{S}|n\rangle$ is also an energy eigenket with the same energy, because

$$H(\mathscr{S}|n\rangle) = \mathscr{S}H|n\rangle = E_n(\mathscr{S}|n\rangle). \tag{4.14}$$

Suppose $|n\rangle$ and $\mathscr{S}|n\rangle$ represent different states. Then these are two states with the same energy, that is, they are degenerate. Quite often $\mathscr{S}$ is characterized by continuous parameters, say $\lambda$, in which case all states of the form $\mathscr{S}(\lambda)|n\rangle$ have the same energy.

We now consider rotation specifically. Suppose the Hamiltonian is rotationally invariant, so

$$[\mathscr{D}(R), H] = 0, \tag{4.15}$$

which necessarily implies that

$$[\mathbf{J}, H] = 0, \quad [\mathbf{J}^2, H] = 0. \tag{4.16}$$

We can then form simultaneous eigenkets of $H$, $\mathbf{J}^2$, and $J_z$, denoted by $|n; j, m\rangle$. The argument just given implies that all states of the form

$$\mathscr{D}(R)|n; j, m\rangle \tag{4.17}$$

have the same energy. We saw in Chapter 3 that under rotation different $m$-values are mixed up. In general, $\mathscr{D}(R)|n; j, m\rangle$ is a linear combination of $2j + 1$ independent states. Explicitly,

$$\mathscr{D}(R)|n; j, m\rangle = \sum_{m'} |n; j, m'\rangle \mathscr{D}_{m'm}^{(j)}(R), \tag{4.18}$$

and by changing the continuous parameter that characterizes the rotation operator $\mathscr{D}(R)$, we can get different linear combinations of $|n; j, m'\rangle$. If all states of form $\mathscr{D}(R)|n; j, m\rangle$ with arbitrary $\mathscr{D}(R)$ are to have the same energy, it is then essential that each of $|n; j, m\rangle$ with different $m$ must have the same energy. So the degeneracy here is $(2j + 1)$-fold, just equal to the number of possible $m$-values. This point is also evident from the fact that all states obtained by successively applying $J_\pm$, which commutes with $H$, to $|n; jm\rangle$ have the same energy.

As an application, consider an atomic electron whose potential is written as $V(r) + V_{LS}(r)\mathbf{L} \cdot \mathbf{S}$. Because $r$ and $\mathbf{L} \cdot \mathbf{S}$ are both rotationally invariant, we expect a $(2j + 1)$-fold degeneracy for each atomic level. On the other hand, suppose there is an external electric or magnetic field, say in the $z$-direction. The rotational symmetry is now manifestly broken; as a result, the $(2j + 1)$-fold degeneracy is no longer expected and states characterized by different $m$-values no longer have the same energy. We will examine how this splitting arises in Chapter 5.

## 4.1.4  SO(4) Symmetry in the Coulomb Potential

A fine example of continuous symmetry in quantum mechanics is afforded by the hydrogen atom problem and the solution for the Coulomb potential. We carried out the solution to this problem in Section 3.7 where we discovered that the energy eigenvalues in (3.315) show the striking degeneracy summarized in (3.318). It would be even more striking if this degeneracy were just an accident, but indeed, it is the result of an additional symmetry that is particular to the problem of bound states of $1/r$ potentials.

The classical problem of orbits in such potentials, the Kepler problem, was of course well studied long prior to quantum mechanics. The fact that the solution leads to elliptical orbits that are *closed* means that there should be some (vector) constant of the motion which maintains the orientation of the major axis of the ellipse. We know that even a small deviation from a $1/r$ potential leads to precession of this axis, so we expect that the constant of the motion we seek is in fact particular to $1/r$ potentials.

Classically, this new constant of the motion is

$$\mathbf{M} = \frac{\mathbf{p} \times \mathbf{L}}{m} - \frac{Ze^2}{r}\mathbf{x} \tag{4.19}$$

where we refer back to the notation used in Section 3.7. This quantity is generally known as the *Lenz vector* or sometimes the *Runge–Lenz vector*. Rather than belabor the classical treatment here, we will move on to the quantum-mechanical treatment in terms of the symmetry responsible for this constant of the motion.

This new symmetry is called SO(4), completely analogous to the symmetry SO(3) studied in Section 3.3. That is, SO(4) is the group of rotation operators in *four* spatial dimensions. Equivalently, it is the group of orthogonal $4 \times 4$ matrices with unit determinant. Let us build up the properties of the symmetry that leads to the Lenz vector as a constant of the motion, and then we will see that these properties of those we expect from SO(4).

Our approach[1] closely follows that given by Schiff (1968), pp. 235–239. We first need to modify (4.19) to construct a Hermitian operator. For two Hermitian vector operators $\mathbf{A}$ and $\mathbf{B}$, it is easy to show that $(\mathbf{A} \times \mathbf{B})^\dagger = -\mathbf{B} \times \mathbf{A}$. Therefore a Hermitian version of the Lenz vector is

$$\mathbf{M} = \frac{1}{2m}(\mathbf{p} \times \mathbf{L} - \mathbf{L} \times \mathbf{p}) - \frac{Ze^2}{r}\mathbf{x}. \tag{4.20}$$

It can be shown that $\mathbf{M}$ commutes with the Hamiltonian

$$H = \frac{\mathbf{p}^2}{2m} - \frac{Ze^2}{r} \tag{4.21}$$

that is,

$$[\mathbf{M}, H] = 0, \tag{4.22}$$

---

[1] A classical treatment of the Runge–Lenz vector is given in Section 3.9 of Goldstein et al. (2002). The quantum-mechanical treatment was first done by Pauli, *Z. Phys.*, **33** (1925) 879, and an English translation "On the hydrogen spectrum from the standpoint of the new quantum mechanics" is published in Van der Waerden (1967).

so that indeed $\mathbf{M}$ is a (quantum mechanical) constant of the motion. (See Problem 3.31 in Chapter 3.) Other useful relations can be proven, namely

$$\mathbf{L} \cdot \mathbf{M} = 0 = \mathbf{M} \cdot \mathbf{L} \qquad (4.23)$$

$$\mathbf{M}^2 = \frac{2}{m} H \left( \mathbf{L}^2 + \hbar^2 \right) + Z^2 e^4. \qquad (4.24)$$

In order to identify the symmetry responsible for this constant of the motion, it is instructive to review the algebra of the generators of this symmetry. We already know part of this algebra, namely

$$[L_i, L_j] = i\hbar \varepsilon_{ijk} L_k \qquad (4.25)$$

which we wrote earlier as (3.211) in a notation where repeated indices ($k$ in this case) are automatically summed over components. One can also show that

$$[M_i, L_j] = i\hbar \varepsilon_{ijk} M_k \qquad (4.26)$$

which in fact establish $\mathbf{M}$ as a vector operator in the sense of (3.451). Finally, it is possible to derive

$$[M_i, M_j] = -i\hbar \varepsilon_{ijk} \frac{2}{m} H L_k. \qquad (4.27)$$

To be sure, (4.25), (4.26), and (4.27) do not form a closed algebra, due to the presence of $H$ in (4.27), and that makes it difficult to identify these operators as generators of a continuous symmetry. However, we can consider the problem of specific bound states. In this case, the vector space is truncated only to those that are eigenstates of $H$, with eigenvalue $E < 0$. In that case, we replace $H$ with $E$ in (4.27) and the algebra is closed. It is instructive to replace $\mathbf{M}$ with the scaled vector operator

$$\mathbf{N} \equiv \left( -\frac{m}{2E} \right)^{1/2} \mathbf{M}. \qquad (4.28)$$

In this case we have the closed algebra

$$[L_i, L_j] = i\hbar \varepsilon_{ijk} L_k \qquad (4.29a)$$

$$[N_i, L_j] = i\hbar \varepsilon_{ijk} N_k \qquad (4.29b)$$

$$[N_i, N_j] = i\hbar \varepsilon_{ijk} L_k. \qquad (4.29c)$$

So what is the symmetry operation generated by the operators $\mathbf{L}$ and $\mathbf{N}$ in (4.29)? Although it is far from obvious, the answer is "rotation in *four* spatial dimensions." The first clue is in the number of generators, namely six, each of which should correspond to rotation about some axis. Think of a rotation as an operation which mixes two orthogonal axes. Then, the number of generators for rotations in $n$ spatial dimensions should be the number of combinations of $n$ things taken two at a time, namely $n(n-1)/2$. Consequently, rotations in two dimensions require one generator, that is $L_z$. Rotations in three dimensions require three, namely $\mathbf{L}$, and four-dimensional rotations require six generators.

It is harder to see that (4.29) is the appropriate algebra for this kind of rotation, but we proceed as follows. In three spatial dimensions, the orbital angular-momentum operator

(3.210) generates rotations. We saw this clearly in (3.215) where an infinitesimal $z$-axis rotation on a state $|\alpha\rangle$ is represented in a rotated version of the $|x,y,z\rangle$ basis. This was just a consequence of the momentum operator being the generator of translations in space. In fact, a combination like $L_z = xp_y - yp_x$ indeed mixes the $x$-axis and $y$-axis, just as one would expect from the generator of rotations about the $z$-axis.

To generalize this to four spatial dimensions, we first associate $(x,y,z)$ and $(p_x,p_y,p_z)$ with $(x_1,x_2,x_3)$ and $(p_1,p_2,p_3)$. We are led to rewrite the generators as $L_3 = \tilde{L}_{12} = x_1p_2 - x_2p_1$, $L_1 = \tilde{L}_{23}$, and $L_2 = \tilde{L}_{31}$. If we then invent a new spatial dimension $x_4$ and its conjugate momentum $p_4$ (with the usual commutation relations) we can define

$$\tilde{L}_{14} = x_1p_4 - x_4p_1 \equiv N_1 \tag{4.30a}$$

$$\tilde{L}_{24} = x_2p_4 - x_4p_2 \equiv N_2 \tag{4.30b}$$

$$\tilde{L}_{34} = x_3p_4 - x_4p_3 \equiv N_3. \tag{4.30c}$$

It is easy to show that these operators $N_i$ obey the algebra (4.29). For example

$$[N_1,L_2] = [x_1p_4 - x_4p_1, x_3p_1 - x_1p_3]$$
$$= p_4[x_1,p_1]x_3 + x_4[p_1,x_1]p_3$$
$$= i\hbar(x_3p_4 - x_4p_3) = i\hbar N_3. \tag{4.31}$$

In other words, this is the algebra of four spatial dimensions. We will return to this notion in a moment, but for now we will press on with the degeneracies in the Coulomb potential that are implied by (4.14).

Defining the operators

$$\mathbf{I} \equiv (\mathbf{L} + \mathbf{N})/2 \tag{4.32}$$

$$\mathbf{K} \equiv (\mathbf{L} - \mathbf{N})/2 \tag{4.33}$$

we easily can prove the following algebra:

$$[I_i,I_j] = i\hbar\varepsilon_{ijk}I_k \tag{4.34a}$$

$$[K_i,K_j] = i\hbar\varepsilon_{ijk}K_k \tag{4.34b}$$

$$[I_i,K_j] = 0. \tag{4.34c}$$

Therefore, these operators obey independent angular momentum algebras. It is also evident that $[\mathbf{I},H] = [\mathbf{K},H] = 0$. Thus, these "angular momenta" are conserved quantities, and we denote the eigenvalues of the operators $\mathbf{I}^2$ and $\mathbf{K}^2$ by $i(i+1)\hbar^2$ and $k(k+1)\hbar^2$ respectively, with $i,k = 0,\frac{1}{2},1,\frac{3}{2},\ldots$.

Since $\mathbf{I}^2 - \mathbf{K}^2 = \mathbf{L}\cdot\mathbf{N} = 0$ by (4.23) and (4.28), we must have $i = k$. On the other hand, the operator

$$\mathbf{I}^2 + \mathbf{K}^2 = \frac{1}{2}\left(\mathbf{L}^2 + \mathbf{N}^2\right) = \frac{1}{2}\left(\mathbf{L}^2 - \frac{m}{2E}\mathbf{M}^2\right) \tag{4.35}$$

leads, with (4.24), to the numerical relation

$$2k(k+1)\hbar^2 = \frac{1}{2}\left(-\hbar^2 - \frac{m}{2E}Z^2e^4\right). \tag{4.36}$$

Solving for $E$ we find

$$E = -\frac{mZ^2 e^4}{2\hbar^2} \frac{1}{(2k+1)^2}.$$  (4.37)

This is the same as (3.315) with the principal quantum number $n$ replaced by $2k+1$. We now see that the degeneracy in the Coulomb problem arises from the two "rotational" symmetries represented by the operators $\mathbf{I}$ and $\mathbf{K}$. The degree of degeneracy, in fact, is $(2i+1)(2k+1) = (2k+1)^2 = n^2$. This is exactly what we arrived at in (3.318) except it is now clear that the degeneracy is no accident.

It is worth noting that we have just solved for the eigenvalues of the hydrogen atom without ever resorting to solving the Schrödinger equation. Instead, we exploited the inherent symmetries to arrive at the same answer. This solution was apparently first carried out by Pauli.

In the language of the theory of continuous groups, which we started to develop in Section 3.3, we see that the algebra (4.29) corresponds to the group SO(4). Furthermore, rewriting this algebra as (4.34) shows that this can also be thought of as two independent groups SU(2), that is SU(2)×SU(2). Although it is not the purpose of this book to include an introduction to group theory, we will carry this a little further to show how one formally carries out rotations in $n$ spatial dimensions, that is, the group SO($n$).

Generalizing the discussion in Section 3.3, consider the group of $n \times n$ orthogonal matrices $R$ which carry out rotations in $n$ dimensions. They can be parameterized as

$$R = \exp\left( i \sum_{q=1}^{n(n-1)/2} \phi^q \tau^q \right)$$  (4.38)

where the $\tau^q$ are purely imaginary, antisymmetric $n \times n$ matrices, that is $(\tau^q)^T = -\tau^q$ and the $\phi^q$ are generalized rotation angles. The antisymmetry condition ensures that $R$ is orthogonal. The overall factor of $i$ implies that the imaginary matrices $\tau^q$ are also Hermitian.

The $\tau^q$ are obviously related to the generators of the rotation operator. In fact, it is their commutation relations which should be parroted by the commutation relations of these generators. Following along as in Section 3.1, we compare the action of performing an infinitesimal rotation first about axis $q$ and then about axis $p$ with the rotation carried out in reverse order. Then,

$$\left(1 + i\phi^p \tau^p\right)\left(1 + i\phi^q \tau^q\right) - \left(1 + i\phi^q \tau^q\right)\left(1 + i\phi^p \tau^p\right)$$
$$= -\phi^p \phi^q \left[\tau^p, \tau^q\right]$$
$$= 1 - \left(1 + i\phi^p \phi^q \sum_r f_r^{pq} \tau^r\right)$$  (4.39)

where the last line of (4.39) recognizes that the result must be a second-order rotation about the two axes with some linear combination of generators. The $f_r^{pq}$ are called *structure constants* for this group of rotations. This gives us the commutation relations

$$[\tau^p, \tau^q] = i \sum_r f_r^{pq} \tau^r.$$  (4.40)

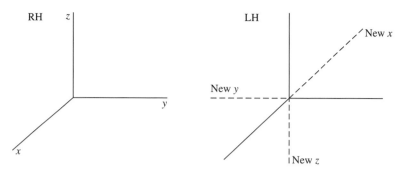

**Fig. 4.1**    Right-handed (RH) and left-handed (LH) systems.

To go further, one would need to determine the structure constants $f_r^{pq}$, and we leave these details to textbooks devoted to group theory. It is not hard to show, however, that in three dimensions, $f_r^{pq} = \varepsilon_{pqr}$ as expected.

## 4.2  Discrete Symmetries, Parity, or Space Inversion

So far we have considered continuous symmetry operators, that is, operations that can be obtained by applying successively infinitesimal symmetry operations. All symmetry operations useful in quantum mechanics are not necessarily of this form. In this chapter we consider three symmetry operations that can be considered to be discrete, as opposed to continuous – parity, lattice translation, and time reversal.

The first operation we consider is **parity**, or space inversion. The parity operation, as applied to transformation on the coordinate system, changes a right-handed (RH) system into a left-handed (LH) system, as shown in Figure 4.1. However, in this book we consider a transformation on state kets rather than on the coordinate system. Given $|\alpha\rangle$, we consider a space-inverted state, assumed to be obtained by applying a unitary operator $\pi$ known as the **parity operator**, as follows:

$$|\alpha\rangle \to \pi|\alpha\rangle. \tag{4.41}$$

We require the expectation value of **x** taken with respect to the space-inverted state to be opposite in sign:

$$\langle\alpha|\pi^\dagger \mathbf{x}\pi|\alpha\rangle = -\langle\alpha|\mathbf{x}|\alpha\rangle, \tag{4.42}$$

a very reasonable requirement. This is accomplished if

$$\pi^\dagger \mathbf{x}\pi = -\mathbf{x} \tag{4.43}$$

or

$$\mathbf{x}\pi = -\pi\mathbf{x}, \tag{4.44}$$

where we have used the fact that $\pi$ is unitary. In other words, **x** and $\pi$ must *anti*commute.

How does an eigenket of the position operator transform under parity? We claim that

$$\pi|\mathbf{x}'\rangle = e^{i\delta}|-\mathbf{x}'\rangle \tag{4.45}$$

where $e^{i\delta}$ is a phase factor ($\delta$ real). To prove this assertion let us note that

$$\mathbf{x}\pi|\mathbf{x}'\rangle = -\pi\mathbf{x}|\mathbf{x}'\rangle = (-\mathbf{x}')\pi|\mathbf{x}'\rangle. \tag{4.46}$$

This equation says that $\pi|\mathbf{x}'\rangle$ is an eigenket of $\mathbf{x}$ with eigenvalue $-\mathbf{x}'$, so it must be the same as a position eigenket $|-\mathbf{x}'\rangle$ up to a phase factor.

It is customary to take $e^{i\delta} = 1$ by convention. Substituting this in (4.45), we have $\pi^2|\mathbf{x}'\rangle = |\mathbf{x}'\rangle$; hence, $\pi^2 = 1$. That is, we come back to the same state by applying $\pi$ twice. We easily see from (4.45) that $\pi$ is now not only unitary but also Hermitian:

$$\pi^{-1} = \pi^{\dagger} = \pi. \tag{4.47}$$

Its eigenvalue can be only $+1$ or $-1$.

What about the momentum operator? The momentum $\mathbf{p}$ is like $m d\mathbf{x}/dt$, so it is natural to expect it to be odd under parity, like $\mathbf{x}$. A more satisfactory argument considers the momentum operator as the generator of translation. Since translation followed by parity is equivalent to parity followed by translation in the *opposite* direction, as can be seen from Figure 4.2, then

$$\pi\mathcal{T}(d\mathbf{x}') = \mathcal{T}(-d\mathbf{x}')\pi \tag{4.48}$$

$$\pi\left(1 - \frac{i\mathbf{p}\cdot d\mathbf{x}'}{\hbar}\right)\pi^{\dagger} = 1 + \frac{i\mathbf{p}\cdot d\mathbf{x}'}{\hbar}, \tag{4.49}$$

from which follows

$$\{\pi, \mathbf{p}\} = 0 \quad \text{or} \quad \pi^{\dagger}\mathbf{p}\pi = -\mathbf{p}. \tag{4.50}$$

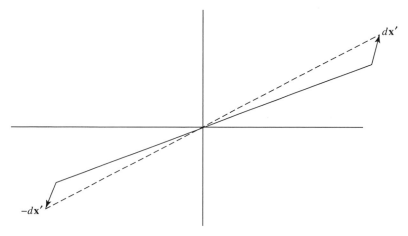

**Fig. 4.2**      Translation followed by parity, and vice versa.

We can now discuss the behavior of $\mathbf{J}$ under parity. First, for orbital angular momentum we clearly have

$$[\pi, \mathbf{L}] = 0 \tag{4.51}$$

because

$$\mathbf{L} = \mathbf{x} \times \mathbf{p}, \tag{4.52}$$

and both $\mathbf{x}$ and $\mathbf{p}$ are odd under parity. However, to show that this property also holds for spin, it is best to use the fact that $\mathbf{J}$ is the generator of rotation. For $3 \times 3$ orthogonal matrices, we have

$$R^{(\text{parity})} R^{(\text{rotation})} = R^{(\text{rotation})} R^{(\text{parity})}, \tag{4.53}$$

where explicitly

$$R^{(\text{parity})} = \begin{pmatrix} -1 & & 0 \\ & -1 & \\ 0 & & -1 \end{pmatrix}; \tag{4.54}$$

that is, the parity and rotation operations commute. In quantum mechanics, it is natural to postulate the corresponding relation for the unitary operators, so

$$\pi \mathscr{D}(R) = \mathscr{D}(R) \pi, \tag{4.55}$$

where $\mathscr{D}(R) = 1 - i \mathbf{J} \cdot \hat{\mathbf{n}} \varepsilon / \hbar$. From (4.55) it follows that

$$[\pi, \mathbf{J}] = 0 \quad \text{or} \quad \pi^\dagger \mathbf{J} \pi = \mathbf{J}. \tag{4.56}$$

This together with (4.51) means that the spin operator $\mathbf{S}$ (leading to the total angular momentum $\mathbf{J} = \mathbf{L} + \mathbf{S}$) also transforms in the same way as $\mathbf{L}$.

Under rotations, $\mathbf{x}$ and $\mathbf{J}$ transform in the same way, so they are both vectors, or spherical tensors, of rank 1. However, $\mathbf{x}$ (or $\mathbf{p}$) is odd under parity [see (4.43) and (4.50)], while $\mathbf{J}$ is even under parity [see (4.56)]. Vectors that are odd under parity are called **polar vectors**, while vectors that are even under parity are called **axial vectors**, or **pseudovectors**.

Let us now consider operators like $\mathbf{S} \cdot \mathbf{x}$. Under rotations they transform like ordinary scalars, such as $\mathbf{S} \cdot \mathbf{L}$ or $\mathbf{x} \cdot \mathbf{p}$. Yet under space inversion we have

$$\pi^{-1} \mathbf{S} \cdot \mathbf{x} \pi = -\mathbf{S} \cdot \mathbf{x}, \tag{4.57}$$

while for ordinary scalars we have

$$\pi^{-1} \mathbf{L} \cdot \mathbf{S} \pi = \mathbf{L} \cdot \mathbf{S} \tag{4.58}$$

and so on. The operator $\mathbf{S} \cdot \mathbf{x}$ is an example of a **pseudoscalar**.

## 4.2.1 Wave Functions under Parity

Let us now look at the parity property of wave functions. First, let $\psi$ be the wave function of a spinless particle whose state ket is $|\alpha\rangle$:

$$\psi(\mathbf{x}') = \langle \mathbf{x}' | \alpha \rangle. \tag{4.59}$$

The wave function of the space-inverted state, represented by the state ket $\pi|\alpha\rangle$, is

$$\langle \mathbf{x}'|\pi|\alpha\rangle = \langle -\mathbf{x}'|\alpha\rangle = \psi(-\mathbf{x}'). \tag{4.60}$$

Suppose $|\alpha\rangle$ is an eigenket of parity. We have already seen that the eigenvalue of parity must be $\pm 1$, so

$$\pi|\alpha\rangle = \pm|\alpha\rangle. \tag{4.61}$$

Let us look at its corresponding wave function,

$$\langle \mathbf{x}'|\pi|\alpha\rangle = \pm\langle \mathbf{x}'|\alpha\rangle. \tag{4.62}$$

But we also have

$$\langle \mathbf{x}'|\pi|\alpha\rangle = \langle -\mathbf{x}'|\alpha\rangle, \tag{4.63}$$

so the state $|\alpha\rangle$ is even or odd under parity depending on whether the corresponding wave function satisfies

$$\psi(-\mathbf{x}') = \pm\psi(\mathbf{x}') \begin{cases} \text{even parity} \\ \text{odd parity.} \end{cases} \tag{4.64}$$

Not all wave functions of physical interest have definite parities in the sense of (4.64). Consider, for instance, the momentum eigenket. The momentum operator anticommutes with the parity operator, so the momentum eigenket is not expected to be a parity eigenket. Indeed, it is easy to see that the plane wave, which is the wave function for a momentum eigenket, does not satisfy (4.64).

An eigenket of orbital angular momentum is expected to be a parity eigenket because $\mathbf{L}$ and $\pi$ commute [see (4.51)]. To see how an eigenket of $\mathbf{L}^2$ and $L_z$ behaves under parity, let us examine the properties of its wave function under space inversion,

$$\langle \mathbf{x}'|\alpha, lm\rangle = R_\alpha(r)Y_l^m(\theta,\phi). \tag{4.65}$$

The transformation $\mathbf{x}' \to -\mathbf{x}'$ is accomplished by letting

$$\begin{aligned} r &\to r \\ \theta &\to \pi - \theta \qquad (\cos\theta \to -\cos\theta) \\ \phi &\to \phi + \pi \qquad (e^{im\phi} \to (-1)^m e^{im\phi}). \end{aligned} \tag{4.66}$$

Using the explicit form of

$$Y_l^m = (-1)^m \sqrt{\frac{(2l+1)(l-m)!}{4\pi(l+m)!}} P_l^m(\cos\theta)e^{im\phi} \tag{4.67}$$

for positive $m$, with (3.247), where

$$P_l^{|m|}(\cos\theta) = \frac{(-1)^{m+l}}{2^l l!}\frac{(l+|m|)!}{(l-|m|)!}\sin^{-|m|}\theta\left(\frac{d}{d(\cos\theta)}\right)^{l-|m|}\sin^{2l}\theta, \tag{4.68}$$

we can readily show that

$$Y_l^m \to (-1)^l Y_l^m \tag{4.69}$$

as $\theta$ and $\phi$ are changed, as in (4.66). Therefore, we can conclude that

$$\pi|\alpha,lm\rangle = (-1)^l|\alpha,lm\rangle. \qquad (4.70)$$

It is actually not necessary to look at $Y_l^m$; an easier way to obtain the same result is to work with $m = 0$ and note that $L_{\pm}^r|l,m=0\rangle\,(r=0,1,\ldots,l)$ must have the same parity because $\pi$ and $(L_{\pm})^r$ commute.

Let us now look at the parity properties of energy eigenstates. We begin by stating a very important theorem.

**Theorem 6** *Suppose*

$$[H,\pi] = 0 \qquad (4.71)$$

*and $|n\rangle$ is a nondegenerate eigenket of $H$ with eigenvalue $E_n$:*

$$H|n\rangle = E_n|n\rangle; \qquad (4.72)$$

*then $|n\rangle$ is also a parity eigenket.*

**Proof** We prove this theorem by first considering the state

$$|\alpha\rangle \equiv \frac{1}{2}(1 \pm \pi)|n\rangle. \qquad (4.73)$$

Since $\pi^2 = 1$, it is simple to show that $\pi|\alpha\rangle = \pm|\alpha\rangle$, so $|\alpha\rangle$ is a parity eigenket. Also, since $H\pi = \pi H$, it is simple to show that $H|\alpha\rangle = E_n|\alpha\rangle$, so $|\alpha\rangle$ is also an energy eigenket with eigenvalue $E_n$. However, we assumed that $|n\rangle$ was nondegenerate. Therefore $|n\rangle$ and $|\alpha\rangle$ must be the same, to within a multiplicative constant, and $|n\rangle$ must be a parity eigenket with parity $\pm 1$, that is $\pi|n\rangle = \pm|n\rangle$. (It is worth noting that $|\alpha\rangle = 0$ if we choose the "wrong" sign in (4.73). That is, despite the $\pm$, (4.73) defines just one state.)  $\square$

As an example, let us look at the simple harmonic oscillator. The ground state $|0\rangle$ has even parity because its wave function, being Gaussian, is even under $\mathbf{x}' \rightarrow -\mathbf{x}'$. The first excited state,

$$|1\rangle = a^{\dagger}|0\rangle, \qquad (4.74)$$

must have odd parity because $a^{\dagger}$ is linear in $x$ and $p$, which are both odd [see (2.123)]. In general, the parity of the $n$th excited state of the simple harmonic operator is given by $(-1)^n$.

It is important to note that the nondegenerate assumption is essential here. For instance, consider the hydrogen atom in nonrelativistic quantum mechanics. As is well known, the energy eigenvalues depend only on the principal quantum number $n$ (for example, $2p$ and $2s$ states are degenerate) – the Coulomb potential is obviously invariant under parity – yet an energy eigenket

$$c_p|2p\rangle + c_s|2s\rangle \qquad (4.75)$$

is obviously not a parity eigenket.

As another example, consider a momentum eigenket. Momentum anticommutes with parity, so – even though the free-particle Hamiltonian $H$ is invariant under parity – the momentum eigenket (though obviously an energy eigenket) is not a parity eigenket. Our theorem remains intact because we have here a degeneracy between $|\mathbf{p}'\rangle$ and $|-\mathbf{p}'\rangle$, which have the same energy. In fact, we can easily construct linear combinations $(1/\sqrt{2})(|\mathbf{p}'\rangle \pm |-\mathbf{p}'\rangle)$, which are parity eigenkets with eigenvalues $\pm 1$. In terms of wave function language, $e^{i\mathbf{p}'\cdot\mathbf{x}'/\hbar}$ does not have a definite parity, but $\cos \mathbf{p}'\cdot\mathbf{x}'/\hbar$ and $\sin \mathbf{p}'\cdot\mathbf{x}'/\hbar$ do.

### 4.2.2 Symmetrical Double-Well Potential

As an elementary but instructive example, we consider a symmetrical double-well potential; see Figure 4.3. The Hamiltonian is obviously invariant under parity. In fact, the two lowest-lying states are as shown in Figure 4.3, as we can see by working out the explicit solutions involving sine and cosine in classically allowed regions and sinh and cosh in the classically forbidden region. The solutions are matched where the potential is discontinuous; we call them the **symmetrical state** $|S\rangle$ and the **antisymmetrical state** $|A\rangle$. Of course, they are simultaneous eigenkets of $H$ and $\pi$. Calculation also shows that

$$E_A > E_S, \tag{4.76}$$

which we can infer from Figure 4.3 by noting that the wave function of the antisymmetrical state has a greater curvature. The energy difference is very tiny if the middle barrier is high, a point which we will discuss later.

We can form

$$|R\rangle = \frac{1}{\sqrt{2}}(|S\rangle + |A\rangle) \tag{4.77a}$$

and

$$|L\rangle = \frac{1}{\sqrt{2}}(|S\rangle - |A\rangle). \tag{4.77b}$$

The wave functions of (4.77a) and (4.77b) are largely concentrated in the right-hand side and the left-hand side, respectively. They are obviously not parity eigenstates; in fact, under parity $|R\rangle$ and $|L\rangle$ are interchanged. Note that they are not energy eigenstates either. Indeed, they are typical examples of **nonstationary states**. To be precise, let us assume that the system is represented by $|R\rangle$ at $t=0$. At a later time, we have

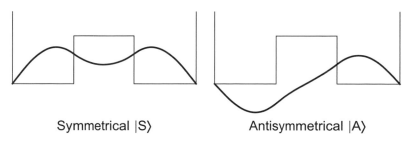

Symmetrical $|S\rangle$          Antisymmetrical $|A\rangle$

**Fig. 4.3**  The symmetrical double well with the two lowest-lying states $|S\rangle$ (symmetrical) and $|A\rangle$ (antisymmetrical) shown.

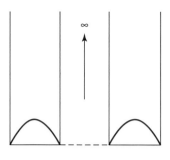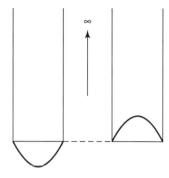

**Fig. 4.4**    The symmetrical double well with an infinitely high middle barrier.

$$|R, t_0 = 0; t\rangle = \frac{1}{\sqrt{2}}\left(e^{-iE_S t/\hbar}|S\rangle + e^{iE_A t/\hbar}|A\rangle\right)$$

$$= \frac{1}{\sqrt{2}} e^{-iE_S t/\hbar}\left(|S\rangle + e^{i(E_A - E_S)t/\hbar}|A\rangle\right). \tag{4.78}$$

At time $t = T/2 \equiv 2\pi\hbar/2(E_A - E_S)$, the system is found in pure $|L\rangle$. At $t = T$, we are back to pure $|R\rangle$, and so forth. Thus, in general, we have an oscillation between $|R\rangle$ and $|L\rangle$ with angular frequency

$$\omega = \frac{(E_A - E_S)}{\hbar}. \tag{4.79}$$

This oscillatory behavior can also be considered from the viewpoint of tunneling in quantum mechanics. A particle initially confined to the right-hand side can tunnel through the classically forbidden region (the middle barrier) into the left-hand side, then back to the right-hand side, and so on. But now let the middle barrier become infinitely high; see Figure 4.4. The $|S\rangle$ and $|A\rangle$ states are now degenerate, so (4.77a) and (4.77b) are also energy eigenkets even though they are not parity eigenkets. Once the system is found in $|R\rangle$, it remains so forever (oscillation time between $|S\rangle$ and $|A\rangle$ is now $\infty$). Because the middle barrier is infinitely high, there is no possibility for tunneling. Thus when there is degeneracy, the physically realizable energy eigenkets need not be parity eigenkets. We have a ground state which is asymmetrical despite the fact that the Hamiltonian itself is symmetrical under space inversion, so with degeneracy the symmetry of $H$ is not necessarily obeyed by energy eigenstates $|S\rangle$ and $|A\rangle$.

This is a very simple example of broken symmetry and degeneracy. Nature is full of situations analogous to this. Consider a ferromagnet. The basic Hamiltonian for iron atoms is rotationally invariant, but the ferromagnet clearly has a definite direction in space; hence, the (infinite) number of ground states is *not* rotationally invariant, since the spins are all aligned along some definite (but arbitrary) direction.

A textbook example of a system that illustrates the actual importance of the symmetrical double well is an ammonia molecule, $NH_3$; see Figure 4.5. We imagine that the three H atoms form the three corners of an equilateral triangle. The N atom can be up or down, where the directions up and down are defined because the molecule is rotating around the axis as shown in Figure 4.5. The up and down positions for the N atom are analogous to

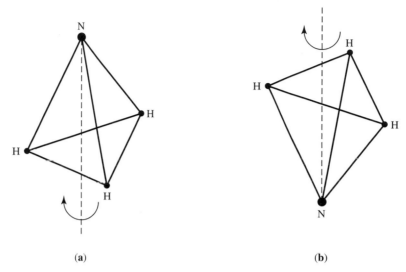

(a)                                              (b)

**Fig. 4.5**    An ammonia molecule, $NH_3$, where the three H atoms form the three corners of an equilateral triangle.

$R$ and $L$ of the double-well potential. The parity and energy eigenstates are *superpositions* of Figure 4.5a and Figure 4.5b in the sense of (4.77a) and (4.77b), respectively, and the energy difference between the simultaneous eigenstates of energy and parity corresponds to an oscillation frequency of 24,000 MHz, a wavelength of about 1 cm, which is in the microwave region. In fact, $NH_3$ is of fundamental importance in maser physics.

There are naturally occurring organic molecules, such as sugars or amino acids, which are of the $R$-type (or $L$-type) only. Such molecules which have definite handedness are called **optical isomers**. In many cases the oscillation time is practically infinite – on the order of $10^4$–$10^6$ years – so $R$-type molecules remain right-handed for all practical purposes. It is amusing that if we attempt to synthesize such organic molecules in the laboratory, we find equal mixtures of $R$ and $L$. Why we have a preponderance of one type is nature's deepest mystery. Is it due to a genetic accident, like the spiral shell of a snail or the fact that our hearts are on the left-hand side?[2]

## 4.2.3 Parity-Selection Rule

Suppose $|\alpha\rangle$ and $|\beta\rangle$ are parity eigenstates:

$$\pi|\alpha\rangle = \varepsilon_\alpha|\alpha\rangle \qquad\qquad (4.80a)$$

and

$$\pi|\beta\rangle = \varepsilon_\beta|\beta\rangle, \qquad\qquad (4.80b)$$

[2]  It has been suggested that parity violation in nuclear processes active during the formation of life, may have contributed to this handedness. See Bonner, *Chirality*, **12** (2000) 114.

where $\varepsilon_\alpha$, $\varepsilon_\beta$ are the parity eigenvalues ($\pm 1$). We can show that

$$\langle \beta | \mathbf{x} | \alpha \rangle = 0 \tag{4.81}$$

unless $\varepsilon_\alpha = -\varepsilon_\beta$. In other words, the parity-odd operator $\mathbf{x}$ connects states of opposite parity. The proof of this follows:

$$\langle \beta | \mathbf{x} | \alpha \rangle = \langle \beta | \pi^{-1} \pi \mathbf{x} \pi^{-1} \pi | \alpha \rangle = \varepsilon_\alpha \varepsilon_\beta (-\langle \beta | \mathbf{x} | \alpha \rangle), \tag{4.82}$$

which is impossible for a finite nonzero $\langle \beta | \mathbf{x} | \alpha \rangle$ unless $\varepsilon_\alpha$ and $\varepsilon_\beta$ are opposite in sign. Perhaps the reader is familiar with this argument from

$$\int \psi_\beta^* \mathbf{x} \psi_\alpha d\tau = 0 \tag{4.83}$$

if $\psi_\beta$ and $\psi_\alpha$ have the same parity. This selection rule, due to Wigner, is of importance in discussing radiative transitions between atomic states. As we will discuss in greater detail later, radiative transitions take place between states of opposite parity as a consequence of multipole expansion formalism. This rule was known phenomenologically from analysis of spectral lines, before the birth of quantum mechanics, as **Laporte's rule**. It was Wigner who showed that Laporte's rule is a consequence of the parity-selection rule.

If the basic Hamiltonian $H$ is invariant under parity, nondegenerate energy eigenstates [as a corollary of (4.83)] cannot possess a permanent electric dipole moment:

$$\langle n | \mathbf{x} | n \rangle = 0. \tag{4.84}$$

This follows trivially from (4.83), because with the nondegenerate assumption, energy eigenstates are also parity eigenstates [see (4.72) and (4.73)]. For a degenerate state, it is perfectly all right to have an electric dipole moment. We will see an example of this when we discuss the linear Stark effect in Chapter 5.

Our considerations can be generalized: Operators that are odd under parity, like $\mathbf{p}$ or $\mathbf{S} \cdot \mathbf{x}$, have nonvanishing matrix elements only between states of opposite parity. In contrast, operators that are even under parity connect states of the same parity.

### 4.2.4  Parity Nonconservation

The basic Hamiltonian responsible for the so-called weak interaction of elementary particles is not invariant under parity. In decay processes we can have final states which are superpositions of opposite parity states. Observable quantities like the angular distribution of decay products can depend on pseudoscalars such as $\langle \mathbf{S} \rangle \cdot \mathbf{p}$. It is remarkable that parity conservation was believed to be a sacred principle until 1956, when Lee and Yang speculated that parity is not conserved in weak interactions and proposed crucial experiments to test the validity of parity conservation. Subsequent experiments indeed showed that observable effects do depend on pseudoscalar quantities such as correlation between $\langle \mathbf{S} \rangle$ and $\mathbf{p}$.

To this day, one of the clearest demonstrations of parity nonconservation is the experiment which first discovered it. This result, from Wu et al., *Phys. Rev.*, **105** (1957) 1413, shows a decay rate that depends on $\langle \mathbf{S} \rangle \cdot \mathbf{p}$. The decay observed is $^{60}\text{Co} \rightarrow {}^{60}\text{Ni} + e^- + \bar{\nu}_e$

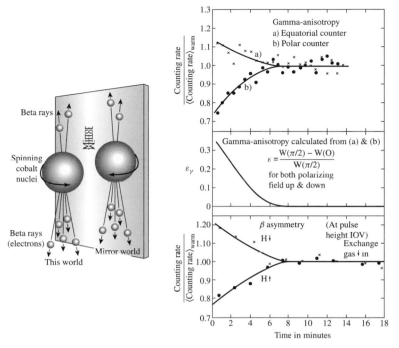

**Fig. 4.6** Experimental demonstration of parity nonconservation. The key observation, shown on the left, is that radioactive cobalt nuclei, oriented according to their nuclear spin, emit "beta rays" (i.e. electrons) preferentially in the the opposite direction. The experiment data, shown on the right, show how the "up/down" beta decay asymmetry (bottom panel) correlates perfectly with the signal which indicates the degree of nuclear polarization (upper panel). As time goes on, the sample warms up and the cobalt nuclei depolarize.

where $S$ is the spin of the $^{60}$Co nucleus and the momentum of the emitted $e^-$ is $\mathbf{p}$. A sample of spin-polarized radioactive $^{60}$Co nuclei is prepared at low temperature, and the decay $e^-$ are detected in the direction parallel or antiparallel to the spin, depending on the sign of the polarizing magnetic field. The polarization of the sample is monitored by observing the anisotropy of the $\gamma$-rays in the decay of the excited $^{60}$Ni daughter nuclei, a parity conserving effect. The results are shown in Figure 4.6. Over a period of several minutes, the sample warms up, and the $\beta$-decay asymmetry disappears at exactly the same rate as the $\gamma$-ray anisotropy.

Because parity is not conserved in weak interactions, previously thought "pure" nuclear and atomic states are, in fact, parity mixtures. These subtle effects have also been found experimentally.

## 4.3 Lattice Translation as a Discrete Symmetry

We now consider another kind of discrete symmetry operation, namely, lattice translation. This subject has extremely important applications in solid-state physics.

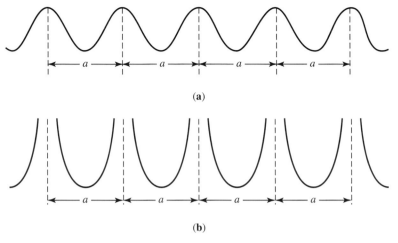

**Fig. 4.7** (a) Periodic potential in one dimension with periodicity $a$. (b) The periodic potential when the barrier height between two adjacent lattice sites becomes infinite.

Consider a periodic potential in one dimension, where $V(x \pm a) = V(x)$, as depicted in Figure 4.7. Realistically, we may consider the motion of an electron in a chain of regularly spaced positive ions. In general, the Hamiltonian is not invariant under a translation represented by $\tau(l)$ with $l$ arbitrary, where $\tau(l)$ has the property (see Section 1.6)

$$\tau^{\dagger}(l)x\tau(l) = x+l, \quad \tau(l)|x'\rangle = |x'+l\rangle. \tag{4.85}$$

However, when $l$ coincides with the lattice spacing $a$, we do have

$$\tau^{\dagger}(a)V(x)\tau(a) = V(x+a) = V(x). \tag{4.86}$$

Because the kinetic-energy part of the Hamiltonian $H$ is invariant under the translation with any displacement, the entire Hamiltonian satisfies

$$\tau^{\dagger}(a)H\tau(a) = H. \tag{4.87}$$

Because $\tau(a)$ is unitary, we have [from (4.87)]

$$[H, \tau(a)] = 0, \tag{4.88}$$

so the Hamiltonian and $\tau(a)$ can be simultaneously diagonalized. Although $\tau(a)$ is unitary, it is not Hermitian, so we expect the eigenvalue to be a *complex* number of modulus 1.

Before we determine the eigenkets and eigenvalues of $\tau(a)$ and examine their physical significance, it is instructive to look at a special case of periodic potential when the barrier height between two adjacent lattice sites is made to go to infinity, as in Figure 4.7b. What is the ground state for the potential of Figure 4.7b? Clearly, a state in which the particle is completely localized in one of the lattice sites can be a candidate for the ground state. To be specific let us assume that the particle is localized at the $n$th site and denote the corresponding ket by $|n\rangle$. This is an energy eigenket with energy eigenvalue $E_0$, namely, $H|n\rangle = E_0|n\rangle$. Its wave function $\langle x'|n\rangle$ is finite only in the $n$th site. However, we note that a similar state localized at some other site also has the same energy $E_0$, so actually there are denumerably infinite ground states $n$, where $n$ runs from $-\infty$ to $+\infty$.

Now $|n\rangle$ is obviously not an eigenket of the lattice translation operator, because when the lattice translation operator is applied to it, we obtain $|n+1\rangle$:

$$\tau(a)|n\rangle = |n+1\rangle. \tag{4.89}$$

So despite the fact that $\tau(a)$ commutes with $H, |n\rangle$ – which is an eigenket of $H$ – is not an eigenket of $\tau(a)$. This is quite consistent with our earlier theorem on symmetry because we have an infinitefold degeneracy. When there is such degeneracy, the symmetry of the world need not be the symmetry of energy eigenkets. Our task is to find a *simultaneous* eigenket of $H$ and $\tau(a)$.

Here we may recall how we handled a somewhat similar situation with the symmetrical double-well potential of the previous section. We noted that even though neither $|R\rangle$ nor $|L\rangle$ is an eigenket of $\pi$, we could easily form a symmetrical and an antisymmetrical combination of $|R\rangle$ and $|L\rangle$ that are parity eigenkets. The case is analogous here. Let us specifically form a linear combination[3]

$$|\theta\rangle \equiv \sum_{n=-\infty}^{\infty} e^{in\theta}|n\rangle, \tag{4.90}$$

where $\theta$ is a real parameter with $-\pi \le \theta \le \pi$. We assert that $|\theta\rangle$ is a simultaneous eigenket of $H$ and $\tau(a)$. That it is an $H$ eigenket is obvious because $|n\rangle$ is an energy eigenket with eigenvalue $E_0$, independent of $n$. To show that it is also an eigenket of the lattice translation operator we apply $\tau(a)$ as follows:

$$\tau(a)|\theta\rangle = \sum_{n=-\infty}^{\infty} e^{in\theta}|n+1\rangle = \sum_{n=-\infty}^{\infty} e^{i(n-1)\theta}|n\rangle$$
$$- e^{-i\theta}|\theta\rangle. \tag{4.91}$$

Note that this simultaneous eigenket of $H$ and $\tau(a)$ is parameterized by a continuous parameter $\theta$. Furthermore, the energy eigenvalue $E_0$ is independent of $\theta$.

Let us now return to the more realistic situation of Figure 4.7a, where the barrier between two adjacent lattice sites is not infinitely high. We can construct a localized ket $|n\rangle$ just as before with the property $\tau(a)|n\rangle = |n+1\rangle$. However, this time we expect that there is some leakage possible into neighboring lattice sites due to quantum-mechanical tunneling. In other words, the wave function $\langle x'|n\rangle$ has a tail extending to sites other than the $n$th site. The diagonal elements of $H$ in the $\{|n\rangle\}$ basis are all equal because of translation invariance, that is,

$$\langle n|H|n\rangle = E_0, \tag{4.92}$$

independent of $n$, as before. However we suspect that $H$ is not completely diagonal in the $\{|n\rangle\}$ basis due to leakage. Now, suppose the barriers between adjacent sites are high (but not infinite). We then expect matrix elements of $H$ between distant sites to be

---

[3]  Beware that the states $|\theta\rangle$ are not orthonormal. That is, the inner product $\langle\theta'|\theta\rangle$ is infinite when $\theta = \theta'$, but is not necessarily zero otherwise.

completely negligible. Let us assume that the only nondiagonal elements of importance connect immediate neighbors. That is,

$$\langle n'|H|n\rangle \neq 0 \quad \text{only if} \quad n' = n \quad \text{or} \quad n' = n \pm 1. \tag{4.93}$$

In solid-state physics this assumption is known as the **tight-binding approximation.** Let us define

$$\langle n \pm 1|H|n\rangle = -\Delta. \tag{4.94}$$

Clearly, $\Delta$ is again independent of $n$ due to translation invariance of the Hamiltonian. To the extent that $|n\rangle$ and $|n'\rangle$ are orthogonal when $n \neq n'$, we obtain

$$H|n\rangle = E_0|n\rangle - \Delta|n+1\rangle - \Delta|n-1\rangle. \tag{4.95}$$

Note that $|n\rangle$ is no longer an energy eigenket.

As we have done with the potential of Figure 4.7b, let us form a linear combination

$$|\theta\rangle = \sum_{n=-\infty}^{\infty} e^{in\theta}|n\rangle. \tag{4.96}$$

Clearly, $|\theta\rangle$ is an eigenket of translation operator $\tau(a)$ because the steps in (4.91) still hold. A natural question is, is $|\theta\rangle$ an energy eigenket? To answer this question, we apply $H$:

$$
\begin{aligned}
H\sum e^{in\theta}|n\rangle &= E_0 \sum e^{in\theta}|n\rangle - \Delta \sum e^{in\theta}|n+1\rangle - \Delta \sum e^{in\theta}|n-1\rangle \\
&= E_0 \sum e^{in\theta}|n\rangle - \Delta \sum (e^{in\theta-i\theta} + e^{in\theta+i\theta})|n\rangle \\
&= (E_0 - 2\Delta\cos\theta)\sum e^{in\theta}|n\rangle.
\end{aligned} \tag{4.97}
$$

The big difference between this and the previous situation is that the energy eigenvalue now depends on the continuous real parameter $\theta$. The degeneracy is lifted as $\Delta$ becomes finite, and we have a continuous distribution of energy eigenvalues between $E_0 - 2\Delta$ and $E_0 + 2\Delta$. See Figure 4.8, where we visualize how the energy levels start forming a continuous energy band as $\Delta$ is increased from zero.

To see the physical meaning of the parameter $\theta$ let us study the wave function $\langle x'|\theta\rangle$. For the wave function of the lattice translated state $\tau(a)|\theta\rangle$, we obtain

$$\langle x'|\tau(a)|\theta\rangle = \langle x' - a|\theta\rangle \tag{4.98}$$

by letting $\tau(a)$ act on $\langle x'|$. But we can also let $\tau(a)$ operate on $|\theta\rangle$ and use (4.91). Thus

$$\langle x'|\tau(a)|\theta\rangle = e^{-i\theta}\langle x'|\theta\rangle, \tag{4.99}$$

so

$$\langle x' - a|\theta\rangle = \langle x'|\theta\rangle e^{-i\theta}. \tag{4.100}$$

We solve this equation by setting

$$\langle x'|\theta\rangle = e^{ikx'}u_k(x'), \tag{4.101}$$

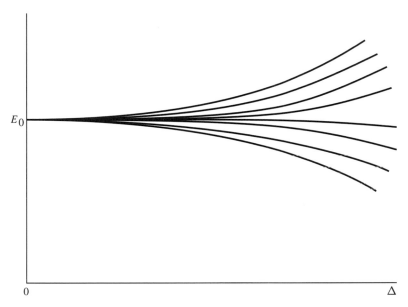

**Fig. 4.8**    Energy levels forming a continuous energy band as $\Delta$ is increased from zero.

with $\theta = ka$, where $u_k(x')$ is a periodic function with period $a$, as we can easily verify by explicit substitutions, namely,

$$e^{ik(x'-a)}u_k(x'-a) = e^{ikx'}u_k(x')e^{-ika}. \tag{4.102}$$

Thus we get the important condition known as **Bloch's theorem**: The wave function of $|\theta\rangle$, which is an eigenket of $\tau(a)$, can be written as a plane wave $e^{ikx'}$ times a periodic function with periodicity $a$. Notice that the only fact we used was that $|\theta\rangle$ is an eigenket of $\tau(a)$ with eigenvalue $e^{-i\theta}$ [see (4.91)]. In particular, the theorem holds even if the tight-binding approximation (4.93) breaks down.

We are now in a position to interpret our earlier result (4.97) for $|\theta\rangle$ given by (4.96). We know that the wave function is a plane wave characterized by the propagation wave vector $k$ modulated by a periodic function $u_k(x')$ [see (4.101)]. As $\theta$ varies from $-\pi$ to $\pi$, the wave vector $k$ varies from $-\pi/a$ to $\pi/a$. The energy eigenvalue $E$ now depends on $k$ as follows:

$$E(k) = E_0 - 2\Delta\cos ka. \tag{4.103}$$

Notice that this energy eigenvalue equation is independent of the detailed shape of the potential as long as the tight-binding approximation is valid. Note also that there is a cutoff in the wave vector $k$ of the Bloch wave function (4.101) given by $|k| = \pi/a$. Equation (4.103) defines a dispersion curve, as shown in Figure 4.9. As a result of tunneling, the denumerably infinitefold degeneracy is now completely lifted, and the allowed energy values form a continuous band between $E_0 - 2\Delta$ and $E_0 + 2\Delta$, known as the **Brillouin zone**.

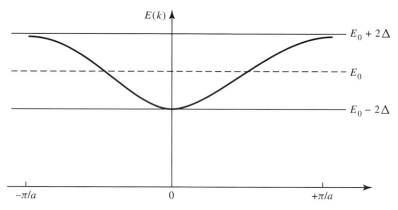

**Fig. 4.9**    Dispersion curve for $E(k)$ versus $k$ in the Brillouin zone $|k| \leq \pi/a$.

So far we have considered only one particle moving in a periodic potential. In a more realistic situation we must look at many electrons moving in such a potential. Actually the electrons satisfy the Pauli exclusion principle, as we will discuss more systematically in Chapter 7, and they start filling the band. In this way, the main qualitative features of metals, semiconductors, and the like can be understood as a consequence of translation invariance supplemented by the exclusion principle.

The reader may have noted the similarity between the symmetrical double-well problem of Section 4.2 and the periodic potential of this section. Comparing Figures 4.3 and 4.7, we note that they can be regarded as opposite extremes (two versus infinite) of potentials with a finite number of troughs.

## 4.4  The Time-Reversal Discrete Symmetry

In this section we study another discrete symmetry operator, called **time reversal**. This is a difficult topic for the novice, partly because the term *time reversal* is a misnomer; it reminds us of science fiction. Actually what we do in this section can be more appropriately characterized by the term *reversal of motion*. Indeed, that is the terminology used by E. Wigner, who formulated time reversal in a very fundamental paper written in 1932.

For orientation purposes let us look at classical mechanics. Suppose there is a trajectory of a particle subject to a certain force field; see Figure 4.10. At $t = 0$, let the particle stop and reverse its motion: $\mathbf{p}|_{t=0} \to -\mathbf{p}|_{t=0}$. The particle traverses backward along the same trajectory. If you run the motion picture of trajectory (a) backward as in (b), you may have a hard time telling whether this is the correct sequence.

More formally, if $\mathbf{x}(t)$ is a solution to

$$m\ddot{\mathbf{x}} = -\nabla V(\mathbf{x}), \tag{4.104}$$

then $\mathbf{x}(-t)$ is also a possible solution in the same force field derivable from $V$. It is, of course, important to note that we do not have a dissipative force here. A block sliding on a

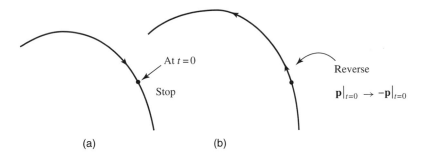

**Fig. 4.10** Classical trajectory which (a) stops at $t = 0$ and (b) reverses its motion $\mathbf{p}|_{t=0} \to -\mathbf{p}|_{t=0}$.

table decelerates (due to friction) and eventually stops. But have you ever seen a block on a table spontaneously start to move and accelerate?

With a magnetic field you may be able to tell the difference. Imagine that you are taking the motion picture of a spiraling electron trajectory in a magnetic field. You may be able to tell whether the motion picture is run forward or backward by comparing the sense of rotation with the magnetic pole labeling N and S. However, from a microscopic point of view, $\mathbf{B}$ is produced by moving charges via an electric current; if you could reverse the current that causes $\mathbf{B}$, then the situation would be quite symmetrical. In terms of the picture shown in Figure 4.11, you may have figured out that N and S are mislabeled! Another more formal way of saying all this is that the Maxwell equations, for example,

$$\nabla \cdot \mathbf{E} = 4\pi\rho, \quad \nabla \times \mathbf{B} - \frac{1}{c}\frac{\partial \mathbf{E}}{\partial t} = \frac{4\pi\mathbf{j}}{c}, \quad \nabla \times \mathbf{E} = -\frac{1}{c}\frac{\partial \mathbf{B}}{\partial t}, \tag{4.105}$$

and the Lorentz force equation $\mathbf{F} = e[\mathbf{E} + (1/c)(\mathbf{v} \times \mathbf{B})]$ are invariant under $t \to -t$ provided we also let

$$\mathbf{E} \to \mathbf{E}, \quad \mathbf{B} \to -\mathbf{B}, \quad \rho \to \rho, \quad \mathbf{j} \to -\mathbf{j}, \quad \mathbf{v} \to -\mathbf{v}. \tag{4.106}$$

Let us now look at wave mechanics, where the basic equation of the Schrödinger wave equation is

$$i\hbar\frac{\partial \psi}{\partial t} = \left(-\frac{\hbar^2}{2m}\nabla^2 + V\right)\psi. \tag{4.107}$$

Suppose $\psi(\mathbf{x}, t)$ is a solution. We can easily verify that $\psi(\mathbf{x}, -t)$ is not a solution, because of the appearance of the first-order time derivative. However, $\psi^*(\mathbf{x}, -t)$ is a solution, as you may verify by complex conjugation of (4.107). It is instructive to convince ourselves of this point for an energy eigenstate, that is, by substituting

$$\psi(\mathbf{x}, t) = u_n(\mathbf{x})e^{-iE_n t/\hbar}, \quad \psi^*(\mathbf{x}, -t) = u_n^*(\mathbf{x})e^{-iE_n t/\hbar} \tag{4.108}$$

into the Schrödinger equation (4.107). Thus we conjecture that time reversal must have something to do with complex conjugation. If at $t = 0$ the wave function is given by

$$\psi = \langle \mathbf{x}|\alpha\rangle, \tag{4.109}$$

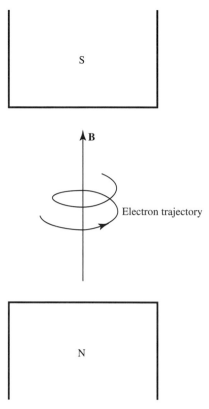

**Fig. 4.11**    Electron trajectory between the north and south poles of a magnet.

then the wave function for the corresponding time-reversed state is given by $\langle \mathbf{x}|\alpha\rangle^*$. We will later show that this is indeed the case for the wave function of a spinless system. As an example, you may easily check this point for the wave function of a plane wave; see Problem 4.8 of this chapter.

### 4.4.1 Digression on Symmetry Operations

Before we begin a systematic treatment of the time-reversal operator, some general remarks on symmetry operations are in order. Consider a symmetry operation

$$|\alpha\rangle \to |\tilde{\alpha}\rangle, \quad |\beta\rangle \to |\tilde{\beta}\rangle. \tag{4.110}$$

One may argue that it is natural to require the inner product $\langle\beta|\alpha\rangle$ to be preserved, that is,

$$\langle\tilde{\beta}|\tilde{\alpha}\rangle = \langle\beta|\alpha\rangle. \tag{4.111}$$

Indeed, for symmetry operations such as rotations, translations, and even parity, this is the case. If $|\alpha\rangle$ is rotated and $|\beta\rangle$ is also rotated in the same manner, $\langle\beta|\alpha\rangle$ is unchanged. Formally this arises from the fact that, for the symmetry operations considered in the previous sections, the corresponding symmetry operator is unitary, so

$$\langle \beta | \alpha \rangle \to \langle \beta | U^\dagger U | \alpha \rangle = \langle \beta | \alpha \rangle. \tag{4.112}$$

However, in discussing time reversal, we see that requirement (4.111) turns out to be too restrictive. Instead, we merely impose the weaker requirement that

$$|\langle \tilde{\beta} | \tilde{\alpha} \rangle| = |\langle \beta | \alpha \rangle|. \tag{4.113}$$

Requirement (4.111) obviously satisfies (4.113). But this is not the only way;

$$\langle \tilde{\beta} | \tilde{\alpha} \rangle = \langle \beta | \alpha \rangle^* = \langle \alpha | \beta \rangle \tag{4.114}$$

works equally well. We pursue the latter possibility in this section because from our earlier discussion based on the Schrödinger equation we inferred that time reversal has something to do with complex conjugation.

**Definition 1**   *The transformation*

$$|\alpha\rangle \to |\tilde{\alpha}\rangle = \theta|\alpha\rangle, \quad |\beta\rangle \to |\tilde{\beta}\rangle = \theta|\beta\rangle \tag{4.115}$$

*is said to be **antiunitary** if*

$$\langle \tilde{\beta} | \tilde{\alpha} \rangle - \langle \beta | \alpha \rangle^*, \tag{4.116a}$$

$$\theta(c_1|\alpha\rangle + c_2|\beta\rangle) = c_1^* \theta|\alpha\rangle + c_2^* \theta|\beta\rangle. \tag{4.116b}$$

In such a case the operator $\theta$ is an antiunitary operator. Relation (4.116b) alone defines an **antilinear** operator.

We now claim that an antiunitary operator can be written as

$$\theta = UK, \tag{4.117}$$

where $U$ is a unitary operator and $K$ is the complex conjugate operator that forms the complex conjugate of any coefficient that multiplies a ket (and stands on the right of $K$). Before checking (4.116) let us examine the property of the $K$ operator. Suppose we have a ket multiplied by a complex number $c$. We then have

$$Kc|\alpha\rangle = c^* K|\alpha\rangle. \tag{4.118}$$

One may further ask: What happens if $|\alpha\rangle$ is expanded in terms of base kets $\{|a'\rangle\}$? Under the action $K$ we have

$$|\alpha\rangle = \sum_{a'} |a'\rangle\langle a'|\alpha\rangle \xrightarrow{K} |\tilde{\alpha}\rangle = \sum_{a'} \langle a'|\alpha\rangle^* K|a'\rangle$$
$$= \sum_{a'} \langle a'|\alpha\rangle^* |a'\rangle. \tag{4.119}$$

Notice that $K$ acting on the base ket does not change the base ket. The explicit representation of $|a'\rangle$ is

$$|a'\rangle = \begin{pmatrix} 0 \\ 0 \\ \vdots \\ 0 \\ 1 \\ 0 \\ \vdots \\ 0 \end{pmatrix}, \tag{4.120}$$

and there is nothing to be changed by $K$. The reader may wonder, for instance, whether the $S_y$ eigenkets for a spin $\frac{1}{2}$ system change under $K$. The answer is that if the $S_z$ eigenkets are used as base kets, we must change the $S_y$ eigenkets because the $S_y$ eigenkets (1.14) undergo under $K$

$$K\left(\frac{1}{\sqrt{2}}|+\rangle \pm \frac{i}{\sqrt{2}}|-\rangle\right) \to \frac{1}{\sqrt{2}}|+\rangle \mp \frac{i}{\sqrt{2}}|-\rangle. \tag{4.121}$$

On the other hand, if the $S_y$ eigenkets themselves are used as the base kets, we do not change the $S_y$ eigenkets under the action of $K$. Thus the effect of $K$ changes with the basis. As a result, the form of $U$ in (4.117) also depends on the particular representation (that is, the choice of base kets) used.

Returning to $\theta = UK$ and (4.116), let us first check property (4.116b). We have

$$\theta(c_1|\alpha\rangle + c_2|\beta\rangle) = UK(c_1|\alpha\rangle + c_2|\beta\rangle)$$
$$= c_1^* UK|\alpha\rangle + c_2^* UK|\beta\rangle$$
$$= c_1^* \theta|\alpha\rangle + c_2^* \theta|\beta\rangle, \tag{4.122}$$

so (4.116b) indeed holds. Before checking (4.116a), we assert that it is always safer to work with the action of $\theta$ on kets only. We can figure out how the bras change just by looking at the corresponding kets. In particular, it is not necessary to consider $\theta$ acting on bras from the right, nor is it necessary to define $\theta^\dagger$. We have

$$|\alpha\rangle \xrightarrow{\theta} |\tilde{\alpha}\rangle = \sum_{a'} \langle a'|\alpha\rangle^* UK|a'\rangle$$
$$= \sum_{a'} \langle a'|\alpha\rangle^* U|a'\rangle$$
$$= \sum_{a'} \langle \alpha|a'\rangle U|a'\rangle. \tag{4.123}$$

As for $|\beta\rangle$, we have

$$|\tilde{\beta}\rangle = \sum_{a'} \langle a'|\beta\rangle^* U|a'\rangle \overset{\text{DC}}{\leftrightarrow} \langle \tilde{\beta}| = \sum_{a'} \langle a'|\beta\rangle \langle a'|U^\dagger$$
$$\langle \tilde{\beta}|\tilde{\alpha}\rangle = \sum_{a''} \sum_{a'} \langle a''|\beta\rangle \langle a''|U^\dagger U|a'\rangle \langle \alpha|a'\rangle$$
$$= \sum_{a'} \langle \alpha|a'\rangle \langle a'|\beta\rangle = \langle \alpha|\beta\rangle$$
$$= \langle \beta|\alpha\rangle^*, \tag{4.124}$$

so this checks. (Recall the notion of dual correspondence, or DC, from Section 1.2.)

In order for (4.113) to be satisfied, it is of physical interest to consider just two types of transformation, unitary and antiunitary. Other possibilities are related to either of the preceding via trivial phase changes. The proof of this assertion is actually very difficult and will not be discussed further here. See, however, Gottfried and Yan (2003), Section 7.1.

## 4.4.2  Time-Reversal Operator

We are finally in a position to present a formal theory of time reversal. Let us denote the time-reversal operator by $\Theta$, to be distinguished from $\theta$, a general antiunitary operator. Consider

$$|\alpha\rangle \rightarrow \Theta|\alpha\rangle, \tag{4.125}$$

where $\Theta|\alpha\rangle$ is the time-reversed state. More appropriately, $\Theta|\alpha\rangle$ should be called the motion-reversed state. If $|\alpha\rangle$ is a momentum eigenstate $|\mathbf{p}'\rangle$, we expect $\Theta|\alpha\rangle$ to be $|-\mathbf{p}'\rangle$ up to a possible phase. Likewise, $\mathbf{J}$ is to be reversed under time reversal.

We now deduce the fundamental property of the time-reversal operator by looking at the time evolution of the time-reversed state. Consider a physical system represented by a ket $|\alpha\rangle$, say at $t = 0$. Then at a slightly later time $t = \delta t$, the system is found in

$$|\alpha, t_0 = 0; t = \delta t\rangle = \left(1 - \frac{iH}{\hbar}\delta t\right)|\alpha\rangle, \tag{4.126}$$

where $H$ is the Hamiltonian that characterizes the time evolution. Instead of the preceding equation, suppose we first apply $\Theta$, say at $t = 0$, and then let the system evolve under the influence of the Hamiltonian $H$. We then have at $\delta t$

$$\left(1 - \frac{iH\delta t}{\hbar}\right)\Theta|\alpha\rangle. \tag{4.127a}$$

If motion obeys symmetry under time reversal, we expect the preceding state ket to be the same as

$$\Theta|\alpha, t_0 = 0; t = -\delta t\rangle \tag{4.127b}$$

that is, first consider a state ket at *earlier time* $t = -\delta t$, and then reverse $\mathbf{p}$ and $\mathbf{J}$; see Figure 4.12. Mathematically,

$$\left(1 - \frac{iH}{\hbar}\delta t\right)\Theta|\alpha\rangle = \Theta\left(1 - \frac{iH}{\hbar}(-\delta t)\right)|\alpha\rangle. \tag{4.128}$$

If the preceding relation is to be true for any ket, we must have

$$-iH\Theta|\rangle = \Theta iH|\rangle, \tag{4.129}$$

where the blank ket $|\rangle$ emphasizes that (4.129) is to be true for any ket.

We now argue that $\Theta$ *cannot* be unitary if the motion of time reversal is to make sense. Suppose $\Theta$ were unitary. It would then be legitimate to cancel the $i$ in (4.129), and we would have the operator equation

$$-H\Theta = \Theta H. \tag{4.130}$$

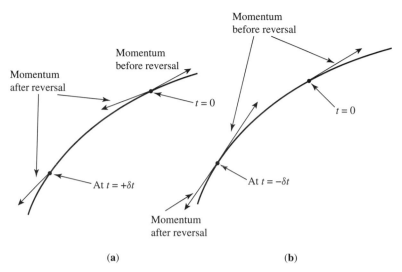

Fig. 4.12 Momentum before and after time reversal at time $t = 0$ and $t = \pm \delta t$.

Consider an energy eigenket $|n\rangle$ with energy eigenvalue $E_n$. The corresponding time-reversed state would be $\Theta|n\rangle$, and we would have, because of (4.130),

$$H\Theta|n\rangle = -\Theta H|n\rangle = (-E_n)\Theta|n\rangle. \tag{4.131}$$

This equation says that $\Theta|n\rangle$ is an eigenket of the Hamiltonian with energy eigenvalues $-E_n$. But this is nonsensical even in the very elementary case of a free particle. We know that the energy spectrum of the free particle is positive semidefinite, from 0 to $+\infty$. There is no state lower than a particle at rest (momentum eigenstate with momentum eigenvalue zero); the energy spectrum ranging from $-\infty$ to 0 would be completely unacceptable. We can also see this by looking at the structure of the free-particle Hamiltonian. We expect $\mathbf{p}$ to change sign but not $\mathbf{p}^2$; yet (4.130) would imply that

$$\Theta^{-1} \frac{\mathbf{p}^2}{2m} \Theta = \frac{-\mathbf{p}^2}{2m}. \tag{4.132}$$

All these arguments strongly suggest that if time reversal is to be a useful symmetry at all, we are not allowed to cancel the $i$ in (4.129); hence, $\Theta$ had better be antiunitary. In this case the right-hand side of (4.129) becomes

$$\Theta i H|\rangle = -i\Theta H|\rangle \tag{4.133}$$

by antilinear property (4.116b). Now at last we can cancel the $i$ in (4.129) leading, finally, via (4.133) to

$$\Theta H = H\Theta. \tag{4.134}$$

Equation (4.134) expresses the fundamental property of the Hamiltonian under time reversal. With this equation the difficulties mentioned earlier [see (4.130) to (4.132)] are absent, and we obtain physically sensible results. From now on we will always take $\Theta$ to be antiunitary.

We mentioned earlier that it is best to avoid an antiunitary operator acting on bras from the right. Nevertheless, we may use

$$\langle \beta | \Theta | \alpha \rangle, \tag{4.135}$$

which is always to be understood as

$$(\langle \beta |) \cdot (\Theta | \alpha \rangle) \tag{4.136}$$

and never as

$$(\langle \beta | \Theta) \cdot | \alpha \rangle. \tag{4.137}$$

In fact, we do not even attempt to define $\langle \beta | \Theta$. This is one place where the Dirac bra-ket notation is a little confusing. After all, that notation was invented to handle linear operators, not antilinear operators.

With this cautionary remark, we are in a position to discuss the behavior of operators under time reversal. We continue to take the point of view that the $\Theta$ operator is to act on kets

$$|\tilde{\alpha}\rangle = \Theta | \alpha \rangle, \quad |\tilde{\beta}\rangle = \Theta | \beta \rangle, \tag{4.138}$$

yet it is often convenient to talk about operators – in particular, observables – which are odd or even under time reversal. We start with an important identity, namely,

$$\langle \beta | X | \alpha \rangle = \langle \tilde{\alpha} | \Theta X^\dagger \Theta^{-1} | \tilde{\beta} \rangle, \tag{4.139}$$

where $X$ is a linear operator. This identity follows solely from the antiunitary nature of $\Theta$. To prove this let us define

$$|\gamma\rangle \equiv X^\dagger | \beta \rangle. \tag{4.140}$$

By dual correspondence we have

$$|\gamma\rangle \overset{\text{DC}}{\leftrightarrow} \langle \beta | X = \langle \gamma |. \tag{4.141}$$

Hence,

$$
\begin{aligned}
\langle \beta | X | \alpha \rangle &= \langle \gamma | \alpha \rangle = \langle \tilde{\alpha} | \tilde{\gamma} \rangle \\
&= \langle \tilde{\alpha} | \Theta X^\dagger | \beta \rangle = \langle \tilde{\alpha} | \Theta X^\dagger \Theta^{-1} \Theta | \beta \rangle \\
&= \langle \tilde{\alpha} | \Theta X^\dagger \Theta^{-1} | \tilde{\beta} \rangle,
\end{aligned}
\tag{4.142}
$$

which proves the identity. In particular, for *Hermitian* observables $A$ we get

$$\langle \beta | A | \alpha \rangle = \langle \tilde{\alpha} | \Theta A \Theta^{-1} | \tilde{\beta} \rangle. \tag{4.143}$$

We say that observables are even or odd under time reversal according to whether we have the upper or lower sign in

$$\Theta A \Theta^{-1} = \pm A. \tag{4.144}$$

Note that this equation, together with (4.143), gives a phase restriction on the matrix elements of $A$ taken with respect to time reversed states as follows:

$$\langle\beta|A|\alpha\rangle = \pm\langle\tilde{\beta}|A|\tilde{\alpha}\rangle^*. \tag{4.145}$$

If $|\beta\rangle$ is identical to $|\alpha\rangle$, so that we are talking about expectation values, we have

$$\langle\alpha|A|\alpha\rangle = \pm\langle\tilde{\alpha}|A|\tilde{\alpha}\rangle, \tag{4.146}$$

where $\langle\tilde{\alpha}|A|\tilde{\alpha}\rangle$ is the expectation value taken with respect to the time-reversed state.

As an example, let us look at the expectation value of $\mathbf{p}$. It is reasonable to expect that the expectation value of $\mathbf{p}$ taken with respect to the time-reversed state be of opposite sign. Thus

$$\langle\alpha|\mathbf{p}|\alpha\rangle = -\langle\tilde{\alpha}|\mathbf{p}|\tilde{\alpha}\rangle, \tag{4.147}$$

so we take $\mathbf{p}$ to be an odd operator, namely,

$$\Theta\mathbf{p}\Theta^{-1} = -\mathbf{p}. \tag{4.148}$$

This implies that

$$\mathbf{p}\Theta|\mathbf{p}'\rangle = -\Theta\mathbf{p}\Theta^{-1}\Theta|\mathbf{p}'\rangle$$
$$= (-\mathbf{p}')\Theta|\mathbf{p}'\rangle. \tag{4.149}$$

Equation (4.149) agrees with our earlier assertion that $\Theta|\mathbf{p}'\rangle$ is a momentum eigenket with eigenvalue $-\mathbf{p}'$. It can be identified with $|-\mathbf{p}'\rangle$ itself with a suitable choice of phase. Likewise, we obtain

$$\Theta\mathbf{x}\Theta^{-1} = \mathbf{x}$$
$$\Theta|\mathbf{x}'\rangle = |\mathbf{x}'\rangle \quad \text{(up to a phase)} \tag{4.150}$$

from the (eminently reasonable) requirement

$$\langle\alpha|\mathbf{x}|\alpha\rangle = \langle\tilde{\alpha}|\mathbf{x}|\tilde{\alpha}\rangle. \tag{4.151}$$

We can now check the invariance of the fundamental commutation relation

$$[x_i, p_j]|\rangle = i\hbar\delta_{ij}|\rangle, \tag{4.152}$$

where the blank ket $|\rangle$ stands for any ket. Applying $\Theta$ to both sides of (4.152), we have

$$\Theta\,[x_i, p_j]\Theta^{-1}\Theta|\rangle = \Theta i\hbar\delta_{ij}|\rangle, \tag{4.153}$$

which leads to, after passing $\Theta$ through $i\hbar$,

$$[x_i, (-p_j)]\Theta|\rangle = -i\hbar\delta_{ij}\Theta|\rangle. \tag{4.154}$$

Note that the fundamental commutation relation $[x_i, p_j] = i\hbar\delta_{ij}$ is preserved by virtue of the fact that $\Theta$ is antiunitary. This can be given as yet another reason for taking $\Theta$ to be antiunitary; otherwise, we will be forced to abandon either (4.148) or (4.150)! Similarly, to preserve

$$[J_i, J_j] = i\hbar\varepsilon_{ijk}J_k, \tag{4.155}$$

the angular-momentum operator must be odd under time reversal, that is,

$$\Theta \mathbf{J} \Theta^{-1} = -\mathbf{J}. \tag{4.156}$$

This is consistent for a spinless system where $\mathbf{J}$ is just $\mathbf{x} \times \mathbf{p}$. Alternatively, we could have deduced this relation by noting that the rotational operator and the time-reversal operator commute (note the extra $i$!).

### 4.4.3 Wave Function

Suppose at some given time, say at $t = 0$, a spinless single-particle system is found in a state represented by $|\alpha\rangle$. Its wave function $\langle \mathbf{x}' | \alpha \rangle$ appears as the expansion coefficient in the position representation

$$|\alpha\rangle = \int d^3 x' |\mathbf{x}'\rangle\langle \mathbf{x}' | \alpha \rangle. \tag{4.157}$$

Applying the time-reversal operator

$$\Theta|\alpha\rangle = \int d^3 x' \Theta|\mathbf{x}'\rangle\langle \mathbf{x}' | \alpha \rangle^*$$
$$= \int d^3 x' |\mathbf{x}'\rangle\langle \mathbf{x}' | \alpha \rangle^*, \tag{4.158}$$

where we have chosen the phase convention so that $\Theta|\mathbf{x}'\rangle$ is $|\mathbf{x}'\rangle$ itself. We then recover the rule

$$\psi(\mathbf{x}') \to \psi^*(\mathbf{x}') \tag{4.159}$$

inferred earlier by looking at the Schrödinger wave equation [see (4.108)]. The angular part of the wave function is given by a spherical harmonic $Y_l^m$. With the usual phase convention we have

$$Y_l^m(\theta, \phi) \to Y_l^{m*}(\theta, \phi) = (-1)^m Y_l^{-m}(\theta, \phi). \tag{4.160}$$

Now $Y_l^m(\theta, \phi)$ is the wave function for $|l, m\rangle$ [see (3.232)]; therefore, from (4.159) we deduce

$$\Theta|l, m\rangle = (-1)^m |l, -m\rangle. \tag{4.161}$$

If we study the probability current density (2.191) for a wave function of type (3.231) going like $R(r) Y_l^m$, we shall conclude that for $m > 0$ the current flows in the counterclockwise direction, as seen from the positive $z$-axis. The wave function for the corresponding time-reversed state has its probability current flowing in the opposite direction because the sign of $m$ is reversed. All this is very reasonable.

As a nontrivial consequence of time-reversal invariance, we state an important theorem on the reality of the energy eigenfunction of a spinless particle.

**Theorem 7**  *Suppose the Hamiltonian is invariant under time reversal and the energy eigenket $|n\rangle$ is nondegenerate; then the corresponding energy eigenfunction is real (or, more generally, a real function times a phase factor independent of $\mathbf{x}$).*

**Proof** To prove this, first note that

$$H\Theta|n\rangle = \Theta H|n\rangle = E_n\Theta|n\rangle, \tag{4.162}$$

so $|n\rangle$ and $\Theta|n\rangle$ have the same energy. The nondegeneracy assumption prompts us to conclude that $|n\rangle$ and $\Theta|n\rangle$ must represent the *same* state; otherwise there would be two different states with the same energy $E_n$, an obvious contradiction! Let us recall that the wave functions for $|n\rangle$ and $\Theta|n\rangle$ are $\langle\mathbf{x}'|n\rangle$ and $\langle\mathbf{x}'|n\rangle^*$, respectively. They must be the same, that is,

$$\langle\mathbf{x}'|n\rangle = \langle\mathbf{x}'|n\rangle^* \tag{4.163}$$

for all practical purposes, or, more precisely, they can differ at most by a phase factor independent of $\mathbf{x}$. $\qquad\square$

Thus if we have, for instance, a nondegenerate bound state, its wave function is always real. On the other hand, in the hydrogen atom with $l \neq 0$, $m \neq 0$, the energy eigenfunction characterized by definite $(n,l,m)$ quantum numbers is complex because $Y_l^m$ is complex; this does not contradict the theorem because $|n,l,m\rangle$ and $|n,l,-m\rangle$ are degenerate. Similarly, the wave function of a plane wave $e^{i\mathbf{p}\cdot\mathbf{x}/\hbar}$ is complex, but it is degenerate with $e^{-i\mathbf{p}\cdot\mathbf{x}/\hbar}$.

We see that for a spinless system, the wave function for the time-reversed state, say at $t = 0$, is simply obtained by complex conjugation. In terms of ket $|\alpha\rangle$ written as in (4.119) or in (4.157), the $\Theta$ operator is the complex conjugate operator $K$ itself because $K$ and $\Theta$ have the same effect when acting on the base ket $|a'\rangle$ (or $|\mathbf{x}'\rangle$). We may note, however, that the situation is quite different when the ket $|\alpha\rangle$ is expanded in terms of the momentum eigenket because $\Theta$ must change $|\mathbf{p}'\rangle$ into $|-\mathbf{p}'\rangle$ as follows:

$$\Theta|\alpha\rangle = \int d^3p'|-\mathbf{p}'\rangle\langle\mathbf{p}'|\alpha\rangle^* = \int d^3p'|\mathbf{p}'\rangle\langle-\mathbf{p}'|\alpha\rangle^*. \tag{4.164}$$

It is apparent that the momentum-space wave function of the time-reversed state is not just the complex conjugate of the original momentum-space wave function; rather, we must identify $\phi^*(-\mathbf{p}')$ as the momentum-space wave function for the time-reversed state. This situation once again illustrates the basic point that the particular form of $\Theta$ depends on the particular representation used.

## 4.4.4 Time Reversal for a Spin $\frac{1}{2}$ System

The situation is even more interesting for a particle with spin, and for spin $\frac{1}{2}$, in particular. We recall from Section 3.2 that the eigenket of $\mathbf{S}\cdot\hat{\mathbf{n}}$ with eigenvalue $\hbar/2$ can be written as

$$|\hat{\mathbf{n}};+\rangle = e^{-iS_z\alpha/\hbar}e^{-iS_y\beta/\hbar}|+\rangle, \tag{4.165}$$

where $\hat{\mathbf{n}}$ is characterized by the polar and azimuthal angles $\beta$ and $\alpha$, respectively. Noting (4.156) we have

$$\Theta|\hat{\mathbf{n}};+\rangle = e^{-iS_z\alpha/\hbar}e^{-iS_y\beta/\hbar}\Theta|+\rangle = \eta|\hat{\mathbf{n}};-\rangle. \tag{4.166}$$

On the other hand, we can easily verify that

$$|\hat{\mathbf{n}};-\rangle = e^{-i\alpha S_z/\hbar}e^{-i(\pi+\beta)S_y/\hbar}|+\rangle. \tag{4.167}$$

In general, we saw earlier that the product $UK$ is an antiunitary operator. Comparing (4.166) and (4.167) with $\Theta$ set equal to $UK$, and noting that $K$ acting on the base ket $|+\rangle$ gives just $|+\rangle$, we see that

$$\Theta = \eta e^{-i\pi S_y/\hbar} K = -i\eta \left(\frac{2S_y}{\hbar}\right) K, \tag{4.168}$$

where $\eta$ stands for an arbitrary phase (a complex number of modulus unity). Another way to be convinced of (4.168) is to verify that if $\chi(\hat{\mathbf{n}}; +)$ is the two-component eigenspinor corresponding to $|\hat{\mathbf{n}}; +\rangle$ [in the sense that $\sigma \cdot \hat{\mathbf{n}} \chi(\hat{\mathbf{n}}; +) = \chi(\hat{\mathbf{n}}; +)$], then

$$-i\sigma_y \chi^*(\hat{\mathbf{n}}; +) \tag{4.169}$$

(note the complex conjugation!) is the eigenspinor corresponding to $|\hat{\mathbf{n}}; -\rangle$, again up to an arbitrary phase, see Problem 4.7 of this chapter. The appearance of $S_y$ or $\sigma_y$ can be traced to the fact that we are using the representation in which $S_z$ is diagonal and the nonvanishing matrix elements of $S_y$ are purely imaginary.

Let us now note

$$e^{-i\pi S_y/\hbar}|+\rangle = +|-\rangle, \qquad e^{-i\pi S_y/\hbar}|-\rangle = -|+\rangle. \tag{4.170}$$

Using (4.170), we are in a position to work out the effect of $\Theta$, written as (4.168), on the most general spin $\frac{1}{2}$ ket:

$$\Theta(c_+|+\rangle + c_-|-\rangle) = +\eta c_+^*|-\rangle - \eta c_-^*|+\rangle. \tag{4.171}$$

Let us apply $\Theta$ once again:

$$\Theta^2(c_+|+\rangle + c_-|-\rangle) = -|\eta|^2 c_+|+\rangle - |\eta|^2 c_-|-\rangle$$
$$= -(c_+|+\rangle + c_-|-\rangle) \tag{4.172}$$

or

$$\Theta^2 = -1 \tag{4.173}$$

(where $-1$ is to be understood as $-1$ times the identity operator), for *any* spin orientation. This is an extraordinary result. It is crucial to note here that our conclusion is completely independent of the choice of phase; (4.173) holds no matter what phase convention we may use for $\eta$. In contrast, we may note that two successive applications of $\Theta$ to a spinless state give

$$\Theta^2 = +1 \tag{4.174}$$

as is evident from, say, (4.161).

More generally, we now prove

$$\Theta^2|j\,\text{half-integer}\rangle = -|j\,\text{half-integer}\rangle \tag{4.175a}$$

$$\Theta^2|j\,\text{integer}\rangle = +|j\,\text{integer}\rangle. \tag{4.175b}$$

Thus the eigenvalue of $\Theta^2$ is given by $(-1)^{2j}$. We first note that (4.168) generalizes for arbitrary $j$ to

$$\Theta = \eta e^{-i\pi J_y/\hbar} K. \tag{4.176}$$

For a ket $|\alpha\rangle$ expanded in terms of $|j,m\rangle$ base eigenkets, we have

$$\Theta\left(\Theta \sum |jm\rangle\langle jm|\alpha\rangle\right) = \Theta\left(\eta \sum e^{-i\pi J_y/\hbar}|jm\rangle\langle jm|\alpha\rangle^*\right)$$
$$= |\eta|^2 e^{-2i\pi J_y/\hbar} \sum |jm\rangle\langle jm|\alpha\rangle. \tag{4.177}$$

But

$$e^{-2i\pi J_y/\hbar}|jm\rangle = (-1)^{2j}|jm\rangle, \tag{4.178}$$

as is evident from the properties of angular-momentum eigenstates under rotation by $2\pi$.

In (4.175b), $|j\,\text{integer}\rangle$ may stand for the spin state

$$\frac{1}{\sqrt{2}}(|+-\rangle \pm |-+\rangle) \tag{4.179}$$

of a two-electron system or the orbital state $|l,m\rangle$ of a spinless particle. It is important only that $j$ is an integer. Likewise, $|j\,\text{half-integer}\rangle$ may stand, for example, for a three-electron system in any configuration. Actually, for a system made up exclusively of electrons, any system with an odd (even) number of electrons – regardless of their spatial orientation (for example, relative orbital angular momentum) – is odd (even) under $\Theta^2$; they need not even be $\mathbf{J}^2$ eigenstates!

We make a parenthetical remark on the phase convention. In our earlier discussion based on the position representation, we saw that with the usual convention for spherical harmonics it is natural to choose the arbitrary phase for $|l,m\rangle$ under time reversal so that

$$\Theta|l,m\rangle = (-1)^m|l,-m\rangle. \tag{4.180}$$

Some authors find it attractive to generalize this to obtain

$$\Theta|j,m\rangle = (-1)^m|j,-m\rangle \quad (j \text{ an integer}) \tag{4.181}$$

regardless of whether $j$ refers to $l$ or $s$ (for an integer spin system). We may naturally ask whether this is compatible with (4.175a) for a spin $\frac{1}{2}$ system when we visualize $|j,m\rangle$ as being built up of "primitive" spin $\frac{1}{2}$ objects according to Wigner and Schwinger. It is easy to see that (4.175a) is indeed consistent provided we choose $\eta$ in (4.176) to be $+i$. In fact, in general, we can take

$$\Theta|j,m\rangle = i^{2m}|j,-m\rangle \tag{4.182}$$

for any $j$, either a half-integer $j$ or an integer $j$; see Problem 4.10 of this chapter. The reader should be warned, however, that this is not the only convention found in the literature. See, for instance, Henley and Garcia (2007). For some physical applications, it is more convenient to use other choices; for instance, the phase convention that makes the $\mathbf{J}_\pm$ operator matrix elements simple is *not* the phase convention that makes the time-reversal operator properties simple. We emphasize once again that (4.175) is completely independent of phase convention.

Having worked out the behavior of angular-momentum eigenstates under time reversal, we are in a position to study once again the expectation values of a Hermitian operator. Recalling (4.146), we obtain under time reversal (canceling the $i^{2m}$ factors)

$$\langle \alpha,j,m|A|\alpha,j,m\rangle = \pm\langle \alpha,j,-m|A|\alpha,j,-m\rangle. \tag{4.183}$$

Now suppose $A$ is a component of a spherical tensor $T_q^{(k)}$. Because of the Wigner–Eckart theorem, it is sufficient to examine just the matrix element of the $q=0$ component. In general, $T^{(k)}$ (assumed to be Hermitian) is said to be even or odd under time reversal depending on how its $q=0$ component satisfies the upper or lower sign in

$$\Theta T_{q=0}^{(k)}\Theta^{-1} = \pm T_{q=0}^{(k)}. \tag{4.184}$$

Equation (4.183) for $A = T_0^{(k)}$ becomes

$$\langle \alpha,j,m|T_0^{(k)}|\alpha,j,m\rangle = \pm\langle \alpha,j,-m|T_0^{(k)}|\alpha,j,-m\rangle. \tag{4.185}$$

Due to (3.255)–(3.258), we expect $|\alpha,j,-m\rangle = \mathscr{D}(0,\pi,0)|\alpha,j,m\rangle$ up to a phase. We next use (3.465) for $T_0^{(k)}$, which leads to

$$\mathscr{D}^\dagger(0,\pi,0)T_0^{(k)}\mathscr{D}(0,\pi,0) = (-1)^k T_0^{(k)} + (q \neq 0 \text{ components}), \tag{4.186}$$

where we have used $\mathscr{D}_{00}^{(k)}(0,\pi,0) = P_k(\cos\pi) = (-1)^k$, and the $q \neq 0$ components give vanishing contributions when sandwiched between $\langle \alpha,j,m|$ and $|\alpha,j,m\rangle$. The net result is

$$\langle \alpha,j,m|T_0^{(k)}|\alpha,j,m\rangle = \pm(-1)^k\langle \alpha,j,m|T_0^{(k)}|\alpha,j,m\rangle. \tag{4.187}$$

As an example, taking $k=1$, the expectation value $\langle \mathbf{x}\rangle$ taken with respect to eigenstates of $j$, $m$ vanishes. We may argue that we already know $\langle \mathbf{x}\rangle = 0$ from parity inversion if the expectation value is taken with respect to parity eigenstates [see (4.81)]. But note that here $|\alpha,j,m\rangle$ need not be parity eigenkets! For example, the $|j,m\rangle$ for spin $\frac{1}{2}$ particles could be $c_s|s_{1/2}\rangle + c_p|p_{1/2}\rangle$.

## 4.4.5 Interactions with Electric and Magnetic Fields; Kramers Degeneracy

Consider charged particles in an external electric or magnetic field. If we have only a static electric field interacting with the electric charge, the interaction part of the Hamiltonian is just

$$V(\mathbf{x}) = e\phi(\mathbf{x}), \tag{4.188}$$

where $\phi(\mathbf{x})$ is the electrostatic potential. Because $\phi(\mathbf{x})$ is a real function of the time-reversal even operator $\mathbf{x}$, we have

$$[\Theta,H] = 0. \tag{4.189}$$

Unlike the parity case, (4.189) does not lead to an interesting conservation law. The reason is that

$$\Theta U(t,t_0) \neq U(t,t_0)\Theta \tag{4.190}$$

even if (4.189) holds, so our discussion following (4.9) of Section 4.1 breaks down. As a result, there is no such thing as the "conservation of time-reversal quantum number." As we already mentioned, requirement (4.189) does, however, lead to a nontrivial phase restriction – the reality of a nondegenerate wave function for a spinless system [see (4.162) and (4.163)].

Another far-reaching consequence of time-reversal invariance is the **Kramers degeneracy**. Suppose $H$ and $\Theta$ commute, and let $|n\rangle$ and $\Theta|n\rangle$ be the energy eigenket and its time-reversed state, respectively. It is evident from (4.189) that $|n\rangle$ and $\Theta|n\rangle$ belong to the same energy eigenvalue $E_n$ ($H\Theta|n\rangle = \Theta H|n\rangle = E_n\Theta|n\rangle$). The question is, does $|n\rangle$ represent the same state as $\Theta|n\rangle$? If it does, $|n\rangle$ and $\Theta|n\rangle$ can differ at most by a phase factor. Hence,

$$\Theta|n\rangle = e^{i\delta}|n\rangle. \tag{4.191}$$

Applying $\Theta$ again to (4.191), we have $\Theta^2|n\rangle = \Theta e^{i\delta}|n\rangle = e^{-i\delta}\Theta|n\rangle = e^{-i\delta}e^{+i\delta}|n\rangle$; hence,

$$\Theta^2|n\rangle = +|n\rangle. \tag{4.192}$$

But this relation is impossible for half-integer $j$ systems, for which $\Theta^2$ is always $-1$, so we are led to conclude that $|n\rangle$ and $\Theta|n\rangle$, which have the same energy, must correspond to distinct states, that is, there must be a degeneracy. This means, for instance, that for a system composed of an odd number of electrons in an external electric field $\mathbf{E}$, each energy level must be at least twofold degenerate no matter how complicated $\mathbf{E}$ may be. Considerations along this line have interesting applications to electrons in crystals where odd-electron and even-electron systems exhibit very different behaviors. Historically, Kramers inferred degeneracy of this kind by looking at explicit solutions of the Schrödinger equation; subsequently, Wigner pointed out that Kramers degeneracy is a consequence of time-reversal invariance.

Let us now turn to interactions with an external magnetic field. The Hamiltonian $H$ may then contain terms like

$$\mathbf{S}\cdot\mathbf{B}, \qquad \mathbf{p}\cdot\mathbf{A} + \mathbf{A}\cdot\mathbf{p} \qquad (\mathbf{B} = \nabla \times \mathbf{A}), \tag{4.193}$$

where the magnetic field is to be regarded as external. The operators $\mathbf{S}$ and $\mathbf{p}$ are odd under time reversal; these interaction terms therefore do lead to

$$\Theta H \neq H\Theta. \tag{4.194}$$

As a trivial example, for a spin $\frac{1}{2}$ system the spin-up state $|+\rangle$ and its time-reversed state $|-\rangle$ no longer have the same energy in the presence of an external magnetic field. In general, Kramers degeneracy in a system containing an odd number of electrons can be lifted by applying an external magnetic field.

Notice that when we treat $\mathbf{B}$ as external, we do not change $\mathbf{B}$ under time reversal; this is because the atomic electron is viewed as a closed quantum-mechanical system to which we apply the time-reversal operator. This should not be confused with our earlier remarks concerning the invariance of the Maxwell equations (4.105) and the Lorentz force equation under $t \to -t$ and (4.106). There we were to apply time reversal to the *whole world*, for example, even to the currents in the wire that produces the $\mathbf{B}$ field!

# Problems

**4.1**  Calculate the *three lowest* energy levels, together with their degeneracies, for the following systems (assume equal mass *distinguishable* particles).

a. Three noninteracting spin $\frac{1}{2}$ particles in a cubic box of side length $L$.

b. Four noninteracting spin $\frac{1}{2}$ particles in a cubic box of side length $L$.

**4.2**  Let $\mathcal{T}_{\mathbf{d}}$ denote the translation operator (displacement vector $\mathbf{d}$), $\mathscr{D}(\hat{\mathbf{n}}, \phi)$ the rotation operator ($\hat{\mathbf{n}}$ and $\phi$ are the axis and angle of rotation, respectively), and $\pi$ the parity operator. Which, if any, of the following pairs commute? Why?

a. $\mathcal{T}_{\mathbf{d}}$ and $\mathcal{T}_{\mathbf{d}'}$ ($\mathbf{d}$ and $\mathbf{d}'$ in different directions).

b. $\mathscr{D}(\hat{\mathbf{n}}, \phi)$ and $\mathscr{D}(\hat{\mathbf{n}}', \phi')$ ($\hat{\mathbf{n}}$ and $\hat{\mathbf{n}}'$ in different directions).

c. $\mathcal{T}_{\mathbf{d}}$ and $\pi$.

d. $\mathscr{D}(\hat{\mathbf{n}}, \phi)$ and $\pi$.

**4.3**  A quantum-mechanical state $|\Psi\rangle$ is known to be a simultaneous eigenstate of two Hermitian operators $A$ and $B$ which *anticommute*,

$$AB + BA = 0.$$

What can you say about the eigenvalues of $A$ and $B$ for state $|\Psi\rangle$? Illustrate your point using the parity operator (which can be chosen to satisfy $\pi = \pi^{-1} = \pi^{\dagger}$) and the momentum operator.

**4.4**  A spin $\frac{1}{2}$ particle is bound to a fixed center by a spherically symmetrical potential.

a. Write down the spin angular function $\mathscr{Y}_{l=0}^{j=1/2, m=1/2}$.

b. Express $(\boldsymbol{\sigma} \cdot \mathbf{x}) \, \mathscr{Y}_{l=0}^{j=1/2, m=1/2}$ in terms of some other $\mathscr{Y}_{l}^{j,m}$.

c. Show that your result in (b) is understandable in view of the transformation properties of the operator $\mathbf{S} \cdot \mathbf{x}$ under rotations and under space inversion (parity).

**4.5**  Because of weak (neutral-current) interactions there is a parity-violating potential between the atomic electron and the nucleus as follows:

$$V = \lambda [\delta^{(3)}(\mathbf{x}) \mathbf{S} \cdot \mathbf{p} + \mathbf{S} \cdot \mathbf{p} \delta^{(3)}(\mathbf{x})],$$

where $\mathbf{S}$ and $\mathbf{p}$ are the spin and momentum operators of the electron, and the nucleus is assumed to be situated at the origin. As a result, the ground state of an alkali atom, usually characterized by $|n, l, j, m\rangle$, actually contains very tiny contributions from other eigenstates as follows:

$$|n, l, j, m\rangle \rightarrow |n, l, j, m\rangle + \sum_{n'l'j'm'} C_{n'l'j'm'} |n', l', j', m'\rangle.$$

On the basis of symmetry considerations *alone*, what can you say about $(n', l', j', m')$, which give rise to nonvanishing contributions? Suppose the radial wave functions and the energy levels are all known. Indicate how you may calculate $C_{n'l'j'm'}$. Do we get further restrictions on $(n', l', j', m')$?

**4.6**    Consider a symmetric rectangular double-well potential:

$$V = \begin{cases} \infty & \text{for } |x| > a+b, \\ 0 & \text{for } a < |x| < a+b, \\ V_0 > 0 & \text{for } |x| < a. \end{cases}$$

Assuming that $V_0$ is very high compared to the quantized energies of low-lying states, obtain an approximate expression for the energy splitting between the two lowest-lying states.

**4.7**    a. Let $\psi(\mathbf{x},t)$ be the wave function of a spinless particle corresponding to a plane wave in three dimensions. Show that $\psi^*(\mathbf{x},-t)$ is the wave function for the plane wave with the momentum direction reversed.

b. Let $\chi(\hat{\mathbf{n}})$ be the two-component eigenspinor of $\boldsymbol{\sigma} \cdot \hat{\mathbf{n}}$ with eigenvalue $+1$. *Using the explicit form of* $\chi(\hat{\mathbf{n}})$ (in terms of the polar and azimuthal angles $\beta$ and $\gamma$ that characterize $\hat{\mathbf{n}}$) verify that $-i\sigma_2\chi^*(\hat{\mathbf{n}})$ is the two-component eigenspinor with the spin direction reversed.

**4.8**    a. Assuming that the Hamiltonian is invariant under time reversal, prove that the wave function for a spinless nondegenerate system at any given instant of time can always be chosen to be real.

b. The wave function for a plane wave state at $t = 0$ is given by a complex function $e^{i\mathbf{p}\cdot\mathbf{x}/\hbar}$. Why does this not violate time-reversal invariance?

**4.9**    Let $\phi(\mathbf{p}')$ be the momentum-space wave function for state $|\alpha\rangle$, that is, $\phi(\mathbf{p}') = \langle\mathbf{p}'|\alpha\rangle$. Is the momentum-space wave function for the time-reversed state $\Theta|\alpha\rangle$ given by $\phi(\mathbf{p}')$, $\phi(-\mathbf{p}')$, $\phi^*(\mathbf{p}')$, or $\phi^*(-\mathbf{p}')$? Justify your answer.

**4.10**    a. For the time-reversal operator $\Theta$, use (4.156) to show that $\Theta|jm\rangle$ equals $|j,-m\rangle$ up to some phase that includes the factor $(-1)^m$. In other words, show that $\Theta|jm\rangle = e^{i\delta}(-1)^m|j,-m\rangle$, where $\delta$ is independent of $m$.

b. Using the same phase convention, find the time-reversed state corresponding to $\mathscr{D}(R)|jm\rangle$. Consider using the infinitesimal form $\mathscr{D}(\hat{\mathbf{n}},d\phi)$ and then generalize to finite rotations.

c. From these results, prove that, independent of $\delta$, one finds

$$\mathscr{D}^{(j)*}_{m'm}(R) = (-1)^{m-m'}\mathscr{D}^{(j)}_{-m',-m}(R).$$

d. Conclude that we are free to choose $\delta = 0$, and $\Theta|jm\rangle = (-1)^m|j,-m\rangle = i^{2m}|j,-m\rangle$.

**4.11**    Suppose a spinless particle is bound to a fixed center by a potential $V(\mathbf{x})$ so asymmetrical that no energy level is degenerate. Using time-reversal invariance prove

$$\langle\mathbf{L}\rangle = 0$$

for any energy eigenstate. (This is known as **quenching** of orbital angular momentum.) If the wave function of such a nondegenerate eigenstate is expanded as

$$\sum_l \sum_m F_{lm}(r) Y_l^m(\theta, \phi),$$

what kind of phase restrictions do we obtain on $F_{lm}(r)$? It is useful to review Theorem 7 on p. 279.

**4.12** The Hamiltonian for a spin 1 system is given by

$$H = AS_z^2 + B(S_x^2 - S_y^2).$$

Solve this problem *exactly* to find the normalized energy eigenstates and eigenvalues. (A spin-dependent Hamiltonian of this kind actually appears in crystal physics.) Is this Hamiltonian invariant under time reversal? How do the normalized eigenstates you obtained transform under time reversal?

# 5

# Approximation Methods

Few problems in quantum mechanics – with either time-independent or time-dependent Hamiltonians – can be solved exactly. Inevitably we are forced to resort to some form of approximation method. One may argue that with the advent of high-speed computers it is always possible to obtain the desired solution numerically to the requisite degree of accuracy; nevertheless, it remains important to understand the basic physics of the approximate solutions even before we embark on ambitious computer calculations. This chapter is devoted to a fairly systematic discussion of approximate solutions to bound-state problems.

## 5.1 Time-Independent Perturbation Theory: Nondegenerate Case

### 5.1.1 Statement of the Problem

The approximation method we consider here is time-independent perturbation theory, sometimes known as the Rayleigh–Schrödinger perturbation theory. We consider a time-independent Hamiltonian $H$ such that it can be split into two parts, namely,

$$H = H_0 + V, \tag{5.1}$$

where the $V = 0$ problem is assumed to have been solved in the sense that both the exact energy eigenkets $|n^{(0)}\rangle$ and the exact energy eigenvalues $E_n^{(0)}$ are known:

$$H_0|n^{(0)}\rangle = E_n^{(0)}|n^{(0)}\rangle. \tag{5.2}$$

We are required to find approximate eigenkets and eigenvalues for the *full* Hamiltonian problem

$$(H_0 + V)|n\rangle = E_n|n\rangle, \tag{5.3}$$

where $V$ is known as the **perturbation**; it is not, in general, the full-potential operator. For example, suppose we consider the hydrogen atom in an external electric or magnetic field. The unperturbed Hamiltonian $H_0$ is taken to be the kinetic energy $\mathbf{p}^2/2m$ *and* the Coulomb potential due to the presence of the proton nucleus $-e^2/r$. Only that part of the potential due to the interaction with the external $\mathbf{E}$ or $\mathbf{B}$ field is represented by the perturbation $V$.

Instead of (5.3) it is customary to solve

$$(H_0 + \lambda V)|n\rangle = E_n|n\rangle, \tag{5.4}$$

where $\lambda$ is a continuous real parameter. This parameter is introduced to keep track of the number of times the perturbation enters. At the end of the calculation we may set $\lambda \to 1$ to get back to the full-strength case. In other words, we assume that the strength of the perturbation can be controlled. The parameter $\lambda$ can be visualized to vary continuously from 0 to 1, the $\lambda = 0$ case corresponding to the unperturbed problem and $\lambda = 1$ corresponding to the full-strength problem of (5.3). In physical situations where this approximation method is applicable, we expect to see a smooth transition of $|n^0\rangle$ into $|n\rangle$ and $E_n^{(0)}$ into $E_n$ as $\lambda$ is "dialed" from 0 to 1.

The method rests on the expansion of the energy eigenvalues and energy eigenkets in powers of $\lambda$. This means that we implicitly assume the analyticity of the energy eigenvalues and eigenkets in a complex $\lambda$-plane around $\lambda = 0$. Of course, if our method is to be of practical interest, good approximations can better be obtained by taking only one or two terms in the expansion.

## 5.1.2  The Two-State Problem

Before we embark on a systematic presentation of the basic method, let us see how the expansion in $\lambda$ might indeed be valid in the exactly soluble two-state problem we have encountered many times already. Suppose we have a Hamiltonian that can be written as

$$H = E_1^{(0)}|1^{(0)}\rangle\langle 1^{(0)}| + E_2^{(0)}|2^{(0)}\rangle\langle 2^{(0)}| + \lambda V_{12}|1^{(0)}\rangle\langle 2^{(0)}| + \lambda V_{21}|2^{(0)}\rangle\langle 1^{(0)}|, \tag{5.5}$$

where $|1^{(0)}\rangle$ and $|2^{(0)}\rangle$ are the energy eigenkets for the $\lambda = 0$ problem, and we consider the case $V_{11} = V_{22} = 0$. In this representation the $H$ may be represented by a square matrix as follows:

$$H = \begin{pmatrix} E_1^{(0)} & \lambda V_{12} \\ \lambda V_{21} & E_2^{(0)} \end{pmatrix}, \tag{5.6}$$

where we have used the basis formed by the unperturbed energy eigenkets. The $V$ matrix must, of course, be Hermitian; let us solve the case when $V_{12}$ and $V_{21}$ are real:

$$V_{12} = V_{12}^*, \quad V_{21} = V_{21}^*; \tag{5.7}$$

hence, by Hermiticity

$$V_{12} = V_{21}. \tag{5.8}$$

This can always be done by adjusting the phase of $|2^{(0)}\rangle$ relative to that of $|1^{(0)}\rangle$. The problem of obtaining the energy eigenvalues here is completely analogous to that of solving the spin-orientation problem, where the analogue of (5.6) is

$$H = a_0 + \sigma \cdot \mathbf{a} = \begin{pmatrix} a_0 + a_3 & a_1 \\ a_1 & a_0 - a_3 \end{pmatrix}, \tag{5.9}$$

where we assume $\mathbf{a} = (a_1, 0, a_3)$ is small and $a_0, a_1, a_3$ are all real. The eigenvalues for this problem are known to be just

$$E = a_0 \pm \sqrt{a_1^2 + a_3^2}. \tag{5.10}$$

By analogy the corresponding eigenvalues for (5.6) are

$$\left\{ \begin{matrix} E_1 \\ E_2 \end{matrix} \right\} = \frac{\left( E_1^{(0)} + E_2^{(0)} \right)}{2} \pm \sqrt{\left[ \frac{\left( E_1^{(0)} - E_2^{(0)} \right)^2}{4} + \lambda^2 |V_{12}|^2 \right]}. \tag{5.11}$$

Let us suppose $\lambda |V_{12}|$ is small compared with the relevant energy scale, the difference of the energy eigenvalues of the unperturbed problem:

$$\lambda |V_{12}| \ll |E_1^{(0)} - E_2^{(0)}|. \tag{5.12}$$

We can then use

$$\sqrt{1 + \varepsilon} = 1 + \frac{1}{2}\varepsilon - \frac{\varepsilon^2}{8} + \cdots \tag{5.13}$$

to obtain the expansion of the energy eigenvalues in the presence of perturbation $\lambda |V_{12}|$, namely,

$$E_1 = E_1^{(0)} + \frac{\lambda^2 |V_{12}|^2}{\left( E_1^{(0)} - E_2^{(0)} \right)} + \cdots$$

$$E_2 = E_2^{(0)} + \frac{\lambda^2 |V_{12}|^2}{\left( E_2^{(0)} - E_1^{(0)} \right)} + \cdots. \tag{5.14}$$

These are expressions that we can readily obtain using the general formalism to be developed shortly. It is also possible to write down the energy eigenkets in analogy with the spin-orientation problem.

The reader might be led to believe that a perturbation expansion always exists for a sufficiently weak perturbation. Unfortunately this is not necessarily the case. As an elementary example, consider a one-dimensional problem involving a particle of mass $m$ in a very weak square-well potential of depth $V_0$ (i.e. $V = -V_0$ for $-a < x < a$ and $V = 0$ for $|x| > a$). This problem admits one bound state of energy,

$$E = -(2ma^2/\hbar^2)|\lambda V|^2, \ \lambda > 0 \text{ for attraction.} \tag{5.15}$$

We might regard the square well as a very weak perturbation to be added to the free-particle Hamiltonian and interpret result (5.15) as the energy shift in the ground state from zero to $|\lambda V|^2$. Specifically, because (5.15) is quadratic in $V$, we might be tempted to associate this as the energy shift of the ground state computed according to second-order perturbation theory. However, this view is false because if this were the case, the system would also admit an $E < 0$ state for a repulsive potential case with $\lambda$ negative, which would be sheer nonsense.

Let us now examine the radius of convergence of series expansion (5.14). If we go back to the exact expression of (5.11) and regard it as a function of a *complex* variable $\lambda$, we see that as $|\lambda|$ is increased from zero, branch points are encountered at

$$\lambda|V_{12}| = \frac{\pm i(E_1^{(0)} - E_2^{(0)})}{2}. \tag{5.16}$$

The condition for the convergence of the series expansion for the $\lambda = 1$ full-strength case is

$$|V_{12}| < \frac{|E_1^{(0)} - E_2^{(0)}|}{2}. \tag{5.17}$$

If this condition is not met, perturbation expansion (5.14) is meaningless.[1]

### 5.1.3 Formal Development of Perturbation Expansion

We now state in more precise terms the basic problem we wish to solve. Suppose we know completely and exactly the energy eigenkets and energy eigenvalues of the unperturbed Hamiltonian $H_0$, that is

$$H_0|n^{(0)}\rangle = E_n^{(0)}|n^{(0)}\rangle. \tag{5.18}$$

The set $\{|n^{(0)}\rangle\}$ is complete in the sense that the closure relation $1 = \sum_n |n^{(0)}\rangle\langle n^{(0)}|$ holds. Furthermore, we assume here that the energy spectrum is nondegenerate; in the next section we will relax this assumption. We are interested in obtaining the energy eigenvalues and eigenkets for the problem defined by (5.4). To be consistent with (5.18) we should write (5.4) as

$$(H_0 + \lambda V)|n\rangle_\lambda = E_n^{(\lambda)}|n\rangle_\lambda \tag{5.19}$$

to denote the fact that the energy eigenvalues $E_n^{(\lambda)}$ and energy eigenkets $|n\rangle_\lambda$ are functions of the continuous parameter $\lambda$; however, we will usually dispense with this correct but more cumbersome notation.

As the continuous parameter $\lambda$ is increased from zero, we expect the energy eigenvalue $E_n$ for the $n$th eigenket to depart from its unperturbed value $E_n^{(0)}$, so we define the energy shift for the $n$th level as follows:

$$\Delta_n \equiv E_n - E_n^{(0)}. \tag{5.20}$$

The basic Schrödinger equation to be solved (approximately) is

$$(E_n^{(0)} - H_0)|n\rangle = (\lambda V - \Delta_n)|n\rangle. \tag{5.21}$$

We may be tempted to invert the operator $E_n^{(0)} - H_0$; however, in general, the inverse operator $1/(E_n^{(0)} - H_0)$ is ill defined because it may act on $|n^{(0)}\rangle$. Fortunately in our case $(\lambda V - \Delta_n)|n\rangle$ has no component along $|n^{(0)}\rangle$, as can easily be seen by multiplying both sides of (5.21) by $\langle n^{(0)}|$ on the left:

$$\langle n^{(0)}|(\lambda V - \Delta_n)|n\rangle = 0. \tag{5.22}$$

---

[1] See the discussion on convergence following (5.44), under general remarks.

Suppose we define the complementary projection operator

$$\phi_n \equiv 1 - |n^{(0)}\rangle\langle n^{(0)}| = \sum_{k \neq n} |k^{(0)}\rangle\langle k^{(0)}|. \tag{5.23}$$

The inverse operator $1/(E_n^{(0)} - H_0)$ is well defined when it multiplies $\phi_n$ on the right. Explicitly,

$$\frac{1}{E_n^{(0)} - H_0}\phi_n = \sum_{k \neq n} \frac{1}{E_n^{(0)} - E_k^{(0)}}|k^{(0)}\rangle\langle k^{(0)}|. \tag{5.24}$$

Also from (5.22) and (5.23), it is evident that

$$(\lambda V - \Delta_n)|n\rangle = \phi_n(\lambda V - \Delta_n)|n\rangle. \tag{5.25}$$

We may therefore be tempted to rewrite (5.21) as

$$|n\rangle \overset{?}{=} \frac{1}{E_n^{(0)} - H_0}\phi_n(\lambda V - \Delta_n)|n\rangle. \tag{5.26}$$

However, this cannot be correct because as $\lambda \to 0$, we must have $|n\rangle \to |n^{(0)}\rangle$ and $\Delta_n \to 0$. Nevertheless, even for $\lambda \neq 0$, we can always add to $|n\rangle$ a solution to the homogeneous equation (5.18), namely, $c_n|n^{(0)}\rangle$, so a suitable final form is

$$|n\rangle = c_n(\lambda)|n^{(0)}\rangle + \frac{1}{E_n^{(0)} - H_0}\phi_n(\lambda V - \Delta_n)|n\rangle, \tag{5.27}$$

where

$$\lim_{\lambda \to 0} c_n(\lambda) = 1. \tag{5.28}$$

Note that

$$c_n(\lambda) = \langle n^{(0)}|n\rangle. \tag{5.29}$$

For reasons we will see later, it is convenient to depart from the usual normalization convention

$$\langle n|n\rangle = 1. \tag{5.30}$$

Rather, we set

$$\langle n^{(0)}|n\rangle = c_n(\lambda) = 1, \tag{5.31}$$

even for $\lambda \neq 0$. We can always do this if we are not worried about the overall normalization because the only effect of setting $c_n \neq 1$ is to introduce a common multiplicative factor. Thus, if desired, we can always normalize the ket at the very end of the calculation. It is also customary to write

$$\frac{1}{E_n^{(0)} - H_0}\phi_n \to \frac{\phi_n}{E_n^{(0)} - H_0} \tag{5.32}$$

and similarly

$$\frac{1}{E_n^{(0)} - H_0}\phi_n = \phi_n\frac{1}{E_n^{(0)} - H_0} = \phi_n\frac{1}{E_n^{(0)} - H_0}\phi_n, \tag{5.33}$$

so we have

$$|n\rangle = |n^{(0)}\rangle + \frac{\phi_n}{E_n^{(0)} - H_0}(\lambda V - \Delta_n)|n\rangle. \tag{5.34}$$

We also note from (5.22) and (5.31) that

$$\Delta_n = \lambda\langle n^{(0)}|V|n\rangle. \tag{5.35}$$

Everything depends on the two equations in (5.34) and (5.35). Our basic strategy is to expand $|n\rangle$ and $\Delta_n$ in the powers of $\lambda$ and then match the appropriate coefficients. This is justified because (5.34) and (5.35) are identities which hold for all values of $\lambda$ between 0 and 1. We begin by writing

$$|n\rangle = |n^{(0)}\rangle + \lambda|n^{(1)}\rangle + \lambda^2|n^{(2)}\rangle + \cdots$$
$$\Delta_n = \lambda\Delta_n^{(1)} + \lambda^2\Delta_n^{(2)} + \cdots. \tag{5.36}$$

Substituting (5.36) into (5.35) and equating the coefficient of various powers of $\lambda$, we obtain

$$\begin{array}{lll}
O(\lambda^1): & \Delta_n^{(1)} = & \langle n^{(0)}|V|n^{(0)}\rangle \\
O(\lambda^2): & \Delta_n^{(2)} = & \langle n^{(0)}|V|n^{(1)}\rangle \\
\vdots & \vdots & \\
O(\lambda^N): & \Delta_n^{(N)} = & \langle n^{(0)}|V|n^{(N-1)}\rangle \\
\vdots & \vdots &
\end{array} \tag{5.37}$$

so to evaluate the energy shift up to order $\lambda^N$ it is sufficient to know $|n\rangle$ only up to order $\lambda^{N-1}$. We now look at (5.34); when it is expanded using (5.36), we get

$$|n^{(0)}\rangle + \lambda|n^{(1)}\rangle + \lambda^2|n^{(2)}\rangle + \cdots$$
$$= |n^{(0)}\rangle + \frac{\phi_n}{E_n^{(0)} - H_0}(\lambda V - \lambda\Delta_n^{(1)} - \lambda^2\Delta_n^2 - \cdots)$$
$$\times (|n^{(0)}\rangle + \lambda|n^{(1)}\rangle + \cdots). \tag{5.38}$$

Equating the coefficient of powers of $\lambda$, we have

$$O(\lambda): \quad |n^{(1)}\rangle = \frac{\phi_n}{E_n^{(0)} - H_0}V|n^{(0)}\rangle, \tag{5.39}$$

where we have used $\phi_n\Delta_n^{(1)}|n^{(0)}\rangle = 0$. Armed with $|n^{(1)}\rangle$, it is now profitable for us to go back to our earlier expression for $\Delta_n^{(2)}$ [see (5.37)]:

$$\Delta_n^{(2)} = \langle n^{(0)}|V\frac{\phi_n}{E_n^{(0)} - H_0}V|n^{(0)}\rangle. \tag{5.40}$$

Knowing $\Delta_n^{(2)}$, we can work out the $\lambda^2$-term in ket equation (5.38) also using (5.39) as follows:

$$O(\lambda^2): \quad |n^{(2)}\rangle = \frac{\phi_n}{E_n^{(0)} - H_0} V \frac{\phi_n}{E_n^{(0)} - H_0} V |n^{(0)}\rangle$$

$$- \frac{\phi_n}{E_n^{(0)} - H_0} \langle n^{(0)}|V|n^{(0)}\rangle \frac{\phi_n}{E_n^{(0)} - H_0} V |n^{(0)}\rangle. \tag{5.41}$$

Clearly, we can continue in this fashion as long as we wish. Our operator method is very compact; it is not necessary to write down the indices each time. Of course, to do practical calculations we must use at the end the explicit form of $\phi_n$ as given by (5.23).

To see how all this works, we write down the explicit expansion for the energy shift

$$\Delta_n \equiv E_n - E_n^{(0)}$$

$$= \lambda V_{nn} + \lambda^2 \sum_{k \neq n} \frac{|V_{nk}|^2}{E_n^{(0)} - E_k^{(0)}} + \cdots, \tag{5.42}$$

where

$$V_{nk} \equiv \langle n^{(0)}|V|k^{(0)}\rangle \neq \langle n|V|k\rangle, \tag{5.43}$$

that is, the matrix elements are taken with respect to *unperturbed* kets. Notice that when we apply the expansion to the two-state problem we recover the earlier expression (5.14). The expansion for the perturbed ket goes as follows:

$$|n\rangle = |n^{(0)}\rangle + \lambda \sum_{k \neq n} |k^{(0)}\rangle \frac{V_{kn}}{E_n^{(0)} - E_k^{(0)}}$$

$$+ \lambda^2 \left( \sum_{k \neq n} \sum_{l \neq n} \frac{|k^{(0)}\rangle V_{kl} V_{ln}}{(E_n^{(0)} - E_k^{(0)})(E_n^{(0)} - E_l^{(0)})} - \sum_{k \neq n} \frac{|k^{(0)}\rangle V_{nn} V_{kn}}{(E_n^{(0)} - E_k^{(0)})^2} \right)$$

$$+ \cdots. \tag{5.44}$$

Equation (5.44) says that the $n$th level is no longer proportional to the unperturbed ket $|n^{(0)}\rangle$ but acquires components along other unperturbed energy kets; stated another way, the perturbation $V$ mixes various unperturbed energy eigenkets.

A few general remarks are in order. First, to obtain the first-order energy shift it is sufficient to evaluate the expectation value of $V$ with respect to the unperturbed kets. Second, it is evident from the expression of the second-order energy shift (5.42) that two energy levels, say the $i$th level and the $j$th level, when connected by $V_{ij}$ tend to repel each other; the lower one, say the $i$th level, tends to get depressed as a result of mixing with the higher $j$th level by $|V_{ij}|^2/(E_j^{(0)} - E_i^{(0)})$, while the energy of the $j$th level goes up by the same amount. This is a special case of the **no-level crossing theorem**, which states that a pair of energy levels connected by perturbation do not cross as the strength of the perturbation is varied.

Suppose there is more than one pair of levels with appreciable matrix elements but the ket $|n\rangle$, whose energy we are concerned with, refers to the ground state; then each term in (5.42) for the second-order energy shift is negative. This means that the second-order energy shift is always negative for the ground state; the lowest state tends to get even lower as a result of mixing.

It is clear that perturbation expansions (5.42) and (5.44) will converge if $|V_{il}/(E_i^{(0)} - E_l^{(0)})|$ is sufficiently "small." A more specific criterion can be given for the case in which $H_0$ is simply the kinetic-energy operator (then this Rayleigh–Schrödinger perturbation expansion is just the Born series): At an energy $E_0 < 0$, the Born series converges if and only if neither $H_0 + V$ nor $H_0 - V$ has bound states of energy $E \leq E_0$. See Newton (1982), p. 233.

## 5.1.4  Wave Function Renormalization

We are in a position to look at the normalization of the perturbed ket. Recalling the normalization convention we use, (5.31), we see that the perturbed ket $|n\rangle$ is not normalized in the usual manner. We can renormalize the perturbed ket by defining

$$|n\rangle_N = Z_n^{1/2}|n\rangle, \tag{5.45}$$

where $Z_n$ is simply a constant with $_N\langle n|n\rangle_N = 1$. Multiplying $\langle n^{(0)}|$ on the left we obtain [because of (5.31)]

$$Z_n^{1/2} = \langle n^0|n\rangle_N. \tag{5.46}$$

What is the physical meaning of $Z_n$? Because $|n\rangle_N$ satisfies the usual normalization requirement (5.30), $Z_n$ can be regarded as the probability for the perturbed energy eigenstate to be found in the corresponding unperturbed energy eigenstate. Noting

$$_N\langle n|n\rangle_N = Z_n\langle n|n\rangle = 1, \tag{5.47}$$

we have

$$\begin{aligned}
Z_n^{-1} = \langle n|n\rangle &= (\langle n^{(0)}| + \lambda\langle n^{(1)}| + \lambda^2\langle n^{(2)}| + \cdots) \\
&\quad \times (|n^{(0)}\rangle + \lambda|n^{(1)}\rangle + \lambda^2|n^{(2)}\rangle + \cdots) \\
&= 1 + \lambda^2\langle n^{(1)}|n^{(1)}\rangle + 0(\lambda^3) \\
&= 1 + \lambda^2\sum_{k\neq n}\frac{|V_{kn}|^2}{(E_n^{(0)} - E_k^{(0)})^2} + 0(\lambda^3),
\end{aligned} \tag{5.48a}$$

so up to order $\lambda^2$, we get for the probability of the perturbed state to be found in the corresponding unperturbed state

$$Z_n \simeq 1 - \lambda^2\sum_{k\neq n}\frac{|V_{kn}|^2}{(E_n^0 - E_k^0)^2}. \tag{5.48b}$$

The second term in (5.48b) is to be understood as the probability for "leakage" to states other than $|n^{(0)}\rangle$. Notice that $Z_n$ is less than 1, as expected on the basis of the probability interpretation for $Z$.

It is also amusing to note from (5.42) that to order $\lambda^2$, $Z$ is related to the derivative of $E_n$ with respect to $E_n^{(0)}$ as follows:

$$Z_n = \frac{\partial E_n}{\partial E_n^{(0)}}. \tag{5.49}$$

We understand, of course, that in taking the partial derivative of $E_n$ with respect to $E_n^{(0)}$, we must regard the matrix elements of $V$ as fixed quantities. Result (5.49) is actually quite general and not restricted to second-order perturbation theory.

### 5.1.5 Elementary Examples

To illustrate the perturbation method we have developed, let us look at two examples. The first one concerns a simple harmonic oscillator whose unperturbed Hamiltonian is the usual one:

$$H_0 = \frac{p^2}{2m} + \frac{1}{2}m\omega^2 x^2. \tag{5.50}$$

Suppose the spring constant $k = m\omega^2$ is changed slightly. We may represent the modification by adding an extra potential

$$V = \frac{1}{2}\varepsilon m\omega^2 x^2, \tag{5.51}$$

where $\varepsilon$ is a dimensionless parameter such that $\varepsilon \ll 1$. From a certain point of view this is the silliest problem in the world to which to apply perturbation theory; the exact solution is immediately obtained just by changing $\omega$ as follows

$$\omega \to \sqrt{1+\varepsilon}\,\omega, \tag{5.52}$$

yet this is an instructive example because it affords a comparison between the perturbation approximation and the exact approach.

We are concerned here with the new ground-state ket $|0\rangle$ in the presence of $V$ and the ground-state energy shift $\Delta_0$:

$$|0\rangle = |0^{(0)}\rangle + \sum_{k\neq 0} |k^{(0)}\rangle \frac{V_{k0}}{E_0^{(0)} - E_k^{(0)}} + \cdots \tag{5.53a}$$

and

$$\Delta_0 = V_{00} + \sum_{k\neq 0} \frac{|V_{k0}|^2}{E_0^{(0)} - E_k^{(0)}} + \cdots. \tag{5.53b}$$

The relevant matrix elements are (see Problem 5.8 in this chapter)

$$\begin{aligned}
V_{00} &= \left(\frac{\varepsilon m\omega^2}{2}\right)\langle 0^{(0)}|x^2|0^{(0)}\rangle = \frac{\varepsilon\hbar\omega}{4} \\
V_{20} &= \left(\frac{\varepsilon m\omega^2}{2}\right)\langle 2^{(0)}|x^2|0^{(0)}\rangle = \frac{\varepsilon\hbar\omega}{2\sqrt{2}}.
\end{aligned} \tag{5.54}$$

All other matrix elements of form $V_{k0}$ vanish. Noting that the nonvanishing energy denominators in (5.53a) and (5.53b) are $-2\hbar\omega$, we can combine everything to obtain

$$|0\rangle = |0^{(0)}\rangle - \frac{\varepsilon}{4\sqrt{2}}|2^{(0)}\rangle + 0(\varepsilon^2) \qquad (5.55a)$$

and

$$\Delta_0 = E_0 - E_0^{(0)} = \hbar\omega\left[\frac{\varepsilon}{4} - \frac{\varepsilon^2}{16} + 0(\varepsilon^3)\right]. \qquad (5.55b)$$

Notice that as a result of perturbation, the ground-state ket, when expanded in terms of original unperturbed energy eigenkets $\{|n^{(0)}\rangle\}$, acquires a component along the second excited state. The absence of a component along the first excited state is not surprising because our *total H* is invariant under parity; hence, an energy eigenstate is expected to be a parity eigenstate.

A comparison with the exact method can easily be made for the energy shift as follows:

$$\frac{\hbar\omega}{2} \to \left(\frac{\hbar\omega}{2}\right)\sqrt{1+\varepsilon} = \left(\frac{\hbar\omega}{2}\right)\left[1 + \frac{\varepsilon}{2} - \frac{\varepsilon^2}{8} + \cdots\right], \qquad (5.56)$$

in complete agreement with (5.55b). As for the perturbed ket, we look at the change in the wave function. In the absence of $V$ the ground-state wave function is

$$\langle x|0^{(0)}\rangle = \frac{1}{\pi^{1/4}}\frac{1}{\sqrt{x_0}}e^{-x^2/2x_0^2}, \qquad (5.57)$$

where

$$x_0 \equiv \sqrt{\frac{\hbar}{m\omega}}. \qquad (5.58)$$

Substitution (5.52) leads to

$$x_0 \to \frac{x_0}{(1+\varepsilon)^{1/4}}; \qquad (5.59)$$

hence,

$$\langle x|0^{(0)}\rangle \to \frac{1}{\pi^{1/4}\sqrt{x_0}}(1+\varepsilon)^{1/8}\exp\left[-\left(\frac{x^2}{2x_0^2}\right)(1+\varepsilon)^{1/2}\right]$$

$$\simeq \frac{1}{\pi^{1/4}}\frac{1}{\sqrt{x_0}}e^{-x^2/2x_0^2} + \frac{\varepsilon}{\pi^{1/4}\sqrt{x_0}}e^{-x^2/2x_0^2}\left[\frac{1}{8} - \frac{1}{4}\frac{x^2}{x_0^2}\right]$$

$$= \langle x|0^{(0)}\rangle - \frac{\varepsilon}{4\sqrt{2}}\langle x|2^{(0)}\rangle, \qquad (5.60)$$

where we have used

$$\langle x|2^{(0)}\rangle = \frac{1}{2\sqrt{2}}\langle x|0^{(0)}\rangle H_2\left(\frac{x}{x_0}\right)$$

$$= \frac{1}{2\sqrt{2}}\frac{1}{\pi^{1/4}}\frac{1}{\sqrt{x_0}}e^{-x^2/2x_0^2}\left[-2 + 4\left(\frac{x}{x_0}\right)^2\right], \qquad (5.61)$$

and $H_2(x/x_0)$ is a Hermite polynomial of order 2.

As another illustration of nondegenerate perturbation theory, we discuss the **quadratic Stark effect**. A one-electron atom – the hydrogen atom or a hydrogenlike atom with one valence electron outside the closed (spherically symmetrical) shell – is subjected to a uniform electric field in the positive $z$-direction. The Hamiltonian $H$ is split into two parts,

$$H_0 = \frac{\mathbf{p}^2}{2m} + V_0(r) \quad \text{and} \quad V = -e|\mathbf{E}|z \quad (e < 0 \text{ for the electron}). \tag{5.62}$$

[*Editor's Note*: Since the perturbation $V \to -\infty$ as $z \to -\infty$, particles bound by $H_0$ can, of course, escape now, and all formerly bound states acquire a finite lifetime. However, we can still formally use perturbation theory to calculate the shift in the energy. (The imaginary part of this shift, which we shall ignore here, would give us the lifetime of the state or the width of the corresponding resonance.)]

It is assumed that the energy eigenkets and the energy spectrum for the unperturbed problem ($H_0$ only) are completely known. The electron spin turns out to be irrelevant in this problem, and we assume that with spin degrees of freedom ignored, no energy level is degenerate. This assumption does not hold for $n \neq 1$ levels of the hydrogen atoms, where $V_0$ is the pure Coulomb potential; we will treat such cases later. The energy shift is given by

$$\Delta_k = -e|\mathbf{E}|z_{kk} + e^2|\mathbf{E}|^2 \sum_{j \neq k} \frac{|z_{kj}|^2}{E_k^{(0)} - E_j^{(0)}} + \cdots, \tag{5.63}$$

where we have used $k$ rather than $n$ to avoid confusion with the principal quantum number $n$. With no degeneracy, $|k^{(0)}\rangle$ is expected to be a parity eigenstate; hence,

$$z_{kk} = 0, \tag{5.64}$$

as we saw in Section 4.2. Physically speaking, there can be no linear Stark effect, that is, there is no term in the energy shift proportional to $|\mathbf{E}|$ because the atom possesses a vanishing permanent electric dipole, so the energy shift is *quadratic* in $|\mathbf{E}|$ if terms of order $e^3|\mathbf{E}|^3$ or higher are ignored.

Let us now look at $z_{kj}$, which appears in (5.63), where $k$ (or $j$) is the **collective index** that stands for $(n, l, m)$ and $(n', l', m')$. First, we recall the selection rule [see (3.482)]

$$\langle n', l'm'|z|n, lm \rangle = 0, \quad \text{unless} \begin{cases} l' = l \pm 1 \\ m' = m \end{cases} \tag{5.65}$$

that follows from angular momentum (the Wigner–Eckart theorem with $T_{q=0}^{(1)}$) and parity considerations.

There is another way to look at the $m$-selection rule. In the presence of $V$, the full spherical symmetry of the Hamiltonian is destroyed by the external electric field that selects the positive $z$-direction, but $V$ (hence the total $H$) is still invariant under rotation around the $z$-axis; in other words, we still have a cylindrical symmetry. Formally this is reflected by the fact that

$$[V, L_z] = 0. \tag{5.66}$$

This means that $L_z$ is still a good quantum number even in the presence of $V$. As a result, the perturbation can be written as a superposition of eigenkets of $L_z$ with the same $m$, where $m = 0$ in our case. This statement is true for all orders, in particular, for the first-order ket. Also, because the second-order energy shift is obtained from the first-order ket [see (5.40)], we can understand why only the $m = 0$ terms contribute to the sum.

The polarizability $\alpha$ of an atom is defined in terms of the energy shift of the atomic state as follows:

$$\Delta = -\frac{1}{2}\alpha|\mathbf{E}|^2. \tag{5.67}$$

Let us consider the special case of the ground state of the hydrogen atom. Even though the spectrum of the hydrogen atom is degenerate for excited states, the ground state (with spin ignored) is nondegenerate, so the formalism of nondegenerate perturbation theory can be applied. The ground state $|0^{(0)}\rangle$ is denoted in the $(n, l, m)$ notation by $(1, 0, 0)$, so

$$\alpha = -2e^2 \sum_{k \neq 0}^{\infty} \frac{|\langle k^{(0)}|z|1,0,0\rangle|^2}{\left[E_0^{(0)} - E_k^{(0)}\right]}, \tag{5.68}$$

where the sum over $k$ includes not only all bound states $|n, l, m\rangle$ (for $n > 1$) but also the positive-energy continuum states of hydrogen.

There are many ways to estimate approximately or evaluate exactly the sum in (5.68) with various degrees of sophistication. We present here the simplest of all the approaches. Suppose the denominator in (5.68) were constant. Then we could obtain the sum by considering

$$\sum_{k \neq 0}|\langle k^{(0)}|z|1,0,0\rangle|^2 = \sum_{\text{all } k}|\langle k^{(0)}|z|1,0,0\rangle|^2$$

$$= \langle 1,0,0|z^2|1,0,0\rangle, \tag{5.69}$$

where we have used the completeness relation in the last step. But we can easily evaluate $\langle z^2 \rangle$ for the ground state as follows:

$$\langle z^2 \rangle = \langle x^2 \rangle = \langle y^2 \rangle = \frac{1}{3}\langle r^2 \rangle, \tag{5.70}$$

and using the explicit form for the wave function we obtain

$$\langle r^2 \rangle = 3a_0^2,$$

where $a_0$ stands for the Bohr radius. Unfortunately the expression for polarizability $\alpha$ involves the energy denominator that depends on $E_k^{(0)}$, but we know that the inequality

$$-E_0^{(0)} + E_k^{(0)} \geq -E_0^{(0)} + E_1^{(0)} = \frac{e^2}{2a_0}\left[1 - \frac{1}{4}\right] \tag{5.71}$$

holds for every energy denominator in (5.68). As a result, we can obtain an upper limit for the polarizability of the ground state of the hydrogen atom, namely,

$$\alpha < \frac{16a_0^3}{3} \simeq 5.3a_0^3. \tag{5.72}$$

It turns out that we can evaluate exactly the sum in (5.68) using a method due to A. Dalgarno and J. T. Lewis (Merzbacher (1970), p. 424, for example), which also agrees with the experimentally measured value. This gives

$$\alpha = \frac{9a_0^3}{2} = 4.5a_0^3. \tag{5.73}$$

We obtain the same result (without using perturbation theory) by solving the Schrödinger equation exactly using parabolic coordinates.

## 5.2 Time-Independent Perturbation Theory: The Degenerate Case

The perturbation method we developed in the previous section fails when the unperturbed energy eigenkets are degenerate. The method of the previous section assumes that there is a unique and well-defined unperturbed ket of energy $E_n^{(0)}$ which the perturbed ket approaches as $\lambda \to 0$. With degeneracy present, however, any linear combination of unperturbed kets has the same unperturbed energy; in such a case it is not a priori obvious to what linear combination of the unperturbed kets the perturbed ket is reduced in the limit $\lambda \to 0$. Here specifying just the energy eigenvalue is not enough; some other observable is needed to complete the picture. To be more specific, with degeneracy we can take as our base kets simultaneous eigenkets of $H_0$ and some other observable $A$, and we can continue labeling the unperturbed energy eigenket by $|k^{(0)}\rangle$, where $k$ now symbolizes a collective index that stands for both the energy eigenvalue *and* the $A$ eigenvalue. When the perturbation operator $V$ does not commute with $A$, the **zeroth-order** eigenkets for $H$ (including the perturbation) are in fact *not* $A$ eigenkets.

From a more practical point of view, a blind application of formulas like (5.42) and (5.44) obviously runs into difficulty because

$$\frac{V_{nk}}{E_n^{(0)} - E_k^{(0)}} \tag{5.74}$$

becomes singular if $V_{nk}$ is nonvanishing and $E_n^{(0)}$ and $E_k^{(0)}$ are equal. We must modify the method of the previous section to accommodate such a situation.

Whenever there is degeneracy we are free to choose our base set of unperturbed kets. We should, by all means, exploit this freedom. Intuitively we suspect that the catastrophe of vanishing denominators may be avoided by choosing our base kets in such a way that $V$ has no off-diagonal matrix elements [such as $V_{nk} = 0$ in (5.74)]. In other words, we should use the linear combinations of the degenerate unperturbed kets that diagonalize $H$ in the subspace spanned by the degenerate unperturbed kets. This is indeed the correct procedure to use.

Suppose there is a $g$-fold degeneracy before the perturbation $V$ is switched on. This means that there are $g$ different eigenkets all with the same unperturbed energy $E_D^{(0)}$. Let us denote these kets by $\{|m^{(0)}\rangle\}$. In general, the perturbation removes the degeneracy in the sense that there will be $g$ perturbed eigenkets all with different energies. Let them form a set $\{|l\rangle\}$. As $\lambda$ goes to zero $|l\rangle \to |l^{(0)}\rangle$, and various $|l^{(0)}\rangle$ are eigenkets of $H_0$ all with the same energy $E_m^{(0)}$. However, the set $|l^{(0)}\rangle$ need not coincide with $\{|m^{(0)}\rangle\}$ even though the two sets of unperturbed eigenkets span the same degenerate subspace, which we call $D$. We can write

$$|l^{(0)}\rangle = \sum_{m \in D} \langle m^{(0)}|l^{(0)}\rangle |m^{(0)}\rangle,$$

where the sum is over the energy eigenkets in the degenerate subspace.

Before expanding in $\lambda$, there is a rearrangement of the Schrödinger equation that will make it much easier to carry out the expansion. Let $P_0$ be a projection operator onto the space defined by $\{|m^{(0)}\rangle\}$. (Recall the discussion of projection operators in Section 1.3.) We define $P_1 = 1 - P_0$ to be the projection onto the remaining states. We shall then write the Schrödinger equation for the states $|l\rangle$ as

$$0 = (E - H_0 - \lambda V)|l\rangle$$
$$= (E - E_D^{(0)} - \lambda V)P_0|l\rangle + (E - H_0 - \lambda V)P_1|l\rangle. \tag{5.75}$$

We next separate (5.75) into two equations by projecting from the left on (5.75) with $P_0$ and $P_1$,

$$(E - E_D^{(0)} - \lambda P_0 V)P_0|l\rangle - \lambda P_0 V P_1|l\rangle = 0 \tag{5.76}$$

$$-\lambda P_1 V P_0|l\rangle + (E - H_0 - \lambda P_1 V)P_1|l\rangle = 0. \tag{5.77}$$

We can solve (5.77) in the $P_1$ subspace because $P_1(E - H_0 - \lambda P_1 V P_1)$ is not singular in this subspace since $E$ is close to $E_D^{(0)}$ and the eigenvalues of $P_1 H_0 P_1$ are all different from $E_D^{(0)}$. Hence we can write

$$P_1|l\rangle = P_1 \frac{\lambda}{E - H_0 - \lambda P_1 V P_1} P_1 V P_0|l\rangle \tag{5.78}$$

or written out explicitly to order $\lambda$ when $|l\rangle$ is expanded as $|l\rangle = |l^{(0)}\rangle + \lambda |l^{(1)}\rangle + \cdots$

$$P_1|l^{(1)}\rangle = \sum_{k \notin D} \frac{|k^{(0)}\rangle V_{kl}}{E_D^{(0)} - E_k^{(0)}}. \tag{5.79}$$

To calculate $P_0|l\rangle$, we substitute (5.78) into (5.76) to obtain

$$\left(E - E_D^{(0)} - \lambda P_0 V P_0 - \lambda^2 P_0 V P_1 \frac{1}{E - H_0 - \lambda V} P_1 V P_0\right) P_0|l\rangle = 0. \tag{5.80}$$

Although there is a term of order $\lambda^2$ in (5.80) that results from the substitution, we shall find that it produces a term of order $\lambda$ in the state $P_0|l\rangle$. So we obtain the equation for the energies to order $\lambda$ and eigenfunctions to order zero,

$$(E - E_D^{(0)} - \lambda P_0 V P_0)(P_0|l^{(0)}\rangle) = 0. \tag{5.81}$$

This is an equation in the $g$-dimensional degenerate subspace and clearly means that the eigenvectors are just the eigenvectors of the $g \times g$ matrix $P_0 V P_0$ and the eigenvalues $E^{(1)}$ are just the roots of the secular equation

$$\det[V - (E - E_D^{(0)})] = 0 \tag{5.82}$$

where $V = $ matrix of $P_0 V P_0$ with matrix elements $\langle m^{(0)}|V|m'^{(0)}\rangle$. Explicitly in matrix form we have

$$
\begin{pmatrix} V_{11} & V_{12} & \cdots \\ V_{21} & V_{22} & \cdots \\ \vdots & \vdots & \ddots \end{pmatrix} \begin{pmatrix} \langle 1^{(0)}|l^{(0)}\rangle \\ \langle 2^{(0)}|l^{(0)}\rangle \\ \vdots \end{pmatrix} = \Delta_l^{(1)} \begin{pmatrix} \langle 1^{(0)}|l^{(0)}\rangle \\ \langle 2^{(0)}|l^{(0)}\rangle \\ \vdots \end{pmatrix}.
\tag{5.83}
$$

The roots determine the eigenvalues $\Delta_l^{(1)}$, there are $g$ altogether, and by substituting them into (5.83), we can solve for $\langle m^{(0)}|l^{(0)}\rangle$ for each $l$ up to an overall normalization constant. Thus by solving the eigenvalue problem, we obtain in one stroke both the first-order energy shifts and the correct zeroth-order eigenkets. Notice that the zeroth-order kets we obtain as $\lambda \to 0$ are just the linear combinations of the various $|m^{(0)}\rangle$ that diagonalize the perturbation $V$, the diagonal elements immediately giving the first-order shift

$$
\Delta_l^{(1)} = \langle l^{(0)}|V|l^{(0)}\rangle.
\tag{5.84}
$$

Note also that if the degenerate subspace were the whole space, we would have solved the problem exactly in this manner. The presence of unperturbed "distant" eigenkets not belonging to the degenerate subspace will show up only in higher orders – first order and higher for the energy eigenkets and second order and higher for the energy eigenvalues.

Expression (5.84) looks just like the first-order energy shift [see (5.37)] in the nondegenerate case except that here we have to make sure that the base kets used are such that $V$ does not have nonvanishing off-diagonal matrix elements in the subspace spanned by the degenerate unperturbed eigenkets. If the $V$ operator is already diagonal in the base ket representation we are using, we can immediately write down the first-order shift by taking the expectation value of $V$, just as in the nondegenerate case.

Let us now look at (5.80). To be safe we keep all terms in the $g \times g$ effective Hamiltonian that appears in (5.80) to order $\lambda^2$ although we want $P_0|l\rangle$ only to order $\lambda$. We find

$$
\left( E - E_D^{(0)} - \lambda P_0 V P_0 - \lambda^2 P_0 V P_1 \frac{1}{E_D^{(0)} - H_0} P_1 V P_0 \right) P_0|l\rangle = 0.
\tag{5.85}
$$

For the $g \times g$ matrix $P_0 V P_0$, let us call the eigenvalues $v_i$ and the eigenvectors $P_0|l_i^{(0)}\rangle$. The eigen energies to first order are $E_i^{(1)} = E_D^{(0)} + \lambda v_i$. We assume that the degeneracy is completely resolved so that $E_i^{(1)} - E_j^{(1)} = \lambda(v_i - v_j)$ are all nonzero. We can now apply nondegenerate perturbation theory (5.39) to the $g \times g$ dimensional Hamiltonian that appears in (5.85). The resulting correction to the eigenvectors $P_0|l_i^{(0)}\rangle$ is

$$
P_0|l_i^{(1)}\rangle = \sum_{j \neq i} \frac{P_0|l_j^{(0)}\rangle}{v_i - v_j} \langle l_j^{(0)}|V P_1 \frac{1}{E_D^{(0)} - H_0} P_1 V|l_i^{(0)}\rangle
\tag{5.86}
$$

or more explicitly

$$
P_0|l_i^{(1)}\rangle = \sum_{j \neq i} \frac{P_0|l_j^{(0)}\rangle}{v_i - v_j} \sum_{k \notin D} \langle l_j^{(0)}|V|k\rangle \frac{1}{E_D^{(0)} - E_k^{(0)}} \langle k|V|l_i^{(0)}\rangle.
\tag{5.87}
$$

Thus, although the third term in the effective Hamiltonian that appears in (5.85) is of order $\lambda^2$, it is divided by energy denominators of order $\lambda$ in forming the correction to the eigenvector, which then gives terms of order $\lambda$ in the vector. If we add together (5.79) and (5.87), we get the eigenvector accurate to order $\lambda$.

As in the nondegenerate case, it is convenient to adopt the normalization convention $\langle l^{(0)}|l\rangle = 1$. We then have, from (5.76) and (5.77), $\lambda\langle l^{(0)}|V|l\rangle = \Delta_l = \lambda\Delta_l^{(1)} + \lambda^2\Delta_l^{(2)} + \cdots$. The $\lambda$-term just reproduces (5.84). As for the $\lambda^2$-term, we obtain $\Delta_l^{(2)} = \langle l^{(0)}|V|l^{(1)}\rangle = \langle l^{(0)}|V|P_1 l^{(1)}\rangle + \langle l^{(0)}|V|P_0 l_i^{(1)}\rangle$. Since the vectors $P_0|l_j^{(0)}\rangle$ are eigenvectors of $V$, the correction to the vector, (5.87), gives no contribution to the second-order energy shift, so we find using (5.79)

$$\Delta_l^{(2)} = \sum_{k\notin D}\frac{|V_{kl}|^2}{E_D^{(0)} - E_k^{(0)}}. \tag{5.88}$$

Our procedure works provided that there is no degeneracy in the roots of secular equation (5.82). Otherwise we still have an ambiguity as to which linear combination of the degenerate unperturbed kets the perturbed kets are reduced in the limit $\lambda \to 0$. Put in another way, if our method is to work, the degeneracy should be removed completely in first order. A challenge for the experts: How must we proceed if the degeneracy is not removed in first order, that is, if some of the roots of the secular equation are equal? (See Problem 5.15 of this chapter.)

Let us now summarize the basic procedure of degenerate perturbation theory.

1. Identify degenerate unperturbed eigenkets and construct the perturbation matrix $V$, a $g \times g$ matrix if the degeneracy is $g$-fold.
2. Diagonalize the perturbation matrix by solving, as usual, the appropriate secular equation.
3. Identify the roots of the secular equation with the first-order energy shifts; the base kets that diagonalize the $V$ matrix are the correct zeroth-order kets to which the perturbed kets approach in the limit $\lambda \to 0$.
4. For higher orders use the formulas of the corresponding nondegenerate perturbation theory except in the summations, where we exclude all contributions from the unperturbed kets in the degenerate subspace $D$.

## 5.2.1 Linear Stark Effect

As an example of degenerate perturbation theory, let us study the effect of a uniform electric field on excited states of the hydrogen atom. As is well known, in the Schrödinger theory with a pure Coulomb potential with no spin dependence, the bound-state energy of the hydrogen atom depends only on the principal quantum number $n$. This leads to degeneracy for all but the ground state because the allowed values of $l$ for a given $n$ satisfy

$$0 \leq l < n. \tag{5.89}$$

To be specific, for the $n = 2$ level, there is an $l = 0$ state called $2s$ and three $l = 1$ ($m = \pm 1, 0$) states called $2p$, all with the same energy, $-e^2/8a_0$. As we apply a uniform electric field in the $z$-direction, the appropriate perturbation operator is given by

$$V = -ez|\mathbf{E}|, \tag{5.90}$$

which we must now diagonalize. Before we evaluate the matrix elements in detail using the usual ($nlm$) basis, let us note that the perturbation (5.90) has nonvanishing matrix elements only between states of opposite parity, that is, between $l = 1$ and $l = 0$ in our case. Furthermore, in order for the matrix element to be nonvanishing, the $m$-values must be the same because $z$ behaves like a spherical tensor of rank one with spherical component (*magnetic quantum number*) zero. So the only nonvanishing matrix elements are between $2s$ ($m = 0$ necessarily) and $2p$ with $m = 0$. Thus

$$V = \begin{pmatrix} 2s & 2p\ m=0 & 2p\ m=1 & 2p\ m=-1 \\ 0 & \langle 2s|V|2p, m=0\rangle & 0 & 0 \\ \langle 2p, m=0|V|2s\rangle & 0 & 0 & 0 \\ 0 & 0 & 0 & 0 \\ 0 & 0 & 0 & 0 \end{pmatrix}. \tag{5.91}$$

Explicitly,

$$\langle 2s|V|2p, m=0\rangle = \langle 2p, m=0|V|2s\rangle$$
$$= 3ea_0|\mathbf{E}|. \tag{5.92}$$

It is sufficient to concentrate our attention on the upper left-hand corner of the square matrix. It then looks very much like the $\sigma_x$ matrix, and we can immediately write down the answer. For the energy shifts we get

$$\Delta_\pm^{(1)} = \pm 3ea_0|\mathbf{E}|, \tag{5.93}$$

where the subscripts $\pm$ refer to the zeroth-order kets that diagonalize $V$:

$$|\pm\rangle = \frac{1}{\sqrt{2}}(|2s, m=0\rangle \pm |2p, m=0\rangle). \tag{5.94}$$

Schematically the energy levels are as shown in Figure 5.1.

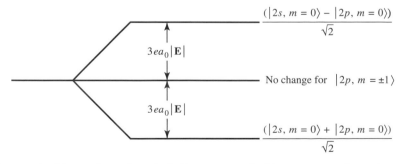

**Fig. 5.1**    Schematic energy-level diagram for the linear Stark effect as an example of degenerate perturbation theory.

Notice that the shift is *linear* in the applied electric field strength, hence the term the **linear Stark effect**. One way we can visualize the existence of this effect is to note that the energy eigenkets (5.94) are not parity eigenstates and are therefore allowed to have nonvanishing electric permanent dipole moments, as we can easily see by explicitly evaluating $\langle z \rangle$. Quite generally, for an energy state that we can write as a superposition of opposite parity states, it is permissible to have a nonvanishing permanent electric dipole moment, which gives rise to the linear Stark effect.

An interesting question can now be asked. If we look at the "real" hydrogen atom, the $2s$ level and $2p$ level are not really degenerate. Due to the spin-orbit force, $2p_{3/2}$ is separated from $2p_{1/2}$, as we will show in the next section, and even the degeneracy between the $2s_{1/2}$ and $2p_{1/2}$ levels that persists in the single-particle Dirac theory is removed by quantum electrodynamics effects (the *Lamb shift*). We might therefore ask: Is it realistic to apply degenerate perturbation theory to this problem? A comparison with the exact result shows that if the perturbation matrix elements are much larger when compared to the Lamb shift splitting, then the energy shift is linear in $|\mathbf{E}|$ for all practical purposes and the formalism of degenerate perturbation theory is applicable. On the opposite extreme, if the perturbation matrix elements are small compared to the Lamb shift splitting, then the energy shift is quadratic and we can apply nondegenerate perturbation theory; see Problem 5.19 of this chapter. This incidentally shows that the formalism of degenerate perturbation theory is still useful when the energy levels are almost degenerate compared to the energy scale defined by the perturbation matrix element. In intermediate cases we must work harder; it is safer to attempt to diagonalize the Hamiltonian exactly in the space spanned by all the nearby levels.

## 5.3  Hydrogenlike Atoms: Fine Structure and the Zeeman Effect

### 5.3.1  The Relativistic Correction to the Kinetic Energy

A hydrogen like atom with a single electron has the potential energy function (3.305) giving the Hamiltonian

$$H_0 = \frac{\mathbf{p}^2}{2m_e} - \frac{Ze^2}{r} \tag{5.95}$$

where the first term is the nonrelativistic kinetic-energy operator. However, the relativistically correct kinetic energy is

$$K = \sqrt{\mathbf{p}^2 c^2 + m_e^2 c^4} - m_e c^2$$

$$\approx \frac{\mathbf{p}^2}{2m_e} - \frac{(\mathbf{p}^2)^2}{8m_e^3 c^2}. \tag{5.96}$$

Therefore, following the notation in (5.1), we can treat this problem in perturbation theory where $H_0$ is given by (5.95) and the perturbation is

$$V = -\frac{(\mathbf{p}^2)^2}{8m_e^3 c^2}. \tag{5.97}$$

Now in principle, this is a complicated problem because of the highly degenerate eigenstates $|nlm\rangle$ of the hydrogen atom. However, since $\mathbf{L}$ commutes with $\mathbf{p}^2$ as we noted in (3.264), we also have

$$[\mathbf{L}, V] = 0. \tag{5.98}$$

In other words, $V$ is rotationally symmetric, and is therefore already diagonal in the $|nlm\rangle$ basis. Therefore the first-order energy shifts due to $V$ are just equal to the expectation values in these basis states. Following (5.37) we write

$$\Delta_{nl}^{(1)} = \langle nlm|V|nlm\rangle = -\langle nlm|\frac{(\mathbf{p}^2)^2}{8m_e^3 c^2}|nlm\rangle \tag{5.99}$$

where the rotational symmetry assures us that the first-order energy shifts cannot depend on $m$.

In principle, (5.99) could be evaluated by brute force, but there is a more elegant way. Since

$$\frac{(\mathbf{p}^2)^2}{8m_e^3 c^2} = \frac{1}{2m_e c^2}\left(\frac{\mathbf{p}^2}{2m_e}\right)^2 = \frac{1}{2m_e c^2}\left(H_0 + \frac{Ze^2}{r}\right)^2 \tag{5.100}$$

we immediately see that

$$\Delta_{nl}^{(1)} = -\frac{1}{2m_e c^2}\left[\left(E_n^{(0)}\right)^2 + 2E_n^{(0)}\langle nlm|\frac{Ze^2}{r}|nlm\rangle + \langle nlm|\frac{(Ze^2)^2}{r^2}|nlm\rangle\right]. \tag{5.101}$$

The problem is therefore reduced to calculating the expectation values for $Ze^2/r$ and $(Ze^2)^2/r^2$. In fact, both of these expectation values can be evaluated using some clever tricks. We simply outline the approach here, but the interested reader is referred to Shankar (1994) or Townsend (2000) for more details.

If one imagines a hydrogen atom with a "perturbation" $V_\gamma = \gamma/r$, then the expectation value in the second term of (5.101) is simply the first-order correction to the energy with $\gamma = Ze^2$. On the other hand, it is simple to solve this problem exactly since it corresponds to the hydrogen atom with $Ze^2 \to Ze^2 - \gamma$, and it is straightforward to find the first-order correction from the exact solution. One finds that

$$\langle nlm|\frac{Ze^2}{r}|nlm\rangle = -2E_n^{(0)}. \tag{5.102}$$

Indeed, this is actually a statement of the virial theorem for the Coulomb potential.

A similar approach can be taken for the third term in (5.101). In this case, imagine a perturbation $V_\gamma = \gamma/r^2$ which modifies the centrifugal barrier term in the effective potential. That is, it takes $l$ to a form that includes $\gamma$, which can again be used to write down the first-order correction. One finds that

$$\langle nlm|\frac{(Ze^2)^2}{r^2}|nlm\rangle = \frac{4n}{l+\frac{1}{2}}\left(E_n^{(0)}\right)^2. \tag{5.103}$$

So, using (5.102) and (5.103), along with $E_n^{(0)}$ from (3.315), we rewrite (5.101) as

$$\Delta_{nl}^{(1)} = E_n^{(0)} \left[ \frac{Z^2 \alpha^2}{n^2} \left( -\frac{3}{4} + \frac{n}{l + \frac{1}{2}} \right) \right] \tag{5.104a}$$

$$= -\frac{1}{2} m_e c^2 Z^4 \alpha^4 \left[ -\frac{3}{4n^4} + \frac{1}{n^3 \left( l + \frac{1}{2} \right)} \right]. \tag{5.104b}$$

Not unexpectedly, the relative size of the first-order correction is proportional to $Z^2 \alpha^2$, the square of the classical electron orbital velocity (in units of $c$).

## 5.3.2 Spin-Orbit Interaction and Fine Structure

Now let us move on to the study of the atomic levels of general hydrogenlike atoms, that is, atoms with one valence electron outside the closed shell. Alkali atoms such as sodium (Na) and potassium (K) belong to this category.

The central (spin-independent) potential $V_c(r)$ appropriate for the valence electron is no longer of the pure Coulomb form. This is because the electrostatic potential $\phi(r)$ that appears in

$$V_c(r) = e\phi(r) \tag{5.105}$$

is no longer due just to the nucleus of electric charge $|e|Z$; we must take into account the cloud of negatively charged electrons in the inner shells. A precise form of $\phi(r)$ does not concern us here. We simply remark that the degeneracy characteristics of the pure Coulomb potential are now removed in such a way that the higher $l$ states lie higher for a given $n$. Physically this arises from the fact that the higher $l$ states are more susceptible to the repulsion due to the electron cloud.

Instead of studying the details of $V_c(r)$, which determines the gross structure of hydrogenlike atoms, we discuss the effect of the spin-orbit $(\mathbf{L} \cdot \mathbf{S})$ interaction that gives rise to *fine structure*. We can understand the existence of this interaction in a qualitative fashion as follows. Because of the central force part (5.105), the valence electron experiences the electric field

$$\mathbf{E} = -\left( \frac{1}{e} \right) \nabla V_c(r). \tag{5.106}$$

But whenever a moving charge is subjected to an electric field, it "feels" an effective magnetic field given by

$$\mathbf{B}_{\text{eff}} = -\left( \frac{\mathbf{v}}{c} \right) \times \mathbf{E}. \tag{5.107}$$

Because the electron has a magnetic moment $\mu$ given by

$$\mu = \frac{e\mathbf{S}}{m_e c}, \tag{5.108}$$

we suspect a spin-orbit potential $V_{LS}$ contribution to $H$ as follows:

$$H_{LS} \overset{?}{=} -\boldsymbol{\mu} \cdot \mathbf{B}_{\text{eff}}$$

$$= \boldsymbol{\mu} \cdot \left( \frac{\mathbf{v}}{c} \times \mathbf{E} \right)$$

$$= \left( \frac{e\mathbf{S}}{m_e c} \right) \cdot \left[ \frac{\mathbf{p}}{m_e c} \times \left( \frac{\mathbf{x}}{r} \right) \frac{1}{(-e)} \frac{dV_c}{dr} \right]$$

$$= \frac{1}{m_e^2 c^2} \frac{1}{r} \frac{dV_c}{dr} (\mathbf{L} \cdot \mathbf{S}). \tag{5.109}$$

When this expression is compared with the observed spin-orbit interaction, it is seen to have the correct sign, but the magnitude turns out to be too large by a factor of two.[2] There is a classical explanation for this due to spin precession (*Thomas precession* after L. H. Thomas), but we shall not bother with that. See Jackson (1975), for example. We simply treat the spin-orbit interaction phenomenologically and take $V_{LS}$ to be one-half of (5.109). The correct quantum-mechanical explanation for this discrepancy must await the Dirac (relativistic) theory of the electron discussed in the last chapter of this textbook.

We are now in a position to apply perturbation theory to hydrogen like atoms using $V_{LS}$ as the perturbation ($V$ of Sections 5.1 and 5.2). The unperturbed Hamiltonian $H_0$ is taken to be

$$H_0 = \frac{\mathbf{p}^2}{2m} + V_c(r), \tag{5.110}$$

where the central potential $V_c$ is no longer of the pure Coulomb form for alkali atoms. With just $H_0$ we have freedom in choosing the base kets.

Set 1:   The eigenkets of $\mathbf{L}^2, L_z, \mathbf{S}^2, S_z$.

Set 2:   The eigenkets of $\mathbf{L}^2, \mathbf{S}^2, \mathbf{J}^2, J_z$.

$$\tag{5.111}$$

Without $V_{LS}$ (or $H_{LS}$) either set is satisfactory in the sense that the base kets are also energy eigenkets. With $H_{LS}$ added it is far superior to use set 2 of (5.111) because $\mathbf{L} \cdot \mathbf{S}$ does not commute with $L_z$ and $S_z$, while it does commute with $\mathbf{J}^2$ and $J_z$. Remember the cardinal rule: choose unperturbed kets that diagonalize the perturbation. You have to be either a fool or a masochist to use the $L_z, S_z$ eigenkets [set 1 of (5.111)] as the base kets for this problem; if we proceeded to apply blindly the method of degenerate perturbation theory starting with set 1 as our base kets, we would be forced to diagonalize the $V_{LS}(H_{LS})$ matrix written in the $L_z, S_z$ representation. The results of this, after a lot of hard algebra, give us just the $\mathbf{J}^2, J_z$ eigenkets as the zeroth-order unperturbed kets to be used!

In degenerate perturbation theory, if the perturbation is already diagonal in the representation we are using, all we need to do for the first-order energy shift is to take the expectation value. The wave function in the two-component form is explicitly written as

$$\psi_{nlm} = R_{nl}(r) \mathscr{Y}_l^{j=l\pm 1/2, m} \tag{5.112}$$

---

[2] Indeed, this is a misleading statement, although it appears often in physics textbooks. See Haar and Curtis, *Am. J. Phys.*, **55** (1987) 1044.

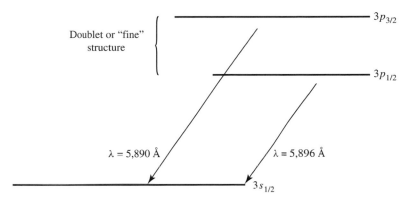

Doublet or "fine" structure

$\lambda = 5{,}890 \text{ Å}$  $\lambda = 5{,}896 \text{ Å}$

$3p_{3/2}$

$3p_{1/2}$

$3s_{1/2}$

**Fig. 5.2** Schematic diagram of $3s$ and $3p$ lines. The $3s$ and $3p$ degeneracy is lifted because $V_c(r)$ is now the screened Coulomb potential due to core electrons rather than pure Coulombic potential; $V_{LS}$ then removes the $3p_{1/2}$ and $3p_{3/2}$ degeneracy.

where $\mathscr{Y}_l^{\,j=l\pm1/2,m}$ is the spin-angular function of Section 3.8 [see (3.384)]. For the first-order shift, we obtain

$$\Delta_{nlj} = \frac{1}{2m_e^2 c^2} \left\langle \frac{1}{r}\frac{dV_c}{dr} \right\rangle_{nl} \frac{\hbar^2}{2} \left\{ \begin{array}{l} l \\ -(l+1) \end{array} \right\} \begin{array}{l} j = l + \dfrac{1}{2} \\[2mm] j = l - \dfrac{1}{2} \end{array} \tag{5.113}$$

$$\left\langle \frac{1}{r}\frac{dV_c}{dr} \right\rangle_{nl} \equiv \int_0^\infty R_{nl}\frac{1}{r}\frac{dV_c}{dr}R_{nl}r^2\,dr$$

where we have used the $m$-independent identity [see (3.386)]

$$\int \mathscr{Y}^\dagger \mathbf{S}\cdot\mathbf{L}\,\mathscr{Y}\,d\Omega = \frac{1}{2}\left[j(j+1)-l(l+1)-\frac{3}{4}\right]\hbar^2 = \frac{\hbar^2}{2}\left\{ \begin{array}{l} l \\ -(l+1) \end{array} \right\} \begin{array}{l} j = l + \dfrac{1}{2} \\[2mm] j = l - \dfrac{1}{2}. \end{array} \tag{5.114}$$

Equation (5.113) is known as **Lande's interval rule**.

To be specific, consider a sodium atom. From standard atomic spectroscopy notation, the ground-state configuration is

$$(1s)^2(2s)^2(2p)^6(3s). \tag{5.115}$$

The inner 10 electrons can be visualized to form a spherically symmetrical electron cloud. We are interested in the excitation of the eleventh electron from $3s$ to a possible higher state. The nearest possibility is excitation to $3p$. Because the central potential is no longer of the pure Coulomb form, $3s$ and $3p$ are now split. The fine structure brought about by $V_{LS}$ refers to an even finer split within $3p$, between $3p_{1/2}$ and $3p_{3/2}$, where the subscript refers to $j$. Experimentally, we observe two closely separated yellow lines, known as the **sodium D lines**, one at 5896 Å and the other at 5890 Å; see Figure 5.2. Notice that $3p_{3/2}$ lies higher because the radial integral in (5.113) is positive.

To appreciate the order of magnitude of the fine-structure splitting, let us note that for $Z \simeq 1$,

$$\left\langle \frac{1}{r} \frac{dV_c}{dr} \right\rangle_{nl} \sim \frac{e^2}{a_0^3} \tag{5.116}$$

just on the basis of dimensional considerations. So the fine-structure splitting is of order $(e^2/a_0^3)(\hbar/m_e c)^2$, which is to be compared with Balmer splittings of order $e^2/a_0$. It is useful to recall here that the classical radius of the electron, the Compton wavelength of the electron, and the Bohr radius are related in the following way:

$$\frac{e^2}{m_e c^2} : \frac{\hbar}{m_e c} : a_0 :: 1 : 137 : (137)^2, \tag{5.117}$$

where we have used

$$\frac{e^2}{\hbar c} = \frac{1}{137}. \tag{5.118}$$

Typically, fine-structure splittings are then related to typical Balmer splittings via

$$\left( \frac{e^2}{a_0^3} \frac{\hbar^2}{m_e^2 c^2} \right) : \left( \frac{e^2}{a_0} \right) :: \left( \frac{1}{137} \right)^2 : 1, \tag{5.119}$$

which explains the origin of the term *fine structure*. There are other effects of similar orders of magnitude, for example, the relativistic correction to kinetic energy discussed earlier in this section.

Before leaving this discussion, let us calculate out (5.113) for the case of the Coulomb potential, that is, a hydrogen atom or one-electron ion with $Z$ protons. In this case

$$\left\langle \frac{1}{r} \frac{dV_c}{dr} \right\rangle_{nl} = \left\langle \frac{Ze^2}{r^3} \right\rangle_{nl}. \tag{5.120}$$

We can evaluate this expectation value with the help of yet another trick. First we note that with $H_0$ given by (5.95) we have

$$\langle nlm | [H_0, A] | nlm \rangle = 0 \tag{5.121}$$

for *any* operator $A$, since $H_0$ acting to the right or left just gives $E_n^{(0)}$. If we let $A = p_r$, the radial momentum operator, then it obviously commutes with the radial part of the kinetic-energy term in $H_0$. Hence we are left with

$$\langle nlm | \left[ \frac{l(l+1)\hbar^2}{2m_e r^2} - \frac{Ze^2}{r}, p_r \right] | nlm \rangle = 0. \tag{5.122}$$

Now in coordinate space, $p_r$ does not commute with functions of the coordinate $r$ because of the presence of the derivative $\partial/\partial r$. Therefore we can explicitly carry out the commutator in (5.122) to arrive at

$$\langle nlm | \left[ -\frac{l(l+1)\hbar^2}{m_e r^3} + \frac{Ze^2}{r^2} \right] | nlm \rangle = 0. \tag{5.123}$$

Finally, then, we make use of (5.103) and (3.315) to write

$$\left\langle \frac{Ze^2}{r^3} \right\rangle_{nl} = \frac{m_e}{l(l+1)\hbar^2} \left\langle \frac{(Ze^2)^2}{r^2} \right\rangle_{nl}$$

$$= -\frac{2m_e^2 c^2 Z^2 \alpha^2}{nl(l+1)(l+1/2)\hbar^2} E_n^{(0)}. \tag{5.124}$$

We therefore have the spin-orbit correction to the energy eigenstates of the hydrogen atom from (5.113) as

$$\Delta_{nlj} = -\frac{Z^2 \alpha^2}{2nl(l+1)(l+1/2)} E_n^{(0)} \begin{Bmatrix} l \\ -(l+1) \end{Bmatrix} \begin{array}{l} j = l + \dfrac{1}{2} \\[2mm] j = l - \dfrac{1}{2}. \end{array} \tag{5.125}$$

Interestingly, this expression is nonzero for $l = 0$. Nevertheless, it gives the correct answer for the energy eigenvalues of the Dirac equation, as we shall see later in this textbook. The origin of this shift, attributed to something called the Darwin term, is discussed elsewhere. See, for example, Townsend (2000).

### 5.3.3 The Zeeman Effect

We now discuss hydrogen or hydrogenlike (one-electron) atoms in a uniform magnetic field – the **Zeeman effect**, sometimes called the *anomalous Zeeman effect* with the electron spin taken into account. Recall that a uniform magnetic field $B$ is derivable from a vector potential

$$\mathbf{A} = \tfrac{1}{2}(\mathbf{B} \times \mathbf{x}). \tag{5.126}$$

For $\mathbf{B}$ in the positive $z$-direction, that is $\mathbf{B} = B\hat{\mathbf{z}}$, this becomes

$$\mathbf{A} = -\tfrac{1}{2}(By\hat{\mathbf{x}} - Bx\hat{\mathbf{y}}). \tag{5.127}$$

Apart from the spin term, the interaction Hamiltonian is generated by the substitution

$$\mathbf{p} \to \mathbf{p} - \frac{e\mathbf{A}}{c}. \tag{5.128}$$

We therefore have

$$H = \frac{\mathbf{p}^2}{2m_e} + V_c(r) - \frac{e}{2m_e c}(\mathbf{p} \cdot \mathbf{A} + \mathbf{A} \cdot \mathbf{p}) + \frac{e^2 \mathbf{A}^2}{2m_e c^2}. \tag{5.129}$$

Because

$$\langle \mathbf{x}' | \mathbf{p} \cdot \mathbf{A}(\mathbf{x}) | \rangle = -i\hbar \nabla' \cdot [\mathbf{A}(\mathbf{x}') \langle \mathbf{x}' | \rangle]$$

$$= \langle \mathbf{x}' | \mathbf{A}(\vec{\mathbf{x}}) \cdot \mathbf{p} | \rangle + \langle \mathbf{x}' | \rangle [-i\hbar \nabla' \cdot \mathbf{A}(\mathbf{x}')], \tag{5.130}$$

it is legitimate to replace $\mathbf{p} \cdot \mathbf{A}$ by $\mathbf{A} \cdot \mathbf{p}$ whenever

$$\nabla \cdot \mathbf{A}(\mathbf{x}) = 0, \tag{5.131}$$

which is the case for the vector potential of (5.127). Noting

$$\mathbf{A} \cdot \mathbf{p} = B(-\tfrac{1}{2}yp_x + \tfrac{1}{2}xp_y)$$

$$= \tfrac{1}{2}BL_z \tag{5.132}$$

and

$$\mathbf{A}^2 = \tfrac{1}{4}B^2(x^2 + y^2), \tag{5.133}$$

we obtain for (5.129)

$$H = \frac{\mathbf{p}^2}{2m_e} + V_c(r) - \frac{e}{2m_ec}BL_z + \frac{e^2}{8m_ec^2}B^2(x^2 + y^2). \tag{5.134}$$

To this we may add the spin magnetic-moment interaction

$$-\boldsymbol{\mu} \cdot \mathbf{B} = \frac{-e}{m_ec}\mathbf{S} \cdot \mathbf{B} = \frac{-e}{m_ec}BS_z. \tag{5.135}$$

The quadratic $B^2(x^2 + y^2)$ is unimportant for a one-electron atom; the analogous term is important for the ground state of the helium atom where $L_z^{(tot)}$ and $S_z^{(tot)}$ both vanish. The reader may come back to this problem when he or she computes diamagnetic susceptibilities in Problems 5.25 and 5.26 of this chapter.

To summarize, omitting the quadratic term, the total Hamiltonian is made up of the following three terms:

$$H_0 = \frac{\mathbf{p}^2}{2m_e} + V_c(r) \tag{5.136a}$$

$$H_{LS} = \frac{1}{2m_e^2c^2}\frac{1}{r}\frac{dV_c(r)}{dr}\mathbf{L} \cdot \mathbf{S} \tag{5.136b}$$

$$H_B = \frac{-eB}{2m_ec}(L_z + 2S_z). \tag{5.136c}$$

Notice the factor 2 in front of $S_z$; this reflects the fact that the $g$-factor of the electron is 2.

Suppose $H_B$ is treated as a small perturbation. We can study the effect of $H_B$ using the eigenkets of $H_0 + H_{LS}$, that is, the $\mathbf{J}^2, J_z$ eigenkets, as our base kets. Noting

$$L_z + 2S_z = J_z + S_z, \tag{5.137}$$

the first-order shift can be written as

$$\frac{-eB}{2m_ec}\langle J_z + S_z \rangle_{j=l\pm 1/2,m}. \tag{5.138}$$

The expectation value of $J_z$ immediately gives $m\hbar$. As for $\langle S_z \rangle$, we first recall

$$\left| j = l \pm \frac{1}{2}, m \right\rangle = \pm\sqrt{\frac{l \pm m + \frac{1}{2}}{2l+1}} \times \left| m_l = m - \frac{1}{2}, m_s = \frac{1}{2} \right\rangle$$

$$+ \sqrt{\frac{l \mp m + \frac{1}{2}}{2l+1}} \left| m_l = m + \frac{1}{2}, m_s = -\frac{1}{2} \right\rangle. \tag{5.139}$$

The expectation value of $S_z$ can then easily be computed:

$$\langle S_z \rangle_{j=l\pm 1/2, m} = \frac{\hbar}{2}(|c_+|^2 - |c_-|^2)$$

$$= \frac{\hbar}{2}\frac{1}{(2l+1)}\left[\left(l\pm m+\frac{1}{2}\right) - \left(l\mp m+\frac{1}{2}\right)\right] = \pm\frac{m\hbar}{(2l+1)}. \qquad (5.140)$$

In this manner we obtain Lande's formula for the energy shift (due to **B** field),

$$\Delta E_B = \frac{-e\hbar B}{2m_e c}m\left[1\pm\frac{1}{(2l+1)}\right]. \qquad (5.141)$$

We see that the energy shift of (5.141) is proportional to $m$. To understand the physical origin for this, we present another method for deriving (5.140). We recall that the expectation value of $S_z$ can also be obtained using the projection theorem of Section 3.11. We get [see (3.488)]

$$\langle S_z \rangle_{j=l\pm 1/2, m} = [\langle \mathbf{S}\cdot\mathbf{J}\rangle_{j=l\pm 1/2}]\frac{m\hbar}{\hbar^2 j(j+1)}$$

$$= \frac{m\langle \mathbf{J}^2 + \mathbf{S}^2 - \mathbf{L}^2\rangle_{j=l\pm 1/2}}{2\hbar j(j+1)}$$

$$= m\hbar\left[\frac{\left(l\pm\frac{1}{2}\right)\left(l\pm\frac{1}{2}+1\right)+\frac{3}{4}-l(l+1)}{2\left(l\pm\frac{1}{2}\right)\left(l\pm\frac{1}{2}+1\right)}\right]$$

$$= \pm\frac{m\hbar}{(2l+1)}, \qquad (5.142)$$

which is in complete agreement with (5.140).

In the foregoing discussion the magnetic field is treated as a small perturbation. We now consider the opposite extreme – **the Paschen–Back limit** – with a magnetic field so intense that the effect of $H_B$ is far more important than that of $H_{LS}$, which we later add as a small perturbation. With $H_0 + H_B$ only, the good quantum numbers are $L_z$ and $S_z$. Even $\mathbf{J}^2$ is no good because spherical symmetry is completely destroyed by the strong **B** field that selects a particular direction in space, the $z$-direction. We are left with cylindrical symmetry only, that is, invariance under rotation around the $z$-axis. So the $L_z, S_z$ eigenkets $|l, s = \frac{1}{2}, m_l, m_s\rangle$ are to be used as our base kets. The effect of the main term $H_B$ can easily be computed:

$$\langle H_B \rangle_{m_l m_s} = \frac{-eB\hbar}{2m_e c}(m_l + 2m_s). \qquad (5.143)$$

The $2(2l+1)$ degeneracy in $m_l$ and $m_s$ we originally had with $H_0$ [see (5.136a)] is now reduced by $H_B$ to states with the same $(m_l) + (2m_s)$, namely, $(m_l) + (1)$ and $(m_l + 2) + (-1)$. Clearly we must evaluate the expectation value of $\mathbf{L}\cdot\mathbf{S}$ with respect to $|m_l, m_s\rangle$:

$$\langle \mathbf{L}\cdot\mathbf{S}\rangle = \langle L_z S_z + \frac{1}{2}(L_+ S_- + L_- S_+)\rangle_{m_l m_s}$$

$$= \hbar^2 m_l m_s, \qquad (5.144)$$

| **Table 5.1** Quantum Numbers in Weak and Strong Magnetic Fields | | | | |
|---|---|---|---|---|
| | Dominant interaction | Almost good | No good | Always good |
| Weak **B**   $H_{LS}$ | | $\mathbf{J}^2 \, (\text{or} \, \mathbf{L} \cdot \mathbf{S})$ | $L_z, S_z{}^*$ | |
| | | | | $\mathbf{L}^2, \mathbf{S}^2, J_z$ |
| Strong **B**   $H_B$ | | $L_z, S_z$ | $\mathbf{J}^2 \, (\text{or} \, \mathbf{L} \cdot \mathbf{S})$ | |

*The exception is the stretched configuration, e.g., $p_{3/2}$ with $m = \pm\frac{3}{2}$. Here $L_z$ and $S_z$ are both good; this is because the magnetic quantum number of $J_z$, $m = m_l + m_s$ can be satisfied in only one way.

where we have used

$$\langle L_\pm \rangle_{m_l} = 0, \qquad \langle S_\pm \rangle_{m_s} = 0. \tag{5.145}$$

Hence,

$$\langle H_{LS} \rangle_{m_l m_s} = \frac{\hbar^2 m_l m_s}{2 m_e^2 c^2} \left\langle \frac{1}{r} \frac{dV_c}{dr} \right\rangle. \tag{5.146}$$

We summarize our results in Table 5.1, where weak and strong **B** fields are "calibrated" by comparing their magnitudes $e\hbar B/2m_e c$ with $(1/137)^2 e^2/a_0$. In this table *almost good* simply means good to the extent that the less dominant interaction could be ignored.

Specifically, let us look at the level scheme of a $p$ electron $l = 1 (p_{3/2}, p_{1/2})$. In the weak **B** case the energy shifts are linear in **B**, with slopes determined by

$$m \left[ 1 \pm \left( \frac{1}{2l+1} \right) \right].$$

As we now increase **B**, mixing becomes possible between states with the same $m$-value, for example, $p_{3/2}$ with $m = \pm\frac{1}{2}$ and $p_{1/2}$ with $m = \pm\frac{1}{2}$; in this connection note the operator $L_z + 2S_z$ that appears in $H_B$ [(5.136c)] is a rank 1 tensor operator $T_{q=0}^{(k=1)}$ with spherical component $q = 0$.

It is instructive to solve the full problem, including fine structure along with the Zeeman effect for arbitrary field strength, and see how the behavior transitions from weak to strong magnetic fields. For the $n = 2$ states, one diagonalizes the $8 \times 8$ matrix of the perturbations in the degenerate subspace. This is not as difficult as it sounds, and is the object of Problem 5.24 at the end of this chapter. The resulting behavior is plotted in Figure 5.3.

## 5.3.4 Van der Waals' Interaction

An important, nice application of the Rayleigh–Schrödinger perturbation theory is to calculate the long range interaction, or **van der Waals' force**, between two hydrogen atoms in their ground states. It is easy to show that the energy between the two atoms for large separation $r$ is attractive and varies as $r^{-6}$.

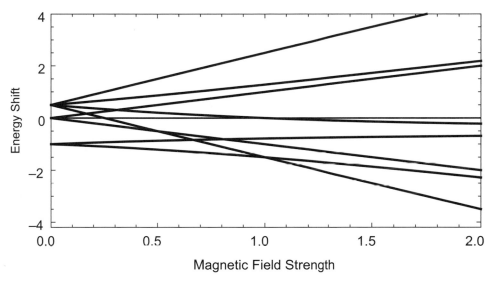

Fig. 5.3 First-order energy shifts for the $n = 2$ states in hydrogen, as a function of the applied magnetic field strength. See Problem 5.24 at the end of this chapter. Note that there is a horizontal line at zero, in addition to the eight curves for the 2s and 2p states. The evolution from eight states at low field to five states at high field is evident.

Fig. 5.4 Two hydrogen atoms with their protons (+) separated by a fixed distance $r$ and their electrons (−) at displacements $r_i$ from them.

Consider the two protons of the hydrogen atoms to be *fixed* at a distance $r$ (along the $z$-axis) with $\mathbf{r}_1$ the vector from the first proton to its electron and $\mathbf{r}_2$ the vector from the second proton to its electron; see Figure 5.4. Then the Hamiltonian $H$ can be written as

$$H = H_0 + V$$

$$H_0 = -\frac{\hbar^2}{2m}(\nabla_1^2 + \nabla_2^2) - \frac{e^2}{r_1} - \frac{e^2}{r_2}$$

$$V = \frac{e^2}{r} + \frac{e^2}{|\mathbf{r} + \mathbf{r}_2 - \mathbf{r}_1|} - \frac{e^2}{|\mathbf{r} + \mathbf{r}_2|} - \frac{e^2}{|\mathbf{r} - \mathbf{r}_1|}. \tag{5.147}$$

The lowest-energy solution of $H_0$ is simply the product of the ground-state wave functions of the noninteracting hydrogen atoms

$$U_0^{(0)} = U_{100}^{(0)}(\mathbf{r}_1) U_{100}^{(0)}(\mathbf{r}_2). \tag{5.148}$$

Now for large $r$ ($\gg$ the Bohr radius $a_0$) expand the perturbation $V$ in powers of $\mathbf{r}_i/\mathbf{r}$ to obtain

$$V = \frac{e^2}{r^3}(x_1 x_2 + y_1 y_2 - 2z_1 z_2) + 0\left(\frac{1}{r^4}\right) + \cdots. \qquad (5.149)$$

The lowest-order $r^{-3}$ term in (5.149) corresponds to the interaction of two electric dipoles $e\mathbf{r}_1$ and $e\mathbf{r}_2$ separated by $\mathbf{r}$. The higher-order terms represent higher-order multipole interactions, and thus every term in $V$ involves spherical harmonics $Y_l^m$ with $l_i > 0$ for each hydrogen atom. Hence, for each term in (5.149) the first-order perturbation energy matrix element $V_{00} \simeq 0$, since the ground-state $U_0^{(0)}$ wave function (5.148) has $l_i = 0$ and $\int d\Omega Y_l^m(\Omega) = 0$ for $l$ and $m \neq 0$. The second-order perturbation

$$E^{(2)}(r) = \frac{e^4}{r^6} \sum_{k \neq 0} \frac{|\langle k^{(0)}| x_1 x_2 + y_1 y_2 - 2z_1 z_2 |0^{(0)}\rangle|^2}{E_0^{(0)} - E_k^{(0)}} \qquad (5.150)$$

will be nonvanishing. We immediately see that this interaction varies as $1/r^6$; since $E_k^{(0)} > E_0^{(0)}$, it is negative. This $1/r^6$ long range attractive van der Waals' potential is a general property of the interaction between two atoms in their ground state.[3]

# 5.4  Variational Methods

The perturbation theory developed in the previous sections is, of course, of no help unless we already know exact solutions to a problem whose Hamiltonian is sufficiently similar. The variational method we now discuss is very useful for estimating the ground-state energy $E_0$ when such exact solutions are not available.

We attempt to guess the ground-state energy $E_0$ by considering a "trial ket" $|\tilde{0}\rangle$, which tries to imitate the true ground-state ket $|0\rangle$. To this end we first obtain a theorem of great practical importance. We define $\overline{H}$ such that

$$\overline{H} \equiv \frac{\langle \tilde{0}|H|\tilde{0}\rangle}{\langle \tilde{0}|\tilde{0}\rangle}, \qquad (5.151)$$

where we have accommodated the possibility that $|\tilde{0}\rangle$ might not be normalized. We can then prove the following.

**Theorem 8**

$$\overline{H} \geq E_0. \qquad (5.152)$$

This means that we can obtain an *upper bound* to $E_0$ by considering various kinds of $|\tilde{0}\rangle$. The proof of this is very straightforward.

---

[3] See the treatment in Schiff (1968), pp. 261–263, which gives a lower and upper bound on the magnitude of the van der Waals' potential from (5.150) and from a variational calculation. Also note the first footnote on p. 263 of Schiff concerning retardation effects.

**Proof**  Even though we do not know the energy eigenket of the Hamiltonian $H$, we can imagine that $|\tilde{0}\rangle$ can be expanded as

$$|\tilde{0}\rangle = \sum_{k=0}^{\infty} |k\rangle\langle k|\tilde{0}\rangle \qquad (5.153)$$

where $|k\rangle$ is an *exact* energy eigenket of $H$:

$$H|k\rangle = E_k|k\rangle. \qquad (5.154)$$

The theorem (5.152) follows when we use $E_k = E_k - E_0 + E_0$ to evaluate $\overline{H}$ in (5.151). We have

$$\overline{H} = \frac{\displaystyle\sum_{k=0} |\langle k|\tilde{0}\rangle|^2 E_k}{\displaystyle\sum_{k=0} |\langle k|\tilde{0}\rangle|^2} \qquad (5.155a)$$

$$= \frac{\displaystyle\sum_{k=1}^{\infty} |\langle k|\tilde{0}\rangle|^2 (E_k - E_0)}{\displaystyle\sum_{k=0} |\langle k|\tilde{0}\rangle|^2} + E_0 \qquad (5.155b)$$

$$\geq E_0, \qquad (5.155c)$$

where we have used the fact that $E_k - E_0$ in the first sum of (5.155b) is necessarily positive. It is also obvious from this proof that the equality sign in (5.152) holds only if $|\tilde{0}\rangle$ coincides exactly with $|0\rangle$, that is, if the coefficients $\langle k|\tilde{0}\rangle$ all vanish for $k \neq 0$.    □

The theorem (5.152) is quite powerful because $\overline{H}$ provides an upper bound to the true ground-state energy. Furthermore, a relatively poor trial ket can give a fairly good energy estimate for the ground state because if

$$\langle k|\tilde{0}\rangle \sim 0(\varepsilon) \quad \text{for } k \neq 0, \qquad (5.156)$$

then from (5.155) we have

$$\overline{H} - E_0 \sim 0(\varepsilon^2). \qquad (5.157)$$

We see an example of this in a moment. Of course, the method does not say anything about the discrepancy between $\overline{H}$ and $E_0$; all we know is that $\overline{H}$ is larger than (or equal to) $E_0$.

Another way to state the theorem is to assert that $\overline{H}$ is stationary with respect to the variation

$$|\tilde{0}\rangle \rightarrow |\tilde{0}\rangle + \delta|\tilde{0}\rangle; \qquad (5.158)$$

that is, $\delta\overline{H} = 0$ when $|\tilde{0}\rangle$ coincides with $|0\rangle$. By this we mean that if $|0\rangle + \delta|\tilde{0}\rangle$ is used in place of $|\tilde{0}\rangle$ in (5.155) and we calculate $\overline{H}$, then the error we commit in estimating the true ground-state energy involves $|\tilde{0}\rangle$ to order $(\delta|\tilde{0}\rangle)^2$.

The variational method per se does not tell us what kind of trial kets are to be used to estimate the ground-state energy. Quite often we must appeal to physical intuition, for example, the asymptotic behavior of a wave function at large distances. What we do in practice is to characterize trial kets by one or more parameters $\lambda_1, \lambda_2, \ldots$, and compute $\overline{H}$

as a function of $\lambda_1, \lambda_2, \ldots$. We then minimize $\overline{H}$ by (1) setting the derivative with respect to the parameters all zero, namely,

$$\frac{\partial \overline{H}}{\partial \lambda_1} = 0, \qquad \frac{\partial \overline{H}}{\partial \lambda_2} = 0, \ldots, \tag{5.159}$$

(2) determining the optimum values of $\lambda_1, \lambda_2, \ldots$, and (3) substituting them back to the expression for $\overline{H}$.

If the wave function for the trial ket already has a functional form of the exact ground-state energy eigenfunction, we of course obtain the true ground-state energy function by this method. For example, suppose somebody has the foresight to guess that the wave function for the ground state of the hydrogen atom must be of the form

$$\langle \mathbf{x}|0 \rangle \propto e^{-r/a}, \tag{5.160}$$

where $a$ is regarded as a parameter to be varied. We then find, upon minimizing $\overline{H}$ with (5.160), the correct ground-state energy $-e^2/2a_0$. Not surprisingly, the minimum is achieved when $a$ coincides with the Bohr radius $a_0$.

As a second example, we attempt to estimate the ground state of the infinite-well (one-dimensional box) problem defined by

$$V = \begin{cases} 0 & \text{for } |x| < a \\ \infty & \text{for } |x| > a. \end{cases} \tag{5.161}$$

The exact solutions are, of course, well known:

$$\langle x|0 \rangle = \frac{1}{\sqrt{a}} \cos\left(\frac{\pi x}{2a}\right),$$
$$E_0 = \left(\frac{\hbar^2}{2m}\right)\left(\frac{\pi^2}{4a^2}\right). \tag{5.162}$$

But suppose we did not know these. Evidently the wave function must vanish at $x = \pm a$; furthermore, for the ground state the wave function cannot have any wiggles. The simplest analytic function that satisfies both requirements is just a parabola going through $x = \pm a$:

$$\langle x|\tilde{0} \rangle = a^2 - x^2, \tag{5.163}$$

where we have not bothered to normalize $|\tilde{0}\rangle$. Here there is no variational parameter. We can compute $\overline{H}$ as follows:

$$\overline{H} = \frac{\left(\frac{-\hbar^2}{2m}\right) \int_{-a}^{a} (a^2 - x^2) \frac{d^2}{dx^2} (a^2 - x^2) dx}{\int_{-a}^{a} (a^2 - x^2)^2 dx}$$

$$= \left(\frac{10}{\pi^2}\right)\left(\frac{\pi^2 \hbar^2}{8a^2 m}\right) \simeq 1.0132 E_0. \tag{5.164}$$

It is remarkable that with such a simple trial function we can come within 1.3% of the true ground-state energy.

A much better result can be obtained if we use a more sophisticated trial function. We try

$$\langle x|\tilde{0}\rangle = |a|^{\lambda} - |x|^{\lambda}, \tag{5.165}$$

where $\lambda$ is now regarded as a variational parameter. Straightforward algebra gives

$$\overline{H} = \left[\frac{(\lambda+1)(2\lambda+1)}{(2\lambda-1)}\right]\left(\frac{\hbar^2}{4ma^2}\right), \tag{5.166}$$

which has a minimum at

$$\lambda = \frac{(1+\sqrt{6})}{2} \simeq 1.72, \tag{5.167}$$

not far from $\lambda = 2$ (a parabola) considered earlier. This gives

$$\overline{H}_{\min} = \left(\frac{5+2\sqrt{6}}{\pi^2}\right)E_0 \simeq 1.00298E_0. \tag{5.168}$$

So the variational method with (5.165) gives the correct ground-state energy within 0.3% – a fantastic result considering the simplicity of the trial function used.

How well does this trial function imitate the true ground-state wave function? It is amusing that we can answer this question without explicitly evaluating the overlap integral $\langle 0|\tilde{0}\rangle$. Assuming that $|\tilde{0}\rangle$ is normalized, we have [from (5.151)–(5.154)]

$$\overline{H}_{\min} = \sum_{k=0}^{\infty} |\langle k|\tilde{0}\rangle|^2 E_k$$

$$\geq |\langle 0|\tilde{0}\rangle|^2 E_0 + 9E_0(1 - |\langle 0|\tilde{0}\rangle|^2) \tag{5.169}$$

where $9E_0$ is the energy of the second excited state; the first excited state ($k = 1$) gives no contribution by parity conservation. Solving for $|\langle 0|\tilde{0}\rangle|$ and using (5.168), we have

$$|\langle 0|\tilde{0}\rangle|^2 \geq \frac{9E_0 - \overline{H}_{\min}}{8E_0} = 0.99963. \tag{5.170}$$

Departure from unity characterizes a component of $|\tilde{0}\rangle$ in a direction orthogonal to $|0\rangle$. If we are talking about the "angle" $\theta$ defined by

$$\langle 0|\tilde{0}\rangle = \cos\theta, \tag{5.171}$$

then (5.170) corresponds to

$$\theta \underset{\sim}{<} 1.1°, \tag{5.172}$$

so $|0\rangle$ and $|\tilde{0}\rangle$ are nearly "parallel."

One of the earliest applications of the variational method involved the ground-state energy of the helium atom, which we will discuss in Section 7.4. We can also use the variational method to estimate the energies of first excited states; all we need to do is work with a trial ket orthogonal to the ground-state wave function, either exact, if known, or an approximate one obtained by the variational method.

## 5.5 Time-Dependent Potentials: The Interaction Picture

### 5.5.1 Statement of the Problem

So far in this book we have been concerned with Hamiltonians that do not contain time explicitly. In nature, however, there are many quantum-mechanical systems of importance with time dependence. In the remaining part of this chapter we show how to deal with situations with time-dependent potentials.

We consider a Hamiltonian $H$ such that it can be split into two parts,

$$H = H_0 + V(t), \tag{5.173}$$

where $H_0$ does not contain time explicitly. The problem $V(t) = 0$ is assumed to be solved in the sense that the energy eigenkets $|n\rangle$ and the energy eigenvalues $E_n$ defined by

$$H_0|n\rangle = E_n|n\rangle \tag{5.174}$$

are completely known.[4] We may be interested in situations where initially only one of the energy eigenstates of $H_0$, for example, $|i\rangle$ is populated. As time goes on, however, states other than $|i\rangle$ are populated because with $V(t) \neq 0$ we are no longer dealing with "stationary" problems; the time-evolution operator is no longer as simple as $e^{-iHt/\hbar}$ when $H$ itself involves time. Quite generally the time-dependent potential $V(t)$ can cause transitions to states other than $|i\rangle$. The basic question we address is, what is the probability as a function of time for the system to be found in $|n\rangle$, with $n \neq i$?

More generally, we may be interested in how an arbitrary state ket changes as time goes on, where the total Hamiltonian is the sum of $H_0$ and $V(t)$. Suppose at $t = 0$, the state ket of a physical system is given by

$$|\alpha\rangle = \sum_n c_n(0)|n\rangle. \tag{5.175}$$

We wish to find $c_n(t)$ for $t > 0$ such that

$$|\alpha, t_0 = 0; t\rangle = \sum_n c_n(t)e^{-iE_n t/\hbar}|n\rangle \tag{5.176}$$

where the ket on the left side stands for the state ket in the Schrödinger picture at $t$ of a physical system whose state ket at $t = 0$ was found to be $|\alpha\rangle$.

The astute reader may have noticed the manner in which we have separated the time dependence of the coefficient of $|n\rangle$ in (5.176). The factor $e^{-iE_n t/\hbar}$ is present even if $V$ is absent. This way of writing the time dependence makes it clear that the time evolution of $c_n(t)$ is due solely to the presence of $V(t)$; $c_n(t)$ would be identically equal to $c_n(0)$ and hence independent of $t$ if $V$ were zero. As we shall see in a moment, this separation is convenient because $c_n(t)$ satisfies a relatively simple differential equation. The probability of finding $|n\rangle$ is found by evaluating $|c_n(t)|^2$.

---

[4] In (5.174) we no longer use the notation $|n^{(0)}\rangle$, $E_n^{(0)}$.

## 5.5.2 The Interaction Picture

Before we discuss the differential equation for $c_n(t)$, we discuss the interaction picture. Suppose we have a physical system such that its state ket coincides with $|\alpha\rangle$ at $t = t_0$, where $t_0$ is often taken to be zero. At a later time, we denote the state ket in the Schrödinger picture by $|\alpha, t_0; t\rangle_S$, where the subscript $S$ reminds us that we are dealing with the state ket of the Schrödinger picture.

We now *define*

$$|\alpha, t_0; t\rangle_I = e^{iH_0 t/\hbar}|\alpha, t_0; t\rangle_S, \tag{5.177}$$

where $|\ \rangle_I$ stands for a state ket that represents the same physical situation in the *interaction picture*. At $t = 0$, $|\ \rangle_I$ evidently coincides with $|\ \rangle_S$. For operators (representing observables) we define observables in the interaction picture as

$$A_I \equiv e^{iH_0 t/\hbar} A_S e^{-iH_0 t/\hbar}. \tag{5.178}$$

In particular,

$$V_I = e^{iH_0 t/\hbar} V e^{-iH_0 t/\hbar} \tag{5.179}$$

where $V$ without a subscript is understood to be the time-dependent potential in the Schrödinger picture. The reader may recall here the connection between the Schrödinger picture and the Heisenberg picture:

$$|\alpha\rangle_H = e^{+iHt/\hbar}|\alpha, t_0 = 0; t\rangle_S \tag{5.180}$$

$$A_H = e^{iHt/\hbar} A_S e^{-iHt/\hbar}. \tag{5.181}$$

The basic difference between (5.180) and (5.181) on the one hand and (5.177) and (5.178) on the other is that $H$ rather than $H_0$ appears in the exponential.

We now derive the fundamental differential equation that characterizes the time evolution of a state ket in the interaction picture. Let us take the time derivative of (5.177) with the full $H$ given by (5.173):

$$i\hbar \frac{\partial}{\partial t}|\alpha, t_0; t\rangle_I = i\hbar \frac{\partial}{\partial t}(e^{iH_0 t/\hbar}|\alpha, t_0; t\rangle_S)$$

$$= -H_0 e^{iH_0 t/\hbar}|\alpha, t_0; t\rangle_S + e^{iH_0 t/\hbar}(H_0 + V)|\alpha, t_0; t\rangle_S$$

$$= e^{iH_0 t/\hbar} V e^{-iH_0 t/\hbar} e^{iH_0 t/\hbar}|\alpha, t_0; t\rangle_S. \tag{5.182}$$

We thus see

$$i\hbar \frac{\partial}{\partial t}|\alpha, t_0; t\rangle_I = V_I|\alpha, t_0; t\rangle_I, \tag{5.183}$$

which is a Schrödinger-like equation with the total $H$ replaced by $V_I$. In other words $|\alpha, t_0; t\rangle_I$ would be a ket fixed in time if $V_I$ were absent. We can also show for an observable $A$ (that does not contain time $t$ explicitly in the Schrödinger picture) that

$$\frac{dA_I}{dt} = \frac{1}{i\hbar}[A_I, H_0], \tag{5.184}$$

which is a Heisenberg-like equation with $H$ replaced by $H_0$.

| | Heisenberg picture | Interaction picture | Schrödinger picture |
|---|---|---|---|
| **Table 5.2** Heisenberg, Interaction, and Schrödinger Pictures | | | |
| State ket | No change | Evolution determined by $V_I$ | Evolution determined by $H$ |
| Observable | Evolution determined by $H$ | Evolution determined by $H_0$ | No change |

In many respects, the interaction picture, or *Dirac picture*, is intermediate between the Schrödinger picture and the Heisenberg picture. This should be evident from Table 5.2.

In the interaction picture we continue using $|n\rangle$ as our base ket. Thus we expand $|\ \rangle_I$ as follows:

$$|\alpha, t_0; t\rangle_I = \sum_n c_n(t)|n\rangle. \tag{5.185}$$

With $t_0$ set equal to 0, we see that the $c_n(t)$ appearing here are the same as the $c_n(t)$ introduced earlier in (5.176), as can easily be verified by multiplying both sides of (5.176) by $e^{iH_0 t/\hbar}$ using (5.174).

We are finally in a position to write the differential equation for $c_n(t)$. Multiplying both sides of (5.183) by $\langle n|$ from the left, we obtain

$$i\hbar \frac{\partial}{\partial t}\langle n|\alpha, t_0; t\rangle_I = \sum_m \langle n|V_I|m\rangle\langle m|\alpha, t_0; t\rangle_I. \tag{5.186}$$

This can also be written using

$$\langle n|e^{iH_0 t/\hbar}V(t)e^{-iH_0 t/\hbar}|m\rangle = V_{nm}(t)e^{i(E_n - E_m)t/\hbar}$$

and

$$c_n(t) = \langle n|\alpha, t_0; t\rangle_I$$

[from (5.185)] as

$$i\hbar \frac{d}{dt}c_n(t) = \sum_m V_{nm}e^{i\omega_{nm}t}c_m(t), \tag{5.187}$$

where

$$\omega_{nm} \equiv \frac{(E_n - E_m)}{\hbar} = -\omega_{mn}. \tag{5.188}$$

Explicitly,

$$i\hbar \begin{pmatrix} \dot{c}_1 \\ \dot{c}_2 \\ \dot{c}_3 \\ \vdots \end{pmatrix} = \begin{pmatrix} V_{11} & V_{12}e^{i\omega_{12}t} & \cdots \\ V_{21}e^{i\omega_{21}t} & V_{22} & \cdots \\ & & V_{33} & \cdots \\ \vdots & \vdots & \vdots & \ddots \end{pmatrix} \begin{pmatrix} c_1 \\ c_2 \\ c_3 \\ \vdots \end{pmatrix}. \tag{5.189}$$

This is the basic coupled differential equation that must be solved to obtain the probability of finding $|n\rangle$ as a function of $t$.

### 5.5.3  Time-Dependent Two-State Problems: Nuclear Magnetic Resonance, Masers, and So Forth

Exact soluble problems with time-dependent potentials are rather rare. In most cases we have to resort to perturbation expansion to solve the coupled differential equations (5.189), as we will discuss in the next section. There is, however, a problem of enormous practical importance, which can be solved exactly – a two-state problem with a sinusoidal oscillating potential.

The problem is defined by

$$H_0 = E_1|1\rangle\langle 1| + E_2|2\rangle\langle 2| \quad (E_2 > E_1)$$
$$V(t) = \gamma e^{i\omega t}|1\rangle\langle 2| + \gamma e^{-i\omega t}|2\rangle\langle 1|, \tag{5.190}$$

where $\gamma$ and $\omega$ are real and positive. In the language of (5.186) and (5.187), we have

$$V_{12} = V_{21}^* = \gamma e^{i\omega t}$$
$$V_{11} = V_{22} = 0. \tag{5.191}$$

We thus have a time-dependent potential that connects the two energy eigenstates of $H_0$. In other words, we can have a transition between the two states $|1\rangle \underset{\leftarrow}{\overset{\rightarrow}{}} |2\rangle$.

An exact solution to this problem is available. If initially, at $t = 0$, only the lower level is populated so that [see (5.175)]

$$c_1(0) = 1, \qquad c_2(0) = 0, \tag{5.192}$$

then the probability for being found in each of the two states is given by **Rabi's formula** (after I. I. Rabi, who is the father of molecular beam techniques)

$$|c_2(t)|^2 = \frac{\gamma^2/\hbar^2}{\gamma^2/\hbar^2 + (\omega - \omega_{21})^2/4} \sin^2\left\{\left[\frac{\gamma^2}{\hbar^2} + \frac{(\omega - \omega_{21})^2}{4}\right]^{1/2} t\right\} \tag{5.193a}$$

$$|c_1(t)|^2 = 1 - |c_2(t)|^2, \tag{5.193b}$$

where

$$\omega_{21} \equiv \frac{(E_2 - E_1)}{\hbar}, \tag{5.194}$$

as the reader may verify by working out Problem 5.37 of this chapter.

Let us now look at $|c_2|^2$ a little more closely. We see that the probability for finding the upper state $E_2$ exhibits an oscillatory time dependence with angular frequency, two times that of

$$\Omega = \sqrt{\left(\frac{\gamma^2}{\hbar^2}\right) + \frac{(\omega - \omega_{21})^2}{4}}. \tag{5.195}$$

The amplitude of oscillation is very large when

$$\omega \simeq \omega_{21} = \frac{(E_2 - E_1)}{\hbar}, \tag{5.196}$$

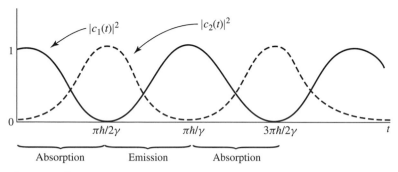

**Fig. 5.5** Plot of $|c_1(t)|^2$ and $|c_2(t)|^2$ against time $t$ exactly at resonance $\omega = \omega_{21}$ and $\Omega = \gamma/\hbar$. The graph also illustrates the back-and-forth behavior between $|1\rangle$ and $|2\rangle$.

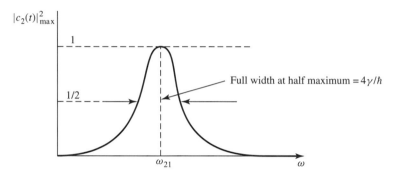

**Fig. 5.6** Graph of $|c_2(t)|^2_{\max}$ as a function of $\omega$, where $\omega = \omega_{21}$ corresponds to the resonant frequency.

that is, when the angular frequency of the potential, usually due to an externally applied electric or magnetic field, is nearly equal to the angular frequency characteristic of the two-state system. Equation (5.196) is therefore known as the **resonance condition**.

It is instructive to look at (5.193a) and (5.193b) a little closely exactly at resonance:

$$\omega = \omega_{21}, \quad \Omega = \frac{\gamma}{\hbar}. \tag{5.197}$$

We can plot $|c_1(t)|^2$ and $|c_2(t)|^2$ as a function of $t$; see Figure 5.5. From $t = 0$ to $t = \pi\hbar/2\gamma$, the two-level system absorbs energy from the time-dependent potential $V(t)$; $|c_1(t)|^2$ decreases from unity as $|c_2(t)|^2$ grows. At $t = \pi\hbar/2\gamma$, only the upper state is populated. From $t = \pi\hbar/2\gamma$ to $t = \pi\hbar/\gamma$, the system gives up its excess energy [of the excited (upper) state] to $V(t)$; $|c_2|^2$ decreases and $|c_1|^2$ increases. This *absorption-emission cycle* is repeated indefinitely, as is also shown in Figure 5.5, so $V(t)$ can be regarded as a source or sink of energy; put in another way, $V(t)$ can cause a transition from $|1\rangle$ to $|2\rangle$ (absorption) or from $|2\rangle$ to $|1\rangle$ (emission). We will come back to this point of view when we discuss emission and absorption of radiation.

The absorption-emission cycle takes place even away from resonance. However, the amplitude of oscillation for $|2\rangle$ is now reduced; $|c_2(t)|^2_{\max}$ is no longer 1 and $|c_1(t)|^2$ does not go down all the way to 0. In Figure 5.6 we plot $|c_2(t)|^2_{\max}$ as a function of $\omega$. This curve has a resonance peak centered around $\omega = \omega_{21}$, and the full width at half maximum is given by $4\gamma/\hbar$. It is worth noting that the weaker the time-dependent potential ($\gamma$ small), the narrower the resonance peak.

## 5.5.4 Spin Magnetic Resonance

The two-state problem defined by (5.190) has many physical applications. As a first example, consider a spin $\frac{1}{2}$ system, say a bound electron, subjected to a $t$-independent uniform magnetic field in the $z$-direction *and*, in addition, a $t$-dependent magnetic field rotating in the $xy$-plane:

$$\mathbf{B} = B_0\hat{\mathbf{z}} + B_1(\hat{\mathbf{x}}\cos\omega t + \hat{\mathbf{y}}\sin\omega t) \qquad (5.198)$$

with $B_0$ and $B_1$ constant. We can treat the effect of the uniform $t$-independent field as $H_0$ and the effect of the rotating field as $V$. For

$$\boldsymbol{\mu} - \frac{e}{m_e c}\mathbf{S} \qquad (5.199)$$

we have

$$H_0 = -\left(\frac{e\hbar B_0}{2m_e c}\right)(|+\rangle\langle+| - |-\rangle\langle-|)$$

$$V(t) = -\left(\frac{e\hbar B_1}{2m_e c}\right)[\cos\omega t(|+\rangle\langle-| + |-\rangle\langle+|) \qquad (5.200)$$

$$+ \sin\omega t(-i|+\rangle\langle-| + i|-\rangle\langle+|)],$$

where we have used the ket-bra forms of $2S_j/\hbar$ [see (3.25)]. With $e < 0$, $E_+$ has a higher energy than $E_-$, and we can identify

$$\begin{aligned}|+\rangle &\to |2\rangle \quad \text{(upper level)} \\ |-\rangle &\to |1\rangle \quad \text{(lower level)}\end{aligned} \qquad (5.201)$$

to make correspondence with the notation of (5.190). The angular frequency characteristic of the two-state system is

$$\omega_{21} = \frac{|e|B_0}{m_e c}, \qquad (5.202)$$

which is just the spin-precession frequency for the $B_0 \neq 0$, $B_1 = 0$ problem already treated in Section 2.1. Even though the expectation values of $\langle S_{x,y}\rangle$ change due to spin precession in the counterclockwise direction (seen from the positive $z$-side), $|c_+|^2$ and $|c_-|^2$ remain unchanged in the absence of the rotating field. We now add a new feature as a result of the rotating field: $|c_+|^2$ and $|c_-|^2$ do change as a function of time. This can be seen by identifying

$$\frac{-e\hbar B_1}{2m_e c} \to \gamma, \quad \omega \to \omega \qquad (5.203)$$

to make correspondence with the notation of (5.190); our time-dependent interaction (5.200) is precisely of form (5.190). The fact that $|c_+(t)|^2$ and $|c_-(t)|^2$ vary in the manner indicated by Figure 5.5 for $\omega = \omega_{21}$ and the correspondence (5.201), for example, implies that the spin $\frac{1}{2}$ system undergoes a succession of spin-flips, $|+\rangle \overset{\rightarrow}{\underset{\leftarrow}{}} |-\rangle$, in addition to spin precession. Semiclassically, spin-flips of this kind can be interpreted as being due to the driving torque exerted by rotating $\mathbf{B}$.

The resonance condition is satisfied whenever the frequency of the rotating magnetic field coincides with the frequency of spin precession determined by the strength of the uniform magnetic field. We see that the probability of spin-flips is particularly large.

In practice, a rotating magnetic field may be difficult to produce experimentally. Fortunately, a horizontally oscillating magnetic field, for instance, in the $x$-direction, is just as good. To see this, we first note that such an oscillating field can be decomposed into a counterclockwise component and a clockwise component as follows:

$$2B_1\hat{\mathbf{x}}\cos\omega t = B_1(\hat{\mathbf{x}}\cos\omega t + \hat{\mathbf{y}}\sin\omega t) + B_1(\hat{\mathbf{x}}\cos\omega t - \hat{\mathbf{y}}\sin\omega t). \qquad (5.204)$$

We can obtain the effect of the counterclockwise component simply by reversing the sign of $\omega$. Suppose the resonance condition is met for the counterclockwise component

$$\omega \simeq \omega_{21}. \qquad (5.205)$$

Under a typical experimental condition,

$$\frac{B_1}{B_0} \ll 1, \qquad (5.206)$$

which implies from (5.202) and (5.203) that

$$\frac{\gamma}{\hbar} \ll \omega_{21}; \qquad (5.207)$$

as a result, whenever the resonance condition is met for the counterclockwise component, the effect of the clockwise component becomes completely negligible, since it amounts to $\omega \to -\omega$, and the amplitude becomes small in magnitude as well as very rapidly oscillating.

The resonance problem we have solved is of fundamental importance in interpreting atomic molecular beam and nuclear magnetic resonance experiments. By varying the frequency of oscillating field, it is possible to make a very precise measurement of magnetic moment. We have based our discussion on the solution to differential equations (5.189); this problem can also be solved, perhaps more elegantly, by introducing the *rotating axis representation* of Rabi, Schwinger, and Van Vleck.

### 5.5.5  Maser

As another application of the time-dependent two-state problem, let us consider a **maser**. Specifically, we consider an ammonia molecule $NH_3$, which – as we may recall from Section 4.2 – has two parity eigenstates $|S\rangle$ and $|A\rangle$ lying close together such that $|A\rangle$ is slightly higher. Let $\mu_{el}$ be the electric dipole operator of the molecule. From symmetry considerations we expect that $\mu_{el}$ is proportional to $\mathbf{x}$, the position operator for the N atom. The basic interaction is like $-\mu_{el} \cdot \mathbf{E}$, where for a maser $\mathbf{E}$ is a time-dependent electric field in a microwave cavity:

$$\mathbf{E} = |\mathbf{E}|_{max}\hat{\mathbf{z}}\cos\omega t. \qquad (5.208)$$

It is legitimate to ignore the spatial variation of **E** because the wavelength in the microwave region is far larger than the molecular dimension. The frequency $\omega$ is tuned to the energy difference between $|A\rangle$ and $|S\rangle$:

$$\omega \simeq \frac{(E_A - E_S)}{\hbar}. \tag{5.209}$$

The diagonal matrix elements of the dipole operator vanish by parity,

$$\langle A|\mu_{el}|A\rangle = \langle S|\mu_{el}|S\rangle = 0, \tag{5.210}$$

but the off-diagonal elements are, in general, nonvanishing:

$$\langle S|\mathbf{x}|A\rangle = \langle A|\mathbf{x}|S\rangle \neq 0. \tag{5.211}$$

This means that there is a time-dependent potential that connects $|S\rangle$ and $|A\rangle$, and the general two-state problem we discussed earlier is now applicable.

We are now in a position to discuss how masers work. Given a molecular beam of $NH_3$ containing both $|S\rangle$ and $|A\rangle$, we first eliminate the $|S\rangle$ component by letting the beam go through a region of time-independent inhomogeneous electric field. Such an electric field separates $|S\rangle$ from $|A\rangle$ in much the same way as the inhomogeneous magnetic field in the Stern–Gerlach experiment separates $|+\rangle$ from $|-\rangle$. A pure beam of $|A\rangle$ then enters a microwave cavity tuned to the energy difference $E_A - E_S$. The dimension of the cavity is such that the time spent by the molecule is just $(\pi/2)\hbar/\gamma$. As a result we stay in the first emission phase of Figure 5.5; we have $|A\rangle$ in and $|S\rangle$ out. The excess energy of $|A\rangle$ is given up to the time-dependent potential as $|A\rangle$ turns into $|S\rangle$ and the radiation (microwave) field gains energy. In this way we obtain Microwave Amplification by Stimulated Emission of Radiation, or MASER.

There are many other applications of the general time-dependent two-state problem, such as the atomic clock and optical pumping. In fact, it is amusing to see that as many as four Nobel Prizes in physics have been awarded to those who exploited time-dependent two-state systems of some form.[5]

## 5.6  Hamiltonians with Extreme Time Dependence

This section is devoted to time-dependent Hamiltonians, with some "obvious" approximations in the case of very fast or very slow time dependence. A careful look, though, points out some interesting phenomena, one of which did not come to light until very late in the twentieth century.

Our treatment here is limited to a discussion of the basics, followed by some specific examples. An instructive example which we do not discuss is that of the square well with

---

[5] *Nobel Prize winners* who took advantage of resonance in the two-level systems are Rabi (1944) on molecular beams and nuclear magnetic resonance; Bloch and Purecell (1952) on **B** field in atomic nuclei and nuclear magnetic moments; Townes, Basov, and Prochorov (1964) on masers, lasers, and quantum optics; and Kastler (1966) on optical pumping.

contracting or expanding walls, where the walls may move quickly or slowly. For these cases, we refer the interested reader to Pinder, *Am. J. Phys.*, **58** (1990) 54, and D. W. Schlitt and Stutz, *Am. J. Phys.*, **38** (1970) 70.

## 5.6.1 Sudden Approximation

If a Hamiltonian changes very quickly, then the system "doesn't have time" to adjust to the change. This leaves the system in the same state it was before the change, and is the essence of the so-called "sudden approximation."

Of course, even though it may have been in an eigenstate beforehand, there is no reason to believe that this is an eigenstate of the transformed Hamiltonian. Therein lie opportunities for interesting physics. One classic example is the calculation of the population of electronic final states in the $^3\text{He}^+$ ion following beta decay of the tritium atom.[6] See Problem 5.42 at the end of this chapter.

Let us consider a more precise statement of the sudden approximation, and work through some of the consequences. Rewrite the Schrödinger equation for the time-evolution operator (2.25) as

$$i\frac{\partial}{\partial s}\mathscr{U}(t,t_0) = \frac{H}{\hbar/T}\mathscr{U}(t,t_0) = \frac{H}{\hbar\Omega}\mathscr{U}(t,t_0), \tag{5.212}$$

where we have written time $t = sT$ in terms of a dimensionless parameter $s$ and a time scale $T$, and defined $\Omega \equiv 1/T$. In the sudden approximation, the time scale $T \to 0$ which means that $\hbar\Omega$ will be much larger than the energy scale represented by $H$. Assuming we can redefine $H$ by adding or subtracting an arbitrary constant, introducing some overall phase factor in the state vectors, we see that

$$\mathscr{U}(t,t_0) \to 1 \qquad \text{as} \qquad T \to 0. \tag{5.213}$$

This proves the validity of the sudden approximation. It should be appropriate if $T$ is small compared to times $2\pi/\omega_{ab}$ where $E_{ab} = \hbar\omega_{ab}$ is the difference between two relevant eigenvalues of the Hamiltonian $H$.

## 5.6.2 Adiabatic Approximation

We tend to take the adiabatic approximation for granted. Given a Hamiltonian which depends on some set of parameters, we will find energy eigenvalues which depend on the values of those parameters. If the parameters vary "slowly" with time, then the energy eigenvalues should just follow the values one gets as the parameters themselves change. The key is what we mean by "slowly." Quantum mechanically or otherwise, presumably we mean that the parameters change on a time scale $T$ that is much larger than $2\pi/\omega_{ab} = 2\pi\hbar/E_{ab}$ for some difference $E_{ab}$ in energy eigenvalues.

---

[6] This has important implications for modern experiments which try to infer a nonzero neutrino mass from beta decay measurements. The Karlsruhe Tritium Neutrino Experiment (KATRIN), for example, has just published first results at the time of this writing. See Aker et al., *Phys. Rev. Lett.*, **123** (2019) 221802.

An obvious classical example is a pendulum which is transported around near the surface of the Earth. The pendulum will behave normally as you climb a mountain, with only the period slowly lengthening as the force of gravity decreases, so long as the time over which the height is changed is long compared to pendulum period. If one slowly changes the electric field which permeates a hydrogen atom, the energy levels will change in pace according to the Stark effect calculation in Section 5.2.

Let us consider the mathematics of adiabatic change from a quantum-mechanical point of view. We follow the treatment given in Griffiths (2005), and pay particular attention to the phase change as a function of time. We number the states in order using the index $n$ and assume no degeneracy[7] so there is no confusion with the ordering of states crossing as time changes. Our starting point is essentially (2.27) but we will take $t_0 = 0$ and suppress the initial time in our notation.

Begin with the eigenvalue equation using the notation

$$H(t)|n; t\rangle = E_n(t)|n; t\rangle, \tag{5.214}$$

simply noting that at any particular time $t$, the states and eigenvalues may change. If we now look for general solutions to the Schrödinger equation of the form

$$i\hbar \frac{\partial}{\partial t}|\alpha; t\rangle = H(t)|\alpha; t\rangle \tag{5.215}$$

then we can write

$$|\alpha; t\rangle = \sum_n c_n(t)e^{i\theta_n(t)}|n; t\rangle \tag{5.216}$$

where

$$\theta_n(t) \equiv -\frac{1}{\hbar}\int_0^t E_n(t')dt'. \tag{5.217}$$

The separation of the expansion coefficient into the factors $c_n(t)$ and $\exp(i\theta_n(t))$ will prove useful in a moment. Substituting (5.216) into (5.215) and using (5.214) we find

$$\sum_n e^{i\theta_n(t)}\left[\dot{c}_n(t)|n; t\rangle + c_n(t)\frac{\partial}{\partial t}|n; t\rangle\right] = 0. \tag{5.218}$$

Now, taking the inner product with $\langle m; t|$ and invoking orthonormality of the eigenstates at equal times, we arrive at a differential equation for the $c_n(t)$, namely

$$\dot{c}_m(t) = -\sum_n c_n(t)e^{i[\theta_n(t)-\theta_m(t)]}\langle m; t|\left[\frac{\partial}{\partial t}|n; t\rangle\right]. \tag{5.219}$$

The inner product $\langle m; t|(\partial/\partial t)|n; t\rangle$ is a new feature. If $H$ were not time dependent, then the $|n; t\rangle$ would be stationary states, and the usual exponential time dependence would emerge. In order to treat this in the general case, we can go back to (5.214) and take the time derivative of both sides. For the case where $m \neq n$ we find

$$\langle m; t|\dot{H}|n; t\rangle = [E_n(t) - E_m(t)]\langle m; t|\left[\frac{\partial}{\partial t}|n; t\rangle\right]. \tag{5.220}$$

---

[7] This is not a significant constraint. If the degeneracy is broken by $H(t)$ after some time, we can just "start" there. If the degeneracy is never broken by $H(t)$ then it is irrelevant.

This finally allows us to rewrite (5.219) as

$$\dot{c}_m(t) = -c_m(t)\langle m; t| \left[\frac{\partial}{\partial t}|m; t\rangle\right] - \sum_n c_n(t) e^{i(\theta_n - \theta_m)} \frac{\langle m; t|\dot{H}|n; t\rangle}{E_n - E_m} \tag{5.221}$$

which is a formal solution to the general time-dependent problem. Equation (5.221) demonstrates that as time goes on, states with $n \neq m$ will mix with $|m; t\rangle$ due to the time dependence of the Hamiltonian $H$, by virtue of the second term.

Now we can apply the adiabatic approximation, which amounts to neglecting the second term in (5.221). Roughly, this means that

$$\frac{|\langle m; t|\dot{H}|n; t\rangle|}{E_{nm}} \equiv \frac{1}{\tau} \ll \langle m; t| \left[\frac{\partial}{\partial t}|m; t\rangle\right] \sim \frac{E_m}{\hbar}. \tag{5.222}$$

In other words, the time scale $\tau$ for changes in the Hamiltonian must be very large compared to the inverse natural frequency of the state phase factor. That is, just as for the pendulum being carried around the Earth, the Hamiltonian changes much more slowly than the oscillation frequency of the system. Consequently, we have

$$c_n(t) = e^{i\gamma_n(t)} c_n(0) \tag{5.223}$$

where

$$\gamma_n(t) \equiv i \int_0^t \langle n; t'| \left[\frac{\partial}{\partial t'}|n; t'\rangle\right] dt'. \tag{5.224}$$

Note that by this definition, $\gamma_n(t)$ is real, since

$$0 = \frac{\partial}{\partial t}\langle n; t|n; t\rangle = \left[\frac{\partial}{\partial t}\langle n; t|\right]|n; t\rangle + \langle n; t|\left[\frac{\partial}{\partial t}|n; t\rangle\right] \tag{5.225}$$

or, in other words,

$$\left(\langle n; t|\left[\frac{\partial}{\partial t}|n; t\rangle\right]\right)^* = -\langle n; t|\left[\frac{\partial}{\partial t}|n; t\rangle\right] \tag{5.226}$$

in which case the integrand in (5.224) is purely imaginary.

Therefore, in the adiabatic approximation, if a system starts out in an eigenstate $|n\rangle$ of $H(0)$, then it remains in the eigenstate $|n; t\rangle$ of $H(t)$, since $c_i(0) = 0$ unless $i = n$ in which case $c_n(0) = 1$. Using (5.216) with (5.223) we have, in an obvious notation,

$$|\alpha^{(n)}; t\rangle = e^{i\gamma_n(t)} e^{i\theta_n(t)}|n; t\rangle. \tag{5.227}$$

It would appear that (5.227) is difficult to use, since the definition (5.224) assumes the time dependence of the state is given, but we will find ways to make good use of this result. In any case, it is easy to see that the result is self consistent. We know that for the case when $H$ is not time dependent, we expect

$$|n; t\rangle = e^{-iE_n t/\hbar}|n\rangle \tag{5.228}$$

and so

$$\langle n; t|\left[\frac{\partial}{\partial t}|n; t\rangle\right] = -i\frac{E_n}{\hbar} \tag{5.229}$$

which, by (5.224) gives $\gamma_n(t) = +E_n t/\hbar$. On the other hand, (5.217) says that $\theta_n(t) = -E_n t/\hbar$. So, the two exponential factors in (5.227) cancel each other, and we find

$$|\alpha^{(n)}; t\rangle = |n; t\rangle \qquad \text{for} \qquad H \neq H(t) \tag{5.230}$$

as we should expect.

The addition of this new phase $\gamma_n(t)$ is the only result of the adiabatic approximation that is less than obvious. It was not considered worth pursuing for many years, until it was discovered that it is in fact measurable. Indeed, it turns out to be the quantum-mechanical manifestation of very many physical phenomena that involve systems that are cyclic in time.

### 5.6.3 Berry's Phase

Excitement in the implications of (5.224) grew dramatically with the publication of "Quantal phase factors accompanying adiabatic changes", by M. V. Berry (*Proc. R. Soc. London, Ser. A*, **392** (1984) 45). Indeed, the accumulated phase for systems which travel in a closed loop is generally called "Berry's phase," although Berry himself refers to this as a "geometric phase."

Berry's paper is widely cited, and the interested reader will have no difficulty finding many references. One particular paper that provides a succinct summary and interesting implications is "The adiabatic theorem and Berry's phase" by Holstein (*Am. J. Phys.*, **57** (1989) 1079). Berry, in fact, gives a lovely history of work prior to his own publication, in "Anticipations of the geometric phase" (*Phys. Today*, **43** (1990) 34).

Assume that the time dependence of the Hamiltonian is represented by a "vector of parameters" $\mathbf{R}(t)$. That is, there exists some space in which the components of a vector $\mathbf{R}(t)$ specify the Hamiltonian, and change as a function of time. (In an example below, $\mathbf{R}(t)$ will be the magnetic field.) Therefore, we have $E_n(t) = E_n(\mathbf{R}(t))$ and $|n; t\rangle = |n(\mathbf{R}(t))\rangle$, and also

$$\langle n; t| \left[ \frac{\partial}{\partial t} |n; t\rangle \right] = \langle n; t| \left[ \nabla_{\mathbf{R}} |n; t\rangle \right] \cdot \frac{d\mathbf{R}}{dt} \tag{5.231}$$

where $\nabla_{\mathbf{R}}$ is simply a gradient operator in the space and direction of $\mathbf{R}$. The geometric phase (5.224) then becomes

$$\gamma_n(T) = i \int_0^T \langle n; t| \left[ \nabla_{\mathbf{R}} |n; t\rangle \right] \cdot \frac{d\mathbf{R}}{dt} dt$$

$$= i \int_{\mathbf{R}(0)}^{\mathbf{R}(T)} \langle n; t| \left[ \nabla_{\mathbf{R}} |n; t\rangle \right] \cdot d\mathbf{R}. \tag{5.232}$$

In the case where $T$ represents the period for one full cycle, so that $\mathbf{R}(T) = \mathbf{R}(0)$ where the vector $\mathbf{R}$ traces a curve $C$, we have

$$\gamma_n(C) = i \oint \langle n; t| \left[ \nabla_{\mathbf{R}} |n; t\rangle \right] \cdot d\mathbf{R}. \tag{5.233}$$

With a notation that shows a bias of how we can proceed, define

$$\mathbf{A}_n(\mathbf{R}) \equiv i\langle n; t| \left[\nabla_{\mathbf{R}}|n; t\rangle\right] \tag{5.234}$$

in which case

$$\gamma_n(C) = \oint_C \mathbf{A}_n(\mathbf{R}) \cdot d\mathbf{R} = \int [\nabla_{\mathbf{R}} \times \mathbf{A}_n(\mathbf{R})] \cdot d\mathbf{a} \tag{5.235}$$

using Stokes's theorem, generalized[8] for the dimensionality of $\mathbf{R}$. (The measure $d\mathbf{a}$ is a small area element on some surface bounded by the closed path.) Thus, Berry's phase is determined by the "flux" of a generalized field

$$\mathbf{B}_n(\mathbf{R}) \equiv \nabla_{\mathbf{R}} \times \mathbf{A}_n(\mathbf{R}) \tag{5.236}$$

through a surface $\mathbf{S}$ bounded by the circuit followed by $\mathbf{R}(t)$ over one complete cycle. One obtains the same phase $\gamma_n$ so long as one encounters the same total flux, regardless of the actual path followed by $\mathbf{R}(t)$. Note that, quite similarly to our derivation of the result (5.226), both $\mathbf{A}_n(\mathbf{R})$ and $\mathbf{B}_n(\mathbf{R})$ are purely real quantities. Soon we will be concerning ourselves with sources of the field $\mathbf{B}_n(\mathbf{R})$.

Equation (5.235) has a remarkable property that betrays the notation we have chosen using $\mathbf{A}_n(\mathbf{R})$. Suppose that we multiply $|n; t\rangle$ by an arbitrary phase factor that changes through $\mathbf{R}$-space. That is

$$|n; t\rangle \longrightarrow e^{i\delta(\mathbf{R})}|n; t\rangle. \tag{5.237}$$

Then by (5.234) we have

$$\mathbf{A}_n(\mathbf{R}) \longrightarrow \mathbf{A}_n(\mathbf{R}) - \nabla_{\mathbf{R}}\delta(\mathbf{R}) \tag{5.238}$$

which leaves (5.235) unchanged. In other words, the value of $\gamma_n(C)$ does not depend on the details of the phase behavior along the path, despite our starting point (5.227). Indeed, $\gamma_n(C)$ depends only on the geometry of the path traced out by $\mathbf{R}(t)$, hence the name "geometric" phase. Of course, it remains for us to show that $\gamma_n(C)$ is nonzero, at least under certain conditions. Note also, that (5.237) and (5.238) have exactly the same form as that for gauge transformations in electromagnetism. See (2.356) and (2.369). This analogy will be exploited more fully before we conclude this section.

We now turn to an evaluation of $\gamma_n(C)$. Noting first that since the curl of a gradient vanishes, we can combine (5.234) and (5.236) to get

$$\mathbf{B}_n(\mathbf{R}) = i \left[\nabla_{\mathbf{R}}\langle n; t|\right] \times \left[\nabla_{\mathbf{R}}|n; t\rangle\right]. \tag{5.239}$$

Now insert a complete set of states $|m; t\rangle$ to find

$$\mathbf{B}_n(\mathbf{R}) = i \sum_{m \neq n} \left[\nabla_{\mathbf{R}}\langle n; t|\right] |m; t\rangle \times \langle m; t| \left[\nabla_{\mathbf{R}}|n; t\rangle\right]. \tag{5.240}$$

---

[8] To be sure, generalizing Stokes's theorem for higher dimensionality, is not trivial. See a discussion of this in Berry's original paper. In our case, however, all of our examples will involve only three-dimensional parameter vectors $\mathbf{R}$.

We explicitly discard the term with $m = n$, but it is easily shown to be zero, since $\langle n; t | n; t \rangle = 1$ implies that $[\nabla_{\mathbf{R}} \langle n; t|] | n; t \rangle = -\langle n; t | [\nabla_{\mathbf{R}} | n; t \rangle]$ and so the cross product in (5.240) must be zero. Now, by taking the $\mathbf{R}$-gradient of (5.214) and taking the inner product with $\langle m; t |$ we determine

$$\langle m; t | [\nabla_{\mathbf{R}} | n; t \rangle] = \frac{\langle m; t | [\nabla_{\mathbf{R}} H] | n; t \rangle}{E_n - E_m} \qquad m \neq n. \tag{5.241}$$

This allows us to write, finally,

$$\gamma_n(C) = \int \mathbf{B}_n(\mathbf{R}) \cdot d\mathbf{a} \tag{5.242}$$

where

$$\mathbf{B}_n(\mathbf{R}) = i \sum_{m \neq n} \frac{\langle n; t | [\nabla_{\mathbf{R}} H] | m; t \rangle \times \langle m; t | [\nabla_{\mathbf{R}} H] | n; t \rangle}{(E_m - E_n)^2}. \tag{5.243}$$

As Berry states in his original paper, these last two equations "embody the central results" of his work. Points in $\mathbf{R}$-space where $E_m(\mathbf{R}) = E_n(\mathbf{R})$ will contribute to the surface integral (5.242) even though the path enclosing that surface does not include those points.

It was realized early on that Berry's phase could be observed using photons moving through a twisted optical fiber, and that the geometric character of this phase could be tested experimentally. See Tomita and Chiao, *Phys. Rev. Lett.*, **57** (1986) 937. Indeed, this experiment can be carried out in the student laboratory. A description of the setup can be found in Melissinos and Napolitano (2003).

The next section carries through an example strongly connected to concepts already developed in this textbook, and compared to measurement. However, there are many examples that can be experimentally verified. One is the precise demonstration of Berry's phase in chemical kinetics, described in Yuan et al., *Science*, **362** (2018) 1289.

## 5.6.4   Example: Berry's Phase for Spin $\frac{1}{2}$

Let us now turn to a specific example, and carry through a calculation of $\gamma_n(C)$ from (5.242). We will study the phase motion for a spin $\frac{1}{2}$ particle manipulated slowly through a time varying magnetic field. This particular example has in fact been studied experimentally.

We return to (2.49), our familiar Hamiltonian for a spin $\frac{1}{2}$ particle in a magnetic field, but with some modification for a particle with arbitrary magnetic moment. Since in this case, it is the magnetic field which changes slowly in time, let the magnetic field be given by the three-dimensional vector[9] $\mathbf{R}(t)$. That is, $\mathbf{R}(t)$ is the vector of parameters that we will change slowly. For a magnetic moment $\mu$, our Hamiltonian is written as

$$H(t) = H(\mathbf{R}(t)) = -\frac{2\mu}{\hbar} \mathbf{S} \cdot \mathbf{R}(t) \tag{5.244}$$

where $\mathbf{S}$ is the spin $\frac{1}{2}$ angular-momentum operator. Written in this way, the expectation value for the magnetic moment in the spin-up state is simply $\mu$.

---

[9] We do not use $\mathbf{B}$ to represent the magnetic field, so that we avoid confusions with (5.243).

Now on to the evaluation of $\mathbf{B}(\mathbf{R})$ using (5.243). First, it is simple enough to show, either explicitly (see Problem 3.3 in Chapter 3) or using rotational symmetry to fix $\mathbf{R}$ in the $\hat{\mathbf{z}}$-direction, that the two energy eigenvalues for (5.244) are

$$E_\pm(t) = \mp \mu R(t) \tag{5.245}$$

where $R(t)$ is the magnitude of the magnetic field vector, and the spin-up(down) eigenstates (with respect to the direction of $\mathbf{R}(t)$) are $|\pm; t\rangle$. The summation in (5.243) consists of only one term, with denominator

$$(E_\pm - E_\mp)^2 = 4\mu^2 R^2. \tag{5.246}$$

It is also clear that

$$\nabla_{\mathbf{R}} H = -\frac{2\mu}{\hbar} \mathbf{S} \tag{5.247}$$

leaving us with the need to evaluate the cross product

$$\langle \pm; t|\mathbf{S}|\mp; t\rangle \times \langle \mp; t|\mathbf{S}|\pm; t\rangle = \langle \pm; t|\mathbf{S}|\mp; t\rangle \times \langle \pm; t|\mathbf{S}|\mp; t\rangle^*. \tag{5.248}$$

Evaluating this matrix element would be tedious, except that we can invoke rotational symmetry and define the components of $\mathbf{S}$ relative to the direction of $\mathbf{R}$. That is, $|\pm; t\rangle$ can be taken to be eigenstates of $S_z$. So, using (3.157) to write

$$\mathbf{S} = \frac{1}{2}(S_+ + S_-)\hat{\mathbf{x}} + \frac{1}{2i}(S_+ - S_-)\hat{\mathbf{y}} + S_z\hat{\mathbf{z}}, \tag{5.249}$$

we invoke (3.191) and (3.192) to find

$$\langle \pm; t|\mathbf{S}|\mp; t\rangle = \frac{\hbar}{2}(\hat{\mathbf{x}} \mp i\hat{\mathbf{y}}). \tag{5.250}$$

Combining (5.246), (5.248), and (5.250), we have

$$\mathbf{B}_\pm(\mathbf{R}) = \mp\frac{1}{2R^2(t)}\hat{\mathbf{z}}. \tag{5.251}$$

Of course, this result was derived by taking $|\pm; t\rangle$ to be eigenstates of $S_z$, when in fact they are in the direction of $\mathbf{R}$. Therefore, we actually have

$$\mathbf{B}_\pm(\mathbf{R}) = \mp\frac{1}{2R^2(t)}\hat{\mathbf{R}}. \tag{5.252}$$

Finally, we calculate Berry's phase (5.242) to be

$$\gamma_\pm(C) = \mp\frac{1}{2}\int \frac{\hat{\mathbf{R}} \cdot d\mathbf{a}}{R^2} = \mp\frac{1}{2}\Omega \tag{5.253}$$

where $\Omega$ is the "solid angle" subtended by the path through which the parameter vector $\mathbf{R}(t)$ travels, relative to an origin $\mathbf{R} = 0$ which is the source point for the field $\mathbf{B}$. This emphasizes the "geometric" character of Berry's phase. Specifics of the path do not matter, so long as the solid angle subtended by the path is the same. The result is also independent of the magnetic moment $\mu$.

Soon after Berry's prediction for this effect in spin $\frac{1}{2}$ systems, two groups carried out measurements using neutrons at the Institut Laue-Langevin in Grenoble, France. One of

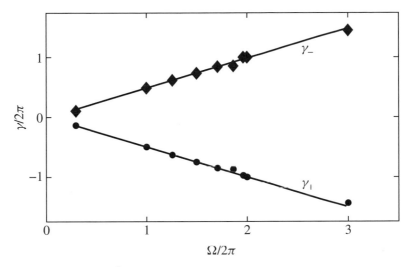

**Fig. 5.7** Observation of Berry's phase for spin $\frac{1}{2}$ particles using ultra cold neutrons, from Richardson et al., *Phys. Rev. Lett.*, **61** (1988) 2030. Data are taken from Table 1 of their paper, and show Berry's phase as a function of "solid angle" for the rotating magnetic field. Both spin-up and spin-down phases were measured. Uncertainties on the data points are about the same size as, or smaller than, the points themselves. The solid lines are taken from (5.253).

these, Bitter and Dubbers, *Phys. Rev. Lett.*, **59** (1987) 251, used a slow (500 m/sec) neutron beam passing through a twisted magnetic field. The second, Richardson et al., *Phys. Rev. Lett.*, **61** (1988) 2030, made use of ultra cold neutrons (UCN) and is more precise. UCN can be stored for long periods of time, so a cycle period $T = 7.387$ sec was used, ensuring the validity of the adiabatic theorem. Initially polarized in the **z**-direction, the neutrons are subjected to a rotating magnetic field component that is switched on at $t = 0$ and switched off at $t = T$. The magnetic field vector traces out a circle (or ellipse, depending on adjustable parameters) in the $yz$-plane, depolarizing the neutrons by an amount depending on the integrated phase. Measuring the final polarization determines the integrated phase, and the dynamical phase (5.217) is subtracted off, leaving Berry's phase.

Figure 5.7 shows the results obtained by Richardson et al. Both spin-up and spin-down phases are measured, and both agree extremely well with Berry's analysis. Even though the value of the magnetic moment does not enter the calculation, its sign determines the direction of rotation, and this experiment confirms that the neutron magnetic moment is indeed negative.

### 5.6.5 Aharonov–Bohm and Magnetic Monopoles Revisited

We have seen that Berry's phase uses a formalism that is closely tied to the formalism of gauge transformations. See (5.237) and (5.238). Let us now makes this connection closer to some physics we have already seen in our study of gauge transformation in Section 2.7.

First we can show that the **Aharonov–Bohm** effect due to the magnetic field can be shown to be just a consequence of a geometrical phase factor. Let a small box confining an

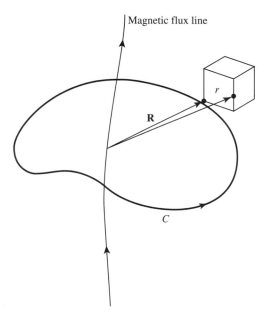

Magnetic flux line

**Fig. 5.8** The Aharonov–Bohm effect as a manifestation of Berry's phase. An electron in a box takes one turn around a magnetic flux line.

electron (charge $e < 0$) make one turn along a closed loop $C$, which surrounds a magnetic flux line $\Phi_B$, as shown in Figure 5.8. Let $\mathbf{R}$ be the vector connecting the origin fixed in the space and a reference point in the box. In this case the vector $\mathbf{R}$ is an external parameter in the real space itself. By using the vector-potential $\mathbf{A}$ to describe the magnetic field $\mathbf{B}$, the $n$th wave function of the electron in the box (with position vector $\mathbf{r}$) is written as

$$\langle \mathbf{r}|n(\mathbf{R})\rangle = \exp\left\{\frac{ie}{\hbar c}\int_{\mathbf{R}}^{\mathbf{r}} \mathbf{A}(\mathbf{r}') \cdot d\mathbf{r}'\right\} \psi_n(\mathbf{r} - \mathbf{R}) \qquad (5.254)$$

where $\psi_n(\mathbf{r}')$ is the wave function of the electron at the $\mathbf{r}'$ position coordinates of the box in the absence of magnetic field.

Now let $\mathbf{R}$ travel around the loop $C$ and calculate Berry's phase. We can easily calculate the derivative of the wave function with respect to the external parameter to obtain

$$\langle n(\mathbf{R}) \,|\, [\nabla_{\mathbf{R}}|n(\mathbf{R})\rangle] = \int d^3x\,\psi_n^*(\mathbf{r} - \mathbf{R})$$

$$\times \left\{-\frac{ie}{\hbar c}A(\mathbf{R})\psi_n(\mathbf{r} - \mathbf{R}) + \nabla_{\mathbf{R}}\psi_n(\mathbf{r} - \mathbf{R})\right\} = -\frac{ieA(\mathbf{R})}{\hbar c}. \quad (5.255)$$

The second term under the integral vanishes for the electron in the box. From (5.232) and (5.255) we see that the geometrical phase is given by

$$\gamma_n(C) = \frac{e}{\hbar c}\oint_C \mathbf{A} \cdot d\mathbf{R} = \frac{e}{\hbar c}\iint_{S(C)} \mathbf{B}(\mathbf{R}) \cdot d\mathbf{S} = \frac{e}{\hbar c}\Phi_B. \qquad (5.256)$$

This result is just the expression (2.390) of the Aharonov–Bohm effect obtained earlier in Section 2.7.

A second example[10] of the physical connection between gauge invariance and Berry's phase, is the Dirac quantization condition (2.405) for **magnetic monopoles**. Consider two surfaces $\mathbf{a}_1$ and $\mathbf{a}_2$ in $\mathbf{R}$-space, each bounded by the same curve $C$. Since the Berry's phase that results from following $C$ is physically measurable, the surface integral (5.242) must be the same for $\mathbf{a}_1$ and $\mathbf{a}_2$ to within a multiple of $2\pi$. That is

$$\int_{\mathbf{a}_1} \mathbf{B} \cdot d\mathbf{a} = \int_{\mathbf{a}_2} \mathbf{B} \cdot d\mathbf{a} + 2N\pi \qquad N = 0, \pm 1, \pm 2, \ldots. \tag{5.257}$$

Now construct a closed surface by putting $\mathbf{a}_1$ "above" $C$ and $\mathbf{a}_2$ below. Assuming that we use something like the right-hand rule to consistently determine the orientation of $d\mathbf{a}$ for each of the two integrals in (5.257), then $d\mathbf{a}$ points inward for one of them. So, reverse the sign of $d\mathbf{a}$ for that integral, and rewrite (5.257) as

$$\oint \mathbf{B} \cdot d\mathbf{a} = 2N\pi. \tag{5.258}$$

Of course, inserting (5.252) into (5.258) yields $\pm 2\pi$, i.e. $N = \pm 1$, but that is an artifact of our choosing a spin $\frac{1}{2}$ system for deriving (5.252). If we carried through that example for a system with arbitrary spin, then $n$ works out to be equal to twice the spin projection quantum number.

Now (5.252) looks suspiciously like the magnetic field from a monopole with charge 1/2, but recall that it is in fact a Berry field. How can we relate this to an actual magnetic field for a monopole? It is easiest to see this when we express Berry's phase in terms of the line integral in (5.235) of the vector potential around the curve $C$. The gauge transformation (2.403) gives the form of the vector potential, and the line integral is just an integral over $\phi$ leading to a factor of $2\pi$. We can then use this to evaluate the left side of (5.258), including the interaction with an single electric charge $e$ to complete the phase, leading to the factor $e/\hbar c$ as in (2.404). Thus, (5.258) becomes

$$\frac{e}{\hbar c}(2e_M)2\pi = 2N\pi$$

or

$$\frac{2ee_M}{\hbar c} = N \tag{5.259}$$

which is the same result (2.405) as we obtained earlier in Section 2.7.

## 5.7 Time-Dependent Perturbation Theory

### 5.7.1 Dyson Series

With the exception of a few problems like the two-level time-dependent problem of the previous section, exact solutions to the differential equation for $c_n(t)$ are usually

---

[10] Our discussion here closely follows that given by Holstein, *Am. J. Phys.*, **57** (1989) 1079.

not available. We must be content with approximate solutions to (5.189) obtained by perturbation expansion:

$$c_n(t) = c_n^{(0)} + c_n^{(1)} + c_n^{(2)} + \cdots, \tag{5.260}$$

where $c_n^{(1)}, c_n^{(2)}, \ldots$ signify amplitudes of first order, second order, and so on in the strength parameter of the time-dependent potential. The iteration method used to solve this problem is similar to what we did in time-independent perturbation theory. If initially only the state $i$ is populated, we approximate $c_n$ on the right-hand side of differential equation (5.189) by $c_n^{(0)} = \delta_{ni}$ (independent of $t$) and relate it to the time derivative of $c_n^{(1)}$, integrate the differential equation to obtain $c_n^{(1)}$, plug $c_n^{(1)}$ into the right-hand side [of (5.189)] again to obtain the differential equation for $c_n^{(2)}$, and so on. This is how Dirac developed time-dependent perturbation theory in 1927.

Instead of working with $c_n(t)$, we propose to look at the time-evolution operator $U_I(t, t_0)$ in the interaction picture, which we will define later. We obtain a perturbation expansion for $U_I(t, t_0)$, and at the very end we relate the matrix elements of $U_I$ to $c_n(t)$. If we are interested only in solving simple problems in nonrelativistic quantum mechanics, all this might look superfluous; however, the operator formalism we develop is very powerful because it can immediately be applied to more advanced problems, such as relativistic quantum field theory and many-body theory.

The time-evolution operator in the interaction picture is defined by

$$|\alpha, t_0; t\rangle_I = U_I(t, t_0)|\alpha, t_0; t_0\rangle_I. \tag{5.261}$$

Differential equation (5.183) for the state ket of the interaction picture is equivalent to

$$i\hbar \frac{d}{dt} U_I(t, t_0) = V_I(t) U_I(t, t_0). \tag{5.262}$$

We must solve this operator differential equation subject to the initial condition

$$U_I(t, t_0)|_{t=t_0} = 1. \tag{5.263}$$

First, let us note that the differential equation together with the initial condition is equivalent to the following integral equation:

$$U_I(t, t_0) = 1 - \frac{i}{\hbar} \int_{t_0}^{t} V_I(t') U_I(t', t_0) dt'. \tag{5.264}$$

We can obtain an approximate solution to this equation by iteration:

$$U_I(t, t_0) = 1 - \frac{i}{\hbar} \int_{t_0}^{t} V_I(t') \left[ 1 - \frac{i}{\hbar} \int_{t_0}^{t'} V_I(t'') U_I(t'', t_0) dt'' \right] dt'$$

$$= 1 - \frac{i}{\hbar} \int_{t_0}^{t} dt' V_I(t') + \left( \frac{-i}{\hbar} \right)^2 \int_{t_0}^{t} dt' \int_{t_0}^{t'} dt'' V_I(t') V_I(t'')$$

$$+ \cdots + \left(\frac{-i}{\hbar}\right)^n \int_{t_0}^{t} dt' \int_{t_0}^{t'} dt'' \cdots$$

$$\times \int_{t_0}^{t^{(n-1)}} dt^{(n)} V_I(t') V_I(t'') \cdots V_I(t^{(n)})$$

$$+ \cdots. \tag{5.265}$$

This series is known as the **Dyson series** after Freeman J. Dyson, who applied this method to covariant quantum electrodynamics (QED).[11] Setting aside the difficult question of convergence, we can compute $U_I(t,t_0)$ to any finite order of perturbation theory.

## 5.7.2 Transition Probability

Once $U_I(t,t_0)$ is given, we can predict the time development of any state ket. For example, if the initial state at $t = 0$ is one of the energy eigenstates of $H_0$, then to obtain the initial state ket at a later time, all we need to do is multiply by $U_I(t,0)$:

$$|i, t_0 = 0; t\rangle_I = U_I(t,0)|i\rangle$$

$$= \sum_n |n\rangle \langle n|U_I(t,0)|i\rangle. \tag{5.266}$$

In fact, $\langle n|U_I(t,0)|i\rangle$ is nothing more than what we called $c_n(t)$ earlier [see (5.185)]. We will say more about this later.

We earlier introduced the time-evolution operator $U(t,t_0)$ in the Schrödinger picture (see Section 2.2). Let us now explore the connection between $U(t,t_0)$ and $U_I(t,t_0)$. We note from (2.87) and (5.177) that

$$|\alpha, t_0; t\rangle_I = e^{iH_0 t/\hbar} |\alpha, t_0; t\rangle_S$$
$$= e^{iH_0 t/\hbar} U(t,t_0) |\alpha, t_0; t_0\rangle_S$$
$$= e^{iH_0 t/\hbar} U(t,t_0) e^{-iH_0 t_0/\hbar} |\alpha, t_0; t_0\rangle_I. \tag{5.267}$$

So we have

$$U_I(t,t_0) = e^{iH_0 t/\hbar} U(t,t_0) e^{-iH_0 t_0/\hbar}. \tag{5.268}$$

Let us now look at the matrix element of $U_I(t,t_0)$ between energy eigenstates of $H_0$:

$$\langle n|U_I(t,t_0)|i\rangle = e^{i(E_n t - E_i t_0)/\hbar} \langle n|U(t,t_0)|i\rangle. \tag{5.269}$$

We recall from Section 2.2 that $\langle n|U(t,t_0)|i\rangle$ is defined to be the transition amplitude. Hence our $\langle n|U_I(t,t_0)|i\rangle$ here is not quite the same as the transition amplitude defined earlier. However, the transition *probability* defined as the square of the modulus of $\langle n|U(t,t_0)|i\rangle$ is the same as the analogous quantity in the interaction picture

$$|\langle n|U_I(t,t_0)|i\rangle|^2 = |\langle n|U(t,t_0)|i\rangle|^2. \tag{5.270}$$

---

[11] Note that in QED, the time-ordered product $(t' > t'' > \cdots)$ is introduced, and then this perturbation series can be summed into an exponential form. This exponential form immediately gives $U(t,t_0) = U(t,t_1)U(t_1,t_0)$ (Bjorken and Drell (1965), pp. 175–178).

Parenthetically, we may remark that if the matrix elements of $U_I$ are taken between initial and final states that are not energy eigenstates, for example, between $|a'\rangle$ and $|b'\rangle$ (eigenkets of $A$ and $B$, respectively), where $[H_0,A] \neq 0$ and/or $[H_0,B] \neq 0$, then we have, in general,

$$|\langle b'|U_I(t,t_0)|a'\rangle| \neq |\langle b'|U(t,t_0)|a'\rangle|,$$

as the reader may easily verify. Fortunately, in problems where the interaction picture is found to be useful, the initial and final states are usually taken to be $H_0$ eigenstates. Otherwise, all that is needed is to expand $|a'\rangle$, $|b'\rangle$, and so on in terms of the energy eigenkets of $H_0$.

Coming back to $\langle n|U_I(t,t_0)|i\rangle$, we illustrate by considering a physical situation where at $t = t_0$, the system is known to be in state $|i\rangle$. The state ket in the Schrödinger picture $|i,t_0;t\rangle_S$ is then equal to $|i\rangle$ up to a phase factor. In applying the interaction picture, it is convenient to choose the phase factor at $t = t_0$ so that

$$|i,t_0;t_0\rangle_S = e^{-iE_i t_0/\hbar}|i\rangle, \tag{5.271}$$

which means that in the interaction picture we have the simple equation

$$|i,t_0;t_0\rangle_I = |i\rangle. \tag{5.272}$$

At a later time we have

$$|i,t_0;t\rangle_I = U_I(t,t_0)|i\rangle. \tag{5.273}$$

Comparing this with the expansion

$$|i,t_0;t\rangle_I = \sum_n c_n(t)|n\rangle, \tag{5.274}$$

we see

$$c_n(t) = \langle n|U_I(t,t_0)|i\rangle. \tag{5.275}$$

We now go back to the perturbation expansion for $U_I(t,t_0)$ [see (5.265)]. We can also expand $c_n(t)$ as in (5.260), where $c_n^{(1)}$ is first order in $V_I(t)$, $c_n^{(2)}$ is second order in $V_I(t)$, and so on. Comparing the expansion of both sides of (5.275), we obtain [using (5.179)]

$$c_n^{(0)}(t) = \delta_{ni} \quad \text{(independent of } t\text{)}$$

$$c_n^{(1)}(t) = \frac{-i}{\hbar} \int_{t_0}^{t} \langle n|V_I(t')|i\rangle dt'$$

$$= \frac{-i}{\hbar} \int_{t_0}^{t} e^{i\omega_{ni}t'} V_{ni}(t') dt' \tag{5.276}$$

$$c_n^{(2)}(t) = \left(\frac{-i}{\hbar}\right)^2 \sum_m \int_{t_0}^{t} dt' \int_{t_0}^{t'} dt'' e^{i\omega_{nm}t'} V_{nm}(t') e^{i\omega_{mi}t''} V_{mi}(t''),$$

where we have used

$$e^{i(E_n - E_i)t/\hbar} = e^{i\omega_{ni}t}. \tag{5.277}$$

The transition probability for $|i\rangle \rightarrow |n\rangle$ with $n \neq i$ is obtained by

$$P(i \rightarrow n) = |c_n^{(1)}(t) + c_n^{(2)}(t) + \cdots|^2. \tag{5.278}$$

### 5.7.3 Constant Perturbation

As an application of (5.276), let us consider a constant perturbation turned on at $t = 0$:

$$V(t) = \begin{cases} 0 & \text{for} \quad t < 0 \\ V \quad \text{(independent of } t) & \text{for} \quad t \ge 0. \end{cases} \tag{5.279}$$

Even though the operator $V$ has no explicit dependence on time, it is, in general, made up of operators like $\mathbf{x}$, $\mathbf{p}$, and $\mathbf{s}$. Now suppose at $t = 0$, we have only $|i\rangle$. With $t_0$ taken to be zero, we obtain

$$c_n^{(0)} = c_n^{(0)}(0) = \delta_{ni},$$

$$c_n^{(1)} = \frac{-i}{\hbar} V_{ni} \int_0^t e^{i\omega_{ni}t'} dt' \tag{5.280}$$

$$= \frac{V_{ni}}{E_n - E_i}(1 - e^{i\omega_{ni}t}),$$

or

$$|c_n^{(1)}|^2 = \frac{|V_{ni}|^2}{|E_n - E_i|^2}(2 - 2\cos\omega_{ni}t)$$

$$= \frac{4|V_{ni}|^2}{|E_n - E_i|^2} \sin^2\left[\frac{(E_n - E_i)t}{2\hbar}\right]. \tag{5.281}$$

The probability of finding $|n\rangle$ depends not only on $|V_{ni}|^2$ but also on the energy difference $E_n - E_i$, so let us try to see how (5.281) looks as a function of $E_n$. In practice, we are interested in this way of looking at (5.281) when there are many states with $E \sim E_n$ so that we can talk about a continuum of final states with nearly the same energy. To this end we define

$$\omega \equiv \frac{E_n - E_i}{\hbar} \tag{5.282}$$

and plot $4\sin^2(\omega t/2)/\omega^2$ as a function of $\omega$ for fixed $t$, the time interval during which the perturbation has been on; see Figure 5.9. We see that the height of the middle peak, centered around $\omega = 0$, is $t^2$ and the width is proportional to $1/t$. As $t$ becomes large, $|c_n^{(1)}(t)|^2$ is appreciable only for those final states that satisfy

$$t \sim \frac{2\pi}{|\omega|} = \frac{2\pi\hbar}{|E_n - E_i|}. \tag{5.283}$$

If we call $\Delta t$ the time interval during which the perturbation has been turned on, a transition with appreciable probability is possible only if

$$\Delta t \Delta E \sim \hbar, \tag{5.284}$$

where by $\Delta E$ we mean the energy change involved in a transition with appreciable probability. If $\Delta t$ is small, we have a broader peak in Figure 5.9, and as a result we can tolerate a fair amount of energy *non*conservation. On the other hand, if the perturbation has been on for a very long time, we have a very narrow peak, and approximate energy

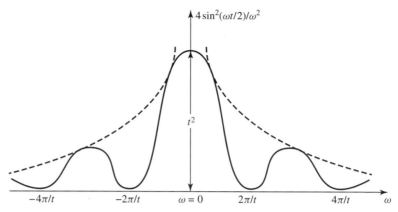

**Fig. 5.9**   Plot of $4\sin^2(\omega t/2)/\omega^2$ versus $\omega$ for a fixed $t$, where in $\omega = (E_n - E_i)/\hbar$ we have regarded $E_n$ as a continuous variable.

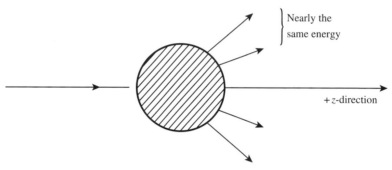

**Fig. 5.10**   Elastic scattering of plane wave by some finite range potential.

conservation is required for a transition with appreciable probability. Note that this "uncertainty relation" is fundamentally different from the $x - p$ uncertainty relation of Section 1.6. There $x$ and $p$ are both observables. In contrast, time in nonrelativistic quantum mechanics is a parameter, not an observable.

For those transitions with exact energy conservation $E_n = E_i$, we have

$$|c_n^{(1)}(t)|^2 = \frac{1}{\hbar^2}|V_{ni}|^2 t^2. \tag{5.285}$$

The probability of finding $|n\rangle$ after a time interval $t$ is quadratic, *not* linear, in the time interval during which $V$ has been on. This may appear intuitively unreasonable. There is no cause for alarm, however. In a realistic situation where our formalism is applicable, there is usually a group of final states, all with nearly the same energy as the energy of the initial state $|i\rangle$. In other words, a final state forms a continuous energy spectrum in the neighborhood of $E_i$. We give two examples along this line. Consider for instance, elastic scattering by some finite range potential (see Figure 5.10), which we will consider in detail in Chapter 6. The initial state is taken to be a plane wave state with its propagation direction oriented in the positive $z$-direction; the final state may also be a plane wave state of the same energy but with its propagation direction, in general, in a direction other than the positive

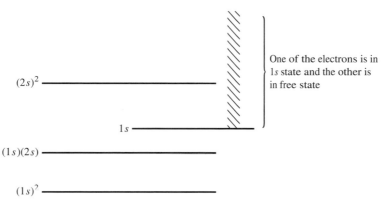

$(2s)^2$ ——————————————

One of the electrons is in $1s$ state and the other is in free state

$1s$ ————————

$(1s)(2s)$ ——————————————

$(1s)^2$ ——————————————

**Fig. 5.11**  Schematic diagram of two electron energy levels of a helium atom.

$z$-direction. Another example of interest is the de-excitation of an excited atomic state via the emission of an Auger electron. The simplest example is a helium atom. The initial state may be $(2s)^2$, where both the electrons are excited; the final state may be $(1s)$ (that is, one of the electrons is still bound) of the $He^+$ ion, while the second electron escapes with a positive energy $E$; see Figure 5.11. In such a case we are interested in the total probability, that is, the transition probabilities summed over final states with $E_n \simeq E_i$:

$$\sum_{n,E_n \simeq E_i} |c_n^{(1)}|^2. \tag{5.286}$$

It is customary to define the density of final states as the number of states within energy interval $(E, E+dE)$ as

$$\rho(E)\, dE. \tag{5.287}$$

We can then write (5.286) as

$$\sum_{n,E_n \simeq E_i} |c_n^{(1)}|^2 \Rightarrow \int dE_n \rho(E_n) |c_n^{(1)}|^2$$

$$= 4 \int \sin^2\left[\frac{(E_n - E_i)t}{2\hbar}\right] \frac{|V_{ni}|^2}{|E_n - E_i|^2} \rho(E_n)\, dE_n. \tag{5.288}$$

As $t \to \infty$, we can take advantage of

$$\lim_{t \to \infty} \frac{1}{|E_n - E_i|^2} \sin^2\left[\frac{(E_n - E_i)t}{2\hbar}\right] = \frac{\pi t}{2\hbar} \delta(E_n - E_i), \tag{5.289}$$

which follows from

$$\lim_{\alpha \to \infty} \frac{1}{\pi} \frac{\sin^2 \alpha x}{\alpha x^2} = \delta(x). \tag{5.290}$$

It is now possible to take the average of $|V_{ni}|^2$ outside the integral sign and perform the integration with the $\delta$-function:

$$\lim_{t \to \infty} \int dE_n \rho(E_n) |c_n^{(1)}(t)|^2 = \left(\frac{2\pi}{\hbar}\right) \overline{|V_{ni}|^2} \rho(E_n) t \bigg|_{E_n \simeq E_i}. \tag{5.291}$$

Thus the total transition probability *is* proportional to $t$ for large values of $t$, which is quite reasonable. Notice that this linearity in $t$ is a consequence of the fact that the total transition probability is proportional to the area under the peak of Figure 5.9, where the height varies as $t^2$ and the width varies as $1/t$.

It is conventional to consider the **transition rate**, that is, the transition probability per unit time. Expression (5.291) tells us that the total transition rate, defined by

$$\frac{d}{dt}\left(\sum_n |c_n^{(1)}|^2\right),\tag{5.292}$$

is constant in $t$ for large $t$. Calling (5.292) $w_{i\to[n]}$, where $[n]$ stands for a group of final states with energy similar to $i$, we obtain

$$w_{i\to[n]} = \frac{2\pi}{\hbar}\overline{|V_{ni}|^2}\rho(E_n)_{E_n\simeq E_i}\tag{5.293}$$

independent of $t$, provided the first-order time-dependent perturbation theory is valid. This formula is of great practical importance; it is called **Fermi's golden rule** even though the basic formalism of $t$-dependent perturbation theory is due to Dirac.[12] We sometimes write (5.293) as

$$w_{i\to n} = \left(\frac{2\pi}{\hbar}\right)|V_{ni}|^2\delta(E_n - E_i),\tag{5.294}$$

where it must be understood that this expression is integrated with $\int dE_n\rho(E_n)$.

We should also understand what is meant by $\overline{|V_{ni}|^2}$. If the final states $|n\rangle$ form a quasi-continuum, the matrix elements $V_{ni}$ are often similar if $|n\rangle$ are similar. However, it may happen that all energy eigenstates with the same $E_n$ do not necessarily have similar matrix elements. Consider, for example, elastic scattering. The $|V_{ni}|^2$ that determines the scattering cross section may depend on the final momentum direction. In such a case the group of final states we should consider must not only have approximately the same energy, but they must also have approximately the same momentum direction. This point becomes clearer when we discuss the photoelectric effect.

Let us now look at the second-order term, still with the constant perturbation of (5.279). From (5.276) we have

$$c_n^{(2)} = \left(\frac{-i}{\hbar}\right)^2 \sum_m V_{nm}V_{mi}\int_0^t dt'\, e^{i\omega_{nm}t'}\int_0^{t'} dt''\, e^{i\omega_{mi}t''}$$

$$= \frac{i}{\hbar}\sum_m \frac{V_{nm}V_{mi}}{E_m - E_i}\int_0^t (e^{i\omega_{ni}t'} - e^{i\omega_{nm}t'})dt'.\tag{5.295}$$

The first term on the right-hand side has the same $t$-dependence as $c_n^{(1)}$ [see (5.280)]. If this were the only term, we could then repeat the same argument as before and conclude that as $t \to \infty$, the only important contribution arises from $E_n \simeq E_i$. Indeed, when $E_m$ differs from $E_n$ and $E_i$, the second contribution gives rise to a rapid oscillation, which does not give a contribution to the transition probability that grows with $t$.

---

[12] For an interesting discussion on the history of Fermi's golden rule, see Visser, "Whose golden rule is it anyway?", *Am. J. Phys.*, **77** (2009) 487.

With $c^{(1)}$ and $c^{(2)}$ together, we have

$$w_{i \to [n]} = \frac{2\pi}{\hbar} \left| V_{ni} + \sum_m \frac{V_{nm} V_{mi}}{E_i - E_m} \right|^2 \rho(E_n) \Bigg|_{E_n \simeq E_i}. \tag{5.296}$$

The formula has the following physical interpretation. We visualize that the transition due to the second-order term takes place in two steps. First, $|i\rangle$ makes an energy nonconserving transition to $|m\rangle$; subsequently, $|m\rangle$ makes an energy nonconserving transition to $|n\rangle$, where between $|n\rangle$ and $|i\rangle$ there is overall energy conservation. Such energy nonconserving transitions are often called *virtual transitions*. Energy need not be conserved for those virtual transitions into (or from) virtual intermediate states. In contrast, the first-order term $V_{ni}$ is often said to represent a direct energy-conserving "real" transition. A special treatment is needed if $V_{nm} V_{mi} \neq 0$ with $E_m \simeq E_i$. The best way to treat this is to use the slow-turn-on method $V \to e^{\eta t} V$, which we will discuss in Section 5.9 and Problem 5.38 of this chapter. The net result is to change the energy denominator in (5.296) as follows:

$$E_i - E_m \to E_i - E_m + i\varepsilon. \tag{5.297}$$

### 5.7.4  Harmonic Perturbation

We now consider a sinusoidally varying time-dependent potential, commonly referred to as **harmonic perturbation**:

$$V(t) = \mathscr{V} e^{i\omega t} + \mathscr{V}^\dagger e^{-i\omega t}, \tag{5.298}$$

where $\mathscr{V}$ may still depend on $\mathbf{x}$, $\mathbf{p}$, $\mathbf{s}$, and so on. Actually, we have already encountered a time-dependent potential of this kind in Section 5.5 in discussing $t$-dependent two-level problems.

Again assume that only one of the eigenstates of $H_0$ is populated initially. Perturbation (5.298) is assumed to be turned on at $t = 0$, so

$$c_n^{(1)} = \frac{-i}{\hbar} \int_0^t (\mathscr{V}_{ni} e^{i\omega t'} + \mathscr{V}_{ni}^\dagger e^{-i\omega t'}) e^{i\omega_{ni} t'} dt'$$

$$= \frac{1}{\hbar} \left[ \frac{1 - e^{i(\omega + \omega_{ni})t}}{\omega + \omega_{ni}} \mathscr{V}_{ni} + \frac{1 - e^{i(\omega_{ni} - \omega)t}}{-\omega + \omega_{ni}} \mathscr{V}_{ni}^\dagger \right] \tag{5.299}$$

where $\mathscr{V}_{ni}^\dagger$ actually stands for $(\mathscr{V}^\dagger)_{ni}$. We see that this formula is similar to the constant perturbation case. The only change needed is

$$\omega_{ni} = \frac{E_n - E_i}{\hbar} \to \omega_{ni} \pm \omega. \tag{5.300}$$

So as $t \to \infty$, $|c_n^{(1)}|^2$ is appreciable only if

$$\omega_{ni} + \omega \simeq 0 \quad \text{or} \quad E_n \simeq E_i - \hbar\omega \tag{5.301a}$$

$$\omega_{ni} - \omega \simeq 0 \quad \text{or} \quad E_n \simeq E_i + \hbar\omega. \tag{5.301b}$$

Clearly, whenever the first term is important because of (5.301a), the second term is unimportant, and vice versa. We see that we have no energy-conservation condition

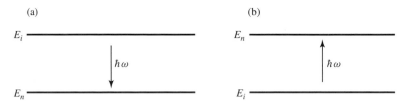

(a)

$E_i$ ——————————————

$\hbar\omega$ ↓

$E_n$ ——————————————

(b)

$E_n$ ——————————————

$\hbar\omega$ ↑

$E_i$ ——————————————

**Fig. 5.12** (a) Stimulated emission: Quantum-mechanical system gives up $\hbar\omega$ to $V$ (possible only if the initial state is excited). (b) Absorption: Quantum-mechanical system receives $\hbar\omega$ from $V$ and ends up as an excited state.

satisfied by the quantum-mechanical system alone; rather the apparent lack of energy conservation is compensated by the energy given out to, or energy taken away from, the "external" potential $V(t)$. Pictorially, we have Figure 5.12. In the first case (*stimulated emission*), the quantum-mechanical system gives up energy $\hbar\omega$ to $V$; this is clearly possible only if the initial state is excited. In the second case (*absorption*), the quantum-mechanical system receives energy $\hbar\omega$ from $V$ and ends up as an excited state. Thus a time-dependent perturbation can be regarded as an inexhaustible source or sink of energy.

In complete analogy with (5.293), we have

$$w_{i\to[n]} = \frac{2\pi}{\hbar}\, \overline{|\mathscr{V}_{ni}|^2}\rho(E_n)\Big|_{E_n\cong E_i-\hbar\omega} \tag{5.302a}$$

$$w_{i\to[n]} = \frac{2\pi}{\hbar}\, \overline{|\mathscr{V}_{ni}^\dagger|^2}\rho(E_n)\Big|_{E_n\cong E_i+\hbar\omega} \tag{5.302b}$$

or, more commonly,

$$w_{i\to n} = \frac{2\pi}{\hbar}\left\{\begin{matrix}|\mathscr{V}_{ni}|^2\\ |\mathscr{V}_{ni}^\dagger|^2\end{matrix}\right\}\delta(E_n-E_i\pm\hbar\omega). \tag{5.303}$$

Note also that

$$|\mathscr{V}_{ni}|^2 = |\mathscr{V}_{in}^\dagger|^2, \tag{5.304}$$

which is a consequence of

$$\langle i|\mathscr{V}^\dagger|n\rangle = \langle n|\mathscr{V}|i\rangle^* \tag{5.305}$$

(remember $\mathscr{V}^\dagger|n\rangle\overset{\mathrm{DC}}{\leftrightarrow}\langle n|\mathscr{V}$). Combining (5.302) and (5.304), we have

$$\frac{\text{emission rate for } i\to[n]}{\text{density of final states for } [n]} = \frac{\text{absorption rate for } n\to[i]}{\text{density of final states for } [i]}, \tag{5.306}$$

where in the absorption case we let $i$ stand for final states. Equation (5.306), which expresses symmetry between emission and absorption, is known as **detailed balancing**.

To summarize, for constant perturbation, we obtain appreciable transition probability for $|i\rangle\to|n\rangle$ only if $E_n\simeq E_i$. In contrast, for harmonic perturbation we have appreciable transition probability only if $E_n\simeq E_i-\hbar\omega$ (stimulated emission) or $E_n\simeq E_i+\hbar\omega$ (absorption).

## 5.8 Applications to Interactions with the Classical Radiation Field

### 5.8.1 Absorption and Stimulated Emission

We apply the formalism of time-dependent perturbation theory to the interactions of atomic electrons with the classical radiation field. By a **classical radiation field** we mean the electric or magnetic field derivable from a classical (as opposed to quantized) radiation field.

For an electron with charge $e$ in an electromagnetic field specified by a vector potential $\mathbf{A}(\mathbf{x})$ and scalar potential $\phi(\mathbf{x})$, the basic Hamiltonian, with $|\mathbf{A}|^2$ omitted, is

$$H = \frac{\mathbf{p}^2}{2m_e} + e\phi(\mathbf{x}) - \frac{e}{m_e c}\mathbf{A} \cdot \mathbf{p}, \tag{5.307}$$

which is justified if

$$\nabla \cdot \mathbf{A} = 0; \tag{5.308}$$

specifically, we work with a monochromatic field of the plane wave for

$$\mathbf{A} = 2A_0 \hat{\boldsymbol{\varepsilon}} \cos\left(\frac{\omega}{c}\hat{\mathbf{n}} \cdot \mathbf{x} - \omega t\right) \tag{5.309}$$

where $\hat{\boldsymbol{\varepsilon}}$ and $\hat{\mathbf{n}}$ are the (linear) polarization and propagation direction. Equation (5.309) obviously satisfies (5.308) because $\hat{\boldsymbol{\varepsilon}}$ is perpendicular to the propagation direction $\hat{\mathbf{n}}$. We write

$$\cos\left(\frac{\omega}{c}\hat{\mathbf{n}} \cdot \mathbf{x} - \omega t\right) = \frac{1}{2}[e^{i(\omega/c)\hat{\mathbf{n}} \cdot \mathbf{x} - i\omega t} + e^{-i(\omega/c)\hat{\mathbf{n}} \cdot \mathbf{x} + i\omega t}] \tag{5.310}$$

and treat $-(e/m_e c)\mathbf{A} \cdot \mathbf{p}$ as a time-dependent potential, where we express $\mathbf{A}$ in (5.309) as

$$\mathbf{A} = A_0 \hat{\boldsymbol{\varepsilon}}[e^{i(\omega/c)\hat{\mathbf{n}} \cdot \mathbf{x} - i\omega t} + e^{-i(\omega/c)\hat{\mathbf{n}} \cdot \mathbf{x} + i\omega t}]. \tag{5.311}$$

Comparing this result with (5.298), we see that the $e^{-i\omega t}$-term in

$$-\left(\frac{e}{m_e c}\right)\mathbf{A} \cdot \mathbf{p} = -\left(\frac{e}{m_e c}\right)A_0 \hat{\boldsymbol{\varepsilon}} \cdot \mathbf{p}[e^{i(\omega/c)\hat{\mathbf{n}} \cdot \mathbf{x} - i\omega t} + e^{-i(\omega/c)\hat{\mathbf{n}} \cdot \mathbf{x} + i\omega t}] \tag{5.312}$$

is responsible for absorption, while the $e^{+i\omega t}$-term is responsible for stimulated emission.

Let us now treat the absorption case in detail. We have

$$\mathscr{V}_{ni}^\dagger = -\frac{eA_0}{m_e c}(e^{i(\omega/c)(\hat{\mathbf{n}} \cdot \mathbf{x})}\hat{\boldsymbol{\varepsilon}} \cdot \mathbf{p})_{ni} \tag{5.313}$$

and

$$w_{i \to n} = \frac{2\pi}{\hbar}\frac{e^2}{m_e^2 c^2}|A_0|^2|\langle n|e^{i(\omega/c)(\hat{\mathbf{n}} \cdot \mathbf{x})}\hat{\boldsymbol{\varepsilon}} \cdot \mathbf{p}|i\rangle|^2 \delta(E_n - E_i - \hbar\omega). \tag{5.314}$$

The meaning of the $\delta$-function is clear. If $|n\rangle$ forms a continuum, we simply integrate with $\rho(E_n)$. But even if $|n\rangle$ is discrete, because $|n\rangle$ cannot be a ground state (albeit a bound-state energy level), its energy is not infinitely sharp; there may be a natural broadening due to a finite lifetime (see Section 5.9); there can also be a mechanism for broadening due to collisions. In such cases, we regard $\delta(\omega - \omega_{ni})$ as

$$\delta(\omega - \omega_{ni}) = \lim_{\gamma \to 0} \left( \frac{\gamma}{2\pi} \right) \frac{1}{([\omega - \omega_{ni}]^2 + \gamma^2/4)}. \tag{5.315}$$

Finally, the incident electromagnetic wave itself is not perfectly monochromatic; in fact, there is always a finite frequency width.

We derive an absorption cross section as

$$\frac{\text{(Energy/unit time) absorbed by the atom}\,(i \to n)}{\text{Energy flux of the radiation field}}. \tag{5.316}$$

For the energy flux (energy per area per unit time), classical electromagnetic theory gives us

$$c\mathscr{U} = \frac{1}{2\pi} \frac{\omega^2}{c} |A_0|^2, \tag{5.317}$$

where we have used

$$\mathscr{U} = \frac{1}{2} \left( \frac{E^2_{\max}}{8\pi} + \frac{B^2_{\max}}{8\pi} \right) \tag{5.318}$$

for energy density (energy per unit volume) with

$$\mathbf{E} = -\frac{1}{c} \frac{\partial}{\partial t} \mathbf{A}, \qquad \mathbf{B} = \nabla \times \mathbf{A}. \tag{5.319}$$

Putting everything together, we get (remembering that $\hbar\omega$ is the energy absorbed by the atom for each absorption process)

$$\sigma_{abs} = \frac{\hbar\omega (2\pi/\hbar)(e^2/m_e^2 c^2)|A_0|^2 |\langle n|e^{i(\omega/c)(\hat{\mathbf{n}} \cdot \mathbf{x})} \hat{\boldsymbol{\varepsilon}} \cdot \mathbf{p}|i\rangle|^2 \delta(E_n - E_i - \hbar\omega)}{(1/2\pi)(\omega^2/c)|A_0|^2}$$

$$= \frac{4\pi^2 \hbar}{m_e^2 \omega} \left( \frac{e^2}{\hbar c} \right) |\langle n|e^{i(\omega/c)(\hat{\mathbf{n}} \cdot \mathbf{x})} \hat{\boldsymbol{\varepsilon}} \cdot \mathbf{p}|i\rangle|^2 \delta(E_n - E_i - \hbar\omega). \tag{5.320}$$

Equation (5.320) has the correct dimension $[1/(M^2/T)](M^2 L^2/T^2)T = L^2$ if we recognize that $\alpha = e^2/\hbar c \simeq 1/137$ (dimensionless) and $\delta(E_n - E_i - \hbar\omega) = (1/\hbar)\delta(\omega_{ni} - \omega)$, where $\delta(\omega_{ni} - \omega)$ has time dimension $T$.

## 5.8.2 Electric Dipole Approximation

The *electric dipole approximation* (E1 approximation) is based on the fact that the wavelength of the radiation field is far longer than the atomic dimension, so that the series (remember $\omega/c = 1/\lambda$)

$$e^{i(\omega/c)\hat{\mathbf{n}} \cdot \mathbf{x}} = 1 + i\frac{\omega}{c} \hat{\mathbf{n}} \cdot \mathbf{x} + \cdots \tag{5.321}$$

can be approximated by its leading term, 1. The validity of this for a light atom is as follows. First, the $\hbar\omega$ of the radiation field must be of order of atomic level spacing, so

$$\hbar\omega \sim \frac{Ze^2}{(a_0/Z)} \simeq \frac{Ze^2}{R_{atom}}. \tag{5.322}$$

This leads to

$$\frac{c}{\omega} = \lambdabar \sim \frac{c\hbar R_{\text{atom}}}{Ze^2} \simeq \frac{137 R_{\text{atom}}}{Z}. \tag{5.323}$$

In other words,

$$\frac{1}{\lambdabar} R_{\text{atom}} \sim \frac{Z}{137} \ll 1 \tag{5.324}$$

for light atoms (small $Z$). Because the matrix element of $x$ is of order $R_{\text{atom}}$, that of $x^2$, of order $R_{\text{atom}}^2$, and so on, we see that the approximation of replacing (5.321) by its leading term is an excellent one.

Now we have

$$\langle n|e^{i(\omega/c)(\hat{\mathbf{n}}\cdot\mathbf{x})}\hat{\boldsymbol{\varepsilon}}\cdot\mathbf{p}|i\rangle \to \hat{\boldsymbol{\varepsilon}}\cdot\langle n|\mathbf{p}|i\rangle. \tag{5.325}$$

In particular, we take $\hat{\boldsymbol{\varepsilon}}$ along the $x$-axis (and $\hat{\mathbf{n}}$ along the $z$-axis). We must calculate $\langle n|p_x|i\rangle$. Using

$$[x, H_0] = \frac{i\hbar p_x}{m}, \tag{5.326}$$

we have

$$\langle n|p_x|i\rangle = \frac{m}{i\hbar}\langle n|[x, H_0]|i\rangle$$

$$= im\omega_{ni}\langle n|x|i\rangle. \tag{5.327}$$

Because of the approximation of the dipole operator, this approximation scheme is called the **electric dipole approximation**. We may here recall [see (3.482)] the selection rule for the dipole matrix element. Since $\mathbf{x}$ is a spherical tensor of rank 1 with $q = \pm 1$, we must have $m' - m = \pm 1, |j' - j| = 0, 1$ (no $0 \to 0$ transition). If $\hat{\boldsymbol{\varepsilon}}$ is along the $y$-axis, the same selection rule applies. On the other hand, if $\hat{\boldsymbol{\varepsilon}}$ is in the $z$-direction, $q = 0$; hence, $m' - m$.

With the electric dipole approximation, the absorption cross section (5.320) now takes a simpler form upon using (5.325) and (5.327) as

$$\sigma_{abs} = 4\pi^2 \alpha \omega_{ni} |\langle n|x|i\rangle|^2 \delta(\omega - \omega_{ni}). \tag{5.328}$$

In other words, $\sigma_{abs}$ treated as a function of $\omega$ exhibits a sharp $\delta$-function-like peak whenever $\hbar\omega$ corresponds to the energy-level spacing at $\omega \simeq (E_n - E_i)/\hbar$. Suppose $|i\rangle$ is the ground state, then $\omega_{ni}$ is necessarily positive; integrating (5.328), we get

$$\int \sigma_{abs}(\omega)\,d\omega = \sum_n 4\pi^2 \alpha \omega_{ni} |\langle n|x|i\rangle|^2. \tag{5.329}$$

In atomic physics we define **oscillator strength**, $f_{ni}$, as

$$f_{ni} \equiv \frac{2m\omega_{ni}}{\hbar} |\langle n|x|i\rangle|^2. \tag{5.330}$$

It is then straightforward (consider $[x, [x, H_0]]$) to establish the **Thomas–Reiche–Kuhn sum rule**,

$$\sum_n f_{ni} = 1. \tag{5.331}$$

In terms of the integration over the absorption cross section, we have

$$\int \sigma_{abs}(\omega)\, d\omega = \frac{4\pi^2\alpha\hbar}{2m_e} = 2\pi^2 c \left(\frac{e^2}{m_e c^2}\right). \tag{5.332}$$

Notice how $\hbar$ has disappeared. Indeed, this is just the oscillation sum rule already known in classical electrodynamics (Jackson (1975), for instance). Historically, this was one of the first examples of how "new quantum mechanics" led to the correct classical result. This sum rule is quite remarkable because we did not specify in detail the form of the Hamiltonian.

### 5.8.3 Photoelectric Effect

We now consider the **photoelectric effect**, that is, the ejection of an electron when an atom is placed in the radiation field. The basic process is considered to be the transition from an atomic (bound) state to a continuum state $E > 0$. Therefore, $|i\rangle$ is the ket for an atomic state, while $|n\rangle$ is the ket for a continuum state, which can be taken to be a plane wave state $|\mathbf{k}_f\rangle$, an approximation that is valid if the final electron is not too slow. Our earlier formula for $\sigma_{abs}(\omega)$ can still be used, except that we must now integrate $\delta(\omega_{ni} - \omega)$ together with the density of final states $\rho(E_n)$.

In fact, we calculated the density of states for a free particle in Section 2.5.1. To review, our basic task is to calculate the number of final states per unit energy interval. As we will see in a moment, this is an example where the matrix element depends not only on the final state energy but also on the momentum *direction*. We must therefore consider a group of final states with both similar momentum directions and similar energies.

To count the number of states it is convenient to use the box normalization convention for plane wave states. We consider a plane wave state to be normalized if, when we integrate the square modulus of its wave function for a cubic box of side $L$, we obtain unity. Furthermore, the state is assumed to satisfy the periodic boundary condition with periodicity of the side of the box. The wave function must then be of form

$$\langle \mathbf{x}|\mathbf{k}_f\rangle = \frac{e^{i\mathbf{k}_f \cdot \mathbf{x}}}{L^{3/2}}, \tag{5.333}$$

where the allowed values of $k_x$ must satisfy

$$k_x = \frac{2\pi n_x}{L}, \dots, \tag{5.334}$$

with $n_x$ a positive or negative integer. Similar restrictions hold for $k_y$ and $k_z$. Notice that as $L \to \infty$, $k_x, k_y$, and $k_z$ become continuous variables.

The problem of counting the number of states is reduced to that of counting the number of dots in three-dimensional lattice space. We define $n$ such that

$$n^2 = n_x^2 + n_y^2 + n_z^2. \tag{5.335}$$

As $L \to \infty$, it is a good approximation to treat $n$ as a continuous variable; in fact it is just the magnitude of the radial vector in the lattice space. Let us consider a small-volume element such that the radial vector falls within $n$ and $n + dn$ and the solid angle element $d\Omega$; clearly, it is of volume $n^2\, dn\, d\Omega$. The energy of the final state plane wave is related to $k_f$ and hence to $n$; we have

$$E = \frac{\hbar^2 k_f^2}{2m_e} = \frac{\hbar^2}{2m_e} \frac{n^2 (2\pi)^2}{L^2}. \tag{5.336}$$

Furthermore, the direction of the radial vector in the lattice space is just the momentum direction of the final state, so the number of states in the interval between $E$ and $E + dE$ with direction into $d\Omega$ being $\mathbf{k}_f$ is (remember $dE = (\hbar^2 k_f/m_e)dk_f$) given by[13]

$$n^2\, d\Omega \frac{dn}{dE}dE = \left(\frac{L}{2\pi}\right)^3 (k_f^2)\frac{dk_f}{dE}d\Omega\, dE$$

$$= \left(\frac{L}{2\pi}\right)^3 \frac{m_e}{\hbar^2}k_f\, dE\, d\Omega. \tag{5.337}$$

We can now put everything together to obtain an expression for the differential cross section for the photoelectric effect:

$$\frac{d\sigma}{d\Omega} = \frac{4\pi^2 \alpha \hbar}{m_e^2 \omega}|\langle \mathbf{k}_f|e^{i(\omega/c)(\hat{\mathbf{n}}\cdot\mathbf{x})}\hat{\varepsilon}\cdot\mathbf{p}|i\rangle|^2 \frac{m_e k_f L^3}{\hbar^2 (2\pi)^3}. \tag{5.338}$$

To be specific, let us consider the ejection of a **K** shell (the innermost shell) electron caused by absorption of light. The initial state wave function is essentially the same as the ground-state hydrogen atom wave function except that the Bohr radius $a_0$ is replaced by $a_0/Z$. Thus

$$\langle \mathbf{k}_f|e^{i(\omega/c)(\hat{\mathbf{n}}\cdot\mathbf{x})}\hat{\varepsilon}\cdot\mathbf{p}|i\rangle = \hat{\varepsilon}\cdot\int d^3x \frac{e^{-i\mathbf{k}_f\cdot\mathbf{x}}}{L^{3/2}}e^{i(\omega/c)(\hat{\mathbf{n}}\cdot\mathbf{x})}$$

$$\times (-i\hbar\nabla)\left[\frac{1}{\sqrt{\pi}}e^{-Zr/a_0}\left(\frac{Z}{a_0}\right)^{3/2}\right]. \tag{5.339}$$

Integrating by parts, we can pass $\nabla$ to the left side. Furthermore,

$$\hat{\varepsilon}\cdot\left[\nabla e^{i(\omega/c)(\hat{\mathbf{n}}\cdot\mathbf{x})}\right] = 0 \tag{5.340}$$

because $\hat{\varepsilon}$ is perpendicular to $\hat{\mathbf{n}}$. On the other hand, $\nabla$ acting on $e^{-i\mathbf{k}_f\cdot\mathbf{x}}$ brings down $-i\mathbf{k}_f$, which can be taken outside the integral. Thus to evaluate (5.339), all we need to do is take the Fourier transform of the atomic wave function with respect to

$$\mathbf{q} \equiv \mathbf{k}_f - \left(\frac{\omega}{c}\right)\hat{\mathbf{n}}. \tag{5.341}$$

The final answer is (see Problem 5.48 of this chapter for the Fourier transform of the hydrogen atom wave function)

$$\frac{d\sigma}{d\Omega} = 32e^2 k_f \frac{(\hat{\varepsilon}\cdot\mathbf{k}_f)^2}{m_e c\omega}\frac{Z^5}{a_0^5}\frac{1}{[(Z^2/a_0^2) + q^2]^4}. \tag{5.342}$$

---

[13]  This is equivalent to taking one state per cube $d^3x\, d^3p/(2\pi\hbar)^3$ in phase space.

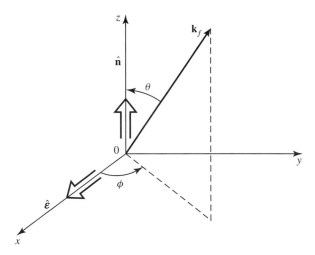

**Fig. 5.13** Polar coordinate system with $\hat{\varepsilon}$ and $\hat{n}$ along the $x$- and $z$-axes, respectively, and $\mathbf{k}_f = (k_f \sin\theta \cos\phi, \, k_f \sin\theta \sin\phi, \, k_f \cos\theta)$.

If we introduce the coordinate system shown in Figure 5.13, we can write the differential cross section in terms of $\theta$ and $\phi$ using

$$(\hat{\varepsilon} \cdot \mathbf{k}_f)^2 = k_f^2 \sin^2\theta \cos^2\phi$$

$$q^2 = k_f^2 - 2k_f \frac{\omega}{c} \cos\theta + \left(\frac{\omega}{c}\right)^2. \tag{5.343}$$

### 5.8.4 Spontaneous Emission

An atom in an excited state will spontaneously emit a photon, making a transition to a lower energy state. We now understand the underlying source of these photons to be electromagnetic energy in the so-called vacuum, but it is possible to accurately calculate this phenomenon using the classical electromagnetic field and the techniques discussed in this section.

Most of the formalism has already been worked out. The Hamiltonian $H = H_0 + V(t)$ is taken to be (5.307) where $H_0 = \mathbf{p}_e^2/2m_e + e\phi(\mathbf{x})$ describes the atomic system. We are considering transitions between eigenstates of $H_0$, and the perturbation $V(t) = -(e/m_e c)\mathbf{A} \cdot \mathbf{p}$ where $\mathbf{A}(\mathbf{x}, t)$ is a plane wave. Thus we once again have a harmonic perturbation, and the transition rate for emission is given by (5.302a). As in (5.309) we write $\mathbf{A}(\mathbf{x}, t)$ as

$$\mathbf{A} = 2A_0 \hat{\varepsilon} \cos(\mathbf{k} \cdot \mathbf{x} - \omega t) \tag{5.344}$$

where $\mathbf{k} \equiv (\omega/c)\hat{n}$ and $\hat{\varepsilon}$ is the polarization unit vector. We contain this field inside a "big box" of side length $L$, large enough to include the frequency $\omega$ that corresponds to the atomic transition, applying periodic boundary conditions, similar to the way we describe the ejected photoelectron in (5.333). This means that the wave vector $\mathbf{k}$ can be written as

$$\mathbf{k} = \frac{2\pi}{L}(n_x\hat{\mathbf{x}} + n_y\hat{\mathbf{y}} + n_z\hat{\mathbf{z}}) \tag{5.345}$$

where $n_x$, $n_y$, and $n_z$ are integers.

In order to determine the amplitude $A_0$, we require that the time-averaged electromagnetic energy in the box equal the transition energy $\hbar\omega$. From Maxwell's equations in free space, the electric and magnetic fields are

$$\mathbf{E} = -\frac{1}{c}\frac{\partial\mathbf{A}}{\partial t} = 2A_0 k\hat{\varepsilon}\sin(\mathbf{k}\cdot\mathbf{x} - \omega t) \tag{5.346a}$$

$$\mathbf{B} = \nabla\times\mathbf{A} = 2A_0(\hat{\varepsilon}\times\mathbf{k})\sin(\mathbf{k}\cdot\mathbf{x} - \omega t). \tag{5.346b}$$

The energy density is $u = (\mathbf{E}^2 + \mathbf{B}^2)/8\pi$, so the time-averaged energy is

$$\int_{L^3} d^3x\,\langle u\rangle = \frac{\omega^2}{2\pi c^2}A_0^2 L^3 = \hbar\omega. \tag{5.347}$$

Therefore, the amplitude of the classical[14] electromagnetic wave is

$$A_0 = \sqrt{\frac{2\pi\hbar c^2}{\omega L^3}}. \tag{5.348}$$

Referring back to (5.302a), we need to find the density of states $\rho(E_n) = \rho(\mathscr{E})$ where $\mathscr{E} = \hbar\omega$ is the energy of the emitted radiation. Writing $\mathscr{E} = 2\pi\hbar cn/L$ where $n \equiv (n_x^2 + n_y^2 + n_z^2)^{1/2}$, we integrate over all directions in $n$-space while keeping $n$ constant to find

$$\rho(\mathscr{E})d\mathscr{E} = 4\pi n^2 dn = 4\pi\left(\frac{\omega L}{2\pi c}\right)^2\frac{L}{2\pi\hbar c}d\mathscr{E} = \frac{\omega^2}{2\pi^2\hbar c^3}L^3 d\mathscr{E}. \tag{5.349}$$

We also need to calculate the transition matrix element

$$\mathscr{V}_{ni} = -\frac{eA_0}{m_e c}\langle n|\hat{\varepsilon}\cdot e^{-i\mathbf{k}\cdot\mathbf{x}}\mathbf{p}|i\rangle. \tag{5.350}$$

The wavelength of emitted radiation in atomic transitions is on the order of 100 nm or longer, three orders of magnitude larger than a typical atom. Therefore the long wavelength approximation (5.321) is valid. In fact, let us go further, and assume the transition is between states that satisfy the electric dipole transition rules (see p. 349). Therefore

$$\mathscr{V}_{ni} = -\frac{eA_0}{m_e c}\langle n|\hat{\varepsilon}\cdot\mathbf{p}|i\rangle = -i\frac{e\omega A_0}{c}\langle n|\hat{\varepsilon}\cdot\mathbf{x}|i\rangle. \tag{5.351}$$

The rate $w$ of a spontaneous transition is usually expressed terms of its inverse $\tau$, called the "lifetime" of the initial state. Then (5.302a) becomes

$$\frac{1}{\tau_{i\to n}} = \frac{2\pi}{\hbar}\left(\frac{e\omega A_0}{c}\right)^2|\langle n|\hat{\varepsilon}\cdot\mathbf{x}|i\rangle|^2\frac{\omega^2}{2\pi^2\hbar c^3}L^3$$

$$= 2\alpha\frac{\omega^3}{c^2}|\langle n|\hat{\varepsilon}\cdot\mathbf{x}|i\rangle|^2 \tag{5.352}$$

---

[14]  See Section 7.8 for a quantum-mechanical treatment of the free electromagnetic field. Note that (5.348) is the same result as (7.180).

where again we make use of $\alpha \equiv e^2/\hbar c \approx 1/137$. Note that this result is free of the box dimension $L$.

Given initial and final state atomic wave functions, we could calculate the rate, perhaps numerically for a complicated system. The rate will depend on the polarization direction $\hat{\varepsilon}$ if the atom itself is somehow polarized. It is useful to write the dipole operator in terms of circular and linear polarization directions as

$$\hat{\varepsilon} \cdot \mathbf{x} = \frac{\varepsilon_x - i\varepsilon_y}{\sqrt{2}} \frac{x+iy}{\sqrt{2}} + \frac{\varepsilon_x + i\varepsilon_y}{\sqrt{2}} \frac{x-iy}{\sqrt{2}} + \varepsilon_z z$$

$$= \sqrt{\frac{4\pi}{3}} \left( -\varepsilon_- Y_1^{+1} r + \varepsilon_- Y_1^{-1} r + \varepsilon_0 Y_1^0 r \right) \tag{5.353}$$

where $\varepsilon_\pm \equiv (\varepsilon_x - i\varepsilon_y)\sqrt{2}$ and $\varepsilon_0 \equiv \varepsilon_z$. Then recall that the $Y_l^m$ are spherical tensors, whose matrix elements are subject to the Wigner–Eckart theorem (3.474).

Let us illustrate using a $p \to s$ transition, where the initial state is unpolarized. That is, the initial atomic $p$-state is an equal mixture of the $m = 0, \pm 1$ states. Now, from (3.471), the matrix element $\langle s | Y_1^q | p, m \rangle = 0$ unless $m = -q$. Therefore, each of the three terms in (5.353) gives a nonzero transition matrix element for exactly one $m$ value for the initial $p$-state. Furthermore, by (3.474),

$$\langle \alpha'; s | Y_1^q r | \alpha; p, m \rangle = \langle 11; -q, q | 11; 00 \rangle \frac{\langle \alpha'; s || Y_1 r || \alpha; p \rangle}{\sqrt{3}} \tag{5.354}$$

where $\alpha$ and $\alpha'$ represent other information needed to specify the state (perhaps a radial quantum number). The reduced matrix element $\langle \alpha; s || Y_1 r || \alpha; p \rangle$ is the same for all three terms in (5.353), so we just need to calculate the matrix element $\langle \alpha'; s | Y_1^q r | \alpha; p, m \rangle$ for one of them. Choosing the $m = 0$ term for this calculation, we find

$$\langle \alpha'; s | Y_1^0 r | \alpha; p, m = 0 \rangle = \frac{1}{\sqrt{4\pi}} \mathscr{R}_{\alpha'\alpha} \tag{5.355}$$

where

$$\mathscr{R}_{\alpha'\alpha} \equiv \int_0^\infty r^2 dr\, R_{\alpha'}^*(r) r R_\alpha(r) \tag{5.356}$$

and so

$$\langle \alpha'; s || Y_1 r || \alpha; p \rangle = -\frac{3}{\sqrt{4\pi}} \mathscr{R}_{\alpha'\alpha}. \tag{5.357}$$

We can then write down the matrix element of (5.353) as

$$\langle \alpha'; s | \hat{\varepsilon} \cdot \mathbf{x} | \alpha; p, m \rangle = -\frac{1}{\sqrt{3}} \left( -\varepsilon_- \delta_{m,-1} + \varepsilon_- \delta_{m,1} - \varepsilon_0 \delta_{m,0} \right) \mathscr{R}_{\alpha'\alpha}. \tag{5.358}$$

We do not know the value of $m$ for the initial state. However, (5.358) shows that each of the three $m$ states gives the same result in (5.352). The initial state is unpolarized, and we are integrating over all directions of the emitted radiation, so we average over the contributions from the three initial states. Since $|\varepsilon_+|^2 + |\varepsilon_-|^2 + |\varepsilon_0|^2 = 1$, this average introduces an additional factor of 1/3. Realizing that we must also multiply (5.352) by two to account for the two orthogonal polarization directions, we have, finally,

$$\frac{1}{\tau_{p \to s}} = \frac{4}{9} \alpha \frac{\omega^3}{c^2} |\mathscr{R}_{\alpha' \alpha}|^2 . \tag{5.359}$$

Before concluding, we note that if the electric dipole selection rules are not satisfied by the initial and final atomic states, we can still use the long wavelength approximation to evaluate (5.350). However, the matrix element will contain higher powers of **x** and is therefore smaller by additional powers of atomic size divided by the wavelength. That is, the lifetimes will be much longer, so if there are final states available which do satisfy the selection rules, these will dominate.

## 5.9 Energy Shift and Decay Width

Our considerations so far have been restricted to the question of how states other than the initial state become populated. In other words, we have been concerned with the time development of the coefficient $c_n(t)$ with $n \neq i$. The question naturally arises, What happens to $c_i(t)$ itself?

To avoid the effect of a sudden change in the Hamiltonian, we propose to increase the perturbation very slowly. In the remote past $(t \to -\infty)$ the time-dependent potential is assumed to be zero. We then *gradually* turn on the perturbation to its full value; specifically,

$$V(t) = e^{\eta t} V \tag{5.360}$$

where $V$ is assumed to be constant and $\eta$ is small and positive. At the end of the calculation, we let $\eta \to 0$ (see Figure 5.14), and the potential then becomes constant at all times.

In the remote past, we take this time to be $-\infty$, so the state ket in the interaction picture is assumed to be $|i\rangle$. Our basic aim is to evaluate $c_i(t)$. However, before we do that, let us make sure that the old formula of the golden rule (see Section 5.7) can be reproduced using this slow-turn-on method. For $c_n(t)$ with $n \neq i$, we have [using (5.276)]

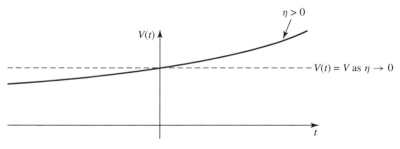

**Fig. 5.14**    Plot of $V(t)$ versus $t$ in the adiabatic (slow-turn-on) picture.

$$c_n^{(0)}(t) = 0$$

$$c_n^{(1)}(t) = \frac{-i}{\hbar} V_{ni} \lim_{t_0 \to -\infty} \int_{t_0}^{t} e^{\eta t'} e^{i\omega_{ni} t'} \, dt'$$

$$= \frac{-i}{\hbar} V_{ni} \frac{e^{\eta t + i\omega_{ni} t}}{\eta + i\omega_{ni}}. \tag{5.361}$$

To lowest nonvanishing order, the transition probability is therefore given by

$$|c_n(t)|^2 \simeq \frac{|V_{ni}|^2}{\hbar^2} \frac{e^{2\eta t}}{\eta^2 + \omega_{ni}^2}, \tag{5.362}$$

or

$$\frac{d}{dt} |c_n(t)|^2 \simeq \frac{2|V_{ni}|^2}{\hbar^2} \left( \frac{\eta e^{2\eta t}}{\eta^2 + \omega_{ni}^2} \right). \tag{5.363}$$

We now let $\eta \to 0$. Clearly, it is all right to replace $e^{\eta t}$ by unity, but note

$$\lim_{\eta \to 0} \frac{\eta}{\eta^2 + \omega_{ni}^2} = \pi \delta(\omega_{ni}) = \pi \hbar \delta(E_n - E_i). \tag{5.364}$$

This leads to the golden rule,

$$w_{i \to n} \simeq \left( \frac{2\pi}{\hbar} \right) |V_{ni}|^2 \delta(E_n - E_i). \tag{5.365}$$

Encouraged by this result, let us calculate $c_i^{(0)}$, $c_i^{(1)}$, and $c_i^{(2)}$, again using (5.276). We have

$$c_i^{(0)} = 1$$

$$c_i^{(1)} = \frac{-i}{\hbar} V_{ii} \lim_{t_0 \to -\infty} \int_{t_0}^{t} e^{\eta t'} \, dt' = \frac{-i}{\hbar \eta} V_{ii} e^{\eta t}$$

$$c_i^{(2)} = \left( \frac{-i}{\hbar} \right)^2 \sum_m |V_{mi}|^2 \lim_{t_0 \to -\infty} \int_{t_0}^{t} dt' e^{i\omega_{im} t' + \eta t'} \frac{e^{i\omega_{mi} t' + \eta t'}}{i(\omega_{mi} - i\eta)} \tag{5.366}$$

$$= \left( \frac{-i}{\hbar} \right)^2 |V_{ii}|^2 \frac{e^{2\eta t}}{2\eta^2} + \left( \frac{-i}{\hbar} \right) \sum_{m \neq i} \frac{|V_{mi}|^2 e^{2\eta t}}{2\eta(E_i - E_m + i\hbar\eta)}.$$

Thus up to second order we have

$$c_i(t) \simeq 1 - \frac{i}{\hbar\eta} V_{ii} e^{\eta t} + \left( \frac{-i}{\hbar} \right)^2 |V_{ii}|^2 \frac{e^{2\eta t}}{2\eta^2} + \left( \frac{-i}{\hbar} \right) \sum_{m \neq i} \frac{|V_{mi}|^2 e^{2\eta t}}{2\eta(E_i - E_m + i\hbar\eta)}. \tag{5.367}$$

Now consider the time derivative of $c_i[dc_i(t)/dt \equiv \dot{c}_i]$, which we have from (5.367). Upon dividing by $c_i$ and letting $\eta \to 0$ (thus replacing $e^{\eta t}$ and $e^{2\eta t}$ by unity), we get

$$\frac{\dot{c}_i}{c_i} \simeq \frac{\dfrac{-i}{\hbar} V_{ii} + \left( \dfrac{-i}{\hbar} \right)^2 \dfrac{|V_{ii}|^2}{\eta} + \left( \dfrac{-i}{\hbar} \right) \displaystyle\sum_{m \neq i} \dfrac{|V_{mi}|^2}{(E_i - E_m + i\hbar\eta)}}{1 - \dfrac{i}{\hbar} \dfrac{V_{ii}}{\eta}}$$

$$\simeq \frac{-i}{\hbar} V_{ii} + \left( \frac{-i}{\hbar} \right) \sum_{m \neq i} \frac{|V_{mi}|^2}{E_i - E_m + i\hbar\eta}. \tag{5.368}$$

Expansion (5.368) is formally correct up to second order in $V$. Note here that $\dot{c}_i(t)/c_i(t)$ is now independent of $t$. Equation (5.368) is a differential equation that is to hold *at all times*. Having obtained this, it is convenient to renormalize $c_i$ so that $c_i(0) = 1$. We now try the ansatz

$$c_i(t) = e^{-i\Delta_i t/\hbar}, \quad \frac{\dot{c}_i(t)}{c_i(t)} = \frac{-i}{\hbar}\Delta_i \tag{5.369}$$

with $\Delta_i$ constant (in time) but not necessarily real. Clearly (5.369) is consistent with (5.368) because the right-hand side of (5.369) is constant. We can see the physical meaning of $\Delta_i$ by noting that $e^{-i\Delta_i t/\hbar}|i\rangle$ in the interaction picture implies $e^{-i\Delta_i t/\hbar - iE_i t/\hbar}|i\rangle$ in the Schrödinger picture. In other words,

$$E_i \rightarrow E_i + \Delta_i \tag{5.370}$$

as a result of perturbation. That is, we have calculated the *level shift* using time-*dependent* perturbation theory. Now expand, as usual,

$$\Delta_i = \Delta_i^{(1)} + \Delta_i^{(2)} + \cdots, \tag{5.371}$$

and compare (5.369) with (5.368); we get to first order

$$\Delta_i^{(1)} = V_{ii}. \tag{5.372}$$

But this is just what we expect from *t-independent perturbation theory*. Before we look at $\Delta_i^{(2)}$, recall

$$\lim_{\varepsilon \to 0} \frac{1}{x + i\varepsilon} = \Pr\frac{1}{x} - i\pi\delta(x). \tag{5.373}$$

Thus

$$\text{Re}(\Delta_i^{(2)}) = \Pr\sum_{m \neq i} \frac{|V_{mi}|^2}{E_i - E_m} \tag{5.374a}$$

$$\text{Im}(\Delta_i^{(2)}) = -\pi \sum_{m \neq i} |V_{mi}|^2 \delta(E_i - E_m). \tag{5.374b}$$

But the right-hand side of (5.374b) is familiar from the golden rule, so we can identify

$$\sum_{m \neq i} w_{i \rightarrow m} = \frac{2\pi}{\hbar} \sum_{m \neq i} |V_{mi}|^2 \delta(E_i - E_m) = -\frac{2}{\hbar}\text{Im}[\Delta_i^{(2)}]. \tag{5.375}$$

Coming back to $c_i(t)$, we can write (5.369) as

$$c_i(t) = e^{-(i/\hbar)[\text{Re}(\Delta_i)t] + (1/\hbar)[\text{Im}(\Delta_i)t]}. \tag{5.376}$$

If we define

$$\frac{\Gamma_i}{\hbar} \equiv -\frac{2}{\hbar}\text{Im}(\Delta_i), \tag{5.377}$$

then

$$|c_i|^2 = e^{2\text{Im}(\Delta_i)t/\hbar} = e^{-\Gamma_i t/\hbar}. \tag{5.378}$$

Therefore, $\Gamma_i$ characterizes the rate at which state $|i\rangle$ disappears.

It is worth checking the probability conservation up to second order in $V$ for small $t$:

$$|c_i|^2 + \sum_{m \neq i} |c_m|^2 = (1 - \Gamma_i t/\hbar) + \sum_{m \neq i} w_{i \to m} t = 1, \tag{5.379}$$

where (5.375) has been used. Thus the probabilities for finding the initial state and all other states add up to 1. Put in another way, the depletion of state $|i\rangle$ is compensated by the growth of states other than $|i\rangle$.

To summarize, the real part of the energy shift is what we usually associate with the level shift. The imaginary part of the energy shift is, apart from $-2$ [see (5.377)], the **decay width**. Note also

$$\frac{\hbar}{\Gamma_i} = \tau_i \tag{5.380}$$

where $\tau_i$ is the mean lifetime of state $|i\rangle$ because

$$|c_i|^2 = e^{-t/\tau_i}. \tag{5.381}$$

To see why $\Gamma_i$ is called *width*, we look at the Fourier decomposition

$$\int f(E) e^{-iEt/\hbar} dE = e^{-i[E_i + \mathrm{Re}(\Delta_i)]t/\hbar - \Gamma_i t/2\hbar}. \tag{5.382}$$

Using the Fourier inversion formula, we get

$$|f(E)|^2 \propto \frac{1}{\{E - [E_i + \mathrm{Re}(\Delta_i)]\}^2 + \Gamma_i^2/4}. \tag{5.383}$$

Therefore, $\Gamma_i$ has the usual meaning of full width at half maximum. Notice that we get the energy-time uncertainty relation from (5.380)

$$\Delta t \Delta E \sim \hbar, \tag{5.384}$$

where we identify the uncertainty in the energy with $\Gamma_i$ and the mean lifetime with $\Delta t$.

Even though we discussed the subject of energy shift and decay width using the constant perturbation $V$ obtained as the limit of (5.360) when $\eta \to 0$, we can easily generalize our considerations to the harmonic perturbation case discussed in Section 5.7. All we must do is to let

$$E_{n(m)} - E_i \to E_{n(m)} - E_i \pm \hbar \omega \tag{5.385}$$

in (5.361), (5.367), and (5.374), and so on. The quantum-mechanical description of unstable states we have developed here is originally due to Wigner and Weisskopf in 1930.

# Problems

**5.1** A simple harmonic oscillator (in one dimension) is subjected to a perturbation

$$H_1 = bx$$

where $b$ is a real constant.

a. Calculate the energy shift of the ground state to *lowest nonvanishing order*.

b. Solve this problem *exactly* and compare with your result obtained in (a).

**5.2**  A one-dimensional potential well has infinite walls at $x = 0$ and $x = L$. The bottom of the well is *not* flat, but rather increases linearly from 0 at $x = 0$ to $V$ at $x = L$. Find the first-order shift in the energy levels as a function of principal quantum number $n$.

**5.3**  A particle of mass $m$ moves in a potential well $V(x) = m\omega^2 x^2/2$. Treating relativistic effects to order $\beta^2 = (p/mc)^2$, find the ground-state energy shift.

**5.4**  A diatomic molecule can be modeled as a rigid rotor with moment of inertia $I$ and an electric dipole moment $d$ along the axis of the rotor. The rotor is constrained to rotate in a plane, and a weak uniform electric field $\mathscr{E}$ lies in the plane. Write the classical Hamiltonian for the rotor, and find the unperturbed energy levels by quantizing the angular-momentum operator. Then treat the electric field as a perturbation, and find the first nonvanishing corrections to the energy levels.

**5.5**  In nondegenerate time-independent perturbation theory, what is the probability of finding in a perturbed energy eigenstate ($|k\rangle$) the corresponding unperturbed eigenstate ($|k^{(0)}\rangle$)? Solve this up to terms of order $\lambda^2$.

**5.6**  Consider a particle in a two-dimensional potential

$$V_0 = \begin{cases} 0 & \text{for } 0 \le x \le L,\ 0 \le y \le L \\ \infty & \text{otherwise.} \end{cases}$$

Write the energy eigenfunctions for the ground and first excited states. We now add a time-independent perturbation of the form

$$V_1 = \begin{cases} \lambda xy & \text{for } 0 \le x \le L,\ 0 \le y \le L \\ 0 & \text{otherwise.} \end{cases}$$

Obtain the zeroth-order energy eigenfunctions and the first-order energy shifts for the ground and first excited states.

**5.7**  Consider an isotropic harmonic oscillator in *two* dimensions. The Hamiltonian is

$$H_0 = \frac{p_x^2}{2m} + \frac{p_y^2}{2m} + \frac{m\omega^2}{2}(x^2 + y^2).$$

a. What are the energies of the three lowest-lying states? Is there any degeneracy?

b. We now apply a perturbation

$$V = \delta m\omega^2 xy,$$

where $\delta$ is a dimensionless real number much smaller than unity. Find the zeroth-order energy eigenket and the corresponding energy to first order [that is, the unperturbed energy obtained in (a) plus the first-order energy shift] for each of the three lowest-lying states.

c. Solve the $H_0 + V$ problem *exactly*. Compare with the perturbation results obtained in (b).

**5.8** Establish (5.54) for the one-dimensional harmonic oscillator given by (5.50) with an additional perturbation $V = \frac{1}{2}\varepsilon m\omega^2 x^2$. Show that all other matrix elements $V_{k0}$ vanish.

**5.9** A slightly anisotropic three-dimensional harmonic oscillator has $\omega_x = \omega_y \equiv \omega$ and $\omega_z = (1+\varepsilon)\omega$ where $\varepsilon \ll 1$. (See Section 3.7.3 for nomenclature and wave functions.) A charged particle moves in the field of this oscillator and is at the same time exposed to a uniform magnetic field in the $x$-direction. Assuming that the Zeeman splitting is comparable to the splitting produced by the anisotropy, but small compared to $\hbar\omega$, calculate to first order the energies of the components of the first excited state. Discuss various limiting cases. (This is taken from Problem 17.7 in Merzbacher (1970). You might find it useful to consult Problem 2.16 in Chapter 2 and Problem 3.29 in Chapter 3.)

**5.10** A one-electron atom whose ground state is nondegenerate is placed in a uniform electric field in the $z$-direction. Obtain an approximate expression for the induced electric dipole moment of the ground state by considering the expectation value of $ez$ with respect to the perturbed state vector computed to first order. Show that the same expression can also be obtained from the energy shift $\Delta = -\alpha|\mathbf{E}|^2/2$ of the ground state computed to second order. (*Note:* $\alpha$ stands for the polarizability.) Ignore spin.

**5.11** Evaluate the matrix elements (or expectation values) given below. If any vanishes, explain why it vanishes using simple symmetry (or other) arguments.
   a. $\langle n=2, l=1, m=0|x|n=2, l=0, m=0\rangle$.
   b. $\langle n=2, l=1, m=0|p_z|n=2, l=0, m=0\rangle$.
   [In (a) and (b), $|nlm\rangle$ stands for the energy eigenket of a nonrelativistic hydrogen atom with spin ignored.]
   c. $\langle L_z\rangle$ for an electron in a central field with $j=\frac{9}{2}$, $m=\frac{7}{2}$, $l=4$.
   d. $\langle \text{singlet}, m_s=0|S_z^{(e-)} - S_z^{(e+)}|\text{triplet}, m_s=0\rangle$ for an $s$-state positronium.
   e. $\langle \mathbf{S}^{(1)} \cdot \mathbf{S}^{(2)}\rangle$ for the ground state of a hydrogen *molecule*.

**5.12** A $p$-orbital electron characterized by $|n, l=1, m=\pm 1, 0\rangle$ (ignore spin) is subjected to a potential

$$V = \lambda(x^2 - y^2) \quad (\lambda = \text{constant}).$$

   a. Obtain the "correct" zeroth-order energy eigenstates that diagonalize the perturbation. You need not evaluate the energy shifts in detail, but show that the original threefold degeneracy is now completely removed.
   b. Because $V$ is invariant under time reversal and because there is no longer any degeneracy, we expect each of the energy eigenstates obtained in (a) to go into itself (up to a phase factor or sign) under time reversal. Check this point explicitly.

**5.13** Consider a spinless particle in a two-dimensional infinite square well:

$$V = \begin{cases} 0 & \text{for } 0 \le x \le a, \ 0 \le y \le a \\ \infty & \text{otherwise.} \end{cases}$$

a. What are the energy eigenvalues for the three lowest states? Is there any degeneracy?

b. We now add a potential

$$V_1 = \lambda xy, \ 0 \le x \le a, \ 0 \le y \le a.$$

Taking this as a weak perturbation, answer the following.

(i) Is the energy shift due to the perturbation linear or quadratic in $\lambda$ for each of the three states?

(ii) Obtain expressions for the energy shifts of the three lowest states accurate to order $\lambda$. (You need not evaluate integrals that may appear.)

(iii) Draw an energy diagram with and without the perturbation for the three energy states. Make sure to specify which unperturbed state is connected to which perturbed state.

**5.14** The Hamiltonian matrix for a two-state system can be written as

$$\mathcal{H} = \begin{pmatrix} E_1^0 & \lambda\Delta \\ \lambda\Delta & E_2^0 \end{pmatrix}.$$

Clearly the energy eigenfunctions for the unperturbed problems ($\lambda = 0$) are given by

$$\phi_1^{(0)} = \begin{pmatrix} 1 \\ 0 \end{pmatrix}, \quad \phi_2^{(0)} = \begin{pmatrix} 0 \\ 1 \end{pmatrix}.$$

a. Solve this problem *exactly* to find the energy eigenfunctions $\psi_1$ and $\psi_2$ and the energy eigenvalues $E_1$ and $E_2$.

b. Assuming that $\lambda|\Delta| \ll |E_1^0 - E_2^0|$, solve the same problem using time-independent perturbation theory up to first order in the energy eigenfunctions and up to second order in the energy eigenvalues. Compare with the exact results obtained in (a).

c. Suppose the two unperturbed energies are "almost degenerate," that is,

$$|E_1^0 - E_2^0| \ll \lambda|\Delta|.$$

Show that the exact results obtained in (a) closely resemble what you would expect by applying *degenerate* perturbation theory to this problem with $E_1^0$ set exactly equal to $E_2^0$.

**5.15** (This is a tricky problem because the degeneracy between the first and the second state is not removed in first order. See also Gottfried (1966), p. 397, Problem 1.) This problem is from Schiff (1968), p. 295, Problem 4. A system that has three unperturbed states can be represented by the perturbed Hamiltonian matrix

$$\begin{pmatrix} E_1 & 0 & a \\ 0 & E_1 & b \\ a^* & b^* & E_2 \end{pmatrix}$$

where $E_2 > E_1$. The quantities $a$ and $b$ are to be regarded as perturbations that are of the same order and are small compared with $E_2 - E_1$. Use the second-order nondegenerate perturbation theory to calculate the perturbed eigenvalues. (Is this procedure correct?) Then diagonalize the matrix to find the exact eigenvalues. Finally, use the second-order degenerate perturbation theory. Compare the three results obtained.

**5.16** Use perturbation theory to calculate the effect of the proton's finite size on the $n = 1$ and $n = 2$ energy levels of the hydrogen atom. Assume the proton is a uniformly charged sphere of radius $R$. Give a physical explanation for why the $\ell = 1$ shifts are so much smaller than for $\ell = 0$.

**5.17** This chapter derived two of the three relativistic corrections to the one-electron atom, namely $\Delta_K^{(1)}$ from "relativistic kinetic energy," and $\Delta_{LS}^{(1)}$ from the spin-orbit interaction. A third term comes from the spread of the electron wave function in the region of changing electric field. The perturbation for this "Darwin term" is

$$V_D = -\frac{1}{8m^2c^2} \sum_{i=1}^{3} [p_i, [p_i, e\phi(r)]]$$

where $\phi(r)$ is the Coulomb potential. Find $\Delta_D^{(1)}$ and show that

$$\Delta_{nj}^{(1)} \equiv \Delta_K^{(1)} + \Delta_{LS}^{(1)} + \Delta_D^{(1)} = \frac{mc^2(Z\alpha)^4}{2n^3}\left[\frac{3}{4n} - \frac{1}{j+1/2}\right].$$

In Section 8.4 we will compare this expression to the result of solving the Dirac equation in the presence of the Coulomb potential.

**5.18** These questions are meant to associate numbers with atomic hydrogen phenomena.
   a. The red $n = 3 \rightarrow 2$ Balmer transition has a wavelength $\lambda \approx 656$ nm. Calculate the wavelength difference $\Delta\lambda$ (in nm) between the $3p_{3/2} \rightarrow 2s_{1/2}$ and $3p_{1/2} \rightarrow 2s_{1/2}$ transitions due to the spin-orbit interaction. Comment on how you might measure this splitting.
   b. How large an electric field $\mathcal{E}$ is needed so that the Stark splitting in the $n = 2$ level is the same as the correction from relativistic kinetic energy between the $2s$ and $2p$ levels? How easy or difficult is it to achieve an electric field of this magnitude in the laboratory?
   c. The Zeeman effect can be calculated with a "weak" or "strong" magnetic field, depending on the size of the energy shift relative to the spin-orbit splitting. Give examples of a weak and a strong field. How easy or difficult is it to achieve such a magnetic field?

**5.19** Compute the Stark effect for the $2s_{1/2}$ and $2p_{1/2}$ levels of hydrogen for a field $\mathcal{E}$ sufficiently weak so that $e\mathcal{E}a_0$ is small compared to the fine structure, but take the Lamb shift $\delta$ ($\delta = 1057$ MHz) into account (that is, ignore $2p_{3/2}$ in this calculation). Show that for $e\mathcal{E}a_0 \ll \delta$, the energy shifts are quadratic in $\mathcal{E}$, whereas for $e\mathcal{E}a_0 \gg \delta$ they are linear in $\mathcal{E}$. Briefly discuss the consequences (if any) of time reversal for this problem. This problem is from Gottfried (1966), Problem 7-3.

**5.20** Work out the Stark effect to lowest nonvanishing order for the $n = 3$ level of the hydrogen atom. Ignoring the spin-orbit force and relativistic correction (Lamb shift), obtain not only the energy shifts to lowest nonvanishing order but also the corresponding zeroth-order eigenket.

**5.21** Suppose the electron had a very small intrinsic *electric* dipole moment analogous to the spin magnetic moment (that is, $\mu_{el}$ proportional to $\sigma$). Treating the hypothetical $-\mu_{el} \cdot \mathbf{E}$ interaction as a small perturbation, discuss qualitatively how the energy levels of the Na atom $(Z = 11)$ would be altered in the absence of any external electromagnetic field. Are the level shifts first order or second order? State explicitly which states get mixed with each other. Obtain an expression for the energy shift of the lowest level that is affected by the perturbation. Assume throughout that only the valence electron is subjected to the hypothetical interaction.

**5.22** Consider a particle bound to a fixed center by a spherically symmetric potential $V(r)$.

a. Prove

$$|\psi(0)|^2 = \left(\frac{m}{2\pi\hbar^2}\right)\left\langle\frac{dV}{dr}\right\rangle$$

for all $s$ states, ground and excited.

b. Check this relation for the ground state of a three-dimensional isotropic oscillator, the hydrogen atom, and so on. (*Note*: This relation has actually been found to be useful in guessing the form of the potential between a quark and an antiquark. See Moxhay and Rosner, *J. Math. Phys.*, **21** (1980) 1688.)

**5.23** a. Suppose the Hamiltonian of a rigid rotator in a magnetic field perpendicular to the axis is of the form (Merzbacher 1970, Problem 17-1)

$$A\mathbf{L}^2 + BL_z + CL_y$$

if terms quadratic in the field are neglected. Assuming $B \gg C$, use perturbation theory to lowest nonvanishing order to get approximate energy eigenvalues.

b. Consider the matrix elements

$$\langle n'l'm'_l m'_s|(3z^2 - r^2)|nlm_l m_s\rangle,$$
$$\langle n'l'm'_l m'_s|xy|nlm_l m_s\rangle$$

of a one-electron (for example, alkali) atom. Write the selection rules for $\Delta l$, $\Delta m_l$, and $\Delta m_s$. Justify your answer.

**5.24** The $n = 2$ state of hydrogen is eightfold degenerate, accounting for both spin and orbital angular momentum. This degeneracy is broken by a perturbation

$$V = \frac{A}{\hbar^2}\mathbf{L}\cdot\mathbf{S} + \frac{B}{\hbar}(L_z + 2S_z)$$

where $\mathbf{L}$ and $\mathbf{S}$ are the orbital and spin angular-momentum operators, $A$ is a constant, and $B$ is proportional (but not equal) to an applied magnetic field in the $z$-direction.

a. Write $V$ in terms of $\mathbf{J}^2$, $\mathbf{L}^2$, $\mathbf{S}^2$, $J_z$, and $S_z$, where $\mathbf{J} = \mathbf{L} + \mathbf{S}$.

b. Find all nonzero matrix elements of $V$ in the basis $|l,s = 1/2, j = l \pm 1/2, m\rangle$ for the eight $n = 2$ states. *Hint:* Show that the $8 \times 8$ matrix decouples into four $2 \times 2$ matrices, two of which are diagonal.

c. Use degenerate perturbation theory to find the first-order energy shifts $\Delta$. *For all eight states*, plot $\Delta/A$ as a function of $B/A$. See Figure 5.3. Explain why the resulting spectrum looks qualitatively different for $B/A \ll 1$ and $B/A \gg 1$.

**5.25** Work out the *quadratic* Zeeman effect for the ground-state hydrogen atom due to the usually neglected $e^2 \mathbf{A}^2/2m_e c^2$-term in the Hamiltonian taken to first order. Write the energy shift as

$$\Delta = -\tfrac{1}{2}\chi \mathbf{B}^2$$

and obtain an expression for *diamagnetic susceptibility*, $\chi$.

**5.26** (Merzbacher (1970), p. 448, Problem 11.) For the He wave function, use

$$\psi(\mathbf{x}_1, \mathbf{x}_2) = (Z_{\mathrm{eff}}^3/\pi a_0^3)\exp\left[\frac{-Z_{\mathrm{eff}}(r_1 + r_2)}{a_0}\right]$$

with $Z_{\mathrm{eff}} = 2 - \tfrac{5}{16}$, as obtained by the variational method. The measured value of the diamagnetic susceptibility is $1.88 \times 10^{-6}\,\mathrm{cm}^3/\mathrm{mole}$.

a. Using the Hamiltonian for an atomic electron in a magnetic field, determine, for a state of zero angular momentum, the energy change to order $B^2$ if the system is in a uniform magnetic field represented by the vector potential $\mathbf{A} = \tfrac{1}{2}\mathbf{B} \times \mathbf{x}$.

b. Defining the atomic diamagnetic susceptibility $\chi$ by $E = -\tfrac{1}{2}\chi B^2$, calculate $\chi$ for a helium atom in the ground state and compare the result with the measured value.

**5.27** Estimate the ground-state energy of a one-dimensional simple harmonic oscillator using

$$\langle x | \tilde{0} \rangle = e^{-\beta|x|}$$

as a trial function with $\beta$ to be varied.

**5.28** Estimate the lowest eigenvalue ($\lambda$) of the differential equation

$$\frac{d^2\psi}{dx^2} + (\lambda - |x|)\psi = 0$$

where $\psi \to 0$ as $|x| \to \infty$ using the variational method with

$$\psi = \begin{cases} c(\alpha - |x|) & \text{for } |x| < \alpha \\ 0 & \text{for } |x| > \alpha \end{cases} \qquad (\alpha \text{ to be varied})$$

as a trial function. (*Caution:* $d\psi/dx$ is discontinuous at $x = 0$.) The *exact* value of the lowest eigenvalue can be shown to be 1.019.

**5.29** Consider a one-dimensional simple harmonic oscillator whose classical angular frequency is $\omega_0$. For $t < 0$ it is known to be in the ground state. For $t > 0$ there is also a time-dependent potential

$$V(t) = F_0 x \cos \omega t$$

where $F_0$ is constant in both space and time. Obtain an expression for the expectation value $\langle x \rangle$ as a function of time using time-dependent perturbation theory to lowest nonvanishing order. Is this procedure valid for $\omega \simeq \omega_0$?

**5.30** A one-dimensional harmonic oscillator is in its ground state for $t < 0$. For $t \geq 0$ it is subjected to a time-dependent but spatially uniform *force* (not potential!) in the $x$-direction,

$$F(t) = F_0 e^{-t/\tau}.$$

a. Using time-dependent perturbation theory to first order, obtain the probability of finding the oscillator in its first excited state for $t > 0$. Show that the $t \to \infty$ ($\tau$ finite) limit of your expression is independent of time. Is this reasonable or surprising?

b. Can we find higher excited states?

**5.31** Consider a particle bound in a simple harmonic oscillator potential. Initially ($t < 0$), it is in the ground state. At $t = 0$ a perturbation of the form

$$H'(x,t) = Ax^2 e^{-t/\tau}$$

is switched on. Using time-dependent perturbation theory, calculate the probability that, after a sufficiently long time ($t \gg \tau$), the system will have made a transition to a given excited state. Consider all final states.

**5.32** The unperturbed Hamiltonian of a two-state system is represented by

$$H_0 = \begin{pmatrix} E_1^0 & 0 \\ 0 & E_2^0 \end{pmatrix}.$$

There is, in addition, a time-dependent perturbation

$$V(t) = \begin{pmatrix} 0 & \lambda \cos \omega t \\ \lambda \cos \omega t & 0 \end{pmatrix} \quad (\lambda \text{ real}).$$

a. At $t = 0$ the system is known to be in the first state, represented by

$$\begin{pmatrix} 1 \\ 0 \end{pmatrix}.$$

Using time-dependent perturbation theory and assuming that $E_1^0 - E_2^0$ is *not* close to $\pm \hbar \omega$, derive an expression for the probability that the system be found in the second state represented by

$$\begin{pmatrix} 0 \\ 1 \end{pmatrix}$$

as a function of $t(t > 0)$.

b. Why is this procedure not valid when $E_1^0 - E_2^0$ is close to $\pm \hbar \omega$?

**5.33** A one-dimensional simple harmonic oscillator of angular frequency $\omega$ is acted upon by a spatially uniform but time-dependent force (*not* potential)

$$F(t) = \frac{(F_0 \tau / \omega)}{(\tau^2 + t^2)}, \quad -\infty < t < \infty.$$

At $t = -\infty$, the oscillator is known to be in the ground state. Using the time-dependent perturbation theory to first order, calculate the probability that the oscillator is found in the first excited state at $t = +\infty$.

Challenge for experts: $F(t)$ is so normalized that the impulse

$$\int F(t)dt$$

imparted to the oscillator is always the same, that is, independent of $\tau$; yet for $\tau \gg 1/\omega$, the probability for excitation is essentially negligible. Is this reasonable?

**5.34** Consider a particle in one dimension moving under the influence of some time-independent potential. The energy levels and the corresponding eigenfunctions for this problem are assumed to be known. We now subject the particle to a traveling pulse represented by a time-dependent potential,

$$V(t) = A\delta(x - ct).$$

a. Suppose at $t = -\infty$ the particle is known to be in the ground state whose energy eigenfunction is $\langle x|i \rangle = u_i(x)$. Obtain the probability for finding the system in some excited state with energy eigenfunction $\langle x|f \rangle = u_f(x)$ at $t = +\infty$.
b. Interpret your result in (a) physically by regarding the $\delta$-function pulse as a superposition of harmonic perturbations; recall

$$\delta(x - ct) = \frac{1}{2\pi c}\int_{-\infty}^{\infty} d\omega e^{i\omega[(x/c)-t]}.$$

Emphasize the role played by energy conservation, which holds even quantum mechanically as long as the perturbation has been on for a very long time.

**5.35** A hydrogen atom in its ground state $[(n,l,m) = (1,0,0)]$ is placed between the plates of a capacitor. A time-dependent but spatial uniform electric field (not potential!) is applied as follows:

$$\mathbf{E} = \begin{cases} 0 & \text{for } t < 0 \\ \mathbf{E}_0 e^{-t/\tau} & \text{for } t > 0 \end{cases} (\mathbf{E}_0 \text{ in the positive } z\text{-direction}).$$

Using first-order time-dependent perturbation theory, compute the probability for the atom to be found at $t \gg \tau$ in each of the three $2p$ states: $(n,l,m) = (2,1,\pm 1 \text{ or } 0)$. Repeat the problem for the $2s$ state: $(n,l,m) = (2,0,0)$. Consider the limit $\tau \to \infty$.

**5.36** Consider a composite system made up of two spin $\frac{1}{2}$ objects. For $t < 0$, the Hamiltonian does not depend on spin and can be taken to be zero by suitably adjusting the energy scale. For $t > 0$, the Hamiltonian is given by

$$H = \left(\frac{4\Delta}{\hbar^2}\right)\mathbf{S}_1 \cdot \mathbf{S}_2.$$

Suppose the system is in $|+-\rangle$ for $t \leq 0$. Find, as a function of time, the probability for being found in each of the following states $|++\rangle$, $|+-\rangle$, $|-+\rangle$, and $|--\rangle$.

a. By solving the problem exactly.
b. By solving the problem assuming the validity of first-order time-dependent perturbation theory with $H$ as a perturbation switched on at $t = 0$. Under what condition does (b) give the correct results?

**5.37** Consider a two-level system with $E_1 < E_2$. There is a time-dependent potential that connects the two levels as follows:

$$V_{11} = V_{22} = 0, \quad V_{12} = \gamma e^{i\omega t}, \quad V_{21} = \gamma e^{-i\omega t} \quad (\gamma \text{ real}).$$

At $t - 0$, it is known that only the lower level is populated, that is, $c_1(0) = 1$, $c_2(0) = 0$.

a. Find $|c_1(t)|^2$ and $|c_2(t)|^2$ for $t > 0$ by *exactly* solving the coupled differential equation

$$i\hbar\dot{c}_k = \sum_{n=1}^{2} V_{kn}(t)e^{i\omega_{kn}t}c_n \quad (k = 1,2).$$

b. Do the same problem using time-dependent perturbation theory to lowest non-vanishing order. Compare the two approaches for small values of $\gamma$. Treat the following two cases separately: (i) $\omega$ very different from $\omega_{21}$ and (ii) $\omega$ close to $\omega_{21}$.

Answer for (a): (Rabi's formula)

$$|c_2(t)|^2 = \frac{\gamma^2/\hbar^2}{\gamma^2/\hbar^2 + (\omega - \omega_{21})^2/4} \sin^2\left\{\left[\frac{\gamma^2}{\hbar^2} + \frac{(\omega - \omega_{21})^2}{4}\right]^{1/2} t\right\},$$

$$|c_1(t)|^2 = 1 - |c_2(t)|^2.$$

**5.38** Show that the slow-turn-on of perturbation $V \to Ve^{\eta t}$ (see Baym (1969), p. 257) can generate contribution from the second term in (5.295).

**5.39** a. Consider the positronium problem solved in Chapter 3, Problem 3.5. In the presence of a uniform and static magnetic field $B$ along the $z$-axis, the Hamiltonian is given by

$$H = A\mathbf{S}_1 \cdot \mathbf{S}_2 + \left(\frac{eB}{m_e c}\right)(S_{1z} - S_{2z}).$$

Solve this problem to obtain the energy levels of all four states *using degenerate time-independent perturbation theory* (instead of diagonalizing the Hamiltonian matrix). Regard the first and the second terms in the expression for $H$ as $H_0$ and $V$, respectively. Compare your results with the exact expressions

$$E = -\left(\frac{\hbar^2 A}{4}\right)\left[1 \pm 2\sqrt{1 + 4\left(\frac{eB}{m_e c\hbar A}\right)^2}\right] \quad \text{for } \begin{cases} \text{singlet } m = 0 \\ \text{triplet } m = 0 \end{cases}$$

$$E = \frac{\hbar^2 A}{4} \quad \text{for triplet } m = \pm 1,$$

where *triplet* (*singlet*) $m = 0$ stands for the state that becomes a pure triplet (singlet) with $m = 0$ as $B \to 0$.

b. We now attempt to cause transitions (via stimulated emission and absorption) between the two $m = 0$ states by introducing an oscillating magnetic field of the "right" frequency. Should we orient the magnetic field along the $z$-axis or along the $x$- (or $y$-) axis? Justify your choice. (The original static field is assumed to be along the $z$-axis throughout.)

c. Calculate the eigenvectors to first order.

**5.40** Repeat Problem 5.39 above, but with the atomic hydrogen Hamiltonian

$$H = A\mathbf{S}_1 \cdot \mathbf{S}_2 + \left( \frac{eB}{m_e c} \right) \mathbf{S}_1 \cdot \mathbf{B}$$

where in the hyperfine term $A\mathbf{S}_1 \cdot \mathbf{S}_2$, $\mathbf{S}_1$ is the electron spin, while $\mathbf{S}_2$ is the proton spin. (Note the problem here has less symmetry than that of the positronium case.)

**5.41** Consider the spontaneous emission of a photon by an excited atom. The process is known to be an $E1$ transition. Suppose the magnetic quantum number of the atom decreases by one unit. What is the angular distribution of the emitted photon? Also discuss the polarization of the photon with attention to angular-momentum conservation for the whole (atom plus photon) system.

**5.42** Consider an atom made up of an electron and a singly charged ($Z = 1$) triton ($^3$H). Initially the system is in its ground state ($n = 1, l = 0$). Suppose the system undergoes beta decay, in which the nuclear charge *suddenly increases* by one unit (realistically by emitting an electron and an antineutrino). This means that the tritium nucleus (called a "triton") turns into a helium ($Z = 2$) nucleus of mass 3 ($^3$He).

a. Obtain the probability for the system to be found in the ground state of the resulting helium ion.

b. The available energy in tritium beta decay is about 18 keV and the size of the $^3$He atom is about 1 Å. Check that the time scale $T$ for the transformation satisfies the criterion of validity for the sudden approximation.

**5.43** Show that $\mathbf{A}_n(\mathbf{R})$ defined in (5.234) is a purely real quantity.

**5.44** Consider a neutron in a magnetic field, fixed at an angle $\theta$ with respect to the $\mathbf{z}$-axis, but rotating slowly in the $\phi$ direction. That is, the tip of the magnetic field traces out a circle on the surface of the sphere, at "latitude" $\pi - \theta$. Explicitly calculate the Berry potential $\mathbf{A}$ for the spin-up state from (5.234), take its curl, and determine Berry's phase $\gamma_+$. Thus, verify (5.253) for this particular example of a curve $C$. (For hints, see "The adiabatic theorem and Berry's phase" by Holstein, *Am. J. Phys.*, **57** (1989) 1079.)

**5.45** The ground state of a hydrogen atom ($n = 1, l = 0$) is subjected to a time-dependent potential as follows:

$$V(\mathbf{x}, t) = V_0 \cos(kz - \omega t).$$

Using time-dependent perturbation theory, obtain an expression for the transition rate at which the electron is emitted with momentum $\mathbf{p}$. Show, in particular, how you may compute the angular distribution of the ejected electron (in terms of $\theta$ and $\phi$ defined with respect to the $z$-axis). Discuss *briefly* the similarities and the differences between this problem and the (more realistic) photoelectric effect. (*Note:* For the initial wave function see Problem 5.42. If you have a normalization problem, the final wave function may be taken to be

$$\psi_f(\mathbf{x}) = \left(\frac{1}{L^{3/2}}\right) e^{i\mathbf{p}\cdot\mathbf{x}/\hbar}$$

with $L$ very large, but you should be able to show that the observable effects are independent of $L$.)

**5.46** A particle of mass $m$ constrained to move in one dimension is confined within $0 < x < L$ by an infinite-wall potential

$$V = \infty \quad \text{for } x < 0, \, x > L,$$
$$V = 0 \quad \text{for } 0 < x < L.$$

Obtain an expression for the density of states (that is, the number of states per unit energy interval) for *high* energies as a function of $E$. (Check your dimension!)

**5.47** Linearly polarized light of angular frequency $\omega$ is incident on a one-electron "atom" whose wave function can be approximated by the ground state of a three-dimensional isotropic harmonic oscillator of angular frequency $\omega_0$. Show that the differential cross section for the ejection of a photoelectron is given by

$$\frac{d\sigma}{d\Omega} = \frac{4\alpha\hbar^2 k_f^3}{m^2\omega\omega_0}\sqrt{\frac{\pi\hbar}{m\omega_0}}\exp\left\{-\frac{\hbar}{m\omega_0}\left[k_f^2 + \left(\frac{\omega}{c}\right)^2\right]\right\}$$
$$\times \sin^2\theta\cos^2\phi\exp\left[\left(\frac{2\hbar k_f\omega}{m\omega_0 c}\right)\cos\theta\right]$$

provided the ejected electron of momentum $\hbar k_f$ can be regarded as being in a plane wave state. (The coordinate system used is shown in Figure 5.13.)

**5.48** Find the probability $|\phi(\mathbf{p}')|^2 d^3 p'$ of the particular momentum $\mathbf{p}'$ for the ground-state hydrogen atom. (This is a nice exercise in three-dimensional Fourier transforms. To perform the angular integration choose the $z$-axis in the direction of $\mathbf{p}$.)

**5.49** Calculate the lifetimes for the $2p \to 1s$ and $3p \to 1s$ transitions in the hydrogen atom. You can find measurements of these lifetimes in Bickel and Goodman, *Phys. Rev.*, **148** (1966) 1.

**5.50** A hydrogen atom is prepared in the $2p, m = +1$ state. Find the rate for the $2p \to 1s$ transition as a function of polar angle $\theta$ with respect to the $z$-axis. Describe the polarization of the emitted radiation as a function of $\theta$ and use this to qualitatively explain the intensity pattern.

**5.51** This problem highlights anomalies in the "exponential" decay of a state. It is inspired by Winter, *Phys. Rev.*, **123** (1961) 1503, but modern computer applications make it straightforward to directly evaluate the integrals numerically.

Consider a particle of mass $m$ that is initially inside a "well" bounded by an infinite wall to the left and a $\delta$-function potential on the right:

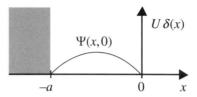

The infinite wall is located at $x = -a$, and the potential at $x = 0$ is $U\delta(x)$ where $U$ is a positive constant. The figure also shows a plausible "ground state" initial wave function $\Psi(x, t = 0) = (2/a)^{1/2} \sin(n\pi x/a)$ with $n = 1$.

a.  Show that the wave function at all times can be written as

$$\Psi(x,t) = 2n \left(\frac{2}{a}\right)^{1/2} \int_0^\infty dq \frac{e^{-iTq^2} q \sin q \left[q \sin(l+1)q + f\right]}{(q^2 - n^2\pi^2)(q^2 + Gq\sin 2q + G^2\sin^2 q)}$$

where $q \equiv [a(2mE)^{1/2}]/\hbar$ for a particle with energy $E$, $T \equiv \hbar t/2ma^2$, $l \equiv x/a$, and $G \equiv 2maU/\hbar^2$ are all dimensionless quantities, and $f = 0$ for $-a \leq x \leq 0$ and $f = G \sin q \sin lq$ for $x \geq 0$. This is most easily done by expanding the wave function in energy eigenstates $|E\rangle$, as

$$\Psi(x,t) = \int_0^\infty dE \phi_E(x) e^{-iEt/\hbar}$$

where $\phi_E(x)$ is an energy eigenfunction and $\langle E|E'\rangle = \delta(E - E')$.

b.  Write a computer program to (numerically) integrate the probability of finding the particle *inside* the well. Carry out the integration for a series of values of $T$ between zero and 12, using the same parameters as Winter, namely $n = 1$ and $G = 6$. Plotting these probabilities as a function of $T$ should resemble Figure 2 of Winter's paper. Fit the points for $2 \leq T \leq 8$ to an exponential, and compare the decay time to Winter's value of 0.644.

c.  Examine the behavior for $T \geq 8$, and compare to the behavior Winter found for the current at $x = 0$. This suggests an experimental measurement. See Norman et al., *Phys. Rev. Lett.*, **60** (1988) 2246.

# 6

# Scattering Theory

This chapter is devoted to the theory of scattering processes. These are processes in which a continuum initial state is transformed into a continuum final state, through the action of some potential which we will treat as a time-dependent perturbation. Such processes are of enormous significance. They are the primary way in which we learn experimentally about distributions in mass, charge, and, in general, potential energy for molecular, atomic, and subatomic systems.

## 6.1 Scattering as a Time-Dependent Perturbation

We assume that the Hamiltonian can be written as

$$H = H_0 + V(\mathbf{x}) \tag{6.1}$$

where

$$H_0 = \frac{\mathbf{p}^2}{2m} \tag{6.2}$$

stands for the kinetic-energy operator, with eigenvalues

$$E_{\mathbf{k}} = \frac{\hbar^2 \mathbf{k}^2}{2m}. \tag{6.3}$$

We denote the plane wave eigenvectors of $H_0$ by $|\mathbf{k}\rangle$ and we assume that the scattering potential $V(\mathbf{r})$ is time independent.

Our treatment realizes that an incoming particle will "see" the scattering potential as a perturbation which is "turned on" only during the time that the particle is in the vicinity of the scatterer. Therefore, we can analyze the problem in terms of time-dependent perturbation theory in the interaction picture.

To review (see Section 5.7), the state $|\alpha, t_0; t_0\rangle_I$ evolves into the state $|\alpha, t_0; t\rangle_I$ according to

$$|\alpha, t_0; t\rangle_I = U_I(t, t_0)|\alpha, t_0; t_0\rangle_I \tag{6.4}$$

where $U_I(t, t_0)$ satisfies the equation

$$i\hbar \frac{\partial}{\partial t} U_I(t, t_0) = V_I(t) U_I(t, t_0) \tag{6.5}$$

with $U_I(t_0, t_0) = 1$ and $V_I(t) = \exp(iH_0 t/\hbar) V \exp(-iH_0 t/\hbar)$. The solution of this equation can be formally written as

$$U_I(t, t_0) = 1 - \frac{i}{\hbar} \int_{t_0}^{t} V_I(t') U_I(t', t_0) dt'. \tag{6.6}$$

Therefore, the "transition amplitude" for an initial state $|i\rangle$ to transform into a final state $|n\rangle$, where both are eigenstates of $H_0$, is given by

$$\langle n | U_I(t, t_0) | i \rangle = \delta_{ni} - \frac{i}{\hbar} \sum_m \langle n | V | m \rangle \int_{t_0}^{t} e^{i\omega_{nm} t'} \langle m | U_I(t', t_0) | i \rangle dt' \tag{6.7}$$

where $\langle n | i \rangle = \delta_{ni}$ and $\hbar \omega_{nm} = E_n - E_m$.

To apply this formalism to scattering theory, we need to make some adjustments. First, there is the normalization of the initial and final states. Equation (6.7) assumes discrete states, but our scattering states are in the continuum. We deal with this by quantizing our scattering states in a "big box," i.e. a cube of side $L$. In the coordinate representation, this gives

$$\langle \mathbf{x} | \mathbf{k} \rangle = \frac{1}{L^{3/2}} e^{i\mathbf{k} \cdot \mathbf{x}} \tag{6.8}$$

in which case $\langle \mathbf{k}' | \mathbf{k} \rangle = \delta_{\mathbf{kk}'}$ where the $\mathbf{k}$ take on discrete values. We will take $L \to \infty$ at the end of any calculation.

We also need to deal with the fact that both the initial and final states exist only asymptotically. That is, we need to work with both $t \to \infty$ and $t_0 \to -\infty$. We can take a hint from a first-order treatment of (6.7) in which case we set $\langle m | U_I(t', t_0) | i \rangle = \delta_{mi}$ inside the integral, that is

$$\langle n | U_I(t, t_0) | i \rangle = \delta_{ni} - \frac{i}{\hbar} \langle n | V | i \rangle \int_{t_0}^{t} e^{i\omega_{ni} t'} dt'. \tag{6.9}$$

In this case, as $t \to \infty$ we saw a "transition rate" emerge as Fermi's golden rule. So, in order to also accommodate $t_0 \to -\infty$, we define a matrix $T$ as follows:

$$\langle n | U_I(t, t_0) | i \rangle = \delta_{ni} - \frac{i}{\hbar} T_{ni} \int_{t_0}^{t} e^{i\omega_{ni} t' + \varepsilon t'} dt' \tag{6.10}$$

where $\varepsilon > 0$ and $t \ll (1/\varepsilon)$. These conditions ensure that $e^{\varepsilon t'}$ is close to unity as $t \to \infty$, and that the integrand goes to zero as $t_0 \to -\infty$. We just need to make sure that we take the limit $\varepsilon \to 0$ first, before we take $t \to +\infty$.

We can now define the scattering (or $S$) matrix in terms of the $T$ matrix:

$$S_{ni} \equiv \lim_{t \to \infty} \left[ \lim_{\varepsilon \to 0} \langle n | U_I(t, -\infty) | i \rangle \right] = \delta_{ni} - \frac{i}{\hbar} T_{ni} \int_{-\infty}^{\infty} e^{i\omega_{ni} t'} dt'$$

$$= \delta_{ni} - 2\pi i \delta(E_n - E_i) T_{ni}. \tag{6.11}$$

Clearly, the $S$ matrix consists of two parts. One part is that in which the final state is the same as the initial state. The second part, governed by the $T$ matrix, is one in which some sort of scattering occurs.

## 6.1.1 Transition Rates and Cross Sections

Proceeding as in Section 5.7, we define the transition rate as

$$w(i \to n) = \frac{d}{dt} |\langle n|U_I(t, -\infty)|i\rangle|^2 \tag{6.12}$$

where for $|i\rangle \neq |n\rangle$ we have

$$\langle n|U_I(t, -\infty)|i\rangle = -\frac{i}{\hbar} T_{ni} \int_{-\infty}^{t} e^{i\omega_{ni}t' + \varepsilon t'} dt' = -\frac{i}{\hbar} T_{ni} \frac{e^{i\omega_{ni}t + \varepsilon t}}{i\omega_{ni} + \varepsilon} \tag{6.13}$$

and therefore

$$w(i \to n) = \frac{d}{dt}\left[ \frac{1}{\hbar^2} |T_{ni}|^2 \frac{e^{2\varepsilon t}}{\omega_{ni}^2 + \varepsilon^2} \right] = \frac{1}{\hbar^2} |T_{ni}|^2 \frac{2\varepsilon \, e^{2\varepsilon t}}{\omega_{ni}^2 + \varepsilon^2}. $$

We need to take $\varepsilon \to 0$ for finite values of $t$, and then $t \to \infty$. Clearly, this will send $w \to 0$ if $\omega_{ni} \neq 0$ so we see something like $\delta(\omega_{ni})$ emerging, which is not unexpected based on (6.11). In fact, since

$$\int_{-\infty}^{\infty} \frac{1}{\omega^2 + \varepsilon^2} d\omega = \frac{\pi}{\varepsilon} \tag{6.14}$$

for $\varepsilon > 0$, we have, for finite $t$,

$$\lim_{\varepsilon \to 0} \frac{\varepsilon \, e^{2\varepsilon t}}{\omega_{ni}^2 + \varepsilon^2} = \pi\delta(\omega_{ni}) = \pi\hbar\delta(E_n - E_i). \tag{6.15}$$

Therefore, the transition rate is

$$w(i \to n) = \frac{2\pi}{\hbar} |T_{ni}|^2 \, \delta(E_n - E_i) \tag{6.16}$$

which is independent of time, so the limit as $t \to \infty$ is trivial. This expression is strikingly similar to Fermi's golden rule (5.294), except that $V_{ni}$ has been replaced by the more general $T_{ni}$. We will see below how to determine the matrix elements $T_{ni}$ in general. First, however, let us continue with this discussion and use the transition rate to express the scattering cross section.

As with Fermi's golden rule, in order to integrate over the final state energy $E_n$, we need to determine the density of final states $\rho(E_n) = \Delta n/\Delta E_n$. We will determine the density of states for elastic scattering, where $|i\rangle = |\mathbf{k}\rangle$ and $|n\rangle = |\mathbf{k}'\rangle$ and $|\mathbf{k}| = |\mathbf{k}'| \equiv k$. (Recall our discussion of "The Free Particle in Three Dimensions" in Section 2.5.1.) For our "big box" normalization, we write

$$E_{\mathbf{n}} = \frac{\hbar^2 \mathbf{k}'^2}{2m} = \frac{\hbar^2}{2m}\left(\frac{2\pi}{L}\right)^2 |\mathbf{n}|^2 \quad \text{so} \quad \Delta E_{\mathbf{n}} = \frac{\hbar^2}{m}\left(\frac{2\pi}{L}\right)^2 |\mathbf{n}|\Delta|\mathbf{n}| \tag{6.17}$$

where $\mathbf{n} = n_x\hat{\mathbf{i}} + n_y\hat{\mathbf{j}} + n_z\hat{\mathbf{k}}$ and $n_{x,y,z}$ are integers. Since $\mathbf{n} = (L/2\pi)|\mathbf{k}'| = (L/2\pi)k$ and $L$ is large, we can think of $|\mathbf{n}|$ as nearly continuous, and the number of states within a spherical shell of radius $|\mathbf{n}|$ and thickness $\Delta|\mathbf{n}|$ is

$$\Delta n = 4\pi|\mathbf{n}|^2\Delta|\mathbf{n}| \times \frac{d\Omega}{4\pi} \tag{6.18}$$

taking into account the fraction of solid angle represented by the final state wave vector $\mathbf{k}$. Therefore

$$\rho(E_n) = \frac{\Delta n}{\Delta E_n} = \frac{m}{\hbar^2}\left(\frac{L}{2\pi}\right)^2 |\mathbf{n}| d\Omega = \frac{mk}{\hbar^2}\left(\frac{L}{2\pi}\right)^3 d\Omega \tag{6.19}$$

and after integrating over final states, the transition rate is given by

$$w(i \to n) = \frac{mkL^3}{(2\pi)^2\hbar^3} |T_{ni}|^2 d\Omega. \tag{6.20}$$

We use the concept of cross section to interpret the transition rate in scattering experiments. That is, we determine the rate at which particles are scattered into a solid angle $d\Omega$ from a "beam" of particles with momentum $\hbar\mathbf{k}$. The speed of these particles is $v = \hbar k/m$ so the time it takes for a particle to cross the "big box" is $L/v$. Therefore the flux in the particle beam is $(1/L^2) \div (L/v) = v/L^3$. Indeed, the probability flux (2.191) for the wave function (6.8) becomes

$$\mathbf{j}(\mathbf{x}, t) = \left(\frac{\hbar}{m}\right)\frac{\mathbf{k}}{L^3} = \frac{v}{L^3}. \tag{6.21}$$

The cross section $d\sigma$ is simply defined as the transition rate divided by the flux. Putting this all together, we have

$$\frac{d\sigma}{d\Omega} = \left(\frac{mL^3}{2\pi\hbar^2}\right)^2 |T_{ni}|^2. \tag{6.22}$$

The job now before us is to relate the matrix elements $T_{ni}$ to the scattering potential distribution $V(\mathbf{r})$.

## 6.1.2 Solving for the $T$ Matrix

We return to the definition of the $T$ matrix. From (6.10) and (6.13) we have

$$\langle n|U_I(t, -\infty)|i\rangle = \delta_{ni} + \frac{1}{\hbar}T_{ni}\frac{e^{i\omega_{ni}t+\varepsilon t}}{-\omega_{ni} + i\varepsilon}. \tag{6.23}$$

We can also return to (6.7). Writing $V_{nm} = \langle n|V|m\rangle$, we have

$$\langle n|U_I(t, -\infty)|i\rangle = \delta_{ni} - \frac{i}{\hbar}\sum_m V_{nm}\int_{-\infty}^t e^{i\omega_{nm}t'}\langle m|U_I(t', -\infty)|i\rangle dt'. \tag{6.24}$$

Now insert (6.23) into the integrand of (6.24). This results in three terms, the first of which is $\delta_{ni}$ and the second looks just like (6.23) but with $T_{ni}$ replaced with $V_{ni}$. The third term is

$$-\frac{i}{\hbar}\frac{1}{\hbar}\sum_m V_{nm}\frac{T_{mi}}{-\omega_{mi} + i\varepsilon}\int_{-\infty}^t e^{i\omega_{nm}t'+i\omega_{mi}t'+\varepsilon t'}dt'. \tag{6.25}$$

The integral is then carried out, and since $\omega_{nm} + \omega_{mi} = \omega_{ni}$, the result can be taken outside the summation. Gathering terms and comparing the result to (6.23), we discover the following relation:

$$T_{ni} = V_{ni} + \frac{1}{\hbar} \sum_m V_{nm} \frac{T_{mi}}{-\omega_{mi} + i\varepsilon} = V_{ni} + \sum_m V_{nm} \frac{T_{mi}}{E_i - E_m + i\hbar\varepsilon}. \tag{6.26}$$

This is an inhomogeneous system of linear equations which can be solved for the values $T_{ni}$, in terms of the known matrix elements $V_{nm}$. It is convenient to define a set of vectors $|\psi^{(+)}\rangle$ in terms of components in some basis $|j\rangle$, so that

$$T_{ni} = \sum_j \langle n|V|j\rangle \langle j|\psi^{(+)}\rangle = \langle n|V|\psi^{(+)}\rangle. \tag{6.27}$$

(The choice of notation will be become apparent shortly.) Therefore (6.26) becomes

$$\langle n|V|\psi^{(+)}\rangle = \langle n|V|i\rangle + \sum_m \langle n|V|m\rangle \frac{\langle m|V|\psi^{(+)}\rangle}{E_i - E_m + i\hbar\varepsilon}. \tag{6.28}$$

Since this must be true for all $|n\rangle$, we have an expression for the $|\psi^{(+)}\rangle$, namely

$$|\psi^{(+)}\rangle = |i\rangle + \sum_m |m\rangle \frac{\langle m|V|\psi^{(+)}\rangle}{E_i - E_m + i\hbar\varepsilon}$$

$$= |i\rangle + \sum_m \frac{1}{E_i - H_0 + i\hbar\varepsilon} |m\rangle \langle m|V|\psi^{(+)}\rangle$$

or

$$|\psi^{(+)}\rangle = |i\rangle + \frac{1}{E_i - H_0 + i\hbar\varepsilon} V|\psi^{(+)}\rangle. \tag{6.29}$$

This is known as the **Lippmann–Schwinger equation**. The physical meaning of $(+)$ is to be discussed in a moment by looking at $\langle \mathbf{x}|\psi^{(+)}\rangle$ at large distances. Clearly, the states $|\psi^{(+)}\rangle$ have a fundamental importance, allowing us to rewrite (6.22) as

$$\frac{d\sigma}{d\Omega} = \left( \frac{mL^3}{2\pi\hbar^2} \right)^2 \left| \langle n|V|\psi^{(+)}\rangle \right|^2. \tag{6.30}$$

We have introduced the matrix elements $T_{ni}$ simply as complex numbers, defined by (6.10). However, we can also define an operator $T$ with matrix elements $\langle n|T|i\rangle = T_{ni}$ by writing $T|i\rangle = V|\psi^{(+)}\rangle$. We can then operate on (6.29) from the left with $V$ which leads to the succinct operator equation

$$T = V + V \frac{1}{E_i - H_0 + i\hbar\varepsilon} T. \tag{6.31}$$

To the extent that the scattering potential $V$ is "weak", an order-by-order approximation scheme presents itself for the transition operator $T$, namely

$$T = V + V \frac{1}{E_i - H_0 + i\hbar\varepsilon} V + V \frac{1}{E_i - H_0 + i\hbar\varepsilon} V \frac{1}{E_i - H_0 + i\hbar\varepsilon} V + \cdots. \tag{6.32}$$

We will return to this approximation scheme in Section 6.3.

### 6.1.3 Scattering from the Future to the Past

We can also picture the scattering process as evolving backwards in time from a plane wave state $|i\rangle$ in the far future, to a state $|n\rangle$ in the distant past. In this case, we would write the formal solution (6.6) as

$$U_I(t, t_0) = 1 + \frac{i}{\hbar} \int_t^{t_0} V_I(t') U_I(t', t_0) dt' \tag{6.33}$$

which is a form suitable for taking $t_0 \to +\infty$. Our $T$ matrix is then defined by regularizing the integral with the opposite sign exponential, namely

$$\langle n|U_I(t, t_0)|i\rangle = \delta_{ni} + \frac{i}{\hbar} T_{ni} \int_t^{t_0} e^{i\omega_{ni}t' - \varepsilon t'} dt'. \tag{6.34}$$

In this case, the $T$ operator is defined through a different set of states $|\psi^{(-)}\rangle$ through $T|i\rangle = V|\psi^{(-)}\rangle$. We are now prepared to study practical solutions to the scattering problem, and gain insight as to the different scattering states $|\psi^{(+)}\rangle$ and $|\psi^{(-)}\rangle$.

## 6.2 The Scattering Amplitude

Let us replace $\hbar\varepsilon$ in the Lippman–Schwinger equation with $\varepsilon$; this will be handy and presents no difficulties since the only constraints on $\varepsilon$ are that it be positive and arbitrarily small. We will also continue to anticipate application to elastic scattering, and use $E$ for the initial (and final) energy. We therefore rewrite (6.29) as

$$|\psi^{(\pm)}\rangle = |i\rangle + \frac{1}{E - H_0 \pm i\varepsilon} V|\psi^{(\pm)}\rangle. \tag{6.35}$$

We now confine ourselves to the position basis by multiplying $\langle x|$ from the left, and inserting a complete set of position basis states. Thus

$$\langle x|\psi^{(\pm)}\rangle = \langle x|i\rangle + \int d^3x' \left\langle x \left| \frac{1}{E - H_0 \pm i\varepsilon} \right| x' \right\rangle \langle x'|V|\psi^{(\pm)}\rangle. \tag{6.36}$$

This is an **integral equation** for scattering because the unknown ket $|\psi^{(\pm)}\rangle$ appears under an integral sign. To make progress we must first evaluate the function

$$G_\pm(x, x') \equiv \frac{\hbar^2}{2m} \left\langle x \left| \frac{1}{E - H_0 \pm i\varepsilon} \right| x' \right\rangle. \tag{6.37}$$

Since the eigenstates of $H_0$ are most easily evaluated in the momentum basis, we proceed by inserting complete sets of states $|k\rangle$. (Recall that these are discrete states in our normalization scheme.) We then write

$$G_\pm(x, x') = \frac{\hbar^2}{2m} \sum_{k'} \sum_{k''} \langle x|k'\rangle \left\langle k' \left| \frac{1}{E - H_0 \pm i\varepsilon} \right| k'' \right\rangle \langle k''|x'\rangle. \tag{6.38}$$

Now let $H_0$ act on $\langle \mathbf{k}'|$, and use

$$\left\langle \mathbf{k}' \left| \frac{1}{E - (\hbar^2 \mathbf{k}'^2/2m) \pm i\varepsilon} \right| \mathbf{k}'' \right\rangle = \frac{\delta_{\mathbf{k}'\mathbf{k}''}}{E - (\hbar^2 \mathbf{k}'^2/2m) \pm i\varepsilon} \tag{6.39}$$

$$\langle \mathbf{x}|\mathbf{k}'\rangle = \frac{e^{i\mathbf{k}'\cdot\mathbf{x}}}{L^{3/2}} \tag{6.40}$$

$$\langle \mathbf{k}''|\mathbf{x}'\rangle = \frac{e^{-i\mathbf{k}''\cdot\mathbf{x}'}}{L^{3/2}} \tag{6.41}$$

and put $E = \hbar^2 k^2/2m$. Equation (6.37) then becomes

$$G_\pm(\mathbf{x},\mathbf{x}') = \frac{1}{L^3} \sum_{\mathbf{k}'} \frac{e^{i\mathbf{k}'\cdot(\mathbf{x}-\mathbf{x}')}}{k^2 - k'^2 \pm i\varepsilon} \tag{6.42}$$

where we have once again redefined $\varepsilon$.

This sum is actually easiest to do if we take $L \to \infty$ and convert it to an integral. Since $k_i = 2\pi n_i/L$ ($i = x,y,z$), the integral measure becomes $d^3 k' = (2\pi)^3/L^3$ and we have

$$G_\pm(\mathbf{x},\mathbf{x}') = \frac{1}{(2\pi)^3} \int d^3 k' \frac{e^{i\mathbf{k}'\cdot(\mathbf{x}-\mathbf{x}')}}{k^2 - k'^2 \pm i\varepsilon}. \tag{6.43}$$

This integral can be done analytically. First, convert to spherical coordinates, with angles measured relative to the $\mathbf{k}'$ direction. The integral over $\phi_{k'}$ just gives a factor of $2\pi$. For the integral over $\sin\theta_{k'} d\theta_{k'}$, put $\mu \equiv \cos\theta_{k'}$ to find

$$G_\pm(\mathbf{x},\mathbf{x}') = \frac{1}{(2\pi)^2} \int_0^\infty k'^2 dk' \int_{-1}^{+1} d\mu \frac{e^{ik'|\mathbf{x}-\mathbf{x}'|\mu}}{k^2 - k'^2 \pm i\varepsilon}$$

$$= \frac{1}{8\pi^2} \frac{1}{i|\mathbf{x}-\mathbf{x}'|} \int_{-\infty}^\infty k' dk' \left[ \frac{e^{ik'|\mathbf{x}-\mathbf{x}'|} - e^{-ik'|\mathbf{x}-\mathbf{x}'|}}{k^2 - k'^2 \pm i\varepsilon} \right] \tag{6.44}$$

where we made use of the fact that the integrand is even in $k'$, allowing us to extend the integration lower limit to $-\infty$ while dividing by two.

We complete the calculation using complex contour integration.[1] This procedure will in fact demonstrate the importance of $\varepsilon$ and its sign. Treating $k'$ as a complex variable, imagine an integration contour running along the $\mathrm{Re}(k')$ axis, and then closed with a semi-circle in either the upper or lower plane. See Figure 6.1.

Now make use of the Cauchy integral formula (F.5), which we rewrite as

$$\oint_C \frac{f(z)}{z - z_0} dz = 2\pi i f(z_0) \tag{6.45}$$

where the contour $C$ is followed counter-clockwise. The denominator of the integrand of (6.44) can be written as $-(k' - k_0)(k' + k_0)$ where $k_0 \equiv k \pm i\varepsilon$. (Once again, $\varepsilon$ is redefined, keeping its sign intact, since $k$ is real and positive.) Therefore, the integral over the complex $k'$ plane has poles at $\pm k_0$, and we can apply (6.45).

---

[1] Integration over the complex plane arises naturally in many subjects, including scattering theory. For a brief summary of the mathematics, see Appendix F of this book. Complex analysis is also covered in just about any textbook on mathematical physics, for example Arfken et al. (2013) or Byron and Fuller (1992).

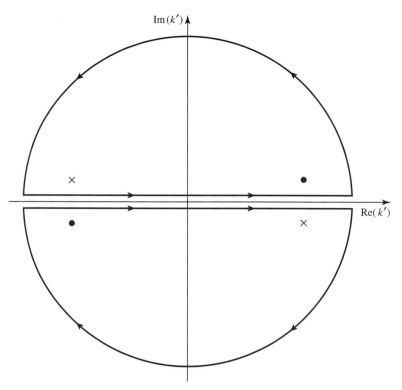

**Fig. 6.1** Integrating the two terms in (6.44) using complex contours. The dots (crosses) mark the positions of the two poles for the $+$ $(-)$ form of $G_\pm(\mathbf{x}, \mathbf{x}')$. We replace the integral over a real-valued $k'$ in (6.44) with one of the two contours in the figure, choosing the one on which the factor $e^{\pm ik' |\mathbf{x}-\mathbf{x}'|}$ tends to zero along the semicircle at large $\mathrm{Im}(k')$. Thus, the only contribution to the contour integral is along the real axis.

Consider separately the two terms in the integrand of (6.44). For the first term, close the contour in the *upper* plane. In this case, the contribution to the integrand along the semicircle goes to zero exponentially with $e^{ik' |\mathbf{x}-\mathbf{x}'|}$ as $\mathrm{Im}(k') \to +\infty$. Closing in the upper plane encloses the pole at $k' = +k + i\varepsilon$ $(k' = -k + i\varepsilon)$ in the calculation of $G_+$ $(G_-)$. Applying (6.45), the integral of the first term in brackets of (6.44) becomes

$$2\pi i(\pm k)\frac{e^{i(\pm k)|\mathbf{x}-\mathbf{x}'|}}{(-)(\pm 2k)} = -\pi i e^{\pm ik|\mathbf{x}-\mathbf{x}'|} \tag{6.46}$$

where we have let $\varepsilon \to 0$. The second term is treated the same way, except that the contour is closed in the *lower* plane, with another overall minus sign because the contour is traced clockwise. The contribution to the integral turns out to be the same as the first-term. We therefore get our final result, namely

$$G_\pm(\mathbf{x}, \mathbf{x}') = -\frac{1}{4\pi}\frac{e^{\pm ik|\mathbf{x}-\mathbf{x}'|}}{|\mathbf{x}-\mathbf{x}'|}. \tag{6.47}$$

The reader may recognize that $G_\pm$ is nothing more than Green's function for the Helmholtz equation,

$$(\nabla^2 + k^2)G_\pm(\mathbf{x}, \mathbf{x}') = \delta^{(3)}(\mathbf{x} - \mathbf{x}'). \tag{6.48}$$

That is, for $\mathbf{x} \neq \mathbf{x}'$, $G_\pm(\mathbf{x}, \mathbf{x}')$ solves the eigenvalue equation $H_0 G_\pm = E G_\pm$.

We can now rewrite (6.36) in a more explicit form, using (6.47), namely

$$\langle \mathbf{x} | \psi^{(\pm)} \rangle = \langle \mathbf{x} | i \rangle - \frac{2m}{\hbar^2} \int d^3 x' \frac{e^{\pm ik|\mathbf{x}-\mathbf{x}'|}}{4\pi|\mathbf{x}-\mathbf{x}'|} \langle \mathbf{x}' | V | \psi^{(\pm)} \rangle. \tag{6.49}$$

Notice that the wave function $\langle \mathbf{x} | \psi^{(\pm)} \rangle$ in the presence of the scatterer is written as the sum of the wave function for the incident wave $\langle \mathbf{x} | i \rangle$ and a term that represents the effect of scattering. As we will see explicitly later, at sufficiently large distances $r$, the spatial dependence of the second term is $e^{\pm ikr}/r$ provided that the potential is of finite range. This means that the positive solution (negative solution) corresponds to the plane wave plus an outgoing (incoming) spherical wave. This is in keeping with the origin of the sign in terms of scattering forward (backward) in time. In most physical problems we are interested in the positive solution because it is difficult to prepare a system satisfying the boundary condition appropriate for the negative solution.

To see the behavior of $\langle \mathbf{x} | \psi^{(\pm)} \rangle$ more explicitly, let us consider the specific case where $V$ is a local potential, that is, a potential diagonal in the $\mathbf{x}$-representation. Potentials that are functions only of the position operator $\mathbf{x}$ belong to this category. In precise terms $V$ is said to be **local** if it can be written as

$$\langle \mathbf{x}' | V | \mathbf{x}'' \rangle = V(\mathbf{x}')\delta^{(3)}(\mathbf{x}' - \mathbf{x}''). \tag{6.50}$$

As a result, we obtain

$$\langle \mathbf{x}' | V | \psi^{(\pm)} \rangle = \int d^3 x'' \langle \mathbf{x}' | V | \mathbf{x}'' \rangle \langle \mathbf{x}'' | \psi^{(\pm)} \rangle = V(\mathbf{x}')\langle \mathbf{x}' | \psi^{(+)} \rangle. \tag{6.51}$$

The integral equation (6.49) now simplifies as

$$\langle \mathbf{x} | \psi^{(\pm)} \rangle = \langle \mathbf{x} | i \rangle - \frac{2m}{\hbar^2} \int d^3 x' \frac{e^{\pm ik|\mathbf{x}-\mathbf{x}'|}}{4\pi|\mathbf{x}-\mathbf{x}'|} V(\mathbf{x}')\langle \mathbf{x}' | \psi^{(\pm)} \rangle. \tag{6.52}$$

Let us attempt to understand the physics contained in this equation. The vector $\mathbf{x}$ is understood to be directed towards the observation point at which the wave function is evaluated. For a finite range potential, the region that gives rise to a nonvanishing contribution is limited in space. In scattering processes we are interested in studying the effect of the scatterer (that is, the finite range potential) at a point far outside the range of the potential. This is quite relevant from a practical point of view because we cannot put a detector at short distance near the scattering center. Observation is always made by a detector placed very far away from the scatterer at $r$ greatly larger than the range of the potential. In other words, we can safely set

$$|\mathbf{x}| \gg |\mathbf{x}'|, \tag{6.53}$$

as depicted in Figure 6.2.

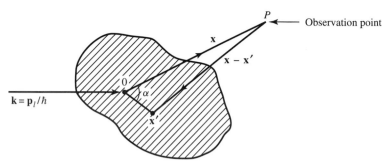

**Fig. 6.2** Finite-range scattering potential. The *observation point* $P$ is where the wave function $\langle \mathbf{x}|\psi^{(\pm)}\rangle$ is to be evaluated, while the contribution to the integral in (6.52) is for $|\mathbf{x}'|$ less than the range of the potential, as depicted by the shaded region of the figure.

Introducing $r = |\mathbf{x}|$, $r' = |\mathbf{x}'|$, and $\alpha = \angle(\mathbf{x},\mathbf{x}')$, we have for $r \gg r'$,

$$|\mathbf{x} - \mathbf{x}'| = \sqrt{r^2 - 2rr'\cos\alpha + r'^2}$$

$$= r\left(1 - \frac{2r'}{r}\cos\alpha + \frac{r'^2}{r^2}\right)^{1/2}$$

$$\approx r - \hat{\mathbf{r}}\cdot\mathbf{x}' \tag{6.54}$$

where

$$\hat{\mathbf{r}} \equiv \frac{\mathbf{x}}{|\mathbf{x}|} \tag{6.55}$$

in which case $\mathbf{k}' \equiv k\hat{\mathbf{r}}$. We then obtain

$$e^{\pm ik|\mathbf{x}-\mathbf{x}'|} \approx e^{\pm ikr}e^{\mp i\mathbf{k}'\cdot\mathbf{x}'} \tag{6.56}$$

for large $r$. It is also legitimate to replace $1/|\mathbf{x} - \mathbf{x}'|$ by just $1/r$.

At this point, we specify the initial state as an eigenstate of the free-particle Hamiltonian $H_0$, that is $|i\rangle = |\mathbf{k}\rangle$. Putting this all together, we have, finally,

$$\langle \mathbf{x}|\psi^{(+)}\rangle \xrightarrow{r\,\text{large}} \langle \mathbf{x}|\mathbf{k}\rangle - \frac{1}{4\pi}\frac{2m}{\hbar^2}\frac{e^{ikr}}{r}\int d^3x' e^{-i\mathbf{k}'\cdot\mathbf{x}'} V(\mathbf{x}')\langle \mathbf{x}'|\psi^{(+)}\rangle$$

$$= \frac{1}{L^{3/2}}\left[e^{i\mathbf{k}\cdot\mathbf{x}} + \frac{e^{ikr}}{r}f(\mathbf{k}',\mathbf{k})\right]. \tag{6.57}$$

This form makes it very clear that we have the original plane wave in propagation direction $\mathbf{k}$ plus an outgoing spherical wave with amplitude $f(\mathbf{k}',\mathbf{k})$ given by

$$f(\mathbf{k}',\mathbf{k}) \equiv -\frac{1}{4\pi}\frac{2m}{\hbar^2}L^3\int d^3x' \frac{e^{-i\mathbf{k}'\cdot\mathbf{x}'}}{L^{3/2}}V(\mathbf{x}')\langle \mathbf{x}'|\psi^{(+)}\rangle$$

$$= -\frac{mL^3}{2\pi\hbar^2}\langle \mathbf{k}'|V|\psi^{(+)}\rangle. \tag{6.58}$$

We can also show from (6.52) and (6.56) that $\langle \mathbf{x}|\psi^{(-)}\rangle$ corresponds to the original plane wave in propagation direction $\mathbf{k}$ plus an incoming spherical wave with spatial dependence $e^{-ikr}/r$ and amplitude $-(mL^3/2\pi\hbar^2)\langle -\mathbf{k}'|V|\psi^{(-)}\rangle$.

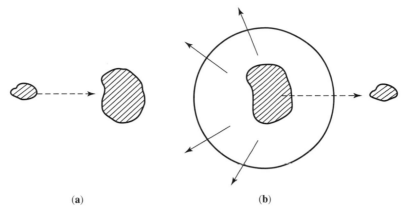

**Fig. 6.3** (a) Incident wave packet approaching scattering center initially. (b) Incident wave packet continuing to move in the original direction plus spherical outgoing wave front (after a long time duration).

We refer to $f(\mathbf{k}',\mathbf{k})$ as the **scattering amplitude**. By comparing (6.58) with (6.30), we see that the differential cross section can be written as

$$\frac{d\sigma}{d\Omega} = |f(\mathbf{k}',\mathbf{k})|^2. \tag{6.59}$$

### 6.2.1 Wave Packet Description

The reader may wonder here whether our formulation of scattering has anything to do with the motion of a particle being bounced by a scattering center. The incident plane wave we have used is infinite in extent in both space and time. In a more realistic situation, we consider a wave packet (a difficult subject!) that approaches the scattering center.[2] After a long time we have both the original wave packet moving in the original direction plus a spherical wave front that moves outward, as in Figure 6.3. Actually the use of a plane wave is satisfactory as long as the dimension of the wave packet is much larger than the size of the scatterer (or range of $V$).

### 6.2.2 The Optical Theorem

There is a fundamental and useful relationship popularly attributed to Bohr, Peierls, and Placzek[3] called the **optical theorem**, which relates the imaginary part of the forward scattering amplitude $f(\theta = 0) \equiv f(\mathbf{k},\mathbf{k})$ to the total cross section $\sigma_{\text{tot}} \equiv \int d\Omega \, (d\sigma/d\Omega)$, as follows:

$$\text{Im} \, f(\theta = 0) = \frac{k\sigma_{\text{tot}}}{4\pi}. \tag{6.60}$$

---

[2] For a fuller account of the wave packet approach, see Chapter 3 in Goldberger and Watson (1964), and Chapter 6 in Newton (1982).

[3] This relationship is in fact due to Eugene Feenberg, *Phys. Rev.*, **40** (1932) 40. See Newton, *Am. J. Phys.*, **44** (1976) 639 for the historical background.

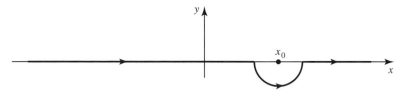

**Fig. 6.4**  Contour used to integrate around a singularity located at $z_0 = x_0 + i\varepsilon$.

To prove this, start with the Lippman–Schwinger Equation (6.35) with $|i\rangle = |\mathbf{k}\rangle$ to write

$$\langle \mathbf{k}|V|\psi^{(+)}\rangle = \left[ \langle \psi^{(+)}| - \langle \psi^{(+)}|V\frac{1}{E - H_0 - i\varepsilon} \right] V|\psi^{(+)}\rangle$$

$$= \langle \psi^{(+)}|V|\psi^{(+)}\rangle - \langle \psi^{(+)}|V\frac{1}{E - H_0 - i\varepsilon}V|\psi^{(+)}\rangle. \tag{6.61}$$

Comparing (6.58) and (6.60) we see that we want to take the imaginary part of both sides of (6.61). The first term on the right side of (6.61) is a real number, since it is the expectation value of a Hermitian operator. Finding the imaginary part of the second term is more difficult, because of singularities along the real axis as $\varepsilon \to 0$. To do this, we use a trick borrowed from the concept of the Cauchy principal value in complex integration.

Figure 6.4 shows a complex integration contour which follows the real axis except for a small semicircle which jumps over a singularity near the real axis. The singularity is located at $z_0 = x_0 + i\varepsilon$, with $\varepsilon > 0$, which is always above the $x$-axis. So, let the semicircle be centered on the real axis at $x_0$, and extend into the lower half of the complex plane with a radius $\delta$. The semicircle is described by $z - x_0 = \delta e^{i\phi}$ with $\phi$ running from $-\pi$ to zero.

Next consider a complex function $f(z)$, with $z = x + iy$. We can write

$$\int_{-\infty}^{\infty} \frac{f(x)}{x - x_0} dx = \int_{-\infty}^{x_0 - \delta} \frac{f(x)}{x - x_0} dx + \int_c \frac{f(z)}{z - z_0} dz + \int_{x_0 + \delta}^{+\infty} \frac{f(x)}{x - x_0} dx = 0$$

$$= \mathscr{P} \int_{-\infty}^{+\infty} \frac{f(x)}{x - x_0} dx + \int_c \frac{f(z)}{z - z_0} dz \tag{6.62}$$

where $c$ denotes the small semicircular contour around the singularity. The Cauchy principal value is defined as

$$\mathscr{P} \int_{-\infty}^{+\infty} \frac{f(x)}{x - x_0} dx = \lim_{\delta \to 0} \left\{ \int_{-\infty}^{x_0 - \delta} \frac{f(x)}{x - x_0} dx + \int_{x_0 + \delta}^{+\infty} \frac{f(x)}{x - x_0} dx \right\}. \tag{6.63}$$

We can evaluate the second term in (6.62) as

$$\int_c \frac{f(z)}{z - z_0} dz = \int_{-\pi}^{0} \frac{f(x_0)}{\delta e^{i\phi}} \left( i\phi \delta e^{i\phi} d\phi \right)$$

$$\to i\pi f(x_0) \qquad \text{as} \qquad \delta \to 0. \tag{6.64}$$

Consequently, we rewrite (6.62) as

$$\int_{-\infty}^{\infty} \frac{f(x)}{x - x_0} dx = \mathscr{P} \int_{-\infty}^{+\infty} \frac{f(x)}{x - x_0} dx + i\pi f(x_0). \tag{6.65}$$

Now we can return to finding the imaginary part of the right side of (6.61). We have

$$\lim_{\varepsilon \to 0} \left( \frac{1}{E - H_0 - i\varepsilon} \right) = \lim_{\varepsilon \to 0} \int_{-\infty}^{+\infty} \frac{\delta(E - E')}{E' - H_0 - i\varepsilon} dE'$$
$$= i\pi \delta(E - H_0) \tag{6.66}$$

where we have made use of (6.65). Therefore

$$\text{Im}\langle \mathbf{k}|V|\psi^{(+)}\rangle = -\pi \langle \psi^{(+)}|V\delta(E - H_0)V|\psi^{(+)}\rangle$$
$$= -\pi \langle \mathbf{k}|T^\dagger \delta(E - H_0)T|\mathbf{k}\rangle \tag{6.67}$$

where we recall that $T$ is defined through $T|\mathbf{k}\rangle = V|\psi^{(+)}\rangle$. Consequently, using (6.58),

$$\text{Im} f(\mathbf{k}, \mathbf{k}) = -\frac{mL^3}{2\pi\hbar^2} \text{Im}\langle \mathbf{k}|V|\psi^{(+)}\rangle$$
$$= \frac{mL^3}{2\hbar^2} \langle \mathbf{k}|T^\dagger \delta(E - H_0)T|\mathbf{k}\rangle$$
$$= \frac{mL^3}{2\hbar^2} \sum_{\mathbf{k}'} \langle \mathbf{k}|T^\dagger \delta(E - H_0)|\mathbf{k}'\rangle \langle \mathbf{k}'|T|\mathbf{k}\rangle$$
$$= \frac{mL^3}{2\hbar^2} \sum_{\mathbf{k}'} |\langle \mathbf{k}'|T|\mathbf{k}\rangle|^2 \delta_{E, \hbar^2 \mathbf{k}'^2/2m} \tag{6.68}$$

where $E = \hbar^2 k^2/2m$.

The optical theorem (6.60) now begins to appear. The factor $|\langle \mathbf{k}'|T|\mathbf{k}\rangle|^2$ is proportional to the differential cross section (6.59). The sum, including the $\delta$ function, is over all scattered momenta which conserve energy; in other words it is over all directions in space. Therefore, the right-hand side of (6.68) is an integral of the differential cross section over all directions and so is proportional to the total cross section.

To carry (6.68) through to the end, we make use of $\langle \mathbf{k}'|\mathbf{T}|\mathbf{k}\rangle = \langle \mathbf{k}'|\mathbf{V}|\psi^{(+)}\rangle$ with (6.58), and converting the sum to an integral as we did to go from (6.42) to (6.44). This gives

$$\text{Im} f(\mathbf{k}, \mathbf{k}) = \frac{mL^3}{2\hbar^2} \left( \frac{2\pi\hbar^2}{mL^3} \right)^2 \sum_{\mathbf{k}'} |f(\mathbf{k}', \mathbf{k})|^2 \delta_{E, \hbar^2 \mathbf{k}'^2/2m}$$
$$\longrightarrow \frac{2\pi^2\hbar^2}{m(2\pi)^3} \int d^3 k' \, |f(\mathbf{k}', \mathbf{k})|^2 \delta\left( E - \frac{\hbar^2 \mathbf{k}'^2}{2m} \right)$$
$$= \frac{\hbar^2}{4\pi m} \frac{1}{\hbar^2 k/m} k^2 \int d\Omega_{k'} \frac{d\sigma}{d\Omega_{k'}}$$
$$= \frac{k}{4\pi} \sigma_{\text{tot}} \tag{6.69}$$

thus proving (6.60).

Section 6.5 will provide some insights as to the physical significance of the optical theorem.

## 6.3  The Born Approximation

Our task now is to calculate the scattering amplitude $f(\mathbf{k}',\mathbf{k})$ for some given potential energy function $V(\mathbf{x})$. This amounts to calculating the matrix element

$$\langle \mathbf{k}'|V|\psi^{(+)}\rangle = \langle \mathbf{k}'|T|\mathbf{k}\rangle. \qquad (6.70)$$

This task is not straightforward, however, since we do not have closed analytic expressions for either $\langle \mathbf{x}'|\psi^{(+)}\rangle$ or $T$. Consequently, one typically resorts to approximations at this point.

We have already alluded to a useful approximation scheme in (6.32). Again replacing $\hbar\varepsilon$ with $\varepsilon$, this is

$$T = V + V\frac{1}{E - H_0 + i\varepsilon}V + V\frac{1}{E - H_0 + i\varepsilon}V\frac{1}{E - H_0 + i\varepsilon}V + \cdots \qquad (6.71)$$

which is an expansion in powers of $V$. We will shortly examine the conditions under which truncations of this expansion should be valid. First, however, we will make use of this scheme and see where it leads us.

Taking the first term in the expansion, i.e. $T = V$ or, equivalently, $|\psi^{(+)}\rangle = |\mathbf{k}\rangle$, is called the **first-order Born approximation**. In this case, the scattering amplitude is denoted by $f^{(1)}$, where

$$f^{(1)}(\mathbf{k}',\mathbf{k}) = -\frac{m}{2\pi\hbar^2}\int d^3x'\, e^{i(\mathbf{k}-\mathbf{k}')\cdot\mathbf{x}'}V(\mathbf{x}') \qquad (6.72)$$

after inserting a complete set of states $|\mathbf{x}'\rangle$ into (6.58). In other words, apart from an overall factor, the first-order amplitude is just the three-dimensional Fourier transform of the potential $V$ with respect to $\mathbf{q} \equiv \mathbf{k} - \mathbf{k}'$.

An important special case is when $V$ is a spherically symmetric potential. This implies that $f^{(1)}(\mathbf{k}',\mathbf{k})$ is a function of $q \equiv |\mathbf{q}|$, which is simply related to kinematic variables easily accessible by experiment. See Figure 6.5. Since $|\mathbf{k}'| = k$ by energy conservation, we have

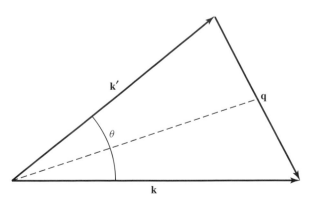

**Fig. 6.5**   Scattering through angle $\theta$, where $\mathbf{q} = \mathbf{k} - \mathbf{k}'$.

$$q = |\mathbf{k} - \mathbf{k}'| = 2k \sin \frac{\theta}{2}. \tag{6.73}$$

We can perform the angular integration in (6.72) explicitly to obtain

$$
\begin{aligned}
f^{(1)}(\theta) &= -\frac{1}{2} \frac{2m}{\hbar^2} \frac{1}{iq} \int_0^\infty \frac{r^2}{r} V(r)(e^{iqr} - e^{-iqr}) \, dr \\
&= -\frac{2m}{\hbar^2} \frac{1}{q} \int_0^\infty rV(r) \sin qr \, dr.
\end{aligned}
\tag{6.74}
$$

A simple but important example is scattering by a finite square well, that is

$$V(r) = \begin{cases} V_0 & r \le a \\ 0 & r > a. \end{cases} \tag{6.75}$$

The integral in (6.74) is readily done, and yields

$$f^{(1)}(\theta) = -\frac{2m}{\hbar^2} \frac{V_0 a^3}{(qa)^2} \left[ \frac{\sin qa}{qa} - \cos qa \right]. \tag{6.76}$$

This function has zeros at $qa = 4.49, 7.73, 10.9\ldots$ and the position of these zeros, along with (6.73), can be used to determine the well radius $a$. Figure 6.6 shows elastic proton scattering from several nuclei, each of which is an isotope of calcium. The nuclear potential is approximated rather nicely by a finite square well, and the differential cross section shows the characteristic minima predicted by (6.76). Furthermore, the data indicate that as neutrons are added to the calcium nucleus, the minima appear at smaller angles, showing that the nuclear radius in fact increases.

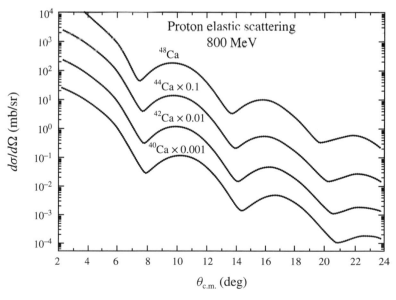

**Fig. 6.6**  Data on elastic scattering of protons from the nuclei of four different isotopes of calcium. The angles at which the cross sections show minima, decrease consistently with increasing neutron number. Therefore, the radius of the calcium nucleus increases as more neutrons are added, as one expects. From Ray et al., *Phys. Rev. C*, **23** (1981) 828.

Another important example is scattering by a Yukawa potential

$$V(r) = \frac{V_0 e^{-\mu r}}{\mu r} \tag{6.77}$$

where $V_0$ is independent of $r$, and $1/\mu$ corresponds, in a certain sense, to the range of the potential. Notice that $V$ goes to zero very rapidly for $r \gg 1/\mu$. For this potential we obtain, from (6.74),

$$f^{(1)}(\theta) = -\left(\frac{2mV_0}{\mu\hbar^2}\right)\frac{1}{q^2+\mu^2}, \tag{6.78}$$

where we note that $\sin qr = \text{Im}(e^{iqr})$ and have used

$$\text{Im}\left[\int_0^\infty e^{-\mu r}e^{iqr}dr\right] = -\text{Im}\left(\frac{1}{-\mu+iq}\right) = \frac{q}{\mu^2+q^2}. \tag{6.79}$$

Notice also that

$$q^2 = 4k^2 \sin^2\frac{\theta}{2} = 2k^2(1-\cos\theta). \tag{6.80}$$

So, in the first Born approximation, the differential cross section for scattering by a Yukawa potential is given by

$$\left(\frac{d\sigma}{d\Omega}\right) \simeq \left(\frac{2mV_0}{\mu\hbar^2}\right)^2 \frac{1}{[2k^2(1-\cos\theta)+\mu^2]^2}. \tag{6.81}$$

It is amusing to observe here that as $\mu \to 0$, the Yukawa potential is reduced to the Coulomb potential, provided the ratio $V_0/\mu$ is fixed, for example, to be $ZZ'e^2$, in the limiting process. We see that the first Born differential cross section obtained in this manner becomes

$$\left(\frac{d\sigma}{d\Omega}\right) \simeq \frac{(2m)^2(ZZ'e^2)^2}{\hbar^4}\frac{1}{16k^4\sin^4(\theta/2)}. \tag{6.82}$$

Even the $\hbar$ disappears if $\hbar k$ is identified as $|\mathbf{p}|$, so

$$\left(\frac{d\sigma}{d\Omega}\right) = \frac{1}{16}\left(\frac{ZZ'e^2}{E_{KE}}\right)^2\frac{1}{\sin^4(\theta/2)}, \tag{6.83}$$

where $E_{KE} = |\mathbf{p}|^2/2m$; this is precisely the Rutherford scattering cross section that can be obtained *classically*.

Coming back to (6.74), the Born amplitude with a spherically symmetric potential, there are several general remarks we can make if $f(\mathbf{k}',\mathbf{k})$ can be approximated by the corresponding first Born amplitude, $f^{(1)}$.

1. $d\sigma/d\Omega$, or $f(\theta)$, is a function of $q$ only; that is, $f(\theta)$ depends on the energy $(\hbar^2 k^2/2m)$ and $\theta$ only through the combination $2k^2(1-\cos\theta)$.
2. $f(\theta)$ is always real.
3. $d\sigma/d\Omega$ is independent of the sign of $V$.

4. For small $k$ ($q$ necessarily small),

$$f^{(1)}(\theta) = -\frac{1}{4\pi}\frac{2m}{\hbar^2}\int V(r)d^3x$$

involving a volume integral independent of $\theta$.

5. $f(\theta)$ is small for large $q$ due to rapid oscillation of the integrand.

In order to study the conditions under which the Born approximation should be valid, let us return to (6.52), slightly rewritten as

$$\langle \mathbf{x}|\psi^{(+)}\rangle = \langle \mathbf{x}|\mathbf{k}\rangle - \frac{2m}{\hbar^2}\int d^3x' \frac{e^{ik|\mathbf{x}-\mathbf{x}'|}}{4\pi|\mathbf{x}-\mathbf{x}'|}V(\mathbf{x}')\langle \mathbf{x}'|\psi^{(+)}\rangle.$$

The approximation is that $T \approx V$, which means that $|\psi^{(+)}\rangle$ can be replaced by $|\mathbf{k}\rangle$. Therefore, the second term on the right-hand side in this equation must be much smaller than the first. Let us assume that a "typical" value for the potential energy $V(\mathbf{x})$ is $V_0$, and that it acts within some "range" $a$. Writing $r' = |\mathbf{x}-\mathbf{x}'|$ and carrying out a rough approximation on the integral, our validity condition becomes

$$\left|\frac{2m}{\hbar^2}\left(\frac{4\pi}{3}a^3\right)\frac{e^{ikr'}}{4\pi a}V_0\frac{e^{i\mathbf{k}\cdot\mathbf{x}'}}{L^{3/2}}\right| \ll \left|\frac{e^{i\mathbf{k}\cdot\mathbf{x}}}{L^{3/2}}\right|.$$

Now for low energies (i.e. $ka \ll 1$), the exponential factors can be replaced by unity. Then, ignoring numerical factors of order unity, the following succinct criterion emerges:

$$\frac{m|V_0|a^2}{\hbar^2} \ll 1. \tag{6.84}$$

Consider the special case of the Yukawa potential in (6.77), in which the range $a = 1/\mu$. The validity criterion becomes $m|V_0|/\hbar^2\mu^2 \ll 1$. This requirement may be compared with the condition for the Yukawa potential to develop a bound state, which we can show to be $2m|V_0|/\hbar^2\mu' \geq 2.7$, with $V_0$ negative. In other words, if the potential is strong enough to develop a bound state, the Born approximation will probably give a misleading result.

At high energies (i.e. $ka \gg 1$), the factors $e^{ikr'}$ and $e^{i\mathbf{k}\cdot\mathbf{x}'}$ oscillate strongly over the region of integration, so they cannot be set equal to unity. Instead, it can be shown that

$$\frac{2m}{\hbar^2}\frac{|V_0|a}{k}\ln(ka) \ll 1. \tag{6.85}$$

As $k$ becomes larger, this inequality is more easily satisfied. Quite generally, the Born approximation tends to get better at higher energies.

## 6.3.1  The Higher-Order Born Approximation

Now, write $T$ to second order in $V$, using (6.71), namely

$$T = V + V\frac{1}{E - H_0 + i\varepsilon}V.$$

It is natural to continue our Born approximation approach and write

$$f(\mathbf{k}',\mathbf{k}) \approx f^{(1)}(\mathbf{k}',\mathbf{k}) + f^{(2)}(\mathbf{k}',\mathbf{k})$$

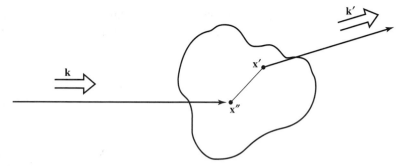

**Fig. 6.7**    Physical interpretation of the higher-order Born term $f^{(2)}(\mathbf{k}',\mathbf{k})$.

where $f^{(1)}(\mathbf{k}',\mathbf{k})$ is given by (6.72) and

$$f^{(2)} = -\frac{1}{4\pi}\frac{2m}{\hbar^2}(2\pi)^3 \int d^3x' \int d^3x'' \langle \mathbf{k}'|\mathbf{x}'\rangle V(\mathbf{x}')$$

$$\times \left\langle \mathbf{x}' \left| \frac{1}{E - H_0 + i\varepsilon} \right| \mathbf{x}'' \right\rangle V(\mathbf{x}'')(\mathbf{x}''|\mathbf{k})$$

$$= -\frac{1}{4\pi}\frac{2m}{\hbar^2} \int d^3x' \int d^3x'' e^{-i\mathbf{k}'\cdot\mathbf{x}'} V(\mathbf{x}')$$

$$\times \left[\frac{2m}{\hbar^2} G_+(\mathbf{x}',\mathbf{x}'')\right] V(\mathbf{x}'')e^{i\mathbf{k}\cdot\mathbf{x}''}. \qquad (6.86)$$

This scheme can obviously be continued to higher orders

A physical interpretation of (6.86) is given in Figure 6.7, where the incident wave interacts at $\mathbf{x}''$, which explains the appearance of $V(\mathbf{x}'')$, and then propagates from $\mathbf{x}''$ to $\mathbf{x}'$ via Green's function for the Helmholtz equation (6.48). Subsequently, a second interaction occurs at $\mathbf{x}'$, thus the appearance of $V(\mathbf{x}')$, and, finally, the wave is scattered into the direction $\mathbf{k}'$. In other words, $f^{(2)}$ corresponds to scattering viewed as a two-step process. Likewise, $f^{(3)}$ can be viewed as a three-step process, and so on.

## 6.4  Phase Shifts and Partial Waves

In considering scattering by a spherically symmetric potential, we often examine how states with definite angular momenta are affected by the scatterer. Such considerations lead to the method of partial waves, which we will discuss shortly. However, before discussing the angular momentum decomposition of scattering states, let us first talk about free-particle states, which are also eigenstates of angular momentum.

### 6.4.1  Free-Particle States

For a free particle the Hamiltonian is just the kinetic-energy operator, which obviously commutes with the momentum operator. We note, however, that the free-particle

Hamiltonian also commutes with $\mathbf{L}^2$ and $L_z$. Thus it is possible to consider a simultaneous eigenket of $H_0, \mathbf{L}^2$, and $L_z$. Ignoring spin, such a state is denoted by $|E, l, m\rangle$, often called a **spherical wave state**.

More generally, the most general free-particle state can be regarded as a superposition of $|E, l, m\rangle$ with various $E$, $l$, and $m$ in much the same way as the most general free-particle state can be regarded as a superposition of $|\mathbf{k}\rangle$ with different $\mathbf{k}$, different in both magnitude and direction. Put in another way, a free-particle state can be analyzed using either the plane wave basis $\{|\mathbf{k}\rangle\}$ or the spherical wave basis $\{|E, l, m\rangle\}$.

We now derive the transformation function $\langle \mathbf{k}|E, l, m\rangle$ that connects the plane wave basis with the spherical wave basis. We can also regard this quantity as the momentum-space wave function for the spherical wave characterized by $E$, $l$, and $m$. We adopt the normalization convention for the spherical wave eigenket as follows:

$$\langle E', l', m'|E, l, m\rangle = \delta_{ll'}\delta_{mm'}\delta(E - E'). \tag{6.87}$$

In analogy with the position-space wave function, we may guess the angular dependence to be

$$\langle \mathbf{k}|E, l, m\rangle = g_{lE}(k)Y_l^m(\hat{\mathbf{k}}), \tag{6.88}$$

where the function $g_{lE}(k)$ will be considered later. To prove this rigorously, we proceed as follows. First, consider the momentum eigenket $|k\hat{\mathbf{z}}\rangle$, that is, a plane wave state whose propagation direction is along the positive $z$-axis. An important property of this state is that it has no orbital angular-momentum component in the $z$-direction:

$$L_z|k\hat{\mathbf{z}}\rangle = (xp_y - yp_x)|k_x = 0, k_y = 0, k_z = k\rangle = 0. \tag{6.89}$$

Actually this is plausible from classical considerations: The angular-momentum component must vanish in the direction of propagation because $\mathbf{L} \cdot \mathbf{p} = (\mathbf{x} \times \mathbf{p}) \cdot \mathbf{p} = 0$. Because of (6.89), and since $\langle E', l', m'|k\hat{\mathbf{z}}\rangle = 0$ for $m' \neq 0$, we must be able to expand $|k\hat{\mathbf{z}}\rangle$ as follows:

$$|k\hat{\mathbf{z}}\rangle = \sum_{l'}\int dE'|E', l', m' = 0\rangle\langle E', l', m' = 0|k\hat{\mathbf{z}}\rangle. \tag{6.90}$$

Notice that there is no $m'$ sum; $m'$ is always zero. We can obtain the most general momentum eigenket, with the direction of $\mathbf{k}$ specified by $\theta$ and $\phi$, from $|k\hat{\mathbf{z}}\rangle$ by just applying the appropriate rotation operator as follows [see Figure 3.3 and (3.256)]:

$$|\mathbf{k}\rangle = \mathscr{D}(\alpha = \phi, \beta = \theta, \gamma = 0)|k\hat{\mathbf{z}}\rangle. \tag{6.91}$$

Multiplying this equation by $\langle E, l, m|$ on the left, we obtain

$$\langle E, l, m|\mathbf{k}\rangle = \sum_{l'}\int dE'\langle E, l, m|\mathscr{D}(\alpha = \phi, \beta = \theta, \gamma = 0)|E', l', m' = 0\rangle$$

$$\times \langle E', l', m' = 0|k\hat{\mathbf{z}}\rangle$$

$$= \sum_{l'}\int dE'\mathscr{D}_{m0}^{(l')}(\alpha = \phi, \beta = \theta, \gamma = 0)$$

$$\times \delta_{ll'}\delta(E - E')\langle E', l', m' = 0|k\hat{\mathbf{z}}\rangle$$

$$= \mathscr{D}_{m0}^{(l)}(\alpha = \phi, \beta = \theta, \gamma = 0)\langle E, l, m = 0|k\hat{\mathbf{z}}\rangle. \tag{6.92}$$

Now $\langle E, l, m = 0 | k\hat{\mathbf{z}} \rangle$ is independent of the orientation of $\mathbf{k}$, that is, independent of $\theta$ and $\phi$, and we may as well call it $\sqrt{\frac{2l+1}{4\pi}} g_{lE}^*(k)$. So we can write, using (3.260),

$$\langle \mathbf{k} | E, l, m \rangle = g_{lE}(k)\, Y_l^m(\hat{\mathbf{k}}). \tag{6.93}$$

Let us determine $g_{lE}(k)$. First, we note that

$$(H_0 - E)|E, l, m\rangle = 0. \tag{6.94}$$

But we also let $H_0 - E$ operate on a momentum eigenbra $\langle \mathbf{k} |$ as follows:

$$\langle \mathbf{k} |(H_0 - E) = \left( \frac{\hbar^2 k^2}{2m} - E \right) \langle \mathbf{k} |. \tag{6.95}$$

Multiplying (6.95) with $|E, l, m\rangle$ on the right, we obtain

$$\left( \frac{\hbar^2 k^2}{2m} - E \right) \langle \mathbf{k} | E, l, m \rangle = 0. \tag{6.96}$$

This means that $\langle \mathbf{k} | E, l, m \rangle$ can be nonvanishing only if $E = \hbar^2 k^2/2m$; so we must be able to write $g_{lE}(k)$ as

$$g_{lE}(k) = N\delta \left( \frac{\hbar^2 k^2}{2m} - E \right). \tag{6.97}$$

To determine $N$ we go back to our normalization convention (6.87). We obtain

$$\langle E', l'm' | E, l, m \rangle = \int d^3 k'' \langle E', l', m' | \mathbf{k}'' \rangle \langle \mathbf{k}'' | E, l, m \rangle$$

$$= \int k''^2 dk'' \int d\Omega_{\mathbf{k}''} |N|^2 \delta \left( \frac{\hbar^2 k''^2}{2m} - E' \right)$$

$$\times \delta \left( \frac{\hbar^2 k''^2}{2m} - E \right) Y_{l'}^{m'*}(\hat{\mathbf{k}}'') Y_l^m(\hat{\mathbf{k}}'')$$

$$= \int \frac{k''^2 dE''}{dE''/dk''} \int d\Omega_{\mathbf{k}''} |N|^2 \delta \left( \frac{\hbar^2 k''^2}{2m} - E' \right) \delta \left( \frac{\hbar^2 k''^2}{2m} - E \right)$$

$$\times Y_{l'}^{m'*}(\hat{\mathbf{k}}'') Y_l^m(\hat{\mathbf{k}}'')$$

$$= |N|^2 \frac{mk'}{\hbar^2} \delta(E - E') \delta_{ll'} \delta_{mm'}, \tag{6.98}$$

where we have defined $E'' = \hbar^2 k''^2/2m$ to change $k''$-integration into $E''$-integration. Comparing this with (6.87), we see that $N = \hbar/\sqrt{mk}$ will suffice. Therefore, we can finally write

$$g_{lE}(k) = \frac{\hbar}{\sqrt{mk}} \delta \left( \frac{\hbar^2 k^2}{2m} - E \right); \tag{6.99}$$

hence

$$\langle \mathbf{k} | E, l, m \rangle = \frac{\hbar}{\sqrt{mk}} \delta \left( \frac{\hbar^2 k^2}{2m} - E \right) Y_l^m(\hat{\mathbf{k}}). \tag{6.100}$$

From (6.100) we infer that the plane wave state $|\mathbf{k}\rangle$ can be expressed as a superposition of free spherical wave states with all possible $l$-values; in particular,

$$
\begin{aligned}
|\mathbf{k}\rangle &= \sum_l \sum_m \int dE |E,l,m\rangle \langle E,l,m|\mathbf{k}\rangle \\
&= \sum_{l=0}^{\infty} \sum_{m=-l}^{l} |E,l,m\rangle \Bigg|_{E=\hbar^2 k^2/2m} \left( \frac{\hbar}{\sqrt{mk}} Y_l^{m*}(\hat{\mathbf{k}}) \right).
\end{aligned}
\tag{6.101}
$$

Because the transverse dimension of the plane wave is infinite, we expect that the plane wave must contain all possible values of impact parameter $b$ (semiclassically, the impact parameter $b \simeq l\hbar/p$). From this point of view it is no surprise that the momentum eigenstates $|\mathbf{k}\rangle$, when analyzed in terms of spherical wave states, contain all possible values of $l$.

We have derived the wave function for $|E,l,m\rangle$ in momentum space. Next, we consider the corresponding wave function in position space. From wave mechanics, the reader should be familiar with the fact that the wave function for a free spherical wave is $j_l(kr)Y_l^m(\hat{\mathbf{r}})$, where $j_l(kr)$ is the spherical Bessel function of order $l$ (see (3.282a) and also Appendix B). The second solution $n_l(kr)$, although it satisfies the appropriate differential equation, is inadmissible because it is singular at the origin. So we can write

$$
\langle \mathbf{x}|E,l,m\rangle = c_l j_l(kr) Y_l^m(\hat{\mathbf{r}}).
\tag{6.102}
$$

To determine $c_l$, all we have to do is compare

$$
\begin{aligned}
\langle \mathbf{x}|\mathbf{k}\rangle &= \frac{e^{i\mathbf{k}\cdot\mathbf{x}}}{(2\pi)^{3/2}} = \sum_l \sum_m \int dE \langle \mathbf{x}|E,l,m\rangle \langle E,l,m|\mathbf{k}\rangle \\
&= \sum_l \sum_m \int dE\, c_l j_l(kr) Y_l^m(\hat{\mathbf{r}}) \frac{\hbar}{\sqrt{mk}} \delta\left( E - \frac{\hbar^2 k^2}{2m} \right) Y_l^{m*}(\hat{\mathbf{k}}) \\
&= \sum_l \frac{(2l+1)}{4\pi} P_l(\hat{\mathbf{k}}\cdot\hat{\mathbf{r}}) \frac{\hbar}{\sqrt{mk}} c_l j_l(kr),
\end{aligned}
\tag{6.103}
$$

where we have used the addition theorem

$$
\sum_m Y_l^m(\hat{\mathbf{r}}) Y_l^{m*}(\hat{\mathbf{k}}) = [(2l+1)/4\pi] P_l(\hat{\mathbf{k}}\cdot\hat{\mathbf{r}})
$$

in the last step. Now $\langle \mathbf{x}|\mathbf{k}\rangle = e^{i\mathbf{k}\cdot\mathbf{x}}/(2\pi)^{3/2}$ can also be written as

$$
\frac{e^{i\mathbf{k}\cdot\mathbf{x}}}{(2\pi)^{3/2}} = \frac{1}{(2\pi)^{3/2}} \sum_l (2l+1) i^l j_l(kr) P_l(\hat{\mathbf{k}}\cdot\hat{\mathbf{r}}),
\tag{6.104}
$$

which can be proved by using the following integral representation for $j_l(kr)$:

$$
j_l(kr) = \frac{1}{2i^l} \int_{-1}^{+1} e^{ikr\cos\theta} P_l(\cos\theta) d(\cos\theta).
\tag{6.105}
$$

Comparing (6.103) with (6.104), we have

$$
c_l = \frac{i^l}{\hbar} \sqrt{\frac{2mk}{\pi}}.
\tag{6.106}
$$

To summarize, we have

$$\langle \mathbf{k}|E,l,m\rangle = \frac{\hbar}{\sqrt{mk}} \delta\left(E - \frac{\hbar^2 k^2}{2m}\right) Y_l^m(\hat{\mathbf{k}}) \tag{6.107a}$$

$$\langle \mathbf{x}|E,l,m\rangle = \frac{i^l}{\hbar}\sqrt{\frac{2mk}{\pi}} j_l(kr) Y_l^m(\hat{\mathbf{r}}). \tag{6.107b}$$

These expressions are extremely useful in developing the partial-wave expansion.

We conclude this section by applying (6.107a) to a decay process. Suppose a parent particle of spin $j$ disintegrates into two spin zero particles $A\,(\text{spin}\,j) \to B\,(\text{spin}\,0) + C\,(\text{spin}\,0)$. The basic Hamiltonian responsible for such a decay process is, in general, very complicated. However, we do know that angular momentum is conserved because the basic Hamiltonian must be rotationally invariant. So the momentum-space wave function for the final state must be of the form (6.107a), with $l$ identified with the spin of the parent particle. This immediately enables us to compute the angular distribution of the decay product because the momentum-space wave function is nothing more than the probability amplitude for finding the decay product with relative momentum direction $\mathbf{k}$.

As a concrete example from nuclear physics, let us consider the decay of an excited nucleus, $\text{Ne}^{20*}$:

$$\text{Ne}^{20^*} \to \text{O}^{16} + \text{He}^4. \tag{6.108}$$

Both $\text{O}^{16}$ and $\text{He}^4$ are known to be spinless particles. Suppose the magnetic quantum number of the parent nucleus is $\pm 1$, relative to some direction $z$. Then the angular distribution of the decay product is proportional to $|Y_1^{\pm 1}(\theta, \phi)|^2 = (3/8\pi)\sin^2\theta$, where $(\theta, \phi)$ are the polar angles defining the relative direction $\mathbf{k}$ of the decay product. On the other hand, if the magnetic quantum number is 0 for a parent nucleus with spin 1, the decay angular distribution varies as $|Y_1^0(\theta, \phi)|^2 = (3/4\pi)\cos^2\theta$.

For a general spin orientation we obtain

$$\sum_{m=-l}^{1} w(m)|Y_{l=1}^m|^2. \tag{6.109}$$

For an unpolarized nucleus the various $w(m)$ are all equal, and we obtain an isotropic distribution; this is not surprising because there is no preferred direction if the parent particle is unpolarized.

For a higher spin object, the angular distribution of the decay is more involved; the higher the spin of the parent decaying system, the greater the complexity of the angular distribution of the decay products. Quite generally, through a study of the angular distribution of the decay products, it is possible to determine the spin of the parent nucleus.

## 6.4.2 Partial-Wave Expansion

Let us now come back to the case $V \neq 0$. We assume that the potential is spherically symmetric, that is, invariant under rotations in three dimensions. It then follows that the transition operator $T$, which is given by (6.71), commutes with $\mathbf{L}^2$ and $\mathbf{L}$. In other words, $T$ is a scalar operator.

It is now useful to use the spherical wave basis because the Wigner–Eckart theorem [see (3.481)], applied to a scalar operator, immediately gives

$$\langle E',l',m'|T|E,l,m\rangle = T_l(E)\delta_{ll'}\delta_{mm'}. \tag{6.110}$$

In other words, $T$ is diagonal both in $l$ and in $m$; furthermore, the (nonvanishing) diagonal element depends on $E$ and $l$ but not on $m$. This leads to an enormous simplification, as we will see shortly.

Let us now look at the scattering amplitude (6.58):

$$f(\mathbf{k}',\mathbf{k}) = -\frac{1}{4\pi}\frac{2m}{\hbar^2}L^3\langle\mathbf{k}'|T|\mathbf{k}\rangle$$

$$\longrightarrow -\frac{1}{4\pi}\frac{2m}{\hbar^2}(2\pi)^3\sum_l\sum_m\sum_{l'}\sum_{m'}\int dE\int dE'\langle\mathbf{k}'|E'l'm'\rangle$$

$$\times\langle E'l'm'|T|Elm\rangle\langle Elm|\mathbf{k}\rangle$$

$$= -\frac{1}{4\pi}\frac{2m}{\hbar^2}(2\pi)^3\frac{\hbar^2}{mk}\sum_l\sum_m T_l(E)\bigg|_{E=\hbar^2k^2/2m}Y_l^m(\hat{\mathbf{k}}')Y_l^{m*}(\hat{\mathbf{k}})$$

$$= -\frac{4\pi^2}{k}\sum_l\sum_m T_l(E)\bigg|_{E=\hbar^2k^2/2m}Y_l^m(\hat{\mathbf{k}}')Y_l^{m*}(\hat{\mathbf{k}}). \tag{6.111}$$

To obtain the angular dependence of the scattering amplitude, let us choose the coordinate system in such a way that $\mathbf{k}$, as usual, is in the positive $z$-direction. We then have [see (3.259)]

$$Y_l^m(\hat{\mathbf{k}}) = \sqrt{\frac{2l+1}{4\pi}}\delta_{m0}, \tag{6.112}$$

where we have used $P_l(1)=1$; hence only the terms $m=0$ contribute. Taking $\theta$ to be the angle between $\mathbf{k}'$ and $\mathbf{k}$, we can write

$$Y_l^0(\hat{\mathbf{k}}') = \sqrt{\frac{2l+1}{4\pi}}P_l(\cos\theta). \tag{6.113}$$

It is customary here to define the **partial-wave amplitude** $f_l(k)$ as follows:

$$f_l(k) \equiv -\frac{\pi T_l(E)}{k}. \tag{6.114}$$

For (6.111) we then have

$$f(\mathbf{k}',\mathbf{k}) = f(\theta) = \sum_{l=0}^{\infty}(2l+1)f_l(k)P_l(\cos\theta), \tag{6.115}$$

where $f(\theta)$ still depends on $k$ (or the incident energy) even though $k$ is suppressed.

To appreciate the physical significance of $f_l(k)$, let us study the large-distance behavior of the wave function $\langle \mathbf{x} | \psi^{(+)} \rangle$ given by (6.57). Using the expansion of a plane wave in terms of spherical waves [(6.104)] and noting that (Appendix B)

$$j_l(kr) \xrightarrow{\text{large } r} \frac{e^{i(kr-(l\pi/2))} - e^{-i(kr-(l\pi/2))}}{2ikr} \qquad (i^l = e^{i(\pi/2)l}) \tag{6.116}$$

and that $f(\theta)$ is given by (6.115), we have

$$\langle \mathbf{x} | \psi^{(+)} \rangle \xrightarrow{\text{large } r} \frac{1}{(2\pi)^{3/2}} \left[ e^{ikz} + f(\theta) \frac{e^{ikr}}{r} \right]$$

$$= \frac{1}{(2\pi)^{3/2}} \left[ \sum_l (2l+1) P_l(\cos\theta) \left( \frac{e^{ikr} - e^{-i(kr-l\pi)}}{2ikr} \right) \right.$$

$$\left. + \sum_l (2l+1) f_l(k) P_l(\cos\theta) \frac{e^{ikr}}{r} \right]$$

$$= \frac{1}{(2\pi)^{3/2}} \sum_l (2l+1) \frac{P_l}{2ik} \left[ [1 + 2ikf_l(k)] \frac{e^{ikr}}{r} - \frac{e^{-i(kr-l\pi)}}{r} \right]. \tag{6.117}$$

The physics of scattering is now clear. When the scatterer is absent, we can analyze the plane wave as the sum of a spherically outgoing wave behaving like $e^{ikr}/r$ and a spherically incoming wave behaving like $-e^{-i(kr-l\pi)}/r$ for each $l$. The presence of the scatterer changes only the coefficient of the outgoing wave, as follows:

$$1 \to 1 + 2ikf_l(k). \tag{6.118}$$

The incoming wave is completely unaffected.

### 6.4.3 Unitarity and Phase Shifts

We now examine the consequences of probability conservation, or unitarity. In a time-independent formulation, the flux current density $\mathbf{j}$ must satisfy

$$\nabla \cdot \mathbf{j} = -\frac{\partial |\psi|^2}{\partial t} = 0. \tag{6.119}$$

Let us now consider a spherical surface of very large radius. By Gauss's theorem, we must have

$$\int_{\text{spherical surface}} \mathbf{j} \cdot d\mathbf{S} = 0. \tag{6.120}$$

Physically (6.119) and (6.120) mean that there is no source or sink of particles. The outgoing flux must equal the incoming flux. Furthermore, because of angular-momentum conservation, this must hold for each partial wave separately. In other words, the coefficient of $e^{ikr}/r$ must be the same in magnitude as the coefficient of $e^{-ikr}/r$. Defining $S_l(k)$ to be

$$S_l(k) \equiv 1 + 2ikf_l(k), \tag{6.121}$$

this means [from (6.118)] that

$$|S_l(k)| = 1, \tag{6.122}$$

that is, the most that can happen is a change in the phase of the outgoing wave. Equation (6.122) is known as the **unitarity relation** for the $l$th partial wave. In a more advanced treatment of scattering, $S_l(k)$ can be regarded as the $l$th diagonal element of the $S$ operator, which is required to be unitary as a consequence of probability conservation.

We thus see that the only change in the wave function at a large distance as a result of scattering is to change the *phase* of the outgoing wave. Calling this phase $2\delta_l$ (the factor of 2 here is conventional), we can write

$$S_l = e^{2i\delta_l}, \tag{6.123}$$

with $\delta_l$ real. It is understood here that $\delta_l$ is a function of $k$ even though we do not explicitly write $\delta_l$ as $\delta_l(k)$. Returning to $f_l$, we can write [from (6.121)]

$$f_l = \frac{(S_l - 1)}{2ik} \tag{6.124}$$

or, explicitly in terms of $\delta_l$,

$$f_l = \frac{e^{2i\delta_l} - 1}{2ik} = \frac{e^{i\delta_l}\sin\delta_l}{k} = \frac{1}{k\cot\delta_l - ik}, \tag{6.125}$$

whichever is convenient. For the full scattering amplitude we have

$$f(\theta) = \sum_{l=0}^{\infty}(2l+1)\left(\frac{e^{2i\delta_l} - 1}{2ik}\right)P_l(\cos\theta)$$

$$= \frac{1}{k}\sum_{l=0}^{\infty}(2l+1)e^{i\delta_l}\sin\delta_l P_l(\cos\theta) \tag{6.126}$$

with $\delta_l$ real. This expression for $f(\theta)$ rests on the twin principles of **rotational invariance** and **probability conservation**. In many books on wave mechanics, (6.126) is obtained by explicitly solving the Schrödinger equation with a real, spherically symmetric potential; our derivation of (6.126) may be of interest because it can be generalized to situations when the potential described in the context of nonrelativistic quantum mechanics may fail.

The differential cross section $d\sigma/d\Omega$ can be obtained by just taking the modulus squared of (6.126). To obtain the total cross section we have

$$\sigma_{\text{tot}} = \int |f(\theta)|^2 d\Omega$$

$$= \frac{1}{k^2}\int_0^{2\pi}d\phi\int_{-1}^{+1}d(\cos\theta)\sum_l\sum_{l'}(2l+1)(2l'+1)$$

$$\times e^{i\delta_l}\sin\delta_l e^{-i\delta_{l'}}\sin\delta_{l'}P_l P_{l'}$$

$$= \frac{4\pi}{k^2}\sum_l(2l+1)\sin^2\delta_l. \tag{6.127}$$

We can check the optical theorem (6.60), which we obtained earlier using a more general argument. All we need to do is note from (6.126) that

$$\text{Im}f(\theta = 0) = \sum_l \frac{(2l+1)\text{Im}[e^{i\delta_l}\sin\delta_l]}{k}P_l(\cos\theta)\bigg|_{\theta=0}$$

$$= \sum_l \frac{(2l+1)}{k}\sin^2\delta_l, \tag{6.128}$$

which is the same as (6.127) except for $4\pi/k$.

As a function of energy, $\delta_l$ changes; hence $f_l(k)$ changes also. The unitarity relation of (6.122) is a restriction on the manner in which $f_l$ can vary. This can be most conveniently seen by drawing an Argand diagram for $kf_l$. We plot $kf_l$ in a complex plane, as shown in Figure 6.8, which is self-explanatory if we note from (6.125) that

$$kf_l = \frac{i}{2} + \frac{1}{2}e^{-(i\pi/2)+2i\delta_l}. \tag{6.129}$$

Notice that there is a circle of radius $\frac{1}{2}$, known as the **unitary circle**, on which $kf_l$ must lie.

We can see many important features from Figure 6.8. Suppose $\delta_l$ is small. Then $f_l$ must stay near the bottom of the circle. It may be positive or negative, but $f_l$ is almost purely real:

$$f_l = \frac{e^{i\delta_l}\sin\delta_l}{k} \simeq \frac{(1+i\delta_l)\delta_l}{k} \simeq \frac{\delta_l}{k}. \tag{6.130}$$

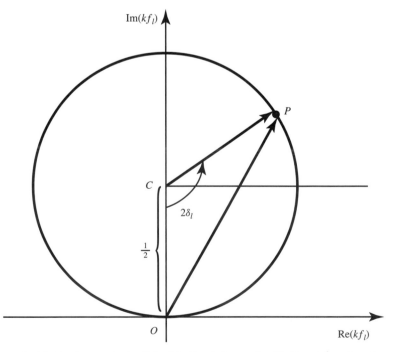

**Fig. 6.8** Argand diagram for $kf_l$. $OP$ is the magnitude of $kf_l$, while $CO$ and $CP$ are each radii of length $\frac{1}{2}$ on the unitary circle; angle $OCP = 2\delta_l$.

On the other hand, if $\delta_l$ is near $\pi/2$, $kf_l$ is almost purely imaginary, and the magnitude of $kf_l$ is maximal. Under such a condition the $l$th partial wave may be in resonance, a concept to be discussed in some detail in Section 6.7. Note that the maximum partial cross section

$$\sigma_{\text{max}}^{(l)} = \frac{4\pi}{k^2}(2l+1) \tag{6.131}$$

is achieved [see (6.127)] when $\sin^2 \delta_l = 1$.

### 6.4.4  Determination of Phase Shifts

Let us now consider how we may actually determine the phase shifts given a potential $V$. We assume that $V$ vanishes for $r > R$, $R$ being the range of the potential. Outside (that is, for $r > R$) the wave function must be that of a free spherical wave. This time, however, there is no reason to exclude $n_l(r)$ because the origin is excluded from our consideration. The wave function is therefore a linear combination of $j_l(kr)P_l(\cos\theta)$ and $n_l(kr)P_l(\cos\theta)$ or, equivalently, $h_l^{(1)}P_l$ and $h_l^{(2)}P_l$, where $h_l^{(1)}$ and $h_l^{(2)}$ are the spherical Hankel functions defined by

$$h_l^{(1)} = j_l + in_l, \quad h_l^{(2)} = j_l - in_l; \tag{6.132}$$

these have the asymptotic behavior (see Appendix B)

$$h_l^{(1)} \overset{r\,\text{large}}{\longrightarrow} \frac{e^{i(kr-(l\pi/2))}}{ikr}, \quad h_l^{(2)} \overset{r\,\text{large}}{\longrightarrow} -\frac{e^{-i(kr-(l\pi/2))}}{ikr}. \tag{6.133}$$

The full-wave function at any $r$ can then be written as:

$$\langle \mathbf{x}|\psi^{(+)}\rangle = \frac{1}{(2\pi)^{3/2}} \sum i^l(2l+1)A_l(r)P_l(\cos\theta) \quad (r > R). \tag{6.134}$$

For $r > R$ we have (for the radial wave function)

$$A_l = c_l^{(1)}h_l^{(1)}(kr) + c_l^{(2)}h_l^{(2)}(kr), \tag{6.135}$$

where the coefficient that multiplies $A_l$ in (6.134) is chosen so that, for $V = 0$, $A_l(r)$ coincides with $j_l(kr)$ everywhere [see (6.104)]. Using (6.133), we can compare the behavior of the wave function for large $r$ given by (6.134) and (6.135) with

$$\frac{1}{(2\pi)^{3/2}} \sum_l (2l+1)P_l \left[\frac{e^{2i\delta_l}e^{ikr}}{2ikr} - \frac{e^{-i(kr-l\pi)}}{2ikr}\right]. \tag{6.136}$$

Clearly, we must have

$$c_l^{(1)} = \tfrac{1}{2}e^{2i\delta_l}, \quad c_l^{(2)} = \tfrac{1}{2}. \tag{6.137}$$

So the radial wave function for $r > R$ is now written as

$$A_l(r) = e^{i\delta_l}[\cos\delta_l j_l(kr) - \sin\delta_l n_l(kr)]. \tag{6.138}$$

Using this, we can evaluate the logarithmic derivative at $r = R$, that is, just outside the range of the potential, as follows:

$$\beta_l \equiv \left( \frac{r}{A_l} \frac{dA_l}{dr} \right)_{r=R}$$

$$= kR \left[ \frac{j'_l(kR) \cos \delta_l - n'_l(kR) \sin \delta_l}{j_l(kR) \cos \delta_l - n_l(kR) \sin \delta_l} \right], \tag{6.139}$$

where $j'_l(kR)$ stands for the derivative of $j_l$ with respect to $kr$ evaluated at $kr = kR$. Conversely, knowing the logarithmic derivative at $R$, we can obtain the phase shift as follows:

$$\tan \delta_l = \frac{kR j'_l(kR) - \beta_l j_l(kR)}{kR n'_l(kR) - \beta_l n_l(kR)}. \tag{6.140}$$

The problem of determining the phase shift is thus reduced to that of obtaining $\beta_l$.

We now look at the solution to the Schrödinger equation for $r < R$, that is, inside the range of the potential. For a spherically symmetric potential, we can solve the Schrödinger equation in three dimensions by looking at the equivalent one-dimensional equation

$$\frac{d^2 u_l}{dr^2} + \left( k^2 - \frac{2m}{\hbar^2} V - \frac{l(l+1)}{r^2} \right) u_l = 0, \tag{6.141}$$

where

$$u_l = r A_l(r) \tag{6.142}$$

subject to the boundary condition

$$u_l|_{r=0} = 0. \tag{6.143}$$

We integrate this one-dimensional Schrödinger equation, if necessary numerically, up to $r = R$, starting at $r = 0$. In this way we obtain the logarithmic derivative at $R$. By continuity we must be able to match the logarithmic derivative for the inside and outside solutions at $r = R$:

$$\beta_l|_{\text{inside solution}} = \beta_l|_{\text{outside solution}}, \tag{6.144}$$

where the left-hand side is obtained by integrating the Schrödinger equation up to $r = R$, while the right-hand side is expressible in terms of the phase shifts that characterize the large-distance behavior of the wave function. This means that the phase shifts are obtained simply by substituting $\beta_l$ for the inside solution into $\tan \delta_l$ [(6.140)]. For an alternative approach it is possible to derive an integral equation for $A_l(r)$, from which we can obtain phase shifts (see Problem 6.9 of this chapter).

## 6.4.5 Hard-Sphere Scattering

Let us work out a specific example. We consider scattering by a hard, or rigid, sphere

$$V = \begin{cases} \infty & \text{for} \quad r < R \\ 0 & \text{for} \quad r > R. \end{cases} \tag{6.145}$$

In this problem we need not even evaluate $\beta_l$ (which is actually $\infty$). All we need to know is that the wave function must vanish at $r = R$ because the sphere is impenetrable. Therefore,

$$A_l(r)|_{r=R} = 0 \tag{6.146}$$

or, from (6.138),

$$j_l(kR)\cos\delta_l - n_l(kR)\sin\delta_l = 0 \tag{6.147}$$

or

$$\tan\delta_l = \frac{j_l(kR)}{n_l(kR)}. \tag{6.148}$$

Thus the phase shifts are now known for any $l$. Notice that no approximations have been made so far.

To appreciate the physical significance of the phase shifts, let us consider the $l = 0$ case (S-wave scattering) specifically. Equation (6.148) becomes, for $l = 0$

$$\tan\delta_0 = \frac{\sin kR/kR}{-\cos kR/kR} = -\tan kR, \tag{6.149}$$

or $\delta_0 = -kR$. The radial wave function (6.138) with $e^{i\delta_0}$ omitted varies as

$$A_{l=0}(r) \propto \frac{\sin kr}{kr}\cos\delta_0 + \frac{\cos kr}{kr}\sin\delta_0 = \frac{1}{kr}\sin(kr + \delta_0). \tag{6.150}$$

Therefore, if we plot $rA_{l=0}(r)$ as a function of distance $r$, we obtain a sinusoidal wave, which is shifted when compared to the free sinusoidal wave by amount $R$; see Figure 6.9.

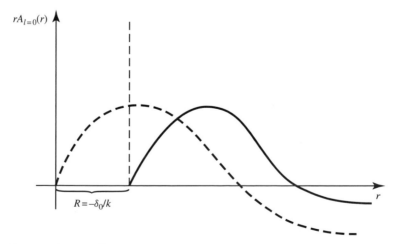

**Fig. 6.9**   Plot of $rA_{l=0}(r)$ versus $r$ (with the $e^{i\delta_0}$ factor removed). The dashed curve for $V = 0$ behaves like $\sin kr$, while the solid curve is for S-wave hard-sphere scattering, shifted by $R = -\delta_0/k$ from the case $V = 0$.

Let us now study the low and high-energy limits of $\tan \delta_l$. Low energy means $kR$ small, $kR \ll 1$. We can then use[4]

$$j_l(kr) \simeq \frac{(kr)^l}{(2l+1)!!}$$

$$n_l(kr) \simeq -\frac{(2l-1)!!}{(kr)^{l+1}}$$

$$\tag{6.151}$$

to obtain

$$\tan \delta_l = \frac{-(kR)^{2l+1}}{\{(2l+1)[(2l-1)!!]^2\}}. \tag{6.152}$$

It is therefore all right to ignore $\delta_l$ with $l \neq 0$. In other words, we have $S$-wave scattering only, which is actually expected for almost any finite-range potential at low energy. Because $\delta_0 = -kR$ regardless of whether $k$ is large or small, we obtain

$$\frac{d\sigma}{d\Omega} = \frac{\sin^2 \delta_0}{k^2} \simeq R^2 \quad \text{for} \quad kR \ll 1. \tag{6.153}$$

It is interesting that the total cross section, given by

$$\sigma_{\text{tot}} = \int \frac{d\sigma}{d\Omega} d\Omega = 4\pi R^2, \tag{6.154}$$

is *four* times the *geometric cross section* $\pi R^2$. By geometric cross section we mean the area of the disc of radius $R$ that blocks the propagation of the plane wave (and has the same cross section area as that of a hard sphere). Low-energy scattering, of course, means a very large wavelength scattering, and we do not necessarily expect a classically reasonable result. We will consider what happens in the high-energy limit when we discuss the eikonal approximation in the next section.

## 6.5 Eikonal Approximation

This approximation covers a situation in which $V(\mathbf{x})$ varies very little over a distance of order of wavelength $\bar{\lambda}$ (which can be regarded as "small"). Note that $V$ itself need not be weak as long as $E \gg |V|$; hence the domain of validity here is different from the Born approximation. Under these conditions, the semiclassical path concept becomes applicable, and we replace the exact wave function $\psi^{(+)}$ by the semiclassical wave function [see (2.193) and (2.197)], namely,

$$\psi^{(+)} \sim e^{iS(\mathbf{x})/\hbar}. \tag{6.155}$$

This leads to the Hamilton–Jacobi equation for $S$,

$$\frac{(\nabla S)^2}{2m} + V = E = \frac{\hbar^2 k^2}{2m}, \tag{6.156}$$

---

[4] Note that $(2n+1)!! \equiv (2n+1)(2n-1)(2n-3) \cdots 1$.

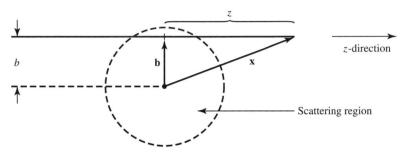

**Fig. 6.10**    Schematic diagram of eikonal approximation scattering where the classical straight-line trajectory is along the $z$-direction, $|\mathbf{x}| = r$, and $b = |\mathbf{b}|$ is the impact parameter.

as discussed in Section 2.4. We propose to compute $S$ from (6.156) by making the further approximation that the classical trajectory is a straight-line path, which should be satisfactory for small deflection at high energy.[5] Consider the situation depicted in Figure 6.10, where the straight-line trajectory is along the $z$-direction. Integrating (6.156) we have

$$\frac{S}{\hbar} = \int_{-\infty}^{z} \left[ k^2 - \frac{2m}{\hbar^2} V\left( \sqrt{b^2 + z'^2} \right) \right]^{1/2} dz' + \text{constant}. \tag{6.157}$$

The additive constant is to be chosen in such a way that

$$\frac{S}{\hbar} \to kz \quad \text{as} \quad V \to 0 \tag{6.158}$$

so that the plane wave form for (6.155) is reproduced in this zero-potential limit. We can then write (6.157) as

$$\frac{S}{\hbar} = kz + \int_{-\infty}^{z} \left[ \sqrt{k^2 - \frac{2m}{\hbar^2} V\left( \sqrt{b^2 + z'^2} \right)} - k \right] dz'$$

$$\cong kz - \frac{m}{\hbar^2 k} \int_{-\infty}^{z} V\left( \sqrt{b^2 + z'^2} \right) dz' \tag{6.159}$$

where for $E \gg V$, we have used

$$\sqrt{k^2 - \frac{2m}{\hbar^2} V\left( \sqrt{b^2 + z'^2} \right)} \sim k - \frac{mV}{\hbar^2 k}$$

at high $E = \hbar^2 k^2 / 2m$. So

$$\psi^{(+)}(\mathbf{x}) = \psi^{(+)}(\mathbf{b} + z\hat{\mathbf{z}}) \simeq \frac{1}{(2\pi)^{3/2}} e^{ikz} \exp\left[ \frac{-im}{\hbar^2 k} \int_{-\infty}^{z} V\left( \sqrt{b^2 + z'^2} \right) dz' \right]. \tag{6.160}$$

Though (6.160) does not have the correct asymptotic form appropriate for an incident plus spherical outgoing wave (that is, it is not of form $e^{i\mathbf{k}\cdot\mathbf{x}} + f(\theta)(e^{ikr}/r)$ and indeed refers only

---

[5] Needless to say, solution of (6.156) to *determine* the classical trajectory would be a forbidding task in general.

to motion along the original direction), it can nevertheless still be used in (6.58) to obtain an approximate expression for $f(\mathbf{k}', \mathbf{k})$, to wit,[6]

$$f(\mathbf{k}', \mathbf{k}) = -\frac{1}{4\pi}\frac{2m}{\hbar^2} \int d^3x' e^{-i\mathbf{k}' \cdot \mathbf{x}'} V\left(\sqrt{b^2 + z'^2}\right) e^{i\mathbf{k} \cdot \mathbf{x}'}$$

$$\times \exp\left[-\frac{im}{\hbar^2 k}\int_{-\infty}^{z'} V\left(\sqrt{b^2 + z''^2}\right) dz''\right]. \tag{6.161}$$

Note that without the last factor, $\exp[\ldots]$, (6.161) is just like the first-order Born amplitude in (6.72). We perform the three-dimensional $(d^3x')$ integration in (6.161) by introducing cylindrical coordinates $d^3x' = b\,db\,d\phi_b\,dz'$ (see Figure 6.10) and noting that

$$(\mathbf{k} - \mathbf{k}') \cdot \mathbf{x}' = (\mathbf{k} - \mathbf{k}') \cdot (\mathbf{b} + z'\hat{\mathbf{z}}) \simeq -\mathbf{k}' \cdot \mathbf{b}, \tag{6.162}$$

where we have used $\mathbf{k} \perp \mathbf{b}$ and $(\mathbf{k} - \mathbf{k}') \cdot \hat{\mathbf{z}} \sim 0(\theta^2)$, which can be ignored for small deflection $\theta$. Without loss of generality we choose scattering to be in the $xz$-plane and write

$$\mathbf{k}' \cdot \mathbf{b} = (k\sin\theta\hat{\mathbf{x}} + k\cos\theta\hat{\mathbf{z}}) \cdot (b\cos\phi_b\hat{\mathbf{x}} + b\sin\phi_b\hat{\mathbf{y}}) \simeq kb\theta\cos\phi_b. \tag{6.163}$$

The expression for $f(\mathbf{k}', \mathbf{k})$ becomes

$$f(\mathbf{k}', \mathbf{k}) = -\frac{1}{4\pi}\frac{2m}{\hbar^2}\int_0^\infty b\,db\int_0^{2\pi} d\phi_b e^{-ikb\theta\cos\phi_b}$$

$$\times \int_{-\infty}^{+\infty} dz\,V\exp\left[\frac{-im}{\hbar^2 k}\int_{-\infty}^z V\,dz'\right]. \tag{6.164}$$

We next use the following identities:

$$\int_0^{2\pi} d\phi_b e^{-ikb\theta\cos\phi_b} = 2\pi J_0(kb\theta) \tag{6.165}$$

and

$$\int_{-\infty}^{+\infty} dz\,V\exp\left[\frac{-im}{\hbar^2 k}\int_{-\infty}^z V\,dz'\right] = \frac{i\hbar^2 k}{m}\exp\left[\frac{-im}{\hbar^2 k}\int_{-\infty}^z V\,dz'\right]\Bigg|_{z=-\infty}^{z=+\infty} \tag{6.166}$$

where, of course, the contribution from $z = -\infty$ on the right-hand side of (6.166) vanishes in the exponent. So, finally

$$f(\mathbf{k}', \mathbf{k}) = -ik\int_0^\infty db\,bJ_0(kb\theta)[e^{2i\Delta(b)} - 1], \tag{6.167}$$

where

$$\Delta(b) \equiv \frac{-m}{2k\hbar^2}\int_{-\infty}^{+\infty} V\left(\sqrt{b^2 + z^2}\right) dz. \tag{6.168}$$

In (6.168) we fix the impact parameter $b$ and integrate along the straight-line path $z$, shown in Figure 6.10. There is no contribution from $[e^{2i\Delta(b)} - 1]$ in (6.167) if $b$ is greater than the range of $V$.

---

[6]  We leave behind the "big box," and write $f(\mathbf{k}', \mathbf{k})$ assuming a continuum normalization.

It can be shown in a straightforward manner that the eikonal approximation satisfies the optical theorem (6.60). This proof plus some interesting applications, for example, when $V$ is a Gaussian potential $\Delta(b)$ becomes Gaussian in $b$-space, are discussed in the literature (Gottfried, 1966). See also Problem 6.8 at the end of this chapter, which includes the case where $V$ is a Yukawa potential.

### 6.5.1 Partial Waves and the Eikonal Approximation

The eikonal approximation is valid at high energies ($\lambda \ll$ range $R$); hence many partial waves contribute. We may regard $l$ as a continuous variable. As an aside we note the semiclassical argument that $l = bk$ (because angular momentum $l\hbar = bp$, where $b$ is the impact parameter and momentum $p = \hbar k$). We take

$$l_{\max} = kR; \tag{6.169}$$

then we make the following substitutions in expression (6.126):

$$\sum_{l}^{l_{\max}=kR} \to k \int db, \quad P_l(\cos\theta) \overset{\substack{\text{large } l \\ \text{small } \theta}}{\simeq} J_0(l\theta) = J_0(kb\theta),$$

$$\delta_l \to \Delta(b)|_{b=l/k}, \tag{6.170}$$

where $l_{\max} = kR$ implies that

$$e^{2i\delta_l} - 1 = e^{2i\Delta(b)} - 1 = 0 \quad \text{for} \quad l > l_{\max}. \tag{6.171}$$

We have

$$f(\theta) \to k \int db \frac{2kb}{2ik}(e^{2i\Delta(b)} - 1)J_0(kb\theta)$$

$$= -ik \int db\, b J_0(kb\theta)[e^{2i\Delta(b)} - 1]. \tag{6.172}$$

The computation of $\delta_l$ can be done by using the explicit form for $\Delta(b)$ given by (6.168) (see Problem 6.8 in this chapter).

Recall now our discussion of partial waves and the "hard sphere" example, from the last section. There, we found that the total cross section was four times the geometric cross section in the low-energy (long wavelength) limit. However, one might conjecture that the geometric cross section is reasonable to expect for high-energy scattering because at high energies the situation might look similar to the semiclassical situation.

At high energies many $l$-values contribute, up to $l_{\max} \simeq kR$, a reasonable assumption. The total cross section is therefore given by

$$\sigma_{\text{tot}} = \frac{4\pi}{k^2} \sum_{l=0}^{l\simeq kR} (2l+1)\sin^2\delta_l. \tag{6.173}$$

But using (6.148), we have

$$\sin^2\delta_l = \frac{\tan^2\delta_l}{1+\tan^2\delta_l} = \frac{[j_l(kR)]^2}{[j_l(kR)]^2 + [n_l(kR)]^2} \simeq \sin^2\left(kR - \frac{\pi l}{2}\right), \tag{6.174}$$

where we have used

$$j_l(kr) \sim \frac{1}{kr}\sin\left(kr - \frac{l\pi}{2}\right)$$

$$n_l(kr) \sim -\frac{1}{kr}\cos\left(kr - \frac{l\pi}{2}\right). \tag{6.175}$$

We see that $\delta_l$ decreases by $90°$ each time $l$ increases by one unit. Thus, for an adjacent pair of partial waves, $\sin^2\delta_l + \sin^2\delta_{l+1} = \sin^2\delta_l + \sin^2(\delta_l - \pi/2) = \sin^2\delta_l + \cos^2\delta_l = 1$, and with so many $l$-values contributing to (6.173), it is legitimate to replace $\sin^2\delta_l$ by its average value, $\frac{1}{2}$. The number of terms in the $l$-sum is roughly $kR$, as is the average of $2l + 1$. Putting all the ingredients together, (6.173) becomes

$$\sigma_{tot} = \frac{4\pi}{k^2}(kR)^2\frac{1}{2} = 2\pi R^2, \tag{6.176}$$

which is not the geometric cross section $\pi R^2$ either! To see the origin of the factor of 2, we may split (6.126) into two parts:

$$f(\theta) = \frac{1}{2ik}\sum_{l=0}^{kR}(2l+1)e^{2i\delta_l}P_l(\cos\theta) + \frac{i}{2k}\sum_{l=0}^{kR}(2l+1)P_l(\cos\theta)$$

$$= f_{reflection} + f_{shadow}. \tag{6.177}$$

In evaluating $\int |f_{refl}|^2 d\Omega$, the orthogonality of the $P_l(\cos\theta)$ ensures that there is no interference amongst contributions from different $l$, and we obtain the sum of the square of partial-wave contributions:

$$\int |f_{refl}|^2 d\Omega = \frac{2\pi}{4k^2}\sum_{l=0}^{l_{max}}\int_{-1}^{+1}(2l+1)^2[P_l(\cos\theta)]^2 d(\cos\theta) = \frac{\pi l_{max}^2}{k^2} = \pi R^2. \tag{6.178}$$

Turning our attention to $f_{shad}$, we note that it is pure imaginary. It is particularly strong in the forward direction because $P_l(\cos\theta) = 1$ for $\theta = 0$, and the contributions from various $l$-values all add up coherently, that is, with the same phase, pure imaginary and positive in our case. We can use the small-angle approximation for $P_l$ to obtain

$$f_{shad} \simeq \frac{i}{2k}\sum(2l+1)J_0(l\theta)$$

$$\simeq ik\int_0^R bdbJ_0(kb\theta)$$

$$= \frac{iRJ_1(kR\theta)}{\theta}. \tag{6.179}$$

But this is just the formula for Fraunhofer diffraction in optics with a strong peaking near $\theta \simeq 0$. Letting $\xi = kR\theta$ and $d\xi/\xi = d\theta/\theta$, we can evaluate

$$\int |f_{shad}|^2 d\Omega = 2\pi\int_{-1}^{+1}\frac{R^2[J_1(kR\theta)]^2}{\theta^2}d(\cos\theta)$$

$$\simeq 2\pi R^2\int_0^\infty\frac{[J_1(\xi)]^2}{\xi}d\xi$$

$$\simeq \pi R^2. \tag{6.180}$$

Finally, the interference between $f_{shad}$ and $f_{refl}$ vanishes:

$$\text{Re}(f^*_{shad}\,f_{refl}) \simeq 0 \tag{6.181}$$

because the phase of $f_{refl}$ oscillates ($2\delta_{l+1} = 2\delta_l - \pi$), approximately averaging to zero, while $f_{shad}$ is pure imaginary. Thus

$$\sigma_{tot} = \underset{\substack{\uparrow \\ \sigma_{refl}}}{\pi R^2} + \underset{\substack{\uparrow \\ \sigma_{shad}}}{\pi R^2}. \tag{6.182}$$

The second term (coherent contribution in the forward direction) is called a *shadow* because for hard-sphere scattering at high energies, waves with impact parameter less than $R$ must be deflected. So, just *behind* the scatterer there must be zero probability for finding the particle and a shadow must be created. In terms of wave mechanics, this shadow is due to destructive interference between the original wave (which would be there even if the scatterer were absent) and the newly scattered wave. Thus we need scattering in order to create a shadow. That this shadow amplitude must be pure imaginary may be seen by recalling from (6.117) that the coefficient of $e^{ikr}/2ikr$ for the $l$th partial wave behaves like $1 + 2ikf_l(k)$, where the 1 would be present even without the scatterer; hence there must be a positive imaginary term in $f_l$ to get cancellation. In fact, this gives a physical interpretation of the optical theorem, which can be checked explicitly. First note that

$$\frac{4\pi}{k}\text{Im}f(0) \simeq \frac{4\pi}{k}\text{Im}[f_{shad}(0)] \tag{6.183}$$

because $\text{Im}[f_{refl}(0)]$ averages to zero due to oscillating phase. Using (6.177), we obtain

$$\frac{4\pi}{k}\text{Im}f_{shad}(0) = \frac{4\pi}{k}\text{Im}\left[\frac{i}{2k}\sum_{l=0}^{kR}(2l+1)P_l(1)\right] = 2\pi R^2 \tag{6.184}$$

which is indeed equal to $\sigma_{tot}$.

## 6.6  Low-Energy Scattering and Bound States

At low energies – or, more precisely, when $\lambdabar = 1/k$ is comparable to or larger than the range $R$ – partial waves for higher $l$ are, in general, unimportant. This point may be obvious classically because the particle cannot penetrate the centrifugal barrier; as a result the potential inside has no effect. In terms of quantum mechanics, the effective potential for the $l$th partial wave is given by

$$V_{eff} = V(r) + \frac{\hbar^2}{2m}\frac{l(l+1)}{r^2}; \tag{6.185}$$

unless the potential is strong enough to accommodate $l \neq 0$ bound states near $E \simeq 0$, the behavior of the radial wave function is largely determined by the centrifugal barrier term, which means that it must resemble $j_l(kr)$. More quantitatively, it is possible to estimate the behavior of the phase shift using the integral equation for the partial wave (see Problem 6.9 of this chapter):

$$\frac{e^{i\delta_l}\sin\delta_l}{k} = -\frac{2m}{\hbar^2}\int_0^\infty j_l(kr)V(r)A_l(r)r^2dr. \tag{6.186}$$

If $A_l(r)$ is not too different from $j_l(kr)$ and $1/k$ is much larger than the range of the potential, the right-hand side would vary as $k^{2l}$; for small $\delta_l$, the left-hand side must vary as $\delta_l/k$. Hence, the phase shift $k$ goes to zero as

$$\delta_l \sim k^{2l+1} \tag{6.187}$$

for small $k$. This is known as **threshold behavior**.

It is therefore clear that at low energies with a finite range potential, $S$-wave scattering is important.

### 6.6.1 Rectangular Well or Barrier

To be specific let us consider $S$-wave scattering by

$$V = \begin{cases} V_0 = \text{constant} & \text{for } r < R \\ 0 & \text{otherwise} \end{cases} \quad \begin{cases} V_0 > 0 & \text{repulsive} \\ V_0 < 0 & \text{attractive.} \end{cases} \tag{6.188}$$

Many of the features we obtain here are common to more complicated finite range potentials.

We have already seen that the outside wave function [see (6.138) and (6.150)] must behave like

$$e^{i\delta_0}[j_0(kr)\cos\delta_0 - n_0(kr)\sin\delta_0] \simeq \frac{e^{i\delta_0}\sin(kr+\delta_0)}{kr}. \tag{6.189}$$

The inside solution can also easily be obtained for $V_0$ a constant:

$$u \equiv rA_{l=0}(r) \propto \sin k'r, \tag{6.190}$$

with $k'$ determined by

$$E - V_0 = \frac{\hbar^2 k'^2}{2m}, \tag{6.191}$$

where we have used the boundary condition $u = 0$ at $r = 0$. In other words, the inside wave function is also sinusoidal as long as $E > V_0$. The curvature of the sinusoidal wave is different than in the free-particle case; as a result the wave function can be pushed in ($\delta_0 > 0$) or pulled out ($\delta_0 < 0$) depending on whether $V_0 < 0$ (attractive) or $V_0 > 0$ (repulsive), as shown in Figure 6.11. Notice also that (6.190) and (6.191) hold even if $V_0 > E$, provided we understand sin to mean sinh, that is, the wave function behaves like

$$u(r) \propto \sinh[\kappa r], \tag{6.190'}$$

where

$$\frac{\hbar^2\kappa^2}{2m} = (V_0 - E). \tag{6.191'}$$

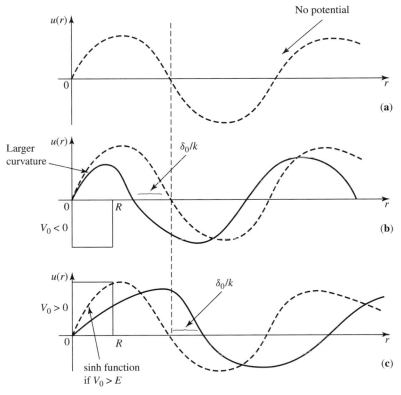

**Fig. 6.11**  Plot of $u(r)$ versus $r$. (a) For $V = 0$ (dashed line). (b) For $V_0 < 0$, $\delta_0 > 0$ with the wave function (solid line) pushed in. (c) For $V_0 > 0$, $\delta_0 < 0$ with the wave function (solid line) pulled out.

We now concentrate on the attractive case and imagine that the magnitude of $V_0$ is increased. Increased attraction will result in a wave function with a larger curvature. Suppose the attraction is such that the interval $[0, R]$ just accommodates one-fourth cycle of the sinusoidal wave. Working in the low-energy $kR \ll 1$ limit, the phase shift is now $\delta_0 = \pi/2$, and this results in a maximal $S$-wave cross section for a given $k$ because $\sin^2 \delta_0$ is unity. Now increase the well depth $V_0$ even further. Eventually the attraction is so strong that one-half cycle of the sinusoidal wave can be fitted within the range of the potential. The phase shift $\delta_0$ is now $\pi$; in other words, the wave function outside $R$ is 180° out of phase compared to the free-particle wave function. What is remarkable is that the partial cross section vanishes ($\sin^2 \delta_0 = 0$),

$$\sigma_{l=0} = 0, \tag{6.192}$$

despite the very strong attraction of the potential. In addition, if the energy is low enough for $l \neq 0$ waves still to be unimportant, we then have an almost perfect transmission of the incident wave. This kind of situation, known as the **Ramsauer–Townsend effect**, is actually observed experimentally for scattering of electrons by such rare gases as argon, krypton, and xenon. This effect was first observed in 1923 prior to the birth of wave mechanics and was considered to be a great mystery. Note the typical parameters here are $R \sim 2 \times 10^{-8}$ cm for electron kinetic energy of order 0.1 eV, leading to $kR \sim 0.324$.

## 6.6.2 Zero-Energy Scattering and Bound States

Let us consider scattering at extremely low energies ($k \simeq 0$). For $r > R$ and for $l = 0$, the outside radial wave function satisfies

$$\frac{d^2 u}{dr^2} = 0. \tag{6.193}$$

The obvious solution to this equation is

$$u(r) = \text{constant}(r - a), \tag{6.194}$$

just a straight line! This can be understood as an infinitely long wavelength limit of the usual expression for the outside wave function [see (6.142) and (6.150)],

$$\lim_{k \to 0} \sin(kr + \delta_0) = \lim_{k \to 0} \sin\left[k\left(r + \frac{\delta_0}{k}\right)\right], \tag{6.195}$$

which looks like (6.194). We have

$$\frac{u'}{u} = k \cot\left[k\left(r + \frac{\delta_0}{k}\right)\right] \overset{k \to 0}{\to} \frac{1}{r - a}. \tag{6.196}$$

Setting $r = 0$ [even though at $r = 0$, (6.194) is not the true wave function], we obtain

$$\lim_{k \to 0} k \cot \delta_0 \overset{k \to 0}{\to} -\frac{1}{a}. \tag{6.197}$$

The quantity $a$ is known as the **scattering length**. The limit of the total cross section as $k \to 0$ is given by [see (6.125)]

$$\sigma_{\text{tot}} = \sigma_{l=0} = 4\pi \lim_{k \to 0} \left| \frac{1}{k \cot \delta_0 - ik} \right|^2 = 4\pi a^2. \tag{6.198}$$

Even though $a$ has the same dimension as the range of the potential $R$, $a$ and $R$ can differ by orders of magnitude. In particular, for an attractive potential, it is possible for the magnitude of the scattering length to be far greater than the range of the potential. To see the physical meaning of $a$, we note that $a$ is nothing more than the intercept of the outside wave function. For a repulsive potential, $a > 0$ and is roughly of order of $R$, as seen in Figure 6.12a. However, for an attractive potential, the intercept is on the negative side (Figure 6.12b). If we *increase* the attraction, the outside wave function can again cross the $r$-axis on the positive side (Figure 6.12c).

The sign change resulting from increased attraction is related to the development of a bound state. To see this point quantitatively, we note from Figure 6.12c that for $a$ very large and positive, the wave function is essentially flat for $r > R$. But (6.194) with $a$ very large is not too different from $e^{-\kappa r}$ with $\kappa$ essentially zero. Now $e^{-\kappa r}$ with $\kappa \simeq 0$ is just a bound-state wave function for $r > R$ with energy $E$ infinitesimally negative. The inside wave functions ($r < R$) for the $E = 0+$ case (scattering with zero kinetic energy) and the $E = 0-$ case (bound state with infinitesimally small binding energy) are essentially the same because in both cases $k'$ in $\sin k'r$ [(6.190)] is determined by

$$\frac{\hbar^2 k'^2}{2m} = E - V_0 \simeq |V_0| \tag{6.199}$$

with $E$ infinitesimal (positive or negative).

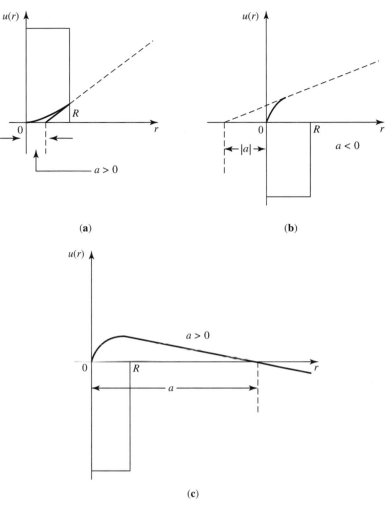

**Fig. 6.12** Plot of $u(r)$ versus $r$ for (a) repulsive potential, (b) attractive potential, and (c) deeper attraction. The intercept $a$ of the zero-energy outside wave function with the $r$-axis is shown for each of three cases.

Because the inside wave functions are the same for the two physical situations ($E = 0+$ and $E = 0-$), we can equate the logarithmic derivative of the bound-state wave function with that of the solution involving zero kinetic-energy scattering,

$$-\frac{\kappa e^{-\kappa r}}{e^{-\kappa r}}\bigg|_{r=R} = \left(\frac{1}{r-a}\right)\bigg|_{r=R}, \tag{6.200}$$

or, if $R \ll a$,

$$\kappa \simeq \frac{1}{a}. \tag{6.201}$$

The binding energy satisfies

$$E_{\mathrm{BE}} = -E_{\mathrm{bound\,state}} = \frac{\hbar^2 \kappa^2}{2m} \simeq \frac{\hbar^2}{2ma^2}, \tag{6.202}$$

and we have a relation between scattering length and bound-state energy. This is a remarkable result. To wit, if there is a loosely bound state, we can infer its binding energy by performing scattering experiments near zero kinetic energy, provided $a$ is measured to be large compared with the range $R$ of the potential. This connection between the scattering length and the bound-state energy was first pointed out by Wigner, who attempted to apply (6.202) to $np$-scattering.

Experimentally, the $^3S_1$-state of the $np$-system has a bound state, that is, the deuteron with

$$E_{\text{BE}} = 2.22 \text{ MeV}. \tag{6.203}$$

The scattering length is measured to be

$$a_{\text{triplet}} = 5.4 \times 10^{-13} \text{ cm}, \tag{6.204}$$

leading to the binding-energy prediction

$$\frac{\hbar^2}{2\mu a^2} = \frac{\hbar^2}{m_N a^2} = m_N c^2 \left(\frac{\hbar}{m_N c a}\right)^2$$

$$= (938 \text{ MeV}) \left(\frac{2.1 \times 10^{-14} \text{ cm}}{5.4 \times 10^{-13} \text{ cm}}\right)^2 = 1.4 \text{ MeV} \tag{6.205}$$

where $\mu$ is the reduced mass approximated by $m_{n,p}/2$. The agreement between experiment and prediction is not too satisfactory. The discrepancy is due to the fact that the inside wave functions are not exactly the same and that $a_{\text{triplet}} \gg R$ is not really such a good approximation for the deuteron. A better result can be obtained by keeping the next term in the expansion of $k \cot \delta$ as a function of $k$,

$$k \cot \delta_0 = -\frac{1}{a} + \frac{1}{2} r_0 k^2, \tag{6.206}$$

where $r_0$ is known as the effective range (see, for example, Preston (1962), p. 23).

### 6.6.3  Bound States as Poles of $S_l(k)$

We conclude this section by studying the analytic properties of the amplitude $S_l(k)$ for $l = 0$. Let us go back to (6.117) and (6.121), where the radial wave function for $l = 0$ at large distance was found to be proportional to

$$S_{l=0}(k) \frac{e^{ikr}}{r} - \frac{e^{-ikr}}{r}. \tag{6.207}$$

Compare this with the wave function for a bound state at large distance,

$$\frac{e^{-\kappa r}}{r}. \tag{6.208}$$

The existence of a bound state implies that a nontrivial solution to the Schrödinger equation with $E < 0$ exists only for a particular (discrete) value of $\kappa$. We may argue that $e^{-\kappa r}/r$ is like $e^{ikr}/r$, except that $k$ is now purely imaginary. Apart from $k$ being imaginary, the important difference between (6.207) and (6.208) is that in the bound-state case, $e^{-\kappa r}/r$ is present even

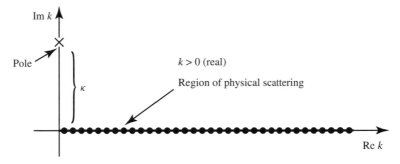

**Fig. 6.13**  The complex $k$-plane with bound-state pole at $k = +i\kappa$.

without the analogue of the incident wave. Quite generally only the ratio of the coefficient of $e^{ikr}/r$ to that of $e^{-ikr}/r$ is of physical interest, and this is given by $S_l(k)$. In the bound-state case we can sustain the outgoing wave (with imaginary $k$) even without an incident wave. So the ratio is $\infty$, which means that $S_{l=0}(k)$, regarded as a function of a complex variable $k$, has a pole at $k = i\kappa$. Thus a bound state implies a pole (which can be shown to be a simple pole) on the positive imaginary axis of the complex $k$-plane; see Figure 6.13. For $k$ real and positive, we have the region of physical scattering. Here we must require [compare with (6.123)]

$$S_{l=0} = e^{2i\delta_0} \tag{6.209}$$

with $\delta_0$ real. Furthermore, as $k \to 0$, $k \cot \delta_0$ has a limiting value $-1/a$ (6.197), which is finite, so $\delta_0$ must behave as follows:

$$\delta_0 \to 0, \pm\pi, \ldots. \tag{6.210}$$

Hence $S_{l=0} = e^{2i\delta_0} \to 1$ as $k \to 0$.

Now let us attempt to construct a simple function satisfying:

1.  Pole at $k = i\kappa$ (existence of bound state)
2.  $|S_{l=0}| = 1$ for $k > 0$ real (unitarity)                                $\qquad$ (6.211)
3.  $S_{l=0} = 1$ at $k = 0$ (threshold behavior).

The simplest function that satisfies all three conditions of (6.211) is

$$S_{l=0}(k) = \frac{-k - i\kappa}{k - i\kappa}. \tag{6.212}$$

[*Editor's Note*: Equation (6.212) is chosen for simplicity rather than as a physically realistic example. For reasonable potentials (not hard spheres!) the phase shift vanishes as $k \to \infty$.]

An assumption implicit in choosing this form is that there is no other singularity that is important apart from the bound-state pole. We can then use (6.124) to obtain, for $f_{l=0}(k)$,

$$f_{l=0} = \frac{S_{l=0} - 1}{2ik} = \frac{1}{-\kappa - ik}. \tag{6.213}$$

Comparing this with (6.125),

$$f_{l=0} = \frac{1}{k \cot \delta_0 - ik}, \tag{6.214}$$

we see that

$$\lim_{k \to 0} k \cot \delta_0 = -\frac{1}{a} = -\kappa, \tag{6.215}$$

precisely the relation between bound state and scattering length (6.201).

It thus appears that by exploiting unitarity and analyticity of $S_l(k)$ in the $k$-plane, we may obtain the kind of information that can be secured by solving the Schrödinger equation explicitly. This kind of technique can be very useful in problems where the details of the potential are not known.

## 6.7 Resonance Scattering

In atomic, nuclear, and particle physics, we often encounter a situation where the scattering cross section for a given partial wave exhibits a pronounced peak. This section is concerned with the dynamics of such a **resonance**.

We continue to consider a finite-ranged potential $V(r)$. The *effective* potential appropriate for the radial wave function of the $l$th partial wave is $V(r)$ plus the centrifugal barrier term as given by (6.185). Suppose $V(r)$ itself is attractive. Because the second term,

$$\frac{\hbar^2}{2m} \frac{l(l+1)}{r^2}$$

is repulsive, we have a situation where the effective potential has an attractive well followed by a repulsive barrier at larger distances, as shown in Figure 6.14.

Suppose the barrier were infinitely high. It would then be possible for particles to be trapped inside, which is another way of saying that we expect bound states, with energy $E > 0$. They are *genuine* bound states in the sense that they are eigenstates of the Hamiltonian with definite values of $E$. In other words, they are *stationary* states with infinite lifetime.

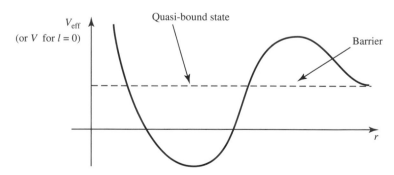

**Fig. 6.14** $V_{\text{eff}} = V(r) + (\hbar^2/2m)[l(l+1)/r^2]$ versus $r$. For $l \neq 0$ the barrier can be due to $(\hbar^2/2m)[l(l+1)/r^2]$; for $l = 0$ barrier must be due to $V$ itself.

In the more realistic case of a finite barrier, the particle can be trapped inside, but it cannot be trapped forever. Such a trapped state has a finite lifetime due to quantum-mechanical tunneling. In other words, a particle leaks through the barrier to the outside region. Let us call such a state a **quasi-bound state** because it would be an honest bound state if the barrier were infinitely high.

The corresponding scattering phase shift $\delta_l$ rises through the value $\pi/2$ as the incident energy rises through that of the quasi-bound state, and at the same time the corresponding partial-wave cross section passes through its maximum possible value (6.131) $4\pi(2l + 1)/k^2$. [*Editor's Note*: Such a sharp rise in the phase shift is, in the time-dependent Schrödinger equation, associated with a delay of the emergence of the trapped particles, rather than an unphysical advance, as would be the case for a sharp decrease through $\pi/2$.]

It is instructive to verify this point with explicit calculations for some known potential. The result of a numerical calculation shows that a resonance behavior is in fact possible for $l \neq 0$ with a spherical-well potential. To be specific we show the results for a spherical well with $2mV_0R^2/\hbar^2 = 5.5^2$ and $l = 3$ in Figure 6.15. The phase shift (Figure 6.15b), which is small at extremely low energies, starts increasing rapidly past $k \approx 1/R$, and goes through $\pi/2$ at $k = 1.41/R$.

Another very instructive example is provided by a repulsive $\delta$-shell potential that is exactly soluble (see Problem 6.10 in this chapter):

$$\frac{2m}{\hbar^2} V(r) = \gamma \delta(r - R). \tag{6.216}$$

Here resonances are possible for $l = 0$ because the $\delta$-shell potential itself can trap the particle in the region $0 < r < R$. For the case $\gamma = \infty$, we expect a series of bound states in the region $r < R$ with

$$kR = \pi, 2\pi, \ldots; \tag{6.217}$$

this is because the radial wave function for $l = 0$ must vanish not only at $r = 0$ but also at $r = R-$ in this case. For the region $r > R$, we simply have hard-sphere scattering with the $S$-wave phase shift, given by

$$\delta_0 = -kR. \tag{6.218}$$

With $\gamma = \infty$, there is no connection between the two problems because the wall at $r = R$ cannot be penetrated.

The situation is more interesting with a finite barrier, as we can show explicitly. The scattering phase shift exhibits a resonance behavior whenever

$$E_{\text{incident}} \simeq E_{\text{quasi-bound state}}. \tag{6.219}$$

Moreover, the larger the $\gamma$, the sharper the resonance peak. However, away from the resonance $\delta_0$ looks very much like the hard-sphere phase shift. Thus we have a situation in which a resonance behavior is superimposed on a smoothly behaving background scattering. This serves as a model for neutron-nucleus scattering, where a series of sharp resonance peaks are observed on top of a smoothly varying cross section.

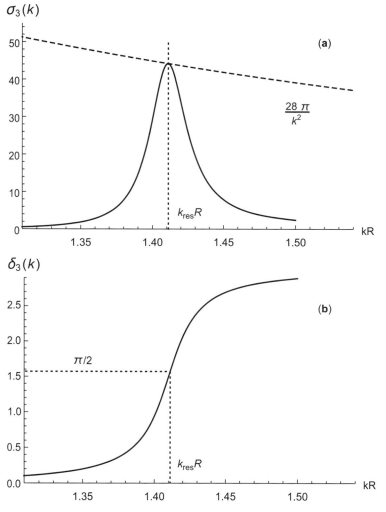

Plots of (a) $\sigma_{l=3}$ versus $k$, where at resonance $\delta_3(k_{res}) = \pi/2$, along with the unitarity limit $4\pi(7)/k^2$ from (6.131); and (b) $\delta_3(k)$ versus $k$. The curves are for a spherical well with $2\,mV_0R^2/\hbar^2 = 5.5^2$.

Coming back to our general discussion of resonance scattering, we ask how the scattering amplitudes vary in the vicinity of the resonance energy. If we are to have any connection between $\sigma_l$ being large and the quasi-bound states, $\delta_l$ *must go through* $\pi/2$ (*or* $3\pi/2,\dots$) *from below*, as discussed above. In other words $\delta_l$ must go through zero from above. Assuming that $\cot\delta_l$ is smoothly varying near the vicinity of resonance, that is,

$$E \simeq E_r, \tag{6.220}$$

we may attempt to expand $\delta_l$ as follows:

$$\cot\delta_l = \underbrace{\cot\delta_l|_{E=E_r}}_{0} - c(E-E_r) + \mathcal{O}\left[(E-E_r)^2\right]. \tag{6.221}$$

This leads to

$$f_l(k) = \frac{1}{k \cot \delta_l - ik} = \frac{1}{k} \frac{1}{[-c(E - E_r) - i]}$$

$$= -\frac{\Gamma/2}{k\left[(E - E_r) + \dfrac{i\Gamma}{2}\right]}, \tag{6.222}$$

where we have defined the *width* $\Gamma$ by

$$\left.\frac{d(\cot \delta_l)}{dE}\right|_{E = E_r} = -c \equiv -\frac{2}{\Gamma}. \tag{6.223}$$

Notice that $\Gamma$ is very small if $\cot \delta_l$ varies rapidly. If a simple resonance dominates the *l*th partial-wave cross section, we obtain a one-level resonance formula (the Breit–Wigner formula):

$$\sigma_l = \frac{4\pi}{k^2} \frac{(2l + 1)(\Gamma/2)^2}{(E - E_r)^2 + \Gamma^2/4}. \tag{6.224}$$

So it is legitimate to regard $\Gamma$ as the full width at half maximum, provided the resonance is reasonably narrow so that variation in $1/k^2$ can be ignored.

It is worthwhile to demonstrate these concepts with actual scattering measurements, but that is typically difficult. For one thing, real world interaction potentials are complicated and not well modeled by something as simple as a $\delta$-function shell or a hard wall spherical well. Furthermore, it usually happens that many different partial waves overlap, so that a sophisticated data analysis is required to separate them.

One good example, however, of an isolated partial wave giving a narrow resonance for an interaction that is reasonably well modeled by a spherical well, is the scattering of positive $\pi$-mesons from protons. Pions and protons are strongly interacting elementary particles, with "sizes" on the order of $1$ fm $\equiv 10^{-15}$ m. The force between them is attractive, and roughly constant when the particles overlap, but falls off rapidly outside this range. It is therefore perhaps not unreasonable that one might observe a relatively narrow resonance in certain partial waves.

Figure 6.16 plots the data[7] with $\pi^+ p$ elastic scattering, for $\pi^+$ center-of-mass momenta in the region of a few hundred MeV/$c$. There is a clear peak in the cross section, wider than, but otherwise similar to, Figure 6.15. One indication that this is in fact a resonance, is the comparison to the unitary limit, but first we need to reconsider the case when the scattering particles are not spinless.

The factor $2l + 1$ in (6.131) comes from summing over the final states with the same $l$ but different $m$. If the scattering particles have spin, then we would replace this factor with $2j + 1$, where $j$ is the total angular-momentum quantum number of the appropriate partial wave. However, we also need to consider the spins $s_1$ and $s_2$ of the scattering particles

---

[7] The data are available from http://pdg.lbl.gov. In fact, this figure plots both the elastic and total cross sections on the same plot. There are many more total cross section data points, but as there are no reactions other than $\pi^+ p \to \pi^+ p$ in this energy range, the cross sections agree.

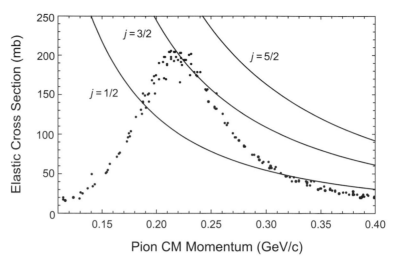

**Fig. 6.16** Elastic cross section data for the elastic scattering reaction $\pi^+ p \rightarrow \pi^+ p$, as a function of momentum in the center-of-mass. Also plotted are the unitary limits for different possibilities of the dominant partial wave. Compare to the calculated cross section for a hard walled square well in Figure 6.15. There is a clear suggestion that this is a resonance with total angular momentum $j = 3/2$.

when considering the initial scattering state. Averaging over all the possibilities gives an additional factor of $1/[(2s_1 + 1)(2s_2 + 1)]$. Therefore (6.131) becomes

$$\sigma_{\max}^{(l)} = \frac{4\pi}{k^2} g \qquad \text{where} \qquad g \equiv \frac{2j+1}{(2s_1 + 1)(2s_2 + 1)} \tag{6.225}$$

is usually called the "statistical factor." More detailed treatments can be found in nearly any textbook on nuclear physics, for example Section VIII.10 in the classic text Blatt and Weisskopf (1952).

For $\pi^+ p$ scattering, the $\pi^+$ is spinless and the proton has spin $\frac{1}{2}$, so $g = (2j + 1)/2$ and $j$ can take on any positive half-integer value. Figure 6.16 includes curves using (6.225) for different values of $j$. The agreement with the maximum cross section for $j = 3/2$ strongly suggests that this is a spin $\frac{3}{2}$ resonance. Indeed, a detailed analysis of the data including relative $l = 0$ and $l = 1$ phase differences shows that $j = 3/2$. This resonances is known as the $\Delta(1232)$, where 1232 MeV/$c^2$ is the invariant mass of the $\pi^+ p$ system at the peak of the cross section.

## 6.8 Symmetry Considerations in Scattering

Let us consider the scattering of two identical spinless charged particles via some central potential, such as the Coulomb potential.[8] The spatial part of the wave function must now be symmetric, so the asymptotic wave function must look like

---

[8] For the student unfamiliar with the elements of permutation symmetry with identical particles, see Chapter 7 of this textbook.

$$e^{i\mathbf{k}\cdot\mathbf{x}} + e^{-i\mathbf{k}\cdot\mathbf{x}} + [f(\theta) + f(\pi - \theta)]\frac{e^{ikr}}{r}, \tag{6.226}$$

where $\mathbf{x} = \mathbf{x}_1 - \mathbf{x}_2$ is the relative position vector between the two particles 1 and 2. This results in a differential cross section,

$$\frac{d\sigma}{d\Omega} = |f(\theta) + f(\pi - \theta)|^2$$

$$= |f(\theta)|^2 + |f(\pi - \theta)|^2 + 2\mathrm{Re}[f(\theta)f^*(\pi - \theta)]. \tag{6.227}$$

The cross section is enhanced through constructive interference at $\theta \simeq \pi/2$.

In contrast, for spin $\frac{1}{2}$– spin $\frac{1}{2}$ scattering with unpolarized beam and $V$ independent of spin, we have the spin-singlet scattering going with space-symmetrical function and the spin-triplet scattering going with space-antisymmetrical wave function (see Section 7.3). If the initial beam is unpolarized, we have the statistical contribution $\frac{1}{4}$ for spin singlet and $\frac{3}{4}$ for spin triplet; hence

$$\frac{d\sigma}{d\Omega} = \frac{1}{4}|f(\theta) + f(\pi - \theta)|^2 + \frac{3}{4}|f(\theta) - f(\pi - \theta)|^2$$

$$= |f(\theta)|^2 + |f(\pi - \theta)|^2 - \mathrm{Re}[f(\theta)f^*(\pi - \theta)]. \tag{6.228}$$

In other words, we expect destructive interference at $\theta \simeq \pi/2$. This has, in fact, been observed.

Now consider symmetries other than exchange symmetry. Suppose $V$ and $H_0$ are both invariant under some symmetry operation. We may ask what this implies for the matrix element of $T$ or for the scattering amplitude $f(\mathbf{k}', \mathbf{k})$.

If the symmetry operator is unitary (for example, rotation and parity), everything is quite straightforward. Using the explicit form of $T$ as given by (6.32), we see that

$$UH_0U^\dagger = H_0, \qquad UVU^\dagger = V \tag{6.229}$$

implies that $T$ is also invariant under $U$, that is,

$$UTU^\dagger = T. \tag{6.230}$$

We define

$$|\tilde{\mathbf{k}}\rangle \equiv U|\mathbf{k}\rangle, \qquad |\tilde{\mathbf{k}}'\rangle \equiv U|\mathbf{k}'\rangle. \tag{6.231}$$

Then

$$\langle \tilde{\mathbf{k}}'|T|\tilde{\mathbf{k}}\rangle = \langle \mathbf{k}'|U^\dagger UTU^\dagger U|\mathbf{k}\rangle$$

$$= \langle \mathbf{k}'|T|\mathbf{k}\rangle. \tag{6.232}$$

As an example, we consider the specific case where $U$ stands for the parity operator

$$\pi|\mathbf{k}\rangle = |-\mathbf{k}\rangle, \qquad \pi|-\mathbf{k}\rangle = |\mathbf{k}\rangle. \tag{6.233}$$

Thus invariance of $H_0$ and $V$ under parity would mean

$$\langle -\mathbf{k}'|T|-\mathbf{k}\rangle = \langle \mathbf{k}'|T|\mathbf{k}\rangle. \tag{6.234}$$

Pictorially, we have the situation illustrated in Figure 6.17a.

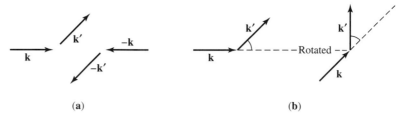

Fig. 6.17 (a) Equality of $T$ matrix elements between $\mathbf{k} \to \mathbf{k}'$ and $-\mathbf{k} \to -\mathbf{k}'$. (b) Equality of $T$ matrix elements under rotation.

We exploited the consequence of angular-momentum conservation when we developed the method of partial waves. The fact that $T$ is diagonal in the $|Elm\rangle$ representation is a direct consequence of $T$ being invariant under rotation. Notice also that $\langle \mathbf{k}'|T|\mathbf{k}\rangle$ depends only on the relative orientation of $\mathbf{k}$ and $\mathbf{k}'$, as depicted in Figure 6.17b.

When the symmetry operation is antiunitary (as in time reversal), we must be more careful. First, we note that the requirement that $V$ as well as $H_0$ be invariant under time reversal invariance requires that

$$\Theta T \Theta^{-1} = T^{\dagger}. \tag{6.235}$$

This is because the antiunitary operator changes

$$\frac{1}{E - H_0 + i\varepsilon} \quad \text{into} \quad \frac{1}{E - H_0 - i\varepsilon} \tag{6.236}$$

in (6.32). We also recall that for an antiunitary operator [see (4.114)],

$$\langle \beta | \alpha \rangle = \langle \tilde{\alpha} | \tilde{\beta} \rangle, \tag{6.237}$$

where

$$|\tilde{\alpha}\rangle \equiv \Theta|\alpha\rangle \quad \text{and} \quad |\tilde{\beta}\rangle \equiv \Theta|\beta\rangle. \tag{6.238}$$

Let us consider

$$|\alpha\rangle = T|\mathbf{k}\rangle, \qquad \langle \beta| = \langle \mathbf{k}'|; \tag{6.239}$$

then

$$|\tilde{\alpha}\rangle = \Theta T|\mathbf{k}\rangle = \Theta T \Theta^{-1}\Theta|\mathbf{k}\rangle = T^{\dagger}|-\mathbf{k}\rangle$$
$$|\tilde{\beta}\rangle = \Theta|\mathbf{k}\rangle = |-\mathbf{k}'\rangle. \tag{6.240}$$

As a result (6.237) becomes

$$\langle \mathbf{k}'|T|\mathbf{k}\rangle = \langle -\mathbf{k}|T|-\mathbf{k}'\rangle. \tag{6.241}$$

Notice that the initial and final momenta are interchanged, in addition to the fact that the directions of the momenta have been reversed.

It is also interesting to combine the requirements of time reversal [(6.241)] and parity [(6.234)]:

$$\langle \mathbf{k}'|T|\mathbf{k}\rangle \overset{\text{under}\,\Theta}{=} \langle -\mathbf{k}|T|-\mathbf{k}'\rangle \overset{\text{under}\,\pi}{=} \langle \mathbf{k}|T|\mathbf{k}'\rangle; \tag{6.242}$$

that is, from (6.58) and (6.70) we have

$$f(\mathbf{k},\mathbf{k}') = f(\mathbf{k}',\mathbf{k}), \tag{6.243}$$

which results in

$$\frac{d\sigma}{d\Omega}(\mathbf{k} \to \mathbf{k}') = \frac{d\sigma}{d\Omega}(\mathbf{k}' \to \mathbf{k}). \tag{6.244}$$

Equation (6.244) is known as **detailed balance**.

It is more interesting to look at the analogue of (6.242) when we have spin. Here we may characterize the initial free-particle ket by $|\mathbf{k}, m_s\rangle$, and we exploit (4.182) for the time reversal portion:

$$\langle \mathbf{k}', m_s'|T|\mathbf{k}, m_s\rangle = i^{-2m_s + 2m_{s'}} \langle -\mathbf{k}, -m_s|T|-\mathbf{k}', -m_s'\rangle$$
$$= i^{-2m_s + 2m_{s'}} \langle \mathbf{k}, -m_s|T|\mathbf{k}', -m_s'\rangle. \tag{6.245}$$

For unpolarized initial states, we sum over the initial spin states and divide by $(2s + 1)$; if the final polarization is not observed, we must sum over final states. We then obtain detailed balance in the form

$$\frac{\overline{d\sigma}}{d\Omega}(\mathbf{k} \to \mathbf{k}') = \frac{\overline{d\sigma}}{d\Omega}(\mathbf{k}' \to \mathbf{k}), \tag{6.246}$$

where we understand the bar on the top of $d\sigma/d\Omega$ in (6.246) to mean that we average over the initial spin states and sum over the final spin states.

## 6.9  Inelastic Electron-Atom Scattering

Let us consider the interactions of electron beams with atoms assumed to be in their ground states. The incident electron may be scattered elastically with final atoms unexcited:

$$e^- + \text{atom (ground state)} \to e^- + \text{atom (ground state)}. \tag{6.247}$$

This is an example of *elastic scattering*. To the extent that the atom can be regarded as infinitely heavy, the kinetic energy of the electron does not change. It is also possible for the target atom to be excited:

$$e^- + \text{atom (ground state)} \to e^- + \text{atom (excited state)}. \tag{6.248}$$

In this case we talk about **inelastic scattering** because the kinetic energy of the final outgoing electron is now less than that of the initial incoming electron, the difference being used to excite the target atom.

The initial ket of the electron plus the atomic system is written as

$$|\mathbf{k}, 0\rangle \tag{6.249}$$

where $\mathbf{k}$ refers to the wave vector of the incident electron and 0 stands for the atomic ground state. Strictly speaking (6.249) should be understood as the direct product of the incident electron ket $|\mathbf{k}\rangle$ and the ground-state atomic ket $|0\rangle$. The corresponding wave function is

$$\frac{1}{L^{3/2}} e^{i\mathbf{k}\cdot\mathbf{x}} \psi_0(\mathbf{x}_1, \mathbf{x}_2, \ldots, \mathbf{x}_z) \tag{6.250}$$

where we use the box normalization for the plane wave.

We may be interested in a final state electron with a definite wave vector $\mathbf{k}'$. The final state ket and the corresponding wave function are

$$|\mathbf{k}', n\rangle \quad \text{and} \quad \frac{1}{L^{3/2}} e^{i\mathbf{k}'\cdot\mathbf{x}} \psi_n(\mathbf{x}_1, \ldots, \mathbf{x}_z), \tag{6.251}$$

where $n = 0$ for elastic scattering and $n \neq 0$ for inelastic scattering.

Assuming that time-dependent perturbation theory is applicable, we can immediately write the differential cross section, as in the previous section:

$$\frac{d\sigma}{d\Omega}(0 \to n) = \frac{1}{(\hbar k/m_e L^3)} \frac{2\pi}{\hbar} |\langle \mathbf{k}'n|V|\mathbf{k}0\rangle|^2 \left(\frac{L}{2\pi}\right)^3 \left(\frac{k'm_e}{\hbar^2}\right)$$

$$= \left(\frac{k'}{k}\right) L^6 \left|\frac{1}{4\pi} \frac{2m_e}{\hbar^2} \langle \mathbf{k}', n|V|\mathbf{k}, 0\rangle\right|^2. \tag{6.252}$$

Everything is similar, including the cancellation of terms such as $L^3$, with one important exception: $k' \equiv |\mathbf{k}'|$ is not, in general, equal to $k \equiv |\mathbf{k}|$ for inelastic scattering.

The next question is, what $V$ is appropriate for this problem? The incident electron can interact with the nucleus, assumed to be situated at the origin; it can also interact with each of the atomic electrons. So $V$ is to be written as

$$V = -\frac{Ze^2}{r} + \sum_i \frac{e^2}{|\mathbf{x} - \mathbf{x}_i|}. \tag{6.253}$$

Here complications may arise because of the identity of the incident electron with one of the atomic electrons; to treat this rigorously is a nontrivial task. Fortunately, for a relatively fast electron we can legitimately ignore the question of identity; this is because there is little overlap between the bound-state electron and the incident electron in *momentum* space. We must evaluate the matrix element $\langle \mathbf{k}', n|V|\mathbf{k}0\rangle$, which, when explicitly written, is

$$\langle \mathbf{k}'n|V|\mathbf{k}0\rangle = \frac{1}{L^3} \int d^3x\, e^{i\mathbf{q}\cdot\mathbf{x}} \langle n| - \frac{Ze^2}{r} + \sum_i \frac{e^2}{|\mathbf{x} - \mathbf{x}_i|} |0\rangle$$

$$= \frac{1}{L^3} \int d^3x\, e^{i\mathbf{q}\cdot\mathbf{x}} \prod_i^z \int d^3x_i \psi_n^*(\mathbf{x}_1, \ldots, \mathbf{x}_z) \left[ -\frac{Ze^2}{r} + \sum_i \frac{e^2}{|\mathbf{x} - \mathbf{x}_i|} \right]$$

$$\times \psi_0(\mathbf{x}_1, \ldots, \mathbf{x}_z) \tag{6.254}$$

with $\mathbf{q} \equiv \mathbf{k} - \mathbf{k}'$.

Let us see how to evaluate the matrix element of the first term, $-Ze^2/r$, where $r$ actually means $|\mathbf{x}|$. First we note that this is a potential between the incident electron and the nucleus, which is independent of the atomic electron coordinates. So it can be taken outside the integration

$$\prod_i^z \int d^3x_i$$

in (6.254); we simply obtain

$$\langle n|0 \rangle = \delta_{n0} \tag{6.255}$$

for the remainder. In other words, this term contributes only to the elastic scattering case, where the target atom remains unexcited. In the elastic case we must still integrate $e^{i\mathbf{q}\cdot\mathbf{x}}/r$ with respect to $\mathbf{x}$, which amounts to taking the Fourier transform of the Coulomb potential. This can readily be done because we already evaluated the Fourier transform of the Yukawa potential; see (6.78). Hence

$$\int d^3x \frac{e^{i\mathbf{q}\cdot\mathbf{x}}}{r} = \lim_{\mu \to 0} \int \frac{d^3x e^{i\mathbf{q}\cdot\mathbf{x}-\mu r}}{r} = \frac{4\pi}{q^2}. \tag{6.256}$$

As for the second term in (6.254), we can evaluate the Fourier transform of $1/|\mathbf{x} - \mathbf{x}_i|$. We can accomplish this by shifting the coordinate variables $\mathbf{x} \to \mathbf{x} + \mathbf{x}_i$:

$$\sum_i \int \frac{d^3x e^{i\mathbf{q}\cdot\mathbf{x}}}{|\mathbf{x} - \mathbf{x}_i|} = \sum_i \int \frac{d^3x e^{i\mathbf{q}\cdot(\mathbf{x}+\mathbf{x}_i)}}{|\mathbf{x}|} = \frac{4\pi}{q^2} \sum_i e^{i\mathbf{q}\cdot\mathbf{x}_i}. \tag{6.257}$$

Notice that this is just the Fourier transform of the Coulomb potential multiplied by the Fourier transform of the electron density due to the atomic electrons situated at $\mathbf{x}_i$:

$$\rho_{\text{atom}}(\mathbf{x}) = \sum_i \delta^{(3)}(\mathbf{x} - \mathbf{x}_i). \tag{6.258}$$

We customarily define the **form factor** $F_n(\mathbf{q})$ for excitation $|0\rangle$ to $|n\rangle$ as follows:

$$ZF_n(\mathbf{q}) \equiv \langle n| \sum_i e^{i\mathbf{q}\cdot\mathbf{x}_i} |0\rangle, \tag{6.259}$$

which is made of coherent, in the sense of definite phase relationships, contributions from the various electrons. Notice that as $q \to 0$, we have

$$\frac{1}{Z} \langle n| \sum_i e^{i\mathbf{q}\cdot\mathbf{x}_i} |0\rangle \to 1$$

for $n = 0$; hence the form factor approaches unity in the elastic scattering case. For $n \neq 0$ (inelastic scattering), $F_n(\mathbf{q}) \to 0$ as $\mathbf{q} \to 0$ by orthogonality between $|n\rangle$ and $|0\rangle$. We can then write the matrix element in (6.254) as

$$\int d^3x e^{i\mathbf{q}\cdot\mathbf{x}} \langle n| \left( -\frac{Ze^2}{r} + \sum_i \frac{e^2}{|\mathbf{x} - \mathbf{x}_i|} \right) |0\rangle = \frac{4\pi Ze^2}{q^2} [-\delta_{n0} + F_n(\mathbf{q})]. \tag{6.260}$$

We are finally in a position to write the differential cross section for inelastic (or elastic) scattering of electrons by atoms:

$$\frac{d\sigma}{d\Omega}(0 \to n) = \left(\frac{k'}{k}\right)\left|\frac{1}{4\pi}\frac{2m_e}{\hbar^2}\frac{4\pi Ze^2}{q^2}[-\delta_{n0} + F_n(\mathbf{q})]\right|^2$$

$$= \frac{4m_e^2}{\hbar^4}\frac{(Ze^2)^2}{q^4}\left(\frac{k'}{k}\right)|-\delta_{n0} + F_n(\mathbf{q})|^2. \tag{6.261}$$

For inelastic scattering the $\delta_{n0}$-term does not contribute, and it is customary to write the differential cross section in terms of the Bohr radius,

$$a_0 = \frac{\hbar^2}{e^2 m_e}, \tag{6.262}$$

as follows:

$$\frac{d\sigma}{d\Omega}(0 \to n) = 4Z^2 a_0^2\left(\frac{k'}{k}\right)\frac{1}{(qa_0)^4}|F_n(\mathbf{q})|^2. \tag{6.263}$$

Quite often $d\sigma/dq$ is used in place of $d\sigma/d\Omega$; using

$$q^2 = |\mathbf{k} - \mathbf{k}'|^2 = k^2 + k'^2 - 2kk'\cos\theta \tag{6.264}$$

and $dq = -d(\cos\theta)kk'/q$, we can write

$$\frac{d\sigma}{dq} = \frac{2\pi q}{kk'}\frac{d\sigma}{d\Omega}. \tag{6.265}$$

The inelastic cross section we have obtained can be used to discuss *stopping power* – the energy loss of a charged particle as it goes through matter. A number of people, including H. A. Bethe and F. Bloch, have discussed the quantum-mechanical derivation of stopping power from the point of view of the inelastic scattering cross section. We are interested in the energy loss of a charged particle per unit length traversed by the incident charged particle. The collision rate per unit length is $N\sigma$, where $N$ is the number of atoms per unit volume; at each collision process the energy lost by the charged particle is $E_n - E_0$. So $dE/dx$ is written as

$$\frac{dE}{dx} = N\sum_n (E_n - E_0)\int \frac{d\sigma}{dq}(0 \to n)\,dq$$

$$= N\sum_n (E_n - E_0)\frac{4Z^2}{a_0^2}\int_{q_{\min}}^{q_{\max}} \frac{k'}{k}\frac{1}{q^4}\frac{2\pi q}{kk'}|F_n(\mathbf{q})|^2\,dq$$

$$= \frac{8\pi N}{k^2 a_0^2}\sum_n (E_n - E_0)\int_{q_{\min}}^{q_{\max}} \left|\langle n|\sum_{i=1}^z e^{i\mathbf{q}\cdot\mathbf{x}_i}|0\rangle\right|^2 \frac{dq}{q^3}. \tag{6.266}$$

There are many papers written on how to evaluate the sum in (6.266).[9] The upshot of all this is to justify quantum mechanically Bohr's 1913 formula for stopping power,

$$\frac{dE}{dx} = \frac{4\pi NZe^4}{m_e v^2}\ln\left(\frac{2m_e v^2}{I}\right), \tag{6.267}$$

[9] For a relatively elementary discussion, see Gottfried (1966) and Bethe and Jackiw (1968).

where $I$ is a semiempirical parameter related to the average excitation energy $\langle E_n - E_0 \rangle$. If the charged particle has electric charge $\pm ze$, we just replace $Ze^4$ by $z^2 Ze^4$. It is also important to note that even if the projectile is not an electron, the $m_e$ that appears in (6.267) is still the electron mass, not the mass of the charged particle. So the energy loss is dependent on the charge and the velocity of the projectile but is independent of the mass of the projectile. This has an important application to the detection of charged particles.

Quantum mechanically, we view the energy loss of a charged particle as a series of inelastic scattering processes. At each interaction between the charged particle and an atom, we may imagine that a "measurement" of the position of the charged particle is made. We may wonder why particle tracks in such media as cloud chambers and nuclear emulsions are nearly straight. The reason is that the differential cross section (6.263) is sharply peaked at small $q$; in an overwhelming number of collisions, the final direction of momentum is nearly the same as the incident electron due to the rapid falloff of $q^{-4}$ and $F_n(\mathbf{q})$ for large $q$.

### 6.9.1  Nuclear Form Factor

The excitation of atoms due to inelastic scattering is important for $q \sim 10^9 \, \text{cm}^{-1}$ to $10^{10} \, \text{cm}^{-1}$. If $q$ is too large, the contributions due to $F_0(\mathbf{q})$ or $F_n(\mathbf{q})$ drop off very rapidly. At extremely high $q$, where $q$ is now of order $1/R_{\text{nucleus}} \sim 10^{12} \, \text{cm}^{-1}$, the structure of the nucleus becomes important. The Coulomb potential due to the point nucleus must now be replaced by a Coulomb potential due to an extended object,

$$-\frac{Ze^2}{r} \to -Ze^2 \int \frac{d^3 x' N(r')}{|\mathbf{x} - \mathbf{x}'|}, \tag{6.268}$$

where $N(r)$ is a nuclear charge distribution, normalized so that

$$\int d^3 x' N(r') = 1. \tag{6.269}$$

The pointlike nucleus can now be regarded as a special case, with

$$N(r') = \delta^{(3)}(r'). \tag{6.270}$$

We can evaluate the Fourier transform of the right-hand side of (6.268) in analogy with (6.256) as follows:

$$Ze^2 \int d^3 x \int \frac{d^3 x' e^{i\mathbf{q}\cdot\mathbf{x}} N(r')}{|\mathbf{x} - \mathbf{x}'|} = Ze^2 \int d^3 x' e^{i\mathbf{q}\cdot\mathbf{x}'} N(r') \int \frac{d^3 x e^{i\mathbf{q}\cdot\mathbf{x}}}{r}$$

$$= Ze^2 \frac{4\pi}{q^2} F_{\text{nucleus}}(\mathbf{q}) \tag{6.271}$$

where we have shifted the coordinates $\mathbf{x} \to \mathbf{x} + \mathbf{x}'$ in the first step and

$$F_{\text{nucleus}} \equiv \int d^3 x e^{i\mathbf{q}\cdot\mathbf{x}} N(r). \tag{6.272}$$

We thus obtain the deviation from the Rutherford formula due to the finite size of the nucleus,

$$\frac{d\sigma}{d\Omega} = \left(\frac{d\sigma}{d\Omega}\right)_{\text{Rutherford}} |F(\mathbf{q})|^2, \tag{6.273}$$

where $(d\sigma/d\Omega)_{\text{Rutherford}}$ is the differential cross section for the electric scattering of electrons by a pointlike nucleus of charge $Z|e|$. For small $q$ we have

$$F_{\text{nucleus}}(\mathbf{q}) = \int d^3x \left(1 + i\mathbf{q}\cdot\mathbf{x} - \frac{1}{2}q^2 r^2 (\hat{\mathbf{q}}\cdot\hat{\mathbf{r}})^2 + \cdots \right) N(r)$$

$$= 1 - \frac{1}{6}q^2 \langle r^2 \rangle_{\text{nucleus}} + \cdots. \tag{6.274}$$

The $\mathbf{q}\cdot\mathbf{x}$-term vanishes because of spherical symmetry, and in the $q^2$-term we have used the fact that the angular average of $\cos^2\theta$ (where $\theta$ is the angle between $\hat{\mathbf{q}}$ and $\hat{\mathbf{r}}$) is just $\frac{1}{3}$:

$$\frac{1}{2}\int_{-1}^{+1} d(\cos\theta)\cos^2\theta = \frac{1}{3}. \tag{6.275}$$

The quantity $\langle r^2 \rangle_{\text{nucleus}}$ is known as the mean square radius of the nucleus. In this way it is possible to "measure" the size of the nucleus and also of the proton, as done by R. Hofstadter and coworkers. In the proton case the spin (magnetic moment) effect is also important.

# Problems

**6.1** Consider scattering in one dimension $x$ from a potential $V(x)$ localized near $x = 0$. The initial state is a plane wave coming from the left, that is $\phi(x) \equiv \langle x|i\rangle = e^{ikx}/\sqrt{2\pi}$.
  a. Find the scattering Green's function $G(x,x')$, defined in one dimension analogously with (6.37), for $G_+(\mathbf{x},\mathbf{x}')$.
  b. For the case of an attractive $\delta$-function potential $V(x) = -\gamma\hbar^2\delta(x)/2m$, with $\gamma > 0$, use the Lippman–Schwinger equation to find the outgoing wave function $\psi(x) \equiv \langle x|\psi^{(+)}\rangle$.
  c. Determine the transmission and reflection coefficients $T(k)$ and $R(k)$, defined as

$$\psi(x) = T(k)\phi(x) \text{ for } x > 0 \quad \text{and} \quad \psi(x) = \phi(x) + R(k)\frac{e^{-ikx}}{\sqrt{2\pi}} \text{ for } x < 0.$$

  Show that $|T|^2 + |R|^2 = 1$, as must be the case.
  d. Confirm that you get the same result by matching right and left going waves on the left with a right going wave on the right at $x = 0$.
  e. We showed in Problem 2.29 that this potential has one, and only one, bound state. Show that your results for $T(k)$ and $R(k)$ have bound-state poles at the expected positions when $k$ is treated as a complex variable.

**6.2**  Prove

$$\sigma_{\text{tot}} \simeq \frac{m^2}{\pi \hbar^4} \int d^3x \int d^3x' \, V(r) V(r') \frac{\sin^2 k|\mathbf{x} - \mathbf{x}'|}{k^2 |\mathbf{x} - \mathbf{x}'|^2}$$

in each of the following ways.

a. By integrating the differential cross section computed using the first-order Born approximation.

b. By applying the optical theorem to the forward-scattering amplitude in the *second*-order Born approximation. [Note that $f(0)$ is real if the first-order Born approximation is used.]

**6.3**  Estimate the radius of the $^{40}$Ca nucleus from the data in Figure 6.6 and compare to that expected from the empirical value $\approx 1.4 A^{1/3}$ fm, where $A$ is the nuclear mass number. Check the validity of using the first-order Born approximation for these data.

**6.4**  Consider a potential

$$V = 0 \quad \text{for} \quad r > R, \qquad V = V_0 = \text{constant} \quad \text{for} \quad r < R,$$

where $V_0$ may be positive or negative. Using the method of partial waves, show that for $|V_0| \ll E = \hbar^2 k^2 / 2m$ and $kR \ll 1$ the differential cross section is isotropic and that the total cross section is given by

$$\sigma_{\text{tot}} = \left( \frac{16\pi}{9} \right) \frac{m^2 V_0^2 R^6}{\hbar^4}.$$

Suppose the energy is raised slightly. Show that the angular distribution can then be written as

$$\frac{d\sigma}{d\Omega} = A + B \cos \theta.$$

Obtain an approximate expression for $B/A$.

**6.5**  A spinless particle is scattered by a weak Yukawa potential

$$V = \frac{V_0 e^{-\mu r}}{\mu r}$$

where $\mu > 0$ but $V_0$ can be positive or negative. It was shown in the text that the first-order Born amplitude is given by

$$f^{(1)}(\theta) = -\frac{2mV_0}{\hbar^2 \mu} \frac{1}{[2k^2(1 - \cos\theta) + \mu^2]}.$$

a. Using $f^{(1)}(\theta)$ and assuming $|\delta_l| \ll 1$, obtain an expression for $\delta_l$ in terms of a Legendre function of the second kind,

$$Q_l(\zeta) = \frac{1}{2} \int_{-1}^{1} \frac{P_l(\zeta')}{\zeta - \zeta'} d\zeta'.$$

b. Use the expansion formula

$$Q_l(\zeta) = \frac{l!}{1 \cdot 3 \cdot 5 \cdots (2l+1)}$$

$$\times \left\{ \frac{1}{\zeta^{l+1}} + \frac{(l+1)(l+2)}{2(2l+3)} \frac{1}{\zeta^{l+3}} \right.$$

$$\left. + \frac{(l+1)(l+2)(l+3)(l+4)}{2 \cdot 4 \cdot (2l+3)(2l+5)} \frac{1}{\zeta^{l+5}} + \cdots \right\} \quad (|\zeta| > 1)$$

to prove each assertion.

(i) $\delta_l$ is negative (positive) when the potential is repulsive (attractive).

(ii) When the de Broglie wavelength is much longer than the range of the potential, $\delta_l$ is proportional to $k^{2l+1}$. Find the proportionality constant.

**6.6** Check explicitly the $x - p_x$ uncertainty relation for the ground state of a particle confined inside a hard sphere: $V = \infty$ for $r > a$, $V = 0$ for $r < a$. (*Hint*: Take advantage of spherical symmetry.)

**6.7** Consider the scattering of a particle by an impenetrable sphere

$$V(r) = \begin{cases} 0 & \text{for } r > a \\ \infty & \text{for } r < a. \end{cases}$$

a. Derive an expression for the $s$-wave ($l = 0$) phase shift. (You need not know the detailed properties of the spherical Bessel functions to be able to do this simple problem!)

b. What is the total cross section $\sigma [\sigma = \int (d\sigma/d\Omega) d\Omega]$ in the extreme low-energy limit $k \to 0$? Compare your answer with the geometric cross section $\pi a^2$. You may assume without proof:

$$\frac{d\sigma}{d\Omega} = |f(\theta)|^2,$$

$$f(\theta) = \left( \frac{1}{k} \right) \sum_{l=0}^{\infty} (2l+1) e^{i\delta_l} \sin \delta_l P_l(\cos\theta).$$

**6.8** Use $\delta_l = \Delta(b)|_{b=l/k}$ to obtain the phase shift $\delta_l$ for scattering at high energies by (a) the Gaussian potential, $V = V_0 \exp(-r^2/a^2)$, and (b) the Yukawa potential, $V = V_0 \exp(-\mu r)/\mu r$. Verify the assertion that $\delta_l$ goes to zero very rapidly with increasing $l$ ($k$ fixed) for $l \gg kR$, where $R$ is the "range" of the potential. [The formula for $\Delta(b)$ is given in (6.168).] It is useful to realize that

$$\int_1^\infty \frac{e^{-ax}}{(x^2 - 1)^{1/2}} dx = K_0(x)$$

where $K_0(x)$ is the modified Bessel function of the second kind, of order zero.

**6.9** a. Prove

$$\frac{\hbar^2}{2m} \langle \mathbf{x} | \frac{1}{E - H_0 + i\varepsilon} | \mathbf{x}' \rangle = -ik \sum_l \sum_m Y_l^m(\hat{\mathbf{r}}) Y_l^{m*}(\hat{\mathbf{r}}') j_l(kr_<) h_l^{(1)}(kr_>)$$

where $r_<$ ($r_>$) stands for the smaller (larger) of $r$ and $r'$.

b. For spherically symmetric potentials, the Lippmann–Schwinger equation can be written for *spherical* waves:

$$|Elm(+)\rangle = |Elm\rangle + \frac{1}{E - H_0 + i\varepsilon} V |Elm(+)\rangle.$$

Using (a), show that this equation, written in the **x**-representation, leads to an equation for the radial function, $A_l(k; r)$, as follows:

$$A_l(k; r) = j_l(kr) - \frac{2mik}{\hbar^2}$$

$$\times \int_0^\infty j_l(kr_<) h_l^{(1)}(kr_>) V(r') A_l(k; r') r'^2 dr'.$$

By taking $r$ very large, also obtain

$$f_l(k) = e^{i\delta_l} \frac{\sin \delta_l}{k}$$

$$= -\left(\frac{2m}{\hbar^2}\right) \int_0^\infty j_l(kr) A_l(k; r) V(r) r^2 dr.$$

**6.10** Consider scattering by a repulsive $\delta$-shell potential:

$$\left(\frac{2m}{\hbar^2}\right) V(r) = \gamma \delta(r - R) \quad (\gamma > 0).$$

a. Set up an equation that determines the $s$-wave phase shift $\delta_0$ as a function of $k$ ($E = \hbar^2 k^2 / 2m$).

b. Assume now that $\gamma$ is very large,

$$\gamma \gg \frac{1}{R}, k.$$

Show that if $\tan kR$ is *not* close to zero, the $s$-wave phase shift resembles the hard-sphere result discussed in the text. Show also that for $\tan kR$ close to (but not exactly equal to) zero, resonance behavior is possible; that is, $\cot \delta_0$ goes through zero from the positive side as $k$ increases. Determine approximately the positions of the resonances keeping terms of order $1/\gamma$; compare them with the bound-state energies for a particle confined *inside* a spherical wall of the same radius,

$$V = 0, \quad r < R; \qquad V = \infty, \quad r > R.$$

Also obtain an approximate expression for the resonance width $\Gamma$ defined by

$$\Gamma = \frac{-2}{[d(\cot \delta_0)/dE]|_{E=E_r}}$$

and notice, in particular, that the resonances become extremely sharp as $\gamma$ becomes large. (*Note*: For a different, more sophisticated approach to this problem see Gottfried (1966), pp. 131–141, who discusses the analytic properties of the $D_l$-function defined by $A_l = j_l/D_l$.)

**6.11** A spinless particle is scattered by a time-dependent potential

$$\mathscr{V}(\mathbf{r},t) = V(\mathbf{r})\cos\omega t.$$

Show that if the potential is treated to first order in the transition amplitude, the energy of the scattered particle is increased or decreased by $\hbar\omega$. Obtain $d\sigma/d\Omega$. Discuss qualitatively what happens if the higher-order terms are taken into account.

**6.12** Show that the differential cross section for the elastic scattering of a fast positron by the ground state of the hydrogen atom is given by

$$\frac{d\sigma}{d\Omega} = \left(\frac{4m^2e^4}{\hbar^4 q^4}\right)\left\{1 - \frac{16}{[4+(qa_0)^2]^2}\right\}^2.$$

**6.13** Write a computer program to reproduce Figure 6.15. Also include a plot showing the relative sizes of the energy, well depth, and effective potential, that is, the analogue of Figure 6.14 for this problem.

**6.14** Let the energy of a particle moving in a central field be $E(J_1 J_2 J_3)$, where $(J_1, J_2, J_3)$ are the three action variables. How does the functional form of $E$ specialize for the Coulomb potential? Using the recipe of the action-angle method, compare the degeneracy of the central field and the Coulomb problems and relate it to the vector $\mathbf{A}$.

If the Hamiltonian is

$$H = \frac{p^2}{2\mu} + V(r) + F(\mathbf{A}^2),$$

how are these statements changed?

Describe the corresponding degeneracies of the central field and Coulomb problems in quantum theory in terms of the usual quantum numbers $(n, l, m)$ and also in terms of the quantum numbers $(k, m, n)$. Here the second set, $(k, m, n)$, labels the wave functions $\mathscr{D}_{mn}^k(\alpha\beta\gamma)$.

How are the wave functions $\mathscr{D}_{mn}^k(\alpha\beta\gamma)$ related to Laguerre times spherical harmonics?

# 7 Identical Particles

This chapter is devoted to a discussion of some striking quantum-mechanical effects arising from the identity of particles. First we present a suitable formalism and the way that nature deals with what appears to be an arbitrary choice. We then consider some applications to atoms more complex than hydrogen like atoms. Next we generalize to systems with many identical particles, and discuss two different formalisms for calculations. Lastly, we will cover one concrete example of a many-particle quantum-mechanical field theory, namely quantizing the electromagnetic field.

## 7.1 Permutation Symmetry

In classical physics it is possible to keep track of individual particles even though they may look alike. When we have particle 1 and particle 2 considered as a system, we can, in principle, follow the trajectory of particle 1 and that of particle 2 separately at each instant of time. For bookkeeping purposes, you may color one of them blue and the other red and then examine how the red particle moves and how the blue particle moves as time passes.

In quantum mechanics, however, identical particles are truly indistinguishable. This is because we cannot specify more than a complete set of commuting observables for each of the particles; in particular, we cannot label the particle by coloring it blue. Nor can we follow the trajectory because that would entail a position measurement at each instant of time, which necessarily disturbs the system; in particular the two situations (a) and (b) shown in Figure 7.1 cannot be distinguished – not even in principle.

For simplicity consider just two particles. Suppose one of the particles, which we call particle 1, is characterized by $|k'\rangle$, where $k'$ is a collective index for a complete set of observables. Likewise, we call the ket of the remaining particle $|k''\rangle$. The state ket for the two particles can be written in product form,

$$|k'\rangle|k''\rangle, \tag{7.1}$$

where it is understood that the first ket refers to particle 1 and the second ket to particle 2. We can also consider

$$|k''\rangle|k'\rangle, \tag{7.2}$$

where particle 1 is characterized by $|k''\rangle$ and particle 2 by $|k'\rangle$. Even though the two particles are indistinguishable, it is worth noting that mathematically (7.1) and (7.2) are *distinct* kets for $k' \neq k''$. In fact, with $k' \neq k''$, they are orthogonal to each other.

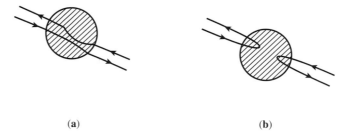

Fig. 7.1 Two different paths, (a) and (b), of a two-electron system, for example, in which we cannot assert even in principle through which of the paths the electrons pass.

Suppose we make a measurement on the two-particle system. We may obtain $k'$ for one particle and $k''$ for the other. However, we do not know a priori whether the state ket is $|k'\rangle|k''\rangle$, $|k''\rangle|k'\rangle$, or, for that matter, any linear combination of the two. Put in another way, all kets of form

$$c_1|k'\rangle|k''\rangle + c_2|k''\rangle|k'\rangle \tag{7.3}$$

lead to an identical set of eigenvalues when measurement is performed. This is known as **exchange degeneracy**. Exchange degeneracy presents a difficulty because, unlike the single-particle case, a specification of the eigenvalue of a complete set of observables does not completely determine the state ket. The way nature avoids this difficulty is quite ingenious. But before proceeding further, let us develop the mathematics of permutation symmetry.

We define the permutation operator $P_{12}$ by

$$P_{12}|k'\rangle|k''\rangle = |k''\rangle|k'\rangle. \tag{7.4}$$

Clearly,

$$P_{21} = P_{12} \quad \text{and} \quad P_{12}^2 = 1. \tag{7.5}$$

Under $P_{12}$, particle 1 having $k'$ becomes particle 1 having $k''$; particle 2 having $k''$ becomes particle 2 having $k'$. In other words, it has the effect of interchanging 1 and 2.

In practice we often encounter an observable that has particle labels. For example in $\mathbf{S_1} \cdot \mathbf{S_2}$ for a two-electron system, $\mathbf{S_1}$ $(\mathbf{S_2})$ stands for the spin operator of particle 1 (2). For simplicity we consider a specific case where the two-particle state ket is completely specified by the eigenvalues of a single observable $A$ for each of the particles:

$$A_1|a'\rangle|a''\rangle = a'|a'\rangle|a''\rangle \tag{7.6a}$$

and

$$A_2|a'\rangle|a''\rangle = a''|a'\rangle|a''\rangle, \tag{7.6b}$$

where the subscripts on $A$ denote the particle labels, and $A_1$ and $A_2$ are thus the observables $A$ for particles 1 and 2, respectively. Applying $P_{12}$ to both sides of (7.6a), and inserting $1 = P_{12}^{-1}P_{12}$, we have

$$P_{12}A_1P_{12}^{-1}P_{12}|a'\rangle|a''\rangle = a'P_{12}|a'\rangle|a''\rangle$$
$$P_{12}A_1P_{12}^{-1}|a''\rangle|a'\rangle = a'|a''\rangle|a'\rangle. \tag{7.7}$$

This is consistent with (7.6b) only if

$$P_{12}A_1P_{12}^{-1} = A_2. \tag{7.8}$$

It follows that $P_{12}$ must change the particle labels of observables.

Let us now consider the Hamiltonian of a system of two identical particles. The observables, such as momentum and position operators, must necessarily appear symmetrically in the Hamiltonian, for example,

$$H = \frac{\mathbf{p}_1^2}{2m} + \frac{\mathbf{p}_2^2}{2m} + V_{\text{pair}}(|\mathbf{x}_1 - \mathbf{x}_2|) + V_{\text{ext}}(\mathbf{x}_1) + V_{\text{ext}}(\mathbf{x}_2). \tag{7.9}$$

Here we have separated the mutual interaction between the two particles from their interaction with some other external potential. Clearly, we have

$$P_{12}HP_{12}^{-1} = H \tag{7.10}$$

for $H$ made up of observables for two identical particles. Because $P_{12}$ commutes with $H$, we can say that $P_{12}$ is a constant of the motion. The eigenvalues of $P_{12}$ allowed are $+1$ and $-1$ because of (7.5). It therefore follows that if the two-particle state ket is symmetric (antisymmetric) to start with, it remains so at all times.

If we insist on eigenkets of $P_{12}$, two particular linear combinations are selected:

$$|k'k''\rangle_+ \equiv \frac{1}{\sqrt{2}}\left(|k'\rangle|k''\rangle + |k''\rangle|k'\rangle\right), \tag{7.11a}$$

and

$$|k'k''\rangle_- \equiv \frac{1}{\sqrt{2}}\left(|k'\rangle|k''\rangle - |k''\rangle|k'\rangle\right). \tag{7.11b}$$

We can define the symmetrizer and antisymmetrizer as follows:

$$S_{12} \equiv \tfrac{1}{2}(1 + P_{12}), \qquad A_{12} \equiv \tfrac{1}{2}(1 - P_{12}). \tag{7.12}$$

We can extend this formalism to include states with more than two identical particles. From (7.12), if we apply $S_{12}(A_{12})$ to an arbitrary linear combination of $|k'\rangle|k''\rangle$ and $|k''\rangle|k'\rangle$, the resulting ket is necessarily symmetric (antisymmetric). This can easily be seen as follows:

$$\begin{Bmatrix} S_{12} \\ A_{12} \end{Bmatrix} [c_1|k'\rangle|k''\rangle + c_2|k''\rangle|k'\rangle]$$
$$= \tfrac{1}{2}\left(c_1|k'\rangle|k''\rangle + c_2|k''\rangle|k'\rangle\right) \pm \tfrac{1}{2}\left(c_1|k''\rangle|k'\rangle + c_2|k'\rangle|k''\rangle\right)$$
$$= \frac{c_1 \pm c_2}{2}\left(|k'\rangle|k''\rangle \pm |k''\rangle|k'\rangle\right). \tag{7.13}$$

In Section 7.5 we will build on this approach.

Before closing this section, we pause to point out that the consequences can be dramatic when permutation symmetry is ignored. Figure 7.2 shows a result which compares two experiments, before and after an error was corrected which ignored permutation symmetry in the analysis.

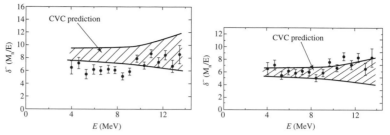

**Fig. 7.2**  Dramatic consequences arise when permutation symmetry is neglected. The data points are from McKeown et al., *Phys. Rev. C*, **22** (1980) 738, who tested a prediction of the CVC hypothesis. The $\beta^{\pm} - \alpha$ correlation $\delta^-$ is plotted against the $\beta^{\pm}$ energy. The prediction comes from a different, previous experiment, which at first neglected permutation symmetry. The corrected plot is on the right, from McKeown et al., *Phys. Rev. C*, **26** (1982) 2336, where the CVC prediction is smaller by a factor of $\sqrt{2}$.

The object of this set of experiments was to test something called the Conserved Vector Current (CVC) hypothesis, based on the assumption of an intimate connection[1] between the electromagnetic and weak interactions. Confirmation, or refutation, of the CVC hypothesis was a high priority, and this experiment was one of the most precise tests. The initial result, shown on the left in Figure 7.2, was less than clear. The corrected result, on the right, was finally a decisive confirmation of CVC.

The data points in Figure 7.2, which are identical for the left-hand and right-hand plots, are from a measurement of the beta decays of $^8$Li and $^8$B, each of which leads to a final state with two (identical) $\alpha$ particles, through an excited state of $^8$Be. That is

$$^8\text{Li} \rightarrow\ ^8\text{Be}^* + e^- + \bar{\nu}_e \tag{7.14a}$$

$$^8\text{B} \rightarrow\ ^8\text{Be}^* + e^+ + \nu_e \tag{7.14b}$$

followed by

$$^8\text{Be}^* \rightarrow \alpha + \alpha. \tag{7.14c}$$

The experiment determines $\delta^-$, the correlation in direction between the $e^{\pm}$ and $\alpha$ directions for the two beta decays, as a function of $e^{\pm}$ energy. The result of this measurement is published as McKeown et al., *Phys. Rev. C*, **22** (1980) 738.

The hatched area shows the CVC prediction derived from an earlier experiment, published as Bowles and Garvey *Phys. Rev. C*, **18** (1978) 1447. This work measured the reaction

$$\alpha + \alpha \rightarrow\ ^8\text{Be}^* \tag{7.15a}$$

followed by

$$^8\text{Be}^* \rightarrow\ ^8\text{Be} + \gamma. \tag{7.15b}$$

---

[1]  The CVC hypothesis predates the unification of electromagnetic and weak interactions in what today is referred to as the "Standard Model."

That is, a process which is rather the inverse of that in (7.14) and which proceeds through an electromagnetic interaction instead of the weak interaction. Deriving the CVC prediction from this result requires that the $\alpha\alpha$ wave function be symmetrized, but this was neglected at first, for the plot shown on the left of Figure 7.2. Some time later, this error was corrected for the missing factor of $\sqrt{2}$, and the plot on the right was published, showing much better agreement between prediction and measurement.

## 7.2 Symmetrization Postulate

So far we have not discussed whether nature takes advantage of totally symmetrical or totally antisymmetrical states. It turns out[2] that systems containing $N$ identical particles are either totally symmetrical under the interchange of any pair, in which case the particles are said to satisfy **Bose–Einstein** (B-E) **statistics**, hence known as **bosons**, or totally antisymmetrical, in which case the particles are said to satisfy **Fermi–Dirac** (F-D) **statistics**, hence known as **fermions**. Thus

$$P_{ij}|N \text{ identical bosons}\rangle = +|N \text{ identical bosons}\rangle \qquad (7.16a)$$

$$P_{ij}|N \text{ identical fermions}\rangle \quad -|N \text{ identical fermions}\rangle, \qquad (7.16b)$$

where $P_{ij}$ is the permutation operator that interchanges the $i$th and the $j$th particle, with $i$ and $j$ arbitrary. It is an empirical fact that a mixed symmetry does not occur.

Even more remarkable is that there is a connection between the spin of a particle and the statistics obeyed by it:

$$\text{Half-integer spin particles are fermions;} \qquad (7.17a)$$

$$\text{Integer spin particles are bosons.} \qquad (7.17b)$$

Here particles can be composite; for example, a $^3$He nucleus is a fermion just as the $e^-$ or the proton; a $^4$He nucleus is a boson just as the $\pi$ meson or $Z^0$ gauge boson.

This spin-statistics connection is, as far as we can tell, an exact law of nature with no known exceptions. In the framework of nonrelativistic quantum mechanics, this principle must be accepted as an empirical postulate. In the relativistic quantum theory, however, it can be proved that half-integer spin particles cannot be bosons and integer spin particles cannot be fermions.

An immediate consequence of the electron being a fermion is that the electron must satisfy the **Pauli exclusion principle**, which states that no two electrons can occupy the same state. This follows because a state like $|k'\rangle|k'\rangle$ is necessarily symmetrical, which is not

---

[2] To be sure, there is an important subtlety that relies on our living in three-dimensional space. It is possible to have objects, called *anyons*, which have a continuum of statistical properties spanning the range between fermions and bosons, if they are constrained to two spatial dimensions. The literature on this subject is fascinating, but scattered. The reader is referred to two early papers, namely "Quantum mechanics of fractional-spin particles," Wilczek, *Phys. Rev. Lett.*, **49** (1982) 957 and "Quantum spectrum of three anyons in an oscillator potential," Murthy et al., *Phys. Rev. Lett.*, **67** (1991) 817.

possible for a fermion. As is well known, the Pauli exclusion principle is the cornerstone of atomic and molecular physics as well as the whole of chemistry. To illustrate the dramatic differences between fermions and bosons, let us consider two particles, each of which can occupy only two states, characterized by $k'$ and $k''$. For a system of two fermions, we have no choice; there is only one possibility:

$$\frac{1}{\sqrt{2}} \left( |k'\rangle |k''\rangle - |k''\rangle |k'\rangle \right). \tag{7.18}$$

For bosons there are three states possible:

$$|k'\rangle |k'\rangle, \qquad |k''\rangle |k''\rangle, \qquad \frac{1}{\sqrt{2}} \left( |k'\rangle |k''\rangle + |k''\rangle |k'\rangle \right). \tag{7.19}$$

In contrast, for "classical" particles satisfying **Maxwell–Boltzmann** (M-B) **statistics** with no restriction on symmetry, we have altogether four independent states:

$$|k'\rangle |k''\rangle, \qquad |k''\rangle |k'\rangle, \qquad |k'\rangle |k'\rangle, \qquad |k''\rangle |k''\rangle. \tag{7.20}$$

We see that in the fermion case it is impossible for both particles to occupy the same state. In the boson case, for two out of the three allowed kets, both particles occupy the same state. In the classical (M-B) statistics case, both particles occupy the same state for two out of the four allowed kets. In this sense fermions are the least sociable; they avoid each other to make sure that they are not in the same state; in contrast, bosons are the most sociable, they really love to be in the same state, even more so than classical particles obeying M-B statistics.

The difference between fermions and bosons shows up most dramatically at low temperatures; a system made up of bosons, such as liquid $^4$He, exhibits a tendency for all particles to get down to the same ground state at extremely low temperatures.[3] This is known as **Bose–Einstein condensation**, a feature not shared by a system made up of fermions.

## 7.3 Two-Electron System

Let us now consider specifically a two-electron system. The eigenvalue of the permutation operator is necessarily $-1$. Suppose the base kets we use may be specified by $\mathbf{x_1}$, $\mathbf{x_2}$, $m_{s1}$, and $m_{s2}$, where $m_{s1}$ and $m_{s2}$ stand for the spin-magnetic quantum numbers of electron 1 and electron 2, respectively.

We can express the wave function for a two-electron system as a linear combination of the state ket with eigenbras of $\mathbf{x_1}$, $\mathbf{x_2}$, $m_{s1}$, and $m_{s2}$ as follows:

$$\psi = \sum_{m_{s1}} \sum_{m_{s2}} C(m_{s1}, m_{s2}) \langle \mathbf{x_1}, m_{s1}; \mathbf{x_2}, m_{s2} | \alpha \rangle. \tag{7.21}$$

---

[3] The visual behavior of liquid helium, as it is cooled past the critical temperature, is striking. Various video examples can be seen at www.youtube.com/ including a classic physics demonstration movie, "Liquid Helium II: The Superfluid" by A. Leitner, from 1963. See also the site http://alfredleitner.com/.

If the Hamiltonian commutes with $\mathbf{S}_{\text{tot}}^2$,

$$\left[\mathbf{S}_{\text{tot}}^2, H\right] = 0, \tag{7.22}$$

then the energy eigenfunction is expected to be an eigenfunction of $\mathbf{S}_{\text{tot}}^2$, and if $\psi$ is written as

$$\psi = \phi(\mathbf{x_1}, \mathbf{x_2})\chi, \tag{7.23}$$

then the spin function $\chi$ is expected to be one of the following:

$$\chi(m_{s1}, m_{s2}) = \begin{cases} \chi_{++} \\ \frac{1}{\sqrt{2}}(\chi_{+-} + \chi_{-+}) \\ \chi_{--} \end{cases} \quad \text{triplet (symmetrical)} \\ \frac{1}{\sqrt{2}}(\chi_{+-} - \chi_{-+}) \quad \text{singlet (antisymmetrical)}, \tag{7.24}$$

where $\chi_{+-}$ corresponds to $\chi(m_{s1} = \frac{1}{2}, m_{s2} = -\frac{1}{2})$. Notice that the triplet spin functions are all symmetrical; this is reasonable because the ladder operator $S_{1-} + S_{2-}$ commutes with $P_{12}$ and the $|+\rangle|+\rangle$ state is even under $P_{12}$.

We note

$$\langle \mathbf{x}_1, m_{s1}; \mathbf{x}_2, m_{s2} | P_{12} | \alpha \rangle = \langle \mathbf{x}_2, m_{s2}; \mathbf{x}_1, m_{s1} | \alpha \rangle. \tag{7.25}$$

Fermi–Dirac statistics thus requires

$$\langle \mathbf{x}_1, m_{s1}; \mathbf{x}_2, m_{s2} | \alpha \rangle = -\langle \mathbf{x}_2, m_{s2}; \mathbf{x}_1, m_{s1} | \alpha \rangle. \tag{7.26}$$

Clearly, $P_{12}$ can be written as

$$P_{12} = P_{12}^{(\text{space})} P_{12}^{(\text{spin})} \tag{7.27}$$

where $P_{12}^{(\text{space})}$ just interchanges the position coordinate, while $P_{12}^{(\text{spin})}$ just interchanges the spin states. It is amusing that we can express $P_{12}^{(\text{spin})}$ as

$$P_{12}^{(\text{spin})} = \frac{1}{2}\left(1 + \frac{4}{\hbar^2}\mathbf{S}_1 \cdot \mathbf{S}_2\right), \tag{7.28}$$

which follows because

$$\mathbf{S}_1 \cdot \mathbf{S}_2 = \begin{cases} \dfrac{\hbar^2}{4} & \text{(triplet)} \\[2mm] \dfrac{-3\hbar^2}{4} & \text{(singlet)}. \end{cases} \tag{7.29}$$

It follows from (7.23) that letting

$$|\alpha\rangle \to P_{12}|\alpha\rangle \tag{7.30}$$

amounts to

$$\phi(\mathbf{x}_1, \mathbf{x}_2) \to \phi(\mathbf{x}_2, \mathbf{x}_1), \qquad \chi(m_{s1}, m_{s2}) \to \chi(m_{s2}, m_{s1}). \tag{7.31}$$

This together with (7.26) implies that if the space part of the wave function is symmetrical (antisymmetrical) the spin part must be antisymmetrical (symmetrical). As a result, the spin triplet state has to be combined with an antisymmetrical space function and the spin-singlet state has to be combined with a space symmetrical function.

The space part of the wave function $\phi(\mathbf{x}_1, \mathbf{x}_2)$ provides the usual probabilistic interpretation. The probability for finding electron 1 in a volume element $d^3x_1$ centered around $\mathbf{x}_1$ and electron 2 in a volume element $d^3x_2$ is

$$|\phi(\mathbf{x}_1, \mathbf{x}_2)|^2 d^3x_1 d^3x_2. \tag{7.32}$$

To see the meaning of this more closely, let us consider the specific case where the mutual interaction between the two electrons [for example, $V_{\text{pair}}(|\mathbf{x}_1 - \mathbf{x}_2|), \mathbf{S}_1 \cdot \mathbf{S}_2$] can be ignored. If there is no spin dependence, the wave equation for the energy eigenfunction $\psi$ [see (7.9)],

$$\left[ \frac{-\hbar^2}{2m} \nabla_1^2 - \frac{\hbar^2}{2m} \nabla_2^2 + V_{\text{ext}}(\mathbf{x}_1) + V_{\text{ext}}(\mathbf{x}_2) \right] \psi = E\psi, \tag{7.33}$$

is now separable. We have a solution of the form $\omega_A(\mathbf{x}_1)\omega_B(\mathbf{x}_2)$ times the spin function. With no spin dependence $S_{\text{tot}}^2$ necessarily (and trivially) commutes with $H$, so the spin part must be a triplet or a singlet, which have definite symmetry properties under $P_{12}^{(\text{spin})}$. The space part must then be written as a symmetrical and antisymmetrical combination of $\omega_A(\mathbf{x}_1)\omega_B(\mathbf{x}_2)$ and $\omega_A(\mathbf{x}_2)\omega_B(\mathbf{x}_1)$:

$$\phi(\mathbf{x}_1, \mathbf{x}_2) = \frac{1}{\sqrt{2}} [\omega_A(\mathbf{x}_1)\omega_B(\mathbf{x}_2) \pm \omega_A(\mathbf{x}_2)\omega_B(\mathbf{x}_1)] \tag{7.34}$$

where the upper sign is for a spin singlet and the lower is for a spin triplet. The probability of observing electron 1 in $d^3x_1$ around $\mathbf{x}_1$ and electron 2 in $d^3x_2$ around $\mathbf{x}_2$ is given by

$$\frac{1}{2} \left\{ |\omega_A(\mathbf{x}_1)|^2 |\omega_B(\mathbf{x}_2)|^2 + |\omega_A(\mathbf{x}_2)|^2 |\omega_B(\mathbf{x}_1)|^2 \right.$$
$$\left. \pm 2\,\text{Re}\left[\omega_A(\mathbf{x}_1)\omega_B(\mathbf{x}_2)\omega_A^*(\mathbf{x}_2)\omega_B^*(\mathbf{x}_1)\right] \right\} d^3x_1 d^3x_2. \tag{7.35}$$

The last term in the curly bracket is known as the **exchange density**.

We immediately see that when the electrons are in a spin-triplet state, the probability of finding the second electron at the same point in space vanishes. Put another way, the electrons tend to avoid each other when their spins are in a triplet state. In contrast, when their spins are in a singlet state, there is enhanced probability of finding them at the same point in space because of the presence of the exchange density.

Clearly, the question of identity is important only when the exchange density is nonnegligible or when there is substantial overlap between function $\omega_A$ and function $\omega_B$. To see this point clearly, let us take the extreme case where $|\omega_A(\mathbf{x})|^2$ (where $\mathbf{x}$ may refer to $\mathbf{x}_1$ or $\mathbf{x}_2$) is big only in region A and $|\omega_B(\mathbf{x})|^2$ is big only in region B such that the two regions are widely separated. Now choose $d^3x_1$ in region A and $d^3x_2$ in region B; see Figure 7.3. The only important term then is just the first term in (7.35),

$$|\omega_A(\mathbf{x}_1)|^2 |\omega_B(\mathbf{x}_2)|^2, \tag{7.36}$$

which is nothing more than the joint probability density expected for classical particles. In this connection, recall that classical particles are necessarily well localized and the question of identity simply does not arise. Thus the exchange density term is unimportant

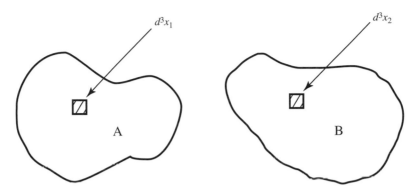

Fig. 7.3 Two widely separated regions A and B; $|\omega_A(\mathbf{x})|^2$ is large in region A while $|\omega_B(\mathbf{x})|^2$ is large in region B.

if regions A and B do not overlap. There is no need to antisymmetrize if the electrons are far apart and the overlap is negligible. This is quite gratifying. We never have to worry about the question of antisymmetrization with 10 billion electrons, nor is it necessary to take into account the antisymmetrization requirement between an electron in New York and an electron in Beijing.

## 7.4 The Helium Atom

A study of the helium atom is rewarding for several reasons. First of all, it is the simplest realistic problem where the question of identity, which we encountered in Section 7.3, plays an important role. Second, even though it is a simple system, the two-particle Schrödinger equation cannot be solved analytically; therefore, this is a nice place to illustrate the use of perturbation theory and also the use of the variational method.

The basic Hamiltonian is given by

$$H = \frac{\mathbf{p}_1^2}{2m} + \frac{\mathbf{p}_2^2}{2m} - \frac{2e^2}{r_1} - \frac{2e^2}{r_2} + \frac{e^2}{r_{12}}, \tag{7.37}$$

where $r_1 \equiv |\mathbf{x}_1|$, $r_2 \equiv |\mathbf{x}_2|$, and $r_{12} \equiv |\mathbf{x}_1 - \mathbf{x}_2|$; see Figure 7.4. Suppose the $e^2/r_{12}$-term were absent. Then, with the identity question ignored, the wave function would be just the product of two hydrogen atom wave functions with $Z = 1$ changed into $Z = 2$. The total spin is a constant of the motion, so the spin state is either singlet or triplet. The space part of the wave function for the important case where one of the electrons is in the ground state and the other in an excited state characterized by $(nlm)$ is

$$\phi(\mathbf{x}_1, \mathbf{x}_2) = \frac{1}{\sqrt{2}} [\psi_{100}(\mathbf{x}_1)\psi_{nlm}(\mathbf{x}_2) \pm \psi_{100}(\mathbf{x}_2)\psi_{nlm}(\mathbf{x}_1)] \tag{7.38}$$

where the upper (lower) sign is for the spin singlet (triplet). We will come back to this general form for an excited state later.

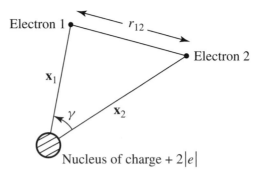

**Fig. 7.4**    Schematic diagram of the helium atom.

For the ground state, we need a special treatment. Here the configuration is characterized by $(1s)^2$, that is, both electrons in $n = 1$, $l = 0$.

The space function must then necessarily be symmetric and only the spin-singlet function is allowed. So we have

$$\psi_{100}(\mathbf{x}_1)\psi_{100}(\mathbf{x}_2)\chi_{\text{singlet}} = \frac{Z^3}{\pi a_0^3}e^{-Z(r_1+r_2)/a_0}\chi \tag{7.39}$$

with $Z = 2$. Not surprisingly, this "unperturbed" wave function gives

$$E = 2 \times 4\left(-\frac{e^2}{2a_0}\right) = -108.8 \text{ eV} \tag{7.40}$$

for the ground-state energy, which is about 30% larger than the experimental value.

This is just the starting point of our investigation because in obtaining the above form (7.39), we have completely ignored the last term in (7.37) that describes the interaction between the two electrons. One way to approach the problem of obtaining a better energy value is to apply first-order perturbation theory using (7.39) as the unperturbed wave function and $e^2/r_{12}$ as the perturbation. We obtain

$$\Delta_{(1s)^2} = \left\langle \frac{e^2}{r_{12}} \right\rangle_{(1s)^2} = \iint \frac{Z^6}{\pi^2 a_0^6}e^{-2Z(r_1+r_2)/a_0}\frac{e^2}{r_{12}}d^3x_1 d^3x_2. \tag{7.41}$$

To carry out the indicated integration we first note

$$\frac{1}{r_{12}} = \frac{1}{\sqrt{r_1^2 + r_2^2 - 2r_1 r_2 \cos\gamma}} = \sum_{l=0}^{\infty} \frac{r_<^l}{r_>^{l+1}}P_l(\cos\gamma), \tag{7.42}$$

where $r_>$ $(r_<)$ is the larger (smaller) of $r_1$ and $r_2$ and $\gamma$ is the angle between $\mathbf{x}_1$ and $\mathbf{x}_2$. The angular integration is easily performed by expressing $P_l(\cos\gamma)$ in terms of $Y_l^m(\theta_1, \phi_1)$ and $Y_l^m(\theta_2, \phi_2)$ using the addition theorem of spherical harmonics. (See, for example, Section 12.8 of Arfken and Weber (1995).) We have

$$P_l(\cos\gamma) = \frac{4\pi}{2l+1}\sum_{m=-l}^{l} Y_l^{m*}(\theta_1, \phi_1)Y_l^m(\theta_2, \phi_2). \tag{7.43}$$

The angular integration is now trivial:

$$\int Y_l^m(\theta_i, \phi_i)\, d\Omega_i = \frac{1}{\sqrt{4\pi}}(4\pi)\delta_{l0}\delta_{m0}. \tag{7.44}$$

The radial integration is elementary (but involves tedious algebra!); it leads to

$$\int_0^\infty \left[ \int_0^{r_1} \frac{1}{r_1} e^{-(2Z/a_0)(r_1+r_2)} r_2^2\, dr_2 + \int_{r_1}^\infty \frac{1}{r_2} e^{-(2Z/a_0)(r_1+r_2)} r_2^2\, dr_2 \right] r_1^2\, dr_1$$
$$= \frac{5}{128} \frac{a_0^5}{Z^5}. \tag{7.45}$$

Combining everything, we have (for $Z = 2$)

$$\Delta_{(1s)^2} = \left( \frac{Z^6 e^2}{\pi^2 a_0^6} \right) 4\pi (\sqrt{4\pi})^2 \left( \frac{5}{128} \right) \left( \frac{a_0^5}{Z^5} \right) = \left( \frac{5}{2} \right) \left( \frac{e^2}{2a_0} \right). \tag{7.46}$$

Adding this energy shift to (7.40), we have

$$E_{\text{cal}} = \left( -8 + \frac{5}{2} \right) \left( \frac{e^2}{2a_0} \right) \simeq -74.8 \text{ eV}. \tag{7.47}$$

Compare this with the experimental value,

$$E_{\text{exp}} = -79.005151042(40) \text{ eV}, \tag{7.48}$$

as determined from the NIST Atomic Spectra Database.

This is not bad, but we can do better! We propose to use the variational method with $Z$, which we call $Z_{\text{eff}}$, as a variational parameter. The physical reason for this choice is that the effective $Z$ seen by one of the electrons is smaller than 2 because the positive charge of 2 units at the origin (see Figure 7.4) is "screened" by the negatively charged cloud of the other electron; in other words, the other electron tends to neutralize the positive charge due to the helium nucleus at the center. For the normalized trial function we use

$$\langle \mathbf{x}_1, \mathbf{x}_2 | \tilde{0} \rangle = \left( \frac{Z_{\text{eff}}^3}{\pi a_0^3} \right) e^{-Z_{\text{eff}}(r_1+r_2)/a_0}. \tag{7.49}$$

From this we obtain

$$\overline{H} = \left\langle \tilde{0} \left| \frac{\mathbf{p}_1^2}{2m} + \frac{\mathbf{p}_2^2}{2m} \right| \tilde{0} \right\rangle - \left\langle \tilde{0} \left| \frac{Ze^2}{r_1} + \frac{Ze^2}{r_2} \right| \tilde{0} \right\rangle + \left\langle \tilde{0} \left| \frac{e^2}{r_{12}} \right| \tilde{0} \right\rangle$$
$$= \left( 2\frac{Z_{\text{eff}}^2}{2} - 2ZZ_{\text{eff}} + \frac{5}{8}Z_{\text{eff}} \right) \left( \frac{e^2}{a_0} \right). \tag{7.50}$$

We easily see that the minimization of $\overline{H}$ is at

$$Z_{\text{eff}} = 2 - \tfrac{5}{16} = 1.6875. \tag{7.51}$$

This is smaller than 2, as anticipated. Using this value for $Z_{\text{eff}}$ we get

$$E_{\text{cal}} = -77.5 \text{ eV}, \tag{7.52}$$

which is already very close considering the crudeness of the trial wave function.

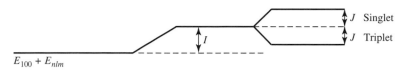

$E_{100} + E_{nlm}$

**Fig. 7.5**    Schematic diagram for the energy-level splittings of $(1s)(nl)$ for the helium atom.

Historically, this achievement was considered to be one of the earliest signs that Schrödinger's wave mechanics was on the right track. We cannot get this kind of number by the purely algebraic (operator) method. The helium calculation was first done by A. Unsöld in 1927.[4]

Let us briefly consider excited states. This is more interesting from the point of view of illustrating quantum-mechanical effects due to identity. We consider just $(1s)(nl)$. We write the energy of this state as

$$E = E_{100} + E_{nlm} + \Delta E. \tag{7.53}$$

In first-order perturbation theory, $\Delta E$ is obtained by evaluating the expectation value of $e^2/r_{12}$. We can write

$$\left\langle \frac{e^2}{r_{12}} \right\rangle = I \pm J, \tag{7.54}$$

where $I$ and $J$, known respectively as the direct integral and the exchange integral, are given by

$$I = \int d^3x_1 \int d^3x_2 |\psi_{100}(\mathbf{x}_1)|^2 |\psi_{nlm}(\mathbf{x}_2)|^2 \frac{e^2}{r_{12}}, \tag{7.55a}$$

$$J = \int d^3x_1 \int d^3x_2 \psi_{100}(\mathbf{x}_1)\psi_{nlm}(\mathbf{x}_2)\frac{e^2}{r_{12}}\psi_{100}^*(\mathbf{x}_2)\psi_{nlm}^*(\mathbf{x}_1). \tag{7.55b}$$

The upper (lower) sign goes with the spin-singlet (spin-triplet) state. Obviously, $I$ is positive; we can also show that $J$ is positive. So the net result is such that for the same configuration, the spin-singlet state lies higher, as shown in Figure 7.5.

The physical interpretation for this is as follows: In the singlet case the space function is symmetric and the electrons have a tendency to come close to each other. Therefore, the effect of the electrostatic repulsion is more serious; hence, a higher energy results. In the triplet case, the space function is antisymmetric and the electrons tend to avoid each other. Helium in spin-singlet states is known as **parahelium**, while helium in spin-triplet states is known as **orthohelium**. Each configuration splits into the para state and the ortho state, the para state lying higher. For the ground state only parahelium is possible. See Figure 7.6 for a schematic energy-level diagram of the helium atom.

It is very important to recall that the original Hamiltonian is spin independent because the potential is made up of just three Coulomb terms. There was no $\mathbf{S}_1 \cdot \mathbf{S}_2$-term whatsoever. Yet there is a spin-dependent effect – the electrons with parallel spins have a lower energy – that arises from Fermi–Dirac statistics.

---

[4] Unsöld, *Ann. Phys.*, **82** (1927) 355.

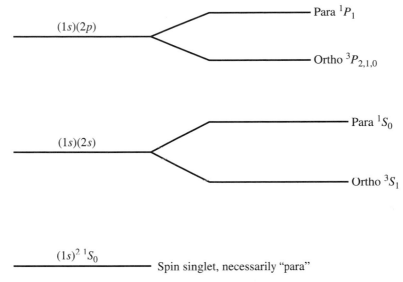

$(1s)(2p)$ — Para $^1P_1$
— Ortho $^3P_{2,1,0}$

$(1s)(2s)$ — Para $^1S_0$
— Ortho $^3S_1$

$(1s)^2\ ^1S_0$ — Spin singlet, necessarily "para"

**Fig. 7.6**   Schematic energy-level diagram for low-lying configurations of the helium atom.

This explanation of the apparent spin dependence of the helium atom energy levels is due to Heisenberg. The physical origin of ferromagnetism – alignment of the electron spins extended over microscopic distances – is also believed to be essentially the same, but the properties of ferromagnets are much harder to calculate quantitatively from first principles.

## 7.5  Multiparticle States

It is not difficult to extend the symmetrization of Section 7.2 to a system made up of many identical particles. Recalling (7.13) we define

$$P_{ij}|k'\rangle|k''\rangle\cdots|k^i\rangle|k^{i+1}\rangle\cdots|k^j\rangle\cdots = |k'\rangle|k''\rangle\cdots|k^j\rangle|k^{i+1}\rangle\cdots|k^i\rangle\cdots. \tag{7.56}$$

Clearly,

$$P_{ij}^2 = 1 \tag{7.57}$$

just as before, and the allowed eigenvalues of $P_{ij}$ are $+1$ and $-1$. It is important to note, however, that in general

$$[P_{ij}, P_{kl}] \neq 0. \tag{7.58}$$

It is worth explicitly working out a system of three identical particles. First, there are $3! = 6$ possible kets of form

$$|k'\rangle|k''\rangle|k'''\rangle \tag{7.59}$$

where $k'$, $k''$, and $k'''$ are all different. Thus there is sixfold exchange degeneracy. Yet if we insist that the state be *totally* symmetrical or *totally* antisymmetrical, we can form only one linear combination each. Explicitly, we have

$$|k'k''k'''\rangle_\pm \equiv \frac{1}{\sqrt{6}} \{|k'\rangle|k''\rangle|k'''\rangle \pm |k''\rangle|k'\rangle|k'''\rangle$$

$$+ |k''\rangle|k'''\rangle|k'\rangle \pm |k'''\rangle|k''\rangle|k'\rangle$$

$$+ |k'''\rangle|k'\rangle|k''\rangle \pm |k'\rangle|k'''\rangle|k''\rangle\}. \tag{7.60}$$

These are both simultaneous eigenkets of $P_{12}$, $P_{23}$, and $P_{13}$. We remarked that there are altogether six independent state kets. It therefore follows that there are four independent kets that are neither totally symmetrical nor totally antisymmetrical. We could also introduce the operator $P_{123}$ by defining

$$P_{123} \left(|k'\rangle|k''\rangle|k'''\rangle\right) = |k''\rangle|k'''\rangle|k'\rangle. \tag{7.61}$$

Note that $P_{123} = P_{12}P_{13}$ because

$$P_{12}P_{13} \left(|k'\rangle|k''\rangle|k'''\rangle\right) = P_{12} \left(|k'''\rangle|k''\rangle|k'\rangle\right) = |k''\rangle|k'''\rangle|k'\rangle. \tag{7.62}$$

In writing (7.60) we assumed that $k'$, $k''$, and $k'''$ are all different. If two of the three indices coincide, it is impossible to have a totally antisymmetrical state. The totally symmetrical state is given by

$$|k'k'k''\rangle_+ = \frac{1}{\sqrt{3}} \left(|k'\rangle|k'\rangle|k''\rangle + |k'\rangle|k''\rangle|k'\rangle + |k''\rangle|k'\rangle|k'\rangle\right), \tag{7.63}$$

where the normalization factor is understood to be $\sqrt{2!/3!}$. For more general cases we have a normalization factor

$$\sqrt{\frac{N_1!N_2!\cdots N_n!}{N!}}, \tag{7.64}$$

where $N$ is the total number of particles and $N_i$ the number of times $|k^{(i)}\rangle$ occurs.

For more general cases of the totally antisymmetrical state, it is helpful to write the antisymmetric case of (7.60) as

$$|k'k''k'''\rangle_- = \frac{1}{\sqrt{3!}} \begin{vmatrix} |k'\rangle & |k''\rangle & |k'''\rangle \\ |k'\rangle & |k''\rangle & |k'''\rangle \\ |k'\rangle & |k''\rangle & |k'''\rangle \end{vmatrix} \tag{7.65}$$

where the ordering of kets in the determinant is set by the row from which they originate. This construction is known as a *Slater determinant*. Its generalization to the $N$-particle case is straightforward.

In principle, we can follow along the lines of Section 7.4 to apply this formalism to $N$-particle systems, but it can quickly become unwieldy, even with modern computational resources. If $N$ is on the order of Avogadro's number, as it is for problems in condensed matter physics or materials chemistry, then this approach is intractable. Other techniques of calculation needed to be invented in order to handle these cases.

The remainder of this chapter develops two such techniques. Section 7.6 covers the basics of density functional theory, a coordinate representation-based approach that reduces the problem to one with only three (spatial) degrees of freedom. Section 7.7 introduces the foundations of quantum field theory using the formalism known as "second quantization." Each of these techniques is both powerful and widely used. Finally, Section 7.8 applies second quantization to the electromagnetic field in free space. (Second quantization is applied to the relativistic Klein–Gordon field in Section 8.1, see p. 485.)

# 7.6  Density Functional Theory

The problem of having many identical particles, interacting with each other and with some external potential, can be written down as an $N$-body Schrödinger equation for a wave function in coordinate space. This equation cannot be solved exactly, however, so approximation schemes must be developed. One particular approach has emerged in the most recent decades as a popular choice, especially for the electronic structure of condensed matter systems.

This approach, known as *density functional theory* (DFT), is based on two seminal papers, namely Hohenberg and Kohn, *Phys. Rev.*, **136** (1964) B864 and Kohn and Sham, *Phys. Rev.*, **140** (1965) A1133. DFT has found wide application in condensed matter and materials physics and chemistry. In fact, the 1998 Nobel Prize in Chemistry was awarded to physicist Walter Kohn and chemist John Pople for the development and application of this formalism.

Our treatment here is strictly an introduction. See Jones and Gunnarsson, *Rev. Mod. Phys.*, **61** (1989) 689 for a thorough review after DFT had matured. Today DFT is used widely in condensed matter physics, and described in modern textbooks on the subject, for example Cohen and Louie (2016). For interesting historical and contextual reading, see Walter Kohn's Nobel Prize lecture, *Rev. Mod. Phys.*, **71** (1999) 1253 and also Perdew et al., *J. Chem. Theory Comput.*, **5** (2009) 902.

## 7.6.1  The Energy Functional for a Single Particle

In Section 5.4 we proved a theorem that was the basis of the variational approximation to the ground-state energy. Cast in terms of wave mechanics, we found that

$$E[\tilde{\psi}] \equiv \int \tilde{\psi}^* H \tilde{\psi} \, d^3x$$

$$= \int \tilde{\psi}^* \left[ -\frac{\hbar^2}{2m} \nabla^2 + V(\mathbf{x}) \right] \tilde{\psi} \, d^3x \geq E_0 \qquad (7.66)$$

for the Hamiltonian $H$ and a (properly normalized) trial wave function $\tilde{\psi}(\mathbf{x})$, where $E_0$ is the true ground-state energy. The notation $E[\tilde{\psi}]$ means that the energy is a *functional* of the trial wave function. That is, given some function $\tilde{\psi}(\mathbf{x})$, we can calculate a value for $E[\tilde{\psi}]$.

We made use of this in the variational approximation by giving $\tilde{\psi}$ free parameters, which we varied to minimize $E[\tilde{\psi}]$, thereby providing an approximation to $E_0$.

*The key idea behind DFT is that the energy can be considered a functional of the density instead of the wave function.* Hohenberg, Kohn, and Sham proved rigorously that this converts the quantum many-body problem into an effective one-body problem, a tractable calculation. Before describing their work, though, let us see how this works for the single-particle case.

We will find it convenient to rewrite (7.66) using the divergence theorem, and the fact that the wave function tends to zero at infinity for bound states. That is

$$\int \tilde{\psi}^* \nabla^2 \tilde{\psi} \, d^3x = \int \left[ \nabla \cdot (\tilde{\psi}^* \nabla \tilde{\psi}) - \nabla \tilde{\psi}^* \cdot \nabla \tilde{\psi} \right] d^3x = -\int \nabla \tilde{\psi}^* \cdot \nabla \tilde{\psi} \, d^3x \qquad (7.67)$$

since the term with the divergence converts to a surface integral at infinity. This transformation is sometimes called a "choice of gauge," for its similarity to the transformations discussed in Section 2.7. The energy functional (7.66) then becomes

$$E[\tilde{\psi}] = \int \left[ \frac{\hbar^2}{2m} \nabla \tilde{\psi}^* \cdot \nabla \tilde{\psi} + V(\mathbf{x}) \tilde{\psi}^* \tilde{\psi} \right] d^3x. \qquad (7.68)$$

Now, the single-particle density for a wave function $\psi(\mathbf{x})$ is $\rho(\mathbf{x}) = \psi^* \psi$. We can take the ground-state trial wave function $\tilde{\psi}(\mathbf{x})$ to be real, and then write the energy as a functional of the density $\tilde{\rho}(\mathbf{x})$. Using $\tilde{\psi} = \sqrt{\tilde{\rho}}$ we find

$$E[\tilde{\rho}] = \int \left[ \frac{\hbar^2}{2m} \left( \nabla \tilde{\rho}^{1/2} \right)^2 + V(\mathbf{x}) \tilde{\rho} \right] d^3x = \int \left[ \frac{\hbar^2}{8m} \frac{(\nabla \tilde{\rho})^2}{\tilde{\rho}} + V(\mathbf{x}) \tilde{\rho} \right] d^3x \qquad (7.69)$$

for the ground-state energy as a functional of density.

At this point, we could make a guess for the function $\tilde{\rho}(\mathbf{x})$, including one or more free parameters, calculate $E[\tilde{\rho}]$, and minimize the result with respect to those parameters. This is directly analogous to the approach we took in Section 5.4.

However, we can take a different approach. Equation (7.69) shows that the energy $E[\tilde{\rho}]$ is a functional of $\tilde{\rho}$ and its first derivatives. Minimizing $E[\tilde{\rho}]$ is equivalent to finding a stationary value of the energy, that is $\delta E[\tilde{\rho}] = 0$, a well-known problem in the calculus of variations.[5] First we write (7.69) as

$$E[\tilde{\rho}] = \int \varepsilon(\tilde{\rho}, \nabla \tilde{\rho}) d^3x = \int \varepsilon(\tilde{\rho}, \tilde{\rho}_x, \tilde{\rho}_y, \tilde{\rho}_z) d^3x \qquad (7.70)$$

where $\tilde{\rho}_x \equiv \partial \tilde{\rho}/\partial x$, $\tilde{\rho}_y \equiv \partial \tilde{\rho}/\partial y$, and $\tilde{\rho}_z \equiv \partial \tilde{\rho}/\partial z$, so that

$$\varepsilon(\tilde{\rho}, \nabla \tilde{\rho}) = \frac{\hbar^2}{8m} \frac{(\nabla \tilde{\rho})^2}{\tilde{\rho}} + V(\mathbf{x}) \tilde{\rho} = \frac{\hbar^2}{8m} \frac{\tilde{\rho}_x^2 + \tilde{\rho}_y^2 + \tilde{\rho}_z^2}{\tilde{\rho}} + V(\mathbf{x}) \tilde{\rho}. \qquad (7.71)$$

---

[5] The calculus of variations is covered in every textbook on classical mechanics, treating the action as a functional of the spatial coordinates and their derivatives, with time as the independent variable. Here we make use of the case of several independent variables, typically covered in textbooks on mathematical physics. See, for example, Arfken, Weber, and Harris (2013), Chapter 22, or Byron and Fuller (1992), Chapter 2.

Before imposing $\delta E[\tilde{\rho}] = 0$, we must include the constraint that the density is normalized, that is

$$\int \tilde{\rho}\, d^3x = 1. \tag{7.72}$$

This is most easily done by multiplying (7.72) by a constant, which we take to be $-\mu$, and adding it to (7.70), creating a new functional $E^{(\mu)}[\tilde{\rho}]$. (This is formally referred to as the method of Lagrange multipliers.) Because of the constraint (7.72), minimizing $E[\tilde{\rho}]$ is equivalent to minimizing $E^{(\mu)}[\tilde{\rho}]$, that is setting $\delta E^{(\mu)}[\tilde{\rho}] = 0$. We have

$$\delta E^{(\mu)}[\tilde{\rho}] = \delta\left[\int \varepsilon(\tilde{\rho}, \mathbf{\nabla}\tilde{\rho})d^3x - \mu\int \tilde{\rho}\, d^3x\right]$$
$$= \int\left[\frac{\partial \varepsilon}{\partial \tilde{\rho}}\delta\tilde{\rho} + \frac{\partial \varepsilon}{\partial \tilde{\rho}_x}\delta\tilde{\rho}_x + \frac{\partial \varepsilon}{\partial \tilde{\rho}_y}\delta\tilde{\rho}_y + \frac{\partial \varepsilon}{\partial \tilde{\rho}_z}\delta\tilde{\rho}_z - \mu\delta\tilde{\rho}\right]d^3x = 0. \tag{7.73}$$

Recognizing that $\delta\tilde{\rho}_x = \partial(\delta\tilde{\rho})/\partial x$, and similarly for $\tilde{\rho}_y$ and $\tilde{\rho}_z$, we integrate those terms by parts with $\delta\tilde{\rho}$ fixed to zero at infinity. Then factor out $\delta\tilde{\rho}$ and set the integrand to zero. The result is

$$\frac{\partial \varepsilon}{\partial \tilde{\rho}} - \frac{\partial}{\partial x}\frac{\partial \varepsilon}{\partial \tilde{\rho}_x} - \frac{\partial}{\partial y}\frac{\partial \varepsilon}{\partial \tilde{\rho}_y} - \frac{\partial}{\partial z}\frac{\partial \varepsilon}{\partial \tilde{\rho}_z} = \mu. \tag{7.74}$$

Noting that

$$\frac{\partial}{\partial x}\frac{\partial \varepsilon}{\partial \tilde{\rho}_x} = \frac{\hbar^2}{4m}\frac{\partial}{\partial x}\frac{\tilde{\rho}_x}{\tilde{\rho}} = \frac{\hbar^2}{4m}\left(\frac{\tilde{\rho}_{xx}}{\tilde{\rho}} - \frac{\tilde{\rho}_x^2}{\tilde{\rho}^2}\right) \tag{7.75}$$

we insert (7.71) into (7.74) to find

$$\frac{\hbar^2}{8m}\frac{(\mathbf{\nabla}\tilde{\rho})^2}{\tilde{\rho}} + V(\mathbf{x})\tilde{\rho} - \frac{\hbar^2}{4m}\mathbf{\nabla}^2\tilde{\rho} = \mu\tilde{\rho}. \tag{7.76}$$

This is a differential equation that could be solved for $\tilde{\rho}(\mathbf{x})$, similar to the way the Schrödinger equation could be solved for the wave function.

Note that integrating both sides of (7.76) proves that $\mu$ equals the ground-state energy, for the exact solution $\tilde{\rho}$. This is because the first two terms on the left are $\varepsilon(\tilde{\rho}, \mathbf{\nabla}\tilde{\rho})$, we have the constraint (7.72), and (similar to the way we dealt with the kinetic-energy term (7.67)) the integral of $\mathbf{\nabla}^2\tilde{\rho}$ becomes a surface integral at infinity and vanishes.

Problem 7.12 at the end of this chapter gives an example of applying these ideas to a single particle in a one-dimensional harmonic oscillator potential. In three dimensions, this could also be demonstrated with the nonisotropic harmonic oscillator or the hydrogen atom.

## 7.6.2 The Hohenberg–Kohn Theorem

Now consider the ground state of a system with $N$ identical particles, with wave function $\Psi(\mathbf{x}_1, \mathbf{x}_2, \ldots, \mathbf{x}_N)$, where the argument position identifies the particle number. (We will abandon the notation $\tilde{\Psi}$.) The wave function must be appropriately symmetrized, that is

$$\Psi(\mathbf{x}_1, \mathbf{x}_2, \ldots, \mathbf{x}_i, \ldots, \mathbf{x}_j, \ldots, \mathbf{x}_N) = \pm\Psi(\mathbf{x}_1, \mathbf{x}_2, \ldots, \mathbf{x}_j, \ldots, \mathbf{x}_i, \ldots, \mathbf{x}_N). \tag{7.77}$$

Other than acknowledging the symmetry of the wave function, we will not be considering spin degrees of freedom in our discussion.

The number density $n(\mathbf{x})$ in this case will be the expectation value of the operator that locates a particular particle at position $\mathbf{x}$, summed over all the particles. That is

$$n(\mathbf{x}) = \sum_{i=1}^{N} \langle \Psi | \delta^3(\mathbf{x} - \mathbf{x}_i) | \Psi \rangle$$

$$= \sum_{i=1}^{N} \int \Psi^*(\mathbf{x}_1, \mathbf{x}_2, \dots, \mathbf{x}_i = \mathbf{x}, \dots, \mathbf{x}_N)$$

$$\times \Psi(\mathbf{x}_1, \mathbf{x}_2, \dots, \mathbf{x}_i = \mathbf{x}, \dots, \mathbf{x}_N) d^3 x_1 \, d^3 x_2 \cdots d^3 x_N \tag{7.78}$$

where the measure of the integral does not include the factor $d^3 x_i$. Invoking the symmetry (7.77), we can exchange $\mathbf{x}_1$ for $\mathbf{x}_i$ for each term in (7.78) to write

$$n(\mathbf{x}) = N \int \Psi^*(\mathbf{x}, \mathbf{x}_2, \dots, \mathbf{x}_N) \Psi(\mathbf{x}, \mathbf{x}_2, \dots, \mathbf{x}_N) d^3 x_2 \cdots d^3 x_N. \tag{7.79}$$

For a properly normalized wave function $\Psi$, we clearly have

$$\int n(\mathbf{x}) d^3 x = N. \tag{7.80}$$

We consider a Hamiltonian that is a straightforward generalization of (7.9), namely

$$H = \sum_i \frac{\mathbf{p}_i^2}{2m} + \frac{1}{2} \sum_i \sum_{j \neq i} V_{\text{pair}}(|\mathbf{x}_i - \mathbf{x}_j|) + \sum_i V_{\text{ext}}(\mathbf{x}_i)$$

$$\equiv T + U_{\text{pair}} + U_{\text{ext}}. \tag{7.81}$$

Our goal is to write the ground-state energy $E = \langle \Psi | H | \Psi \rangle$ as a functional of the density $n(\mathbf{x})$. We will take the point of view that it is the external potential $V_{\text{ext}}(\mathbf{x})$ that defines our problem. For example, if we are trying to find the ground state of a many-electron system, then $T + U_{\text{pair}}$ is universal for all such problems. In other words, the wave function $\Psi$ is determined by $V_{\text{ext}}(\mathbf{x})$. The main point of the Hohenberg–Kohn theorem is that $n(\mathbf{x})$ is also determined by $V_{\text{ext}}(\mathbf{x})$. This was trivial when we wrote $\rho = \psi^2$ for the one-particle system, but it is more subtle for the many-particle case.

First let us find the ground-state expectation value for the $U_{\text{ext}}$ term in (7.81). Returning for a moment to our formal distinction between an operator $\mathbf{x}$ and its eigenvalue $\mathbf{x}'$, we have

$$\langle \Psi | U_{\text{ext}} | \Psi \rangle = \sum_i \int d^3 x_i' \langle \Psi | V_{\text{ext}}(\mathbf{x}_i) | \mathbf{x}_i' \rangle \langle \mathbf{x}_i' | \Psi \rangle = \sum_i \int d^3 x_i' \, V_{\text{ext}}(\mathbf{x}_i') \langle \Psi | \mathbf{x}_i' \rangle \langle \mathbf{x}_i' | \Psi \rangle$$

$$= \sum_i \int d^3 x_1' \, d^3 x_2' \cdots d^3 x_i' \cdots d^3 x_N' \, V_{\text{ext}}(\mathbf{x}_i')$$

$$\times \Psi^*(\mathbf{x}_1', \mathbf{x}_2', \dots, \mathbf{x}_i', \dots, \mathbf{x}_N') \Psi(\mathbf{x}_1', \mathbf{x}_2', \dots, \mathbf{x}_i', \dots, \mathbf{x}_N'). \tag{7.82}$$

We then again change integration variables in each term of the sum, exchanging $\mathbf{x}_i'$ for $\mathbf{x}_1'$, and realizing that each term is the same because of (7.77). Renaming the dummy integration variable $\mathbf{x}_1'$ to $\mathbf{x}$, we use (7.79) to write

$$\langle \Psi | U_{\text{ext}} | \Psi \rangle = \int V_{\text{ext}}(\mathbf{x}) n(\mathbf{x}) d^3 x. \tag{7.83}$$

This shows that this expectation value is a functional of $n$, and is of course completely analogous to the second term in (7.69).

Now we need to show that $\langle\Psi|(T+U_{\mathrm{pair}})|\Psi\rangle$ is a functional of $n(\mathbf{x})$. That is, we need to show that the external potential $V_{\mathrm{ext}}$ uniquely determines $n(\mathbf{x})$, just as it does the ground-state[6] wave function $\Psi$.

The variational theorem of Section 5.4 provides the proof we need. Suppose there is another external potential $V'_{\mathrm{ext}}$, differing by more than an additive constant from $V_{\mathrm{ext}}$, which gives a different wave function $\Psi'$ but the same density $n(\mathbf{x})$. If we consider $\Psi'$ as a trial wave function, then

$$\langle\Psi'|H|\Psi'\rangle > \langle\Psi|H|\Psi\rangle. \tag{7.84}$$

However, since the density $n(\mathbf{x})$ is the same, we invoke (7.83) and write

$$\langle\Psi'|(T+U_{\mathrm{pair}})|\Psi'\rangle > \langle\Psi|(T+U_{\mathrm{pair}})|\Psi\rangle. \tag{7.85}$$

This result is absurd, of course. We could just have easily let $\Psi$ be the trial wave function, and would have found the opposite result. The only way out is that $\Psi'$ is the same as $\Psi$ (to within an overall phase), resulting in the same density $n(\mathbf{x})$.

Using a standard notation, we write

$$F[n] \equiv \langle\Psi|(T+U_{\mathrm{pair}})|\Psi\rangle \tag{7.86}$$

for the "universal" function of a problem involving a specific particle species and two-body interaction. We then have

$$\langle\Psi|H|\Psi\rangle = \int V_{\mathrm{ext}}(\mathbf{x})n(\mathbf{x})\,d^3x + F[n] \equiv E[n] \tag{7.87}$$

which is known as the Hohenberg–Kohn theorem. This result is, of course, a tremendous simplification of the quantum-mechanical many-body problem, reducing it to finding a function $n(\mathbf{x})$ of a single position variable, instead of needing to find the multiparticle wave function $\Psi(\mathbf{x}_1,\mathbf{x}_2,\ldots,\mathbf{x}_N)$.

### 7.6.3 The Kohn–Sham Equations

Even though (7.87) simplifies the approach to the quantum-mechanical many-body problem in principle, it is difficult to directly apply it to practical problems. Hohenberg and Kohn used it to study two limiting cases of density inhomogeneities in an electron gas, but without a prescription for $F[n]$, detailed calculations of the many-body ground state are not possible.

Kohn and Sham found a way to derive a self-consistent approximation scheme, based on single-particle wave functions[7] $\phi_j(\mathbf{x})$ that solve a particular, albeit fictitious, one-body Schrödinger equation. The $\phi_j(\mathbf{x})$ form the multiparticle wave function $\Psi(\mathbf{x}_1,\mathbf{x}_2,\ldots,\mathbf{x}_N)$ through the $N$-body generalizations of (7.60), that is,

---

[6] We are only considering nondegenerate ground states. In fact, the Hohenberg–Kohn theorem can be proved in general, using a constrained search approach.

[7] In much of the literature on DFT, particularly that aimed at chemistry, these wave functions are called "orbitals."

$$\Psi(\mathbf{x}_1, \mathbf{x}_2, \ldots, \mathbf{x}_N) = \phi_1(\mathbf{x}_1)\phi_2(\mathbf{x}_2)\cdots\phi_N(\mathbf{x}_N)$$
$$\pm \phi_2(\mathbf{x}_1)\phi_1(\mathbf{x}_2)\cdots\phi_N(\mathbf{x}_N)$$
$$\pm \ldots . \tag{7.88}$$

Note that although we labeled the $\phi_j(\mathbf{x}_i)$ with $j = 1, 2, \ldots, N$, they do not necessarily have to be distinct functions. For bosons, for example, they could all be the same function, and for fermions, they could be pairwise the same, each with a different spin quantum number.

Next, a constraint is applied so that the $\phi_j(\mathbf{x})$ yield the correct density $n(\mathbf{x})$. Since the $\phi_j(\mathbf{x})$ solve a Schrödinger equation, they are orthonormal, so inserting (7.88) into (7.78), this constraint takes the form

$$n(\mathbf{x}) = \sum_{j=1}^{N} |\phi_j(\mathbf{x})|^2 . \tag{7.89}$$

The $\phi_j(\mathbf{x})$ are used to construct $F[n]$, and then the calculus of variations is applied to minimize the energy functional.

The Kohn–Sham wave functions $\phi_j(\mathbf{x})$ are solutions to Schrödinger equations of the form

$$\langle \mathbf{x}|H_{KS}|\phi_j\rangle = -\frac{\hbar^2}{2m}\nabla^2 \phi_j(\mathbf{x}) + V_{KS}(\mathbf{x})\phi_j(\mathbf{x}) = \varepsilon_j \phi_j(\mathbf{x}) \tag{7.90}$$

where the $\varepsilon_j$ are the energy eigenvalues. To the extent that

$$\langle \Psi|H_{KS}|\Psi\rangle = \int d^3x_1 \cdots d^3x_N \langle \Psi| \{|\mathbf{x}_1\rangle\langle \mathbf{x}_1|\cdots|\mathbf{x}_N\rangle\langle \mathbf{x}_N|H_{KS}|\} |\Psi\rangle$$
$$= \sum_{j=1}^{N} \int d^3x_i \, \phi_j^*(\mathbf{x}_i) \left[ -\frac{\hbar^2}{2m}\nabla_i^2 \phi_j(\mathbf{x}_i) + V_{KS}(\mathbf{x}_i)\phi_j(\mathbf{x}_i) \right] = \sum_{j=1}^{N} \varepsilon_j \tag{7.91}$$

approximates the true ground-state energy $\langle \Psi|H|\Psi\rangle$, we can rely on the kinetic-energy term

$$\langle \Psi|T_{KS}|\Psi\rangle = \sum_{j=1}^{N} \int d^3x_i \, \phi_j^*(\mathbf{x}_i) \left[ -\frac{\hbar^2}{2m}\nabla_i^2 \phi_j(\mathbf{x}_i) \right] \tag{7.92}$$

to approximate $\langle \Psi|T|\Psi\rangle$ in (7.86). In fact, an important underlying notion for the Kohn–Sham formalism is that the constraint (7.89) implies that (7.92) is an exceptionally good approximation for $\langle \Psi|T|\Psi\rangle$.

Unlike the single-particle case, it is not simple to explicitly write $\langle \Psi|T_{KS}|\Psi\rangle$ in terms of the density $n(\mathbf{x})$, but we can make use of the Hohenberg–Kohn theorem to write $\langle \Psi|T_{KS}|\Psi\rangle \equiv T_{KS}[n]$. We then write

$$F[n] = T[n] + U_{\text{pair}}[n] \tag{7.93a}$$
$$= T_{KS}[n] + U_{\text{lr}}[n] + U_{\text{xc}}[n] \tag{7.93b}$$

where $U_{\text{lr}}[n]$ is the long range component of the two-particle interaction, and $U_{\text{xc}}[n]$, called the "exchange-correlation energy," is what remains. As we believe that $T_{KS}[n]$ is a good approximation to the kinetic energy, and that in typical problems we expect the long range interaction to dominate, $U_{\text{xc}}[n]$ turns out to be small in many problems of interest.

Therefore, good approximations to $U_{xc}[n]$ should yield excellent approximations for $E[n]$ in (7.86).

We note that for the **many-electron problem in atomic, molecular, or condensed matter systems**, which is by far the most common application of DFT, the long range two-particle interaction is just the Coulomb potential between pairs of electrons. In this case, we can use a classical approach to writing down $U_{lr}[n] = U_{ee}[n]$. The charge density at any point is just $e\,n(\mathbf{x})$ so the Coulomb energy is

$$U_{ee}[n] = \frac{e^2}{2} \int d^3x \int d^3x' \, \frac{n(\mathbf{x})n(\mathbf{x}')}{|\mathbf{x} - \mathbf{x}'|} \tag{7.94}$$

where the factor of $1/2$ just removes the effect of double-counting the electrons, as in (7.81).

**At this point, we will proceed as we would for a many-electron problem**, that is with $U_{lr}[n] = U_{ee}[n]$. Following (7.73) to minimize the energy functional, and using (7.87) with (7.93b) and applying the constraint (7.80), we have

$$\delta \left[ \int V_{ext}(\mathbf{x})n(\mathbf{x})\,d^3x + T_{KS}[n] + U_{ee}[n] + U_{xc}[n] - \mu \int n(\mathbf{x})\,d^3x \right] = 0 \tag{7.95}$$

where the variation is in terms of the density $n(\mathbf{x})$. The first and last terms are simple, becoming $\int [V_{ext} - \mu]\delta n\,d^3x$. The long range electron-electron interaction is also easy, namely

$$\delta U_{ee}[n] = e^2 \int \left[ \int d^3x' \, \frac{n(\mathbf{x}')}{|\mathbf{x} - \mathbf{x}'|} \right] \delta n(\mathbf{x})\,d^3x. \tag{7.96}$$

This allows us to use (7.95) to write the *chemical potential* $\mu = \delta E[n]/\delta n$ as

$$\mu = \frac{\delta T_{KS}[n]}{\delta n} + V_{ext}(\mathbf{x}) + e^2 \int d^3x' \, \frac{n(\mathbf{x}')}{|\mathbf{x} - \mathbf{x}'|} + \frac{\delta U_{xc}[n]}{\delta n}. \tag{7.97}$$

Now imagine, instead, that the system is governed by a collection of noninteracting electrons with single-particle wave functions $\phi_j(\mathbf{x})$, determined by an effective "external" potential $V_{KS}(\mathbf{r})$. Using (7.90) we would obviously find

$$\mu = \frac{\delta T_{KS}[n]}{\delta n} + V_{KS}(\mathbf{x}) \tag{7.98}$$

which means we can write down the Kohn–Sham potential of (7.90) as

$$V_{KS}(\mathbf{x}) = V_{ext}(\mathbf{x}) + e^2 \int d^3x' \, \frac{n(\mathbf{x}')}{|\mathbf{x} - \mathbf{x}'|} + \frac{\delta U_{xc}[n]}{\delta n}. \tag{7.99}$$

Given some external potential $V_{ext}(\mathbf{x})$ (perhaps a screened atomic or molecular potential, or even a periodic lattice of them) and exchange-correlation functional $U_{xc}[n]$, we now see how we might carry through a self-consistent iterative approximation. First, choose a reasonable approximation for the density $n(\mathbf{x})$. Then use (7.99) to calculate the Kohn–Sham single-particle potential $V_{KS}(\mathbf{x})$. Solve (7.90) using this potential to find the single-particle wave functions $\phi_j(\mathbf{x})$, and redetermine $n(\mathbf{x})$ using (7.89). Then repeat this process using the new density, comparing the function $n(\mathbf{x})$ at the end with what you started out with, until you are satisfied that convergence is reached. When you have your final function $n(\mathbf{x})$, you then calculate the ground-state energy from

$$E[n] = \int V_{\text{ext}}(\mathbf{x}) n(\mathbf{x}) \, d^3x + T_{\text{KS}}[n] + U_{\text{ee}}[n] + U_{\text{xc}}[n] \qquad (7.100)$$

where we understand that the kinetic-energy functional $T_{\text{KS}}[n]$ is calculated from (7.92) using the wave functions determined by the iteration procedure that found $n(\mathbf{x})$. This procedure also gives us a good approximation for the wave function (7.88), from which other ground-state properties can be calculated.

One remaining issue is which eigenfunctions $\phi_j(\mathbf{x})$, obtained by solving (7.90), should be used to construct $n(\mathbf{x})$. Presumably, our iterative procedure will converge most quickly if we use the lowest energy single-particle wave functions to construct the wave function (7.88). For a system of bosons, this would be all $\phi_j(\mathbf{x}) = \phi_0(\mathbf{x})$, the ground-state eigenfunction. For fermions, we would use the lowest energy eigenfunctions consistent with total antisymmetry, including accounting for the spin degrees of freedom.

Rarely is there an analytic solution to (7.90). However, it can be solved numerically, or the Hamiltonian can be represented in a (truncated) basis of known eigenfunctions, for example the hydrogen atom or harmonic oscillator, and its matrix diagonalized.

It is important to realize that the Kohn–Sham procedure yields much more than just the ground-state energy of the system. Having determined the eigenfunctions $\phi_j(\mathbf{x})$ that give the ground-state density, we then have an excellent approximation (7.88) to the ground-state wave function. This wave function can be used to calculate expectation values of any dynamical operator.

For example, in condensed matter physics where DFT finds many applications, results of calculations include electrical and thermal resistivity, and response to electromagnetic fields, mechanical strain, and other perturbations. See Cohen and Louie (2016), Section 7.4 for a detailed discussion.

## 7.6.4 Models of the Exchange-Correlation Energy

Density functional theory is widely used because it can give precise, accurate results for many-electron systems. This is because everything in (7.100) is known from first principles, except for the exchange-correlation energy $U_{\text{xc}}[n]$, which can be approximated well. It contains the corrections to the Kohn–Sham one-body effective potential, as well as the effects of the correlations between identical particles, for example as in (7.35).

Several authors have derived a set of exact constraints that the exchange-correlation functional must satisfy. Model functionals can then be produced which satisfy as many of those constraints as possible. These models can be tested against simple systems like the uniform electron gas, or systems that are somewhat more complicated, but exactly calculable. In this way, different functionals have been derived for systems ranging from single atoms to complex materials.

Functionals $U_{\text{xc}}[n]$ are generally divided into two classes, namely the local density approximation (LDA) and the generalized gradient approximation (GGA). They take the forms

$$U_{\text{xc}}^{\text{LDA}}[n] = \int d^3x \, n(\mathbf{x}) \, \varepsilon(n) \qquad (7.101a)$$

$$U_{\text{xc}}^{\text{GGA}}[n] = \int d^3 x f(n, \nabla n) \tag{7.101b}$$

where $\varepsilon(n)$ and $f(n, \nabla n)$ can be parameterized functions. A goal is to determine those parameters from fundamental first principles, but it is also possible to determine them by fitting to data. In the original Kohn–Sham formalism, $\varepsilon(n)$ in (7.101) is the exchange-correlation energy per electron for an electron gas of uniform density, and $f(n, 0) = n\varepsilon(n)$. A particularly popular form for $f(n, \nabla n)$ which otherwise includes only fundamental constants is described in a paper[8] by Perdew et al., *Phys. Rev. Lett.*, **77** (1996) 3865.

## 7.6.5  Application to the Helium Atom

We conclude this section on DFT with an application to the two-electron system that is the helium atom. It is also a good opportunity to compare to the results in Section 7.4. Our discussion here outlines the solution, with details left to Problem 7.14 at the end of this chapter.

As always, we want to solve differential equations written in terms of dimensionless variables. It is common in DFT calculations to use "atomic units," where $\hbar = e = m = 1$. This means that all distances are measured in units of the Bohr radius (3.317) $a_0 = \hbar^2/me^2 = 0.53$ Å, and energies are measured in units of $e^2/a_0$, called a "Hartree." Recall from (3.315) that the magnitude of the ground-state energy of a hydrogen atom is $E_0 = e^2/2a_0 = 13.6$ eV, so one Hartree equals 27.2 eV. The Kohn–Sham Schrödinger equation (7.90) becomes

$$-\frac{1}{2}\nabla^2 \phi_j(\mathbf{x}) + V_{\text{KS}}(\mathbf{x})\phi_j(\mathbf{x}) = \varepsilon_j \phi_j(\mathbf{x}). \tag{7.102}$$

We will take the external potential $V_{\text{ext}}(\mathbf{x})$ to be Coulomb's law, which in atomic units becomes

$$V_{\text{ext}}(\mathbf{x}) = V_{\text{ext}}(r) = -\frac{Z}{r}, \tag{7.103}$$

where $Z = 2$ for helium.

We need to choose an initial approximation $n^{(0)}(\mathbf{x})$ to the density. We learned in Section 7.4 that a good starting wave function would be proportional to $e^{-Z_{\text{eff}}r}$ where $Z_{\text{eff}} = 2 - \frac{5}{16}$. This suggests a starting density proportional to $e^{-2Z_{\text{eff}}r}$. However, it is frequently the case with DFT calculations, that we do not have such a good guess for $n^{(0)}(\mathbf{x})$. So, we illustrate the convergence process by making a poor guess, leaving off the factor of two and starting with $n^{(0)}(\mathbf{x})$ proportional to $e^{-Z_{\text{eff}}r}$. With the proper normalization, we find

$$n^{(0)}(r) = \frac{Z_{\text{eff}}^3}{4\pi} e^{-Z_{\text{eff}}r}. \tag{7.104}$$

Note that for this, or any other spherically symmetric density $n(r)$, the long range *ee* interaction term in (7.99) can be written as

---

[8] With nearly 100,000 citations, this paper lays claim to the most highly cited article ever published in *Physical Review Letters*.

$$\int d^3x' \frac{n(\mathbf{x}')}{|\mathbf{x} - \mathbf{x}'|} = 8\pi \int_0^\infty r'^2 \, dr' \frac{n(r')}{\sqrt{(r - r')^2 + r + r'}} \tag{7.105}$$

which is a useful expression for the numerical computations that follow.

For the exchange-correlation functional $U_{xc}[n]$, we use a form of (7.101a) derived by Sun et al., *J. Chem. Phys.*, **144** (2016) 191101. This functional was developed specifically for precise calculations in two-electron systems, and uses only fundamental constants. We have

$$\varepsilon_{xc}(n) = \varepsilon_x(n) + \varepsilon_c(n) \tag{7.106}$$

$$\varepsilon_x(n) = -h_x^0 \frac{3}{4\pi} \left(3\pi^2 n\right)^{1/3}$$

$$\varepsilon_c(n) = -\frac{-0.0233504}{1 + 0.1018 r_s^{1/2} + 0.102582 r_s}$$

where $h_x^0 = 1.174$ and $r_s = (4\pi n/3)^{-1/3}$. From here, it is straightforward to write down $\delta U_{xc}[n]/\delta n$ for use in (7.99).

Thus we have all the ingredients we need to build our Kohn–Sham potential (7.99), in particular (7.103) and (7.106). This potential is used to solve the Schrödinger equation (7.102). Note that for the two-electron system, the ground state is a spin-singlet. Therefore the two-particle wave function is symmetric, and we construct it from the ground-state wave function $\phi_0(\mathbf{x})$ that solves (7.90). That is

$$\Psi(\mathbf{x}_1, \mathbf{x}_2) = \phi_0(\mathbf{x}_1)\phi_0(\mathbf{x}_2). \tag{7.107}$$

Therefore, our goal now is to solve (7.102) for the ground-state wave function. We do this first using $n(r) = n^{(0)}(r)$, using the result to find the next iteration of the density $n^{(1)}(r)$ from (7.89). This process is repeated until the result converges, and the ground-state energy is calculated from (7.100).

In principle we could solve (7.102) numerically, but singular potentials like (7.103) will make this difficult. It is easier to diagonalize the Hamiltonian in some basis, and for this we will use the one-electron atomic wave functions based on (7.103). Since the problem has spherical symmetry, only the $s$-states will contribute. We proceed, therefore, using the $1s$, $2s$, and $3s$ wave functions for a one-electron atom with $Z = Z_{eff}$, and check that the contribution from the $3s$ state is small.

At this point, the calculation is numerical and done on a computer. (The results here were obtained using MATHEMATICA, but there are many other options available.) A $3 \times 3$ matrix is constructed using the Hamiltonian from (7.102), and diagonalized to find the ground-state eigenfunction. One finds

$$\phi_0^{(1)}(\mathbf{x}) = \frac{1}{\sqrt{4\pi}} \left[0.99851 R_{1s}(r) + 0.04831 R_{2s}(r) + 0.02519 R_{3s}(r)\right]. \tag{7.108}$$

As expected, the Kohn–Sham wave function is dominated by the ground $1s$ state for $Z = Z_{eff}$. We use this wave function, with (7.89), to get the next iteration of the density, namely

$$n^{(1)}(r) = 2 \left[\phi_0^{(1)}(\mathbf{x})\right]^2. \tag{7.109}$$

The first iteration of the energy functional (7.100) is found using $n^{(1)}(r)$, and also $\phi_0^{(1)}(\mathbf{x})$ for $T_{KS}[n]$, and this is our first estimate of the ground-state energy.

**Table 7.1** Contributions to the Ground-State Energy (7.100) of the Helium Atom for Three Iterations of the Kohn–Sham Equations. The values for the functionals are in Hartree, and their sum for the ground-state energy is given in eV. The numerical calculation is accurate for all significant figures shown, including the value used to convert Hartree to eV.

| Iteration | $U_{\text{ext}}[n]$ | $T_{\text{KS}}[n]$ | $U_{\text{ee}}[n]$ | $U_{\text{xc}}[n]$ | $E$ |
|-----------|---------------------|--------------------|--------------------|--------------------|------|
| 1 | −6.90876 | 2.9882 | 2.17546 | −1.13109 | −78.2649 |
| 2 | −6.48303 | 2.6350 | 1.99548 | −1.04472 | −78.8379 |
| 3 | −6.56468 | 2.6978 | 2.03054 | −1.06127 | −78.8478 |

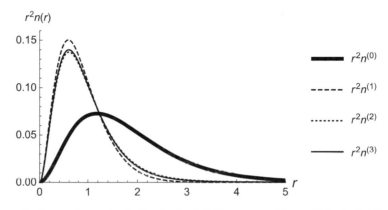

**Fig. 7.7** The initial guess for the two-electron density function for the helium atom, and the results for the first three iterations. Plotted is the density times $r^2$ as a function of $r$, with $r$ in Bohr radii. The first iteration is already close to the final function, and convergence is rapid.

The second iteration finds a new Kohn–Sham potential (7.99), and the process is repeated, yielding a new Kohn–Sham wave function $\phi_0^{(2)}(\mathbf{x})$, a new density $n^{(2)}(r)$, and a new calculation of the ground-state energy. This can be repeated as often as desired, to achieve the desired precision. Of course, the ultimate accuracy of the calculation will only be as good as the form used for the exchange-correlation energy functional.

Table 7.1 shows the results for three iterations of this procedure. Recall from (7.48) that the experimental value is −79.0 eV, and the result (7.52) from the variational principle is −77.5 eV. The first iteration is better than (7.52), and iterations converge rapidly to the measured value.

Note that for this problem, typical of small systems, the exchange-correlation energy is *not* a small contribution. Table 7.1 shows that here it is in fact relatively large, $\approx 40\%$ of the total ground-state energy.

It is instructive to examine the density functions for each iteration. See Figure 7.7. Even though our initial guess is rather different from the function on which the procedure converges, even one iteration gives something very close to the final answer.

We note again that in addition to calculating the ground-state energy, we also now have a good approximation for the two-particle wave function (7.107) that can be used for other calculations of the properties of the helium atom.

# 7.7 Quantum Fields

A different approach to dealing with multiparticle states involves the introduction of quantum-mechanical *fields*. One of the major attractions of quantum field theory is that it can deal with relativistic systems, where particles can be created or destroyed. The techniques are also very useful for nonrelativistic systems, however, so we will take a more general approach here.

## 7.7.1 Second Quantization

This approach known as[9] **second quantization** reexamines the way in which we define the state vector. Define a multiparticle state vector as

$$|n_1, n_2, \ldots, n_i, \ldots\rangle \tag{7.110}$$

where the $n_i$ specify the number of particles with eigenvalue $k_i$ for some operator. Although we take it as a perfectly valid nomenclature for a state vector, it is a member of a new kind of vector space, called "Fock space," in which we need to build in the necessary permutation symmetry.

A word of caution is in order. Our Fock space (or "occupation number") notation for state vectors itself makes an important assumption, namely that there indeed exists a basis of noninteracting states. Interactions between particles can in fact, in principle, affect their very nature. Whether or not we can make a self-consistent theory starting on this assumption, and which in turn accurately describes nature, can only be tested by experiment. See the discussions in Landau (1996) and Merzbacher (1998). We will set this question aside, however, and move ahead at full steam.

Let us now build a framework for a theory of many-particle systems using states in Fock space. We begin this task by recognizing two special cases of states in Fock space. The first of these is

$$|0, 0, \ldots, 0, \ldots\rangle \equiv |0\rangle \tag{7.111}$$

for which there are no particles in any single-particle states. This state is called the "vacuum" and is, as usual, normalized to unity. The second special case is

$$|0, 0, \ldots, n_i = 1, \ldots\rangle \equiv |k_i\rangle \tag{7.112}$$

which is the state in which there is exactly one particle in the state with eigenvalue $k_i$. Of course, this is just the single-particle state which has dominated our discussion of quantum mechanics, prior to this chapter.

Now we need to learn how to build multiparticle states, and then make sure that this building process respects permutation symmetry. In an obvious nod to the creation and

---

[9] The term "second quantization" was apparently coined in the early days of trying to extend quantum mechanics into field theory. The idea was that wave functions were to be turned into operators, which in turn were subject to their own canonical quantization rules. Hence, quantization was enforced a "second" time. See, for example, Section III.12.1 in Heitler (1954).

annihilation operators that we first encountered in Section 2.3, we define a "field operator" $a_i^\dagger$ which increases by one the number of particles in the state with eigenvalue $k_i$, that is

$$a_i^\dagger |n_1, n_2, \ldots, n_i, \ldots\rangle \propto |n_1, n_2, \ldots, n_i + 1, \ldots\rangle \qquad (7.113)$$

where a normalization criterion will be used later to determine the proportionality constant. We *postulate* that the action of the particle creation operator $a_i^\dagger$ on the vacuum is to create a properly normalized single-particle state, namely

$$a_i^\dagger |\mathbf{0}\rangle = |k_i\rangle. \qquad (7.114)$$

This leads us to write

$$1 = \langle k_i | k_i \rangle = [\langle \mathbf{0} | a_i] \left[ a_i^\dagger |\mathbf{0}\rangle \right]$$
$$= \langle \mathbf{0} | \left[ a_i a_i^\dagger |\mathbf{0}\rangle \right] = \langle \mathbf{0} | a_i | k_i \rangle \qquad (7.115)$$

which implies that

$$a_i |k_i\rangle = |\mathbf{0}\rangle \qquad (7.116)$$

so that $a_i$ acts as a particle annihilation operator. We conclude with the following postulates for the particle annihilation operator, namely

$$a_i |n_1, n_2, \ldots, n_i, \ldots\rangle \propto |n_1, n_2, \ldots, n_i - 1, \ldots\rangle \qquad (7.117)$$
$$a_i |\mathbf{0}\rangle = 0 \qquad (7.118)$$
$$a_i |k_j\rangle = 0 \qquad \text{if} \qquad i \neq j \qquad (7.119)$$

where an economy of notation lets us combine (7.116) and (7.119) into

$$a_i |k_j\rangle = \delta_{ij} |\mathbf{0}\rangle. \qquad (7.120)$$

These are enough postulates to fully define the field operators $a_i$, short of actually incorporating permutation symmetry.

The act of permuting two particles, one for the other, is most easily seen by putting the "first" particle in state $|k_i\rangle$ and then the "second" particle in $|k_j\rangle$, and comparing to what happens when we reverse the order in which these states are populated. That is, we expect that, for a two-particle state,

$$a_i^\dagger a_j^\dagger |\mathbf{0}\rangle = \pm a_j^\dagger a_i^\dagger |\mathbf{0}\rangle \qquad (7.121)$$

where the $+$ $(-)$ sign is for bosons (fermions). Applying this same logic to particle exchange in multi-particle states, we are led to

$$a_i^\dagger a_j^\dagger - a_j^\dagger a_i^\dagger = [a_i^\dagger, a_j^\dagger] = 0 \qquad \text{Bosons} \qquad (7.122a)$$
$$a_i^\dagger a_j^\dagger + a_j^\dagger a_i^\dagger = \{a_i^\dagger, a_j^\dagger\} = 0 \qquad \text{Fermions} \qquad (7.122b)$$

where we make use of the "anticommutator" $\{A, B\} \equiv AB + BA$. Simply taking the adjoint of these equations tells us that

$$[a_i, a_j] = 0 \qquad \text{Bosons} \qquad (7.123a)$$
$$\{a_i, a_j\} = 0 \qquad \text{Fermions.} \qquad (7.123b)$$

| Table 7.2 The Algebra for Identical Particles in Second Quantization | |
|---|---|
| Bosons | Fermions |
| $a_i^\dagger a_j^\dagger - a_j^\dagger a_i^\dagger = [a_i^\dagger, a_j^\dagger] = 0$ | $a_i^\dagger a_j^\dagger + a_j^\dagger a_i^\dagger = \{a_i^\dagger, a_j^\dagger\} = 0$ |
| $a_i a_j - a_j a_i = [a_i, a_j] = 0$ | $a_i a_j + a_j a_i = \{a_i, a_j\} = 0$ |
| $a_i a_j^\dagger - a_j^\dagger a_i = [a_i, a_j^\dagger] = \delta_{ij}$ | $a_i a_j^\dagger + a_j^\dagger a_i = \{a_i, a_j^\dagger\} = \delta_{ij}$ |

Note that the Pauli exclusion principle is automatically built into our formalism, since (7.122b) implies that $a_i^\dagger a_i^\dagger = 0$ for some single-particle state $|k_i\rangle$.

Now what about the commutation rules for $a_i$ and $a_j^\dagger$? We would like to define a "number operator" $N_i = a_i^\dagger a_i$ which would count the number of particles in the single-particle state $|k_i\rangle$. Our experience from Section 2.3 argues shows that this is possible if we have $[a_i, a_i^\dagger] = 1$. In fact, a self-consistent picture of both bosons and fermions can be built in just this way, by replacing commutators with anticommutators. The complete algebra is summarized in Table 7.2. For both bosons and fermions, we can define the operator

$$N = \sum_i a_i^\dagger a_i \tag{7.124}$$

which counts the total number of identical particles. (See Problem 7.9 at the end of this chapter.)

We have taken a very ad hoc approach to coming up with the algebra in Table 7.2, rather contrary to the general tone of this book. It is in fact possible to do a somewhat better job in this respect, by postulating, for example, that certain quantities such as the total number of particles be unchanged under a basis change from single-particle states $|k_i\rangle$ to different states $|l_j\rangle$ which are connected by a unitary transformation.[10] Nevertheless, it is not possible to do a fully self-consistent treatment minimizing ad hoc assumptions without developing relativistic quantum field theory, and that is not our mission here.

### 7.7.2 Dynamical Variables in Second Quantization

How do we build operators in second quantization which do more than simply count the number of particles? The answer is straightforward, but once again it is necessary to make some ad hoc assumptions with our current approach.

Suppose the single-particle states $|k_i\rangle$ are eigenstates of some "additive" single-particle operator $K$. Examples might be momentum or kinetic energy. In some multiparticle state

$$|\Psi\rangle = |n_1, n_2, \ldots, n_i, \ldots\rangle \tag{7.125}$$

we expect the eigenvalue of the multiparticle operator $\mathcal{K}$ to be $\sum_i n_i k_i$. This is easy to accomplish if we write

$$\mathcal{K} = \sum_i k_i N_i = \sum_i k_i a_i^\dagger a_i. \tag{7.126}$$

---

[10] This approach, sometimes called the Principle of Unitary Symmetry, is exploited in Merzbacher (1998).

Now suppose that the basis for which the single-particle states are specified is different from the basis in which it is easiest to work. After all, we are used to working with the momentum operator in the coordinate basis, for example. If we use completeness to write

$$|k_i\rangle = \sum_j |l_j\rangle\langle l_j|k_i\rangle \qquad (7.127)$$

then it makes sense to postulate that

$$a_i^\dagger = \sum_j b_j^\dagger \langle l_j|k_i\rangle \qquad (7.128a)$$

which implies that

$$a_i = \sum_j \langle k_i|l_j\rangle b_j \qquad (7.128b)$$

where the operators $b_j^\dagger$ and $b_j$ create and annihilate particles in the single-particle states $|l_j\rangle$. With these assignments, acting on the vacuum state (7.111) with (7.128a) yields (7.127).

Equations (7.128) give us what we need to change the basis for our dynamical single particle operator. We have

$$\mathscr{K} = \sum_i k_i \sum_{mn} b_m^\dagger \langle l_m|k_i\rangle\langle k_i|l_n\rangle b_n$$

$$= \sum_{mn} b_m^\dagger b_n \sum_i \langle l_m|k_i\rangle k_i \langle k_i|l_n\rangle$$

$$= \sum_{mn} b_m^\dagger b_n \langle l_m| \left[ K \sum_i |k_i\rangle\langle k_i| \right] |l_n\rangle$$

or

$$\mathscr{K} = \sum_{mn} b_m^\dagger b_n \langle l_m|K|l_n\rangle. \qquad (7.129)$$

This general form is suitable for writing down a second quantized version of any additive single-particle operator. Examples not only include momentum and kinetic energy, but also any "external" potential energy function which acts individually on each of the particles. All that matters is that the particles do not interact with each other. In the case of bosons, essentially all the particles may find themselves in the lowest energy level of such a potential well, so long as the temperature is low enough. (Experimentally, this phenomenon is referred to as a Bose–Einstein condensate.)

Fermions would behave differently, however. The Pauli exclusion principle will force the particles to populate increasingly higher energy levels in the well. For a system with a very large number of fermions, the total energy of the ground state could be enormous. The highest populated energy level (known as the "Fermi energy") might easily be much larger than the thermal energy $\sim kT$. A classic example is a white dwarf star, a very dense object consisting basically of carbon atoms. The electrons in a white dwarf are, to a good approximation, bound in a potential well. The Fermi level is very high, much larger than the thermal energy for a temperature of tens of millions of degrees Kelvin.

Many-particle systems, however, present a new situation, namely the inevitable possibility that the particles actually interact among themselves. Once again, we postulate an additive operator, that is, one in which the individual two-particle interactions add up independently. Let the symmetric real matrix $V_{ij}$ specify the two-particle eigenvalue for an interaction between particles in single-particle states $|k_i\rangle$ and $|k_j\rangle$. Then the second quantized version of this operator becomes

$$\mathcal{V} = \frac{1}{2}\sum_{i\neq j} V_{ij}N_iN_j + \frac{1}{2}\sum_i V_{ii}N_i(N_i - 1).$$
(7.130)

The first term sums up all of the two-particle interactions, where the factor of 1/2 is necessary because this form double-counts pairs. The second term accounts for all "self interactions" for particles in the same state; there are $n(n-1)/2$ ways to take $n$ things two at a time. The requirement that $V_{ij}$ be real ensures that $\mathcal{V}$ is Hermitian.

The part of the self energy term in (7.130) containing $N_i^2$ exactly represents the parts of the sum in the first term removed by specifying $i \neq j$. Therefore, we can combine this more neatly as

$$\mathcal{V} = \frac{1}{2}\sum_{ij} V_{ij}\left(N_iN_j - N_i\delta_{ij}\right) = \frac{1}{2}\sum_{ij} V_{ij}\Pi_{ij}$$
(7.131)

where $\Pi_{ij} \equiv N_iN_j - N_i\delta_{ij}$ is called the pair distribution operator. Furthermore, we use Table 7.2 to write

$$\Pi_{ij} = a_i^\dagger a_i a_j^\dagger a_j - a_i^\dagger a_i \delta_{ij}$$
$$= a_i^\dagger\left(\delta_{ij} \pm a_j^\dagger a_i\right)a_j - a_i^\dagger a_i\delta_{ij}$$
$$= \pm a_i^\dagger a_j^\dagger a_i a_j$$

or

$$\Pi_{ij} = (\pm)(\pm)a_i^\dagger a_j^\dagger a_j a_i$$
(7.132)

where we used (7.123) to reverse the order of the last two factors. This allows us to rewrite (7.130) as

$$\mathcal{V} = \frac{1}{2}\sum_{ij} V_{ij}a_i^\dagger a_j^\dagger a_j a_i.$$
(7.133)

This sequence of creation and annihilation operators, first one particle is annihilated, then another, and then creating them in reverse order, is called "normal ordering." Note that we see explicitly from (7.122b) or (7.123b) that there is no contribution from diagonal elements of $V$ for fermions.

We can use (7.128) to rewrite (7.133) in a different basis. We have

$$\mathcal{V} = \frac{1}{2}\sum_{mnpq} \langle mn|V|pq\rangle b_m^\dagger b_n^\dagger b_q b_p$$
(7.134)

where

$$\langle mn|V|pq\rangle \equiv \sum_{ij} V_{ij}\langle l_m|k_i\rangle\langle k_i|l_p\rangle\langle l_n|k_j\rangle\langle k_j|l_q\rangle. \tag{7.135}$$

This result provides some insight into the physical meaning of our formalism. Suppose, for example, that the $|k_i\rangle$ are position basis states $|\mathbf{x}\rangle$, and that the $|l_i\rangle$ are momentum basis states $|\mathbf{p} = \hbar\mathbf{k}\rangle$. Then $V_{ij}$ would represent an interaction between two particles, one located at $\mathbf{x}$ and the other at $\mathbf{x}'$. A natural example would be for a collection of particles each with charge $q = -e$ in which case we would write

$$V_{ij} \to V(\mathbf{x},\mathbf{x}') = \frac{e^2}{|\mathbf{x}-\mathbf{x}'|} \tag{7.136}$$

$$\sum_{ij} \to \int d^3x \int d^3x' \tag{7.137}$$

but any mutual interaction between the particles would be treated in a similar way. The quantity $\langle mn|V|pq\rangle$ therefore represents a momentum-space version of the interaction, with $m$ and $p$ following one particle, and $n$ and $q$ following the other. (It is easy to show that $\langle mn|V|pq\rangle = \langle nm|V|qp\rangle$ but interchanging one side and not the other will depend on whether the particles are bosons or fermions.) The four inner products in (7.135) lead to a factor

$$e^{i(\mathbf{k}_m-\mathbf{k}_p)\cdot\mathbf{x} + i(\mathbf{k}_n-\mathbf{k}_q)\cdot\mathbf{x}'}$$

which, after the integrals (7.136) are carried out, results in a $\delta$-function which conserves momentum. One might diagrammatically represent the two-particle interaction as shown in Fig. 7.8.

Clearly we are on our way towards developing a nonrelativistic version of quantum field theory. As a specific example, we will treat the quantum-mechanical version of the

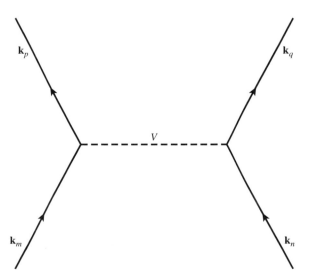

**Fig. 7.8**    Diagrammatic representation of the "momentum-space matrix element" $\langle mn|V|pq\rangle$.

noninteracting electromagnetic field shortly. However, we will not push the general case beyond this point, as it goes beyond the scope of this book and is treated well in any one of several other books. See, for example, Merzbacher (1998), Landau (1996), and Fetter and Walecka (2003).

### 7.7.3 Example: The Degenerate Electron Gas

An excellent example of the principles discussed in this section, is the degenerate electron gas. This is a collection of electrons, interacting with each other through their mutual Coulomb repulsion, bound in some positively charged background medium. Physical examples might include a high temperature plasma, or even, to some approximation, bulk metals.

This problem is treated thoroughly in Chapter One, Section 3 of Fetter and Walecka (2003). We present the problem and outline its solution here, but the interested reader is referred to the original reference to fill in the details.

Our task is to find the eigenvalues of the Hamiltonian

$$H = H_{el} + H_b + H_{el-b} \tag{7.138}$$

for a system of $N$ electrons. The electrons interact among themselves according to

$$H_{el} = \sum_i \frac{\mathbf{p}_i^2}{2m} + \frac{1}{2}e^2 \sum_i \sum_{j \neq i} \frac{e^{-\mu|\mathbf{x}_i - \mathbf{x}_j|}}{|\mathbf{x}_i - \mathbf{x}_j|} \tag{7.139}$$

where we employ a "screened" Coulomb potential but will let $\mu \to 0$ before we finish the calculation. The energy of the positive background is

$$H_b = \frac{1}{2}e^2 \int d^3x' \int d^3x'' \rho(\mathbf{x}')\rho(\mathbf{x}'') \frac{e^{-\mu|\mathbf{x}' - \mathbf{x}''|}}{|\mathbf{x}' - \mathbf{x}''|} \tag{7.140}$$

where $\rho(\mathbf{x})$ is the number density of background particle sites. We will assume a uniform background, with $\rho(\mathbf{x}) = N/V$ for a system of volume $V = L^3$. Then, translating to a variable $\mathbf{x} \equiv \mathbf{x}' - \mathbf{x}''$ (7.140) becomes

$$H_b = \frac{1}{2}e^2 \left(\frac{N}{V}\right)^2 \int d^3x' \int d^3x \frac{e^{-\mu|\mathbf{x}|}}{|\mathbf{x}|} = \frac{1}{2}e^2 \frac{N^2}{V}\frac{4\pi}{\mu^2}. \tag{7.141}$$

Thus $H_b$ contributes simply an additive constant to the energy. The fact that this constant grows without bound as $\mu \to 0$ will not be a problem, as we shall see shortly. The interaction of the electrons with the constant background is

$$H_{el-b} = -e^2 \sum_i \int d^3x \rho(\mathbf{x}) \frac{e^{-\mu|\mathbf{x} - \mathbf{x}_i|}}{|\mathbf{x} - \mathbf{x}_i|}$$

$$= -e^2 \frac{N}{V} \sum_i \int d^3x \frac{e^{-\mu|\mathbf{x} - \mathbf{x}_i|}}{|\mathbf{x} - \mathbf{x}_i|} = -e^2 \frac{N^2}{V}\frac{4\pi}{\mu^2}. \tag{7.142}$$

Therefore (7.138) becomes

$$H = -\frac{1}{2}e^2 \frac{N^2}{V}\frac{4\pi}{\mu^2} + \sum_i \frac{\mathbf{p}_i^2}{2m} + \frac{1}{2}e^2 \sum_i \sum_{j \neq i} \frac{e^{-\mu|\mathbf{x}_i - \mathbf{x}_j|}}{|\mathbf{x}_i - \mathbf{x}_j|}. \tag{7.143}$$

The first term in this equation is just a number. The second is a one-body operator which we will express simply in terms of operators in second quantization in momentum space. The third term is a two-body operator which will involve a bit more work to write in second quantization.

Writing the kinetic-energy term in (7.143) is just a matter of rewriting (7.129) for $K$ being the momentum operator $\mathbf{p}$ and the $|l_n\rangle$ being momentum basis states. Single-particle states are denoted by $i = \{\mathbf{k}, \lambda\}$ where $\lambda = \pm$ indicates electron spin. We know that

$$\langle \mathbf{k}'\lambda'|\mathbf{p}|\mathbf{k}\lambda\rangle = \hbar\mathbf{k}\,\delta_{\mathbf{kk}'}\delta_{\lambda\lambda'} \tag{7.144}$$

so we have

$$\sum_i \frac{\mathbf{p}_i^2}{2m} \Rightarrow \mathscr{K} = \sum_{\mathbf{k}\lambda} \frac{\hbar^2\mathbf{k}^2}{2m} a_{\mathbf{k}\lambda}^\dagger a_{\mathbf{k}\lambda}. \tag{7.145}$$

Now we write the potential energy term in (7.143) in second quantization, using (7.134) and (7.135). Note (7.136) and (7.137). We have

$$\mathscr{V} = \frac{1}{2} \sum_{\mathbf{k}_1\lambda_1} \sum_{\mathbf{k}_2\lambda_2} \sum_{\mathbf{k}_3\lambda_3} \sum_{\mathbf{k}_4\lambda_4} \langle \mathbf{k}_1\lambda_1\mathbf{k}_2\lambda_2|V|\mathbf{k}_3\lambda_3\mathbf{k}_4\lambda_4\rangle a_{\mathbf{k}_1\lambda_1}^\dagger a_{\mathbf{k}_2\lambda_2}^\dagger a_{\mathbf{k}_4\lambda_4} a_{\mathbf{k}_3\lambda_3} \tag{7.146}$$

where

$$\begin{aligned}
\langle &\mathbf{k}_1\lambda_1\mathbf{k}_2\lambda_2|V|\mathbf{k}_4\lambda_4\mathbf{k}_3\lambda_3\rangle \\
&= \int d^3x' \int d^3x'' V(\mathbf{x}',\mathbf{x}'') \langle \mathbf{k}_1\lambda_1|\mathbf{x}'\rangle\langle \mathbf{x}'|\mathbf{k}_3\lambda_3\rangle\langle \mathbf{k}_2\lambda_2|\mathbf{x}''\rangle\langle \mathbf{x}''|\mathbf{k}_4\lambda_4\rangle \\
&= \frac{e^2}{V^2} \int d^3x' \int d^3x'' \frac{e^{-\mu|\mathbf{x}'-\mathbf{x}''|}}{|\mathbf{x}'-\mathbf{x}''|} e^{-i\mathbf{k}_1\cdot\mathbf{x}'}\chi_{\lambda_1}^\dagger e^{i\mathbf{k}_3\cdot\mathbf{x}'}\chi_{\lambda_3} e^{-i\mathbf{k}_2\cdot\mathbf{x}''}\chi_{\lambda_2}^\dagger e^{i\mathbf{k}_4\cdot\mathbf{x}''}\chi_{\lambda_4} \\
&= \frac{e^2}{V^2} \int d^3x \int d^3y \frac{e^{-\mu y}}{y} e^{-i\mathbf{k}_1\cdot\mathbf{x}'}\delta_{\lambda_1\lambda_3} e^{i\mathbf{k}_3\cdot\mathbf{x}'} e^{-i\mathbf{k}_2\cdot\mathbf{x}''}\delta_{\lambda_2\lambda_4} e^{i\mathbf{k}_4\cdot\mathbf{x}''} \\
&= \frac{e^2}{V^2}\delta_{\lambda_1\lambda_4}\delta_{\lambda_2\lambda_3} \int d^3x\, e^{-i(\mathbf{k}_1+\mathbf{k}_2-\mathbf{k}_3-\mathbf{k}_4)\cdot\mathbf{x}} \int d^3y \frac{e^{-\mu y}}{y} e^{-i(\mathbf{k}_1-\mathbf{k}_3)\cdot\mathbf{y}} \\
&= \frac{e^2}{V}\delta_{\lambda_1\lambda_4}\delta_{\lambda_2\lambda_3}\delta_{\mathbf{k}_1+\mathbf{k}_2,\mathbf{k}_3+\mathbf{k}_4} \int d^3y \frac{e^{-\mu y}}{y} e^{-i(\mathbf{k}_1-\mathbf{k}_3)\cdot\mathbf{y}}
\end{aligned} \tag{7.147}$$

where the $\chi_\lambda$ are spinor representations, and we use a change of variables $\mathbf{x} = \mathbf{x}''$ and $\mathbf{y} = \mathbf{x}' - \mathbf{x}''$. Finally, we define the momentum transfer $\mathbf{q} \equiv \mathbf{k}_1 - \mathbf{k}_3$ and find

$$\langle \mathbf{k}_1\lambda_1\mathbf{k}_2\lambda_2|V|\mathbf{k}_3\lambda_3\mathbf{k}_4\lambda_4\rangle = \frac{e^2}{V}\delta_{\lambda_1\lambda_4}\delta_{\lambda_2\lambda_3}\delta_{\mathbf{k}_1+\mathbf{k}_2,\mathbf{k}_3+\mathbf{k}_4}\frac{4\pi}{\mathbf{q}^2+\mu^2}. \tag{7.148}$$

The Kronecker deltas in the spin just insure that that no spins are flipped by this interaction, which we expect since the interaction is spin independent. The Kronecker delta in wave number insures that momentum is conserved. Thus (7.146) becomes

$$\mathscr{V} = \frac{e^2}{2V} \sum_{\mathbf{k}_1\lambda_1} \sum_{\mathbf{k}_2\lambda_2} \sum_{\mathbf{k}_3} \sum_{\mathbf{k}_4} \delta_{\mathbf{k}_1+\mathbf{k}_2,\mathbf{k}_3+\mathbf{k}_4}\frac{4\pi}{\mathbf{q}^2+\mu^2} a_{\mathbf{k}_1\lambda_1}^\dagger a_{\mathbf{k}_2\lambda_2}^\dagger a_{\mathbf{k}_4\lambda_2} a_{\mathbf{k}_3\lambda_1} \tag{7.149}$$

after reducing the summations using the spin-conserving Kronecker deltas.

An important feature of (7.149) becomes apparent if we first redefine $\mathbf{k}_3 \equiv \mathbf{k}$ and $\mathbf{k}_4 \equiv \mathbf{p}$. Then the terms of (7.149) for which $\mathbf{q} = 0$ become

$$\frac{e^2}{2V} \sum_{\mathbf{kp}} \sum_{\lambda_1 \lambda_2} \frac{4\pi}{\mu^2} a^\dagger_{\mathbf{k}\lambda_1} a^\dagger_{\mathbf{p}\lambda_2} a_{\mathbf{p}\lambda_2} a_{\mathbf{k}\lambda_1} = \frac{e^2}{2V} \frac{4\pi}{\mu^2} \sum_{\mathbf{k}\lambda_1} \sum_{\mathbf{p}\lambda_2} a^\dagger_{\mathbf{k}\lambda_1} a_{\mathbf{k}\lambda_1} \left( a^\dagger_{\mathbf{p}\lambda_2} a_{\mathbf{p}\lambda_2} - \delta_{\mathbf{kp}} \delta_{\lambda_1 \lambda_2} \right)$$

$$= \frac{e^2}{2V} \frac{4\pi}{\mu^2} (N^2 - N) \tag{7.150}$$

where we have made use of the fermion anticommutation relations, and the definition of the number operator. The first term in this relation just cancels the first term of (7.143). The second term represents an energy $-2\pi e^2/\mu^2 V$ per particle, but this will vanish in the limit where $V = L^3 \to \infty$ while always keeping $\mu \gg 1/L$. Thus the terms with $\mathbf{q} = 0$ do not contribute, and cancel the rapidly diverging terms in the Hamiltonian. Indeed, this finally allows us to set the screening parameter $\mu = 0$ and write the second quantized Hamiltonian as

$$H = H_0 + H_1 \tag{7.151a}$$

$$H_0 = \sum_{\mathbf{k}\lambda} \frac{\hbar^2 \mathbf{k}^2}{2m} a^\dagger_{\mathbf{k}\lambda} a_{\mathbf{k}\lambda} \tag{7.151b}$$

$$H_1 = \frac{e^2}{2V} {\sum_{\mathbf{kpq}}}' \sum_{\lambda_1 \lambda_2} \frac{4\pi}{q^2} a^\dagger_{\mathbf{k}+\mathbf{q},\lambda_1} a^\dagger_{\mathbf{p}-\mathbf{q},\lambda_2} a_{\mathbf{p}\lambda_2} a_{\mathbf{k}\lambda_1} \tag{7.151c}$$

where the notation $\Sigma'$ indicates that terms with $\mathbf{q} = 0$ are to be omitted. Note that in the limit we have taken, a finite density $n = N/V$ is implicitly assumed.

Finding the eigenvalues of (7.151) is a difficult problem, although solutions are possible. Our approach will be to find the ground-state energy by treating the second term as a perturbation on the first. Although reasonable arguments can be made why this should be a good approximation (see Fetter and Walecka), those arguments only hold in a particular range of densities. Fortunately, that range of densities is relevant to physical systems such as metals, so our approach indeed has practical interest.

This is a good time to introduce some scaling variables. The density is determined by the interatomic spacing $r_0$, that is

$$n = \frac{N}{V} = \left[ \frac{4\pi}{3} r_0^3 \right]^{-1} \tag{7.152}$$

and a natural scale for $r_0$ is the Bohr radius (3.317), i.e. $a_0 = \hbar^2/me^2$. We define a dimensionless distance scale $r_s = r_0/a_0$, called the Wigner–Seitz radius. Our calculation of the ground-state energy will be as a function of $r_s$.

As an introduction to calculating the expectation value $E^{(0)}$ of the operator $H_0$ for the ground state, we discuss the concept of *Fermi energy*. (Recall the discussion on p. 457.) Because of the Pauli exclusion principle, electrons will fill the available energy levels up to some maximum wavenumber $k_F$. We can relate $k_F$ to the total number of electrons by adding up all of the states with $k \leq k_F$, that is

$$N = \sum_{\mathbf{k}\lambda} \theta(k_F - k)$$

$$\rightarrow \frac{V}{(2\pi)^3} \sum_{\lambda} \int d^3k \, \theta(k_F - k) = \frac{V}{3\pi^2} k_F^3 \tag{7.153}$$

where $\theta(x) = 0$ for $x < 0$ and unity otherwise. This implies that

$$k_F = \left(\frac{3\pi^2 N}{V}\right)^{1/3} = \left(\frac{9\pi}{4}\right)^{1/3} \frac{1}{r_0} \tag{7.154}$$

which shows that $k_F$ is about the same size as the inverse interparticle spacing.

Now use the same approach to calculate the unperturbed energy $E^{(0)}$. Denoting the ground state as $|F\rangle$ we have

$$E^{(0)} = \langle F|H_0|F\rangle = \frac{\hbar^2}{2m} \sum_{\mathbf{k}\lambda} k^2 \theta(k_F - k)$$

$$\rightarrow \frac{\hbar^2}{2m} \frac{V}{(2\pi)^3} \sum_{\lambda} \int d^3k \, k^2 \theta(k_F - k) = \frac{e^2}{2a_0} N \frac{3}{5} \left(\frac{9\pi}{4}\right)^{2/3} \frac{1}{r_s^2}. \tag{7.155}$$

Note that $e^2/2a_0 \approx 13.6$ eV, the ground-state energy of the hydrogen atom.

The first-order correction to the ground-state energy is

$$E^{(1)} = \langle F|H_1|F\rangle$$

$$= \frac{e^2}{2V} \sum_{\mathbf{kpq}}' \sum_{\lambda_1\lambda_2} \frac{4\pi}{q^2} \langle F|a_{\mathbf{k}+\mathbf{q},\lambda_1}^\dagger a_{\mathbf{p}-\mathbf{q},\lambda_2}^\dagger a_{\mathbf{p}\lambda_2} a_{\mathbf{k}\lambda_1}|F\rangle. \tag{7.156}$$

The summation is easy to reduce since $|F\rangle$ is a collection of single-particle states with occupation numbers either zero or one. The only way for the matrix element in (7.156) to be nonzero is if the annihilation and creation operators pair up appropriately. Since $\mathbf{q} \neq 0$ in the sum, the only way to pair up the operators is by setting $\{\mathbf{p}-\mathbf{q}, \lambda_2\} = \{\mathbf{k}, \lambda_1\}$ and $\{\mathbf{k}+\mathbf{q}, \lambda_1\} = \{\mathbf{p}, \lambda_2\}$. Therefore

$$E^{(1)} = \frac{e^2}{2V} \sum_{\lambda_1} \sum_{\mathbf{kq}}' \frac{4\pi}{q^2} \langle F|a_{\mathbf{k}+\mathbf{q},\lambda_1}^\dagger a_{\mathbf{k},\lambda_1}^\dagger a_{\mathbf{k}+\mathbf{q},\lambda_1} a_{\mathbf{k}\lambda_1}|F\rangle$$

$$= -\frac{e^2}{2V} \sum_{\lambda_1} \sum_{\mathbf{kq}}' \frac{4\pi}{q^2} \langle F|\left(a_{\mathbf{k}+\mathbf{q},\lambda_1}^\dagger a_{\mathbf{k}+\mathbf{q},\lambda_1}\right)\left(a_{\mathbf{k},\lambda_1}^\dagger a_{\mathbf{k}\lambda_1}\right)|F\rangle$$

$$= -\frac{e^2}{2V} 2 \frac{V^2}{(2\pi)^6} \int d^3k \int d^3q \frac{4\pi}{q^2} \theta(k_F - |\mathbf{k}+\mathbf{q}|)\theta(k_F - k)$$

$$= -e^2 \frac{4\pi V}{(2\pi)^6} \int d^3q \frac{1}{q^2} \int d^3P \, \theta\left(k_F - |\mathbf{P}+\tfrac{1}{2}\mathbf{q}|\right) \theta\left(k_F - |\mathbf{P}-\tfrac{1}{2}\mathbf{q}|\right). \tag{7.157}$$

The integral over $\mathbf{P}$ is just the intersection between two spheres of radius $k_F$ but with centers separated by $\mathbf{q}$, and is easy to evaluate. The result is

$$E^{(1)} = -\frac{e^2}{2a_0} N \frac{3}{2\pi} \left(\frac{9\pi}{4}\right)^{1/3} \frac{1}{r_s}. \tag{7.158}$$

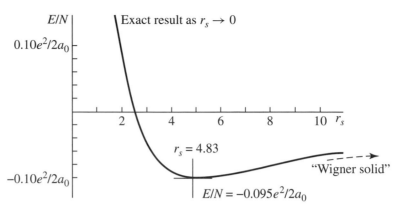

**Fig. 7.9** The ground-state energy, to first order in perturbation theory, for a system of *N* electrons inside a uniform, positively charged background. The energy per electron is plotted as a function of the interparticle spacing in units of the Bohr radius. From Fetter and Walecka (2003).

Therefore, the ground-state energy to first order is given by

$$\frac{E}{N} = \frac{e^2}{2a_0}\left(\frac{9\pi}{4}\right)^{2/3}\left(\frac{3}{5}\frac{1}{r_s^2} - \frac{3}{2\pi}\frac{1}{r_s}\right). \tag{7.159}$$

This is plotted in Figure 7.9. The unperturbed energy decreases monotonically as $r_s \to 0$, but the first-order correction is an attraction which falls more slowly. The result is a minimum at a value $E/N = -0.095e^2/2a_0 = -1.29$ eV where $r_s = 4.83$. Our model is crude, and the solution only approximate, but the agreement with experiment is surprisingly good. For sodium metal, one finds $E/N = -1.13$ eV where $r_s = 3.96$.

## 7.8  Quantization of the Electromagnetic Field

Maxwell's equations form a complete classical description of noninteracting electric and magnetic fields in free space. It is tricky to apply quantum mechanics to that description, but it can be done in a number of ways. In this section, we will once again take a "follow our nose" approach to the problem, based on the many-particle formalism developed in this chapter. The particles, of course, are photons,[11] whose creation and annihilation operators obey Bose–Einstein commutation relations.

We start with a brief summary of Maxwell's equations to establish our notation, and their solution in terms of electromagnetic waves. Then we derive the energy, and associate it with the eigenvalues of a Hamiltonian constructed using bosonic creation and annihilation operators.

---

[11] The concept of photons is not without controversy. See "Anti-photon" by Lamb *Appl. Phys. B*, **60** (1995) 77. The astute reader will note that we derived the cross section for the photoelectric effect in Section 5.8.3 using the classical electromagnetic field.

Including interactions with electromagnetic fields, through the inclusion of spin $\frac{1}{2}$ charged electrons, is the subject of quantum electrodynamics. We do not pursue this subject in this book. (See Section 5.8 for a discussion of a more ad hoc way to apply electromagnetic field interactions to atomic systems.) However, there is a fascinating quantum-mechanical effect observable with free electromagnetic fields, the Casimir effect, and we conclude this section with a description of the calculation and the experimental data.

Our treatment here more or less follows Chapter Four in Loudon (2000), although the approach has become rather standard. See, for example, Chapter 23 in Merzbacher (1998).

## 7.8.1  Maxwell's Equations in Free Space

In the absence of any charges or currents, Maxwell's equations (in Gaussian units; see Appendix A) take the form

$$\nabla \cdot \mathbf{E} = 0 \tag{7.160a}$$

$$\nabla \cdot \mathbf{B} = 0 \tag{7.160b}$$

$$\nabla \times \mathbf{E} + \frac{1}{c}\frac{\partial \mathbf{B}}{\partial t} = 0 \tag{7.160c}$$

$$\nabla \times \mathbf{B} - \frac{1}{c}\frac{\partial \mathbf{E}}{\partial t} = 0. \tag{7.160d}$$

Following standard procedure, we postulate a *vector potential* $\mathbf{A}(\mathbf{x}, t)$ such that

$$\mathbf{B} = \nabla \times \mathbf{A} \tag{7.161}$$

which means that (7.160b) is immediately satisfied. If we impose the further condition

$$\nabla \cdot \mathbf{A} = 0 \tag{7.162}$$

(which is known as "choosing the Coulomb gauge") then

$$\mathbf{E} = -\frac{1}{c}\frac{\partial \mathbf{A}}{\partial t} \tag{7.163}$$

means that (7.160a) and (7.160c) are also satisfied. Therefore, determining $\mathbf{A}(\mathbf{x}, t)$ is equivalent to determining $\mathbf{E}(\mathbf{x}, t)$ and $\mathbf{B}(\mathbf{x}, t)$. A solution for $\mathbf{A}(\mathbf{x}, t)$ is evident, though, by observing that (7.160d) leads directly to

$$\nabla^2 \mathbf{A} - \frac{1}{c^2}\frac{\partial^2 \mathbf{A}}{\partial t^2} = 0. \tag{7.164}$$

That is, $\mathbf{A}(\mathbf{x}, t)$ satisfies the wave equation, with wave speed $c$, just as we might have guessed.

The set of solutions to (7.164) are naturally written as

$$\mathbf{A}(\mathbf{x}, t) = \mathbf{A}(\mathbf{k}) e^{\pm i \mathbf{k} \cdot \mathbf{x}} e^{\pm i \omega t} \tag{7.165}$$

where $\omega = \omega_k \equiv |\mathbf{k}|c = kc$ for the solution to be valid. The Coulomb gauge condition (7.162) implies that $\pm i \mathbf{k} \cdot \mathbf{A}(\mathbf{x}, t) = 0$ or

$$\mathbf{k} \cdot \mathbf{A}(\mathbf{k}) = 0. \tag{7.166}$$

In other words, $\mathbf{A}(\mathbf{x},t)$ is perpendicular to the propagation direction $\mathbf{k}$. For this reason, the Coulomb gauge is frequently referred to as the "transverse gauge." This allows us to write the general solution to (7.164) as

$$\mathbf{A}(\mathbf{x},t) = \sum_{\mathbf{k},\lambda} \hat{\mathbf{e}}_{\mathbf{k}\lambda} A_{\mathbf{k},\lambda}(\mathbf{x},t) \tag{7.167}$$

where $\hat{\mathbf{e}}_{\mathbf{k}\lambda}$ are two unit vectors (corresponding to two values for $\lambda$) perpendicular to $\mathbf{k}$, and where

$$\mathbf{A}_{\mathbf{k},\lambda}(\mathbf{x},t) = \mathbf{A}_{\mathbf{k},\lambda} e^{-i(\omega_k t - \mathbf{k}\cdot\mathbf{x})} + \mathbf{A}_{\mathbf{k},\lambda}^* e^{+i(\omega_k t - \mathbf{k}\cdot\mathbf{x})}. \tag{7.168}$$

Note that in (7.168) the quantities written $\mathbf{A}_{\mathbf{k},\lambda}$ on the right side of the equation are numerical coefficients, not functions of either position or time. Note also that $\mathbf{k}$ and $-\mathbf{k}$ represent different terms in the sum. We write the superposition (7.167) as a sum, not an integral, because we envision quantizing the electromagnetic field inside a "big box" whose dimensions may eventually be taken to grow without bound.

We use the form (7.168) to insure that $\mathbf{A}_{\mathbf{k},\lambda}(\mathbf{x},t)$ is real. When we quantize the electromagnetic field, $\mathbf{A}_{\mathbf{k},\lambda}(\mathbf{x},t)$ will become a Hermitian operator. The coefficients $\mathbf{A}_{\mathbf{k},\lambda}^*$ and $\mathbf{A}_{\mathbf{k},\lambda}$ will become creation and annihilation operators.

As we shall see later, it is useful to take the unit vectors $\hat{\mathbf{e}}_{\mathbf{k}\lambda}$ as directions of *circular* polarization as opposed to linear. That is, if $\hat{\mathbf{e}}_{\mathbf{k}}^{(1)}$ and $\hat{\mathbf{e}}_{\mathbf{k}}^{(2)}$ are the linear unit vectors perpendicular to $\mathbf{k}$, then

$$\hat{\mathbf{e}}_{\mathbf{k}\pm} = \mp\frac{1}{\sqrt{2}}\left(\hat{\mathbf{e}}_{\mathbf{k}}^{(1)} \pm i\hat{\mathbf{e}}_{\mathbf{k}}^{(2)}\right) \tag{7.169}$$

where $\lambda = \pm$ denotes the polarization state. With these definitions, it is easy to show that

$$\hat{\mathbf{e}}_{\mathbf{k}\lambda}^* \cdot \hat{\mathbf{e}}_{\pm\mathbf{k}\lambda'} = \pm\delta_{\lambda\lambda'} \tag{7.170a}$$

$$\hat{\mathbf{e}}_{\mathbf{k}\lambda}^* \times \hat{\mathbf{e}}_{\pm\mathbf{k}\lambda'} = \pm i\lambda\delta_{\lambda\lambda'}\hat{\mathbf{k}} \tag{7.170b}$$

where $\hat{\mathbf{k}}$ is a unit vector in the direction of $\mathbf{k}$. The electric field $\mathbf{E}(\mathbf{x},t)$ can now be written down from (7.163), and similarly for the magnetic field $\mathbf{B}(\mathbf{x},t)$ using (7.161).

The energy $\mathscr{E}$ in the electromagnetic field is given by integrating the energy density over all space, that is

$$\mathscr{E} = \frac{1}{8\pi}\int_V \left[|\mathbf{E}(\mathbf{x},t)|^2 + |\mathbf{B}(\mathbf{x},t)|^2\right]d^3x \tag{7.171}$$

where, as discussed earlier, "all space" is a finite volume $V = L^3$ with periodic boundary conditions. In other words, we are working inside an electromagnetic cavity with conducting walls. This means that

$$\mathbf{k} = (k_x, k_y, k_z) = \frac{2\pi}{L}(n_x, n_y, n_z) \tag{7.172}$$

where $n_x$, $n_y$, and $n_z$ are integers.

Consider first the term dependent on the electric field in (7.171). Using (7.163) with (7.167) and (7.168) we have

$$\mathbf{E} = \frac{i}{c} \sum_{\mathbf{k},\lambda} \omega_k \left[ \mathbf{A}_{\mathbf{k},\lambda} e^{-i(\omega_k t - \mathbf{k}\cdot\mathbf{x})} - \mathbf{A}_{\mathbf{k},\lambda}^* e^{+i(\omega_k t - \mathbf{k}\cdot\mathbf{x})} \right] \hat{\mathbf{e}}_{\mathbf{k}\lambda} \tag{7.173a}$$

and

$$\mathbf{E}^* = -\frac{i}{c} \sum_{\mathbf{k}',\lambda'} \omega_{k'} \left[ \mathbf{A}_{\mathbf{k}',\lambda'}^* e^{+i(\omega_{k'} t - \mathbf{k}'\cdot\mathbf{x})} - \mathbf{A}_{\mathbf{k}',\lambda'} e^{-i(\omega_{k'} t - \mathbf{k}'\cdot\mathbf{x})} \right] \hat{\mathbf{e}}_{\mathbf{k}'\lambda'}^*. \tag{7.173b}$$

Since we have already suggested that the $\mathbf{A}_{\mathbf{k},\lambda}^*$ and $\mathbf{A}_{\mathbf{k},\lambda}$ will become creation and annihilation operators, we need to take care and keep their order intact.

This all leads to an awkward expression for $|\mathbf{E}|^2 = \mathbf{E}^* \cdot \mathbf{E}$, a summation over $\mathbf{k}$, $\lambda$, $\mathbf{k}'$, and $\lambda'$, with four terms inside the summation. However, an important simplification follows from the integral over the spatial volume. Each term inside the sum packs all of its position dependence into an exponential so that the volume integral is of the form

$$\int_V e^{i(\mathbf{k}\mp\mathbf{k}')\cdot\mathbf{x}} d^3 x = V \delta_{\mathbf{k},\pm\mathbf{k}'}. \tag{7.174}$$

Combining this with (7.170a) one finds

$$\int_V |\mathbf{E}(\mathbf{x},t)|^2 d^3 x = \sum_{\mathbf{k},\lambda} \frac{\omega_k^2}{c^2} V \left[ \mathbf{A}_{\mathbf{k},\lambda}^* \mathbf{A}_{\mathbf{k},\lambda} + \mathbf{A}_{\mathbf{k},\lambda} \mathbf{A}_{\mathbf{k},\lambda}^* + \mathbf{A}_{\mathbf{k},\lambda}^* \mathbf{A}_{-\mathbf{k},\lambda}^* e^{2i\omega_k t} + \mathbf{A}_{\mathbf{k},\lambda} \mathbf{A}_{-\mathbf{k},\lambda} e^{-2i\omega_k t} \right]. \tag{7.175}$$

Starting with (7.161), the calculation for $|\mathbf{B}|^2 = \mathbf{B}^* \cdot \mathbf{B}$ is very similar. The curl brings in factors like $\mathbf{k} \times \hat{\mathbf{e}}_{\mathbf{k}\lambda}$ instead of the $\omega_k/c$ in the calculation involving the electric field, but since $\mathbf{k}^2 = \omega_k^2/c^2$ the result is nearly identical. The key difference, though, is that under the change $\mathbf{k} \to -\mathbf{k}$ terms like $\mathbf{k} \times \hat{\mathbf{e}}_{\mathbf{k}\lambda}$ *do not* change sign. This means that the terms analogous to the third and fourth terms in (7.175) appear the same way but with opposite signs. Therefore, they cancel when evaluating (7.171). The result is

$$\mathscr{E} = \frac{1}{4\pi} V \sum_{\mathbf{k},\lambda} \frac{\omega_k^2}{c^2} \left[ \mathbf{A}_{\mathbf{k},\lambda}^* \mathbf{A}_{\mathbf{k},\lambda} + \mathbf{A}_{\mathbf{k},\lambda} \mathbf{A}_{\mathbf{k},\lambda}^* \right]. \tag{7.176}$$

## 7.8.2 Photons and Energy Quantization

Our goal now is to associate (7.176) with the eigenvalues of a Hamiltonian operator. We will do this by hypothesizing that the quantized electromagnetic field is made up of a collection of identical particles called *photons*. An operator $a_\lambda^\dagger(\mathbf{k})$ creates a photon with polarization $\lambda$ and momentum $\hbar\mathbf{k}$, and $a_\lambda(\mathbf{k})$ annihilates this photon. The energy of a photon is $\hbar\omega_k = \hbar c k$, so we will build our Hamiltonian operator based on (7.126) and write

$$\mathscr{H} = \sum_{\mathbf{k},\lambda} \hbar\omega_k a_\lambda^\dagger(\mathbf{k}) a_\lambda(\mathbf{k}) + E_0 \tag{7.177}$$

where we will find it convenient to allow an arbitrary constant $E_0$ to be added to the Hamiltonian. We do not need to consider terms like (7.133) since, by our starting assumption, we are building a noninteracting electromagnetic field.

We are now faced with an important question. Are photons bosons or fermions? That is, what is the "spin" of the photon? We need to know whether it is integer or half-integer, in order to know which algebra is followed by the creation and annihilation operators. A fully relativistic treatment of the photon field demonstrates that the photon has spin one and is therefore a boson, but do we have enough preparation at this point to see that this should be the case?

Yes, we do. We know from Chapter 3 that rotation through an angle $\phi$ about (say) the $z$-axis is carried out by the operator $\exp(-iJ_z\phi/\hbar)$. The possible eigenvalues $m$ of $J_z$ show up explicitly if we rotate a state which happens to be an eigenstate of $J_z$, introducing a phase factor $\exp(-im\phi)$. (This is what gives rise to the "famous" minus sign when a spin $\frac{1}{2}$ state is rotated through $2\pi$. Recall (3.33).)

So, consider what happens if we rotate about the photon direction $\mathbf{k}$ through an angle $\phi$ for a right- or left-handed circularly polarized electromagnetic wave. The polarization directions are the unit vectors $\hat{\mathbf{e}}_{\mathbf{k}\pm}$ given by (7.169). The rotation is equivalent to the transformation

$$\hat{\mathbf{e}}_{\mathbf{k}}^{(1)} \to \hat{\mathbf{e}}_{\mathbf{k}}^{(1)\prime} = \cos\phi\,\hat{\mathbf{e}}_{\mathbf{k}}^{(1)} - \sin\phi\,\hat{\mathbf{e}}_{\mathbf{k}}^{(2)} \tag{7.178a}$$

$$\hat{\mathbf{e}}_{\mathbf{k}}^{(2)} \to \hat{\mathbf{e}}_{\mathbf{k}}^{(2)\prime} = \sin\phi\,\hat{\mathbf{e}}_{\mathbf{k}}^{(1)} + \cos\phi\,\hat{\mathbf{e}}_{\mathbf{k}}^{(2)} \tag{7.178b}$$

which means that the rotation introduces a phase change $\exp(\mp i\phi)$ to the $\hat{\mathbf{e}}_{\mathbf{k}\pm}$. Apparently, right- and left-handed circularly polarized photons correspond to eigenvalues $\pm 1\hbar$ of $J_z$. The photon seems to have spin one.

Consequently, we proceed under the assumption that photons are bosons. We therefore rewrite (7.177) slightly as

$$\mathcal{H} = \sum_{\mathbf{k},\lambda} \hbar\omega_k \frac{1}{2} \left[ a_\lambda^\dagger(\mathbf{k})a_\lambda(\mathbf{k}) + a_\lambda^\dagger(\mathbf{k})a_\lambda(\mathbf{k}) \right] + E_0$$

$$= \sum_{\mathbf{k},\lambda} \hbar\omega_k \frac{1}{2} \left[ a_\lambda^\dagger(\mathbf{k})a_\lambda(\mathbf{k}) + a_\lambda(\mathbf{k})a_\lambda^\dagger(\mathbf{k}) - 1 \right] + E_0 \tag{7.179}$$

and recover the classical energy (7.176) with the definition of the *operator*

$$\mathbf{A}_{\mathbf{k},\lambda} = (4\pi\hbar c^2)^{1/2} \frac{1}{\sqrt{V}} \frac{1}{\sqrt{2\omega_k}} a_\lambda(\mathbf{k}) \tag{7.180}$$

along with the realization of a "zero point" energy

$$E_0 = \frac{1}{2}\sum_{\mathbf{k},\lambda} \hbar\omega_k = \sum_{\mathbf{k}} \hbar\omega_k. \tag{7.181}$$

This is the energy in the electromagnetic field when there are *zero* photons present, and is sometimes called the *vacuum energy*. It is an infinite number, but nevertheless a constant. More important, it has observable consequences.

### 7.8.3  The Casimir Effect

The vacuum energy of the electromagnetic field has a number of physical consequences, but probably the most dramatic is its ability to exert a macroscopic force between conducting surfaces. This is called the Casimir effect, and it has been precisely measured and compared to calculations. Fine accounts by S. Lamoreaux have been published, including a popular article in *Physics Today* (**60** (2007) 40), and a more technical review *Rep. Prog. Phys.*, **68** (2005) 201. See also M. Fortun, "Fluctuating about zero, taking nothing's measure," in Marcus (2000).

Casimir's calculation relies only on the assumption of the vacuum energy (7.181). We reproduce it here, following Lamoreaux's technical review article,[12] above. Two large, parallel, conducting plates are separated by a distance $d$. Define a coordinate system where the $(x, y)$ plane is parallel to the surface of the conducting plates, so $z$ measures the distance perpendicularly away from one surface. This allows us to write down a potential energy function

$$U(d) = E_0(d) - E_0(\infty) \tag{7.182}$$

which just gives the difference in the vacuum energy for plates with a finite and infinite separation. Combining (7.181) with (7.172) (and combining positive and negative integer values) we have

$$E_0(d) = \hbar \sum_{k_x, k_y, n} \omega_k = \hbar c \sum_{k_x, k_y, n} \sqrt{k_x^2 + k_y^2 + \left(\frac{n\pi}{d}\right)^2}. \tag{7.183}$$

(This equation actually is missing a "lost" factor of $1/2$ on the $n = 0$ term. This is because only one polarization state should be counted in (7.181) for $n = 0$, since there is only one purely transverse mode when $k_z = 0$. We will recover this factor below.) Now assume square plates with $x$ and $y$ lengths $L \gg d$. Since $L$ is large, we can replace the summations over $k_x$ and $k_y$ with integrals and write

$$E_0(d) = \hbar c \left(\frac{L}{\pi}\right)^2 \int_0^\infty dk_x \int_0^\infty dk_y \sum_n \sqrt{k_x^2 + k_y^2 + \left(\frac{n\pi}{d}\right)^2}. \tag{7.184}$$

For the limit $d \to \infty$ we can also replace the sum over $n$ with an integral. This gives us all the necessary ingredients to evaluate (7.182).

Unfortunately, however, (7.182) is the difference between two infinite numbers. It is plausible that the difference is finite, since for any particular value of $d$, terms with large enough $n$ will give the same result for different values of $d$. That is, both terms in (7.182) should tend towards infinity in the same way, and these parts will cancel when taking the difference.

This suggests that we can handle the infinities by multiplying the integrand in (7.184) by a function $f(k)$ where $f(k) \to 1$ for $k \to 0$ and $f(k) \to 0$ for $k \to \infty$. This function "cuts off" the integrand before it gets too large, but does so in the same way to both terms in (7.182) so that the contributions from large $k$ still cancel.[13] It is also helpful to introduce the polar

---

[12] We note that Lamoreaux's derivation closely follows that of Itzykson and Zuber (1980), Section 3-2-4. See also Holstein (1992) for a somewhat different approach, and attendant discussion, with a particularly physical perspective.

[13] We can think of many physical reasons why there should be a cutoff at very high frequencies. In general, we expect the main contributions to come from values of $k \sim 1/d$, but there are more specific examples such as the response of electrons in metals to very high-energy photons. It remains an interesting problem, in any case, to see whether the eventual result can be derived even if there is no cutoff frequency.

coordinate $\rho = \sqrt{k_x^2 + k_y^2}$ in which case $dk_x dk_y = 2\pi \rho d\rho$. Note that the integration limits in (7.184) correspond to 1/4 of the $(k_x, k_y)$ plane. Then (7.182) becomes

$$U(d) = 2\pi \hbar c \left(\frac{L}{\pi}\right)^2 \frac{1}{4} \int_0^\infty \rho d\rho \left[\sum_n f\left(\sqrt{\rho^2 + \left(\frac{n\pi}{d}\right)^2}\right) \sqrt{\rho^2 + \left(\frac{n\pi}{d}\right)^2}\right.$$
$$\left. - \frac{d}{\pi} \int_0^\infty dk_z f\left(\sqrt{\rho^2 + k_z^2}\right) \sqrt{\rho^2 + k_z^2}\right]. \tag{7.185}$$

Now define a function $F(\kappa)$ as

$$F(\kappa) = \int_0^\infty dx\, f\left(\frac{\pi}{d}\sqrt{x + \kappa^2}\right) \sqrt{x + \kappa^2} \tag{7.186a}$$

$$= \int_\kappa^\infty 2y^2 f\left(\frac{\pi}{d}y\right) dy. \tag{7.186b}$$

Putting $\rho^2 = (\pi/d)^2 x$ and $k_z = (\pi/d)\kappa$ allows us to write the potential energy more succinctly, reclaiming the lost factor of two, as

$$U(d) = \frac{\pi^2 \hbar c}{4d^3} L^2 \left[\frac{1}{2}F(0) + \sum_{n=1}^\infty F(n) - \int_0^\infty F(\kappa)d\kappa\right]. \tag{7.187}$$

We are therefore left with evaluating the difference between an integral and a sum, both of which are reasonable approximations of each other. Indeed, if a function $F(x)$, defined over range $0 \le x \le N$, is evaluated at integer points $x = i$, then the approximation scheme known as the trapezoidal rule says that

$$\int_0^N F(x)dx \approx \frac{F(0) + F(N)}{2} + \sum_{i=1}^N F(i). \tag{7.188}$$

In our case, $N \to \infty$ with $F(N) \to 0$ (thanks to the cutoff function $f(k)$), and our job is to find the difference between the left- and right-hand sides of (7.188).

Fortunately, there is a theorem which evaluates this difference. It is called the Euler–Maclaurin summation formula and can be written as

$$\frac{F(0)}{2} + \sum_{i=1}^\infty F(i) - \int_0^\infty F(x)dx = -\frac{1}{12}F'(0) + \frac{1}{720}F'''(0) + \cdots. \tag{7.189}$$

The derivatives can be calculated using (7.186b). Since $F(x) \to 0$ as $x \to \infty$, we have

$$F'(y) = -2y^2 f\left(\frac{\pi}{d}y\right) \tag{7.190}$$

which gives $F'(0) = 0$. If we make one further, but natural, assumption about the cutoff function $f(k)$, namely that all of its derivatives go to zero as $k \to 0$, then we are only left with the third derivative term in (7.189). In fact, $F'''(0) = -4$ and

$$U(d) = \frac{\pi^2 \hbar c}{4d^3} L^2 \left[\frac{-4}{720}\right] = -\frac{\pi^2 \hbar c}{720d^3} L^2. \tag{7.191}$$

So, finally, we derive the Casimir force (per unit area) to be

$$\mathscr{F}(d) = \frac{1}{L^2}\left(-\frac{dU}{dd}\right) = -\frac{\pi^2 \hbar c}{240d^4}. \tag{7.192}$$

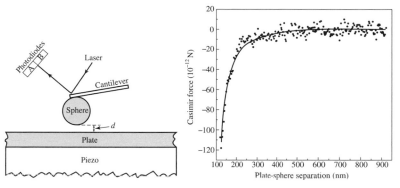

**Fig. 7.10** Experimental verification of the Casimir effect, from Mohideen and Anushree *Phys. Rev. Lett.*, **81** (1998) 4549. For experimental reasons, the force is measured between a metallic sphere and a flat plate, rather than two flat plates. A laser precisely measures the small deflection, from which the force is deduced. The force (measured in $10^{-12}$ N) varies as a function of separation between the sphere and the plate, in excellent agreement with the prediction, shown by the line through the data points, based on a quantized electromagnetic field.

Thus, there is an attractive force between the plates which varies as the inverse fourth power of the separation, due to the reconfiguration of the vacuum energy in the quantized electromagnetic field.

This is one of the examples in nature where a purely quantum-mechanical effect manifests itself in a macroscopic system. Indeed, the Casimir force between conductors has been precisely measured and the result is in excellent agreement with the theory. See Figure 7.10. This experiment makes use of the "atomic force microscope" concept, which relies on the bending of a microscopic cantilever beam in response to a tiny force between nearby surfaces. For this reason, an apparatus is used which suspends a small conducting sphere from the cantilever, and measures the force between the sphere and a flat plate, given by $-(\pi^3 R/360)(\hbar c/d^3)$ where $R$ is the sphere radius. The force deflects the cantilever, and this motion is detected using a laser which reflects from the sphere surface. The measured force as a function of the distance $d$ is shown in the figure as data points, and is compared with the theoretical prediction.

The Casimir effect has also been observed experimentally with parallel conducting surfaces. See, for example, Bressi et al., *Phys. Rev. Lett.*, **88** (2002) 041804.

If the Casimir Effect is due to the presence of electromagnetic fields, and these fields interact only with charges, then why does the electric charge $e$ not appear anywhere in (7.192)? The answer lies in our starting point for the calculation, where we assumed the boundary conditions for conducting plates. These arise from the relative mobility of the electrons in the metal, with which the electromagnetic field interacts. In fact, we made use of a cutoff frequency whose physical basis can lie in the penetrability of electromagnetic radiation at short wavelengths. Indeed, if this penetrability existed for all wavelengths, there would be no Casimir effect.

The Casimir effect has seen renewed interest in recent years, not only for its potential application in nanomechanical devices, but also for its calculation and interpretation[14] using fundamental quantum field theoretical principles. In a formulation in terms of path integrals, the Casimir energy can be written down in terms of the free field propagator with appropriate boundary conditions. The boundary conditions are simply defined by the objects under consideration. The result is an elegant expression for the Casimir energy in terms of the $T$-matrix scattering amplitudes for the free field from the objects, and transformation matrices that express each object's geometry in a natural way with respect to the other. This approach lends itself to a number of insights. For example, it allows one to calculate the Casimir energy for any field which can be expressed in terms of this constrained propagator, such as scalar or fermion fields. It is also clearly amenable to any number of geometries, far beyond simple parallel plates.

## 7.8.4 Concluding Remarks

Before leaving this chapter, we should point out that our treatment in this section only scratches the surface of applications of quantizing the electromagnetic field. Now that we have expression (7.180) (and its adjoint) which is an operator which destroys (or creates) photons of specific wavelength and polarization, we can incorporate it in any number of ways.

For example, we have already seen in (2.343) how we can add the electromagnetic field into the conjugate momentum. This is built into the Hamiltonian in (2.346). Using the quantized version for **A** we have an ad hoc Hamiltonian operator which can create or destroy photons. Terms, then, proportional to $\mathbf{A} \cdot \mathbf{p}$ can be treated as time-dependent perturbations. Thus we can let a photon be absorbed by an atom (the photoelectric effect) or let an excited state of an atom decay spontaneously and emit a photon.

These applications, of course, can be brought to bear just as well in systems covered by nuclear physics or condensed matter physics. These topics are covered in a wide variety of books, some on quantum mechanics in general, but many in books which cover specific research areas.

One particularly fascinating direction, which in fact involves noninteracting electromagnetic fields, is **quantum optics**. This is a field that has come of age in the past few decades, spurred on partly by advances in laser technology and a growing interest in quantum computing and quantum information. A reflective view of the field is given in Roy Glauber's Nobel Prize lecture, *Rev. Mod. Phys.*, **78** (2006) 1267. In the remainder of this section, we give a very brief overview of this large subject.

A hint to the richness of quantum optics is immediately apparent. By virtue of (7.180), the electric field vector (7.173a) becomes an operator which creates and destroys photons. The expectation value of this operator vanishes in any state $|\Psi\rangle$ with a definite number of photons, that is

$$|\Psi\rangle = |\ldots, n_{\mathbf{k}\lambda}, \ldots\rangle. \tag{7.193}$$

This is simple to see, since (7.173a) changes the number of photons $n_{\mathbf{k}\lambda}$ in which case $\langle\Psi|\mathbf{E}|\Psi\rangle$ becomes the inner product between orthogonal states. Therefore *any* physical

---

[14]   There is quite a lot of recent literature. I recommend that the interested reader start with Emig and Jaffe, *J. Phys. A*, **41** (2008) 164001; Emig et al., *Phys. Rev. Lett.*, **99** (2007) 170403; and Jaffe, *Phys. Rev. D*, **72** (2005) 021301.

state needs to be a superposition of states with different number of photons. A wide variety of physical states with different properties can in principle be realized, if one can manipulate this superposition. It is the ability to carry out this manipulation which has given birth to quantum optics. Problem 2.21 of Chapter 2 suggests one possible manipulation leading to something known as a *coherent state*. Coherent states are eigenstates of the annihilation operator $a$, and therefore serve as eigenstates of positive or negative frequency parts of $\mathbf{E}$.

Let us explore one such type of manipulation of single-mode electric field operators, following Chapter Five of Loudon (2000). For a given direction of linear polarization, the electric field is given by

$$E(\chi) = E^+(\chi) + E^-(\chi) = \frac{1}{2}ae^{-i\chi} + \frac{1}{2}a^\dagger e^{i\chi} \tag{7.194}$$

where $\chi \equiv \omega t - kz - \pi/2$. (We absorb a factor of $-(8\pi\hbar\omega_k/V)^{1/2}$ into the definition of the electric field.) The phase angle $\chi$ can be adjusted experimentally. Furthermore, fields with different phase angles generally do not commute. From (2.124) it is easily shown that

$$[E(\chi_1), E(\chi_2)] = -\frac{i}{2}\sin(\chi_1 - \chi_2). \tag{7.195}$$

The uncertainty relation (1.146) therefore implies that

$$\Delta E(\chi_1)\Delta E(\chi_2) \geq \frac{1}{4}|\sin(\chi_1 - \chi_2)| \tag{7.196}$$

where the electric field variance $(\Delta E(\chi))^2$ is defined in the usual way as

$$(\Delta E(\chi))^2 = \left\langle (E(\chi))^2 \right\rangle - \langle E(\chi)\rangle^2$$
$$= \left\langle (E(\chi))^2 \right\rangle \tag{7.197}$$

since $\langle E(\chi)\rangle = 0$ for a state with a single mode. A state $|\zeta\rangle$ for which

$$0 \leq (\Delta E(\chi))^2 < \frac{1}{4} \tag{7.198}$$

is said to be *quadrature squeezed*. It is possible to write $|\zeta\rangle$ as the action of a unitary operator on the vacuum, that is,

$$|\zeta\rangle = \exp\left( \frac{1}{2}\zeta^* a^2 - \frac{1}{2}\zeta(a^\dagger)^2 \right) \tag{7.199}$$

where $\zeta = se^{i\theta}$ is called the *squeeze parameter*. In this state, the electric field variance is

$$(\Delta E(\chi))^2 = \frac{1}{4}\left\{ e^{2s}\sin^2\left( \chi - \frac{1}{2}\theta \right) + e^{-2s}\cos^2\left( \chi - \frac{1}{2}\theta \right) \right\}. \tag{7.200}$$

Thus one can achieve for $\Delta E(\chi)$ a minimum

$$\Delta E_{min} = \frac{1}{2}e^{-s} \quad \text{for} \quad \chi = \frac{\theta}{2} + m\pi \tag{7.201}$$

where $m$ is an integer, and a maximum

$$\Delta E_{max} = \frac{1}{2}e^{s} \quad \text{for} \quad \chi = \frac{\theta}{2} + \left(m + \frac{1}{2}\right)\pi. \tag{7.202}$$

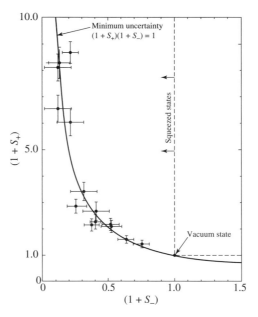

**Fig. 7.11**   Observation of states of "squeezed light," from Wu et al., *J. Opt. Soc. Am. B*, **4** (1987) 1465. (See also Chapter Five in Loudon (2000).) Data are obtained by measuring the electric field variance, that is the noise, for different scans of the phase angle $\chi$. The different points correspond to different squeezed states, formed by selecting different values of the magnitude $s$ of the squeeze parameter $\zeta$. The solid line through the points is given by (7.203).

The resulting uncertainty relation is

$$\Delta E_{\min}\Delta E_{\max} = \frac{1}{4} \qquad (7.203)$$

which satisfies (7.196) as an equality.

The observation of squeezed light is challenging, but such measurements have been carried out. See Figure 7.11. The squeezed states are prepared using an optical technique known as *parametric down conversion* which allows different magnitudes of $\zeta$ to be selected. Each point is the result of sweeping over the phase $\chi$ and measuring the noise spectrum of the electric field.

# Problems

**7.1**   Liquid helium makes a transition to a macroscopic quantum fluid, called superfluid helium, when cooled below a phase transition temperature $T = 2.17$ K. Calculate the de Broglie wavelength $\lambda = h/p$ for helium atoms with average energy at this temperature, and compare it to the size of the atom itself. Use this to predict the superfluid transition temperature for other noble gases, and explain why none of them can form superfluids. (You will need to look up some empirical data for these elements.)

**7.2**    Three identical particles are in a one-dimensional harmonic oscillator potential well with classical angular frequency $\omega$.

   a. Write the complete time-independent Hamiltonian for this system, and express it in coordinate space as a differential equation whose solution is the three-body wave function $\Psi(x_1,x_2,x_3)$.

   b. Assume the particles have zero spin. Use the single-particle wave functions to construct the ground-state wave function $\Psi_0(x_1,x_2,x_3)$, and show that it satisfies the differential equation in (a), and find the ground-state energy.

   c. Assume the particles have spin $\frac{1}{2}$. Repeat (b), and also construct the ground state spin state from single-particle spin eigenstates.

**7.3**    a. $N$ identical spin $\frac{1}{2}$ particles are subjected to a one-dimensional simple harmonic oscillator potential. Ignore any mutual interactions between the particles. What is the ground-state energy? What is the Fermi energy?

   b. What are the ground-state and Fermi energies if we ignore the mutual interactions and assume N to be very large?

**7.4**    It is obvious that two nonidentical spin 1 particles with no orbital angular momenta (that is, $s$-states for both) can form $j = 0$, $j = 1$, and $j = 2$. Suppose, however, that the two particles are *identical*. What restrictions do we get?

**7.5**    Discuss what would happen to the energy levels of a helium atom if the electron were a spinless boson. Be as quantitative as you can.

**7.6**    Three spin 0 particles are situated at the corners of an equilateral triangle. Let us define the $z$-axis to go through the center and in the direction normal to the plane of the triangle. The whole system is free to rotate about the $z$-axis. Using statistics considerations, obtain restrictions on the magnetic quantum numbers corresponding to $J_z$.

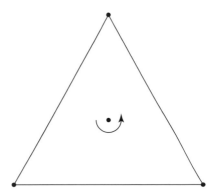

**7.7**    Consider three weakly interacting, identical spin 1 particles.

   a. Suppose the space part of the state vector is known to be symmetric under interchange of *any* pair. Using notation $|+\rangle|0\rangle|+\rangle$ for particle 1 in $m_s = +1$, particle 2 in $m_s = 0$, particle 3 in $m_s = +1$, and so on, construct the normalized spin states in the following three cases.

(i) All three of them in $|+\rangle$.

(ii) Two of them in $|+\rangle$, one in $|0\rangle$.

(iii) All three in different spin states.

What is the total spin in each case?

b. Attempt to do the same problem when the space part is antisymmetric under interchange of any pair.

**7.8** A porphyrin ring is a molecule which is present in chlorophyll, hemoglobin, and other biological compounds. It can be modeled as 18 electrons moving freely along a one-dimensional circular path of radius $R = 0.4\,\text{nm}$.

a. Using a polar angular coordinate $\theta$, write down the appropriately normalized single-particle wave functions $\psi(\theta)$, including periodic boundary conditions. Find an expression for the single-particle energy eigenvalues.

b. Find the electron configurations and energies for the ground state and first excited state of porphyrin.

c. Find a numerical value for the wavelength of electromagnetic radiation that would excite the ground state into the first excited state. This is a very simple model, and porphyrin comes in many varieties, but compare your result to an experimental result.

**7.9** Show that for an operator $a$ which, with its adjoint, obeys the anticommutation relation $\{a, a^\dagger\} = aa^\dagger + a^\dagger a = 1$, then the operator $N = a^\dagger a$ has eigenstates with the eigenvalues 0 and 1.

**7.10** Suppose the electron were a spin $\frac{3}{2}$ particle obeying Fermi–Dirac statistics. Write the configuration of a hypothetical Ne ($Z = 10$) atom made up of such "electrons" [that is, the analogue of $(1s)^2(2s)^2(2p)^6$]. Show that the configuration is highly degenerate. What is the ground state (the lowest term) of the hypothetical Ne atom in spectroscopic notation ($^{2S+1}L_J$, where $S$, $L$, and $J$ stand for the total spin, the total orbital angular momentum, and the total angular momentum, respectively) when exchange splitting and spin-orbit splitting are taken into account?

**7.11** Two identical spin $\frac{1}{2}$ fermions move in one dimension under the influence of the infinite-wall potential $V = \infty$ for $x < 0$, $x > L$, and $V = 0$ for $0 \leq x \leq L$.

a. Write the ground-state wave function and the ground-state energy when the two particles are constrained to a triplet spin state (*ortho* state).

b. Repeat (a) when they are in a singlet spin state (*para* state).

c. Let us now suppose that the two particles interact mutually via a very short range attractive potential that can be approximated by

$$V = -\lambda\delta(x_1 - x_2) \quad (\lambda > 0).$$

Assuming that perturbation theory is valid even with such a singular potential, discuss semiquantitatively what happens to the energy levels obtained in (a) and (b).

**7.12** Consider the case of a single particle of mass $m$ in a one-dimensional simple harmonic oscillator potential $V(x) = m\omega^2 x^2/2$.

    a. Using a form for $\tilde{\rho}(x)$ proportional to $\exp(-ax^2)$, calculate the energy functional (7.69) in terms of $a$. Minimize the result with respect to $a$ and show that you get the correct ground-state energy.

    b. Now show that a form proportional to $\exp(-ax^2)$ is a solution to the differential equation (7.76), for the appropriate value of $a$. Check that (7.69) gives you the correct ground-state energy, and show that it equals the value of $\mu$.

**7.13** Show by explicit construction for $N = 2$ that the normalization condition (7.80) is met for $n(\mathbf{x})$ as defined by (7.89) with the multiparticle wave function (7.88) for the cases when

    a. $\phi_1(\mathbf{x})$ and $\phi_2(\mathbf{x})$ are distinct orthonormal functions,

    b. $\phi_1(\mathbf{x}) = \phi_2(\mathbf{x})$.

**7.14** Using whatever code or application you prefer, fill in the details of the calculation on the helium atom in density functional theory.

    a. Confirm the dimensionless forms (7.102) and (7.103).

    b. Show that (7.104) is properly normalized.

    c. Verify (7.105).

    d. Obtain (7.108), or something very close, based on your own numerical approach.

    e. Complete the calculation and reproduce Table 7.1 and Figure 7.7.

    f. Try repeating the calculation using an initial density $n^{(0)}(r)$ that is smarter than (7.104). In fact, (7.108) shows that the first Kohn–Sham ground-state wave function is very close to the $1s$ state for $Z = Z_{\text{eff}}$. What starting density does this imply?

There are many variations to this problem you might consider, including trying different forms for the exchange-correlation energy, a larger or smaller set of basis states, or more iterations or different starting densities.

**7.15** Prove the relations (7.170), and then carry through the calculation to derive (7.176).

**7.16** A Hamiltonian for a system of bosons has the form

$$\mathcal{H} = \sum_{\mathbf{k}} T(\mathbf{k}) a_{\mathbf{k}}^{\dagger} a_{\mathbf{k}} + \lambda \sum_{\mathbf{l}} \sum_{\mathbf{m}} V(\mathbf{l} + \mathbf{m}) a_{\mathbf{l}}^{\dagger} a_{-\mathbf{l}}^{\dagger} a_{\mathbf{m}} a_{-\mathbf{m}}$$

where $\lambda$ is a constant. Prove that the number operator

$$\mathcal{N} = \sum_{\mathbf{k}} a_{\mathbf{k}}^{\dagger} a_{\mathbf{k}}$$

is a constant of the motion.

# Relativistic Quantum Mechanics

This final chapter provides a succinct look at how one constructs single-particle wave equations that are consistent with special relativity.

To be sure, this effort is ultimately doomed to failure. Special relativity makes it possible to create particles out of energy, but much of our development of quantum mechanics was based on the conservation of probability, so we cannot expect to be entirely successful. The right way to attack this problem is by starting with the postulates of quantum mechanics and building a many-body theory of fields that is relativistically consistent. Nevertheless, at energies low compared to the masses involved, we can expect single-particle quantum mechanics to be a very good approximation to nature. Furthermore, this is a natural way to develop the nomenclature and mathematics of a relativistic field theory.

We will start with the general problem of forming a relativistic wave equation for a free particle. This leads more or less intuitively to the Klein–Gordon equation, which we will discuss in some detail. Along the way, we introduce and continue to use the concepts of natural units, and of relativistically covariant notation. Then, we will go through Dirac's approach to finding a relativistic wave equation that is linear in space-time derivatives, not quadratic. A study of the symmetries of the Dirac equation is presented. The chapter concludes with the solution of the one-electron atom problem and its comparison with data.

This material is of course covered by many other authors, but one nice reference, written at the time when relativistic field theory was emerging out of decades of relativistic quantum mechanics, is "Elementary relativistic wave mechanics of spin 0 and spin 1/2 particles," Feshbach and Villars, *Rev. Mod. Phys.*, **30** (1958) 24.

## 8.1 Paths to Relativistic Quantum Mechanics

The early part of the twentieth century saw the more or less simultaneous developments of both relativity and quantum theory. Therefore, it is not surprising to learn that early attempts to develop wave mechanics produced relativistic wave equations.[1] Although we now understand the many pitfalls which confounded these early pioneers, it took many decades to sort things out.

We begin by focussing on the Hamiltonian operator, the (Hermitian) generator of time translations which led us to the Schrödinger equation (2.25) for the time evolution of a state. That is, a state $|\psi(t)\rangle$ evolves in time according to the equation

[1] See Volume I, Section 1.1 of Weinberg (1995).

$$i\hbar\frac{\partial}{\partial t}|\psi(t)\rangle = H|\psi(t)\rangle. \tag{8.1}$$

We interpret the eigenvalues of the Hamiltonian, of course, as the allowed energies of the system. This is where we can start to incorporate special relativity.

## 8.1.1 Natural Units

This is a good time to graduate to the use of so-called *natural units*, that is, units in which $\hbar = c = 1$. Most people react to this with bewilderment when they first see it, but it is in fact very simple and useful.

First consider the consequences of setting $c = 1$. Then, we measure time ($=$ distance/$c$) in length units, like meters or centimeters. If you really need to know the value of time in seconds, just divide by $c = 3 \times 10^{10}$ cm/sec. Velocity becomes a dimensionless number, which we typically denote by $\beta$.

Setting $c = 1$ means that we also measure both momentum and mass in units of energy, like eV or MeV. Frequently, one puts in the $c$ explicitly and writes momentum units as MeV/$c$ and mass as MeV/$c^2$. Most physicists know that the electron mass, for example, is 0.511 MeV/$c^2$, but very few know this value in kilograms without doing the conversion arithmetic! Just don't be surprised if someone tells you that the mass is 0.511 MeV, and leaves off the $c^2$.

Now consider what happens when we set $\hbar = 1$ as well. This ties together units for length and units for energy. For example, the canonical commutation relation for the momentum and position operators says that their product has the same units as $\hbar$. Therefore, we would measure position in units of MeV$^{-1}$, or some other inverse energy unit.

Remember that you can always put back the $\hbar$ and $c$ in the right places if you need to go back to the old way of doing things. It is not uncommon to do this if you are trying to evaluate the result of some experiment, for example. It is handy to keep in mind that, to a very good approximation, $\hbar c = 200$ MeV·fm for doing these conversions.

As a final note, we point out that in a textbook on statistical mechanics, one would also "naturally" set Boltzmann's constant $k = 1$. That is, temperature would be measured in units of energy as well.

## 8.1.2 The Energy of a Free Relativistic Particle

Consider the energy of a free particle with momentum $p = |\mathbf{p}|$ and mass $m$, namely

$$E_p = +\sqrt{p^2 + m^2}. \tag{8.2}$$

We need to come up with a Hamiltonian which yields this energy eigenvalue for a state $|\mathbf{p}\rangle$ with momentum eigenvalue $\mathbf{p}$. It is the square root, however, which plagued early efforts to come up with a relativistic wave equation, and which we must figure out how to deal with here.

We have faced transcendental functions of operators before, such as $U(t) = \exp(-iHt)$, interpreting them in terms of their Taylor expansions. We could take the same approach here, and write

$$H = \sqrt{p^2 + m^2} = m \left[1 + \frac{p^2}{m^2}\right]^{1/2}$$

$$= m + \frac{p^2}{2m} - \frac{p^4}{8m^3} + \frac{p^6}{16m^5} + \cdots. \tag{8.3}$$

In fact, this would be a viable way to proceed, but it has some serious shortcomings. For one, it would make it impossible to formulate a "covariant" wave equation. That is, if we formed a coordinate space (or momentum space) representation of a state vector $|\psi\rangle$, the resulting wave equation would have one time derivative and an infinite series of increasing spatial derivatives from the momentum operator. There would be no way to put time and space on an "equal footing," so to speak.

This consideration actually leads to a more important problem. Let us go ahead and try to build this wave equation. From (8.1) we have

$$i\frac{\partial}{\partial t}\langle \mathbf{x}|\psi(t)\rangle = \int d^3p \langle \mathbf{x}|\mathbf{p}\rangle\langle \mathbf{p}|H|\psi(t)\rangle$$

$$= \int d^3x' \int d^3p \langle \mathbf{x}|\mathbf{p}\rangle\langle \mathbf{p}|\mathbf{x}'\rangle\langle \mathbf{x}'|E_p|\psi(t)\rangle$$

$$= \int d^3x' \int d^3p \frac{e^{i\mathbf{p}\cdot(\mathbf{x}-\mathbf{x}')}}{(2\pi)^3}\langle \mathbf{x}'|E_p|\psi(t)\rangle \tag{8.4}$$

and (8.3) means that $\langle \mathbf{x}'|E_p|\psi(t)\rangle$ becomes an infinite series of ever higher order derivatives; see (1.252). This renders this wave equation *nonlocal* since it must reach further and further away from the region near $\mathbf{x}'$ in order to evaluate the time derivative. Eventually, causality will be violated for any spatially localized wave function $\langle \mathbf{x}|\psi(t)\rangle$. The loss of covariance costs us a great deal indeed.

We abandon this approach, and instead work with the *square* of the Hamiltonian, instead of the Hamiltonian itself. This removes the problem of the square root, and all of its attendant problems, but it will introduce a different problem. There will be solutions to the wave equation with negative energies, necessary to form a complete set of basis states, but which have no obvious physical meaning. Nevertheless, this approach is more useful than the one we are now leaving.

## 8.1.3 The Klein–Gordon Equation

Start with (8.1) and take the time derivative once more. That is

$$-\frac{\partial^2}{\partial t^2}|\psi(t)\rangle = i\frac{\partial}{\partial t}H|\psi(t)\rangle = H^2|\psi(t)\rangle. \tag{8.5}$$

We can now write down a simple wave equation for $\Psi(\mathbf{x},t) \equiv \langle \mathbf{x}|\psi(t)\rangle$. Taking $H^2 = p^2 + m^2$ and using $\langle \mathbf{x}|p^2|\psi(t)\rangle = -\nabla^2\Psi(\mathbf{x},t)$, we obtain

$$\left[\frac{\partial^2}{\partial t^2} - \nabla^2 + m^2\right]\Psi(\mathbf{x},t) = 0. \tag{8.6}$$

Equation (8.6) is known as the *Klein–Gordon equation*. This looks very much like a classical wave equation, except for the $m^2$ term. Putting back our $\hbar$ and $c$, we see that this term introduces a length scale $\hbar/mc$, called the Compton wavelength.

The Klein–Gordon equation has nearly all the desirable qualities of a relativistic wave equation. Firstly, it is relativistically covariant. You can see that because a Lorentz transformation leaves the square of the space-time interval $ds^2 = dt^2 - d\mathbf{x}^2$ invariant. Therefore, the combination of derivatives in (8.6) is the same if one changes frames from $(\mathbf{x}, t)$ to $(\mathbf{x}', t')$. In other words, $\Psi(\mathbf{x}', t')$ solves the same equation as $\Psi(\mathbf{x}, t)$.

Relativistic covariance is easier to see if one uses relativistic covariant notation. We will use a notation which has become standard. That is, Greek indices run $0, 1, 2, 3$ and Latin indices run $1, 2, 3$. If an index is ever repeated in an expression, summation over that index is implied. A contravariant four-vector $a^\mu \equiv (a^0, \mathbf{a})$ has a dual covariant vector $a_\mu = \eta_{\mu\nu} a^\nu$, where $\eta_{00} = +1$, $\eta_{11} = \eta_{22} = \eta_{33} = -1$, and all other elements are zero. Thus $a_\mu = (a^0, -\mathbf{a})$. Inner products of four-vectors can only be taken between a contravariant vector and a covariant vector, e.g. $a^\mu b_\mu = a^0 b^0 - \mathbf{a} \cdot \mathbf{b}$. In particular, $a^\mu a_\mu = \left(a^0\right)^2 - \mathbf{a}^2$.

A key point of Lorentz transformations is that inner products of four-vectors are invariant. That is $a^\mu b_\mu$ will have the same value in any reference frame. This is the reason that covariant notation is very useful for demonstrating the covariance of a particular expression.

The space-time position four-vector is $x^\mu = (t, \mathbf{x})$. This gives the four-gradient

$$\frac{\partial}{\partial x^\mu} = \left(\frac{\partial}{\partial t}, \mathbf{\nabla}\right) \equiv \partial_\mu \tag{8.7}$$

which is a *covariant* vector operator, despite the positive sign in front of the spacelike part. Now, the covariance of (8.6) is absolutely clear. The Klein–Gordon equation becomes

$$\left[\partial_\mu \partial^\mu + m^2\right] \Psi(\mathbf{x}, t) = 0. \tag{8.8}$$

Sometimes, an even further economy of notation is used by writing $\partial^2 \equiv \partial_\mu \partial^\mu$.

Another desirable property of the Klein–Gordon equation is that it has solutions that are in fact what we expect for a free relativistic particle of mass $m$. We expect the time dependence to be like $\exp(-iEt)$, where $E$ is an eigenvalue of the Hamiltonian. We also expect the spatial dependence to be that of a plane wave, that is $\exp(+i\mathbf{p} \cdot \mathbf{x})$ for momentum $\mathbf{p}$. In other words, our solution should be

$$\Psi(\mathbf{x}, t) = N e^{-i(Et - \mathbf{p} \cdot \mathbf{x})} = N e^{-i p^\mu x_\mu} \tag{8.9}$$

where $p^\mu = (E, \mathbf{p})$. Indeed (8.9) solves (8.8) so long as

$$-p^\mu p_\mu + m^2 = -E^2 + \mathbf{p}^2 + m^2 = 0 \tag{8.10}$$

or $E^2 = E_p^2$. Thus the energy eigenvalues $E = +E_p$ are included, as they should be. On the other hand, the negative energy eigenvalues $E = -E_p$ are also included. This was a serious stumbling block in the historical development of relativistic quantum mechanics, but we will take up a practical explanation of it shortly.

Schrödinger's nonrelativistic wave equation has a very important property, namely that it implies that probability is conserved. The probability density $\rho(\mathbf{x}, t) = \psi^* \psi$ (2.189) is a positive definite quantity, and the probability flux (2.191) obeys a continuity equation (2.190) which proves that the probability density can only be influenced by the flux into or out of a particular region.

One would like to identify analogous expressions using the Klein–Gordon equations, so that the wave function $\Psi(\mathbf{x}, t)$ can be similarly interpreted. The form of the continuity equation strongly suggests that we construct a four-vector current $j^\mu$ with the property $\partial_\mu j^\mu = 0$, with the probability density $\rho \equiv j^0$. In fact, if we follow (2.191) to write

$$j^\mu = \frac{i}{2m} \left[ \Psi^* \partial^\mu \Psi - (\partial^\mu \Psi)^* \Psi \right] \tag{8.11}$$

then it is easy to show that $\partial_\mu j^\mu = 0$. Therefore we calculate a density

$$\rho(\mathbf{x}, t) = j^0(\mathbf{x}, t) = \frac{i}{2m} \left[ \Psi^* \frac{\partial \Psi}{\partial t} - \left( \frac{\partial \Psi}{\partial t} \right)^* \Psi \right]. \tag{8.12}$$

Although this density is conserved, it is *not* positive definite! This was a tremendous problem in the development of relativistic quantum mechanics, since it rendered the standard probabilistic interpretation of the wave function impossible. Eventually a consistent physical interpretation was found. Before discussing this interpretation, though, we need to consider the effect of electromagnetic interactions within the context of our relativistic framework.

The explicitly covariant nature of the Klein–Gordon equation makes it straightforward to add electromagnetic interactions into the Hamiltonian. See Section 2.7, especially (2.343) and (2.346). As before, we assume that the particle has an electric charge $e < 0$. In a classical Hamiltonian, one simply makes the substitutions[2] $E \rightarrow E - e\Phi$ and $\mathbf{p} \rightarrow \mathbf{p} - e\mathbf{A}$, where $\Phi$ is the "scalar" electric potential and $\mathbf{A}$ is the vector potential. In covariant form, this becomes

$$p^\mu \rightarrow p^\mu - eA^\mu \tag{8.13}$$

where $A^\mu = (\Phi, \mathbf{A})$, so $A_\mu = (\Phi, -\mathbf{A})$. This all amounts to rewriting (8.8) as

$$\left[ D_\mu D^\mu + m^2 \right] \Psi(\mathbf{x}, t) = 0 \tag{8.14}$$

where $D_\mu \equiv \partial_\mu + ieA_\mu$. We refer to $D_\mu$ as the covariant derivative.

Unlike the nonrelativistic Schrödinger wave equation, the Klein–Gordon equation is second order in time derivatives, not first order. That implies that not only must one specify $\Psi(\mathbf{x}, t)|_{t=0}$ for its solution, but also $\partial \Psi(\mathbf{x}, t) / \partial t|_{t=0}$. Consequently, more information is necessary than one might have originally expected based on our experience from nonrelativistic quantum mechanics. In fact, this additional "degree of freedom" shows up as the sign of the charge of the particle. This is clear by noting that if $\Psi(\mathbf{x}, t)$ solves (8.14), then $\Psi^*(\mathbf{x}, t)$ solves the same equation, but with $e \rightarrow -e$.

More explicitly, we can reduce the second-order Klein–Gordon equation to two first-order equations, and then interpret the result in terms of the sign of the electric charge.

---

[2] It is worthwhile to take a moment and review the origin of these substitutions. A Lagrangian $L$ is constructed which yields the Lorentz force law, $\mathbf{F} = e[\mathbf{E} + \mathbf{v} \times \mathbf{B}/c]$. For a coordinate $x_i$, the canonical momentum is $p_i \equiv \partial L / \partial \dot{x}_i = m\dot{x}_i + eA_i$. Hence, the kinetic energy uses the "kinematic momentum" $m\dot{x}_i = p_i - eA_i$. See Appendix C of this book, or Taylor (2005), Section 7.9 for more details. Extension to relativistic kinematics is relatively straightforward. The four-momentum $p^\mu$ is replaced by $p^\mu - eA^\mu$; see Jackson (1998), Section 12.1A. When working in coordinate space, the quantum-mechanical operator for the (covariant vector) $p_\mu = (E, -\mathbf{p})$ is $i\partial_\mu = (i\partial_t, i\nabla)$. Therefore, to incorporate electromagnetism, we replace $i\partial_\mu$ with $i\partial_\mu - eA_\mu = i(\partial_\mu + ieA_\mu) \equiv iD_\mu$.

Using a rather obvious notation in which $D_\mu D^\mu = D_t^2 - \mathbf{D}^2$, we can easily write (8.14) as two equations, each first order in time, defining two new functions

$$\phi(\mathbf{x},t) = \frac{1}{2}\left[\Psi(\mathbf{x},t) + \frac{i}{m}D_t\Psi(\mathbf{x},t)\right] \tag{8.15a}$$

$$\chi(\mathbf{x},t) = \frac{1}{2}\left[\Psi(\mathbf{x},t) - \frac{i}{m}D_t\Psi(\mathbf{x},t)\right] \tag{8.15b}$$

so that instead of specifying $\Psi(\mathbf{x},t)|_{t=0}$ and $\partial\Psi(\mathbf{x},t)/\partial t|_{t=0}$, we can specify $\phi(\mathbf{x},t)|_{t=0}$ and $\chi(\mathbf{x},t)|_{t=0}$. Furthermore, $\phi(\mathbf{x},t)$ and $\chi(\mathbf{x},t)$ satisfy the coupled equations

$$iD_t\phi = -\frac{1}{2m}\mathbf{D}^2(\phi+\chi) + m\phi \tag{8.16a}$$

$$iD_t\chi = +\frac{1}{2m}\mathbf{D}^2(\phi+\chi) - m\chi \tag{8.16b}$$

which bear a striking resemblance to the nonrelativistic Schrödinger equation. We can demonstrate this resemblance even more keenly be defining a two-component object $\Upsilon(\mathbf{x},t)$ in terms of the two functions $\phi(\mathbf{x},t)$ and $\chi(\mathbf{x},t)$, and using the Pauli matrices (3.50). That is, for the functions $\phi(\mathbf{x},t)$ and $\chi(\mathbf{x},t)$ which satisfy (8.16), we define a column vector function

$$\Upsilon(\mathbf{x},t) \equiv \left[\begin{array}{c} \phi(\mathbf{x},t) \\ \chi(\mathbf{x},t) \end{array}\right]. \tag{8.17}$$

We now write the Klein–Gordon equation as

$$iD_t\Upsilon = \left[-\frac{1}{2m}\mathbf{D}^2(\tau_3 + i\tau_2) + m\tau_3\right]\Upsilon \tag{8.18}$$

(Note that we use $\tau$ rather than $\sigma$ to denote the Pauli matrices, to avoid any confusion with the concept of spin.) Equation (8.18) is completely equivalent to our formulation in (8.14), but it is a first-order differential equation in time. We have "hidden" the additional degree of freedom in the two-component nature of $\Upsilon(\mathbf{x},t)$.

Now let us return to the question of probability current density. Having rewritten the Klein–Gordon equation using the covariant derivative as (8.14), the correct form of the conserved current is now

$$j^\mu = \frac{i}{2m}\left[\Psi^* D^\mu\Psi - (D^\mu\Psi)^*\Psi\right]. \tag{8.19}$$

The "probability" density (8.12) therefore becomes

$$\rho = j^0 = \frac{i}{2m}\left[\Psi^* D_t\Psi - (D_t\Psi)^*\Psi\right]$$
$$= \phi^*\phi - \chi^*\chi = \Upsilon^\dagger\tau_3\Upsilon. \tag{8.20}$$

This is easy to see by using (8.15) to write $\Psi(\mathbf{x},t)$ and $D_t\Psi$ in terms of $\phi(\mathbf{x},t)$ and $\chi(\mathbf{x},t)$.

We are therefore led to interpret $\rho$ as a probability *charge* density, where $\phi(\mathbf{x},t)$ is the wave function of a positive particle, and $\chi(\mathbf{x},t)$ is the wave function of a negative particle. That is, the Klein–Gordon equation has buried in it simultaneous degrees of freedom for a particle of a certain charge, as well as a particle that behaves identically but with the opposite charge. Before going so far as to refer to these as "particle" and "antiparticle," we should go back and consider the interpretation of the negative energy solutions.

### 8.1.4  An Interpretation of Negative Energies

First consider free particles, in which case $D_\mu = \partial_\mu$, and for which $\Upsilon(\mathbf{x}, t) \propto \exp[-i(Et - \mathbf{p} \cdot \mathbf{x})]$. Inserting this into (8.18) yields the eigenvalues $E = \pm E_p$ as it, of course, should. We find for the eigenfunctions

$$\Upsilon(\mathbf{x}, t) = \frac{1}{2(mE_p)^{1/2}} \begin{pmatrix} E_p + m \\ m - E_p \end{pmatrix} e^{-iE_p t + i\mathbf{p} \cdot \mathbf{x}} \text{ for } E = +E_p \qquad (8.21a)$$

$$\Upsilon(\mathbf{x}, t) = \frac{1}{2(mE_p)^{1/2}} \begin{pmatrix} m - E_p \\ E_p + m \end{pmatrix} e^{+iE_p t + i\mathbf{p} \cdot \mathbf{x}} \text{ for } E = -E_p \qquad (8.21b)$$

with a normalization that leads to a charge density $\rho = \pm 1$ for $E = \pm E_p$. That is, we impose the condition that a free particle with negative charge is to be associated with a particle that has negative total energy. Also, for a particle at rest, $E_p = m$ and the positive energy solution 8.21a has only an upper component (that is, $\chi(\mathbf{x}, t) = 0$), while the negative energy solution 8.21b, has only a lower component (that is, $\phi(\mathbf{x}, t) = 0$). This continues for the nonrelativistic case where $p \ll E_p$ and the positive energy solution is dominated by $\phi(\mathbf{x}, t)$ and the negative energy solution by $\chi(\mathbf{x}, t)$.

More insight to the meaning of negative energies comes from considering the probability current density $\mathbf{j}$. Making use of (3.52), (3.53), and (8.18) we have

$$\partial_t \rho = \partial_t(\Upsilon^\dagger \tau_3 \Upsilon) = (\partial_t \Upsilon^\dagger) \tau_3 \Upsilon + \Upsilon^\dagger \tau_3 (\partial_t \Upsilon)$$

$$= \frac{1}{2im} \left[ (\nabla^2 \Upsilon^\dagger)(1 + \tau_1)\Upsilon - \Upsilon^\dagger(1 + \tau_1)(\nabla^2 \Upsilon) \right]$$

$$= -\nabla \cdot \mathbf{j} \qquad (8.22)$$

where

$$\mathbf{j} = \frac{1}{2im} \left[ \Upsilon^\dagger(1 + \tau_1)(\nabla \Upsilon) - (\nabla \Upsilon^\dagger)(1 + \tau_1)\Upsilon \right].$$

In the case of a free particle, for either positive or negative energies, this reduces to

$$\mathbf{j} = \frac{\mathbf{p}}{m} \Upsilon^\dagger(1 + \tau_1)\Upsilon = \frac{\mathbf{p}}{E_p}. \qquad (8.23)$$

Now this would appear to be quite peculiar. With a normalization that imposes positive and negative charges on positive and negative energy solutions, respectively, we end up with a charge current density that is the same regardless of the sign of the charge and energy. One way to "fix" this would be to recognize that the sign of the momentum vector $\mathbf{p}$ in (8.21b) is "wrong" since we want the exponent to have the form $ip_\mu x^\mu$ in order to be relativistically invariant. We might reverse the sign of $\mathbf{p}$, therefore, for negative energy solutions, in which case (8.23) would carry the "correct" sign to account for the charge of the particle. Another way of "fixing" this problem would be to say that the negative energy particles are moving "backwards in time." This not only reverses the sign of $\mathbf{p}$, but also lets the energy be positive in the exponent of (8.21b)! We have made some contact with the popular lore of particles and antiparticles.

If we like, we can formally associate the positive energy solution $\Psi_{E>0}(\mathbf{x},t)$ to the Klein–Gordon equation with that for a "particle," and the complex conjugate of the negative energy solution $\Psi^*_{E<0}(\mathbf{x},t)$ with an "antiparticle." In this case (8.14) yields us the two equations

$$\left[(\partial_\mu - ieA_\mu)(\partial^\mu - ieA_\mu) + m^2\right]\Psi_{\text{particle}}(\mathbf{x},t) = 0 \qquad (8.24\text{a})$$

$$\left[(\partial_\mu + ieA_\mu)(\partial^\mu + ieA_\mu) + m^2\right]\Psi_{\text{antiparticle}}(\mathbf{x},t) = 0. \qquad (8.24\text{b})$$

This makes it explicitly clear how to split the solutions to the Klein–Gordon equation into two pieces that correspond individually to particles with charge $\pm e$.

It is possible to continue along these lines and solve the Klein–Gordon equation for an atomic system. In fact, the results compare well with experiment, so long as the orbiting charged particle has no spin. (See Problem 8.7 at the end of this chapter.) Nevertheless, the interpretation of the probability density and negative energies is strictly ad hoc, and not founded in the kind of first-principles derivations we espouse in this book.

More importantly, this formalism does not allow for expected relativistic effects, in particular the creation and destruction of massive particles from energy. For this, we must resort to quantum field theory. Of course, Dirac did find a way to write down a single-particle wave equation – with a different, and more palatable interpretation of negative energies – and we take this up in Section 8.2. First, though, we take a small detour to show how the conceptual problems with the Klein–Gordon wave function are solved if we interpret the solution as a quantum field.

## 8.1.5 The Klein–Gordon Field

We now return to finding a solution $\Psi(\mathbf{x},t)$ to the Klein–Gordon equation (8.6). This time, however, we interpret $\Psi(\mathbf{x},t)$ as a second-quantized field, as discussed in Section 7.7.1. We will find it convenient to write $\Psi(\mathbf{x},t)$ as a complex linear combination of two Hermitian fields, so we start by looking for a Hermitian solution $\Phi(\mathbf{x},t)$ of (8.6). We write

$$\Phi(\mathbf{x},t) = \frac{1}{L^{3/2}}\sum_{\mathbf{k}} q_{\mathbf{k}}(t)e^{i\mathbf{k}\cdot\mathbf{x}} \qquad (8.25)$$

where we interpret the $q_{\mathbf{k}}(t)$ as operators in second quantization. Once again, we use a "big box" normalization in which the $\mathbf{k}$ form a complete set of discrete wave vectors. We ensure that $\Phi(\mathbf{x},t)$ is Hermitian by requiring that

$$q^\dagger_{\mathbf{k}}(t) = q_{-\mathbf{k}}(t). \qquad (8.26)$$

Inserting (8.25) into (8.6), we find that

$$\ddot{q}_{\mathbf{k}} + \omega^2_{\mathbf{k}} q_{\mathbf{k}} = 0 \qquad (8.27)$$

where $\omega^2_{\mathbf{k}} \equiv \mathbf{k}^2 + m^2$. (Recall that we are using natural units, in which $\hbar = c = 1$.)

Equation (8.27) suggests how to quantize the Klein–Gordon field. If the $q_{\mathbf{k}}(t)$ were classical quantities, then (8.27) would imply that they behave just like the position coordinate in a simple harmonic oscillator, and we learned how to quantize the simple harmonic oscillator in Chapter 2. However, by virtue of (8.26), unlike the position coordinate, $q_{\mathbf{k}}$ is not a Hermitian operator. Nevertheless, we can derive commutation relations for the $q_{\mathbf{k}}$ from first principles of the quantum harmonic oscillator.

Consider a two-dimensional isotropic harmonic oscillator with position coordinates $x$ and $y$ and natural frequency $\omega$. (The two degrees of freedom will correspond to $q_{\mathbf{k}}$ and $q_{-\mathbf{k}}$. Note that $\omega_{\mathbf{k}} = \omega_{-\mathbf{k}}$.) In terms of their respective creation and annihilation operators, in natural units, the position coordinates are

$$x = \frac{1}{\sqrt{2m\omega}}\left(a_x + a_x^\dagger\right) \quad \text{and} \quad y = \frac{1}{\sqrt{2m\omega}}\left(a_y + a_y^\dagger\right). \tag{8.28}$$

Now define the (non-Hermitian) operators $q_\pm$ as

$$q_\pm = \sqrt{\frac{m}{2}}\left(x \pm iy\right) \tag{8.29}$$

so that $q_+^\dagger = q_-$, which is similar to (8.26). Making the definitions $a_\pm = (a_x \pm ia_y)/\sqrt{2}$, we write

$$q_+ = \frac{1}{\sqrt{2\omega}}\left(a_+ + a_-^\dagger\right). \tag{8.30}$$

Note that $a_\pm$ each satisfy the appropriate commutation relations with their respective creation operator $a_\pm^\dagger$, but also $[a_+, a_-^\dagger] = 0$. That is, the $a_+$ and $a_-$ correspond to independent quanta.

Therefore, following (8.30), the field $\Phi(\mathbf{x}, t)$ is given by (8.25) with

$$q_{\mathbf{k}}(t) = \frac{1}{\sqrt{2\omega_{\mathbf{k}}}}\left[a_{\mathbf{k}}e^{-i\omega_{\mathbf{k}}t} + a_{-\mathbf{k}}^\dagger e^{i\omega_{\mathbf{k}}t}\right] \tag{8.31a}$$

with

$$\left[a_{\mathbf{k}}, a_{\mathbf{k}'}^\dagger\right] = \delta_{\mathbf{k},\mathbf{k}'} \tag{8.31b}$$

and $[a_{\mathbf{k}}, a_{\mathbf{k}'}] = 0 = [a_{\mathbf{k}}^\dagger, a_{\mathbf{k}'}^\dagger]$. Evidently, the Klein–Gordon field specifically applies to bosons.

Of course, (8.31) are contrived. When quantizing the simple harmonic oscillator in Chapter 2, we made use of the definition of momentum as the generator of translations, and this provided the physical basis for deriving the properties of the creation and annihilation operators. We simply used this as a model for building the $q_{\mathbf{k}}$, because (8.27) shows that they behave like a classical oscillator. Therefore, we should check that implications of (8.31) are consistent with the physical interpretation of momentum.

The momentum operator, in second quantization, should be $\mathbf{P} = \sum_{\mathbf{k}} N_{\mathbf{k}}\mathbf{k} = \sum_{\mathbf{k}} \mathbf{k}a_{\mathbf{k}}^\dagger a_{\mathbf{k}}$. Showing that it is indeed the generator of translation, as we defined in Section 1.6, is equivalent to showing that $[\Phi(\mathbf{x}, t), \mathbf{P}] = -i\nabla\Phi(\mathbf{x}, t)$. Writing $\Phi(\mathbf{x}, t)$ explicitly as

$$\Phi(\mathbf{x}, t) = \frac{1}{L^{3/2}} \sum_{\mathbf{k}} \frac{1}{\sqrt{2\omega_{\mathbf{k}}}}\left(a_{\mathbf{k}}e^{-i\omega_{\mathbf{k}}t} + a_{-\mathbf{k}}^\dagger e^{i\omega_{\mathbf{k}}t}\right)e^{i\mathbf{k}\cdot\mathbf{x}} \tag{8.32}$$

we can evaluate the commutators

$$\left[a_{\mathbf{k}}, a_{\mathbf{k}'}^{\dagger} a_{\mathbf{k}'}\right] = a_{\mathbf{k}} a_{\mathbf{k}'}^{\dagger} a_{\mathbf{k}'} - a_{\mathbf{k}'}^{\dagger} a_{\mathbf{k}'} a_{\mathbf{k}} = \left[a_{\mathbf{k}}, a_{\mathbf{k}'}^{\dagger}\right] a_{\mathbf{k}'} = \delta_{\mathbf{k},\mathbf{k}'} a_{\mathbf{k}'} \tag{8.33}$$

$$\left[a_{-\mathbf{k}}^{\dagger}, a_{\mathbf{k}'}^{\dagger} a_{\mathbf{k}'}\right] = a_{-\mathbf{k}}^{\dagger} a_{\mathbf{k}'}^{\dagger} a_{\mathbf{k}'} - a_{\mathbf{k}'}^{\dagger} a_{\mathbf{k}'} a_{-\mathbf{k}}^{\dagger} = a_{\mathbf{k}'}^{\dagger} \left[a_{-\mathbf{k}}^{\dagger}, a_{\mathbf{k}'}\right] = -\delta_{-\mathbf{k},\mathbf{k}'} a_{\mathbf{k}'}^{\dagger} \tag{8.34}$$

to calculate the commutator

$$
\begin{aligned}
[\Phi(\mathbf{x},t), \mathbf{P}] &= \frac{1}{L^{3/2}} \sum_{\mathbf{k}} \sum_{\mathbf{k}'} \frac{1}{\sqrt{2\omega_{\mathbf{k}}}} \left( \left[a_{\mathbf{k}}, a_{\mathbf{k}'}^{\dagger} a_{\mathbf{k}'}\right] e^{-i\omega_{\mathbf{k}} t} + \left[a_{-\mathbf{k}}^{\dagger}, a_{\mathbf{k}'}^{\dagger} a_{\mathbf{k}'}\right] e^{i\omega_{\mathbf{k}} t} \right) \mathbf{k}' e^{i\mathbf{k}\cdot\mathbf{x}} \\
&= \frac{1}{L^{3/2}} \sum_{\mathbf{k}} \frac{1}{\sqrt{2\omega_{\mathbf{k}}}} \left( a_{\mathbf{k}} \mathbf{k} e^{-i\omega_{\mathbf{k}} t} - a_{-\mathbf{k}}^{\dagger}(-\mathbf{k}) e^{i\omega_{\mathbf{k}} t} \right) e^{i\mathbf{k}\cdot\mathbf{x}} \\
&= \frac{1}{L^{3/2}} \sum_{\mathbf{k}} \frac{1}{\sqrt{2\omega_{\mathbf{k}}}} \mathbf{k} \left( a_{\mathbf{k}} e^{-i\omega_{\mathbf{k}} t} + a_{-\mathbf{k}}^{\dagger} e^{i\omega_{\mathbf{k}} t} \right) e^{i\mathbf{k}\cdot\mathbf{x}} \\
&= -i\boldsymbol{\nabla}\Phi(\mathbf{x},t) \tag{8.35}
\end{aligned}
$$

as it should be. This gives us confidence in the physical interpretation of our construction of the Klein–Gordon field (8.32).

The Hamiltonian is simply given by the sum of the single oscillator energies, that is

$$H = \sum_{\mathbf{k}} \left(a_{\mathbf{k}}^{\dagger} a_{\mathbf{k}} + \frac{1}{2}\right) \omega_{\mathbf{k}} = \sum_{\mathbf{k}} N_{\mathbf{k}} \omega_{\mathbf{k}} + \sum_{\mathbf{k}} \frac{1}{2} \omega_{\mathbf{k}}. \tag{8.36}$$

We interpret the energy eigenvalues as the sum over the numbers of quanta, times their respective energies, plus the zero point energy. This is similar to the situation with energy in the electromagnetic field, in terms of photons, discussed in Section 7.8.

The energy eigenvalues are manifestly positive. *There is no longer an issue with negative energies!* We have solved one of the big problems of our relativistic wave equation without even trying. It remains, however, to see if we can resolve the problem of a probabilistic interpretation using our construction.

We saw in (8.20) the suggestion that charge conservation might be associated with breaking up the wave function into two complex components. To that end, let us use the Hermitian field $\Phi(\mathbf{x},t)$ to form a complex (non-Hermitian) field $\Psi(\mathbf{x},t)$ as

$$\Psi(\mathbf{x},t) \equiv \frac{1}{\sqrt{2}} [\Phi_1(\mathbf{x},t) + i\Phi_2(\mathbf{x},t)] \tag{8.37}$$

which now contains two independent sets of field operators $a_{j\mathbf{k}}$ with $j = 1,2$. That is

$$\Phi_j(\mathbf{x},t) = \frac{1}{L^{3/2}} \sum_{\mathbf{k}} \frac{1}{\sqrt{2\omega_{\mathbf{k}}}} \left[a_{j\mathbf{k}} e^{i(\mathbf{k}\cdot\mathbf{x}-\omega_{\mathbf{k}} t)} + a_{j\mathbf{k}}^{\dagger} e^{-i(\mathbf{k}\cdot\mathbf{x}-\omega_{\mathbf{k}} t)}\right] \tag{8.38a}$$

$$\left[a_{j\mathbf{k}}, a_{j'\mathbf{k}'}^{\dagger}\right] = \delta_{jj'} \delta_{\mathbf{k},\mathbf{k}'} \tag{8.38b}$$

$$\left[a_{j\mathbf{k}}, a_{j'\mathbf{k}'}\right] = 0. \tag{8.38c}$$

Note that in deriving (8.38a) from (8.32), we switched the dummy summation index in the second term from $-\mathbf{k}$ to $\mathbf{k}$. Now write $\Psi(\mathbf{x},t)$ in terms of its own field operators $b_{\mathbf{k}}$ and $c_{\mathbf{k}}$

as

$$\Psi(\mathbf{x},t) = \frac{1}{L^{3/2}} \sum_{\mathbf{k}} \frac{1}{\sqrt{2\omega_{\mathbf{k}}}} \left[ b_{\mathbf{k}} e^{i(\mathbf{k}\cdot\mathbf{x} - \omega_{\mathbf{k}}t)} + c_{\mathbf{k}}^{\dagger} e^{-i(\mathbf{k}\cdot\mathbf{x} - \omega_{\mathbf{k}}t)} \right] \tag{8.39a}$$

$$b_{\mathbf{k}} \equiv \frac{1}{\sqrt{2}} \left[ a_{1\mathbf{k}} + i a_{2\mathbf{k}} \right] \tag{8.39b}$$

$$c_{\mathbf{k}} \equiv \frac{1}{\sqrt{2}} \left[ a_{1\mathbf{k}} - i a_{2\mathbf{k}} \right]. \tag{8.39c}$$

Note that $b_{\mathbf{k}}$ and $c_{\mathbf{k}}$ act as an alternative set of field operators, with respect to $a_{1\mathbf{k}}$ and $a_{2\mathbf{k}}$. That is, for example,

$$\begin{aligned}
\left[ b_{\mathbf{k}}, b_{\mathbf{k}}^{\dagger} \right] &= \frac{1}{2} \left[ a_{1\mathbf{k}} + i a_{2\mathbf{k}}, a_{1\mathbf{k}}^{\dagger} - i a_{2\mathbf{k}}^{\dagger} \right] \\
&= \frac{1}{2} \left[ a_{1\mathbf{k}}, a_{1\mathbf{k}}^{\dagger} \right] + \frac{1}{2} \left[ a_{2\mathbf{k}}, a_{2\mathbf{k}}^{\dagger} \right] = 1,
\end{aligned} \tag{8.40a}$$

$$\begin{aligned}
\left[ b_{\mathbf{k}}, c_{\mathbf{k}}^{\dagger} \right] &= \frac{1}{2} \left[ a_{1\mathbf{k}} + i a_{2\mathbf{k}}, a_{1\mathbf{k}}^{\dagger} + i a_{2\mathbf{k}}^{\dagger} \right] \\
&= \frac{1}{2} \left[ a_{1\mathbf{k}}, a_{1\mathbf{k}}^{\dagger} \right] - \frac{1}{2} \left[ a_{2\mathbf{k}}, a_{2\mathbf{k}}^{\dagger} \right] = 0,
\end{aligned} \tag{8.40b}$$

$$b_{\mathbf{k}}^{\dagger} b_{\mathbf{k}} + c_{\mathbf{k}}^{\dagger} c_{\mathbf{k}} = a_{1\mathbf{k}}^{\dagger} a_{1\mathbf{k}} + a_{2\mathbf{k}}^{\dagger} a_{2\mathbf{k}}. \tag{8.40c}$$

In terms of $b_{\mathbf{k}}$ and $c_{\mathbf{k}}$, the Hamiltonian and momentum operators become

$$H = \sum_{\mathbf{k}} \omega_{\mathbf{k}} \left( b_{\mathbf{k}}^{\dagger} b_{\mathbf{k}} + c_{\mathbf{k}}^{\dagger} c_{\mathbf{k}} + 1 \right) \tag{8.41a}$$

$$\mathbf{P} = \sum_{\mathbf{k}} \mathbf{k} \left( b_{\mathbf{k}}^{\dagger} b_{\mathbf{k}} + c_{\mathbf{k}}^{\dagger} c_{\mathbf{k}} \right). \tag{8.41b}$$

It begs the question, why convert to the fields $b_{\mathbf{k}}$ and $c_{\mathbf{k}}$ from $a_{1\mathbf{k}}$ and $a_{2\mathbf{k}}$ if the energy and momentum are equivalently expressed in terms of either sets of quanta? The answer is that, as we will see in a moment, $b_{\mathbf{k}}$ and $c_{\mathbf{k}}$ will naturally take on the role of "particle" and "antiparticle" operators.

In fact, we are now ready to consider the conservation of charge[3] with our free Klein–Gordon field. The field $\Psi(\mathbf{x},t)$ satisfies the Klein–Gordon equation because $\Phi_1(\mathbf{x},t)$ and $\Phi_2(\mathbf{x},t)$ are solutions. Therefore, following (8.11), we know that $\partial_\mu j^\mu = 0$ for the field

$$j^\mu(\mathbf{x},t) = C \times \left[ \Psi(\partial^\mu \Psi^{\dagger}) - \Psi^{\dagger}(\partial^\mu \Psi) \right] \tag{8.42}$$

where $C$ is an arbitrary constant, so the quantity represented by the field

$$Q = \int d^3x\, j^0(\mathbf{x},t) = C \int_{V=L^3} d^3x \left[ \Psi \frac{\partial \Psi^{\dagger}}{\partial t} - \Psi^{\dagger} \frac{\partial \Psi}{\partial t} \right] \tag{8.43}$$

is conserved. (It is worth noting that (8.37) implies that $\Psi(\mathbf{x},t)$ and $\Psi^{\dagger}(\mathbf{x},t)$ commute, since $\Phi_1(\mathbf{x},t)$ and $\Phi_2(\mathbf{x},t)$ are Hermitian and commute with each other.) Inserting (8.39a), we find

---

[3] Since we are in a multiparticle field theory now, the concept of conservation of probability for a single particle is irrelevant.

$$Q = C\frac{1}{L^3} \int_{V=L^3} d^3x \sum_{\mathbf{k}} \sum_{\mathbf{k}'} \frac{1}{\sqrt{2\omega_{\mathbf{k}}}} \frac{1}{\sqrt{2\omega_{\mathbf{k}'}}}$$
$$\times \left[ \left( b_{\mathbf{k}} e^{i(\mathbf{k}\cdot\mathbf{x}-\omega_{\mathbf{k}}t)} + c_{\mathbf{k}}^\dagger e^{-i(\mathbf{k}\cdot\mathbf{x}-\omega_{\mathbf{k}}t)} \right) \left( i\omega_{\mathbf{k}'} b_{\mathbf{k}'}^\dagger e^{-i(\mathbf{k}'\cdot\mathbf{x}-\omega_{\mathbf{k}'}t)} - i\omega_{\mathbf{k}'} c_{\mathbf{k}'} e^{i(\mathbf{k}'\cdot\mathbf{x}-\omega_{\mathbf{k}'}t)} \right) \right.$$
$$\left. - \left( b_{\mathbf{k}}^\dagger e^{-i(\mathbf{k}\cdot\mathbf{x}-\omega_{\mathbf{k}}t)} + c_{\mathbf{k}} e^{i(\mathbf{k}\cdot\mathbf{x}-\omega_{\mathbf{k}}t)} \right) \left( -i\omega_{\mathbf{k}'} b_{\mathbf{k}'} e^{i(\mathbf{k}'\cdot\mathbf{x}-\omega_{\mathbf{k}'}t)} + i\omega_{\mathbf{k}'} c_{\mathbf{k}'}^\dagger e^{-i(\mathbf{k}'\cdot\mathbf{x}-\omega_{\mathbf{k}'}t)} \right) \right].$$

$$\text{(8.44)}$$

In the next step, we carry through the volume integrals. The procedure is reminiscent of our work with the electromagnetic field, and in fact invokes the same result as in (7.174). The two sums in (8.44) collapse to one, with $\mathbf{k}' = \pm\mathbf{k}$, so $\omega_{\mathbf{k}'} = \omega_{\mathbf{k}}$, and an overall factor of $L^3$ appears. We find

$$Q = C\frac{i}{2} \sum_{\mathbf{k}} \left[ b_{\mathbf{k}} b_{\mathbf{k}}^\dagger - b_{\mathbf{k}} c_{-\mathbf{k}} e^{-2i\omega_{\mathbf{k}}t} + c_{\mathbf{k}}^\dagger b_{-\mathbf{k}}^\dagger e^{2i\omega_{\mathbf{k}}t} - c_{\mathbf{k}}^\dagger c_{\mathbf{k}} \right.$$
$$\left. + b_{\mathbf{k}}^\dagger b_{\mathbf{k}} - b_{\mathbf{k}}^\dagger c_{-\mathbf{k}}^\dagger e^{2i\omega_{\mathbf{k}}t} + c_{\mathbf{k}} b_{-\mathbf{k}} e^{-2i\omega_{\mathbf{k}}t} - c_{\mathbf{k}} c_{\mathbf{k}}^\dagger \right]$$
$$= C\frac{i}{2} \sum_{\mathbf{k}} \left[ b_{\mathbf{k}} b_{\mathbf{k}}^\dagger - c_{\mathbf{k}}^\dagger c_{\mathbf{k}} + b_{\mathbf{k}}^\dagger b_{\mathbf{k}} - c_{\mathbf{k}} c_{\mathbf{k}}^\dagger \right]$$

where the cross terms cancel because the $b_{\mathbf{k}}$ and $c_{\mathbf{k}}$ commute, and we switch dummy indices $\mathbf{k}$ for $-\mathbf{k}$ in the second pair of terms. Finally, since $b_{\mathbf{k}} b_{\mathbf{k}}^\dagger = b_{\mathbf{k}}^\dagger b_{\mathbf{k}} + 1$ and $c_{\mathbf{k}} c_{\mathbf{k}}^\dagger = c_{\mathbf{k}}^\dagger c_{\mathbf{k}} + 1$, we have

$$Q = \tilde{C} \sum_{\mathbf{k}} \left[ b_{\mathbf{k}}^\dagger b_{\mathbf{k}} - c_{\mathbf{k}}^\dagger c_{\mathbf{k}} \right] \qquad \text{(8.45)}$$

where $\tilde{C} \equiv iC$ is still an arbitrary constant.

*Equation (8.45) is quite a triumph.* It says that a conserved quantity $Q$, based on the conserved current (8.42), is proportional to the total number of "$b$" quanta, minus the total number of "$c$" quanta. Of course, the physical meaning of $Q$ depends on the particular current in question, but electric charge is a good example if we incorporate Maxwell's equations. It is natural, therefore, to interpret the "$b$" quanta as "particles" and the "$c$" quanta as "antiparticles."

## 8.1.6 Summary: The Klein–Gordon Equation and the Scalar Field

A straightforward pursuit of a relativistic wave equation led us to (8.6), which gives the correct wave function for a free relativistic particle. One big problem, though, is that it implied a full complement of negative total energies, for which there is no good physical interpretation. Another big problem is that we were unable to come up with a conserved current that would provide a (necessarily positive) probability density. By adding electromagnetism in the appropriately covariant fashion, we developed some insight about these problems, but no good answers. It also allows a solution for one-electron atoms (Problem 8.7 at the end of this chapter) but that solution gives fine structure that does not agree with experiment.

We then discovered that interpreting the solution to (8.6) as a field in second quantization, rather than a wave function, led to the implication that the energies of free particles are all positive. We also found a perfectly consistent way to interpret the conserved current, in terms of "particles and antiparticles." The creation and annihilation operators are bosonic, and they have no additional degrees of freedom, so the Klein–Gordon field is necessarily spin zero, that is, a scalar field.

It would seem, then, that we have found our way out of the woods. The downside, though, is that we have traded a problem with a single degree of freedom in the one-particle wave function, to one with an infinite number of degrees of freedom. This presents a major hurdle to solving problems that include realistic interactions. Perturbation theory is generally a viable technique for arriving at a solution, covered in any textbook devoted to quantum field theory.

The problem of getting the wrong fine structure for one-electron atoms remains, however. For this, it took Dirac's insight to find a solution, and we cover that next.

## 8.2  The Dirac Equation

Many of the difficulties with interpreting the results from the Klein–Gordon equation stem from the fact that it is a second-order differential equation in time. These include a nonpositive definite probability density, and additional degrees of freedom, although both of these can be identified to some extent with particles and their oppositely charged antiparticles. Nevertheless, Dirac looked for a way to write a wave equation that is first order in time, and along the way discovered the need for $j = 1/2$ angular momentum states in nature. This also lent itself to a particularly useful interpretation of the negative energy states.

The linear differential equation we seek can be written as

$$(i\gamma^\mu \partial_\mu - m)\Psi(\mathbf{x}, t) = 0 \tag{8.46}$$

where the $\gamma^\mu$ have yet to be determined. (Of course, the constant $m$ is also yet to be determined, but it will turn out to be the mass.) We must still insist that the correct energy eigenvalues (8.10) are obtained for a free particle (8.9), as does (8.8). We can turn (8.46) into (8.8) simply by operating on it with $-i\gamma^\nu \partial_\nu - m$ to get

$$(\gamma^\nu \partial_\nu \gamma^\mu \partial_\mu + m^2)\Psi(\mathbf{x}, t) = 0 \tag{8.47}$$

and then imposing the condition that $\gamma^\nu \gamma^\mu \partial_\nu \partial_\mu = \partial^\mu \partial_\mu = \eta^{\mu\nu} \partial_\nu \partial_\mu$. This condition can be written succinctly, by reversing dummy indices to symmetrize, as

$$\frac{1}{2}\left(\gamma^\mu \gamma^\nu + \gamma^\nu \gamma^\mu\right) \equiv \frac{1}{2}\{\gamma^\mu, \gamma^\nu\} = \eta^{\mu\nu}. \tag{8.48}$$

Thus the four quantities $\gamma^\mu$, $\mu = 0, 1, 2, 3$, are not simply complex numbers, but rather entities that obey something called a Clifford algebra. Clearly this algebra implies that

$$\left(\gamma^0\right)^2 = 1 \tag{8.49a}$$

$$\left(\gamma^i\right)^2 = -1 \qquad i = 1, 2, 3 \tag{8.49b}$$

$$\gamma^\mu \gamma^\nu = -\gamma^\nu \gamma^\mu \qquad \text{if } \mu \neq \nu. \tag{8.49c}$$

Note that the anticommutation property of the $\gamma^\mu$ means that each of these matrices (as well as $\alpha$ and $\beta$, as defined below) are traceless.

Now substitute the free-particle solution (8.9) into (8.46) to find

$$\gamma^\mu p_\mu - m = 0 \tag{8.50}$$

from which we can recover the free-particle energy eigenvalues $E$. Writing (8.50) out in terms of timelike and spacelike parts, and then multiplying through by $\gamma^0$, we obtain

$$E = \gamma^0 \boldsymbol{\gamma} \cdot \mathbf{p} + \gamma^0 m. \tag{8.51}$$

This leads to the Dirac Hamiltonian, written in a traditional form. Making the definitions

$$\alpha_i \equiv \gamma^0 \gamma^i \quad \text{and} \quad \beta \equiv \gamma^0 \tag{8.52}$$

we arrive at

$$H = \boldsymbol{\alpha} \cdot \mathbf{p} + \beta m. \tag{8.53}$$

Note that if we add electromagnetism by making the substitution (8.13) into (8.50), and then setting $\mathbf{A} = 0$ and $A_0 = \Phi$, we have

$$H = \boldsymbol{\alpha} \cdot \mathbf{p} + \beta m + e\Phi \tag{8.54}$$

which controls the motion of a charged particle in an electrostatic potential $\Phi$. We will make use of this when we solve the relativistic one-electron atom in Section 8.4.1.

Which form of the Dirac equation we use, whether it be (8.1) with (8.53) or (8.54), or the covariant form (8.46) perhaps with the substituion (8.13), depends on the specific problem at hand. For example, sometimes it is easiest to use (8.53) when solving problems involving dynamics and the Dirac equation, whereas it is easier to discuss symmetries of the Dirac equation in covariant forms using $\gamma^\mu$.

The algebra (8.49) can be realized with square matrices, so long as they are at least $4 \times 4$. We know that $2 \times 2$ matrices are not large enough, for example, since the Pauli matrices $\sigma$ form a complete set along with the identity matrix. However $\{\sigma_k, 1\} = 2\sigma_k$, so this set is not large enough to realize the Clifford algebra. Therefore $\Psi(\mathbf{x}, t)$ in (8.46) would be a four-dimensional column vector. In order to keep a convention consistent with our matrix representation of states and operators, we insist that $\boldsymbol{\alpha}$ and $\beta$ are Hermitian matrices. Note that this implies that $\gamma^0$ is Hermitian, while $\boldsymbol{\gamma}$ is anti-Hermitian.

We choose to make use of the $2 \times 2$ Pauli spin matrices $\sigma$ (3.50) and we write

$$\alpha = \begin{bmatrix} 0 & \sigma \\ \sigma & 0 \end{bmatrix} \quad \text{and} \quad \beta = \begin{bmatrix} 1 & 0 \\ 0 & -1 \end{bmatrix}. \tag{8.55}$$

That is, we write these $4 \times 4$ matrices as $2 \times 2$ matrices of $2 \times 2$ matrices.

## 8.2.1 The Conserved Current

The Dirac equation immediately solves the problem of the positive definite nature of the probability density. Defining $\Psi^\dagger$ in the usual way, namely as the complex conjugate of the row vector corresponding to the column vector $\Psi$, we can show that the quantity $\rho = \Psi^\dagger \Psi$

is in fact interpretable as a probability density. First, as the sum of the squared magnitudes of all four components of $\Psi(\mathbf{x},t)$, it is positive definite.

Historically, the ability of the Dirac equation to provide a positive definite probability current was one of the main reasons it was adopted as the correct direction for relativistic quantum mechanics. An examination of its free-particle solutions led to an attractive interpretation of negative energies, and in fact to the discovery of the positron.

Second, it satisfies the continuity equation

$$\frac{\partial \rho}{\partial t} + \mathbf{V} \cdot \mathbf{j} = 0 \tag{8.56}$$

for $\mathbf{j} = \Psi^\dagger \boldsymbol{\alpha} \Psi$. (This is simple to prove. Just use the Schrödinger equation and its adjoint. See Problem 8.10 at the end of this chapter.) This means that $\rho$ can only change on the basis of a flow into or out of the immediate region of interest, and is therefore a conserved quantity.

Instead of $\Psi^\dagger$ one usually uses $\overline{\Psi} \equiv \Psi^\dagger \beta = \Psi^\dagger \gamma^0$ in forming the probability density and current. In this case $\rho = \Psi^\dagger \Psi = \Psi^\dagger \gamma^0 \gamma^0 \Psi = \overline{\Psi}\gamma^0\Psi$ and $\mathbf{j} = \Psi^\dagger \gamma^0 \gamma^0 \boldsymbol{\alpha}\Psi = \overline{\Psi}\gamma^0\boldsymbol{\alpha}\Psi$. Since $\gamma^0\boldsymbol{\alpha} = \boldsymbol{\gamma}$ by (8.52) we have

$$\frac{\partial}{\partial t}\left(\overline{\Psi}\gamma^0\Psi\right) + \mathbf{V} \cdot \left(\overline{\Psi}\boldsymbol{\gamma}\Psi\right) = \partial_\mu j^\mu = 0 \tag{8.57}$$

where

$$j^\mu = \overline{\Psi}\gamma^\mu\Psi \tag{8.58}$$

is a four-vector current. Rewriting (8.46) in terms of four-momentum as

$$(\gamma^\mu p_\mu - m)\Psi(\mathbf{x},t) = 0, \tag{8.59}$$

and also taking the adjoint of this equation and using (8.49) to insert a factor of $\gamma^0$,

$$\overline{\Psi}(\mathbf{x},t)(\gamma^\mu p_\mu - m) = 0, \tag{8.60}$$

we come to an insightful interpretation of the conserved current for a free particle. We write

$$\begin{aligned}
j^\mu &= \frac{1}{2}\left\{\left[\overline{\Psi}\gamma^\mu\right]\Psi + \overline{\Psi}\left[\gamma^\mu\Psi\right]\right\} \\
&= \frac{1}{2m}\left\{\left[\overline{\Psi}\gamma^\mu\right]\gamma^\nu p_\nu\Psi + \overline{\Psi}\gamma^\nu p_\nu\left[\gamma^\mu\Psi\right]\right\} \\
&= \frac{1}{2m}\overline{\Psi}\left[\gamma^\mu\gamma^\nu + \gamma^\nu\gamma^\mu\right]p_\nu\Psi \\
&= \frac{p^\mu}{m}\overline{\Psi}\Psi.
\end{aligned} \tag{8.61}$$

Thus, writing the usual Lorentz contraction factor as $\gamma$,

$$j^0 = \frac{E}{m}\overline{\Psi}\Psi = \gamma\left[\Psi^\dagger_{\text{up}}\Psi_{\text{up}} - \Psi^\dagger_{\text{down}}\Psi_{\text{down}}\right] \tag{8.62}$$

$$\mathbf{j} = \frac{\mathbf{p}}{m}\overline{\Psi}\Psi = \gamma\mathbf{v}\left[\Psi^\dagger_{\text{up}}\Psi_{\text{up}} - \Psi^\dagger_{\text{down}}\Psi_{\text{down}}\right]. \tag{8.63}$$

The factor of $\gamma$ is expected because of the Lorentz contraction of the volume element $d^3x$ in the direction of motion. (See Holstein (1992) for an extended discussion.) The meaning of the relative negative sign of the upper and lower components becomes clearer after we study the specific solutions for the free particle.

## 8.2.2 Free-Particle Solutions

We are now in a position to study solutions of the Dirac equation and their symmetry properties. Already we notice, however, that the wave function $\Psi(\mathbf{x},t)$ has four components, whereas the Klein–Gordon wave function $\Upsilon(\mathbf{x},t)$ has two. We will see that the additional degree of freedom in the Dirac equation is the same quantity we called "spin $\frac{1}{2}$" at the very beginning of this book. The four-component object $\Psi(\mathbf{x},t)$ is called a "spinor."

We get immediate insight as to the nature of the solutions of the Dirac equation, just by considering free particles at rest ($\mathbf{p}=\mathbf{0}$). In this case the Dirac equation is simply $i\partial_t\Psi = \beta m\Psi$. Given the diagonal form of $\beta$ (8.55) we see that there are four independent solutions for $\Psi(\mathbf{x},t)$. These are

$$\Psi_1 = e^{-imt}\begin{bmatrix} 1 \\ 0 \\ 0 \\ 0 \end{bmatrix}, \quad \Psi_2 = e^{-imt}\begin{bmatrix} 0 \\ 1 \\ 0 \\ 0 \end{bmatrix}, \quad \Psi_3 = e^{+imt}\begin{bmatrix} 0 \\ 0 \\ 1 \\ 0 \end{bmatrix}, \quad \Psi_4 = e^{+imt}\begin{bmatrix} 0 \\ 0 \\ 0 \\ 1 \end{bmatrix}.$$

(8.64)

Just as in the case of the Klein–Gordon equation, the lower half of the wave function corresponds to negative energy, and we will need to deal with its interpretation later. Both the upper and lower halves of the Dirac wave function, however, have one component that it is tempting to call "spin up" and the other we would call "spin down." This interpretation is in fact correct, but we need to be somewhat more ambitious before we can state this with confidence.

Let us go on and consider free-particle solutions with nonzero momentum $\mathbf{p}=p\hat{\mathbf{z}}$, that is, a particle moving freely in the $z$-direction. In this case, we want to solve the eigenvalue problem $H\Psi = E\Psi$ for $H = \alpha_z p + \beta m$, which is no longer diagonal in spinor space. The eigenvalue equation becomes

$$\begin{bmatrix} m & 0 & p & 0 \\ 0 & m & 0 & -p \\ p & 0 & -m & 0 \\ 0 & -p & 0 & -m \end{bmatrix}\begin{bmatrix} u_1 \\ u_2 \\ u_3 \\ u_4 \end{bmatrix} = E\begin{bmatrix} u_1 \\ u_2 \\ u_3 \\ u_4 \end{bmatrix}.$$

(8.65)

Notice that the equations for $u_1$ and $u_3$ are coupled together, as are the equations for components $u_2$ and $u_4$, but these components are otherwise independent of each other. This makes it simple to find the eigenvalues and eigenfunctions. Details are left as an exercise. (See Problem 8.11 at the end of this chapter.) From the two equations coupling $u_1$ and $u_3$, we find $E = \pm E_p$. We find the same for the two equations coupling $u_2$ and $u_4$. Once again, we find the expected "correct" positive energy eigenvalue, and also the "spurious" negative energy solution. In the case of the Dirac equation, however, a relatively palatable interpretation is forthcoming, as we shall see shortly.

First, however, let us return to the question of spin. Continue to construct the free-particle spinors. For $E = +E_p$ we can either set $u_1 = 1$ (and $u_2 = u_4 = 0$) in which case $u_3 = +p/(E_p + m)$ or $u_2 = 1$ (and $u_1 = u_3 = 0$) in which case $u_4 = -p/(E_p + m)$. In both cases, as for the Klein–Gordon equation, the upper components dominate in the nonrelativistic case. Similarly, for $E = -E_p$ the nonzero components are either $u_3 = 1$ and $u_1 = -p/(E_p + m)$ or $u_4 = 1$ and $u_2 = p/(E_p + m)$ and the lower components dominate nonrelativistically.

Now consider the behavior of the operator $\boldsymbol{\Sigma} \cdot \hat{\mathbf{p}} = \Sigma_z$ where $\boldsymbol{\Sigma}$ is the $4 \times 4$ matrix

$$\boldsymbol{\Sigma} \equiv \begin{bmatrix} \sigma & 0 \\ 0 & \sigma \end{bmatrix}. \tag{8.66}$$

We expect this operator to project out components of spin in the direction of momentum. Indeed, it is simple to see that the spin operator $\mathbf{S} = \frac{\hbar}{2}\boldsymbol{\Sigma}$ projects out positive (negative) helicity for the positive energy solution with $u_1 \neq 0$ ($u_2 \neq 0$). We find the analogous results for the negative energy solutions. In other words, the free-particle solutions do indeed behave appropriately according to the spin-up/down assignment that we have conjectured.

Putting this together, we write the positive energy solutions as

$$u_R^{(+)}(p) = \begin{bmatrix} 1 \\ 0 \\ \frac{p}{E_p + m} \\ 0 \end{bmatrix}, \qquad u_L^{(+)}(p) = \begin{bmatrix} 0 \\ 1 \\ 0 \\ \frac{-p}{E_p + m} \end{bmatrix} \qquad \text{for } E = +E_p \tag{8.67a}$$

where the subscript $R$ ($L$) stands for right (left) handedness, i.e. positive (negative) helicity. For the negative energy solutions, we have

$$u_R^{(-)}(p) = \begin{bmatrix} \frac{-p}{E_p + m} \\ 0 \\ 1 \\ 0 \end{bmatrix}, \qquad u_L^{(-)}(p) = \begin{bmatrix} 0 \\ \frac{p}{E_p + m} \\ 0 \\ 1 \end{bmatrix} \qquad \text{for } E = -E_p. \tag{8.67b}$$

These spinors are normalized to the factor $2E_p/(E_p + m)$. The free-particle wave functions are formed by including the normalization, and also the factor $\exp(-ip_\mu x^\mu)$.

## 8.2.3 Interpretation of Negative Energies

Dirac made use of the Pauli exclusion principle in order to interpret the negative energy solutions. One conjectures a "negative energy sea" that is filled with electrons, as shown in Figure 8.1. (This represents a "background" of infinite energy and infinite charge, but it is possible to imagine that we would be insensitive to both of these.) Since this fills all of the negative energy states, it is not possible for positive energy electrons to fall into negative energies. It would be possible, however, for a high-energy photon to promote electrons out of the sea into positive energies, where it would be observable. The "hole" left in the sea would also be observable, as an object will all the properties of an electron, but with positive charge.

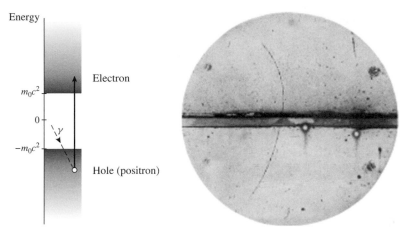

**Fig. 8.1** The figure on the left shows Dirac's interpretation of negative energy states, including the possibility that a negative energy electron can be promoted to positive energy, leaving a positively charge hole, or a "positron." The figure on the right shows the discovery of the positron, Anderson, *Phys. Rev.*, **43** (1933) 491. This cloud chamber photograph shows a particle track moving upward, bending in a known magnetic field. The direction is determined because the curvature above the lead plate is larger than below, since the particle lost energy while traversing it. The amount of energy loss is not consistent with a proton of this momentum, but is consistent with a particle having the mass of an electron.

Figure 8.1 also shows the original discovery of the positron, by Carl Anderson in 1933. After this discovery, the Dirac equation became the standard treatment of relativistic quantum mechanics, explaining the spin $\frac{1}{2}$ electron and (as we will see) its electromagnetic interactions.

### 8.2.4 Electromagnetic Interactions

We introduce electromagnetic interactions in the Dirac equation in the same way that we did for the Klein–Gordon equation, that is through

$$\tilde{\mathbf{p}} \equiv \mathbf{p} - e\mathbf{A} \tag{8.68}$$

and the Dirac equation becomes, in $2 \times 2$ matrix form,

$$\begin{bmatrix} m & \sigma \cdot \tilde{\mathbf{p}} \\ \sigma \cdot \tilde{\mathbf{p}} & -m \end{bmatrix} \begin{bmatrix} u \\ v \end{bmatrix} = E \begin{bmatrix} u \\ v \end{bmatrix} \qquad \text{where} \qquad \Psi = \begin{bmatrix} u \\ v \end{bmatrix}. \tag{8.69}$$

At nonrelativistic (positive) energies $E = K + m$, the kinetic energy $K \ll m$ and the lower equation becomes

$$\sigma \cdot \tilde{\mathbf{p}} u = (E + m)v \approx 2mv \tag{8.70}$$

which lets us write the upper equation as

$$\frac{(\sigma \cdot \tilde{\mathbf{p}})(\sigma \cdot \tilde{\mathbf{p}})}{2m} u = \left[ \frac{\tilde{\mathbf{p}}^2}{2m} + \frac{i\sigma}{2m} \cdot (\tilde{\mathbf{p}} \times \tilde{\mathbf{p}}) \right] u = Ku \tag{8.71}$$

where we have used (3.57). Now in the coordinate representation

$$\tilde{\mathbf{p}} \times \tilde{\mathbf{p}}u = (i\boldsymbol{\nabla} + e\mathbf{A}) \times (i\boldsymbol{\nabla}u + e\mathbf{A}u)$$
$$= ie\left[\boldsymbol{\nabla} \times (\mathbf{A}u) + \mathbf{A} \times \boldsymbol{\nabla}u\right]$$
$$= ie(\boldsymbol{\nabla} \times \mathbf{A})u = ie\mathbf{B}u \tag{8.72}$$

where $\mathbf{B}$ is the magnetic field associated with the vector potential $\mathbf{A}$. Therefore (8.71) becomes

$$\left[\frac{\tilde{\mathbf{p}}^2}{2m} - \boldsymbol{\mu} \cdot \mathbf{B}\right] u = Ku \tag{8.73}$$

where

$$\boldsymbol{\mu} = g\frac{e}{2m}\mathbf{S} \tag{8.74}$$

with

$$\mathbf{S} = \frac{\hbar}{2}\boldsymbol{\sigma} \tag{8.75}$$

and

$$g = 2. \tag{8.76}$$

In other words, the Dirac equation in the presence of an electromagnetic field, reduces nonrelativistically to (8.73), which is just the time-independent Schrödinger equation (with energy eigenvalue $K$) for a particle with magnetic moment $\boldsymbol{\mu}$ in the presence of an external magnetic field. The magnetic moment is derived from the spin operator with a gyromagnetic ratio $g = 2$.

*This brings us full circle.* We began this book by discussing the behavior of particles with magnetic moments in the presence of inhomogeneous magnetic fields, where it appears that they behaved as if they had spin projections quantized in one of two states. We now see that this stems from a consideration of relativity and quantum mechanics, for particles that obey the Dirac equation.

## 8.3 Symmetries of the Dirac Equation

Let us now examine some symmetries inherent in the Dirac equation. We will consider situations where a spin $\frac{1}{2}$ particle sits in some external potential, that is, solutions to the equation

$$i\frac{\partial}{\partial t}\Psi(\mathbf{x},t) = H\Psi(\mathbf{x},t) = E\Psi(\mathbf{x},t) \tag{8.77}$$

where

$$H = \boldsymbol{\alpha} \cdot \mathbf{p} + \beta m + V(\mathbf{x}) \tag{8.78}$$

for some potential energy function $V(\mathbf{x})$. This form, of course, ruins our ability to write a covariant equation, but that is a necessary penalty if we want to talk about potential energy. We note, though, that in the case of electromagnetic interactions, we can end up with exactly this form from the covariant equation if we choose a frame in which the vector potential $\mathbf{A} = 0$.

## 8.3.1 Angular Momentum

Our discussion of rotational invariance for wave mechanics in three dimensions centered on the fact that the orbital angular momentum operator $\mathbf{L} = \mathbf{x} \times \mathbf{p}$ commuted with Hamiltonians with "central potentials." This in turn hinges on the fact that $\mathbf{L}$ commutes with $\mathbf{p}^2$, and so the kinetic-energy operator, and also $\mathbf{x}^2$ (see (3.264)).

Let us now consider the commutator $[H, \mathbf{L}]$ first for the free Dirac Hamiltonian (8.53). It is obvious that $[\beta, \mathbf{L}] = 0$ so, we need to consider the commutator

$$\begin{aligned} [\boldsymbol{\alpha} \cdot \mathbf{p}, L_i] &= [\alpha_\ell p_\ell, \varepsilon_{ijk} x_j p_k] \\ &= \varepsilon_{ijk} \alpha_\ell [p_\ell, x_j] p_k \\ &= -i\varepsilon_{ijk} \alpha_j p_k \neq 0. \end{aligned} \tag{8.79}$$

(Recall our convention of summing over repeated indices.) In other words, the orbital angular momentum operator does not commute with the Dirac Hamiltonian! Therefore $\mathbf{L}$ will not be a conserved quantity for spin $\frac{1}{2}$ particles that are either free or bound in central potentials.

Consider, however, the spin operator (8.66) and its commutator with the Hamiltonian. It is simple to show that $\beta\Sigma_i = \Sigma_i\beta$. It is also easy to use (3.53) to show that $[\alpha_i, \Sigma_j] = 2i\varepsilon_{ijk}\alpha_k$. So, we need to evaluate

$$[\boldsymbol{\alpha} \cdot \mathbf{p}, \Sigma_j] - [\alpha_l, \Sigma_j]p_l = 2i\varepsilon_{ijk}\alpha_k p_l. \tag{8.80}$$

Thus we see that (putting back in $\hbar$ momentarily) even though neither $\mathbf{L}$ nor $\boldsymbol{\Sigma}$ commutes with the free Dirac Hamiltonian, the combined vector operator

$$\mathbf{J} \equiv \mathbf{L} + \frac{\hbar}{2}\boldsymbol{\Sigma} = \mathbf{L} + \mathbf{S} \tag{8.81}$$

*does* commute. That is, the Dirac Hamiltonian conserves *total* angular momentum, but not orbital or spin angular momenta separately.

## 8.3.2 Parity

In the case where $V(\mathbf{x}) = V(|\mathbf{x}|)$ we expect solutions to be parity symmetric. That is, we should have $\Psi(-\mathbf{x}) = \pm\Psi(\mathbf{x})$. It does not appear that this is the case, however, since if $\mathbf{x} \to -\mathbf{x}$, then $\mathbf{p} \to -\mathbf{p}$ in (8.78) and the Hamiltonian changes form. However, this simple look does not take into account the parity transformation on spinors.

Indeed, the parity transformation operator $\pi$, discussed in Section 4.2, only concerns coordinate reflection. That is, $\pi$ is a unitary (and also Hermitian) operator with the properties

$$\pi^\dagger \mathbf{x} \pi = -\mathbf{x} \tag{8.82a}$$

$$\pi^\dagger \mathbf{p} \pi = -\mathbf{p} \tag{8.82b}$$

(See equations (4.43) and (4.50).) The full parity operator, let us call it $\mathscr{P}$, needs to be augmented with a unitary operator $U_P$, which is a $4 \times 4$ matrix in spinor space, and which renders the Hamiltonian (8.78) invariant under a parity transformation. That is

$$\mathscr{P} \equiv \pi U_P \tag{8.83}$$

where the matrix $U_P$ must have the properties

$$U_P^\dagger \boldsymbol{\alpha} U_P = -\boldsymbol{\alpha} \tag{8.84a}$$

$$U_P^\dagger \beta U_P = \beta \tag{8.84b}$$

as well as

$$U_P^2 = 1. \tag{8.84c}$$

Obviously $U_P = \beta = \beta^\dagger$ is consistent with these requirements. Consequently, a parity transformation on a Dirac Hamiltonian consists of taking $\mathbf{x} \to -\mathbf{x}$ and multiplying on both the left and the right by $\beta$. The transformation of a spinor $\Psi(\mathbf{x})$ yields $\beta \Psi(-\mathbf{x})$.

Therefore, in cases where $V(\mathbf{x}) = V(|\mathbf{x}|)$, we expect to find eigenstates of the Dirac Hamiltonian which are simultaneously eigenstates of parity, $\mathbf{J}^2$ and $J_z$. Luckily, we have already constructed angular and spinor parts of these eigenfunctions. These are the two-component **spin-angular functions** $\mathscr{Y}_l^{j=l\pm1/2,m}(\theta,\phi)$ defined in (3.384). We will make use of these as we proceed to solve the Dirac equation for a particular potential energy function of this form in Section 8.4.

### 8.3.3 Charge Conjugation

We saw in (8.24) that for the Klein–Gordon equation, we could split the positive and negative energy solutions into "particle" and "antiparticle" solutions on the basis of the association

$$\Psi_{\text{particle}}(\mathbf{x},t) \equiv \Psi_{E>0}(\mathbf{x},t) \tag{8.85a}$$

$$\Psi_{\text{antiparticle}}(\mathbf{x},t) \equiv \Psi_{E<0}^*(\mathbf{x},t). \tag{8.85b}$$

Let us work towards a similar association for the Dirac equation, and then explore a symmetry operation that connects the two solutions.

For our purposes, an "antiparticle" is an object whose wave function behaves just as that for a "particle" but with opposite electric charge. So, let us return to the covariant form (8.46) of the Dirac equation, and add an electromagnetic field according to our usual prescription (8.13). We have

$$\left(i\gamma^\mu \partial_\mu - e\gamma^\mu A_\mu - m\right) \Psi(\mathbf{x},t) = 0. \tag{8.86}$$

We seek an equation where $e \to -e$, and which relates the new wave function to $\Psi(\mathbf{x},t)$. The key is that the operator in (8.86) has three terms, only two of which contain $\gamma^\mu$, and only one of those two contains $i$. So, take the complex conjugate of this equation to find

$$\left[-i(\gamma^\mu)^* \partial_\mu - e(\gamma^\mu)^* A_\mu - m\right] \Psi^*(\mathbf{x}, t) = 0 \tag{8.87}$$

and the relative sign between the first two terms is reversed. If we can now identify a matrix $\tilde{C}$ such that

$$\tilde{C}(\gamma^\mu)^* \tilde{C}^{-1} = -\gamma^\mu \tag{8.88}$$

we just insert $1 = \tilde{C}^{-1}\tilde{C}$ before the wave function in (8.87), and multiply on the left by $\tilde{C}$. The result is

$$\left[i\gamma^\mu \partial_\mu + e\gamma^\mu A_\mu - m\right] \tilde{C}\Psi^*(\mathbf{x}, t) = 0. \tag{8.89}$$

Therefore, the wave function $\tilde{C}\Psi^*(\mathbf{x}, t)$ satisfies the "positron" equation (8.89) where $\Psi(\mathbf{x}, t)$ satisfies the "electron" equation (8.86).

We need to identify the matrix $\tilde{C}$. From (8.52) and (8.55) we see that $\gamma^0$, $\gamma^1$, and $\gamma^3$ are real matrices, but $(\gamma^2)^* = -\gamma^2$. This makes it possible to realize (8.88) with

$$\tilde{C} = i\gamma^2. \tag{8.90}$$

Therefore, the "positron wave" function corresponding to $\Psi(\mathbf{x}, t)$ is $i\gamma^2\Psi^*(\mathbf{x}, t)$. It is more convenient, in turns out, to write this wave function in terms of $\overline{\Psi} = \Psi^\dagger \gamma^0 = (\Psi^*)^T \gamma^0$. (The superscript $T$ indicates "transpose.") This means that the "positron" wave function can be written as

$$\tilde{C}\Psi^*(\mathbf{x}, t) = i\gamma^2 \left(\overline{\Psi}\gamma^0\right)^T = U_C \left(\overline{\Psi}\right)^T \tag{8.91}$$

where

$$U_C \equiv i\gamma^2\gamma^0. \tag{8.92}$$

Therefore, the *charge conjugation operator* is $\mathscr{C}$ where

$$\mathscr{C}\Psi(\mathbf{x}, t) = U_C \left(\overline{\Psi}\right)^T. \tag{8.93}$$

Note that the change in the space-time part of the free-particle wave function $\Psi(\mathbf{x}, t) \propto \exp(-ip^\mu x_\mu)$ due to $\mathscr{C}$ is to, effectively, take $\mathbf{x} \to -\mathbf{x}$ and $t \to -t$.

### 8.3.4  Time Reversal

Let us now apply the ideas of Section 4.4 to the Dirac equation. First, a brief review. This discussion hinged on the definition (4.117) of an antiunitary operator $\theta$

$$\theta = UK \tag{8.94}$$

where $U$ is a unitary operator and $K$ is an operator which takes the complex conjugate of any complex numbers which follow it. Clearly, $K$ does not affect the kets to the right, and $K^2 = 1$.

Based on this, we defined an antiunitary operator $\Theta$ which takes an arbitrary state $|\alpha\rangle$ to a time-reversed (or, more properly, a motion-reversed) state $|\tilde{\alpha}\rangle$, that is

$$\Theta|\alpha\rangle = |\tilde{\alpha}\rangle. \tag{8.95}$$

We imposed two reasonable requirements on $\Theta$, namely (4.148) and (4.150), that is

$$\Theta \mathbf{p} \Theta^{-1} = -\mathbf{p} \tag{8.96a}$$

$$\Theta \mathbf{x} \Theta^{-1} = \mathbf{x} \tag{8.96b}$$

and so

$$\Theta \mathbf{J} \Theta^{-1} = -\mathbf{J}. \tag{8.96c}$$

For Hamiltonians that commute with $\Theta$, we made the important observation (4.162) that an energy eigenstate $|n\rangle$ has the same energy eigenvalue as its time-reversed counterpart state $\Theta|n\rangle$. We further learned that acting twice on a spin $\frac{1}{2}$ state yields $\Theta^2 = -1$, since the two-component spinor form of (4.168) shows that

$$\Theta = -i\sigma_y K. \tag{8.97}$$

In other words, $U = -i\sigma_y$ in (8.94). Indeed, in this case

$$\Theta^2 = -i\sigma_y K(-i\sigma_y K) = \sigma_y \sigma_y^* K^2 = -1. \tag{8.98}$$

We will see something similar when we apply time reversal to the Dirac equation, which we now take up.

Returning to the Schrödinger equation (8.1) with the Dirac Hamiltonian (8.52), but using the $\gamma$ matrices instead of $\alpha$ and $\beta$, we have

$$i\partial_t \Psi(\mathbf{x}, t) = \left[ -i\gamma^0 \boldsymbol{\gamma} \cdot \boldsymbol{\nabla} + \gamma^0 m \right] \Psi(\mathbf{x}, t). \tag{8.99}$$

We write our time-reversal operator, following the scheme earlier in this Section, as $\mathscr{T}$ instead of $\Theta$, where

$$\mathscr{T} = U_T K \tag{8.100}$$

and $U_T$ is a unitary matrix which we need to identify. As before, insert $\mathscr{T}^{-1}\mathscr{T}$ before the wave function on the left and right sides, and multiply through from the left by $\mathscr{T}$. The left side of (8.99) becomes

$$\mathscr{T}(i\partial_t)\mathscr{T}^{-1}\mathscr{T}\Psi(\mathbf{x}, t) = U_T K(i\partial_t) K U_T^{-1} U_T \Psi^*(\mathbf{x}, t)$$
$$= -i\partial_t U_T \Psi^*(\mathbf{x}, t) = i\partial_{-t}[U_T \Psi^*(\mathbf{x}, t)] \tag{8.101}$$

reversing the sign of $t$ in the derivative, as we need. In order for $[U_T \Psi^*(\mathbf{x}, t)]$ to satisfy the time-reversed form of (8.99), we must insist that

$$\mathscr{T}\left(i\gamma^0\boldsymbol{\gamma}\right)\mathscr{T}^{-1} = i\gamma^0\boldsymbol{\gamma} \tag{8.102a}$$

$$\mathscr{T}\left(\gamma^0\right)\mathscr{T}^{-1} = \gamma^0. \tag{8.102b}$$

These are easily converted to a more convenient form so that we can identify $U_T$. First, do $\mathscr{T}^{-1}$ on the left and $\mathscr{T}$ on the right. Then, do $K$ on the left and right. Finally, insert $U_T U_T^{-1}$ in between the $\gamma$ matrices in (8.102a) and then use the result from (8.102b). Both results then become

$$U_T^{-1}\left(\boldsymbol{\gamma}\right)U_T = -\left(\boldsymbol{\gamma}\right)^* \tag{8.103a}$$

$$U_T^{-1}\left(\gamma^0\right)U_T = \left(\gamma^0\right)^*. \tag{8.103b}$$

We now specialize to our choice (8.55), with (8.52), for the $\gamma$ matrices. Only $\gamma^2$ is imaginary in this representation. Therefore, if we want to build $U_T$ out of $\gamma$ matrices, then (8.103) says that we need a combination that does not change the sign when commuted with $\gamma^0$ and $\gamma^2$, but does change sign with $\gamma^1$ and $\gamma^3$. Rather obviously, this is accomplished by

$$U_T = \gamma^1 \gamma^3 \tag{8.104}$$

up to some arbitrary phase factor. Indeed, this works out to be equivalent to the result $U = i\sigma_y$ in (8.97). See Problem 8.13 at the end of this chapter.

## 8.3.5 CPT

We conclude with a brief look of the operator combination $\mathscr{C}\mathscr{P}\mathscr{T}$. Its action on a Dirac wave function $\Psi(\mathbf{x},t)$ is straightforward to work out, given the above discussion. That is

$$\begin{aligned}\mathscr{C}\mathscr{P}\mathscr{T}\Psi(\mathbf{x},t) &= i\gamma^2 [\mathscr{P}\mathscr{T}\Psi(\mathbf{x},t)]^* \\ &= i\gamma^2\gamma^0 [\mathscr{T}\Psi(-\mathbf{x},t)]^* \\ &= i\gamma^2\gamma^0\gamma^1\gamma^3\Psi(-\mathbf{x},t) = i\gamma^0\gamma^1\gamma^2\gamma^3\Psi(-\mathbf{x},t).\end{aligned} \tag{8.105}$$

This combination of $\gamma$ matrices is well known, and in fact given a special name. We define

$$\gamma^5 \equiv i\gamma^0\gamma^1\gamma^2\gamma^3. \tag{8.106}$$

In our basis (8.55), again writing $4 \times 4$ matrices as $2 \times 2$ matrices of $2 \times 2$ matrices, we find that

$$\gamma^5 = \begin{bmatrix} 0 & 1 \\ 1 & 0 \end{bmatrix}. \tag{8.107}$$

That is, $\gamma^5$, and therefore also $\mathscr{C}\mathscr{P}\mathscr{T}$, reverses the up and down two-component spinors in the Dirac wave function. The net effect of $\mathscr{C}\mathscr{P}\mathscr{T}$ on a free-particle electron wave function is in fact to convert it into the "positron" wave function. See Problem 8.14 at the end of this chapter.

This is a "tip of the iceberg" observation of a profound concept in relativistic quantum field theory. The notion of $\mathscr{C}\mathscr{P}\mathscr{T}$ invariance is equivalent to a total symmetry between matter and antimatter, so long as one integrates over all potential final states. For example, it predicts that the mass of any particle must equal the mass of the corresponding antiparticle.

Indeed, one can show, although it is far from straightforward, that any Lorentz invariant quantum field theory is invariant under $\mathscr{C}\mathscr{P}\mathscr{T}$. The implications are far reaching, particularly in this day and age when string theories offer the possibility that Lorentz invariance is broken at distances approaching the Planck mass. The reader is referred to any one of a number of advanced textbooks, and also the current literature, for more details.

# 8.4 Solving with a Central Potential

Our goal is to solve the eigenvalue problem

$$H\Psi(\mathbf{x}) = E\Psi(\mathbf{x}) \tag{8.108}$$

where

$$H = \boldsymbol{\alpha} \cdot \mathbf{p} + \beta m + V(r) \tag{8.109}$$

and we write the four-component wave function $\Psi(\mathbf{x})$ in terms of two two-component wave functions $\psi_1(\mathbf{x})$ and $\psi_2(\mathbf{x})$ as

$$\Psi(\mathbf{x}) = \left[ \begin{array}{c} \psi_1(\mathbf{x}) \\ \psi_2(\mathbf{x}) \end{array} \right]. \tag{8.110}$$

Based on the symmetries of the Dirac equation that we have already discussed, we expect $\Psi(\mathbf{x})$ to be an eigenfunction of parity, $\mathbf{J}^2$ and $J_z$.

Parity conservation implies that $\beta\Psi(-\mathbf{x}) = \pm\Psi(\mathbf{x})$. Given the form (8.55) of $\beta$, this implies that

$$\left[ \begin{array}{c} \psi_1(-\mathbf{x}) \\ -\psi_2(-\mathbf{x}) \end{array} \right] = \pm \left[ \begin{array}{c} \psi_1(\mathbf{x}) \\ \psi_2(\mathbf{x}) \end{array} \right]. \tag{8.111}$$

This leaves us with two choices, namely

$$\psi_1(-\mathbf{x}) = +\psi_1(\mathbf{x}) \text{ and } \psi_2(-\mathbf{x}) = -\psi_2(\mathbf{x}) \tag{8.112a}$$

$$\psi_1(-\mathbf{x}) = -\psi_1(\mathbf{x}) \text{ and } \psi_2(-\mathbf{x}) = +\psi_2(\mathbf{x}). \tag{8.112b}$$

These conditions are neatly realized by the spinor functions $\mathscr{Y}_l^{jm}(\theta, \phi)$ defined in (3.384), where $l = j \pm (1/2)$. For a given value of $j$, one possible value of $l$ is even and the other is odd. Since the parity of any particular $Y_l^m$ is just $(-1)^l$, then we are presented with two natural choices for the angular and spinor dependences for the conditions (8.112). We write

$$\Psi(\mathbf{x}) = \Psi_A(\mathbf{x}) \equiv \left[ \begin{array}{c} u_A(r)\mathscr{Y}_{j-1/2}^{jm}(\theta, \phi) \\ -iv_A(r)\mathscr{Y}_{j+1/2}^{jm}(\theta, \phi) \end{array} \right] \tag{8.113a}$$

which is an even (odd) parity solution if $j - 1/2$ is even (odd), or

$$\Psi(\mathbf{x}) = \Psi_B(\mathbf{x}) \equiv \left[ \begin{array}{c} u_B(r)\mathscr{Y}_{j+1/2}^{jm}(\theta, \phi) \\ -iv_B(r)\mathscr{Y}_{j-1/2}^{jm}(\theta, \phi) \end{array} \right] \tag{8.113b}$$

which is an odd (even) parity solution if $j - 1/2$ is even (odd). (The factor of $-i$ on the lower spinors is included for later convenience.) Note that although both $\Psi_A(\mathbf{x})$ and $\Psi_B(\mathbf{x})$ have definite parity and quantum numbers $j$ and $m$, they mix values of $l$. Orbital angular momentum is no longer a good quantum number when considering central potentials in the Dirac equation.

We are now ready to turn (8.108) into a differential equation in $r$ for the functions $u_{A(B)}(r)$ and $v_{A(B)}(r)$. First, rewrite the Dirac equation as two coupled equations for the spinors $\psi_1(\mathbf{x})$ and $\psi_2(\mathbf{x})$. Thus

$$[E - m - V(r)]\,\psi_1(\mathbf{x}) - (\boldsymbol{\sigma}\cdot\mathbf{p})\psi_2(\mathbf{x}) = 0 \tag{8.114a}$$

$$[E + m - V(r)]\,\psi_2(\mathbf{x}) - (\boldsymbol{\sigma}\cdot\mathbf{p})\psi_1(\mathbf{x}) = 0. \tag{8.114b}$$

Now make use of (3.57) and (3.59) to write

$$\boldsymbol{\sigma}\cdot\mathbf{p} = \frac{1}{r^2}(\boldsymbol{\sigma}\cdot\mathbf{x})(\boldsymbol{\sigma}\cdot\mathbf{x})(\boldsymbol{\sigma}\cdot\mathbf{p})$$

$$= \frac{1}{r^2}(\boldsymbol{\sigma}\cdot\mathbf{x})[\mathbf{x}\cdot\mathbf{p} + i\boldsymbol{\sigma}\cdot(\mathbf{x}\times\mathbf{p})]$$

$$= (\boldsymbol{\sigma}\cdot\hat{\mathbf{r}})\left[\hat{\mathbf{r}}\cdot\mathbf{p} + i\boldsymbol{\sigma}\cdot\frac{\mathbf{L}}{r}\right]. \tag{8.115}$$

Working in coordinate space, we have

$$\hat{\mathbf{r}}\cdot\mathbf{p} \to \hat{\mathbf{r}}\cdot(-i\boldsymbol{\nabla}) = -i\frac{\partial}{\partial r} \tag{8.116}$$

which will act on the radial part of the wave function only. We also know that

$$\boldsymbol{\sigma}\cdot\mathbf{L} = 2\mathbf{S}\cdot\mathbf{L} = \mathbf{J}^2 - \mathbf{L}^2 - \mathbf{S}^2 \tag{8.117}$$

so that we can write

$$(\boldsymbol{\sigma}\cdot\mathbf{L})\mathscr{Y}_l^{jm} = \left[j(j+1) - l(l+1) - \frac{3}{4}\right]\mathscr{Y}_l^{jm}$$

$$\equiv \kappa(j,l)\mathscr{Y}_l^{jm} \tag{8.118}$$

where

$$\kappa = -j - \frac{3}{2} = -(\lambda + 1) \quad \text{for } l = j + \frac{1}{2} \tag{8.119a}$$

$$\kappa = j - \frac{1}{2} = +(\lambda - 1) \quad \text{for } l = j - \frac{1}{2} \tag{8.119b}$$

where

$$\lambda \equiv j + \frac{1}{2}. \tag{8.120}$$

It is trickier to calculate the effect of the matrix factor

$$\boldsymbol{\sigma}\cdot\hat{\mathbf{r}} = \begin{bmatrix} \cos\theta & e^{-i\phi}\sin\theta \\ e^{i\phi}\sin\theta & -\cos\theta \end{bmatrix} \tag{8.121}$$

on the spinor wave functions. In principle, we can carry out the multiplication on the $\mathscr{Y}_l^{jm}$ as defined in (3.384) and then use the definition (3.246) to evaluate the result. There is an easier way, however.

We expect $\boldsymbol{\sigma}\cdot\hat{\mathbf{r}}$ to behave as a (pseudo)-scalar under rotations, so if we evaluate its effect at one particular $\hat{\mathbf{r}}$ then it should behave this way for all $\hat{\mathbf{r}}$. Choose $\hat{\mathbf{r}} = \hat{\mathbf{z}}$, that is, $\theta = 0$.

Since the $\theta$-dependent part of any $Y_l^m(\theta,\phi)$ contains a factor $[\sin\theta]^{|m|}$ we use (3.248) to write

$$Y_l^m(\theta = 0,\phi) = \sqrt{\frac{2l+1}{4\pi}}\,\delta_{m0} \tag{8.122}$$

in which case

$$\mathscr{Y}_l^{j=l\pm 1/2,m}(\theta = 0,\phi) = \frac{1}{\sqrt{2l+1}}\left[\begin{array}{c} \pm\sqrt{l\pm m+1/2}\,Y_l^{m-1/2}(0,\phi) \\ \sqrt{l\mp m+1/2}\,Y_l^{m+1/2}(0,\phi) \end{array}\right]$$

$$= \frac{1}{\sqrt{4\pi}}\left[\begin{array}{c} \pm\sqrt{l\pm m+1/2}\,\delta_{m,1/2} \\ \sqrt{l\mp m+1/2}\,\delta_{m,-1/2} \end{array}\right]$$

or

$$\mathscr{Y}_{l=j\mp 1/2}^{j,m}(\theta = 0,\phi) = \sqrt{\frac{j+1/2}{4\pi}}\left[\begin{array}{c} \pm\delta_{m,1/2} \\ \delta_{m,-1/2} \end{array}\right]. \tag{8.123}$$

Therefore

$$(\sigma\cdot\hat{\mathbf{z}})\mathscr{Y}_{l=j\mp 1/2}^{j,m}(\theta = 0,\phi) = -\sqrt{\frac{j+1/2}{4\pi}}\left[\begin{array}{c} \mp\delta_{m,1/2} \\ \delta_{m,-1/2} \end{array}\right] = -\mathscr{Y}_{l=j\pm 1/2}^{j,m}(\theta = 0,\phi) \tag{8.124}$$

and so, as we have argued that this result is independent of $\theta$ and $\phi$, we have

$$(\sigma\cdot\hat{\mathbf{r}})\mathscr{Y}_{l=j\pm 1/2}^{j,m}(\theta,\phi) = -\mathscr{Y}_{l=j\mp 1/2}^{j,m}(\theta,\phi) \tag{8.125}$$

where we have used the fact that $(\sigma\cdot\hat{\mathbf{r}})^2 = 1$. In other words, for a given $j$ and $m$, $\mathscr{Y}_{l=j\pm 1/2}^{j,m}(\theta,\phi)$ is an eigenstate of $\sigma\cdot\hat{\mathbf{r}}$ with eigenvalue $-1$ (a consequence of the pseudo-scalar nature of the operator) and changes $l$ to the other allowed value, which naturally has opposite parity.

Now we return to the coupled equations (8.114) with solutions in the form (8.113). We have two choices for $\psi_1(\mathbf{x})$ and $\psi_2(\mathbf{x})$, namely "Choice A"

$$\psi_1(\mathbf{x}) = u_A(r)\mathscr{Y}_{j-1/2}^{jm}(\theta,\phi) \quad\text{and}\quad \psi_2(\mathbf{x}) = -iv_A(r)\mathscr{Y}_{j+1/2}^{jm}(\theta,\phi) \tag{8.126}$$

or "Choice B"

$$\psi_1(\mathbf{x}) = u_B(r)\mathscr{Y}_{j+1/2}^{jm}(\theta,\phi) \quad\text{and}\quad \psi_2(\mathbf{x}) = -iv_B(r)\mathscr{Y}_{j-1/2}^{jm}(\theta,\phi). \tag{8.127}$$

Note that for whichever of these two choices we pick, the effect of the factor $(\sigma\cdot\hat{\mathbf{r}})$, which is part of $(\sigma\cdot\mathbf{p})$ in (8.114) is to exchange $l = j\pm 1/2$ for $l = j\mp 1/2$ in the angular spinor $\mathscr{Y}_l^{jm}$, that is, switching the angular spinor factor of the second term in each of (8.114) so that it is the same as the first term. In other words, the angular factors drop out, and we are left with just the radial equations.

Putting this all together, finally, (8.114) becomes for "Choice A"

$$[E - m - V(r)]u_A(r) - \left[\frac{d}{dr} + \frac{\lambda+1}{r}\right]v_A(r) = 0 \tag{8.128a}$$

$$[E + m - V(r)]v_A(r) + \left[\frac{d}{dr} - \frac{\lambda-1}{r}\right]u_A(r) = 0 \tag{8.128b}$$

and, for "Choice B"

$$[E - m - V(r)]u_B(r) - \left[\frac{d}{dr} - \frac{\lambda - 1}{r}\right]v_B(r) = 0 \tag{8.129a}$$

$$[E + m - V(r)]v_B(r) + \left[\frac{d}{dr} + \frac{\lambda + 1}{r}\right]u_B(r) = 0. \tag{8.129b}$$

However, formally, equations (8.128) become (8.129) with the exchange $\lambda \leftrightarrow -\lambda$. Therefore, we can focus on the solution to (8.128) and drop the subscript $A$.

Equations (8.128) are coupled, first-order ordinary differential equations to be solved for the $u(r)$ and $v(r)$, subject to certain boundary conditions (i.e. normalizability) which will yield eigenvalues $E$. This solution can at least be carried out numerically, which is practical in many situations. We conclude this section, however, with one case that can be solved analytically.

### 8.4.1 The One-Electron Atom

We can now consider atoms with one electron, with the potential energy function

$$V(r) = -\frac{Ze^2}{r}. \tag{8.130}$$

We expect that the "fine structure" of the hydrogen atom, which we studied using perturbation theory in Section 5.3, should emerge naturally with our solution using the Dirac equation.

Start by writing (8.128) in terms of scaled variables. That is

$$\varepsilon \equiv \frac{E}{m} \tag{8.131}$$

$$x \equiv mr \tag{8.132}$$

and recall that we write $\alpha \equiv e^2/(\hbar c) \approx 1/137$. This gives

$$\left[\varepsilon - 1 + \frac{Z\alpha}{x}\right]u(x) - \left[\frac{d}{dx} + \frac{\lambda + 1}{x}\right]v(x) = 0 \tag{8.133a}$$

$$\left[\varepsilon + 1 + \frac{Z\alpha}{x}\right]v(x) + \left[\frac{d}{dx} - \frac{\lambda - 1}{x}\right]u(x) = 0. \tag{8.133b}$$

Next consider the behavior of the solutions as $x \to \infty$. Equation (8.133a) becomes

$$(\varepsilon - 1)u - \frac{dv}{dx} = 0 \tag{8.134}$$

and so (8.133b) implies that

$$(\varepsilon + 1)v + \frac{du}{dx} = (\varepsilon + 1)v + \frac{1}{\varepsilon - 1}\frac{d^2v}{dx^2} = 0 \tag{8.135}$$

which leads to

$$\frac{d^2v}{dx^2} = (1 - \varepsilon^2)v. \tag{8.136}$$

Note that, classically, bound states require that the kinetic energy $E - m - V(r) = 0$ at some distance $r$, and $V(r) < 0$ everywhere, so $E - m < 0$ and $\varepsilon = E/m < 1$. Therefore $1 - \varepsilon^2$ is guaranteed to be positive, and (8.136) implies that

$$v(x) = \exp\left[-(1 - \varepsilon^2)^{1/2}x\right] \qquad \text{for} \quad x \to \infty \tag{8.137}$$

where we require that $v(x)$ be normalizable as $x \to \infty$, but ignore the normalization constant for now. Similarly, (8.134) then implies that

$$u(x) = \exp\left[-(1 - \varepsilon^2)^{1/2}x\right] \qquad \text{for} \quad x \to \infty \tag{8.138}$$

as well.

Now write $u(x)$ and $v(x)$ as power series with their own expansion coefficients, and look for relationships consistent with the differential equations. Using

$$u(x) = e^{-(1-\varepsilon^2)^{1/2}x} x^\gamma \sum_{i=0}^\infty a_i x^i \tag{8.139}$$

$$v(x) = e^{-(1-\varepsilon^2)^{1/2}x} x^\gamma \sum_{i=0}^\infty b_i x^i \tag{8.140}$$

we tacitly assume that we can find series solutions making use of the same overall power $\gamma$ for both $u(x)$ and $v(x)$. Indeed, inserting these expressions into (8.133) and first considering terms proportional to $x^{\gamma-1}$, we find, after a little rearrangement,

$$(Z\alpha)a_0 - (\gamma + \lambda + 1)b_0 = 0 \tag{8.141a}$$

$$(\gamma - \lambda + 1)a_0 + (Z\alpha)b_0 = 0. \tag{8.141b}$$

Our approach will soon yield recursion relations for the coefficients $a_i$ and $b_i$. This means that we need to avoid $a_0 = 0$ and $b_0 = 0$, and the only way to do this is to require that the determinant of (8.141) vanish. That is

$$(Z\alpha)^2 + (\gamma + 1 + \lambda)(\gamma + 1 - \lambda) = 0 \tag{8.142}$$

or, solving for $\gamma$,

$$\gamma = -1 \pm \left[\lambda^2 - (Z\alpha)^2\right]^{1/2}. \tag{8.143}$$

Notice first $\lambda = j + 1/2$ is of order unity, so that things will break down if $Z\alpha \sim 1$. In the strong Coulomb fields for $Z \approx 137$, spontaneous $e^+e^-$ production would occur, and the single-particle nature of the Dirac equation cannot be expected to prevail. Indeed, we have $Z\alpha \ll 1$ in cases of interest here. This also means that the expression within the brackets in (8.143) is of order unity. For the $-$ sign, this gives $\gamma \sim -2$ which would be too singular at the origin. As a result, we choose the $+$ sign and have

$$\gamma = -1 + \left[\left(j + \frac{1}{2}\right)^2 - (Z\alpha)^2\right]^{1/2}. \tag{8.144}$$

Note that for $j = 1/2$ we still have a singularity at the origin, since $\gamma < 0$, but this singularity is very weak, and integrable over space.

Starting with a value for $a_0$ which is determined by normalization, and $b_0 = a_0(Z\alpha)/(\gamma + \lambda + 1)$, we can find the remaining $a_i$ and $b_i$ by going back to the result of inserting (8.139) and (8.140) into (8.133). Collecting powers of $x^\gamma$ and higher, we find

$$(1-\varepsilon)a_{i-1} - Z\alpha a_i - (1-\varepsilon^2)^{1/2}b_{i-1} + (\lambda+1+\gamma+i)b_i = 0 \tag{8.145a}$$

$$(1+\varepsilon)b_{i-1} + Z\alpha b_i - (1-\varepsilon^2)^{1/2}a_{i-1} - (\lambda-1-\gamma-i)a_i = 0. \tag{8.145b}$$

Multiply (8.145a) by $(1+\varepsilon)^{1/2}$ and (8.145b) by $(1-\varepsilon)^{1/2}$ and then add them. This leads to a relationship between the coefficients $a_i$ and $b_i$, namely

$$\frac{b_i}{a_i} = \frac{Z\alpha(1+\varepsilon)^{1/2} + (\lambda-1-\gamma-i)(1-\varepsilon)^{1/2}}{Z\alpha(1-\varepsilon)^{1/2} + (\lambda+1+\gamma+i)(1+\varepsilon)^{1/2}}. \tag{8.146}$$

This relation shows that for large values of $x$, where terms with large $i$ dominate, $a_i$ and $b_i$ are proportional to each other. Furthermore (8.145) also implies that $a_i/a_{i-1} \sim 1/i$ for large $i$. (See Problem 8.15 at the end of this chapter.) In other words, the series (8.139) and (8.140) will grow exponentially, and not be normalizable, unless we force the series to terminate.

If we then assume that $a_i = b_i = 0$ for $i = n' + 1$ then

$$(1-\varepsilon)a_{n'} - (1-\varepsilon^2)^{1/2}b_{n'} = 0 \tag{8.147a}$$

$$(1+\varepsilon)b_{n'} - (1-\varepsilon^2)^{1/2}a_{n'} = 0. \tag{8.147b}$$

Either of these leads to the same condition on the ratio of these terminating coefficients, namely

$$\frac{b_{n'}}{a_{n'}} = \left[\frac{1-\varepsilon}{1+\varepsilon}\right]^{1/2}. \tag{8.148}$$

Combining (8.146) and (8.147) we have

$$(1+\gamma+n')(1-\varepsilon^2)^{1/2} = Z\alpha\varepsilon \tag{8.149}$$

which, finally, can be solved for $\varepsilon$. Putting $c$ back in, we determine the energy eigenvalues

$$E = \frac{mc^2}{\left[1 + \dfrac{(Z\alpha)^2}{\left[\sqrt{(j+1/2)^2 - (Z\alpha)^2} + n'\right]^2}\right]^{1/2}}. \tag{8.150}$$

We emphasize that, for any given quantum number $n'$, the energy eigenvalues depend on the total angular momentum $j$. That is, for example, the energy will be the same for $j = 1/2$, regardless of whether or not it comes from coupling $l = 0$ or $l = 1$ with spin $\frac{1}{2}$.

To lowest order in $Z\alpha$, (8.150) becomes

$$E = mc^2 - \frac{1}{2}\frac{mc^2(Z\alpha)^2}{n^2} \tag{8.151}$$

where $n \equiv j + 1/2 + n'$. Comparison to (3.315) shows that this is simply the familiar Balmer formula, with the addition of rest mass energy, with $n$ being the principal quantum number. Including higher orders of $Z\alpha$ leads to the well-known expressions for the relativistic correction to kinetic energy (5.104) and the spin-orbit interaction (5.125).

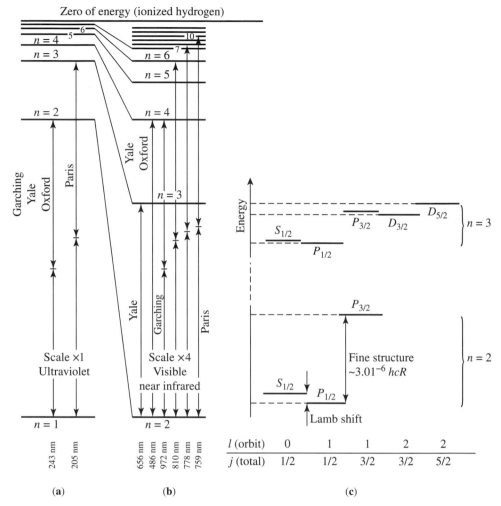

**Fig. 8.2**  The energy levels of the hydrogen atom, with reference to the many high precision experiments which have measured electromagnetic transitions between them. Taken from Cagnac, *Phys. Scr.*, **T70** (1997) 24. (a) The full energy-level diagram, (b) with energy scale multiplied by four, and (c) fine structure of the $n = 2$ and $n = 3$ levels, detailing behavior on the quantum numbers $l$ and $j$.

Figure 8.2 shows the energy levels of the hydrogen atom, and the experiments which made these measurements. A number of ingenious techniques were devised, including two-photon absorption for connecting "forbidden" transitions, in order to obtain these results. The so-called "fine structure" is clear, that is, the relativistic effects that lead to splitting between the $S$ and $P$ levels for the $n = 2$ states, and between the $S$, $P$, and $D$ levels for the $n = 3$ states. Problem 8.16 at the end of this chapter shows that the full relativistic energy levels give the same splittings as obtained using perturbation theory.

There is, of course, a profound discrepancy between the energy levels in Figure 8.2 and the result (8.150) we derived for the energy eigenvalues. According to (8.150) the energy can only depend on $n$ and $j$. However, we see that there is a small splitting between, for example, the $2P_{1/2}$ and $2S_{1/2}$ states. This splitting, called the *Lamb shift* after its discoverer, played a central role in recognizing the importance of relativistic quantum field theory in atomic structure. See Holstein (1992) for a discussion of the history, as well as both formal and "physically intuitive" derivations of the size of the Lamb shift.

Problem 8.17 at the end of this chapter compares energy levels predicted by (8.150) to available high precision data.

## 8.5  Relativistic Quantum Field Theory

We now conclude our coverage of "Modern Quantum Mechanics." The framework outlined in this book continues to be the most fundamental basis by which we understand the physical world. Although the probabilistic interpretation of the concept of measurement is disturbing in some ways, it always prevails when confronted with experiment.

The deficiencies that remain, for example in the Lamb shift, are not the result of problems in the underlying axioms of quantum mechanics. Instead, they are the result of a necessary approximation we make when we try to develop quantum-mechanical wave equations that are consistent with relativity. The ability to create and destroy particles is inconsistent with our "single-particle" approach to writing down dynamics in quantum mechanics. Instead, we would need to re-examine the Hamiltonian formalism, on which much of this book is based, in order address these issues.

Quantum field theory is the correct framework for addressing relativistic quantum mechanics, and multiparticle quantum mechanics in general. There are essentially two ways to approach quantum field theory, neither of which is developed here. We only mention them for the reader interested in going on to the next steps.

One approach is through the method of "second quantization" where operators are introduced which create and destroy particles. These operators commute with each other if they have integer spin, and anticommute for half-integer spin. Work needs to be done in order to build in relativistic covariance, but it is relatively straightforward. It is also, however, not necessary if the problem does not warrant it. This is the case, for example, in a vast number of fascinating problems in condensed matter physics.

Second quantization is discussed in Section 7.5 of this book. For other examples, see *Quantum Mechanics*, by Eugen Merzbacher (1998), and *Quantum Theory of Many-Particle Systems*, by Alexander L. Fetter and John Dirk Walecka (2003).

The second approach is through the path-integral approach to quantum mechanics, famously pioneered by Richard Feynman in his Ph.D. thesis. This conceptually appealing formalism is straightforward to extend from particle quantum mechanics to quantum fields. However, it is not straightforward to use this formalism for calculation of typical problems, until one makes the connection onto the "canonical" formalism that eventually becomes second quantization. Nevertheless, it is a worthwhile subject for students who would like to have a better understanding of the principles that lead to the quantum many-body problem.

Path integrals are not the basis for many books on quantum field theory, but they are beautifully exploited in Zee (2010).

# Problems

**8.1** These exercises are to give you some practice with natural units.

     a. Express the proton mass $m_p = 1.67262158 \times 10^{-27}$ kg in units of GeV.

     b. Assume that a particle with negligible mass is confined to a box the size of the proton, around 1 fm $= 10^{-15}$ m. Use the uncertainty principle to estimate the energy of the confined particle. You might be interested to know that the mass, in natural units, of the pion, the lightest strongly interacting particle, is $m_\pi = 135$ MeV.

     c. String theory concerns the physics at a scale which combines gravity, relativity, and quantum mechanics. Use dimensional analysis to find the "Planck mass" $M_P$ which is formed from $G$, $\hbar$, and $c$, and express the result in GeV.

**8.2** Show that a matrix $\eta^{\mu\nu}$ with the same elements as the metric tensor $\eta_{\mu\nu}$ used in this chapter, has the property that $\eta^{\mu\lambda}\eta_{\lambda\nu} = \delta^\mu_\nu$, the identity matrix. Thus, show that the natural relationship $\eta^{\mu\nu} = \eta^{\mu\lambda}\eta^{\nu\sigma}\eta_{\lambda\sigma}$ in fact holds with this definition. Show also that $a^\mu b_\mu = a_\mu b^\mu$ for two four-vectors $a^\mu$ and $b^\mu$.

**8.3** Show that (8.11) is in fact a conserved current, when $\Psi(\mathbf{x}, t)$ satisfies the Klein–Gordon equation.

**8.4** Show that (8.14) follows from (8.8).

**8.5** Derive (8.16a), (8.16b), and (8.18).

**8.6** Show that the free-particle energy eigenvalues of (8.18) are $E = \pm E_p$ and that the eigenfunctions are indeed given by (8.21), subject to the normalization that $\Upsilon^\dagger \tau_3 \Upsilon = \pm 1$ for $E = \pm E_p$.

**8.7** This problem is taken from Landau (1996). A spinless electron is bound by the Coulomb potential $V(r) = -Ze^2/r$ in a stationary state of total energy $E \le m$. You can incorporate this interaction into the Klein–Gordon equation by using the covariant derivative with $V = -e\Phi$ and $\mathbf{A} = 0$.

     a. Assume that the radial and angular parts of the equation separate, and that the wave function can be written as $e^{-iEt}[u_l(r)/r]Y_{lm}(\theta, \phi)$ and show that the radial equation becomes

$$\frac{d^2 u}{d\rho^2} + \left[ \frac{2EZ\alpha}{\gamma\rho} - \frac{1}{4} - \frac{l(l+1) - (Z\alpha)^2}{\rho^2} \right] u_l(\rho) = 0$$

     where $\alpha = e^2$, $\gamma^2 = 4(m^2 - E^2)$, and $\rho = \gamma r$.

     b. Assume that this equation has a solution of the usual form of a power series times the $\rho \to \infty$ and $\rho \to 0$ solutions, that is

$$u_l(\rho) = \rho^k (1 + c_1\rho + c_2\rho^2 + \cdots)e^{-\rho/2}$$

and show that

$$k = k_\pm = \frac{1}{2} \pm \sqrt{\left(l + \frac{1}{2}\right)^2 - (Z\alpha)^2}$$

and that only for $k_+$ is the expectation value of the kinetic energy finite and that this solution has a nonrelativistic limit which agrees with the solution found for the Schrödinger equation.

c. Determine the recurrence relation among the $c_i$ for this to be a solution of the Klein–Gordon equation, and show that unless the power series terminates, the wave function will have an incorrect asymptotic form.

d. In the case where the series terminates, show that the energy eigenvalue for the $k_+$ solution is

$$E = \frac{m}{\left(1 + (Z\alpha)^2 \left[n - l - \frac{1}{2} + \sqrt{\left(l + \frac{1}{2}\right)^2 - (Z\alpha)^2}\right]^{-2}\right)^{1/2}}$$

where $n$ is the principal quantum number.

e. Expand $E$ in powers of $(Z\alpha)^2$ and show that the first-order term yields the Bohr formula. Connect the higher-order terms with relativistic corrections, and discuss the degree to which the degeneracy in $l$ is removed.

Jenkins and Kunselman, *Phys. Rev. Lett.*, **17** (1966) 1148, report measurements of a large number of transition energies for $\pi^-$ atoms in large $Z$ nuclei. Compare some of these to the calculated energies and discuss the accuracy of the prediction. (For example, consider the $3d \to 2p$ transition in $^{59}$Co which emits a photon with energy $384.6 \pm 1.0$ keV.) You will probably either need a computer to carry out the energy differences with high enough precision, or else expand to higher powers of $(Z\alpha)^2$.

**8.8** Prove that the traces of the $\gamma^\mu$, $\alpha$, and $\beta$ are all zero.

**8.9**  a. Derive the matrices $\gamma^\mu$ from (8.55) and show that they satisfy the Clifford algebra (8.49).

b. Show that

$$\gamma^0 = \begin{pmatrix} I & 0 \\ 0 & -I \end{pmatrix} = I \otimes \tau_3$$

$$\gamma^i = \begin{pmatrix} 0 & \sigma^i \\ -\sigma^i & 0 \end{pmatrix} = \sigma^i \otimes i\tau_2$$

where $I$ is the $2 \times 2$ identity matrix, and $\sigma^i$ and $\tau_i$ are the Pauli matrices. (The $\otimes$ notation is a formal way to write our $4 \times 4$ matrices as $2 \times 2$ matrices of $2 \times 2$ matrices.)

**8.10** Prove the continuity equation (8.56) for the Dirac equation.

**8.11** Find the eigenvalues for the free-particle Dirac equation (8.65).

**8.12** Insert one of the four solutions $u_{R,L}^{(\pm)}(p)$ from (8.67) into the four-vector probability current (8.58) and interpret the result.

**8.13** Make use of Problem 8.9 to show that $U_T$ as defined by (8.104) is just $\sigma^2 \otimes I$, up to a phase factor.

**8.14** Write down the positive helicity, positive energy free-particle Dirac spinor wave function $\Psi(\mathbf{x}, t)$.

    a. Construct the spinors $\mathscr{P}\Psi, \mathscr{C}\Psi, \mathscr{T}\Psi$.

    b. Construct the spinor $\mathscr{C}\mathscr{P}\mathscr{T}\Psi$ and interpret it using the discussion of negative energy solutions to the Dirac equation.

**8.15** Show that (8.145) imply that $u(x)$ and $v(x)$ grow like exponentials if the series (8.139) and (8.140) do not terminate.

**8.16** Expand the energy eigenvalues given by (8.150) in powers of $Z\alpha$ and show that the result is equivalent to including the relativistic correction to kinetic energy (5.104) and the spin-orbit interaction (5.125) to the nonrelativistic energy eigenvalues for the one-electron atom (8.151).

**8.17** The National Institute of Standards and Technology (NIST) maintains a website with up-to-date high precision data on the atomic energy levels of hydrogen and deuterium:

        http://physics.nist.gov/PhysRefData/HDEL/data.html

Following is a table of data obtained from that website. These are the energies of transitions between the $(n, l, j) = (1, 0, 1/2)$ energy level and the energy level indicated by the columns on the left:

| $n$ | $l$ | $j$ | $[E(n,l,j) - E(1,0,1/2)]/hc$ $(\text{cm}^{-1})$ |
|---|---|---|---|
| 2 | 0 | 1/2 | 82 258.954 399 2832(15) |
| 2 | 1 | 1/2 | 82 258.919 113 406(80) |
| 2 | 1 | 3/2 | 82 259.285 001 249(80) |
| 3 | 0 | 1/2 | 97 492.221 724 658(46) |
| 3 | 1 | 1/2 | 97 492.211 221 463(24) |
| 3 | 1 | 3/2 | 97 492.319 632 775(24) |
| 3 | 2 | 3/2 | 97 492.319 454 928(23) |
| 3 | 2 | 5/2 | 97 492.355 591 167(23) |
| 4 | 0 | 1/2 | 102 823.853 020 867(68) |
| 4 | 1 | 1/2 | 102 823.848 581 881(58) |
| 4 | 1 | 3/2 | 102 823.894 317 849(58) |
| 4 | 2 | 3/2 | 102 823.894 241 542(58) |
| 4 | 2 | 5/2 | 102 823.909 486 535(58) |
| 4 | 3 | 5/2 | 102 823.909 459 541(58) |
| 4 | 3 | 7/2 | 102 823.917 081 991(58) |

(The number in parentheses is the numerical value of the standard uncertainty referred to the last figures of the quoted value.) Compare these values to those predicted by (8.150) (you may want to make use of Problem 8.16, in particular the following).

a. Compare fine-structure splitting between the $n = 2$, $j = 1/2$ and $n = 2$, $j = 3/2$ states to (8.150).

b. Compare fine-structure splitting between the $n = 4$, $j = 5/2$ and $n = 4$, $j = 7/2$ states to (8.150).

c. Compare the $1S \rightarrow 2S$ transition energy to the first line in the table. Use as many significant figures as necessary in the values of the fundamental constants, to compare the results within standard uncertainty.

d. How many examples of the Lamb shift are demonstrated in this table? Identify one example near the top and the other near the bottom of the table, and compare their values.

# Appendix A  Electromagnetic Units

Two divergent systems of units established themselves over the course of the twentieth century. One system, known as *SI* (from the French Le Système International d'Unités), is rooted in the laboratory. It gained favor in the engineering community and forms the basis for most undergraduate curricula. The other system, called *Gaussian*, is aesthetically cleaner and is much favored in the theoretical physics community. We use the *Gaussian* system in this book, as do most graduate level physics texts on quantum mechanics and other subjects.

The *SI* system is also known as *MKSA* (for meter, kilogram, second, Ampere), and the *Gaussian* system is sometimes called[1] *CGS* (for centimeter, gram, second). For problems in mechanics, the difference is trivial, amounting only to some powers of ten. The difficulty comes when incorporating electromagnetism. As discussed below, *SI* incorporates a fourth base unit, the Ampere, which implies that charge and other electromagnetic quantities are dimensionally distinct between the *SI* and *Gaussian* systems. This one point is the source of all the confusion.

In other words, although we can write that a meter is equal to 100 centimeters, the *SI* unit of charge, called the Coulomb, is not equal to any number of electrostatic units (esu), the *Gaussian* unit of charge. The two kinds of charges literally have different physical meanings. They should probably have different names, but historically they are both called "charge." Similar comments apply to electric current, electric and magnetic fields, the electric potential, and so on. This is why you see factors like $\varepsilon_0$ and $\mu_0$ appear in *SI* formulas, whereas factors of $c$ are common in *Gaussian* formulas.

This appendix gives an explanation for how these two systems of units diverge when it comes to electromagnetism. It also shows how physical quantities such as force and energy can be used to relate electromagnetic quantities between the two systems. For a more detailed discussion, see "On electric and magnetic units and dimensions," by Birge, *Am. Phys. Teach.*, **2** (1934) 41. (This journal is now called the *American Journal of Physics*.) See also the articles in the *American Journal of Physics*, volume 3, 2005, pp. 90, 102, and 171.

## Electricity, Magnetism, and Electromagnetism: A Review

Electricity and magnetism are not separate phenomena. They are different manifestations of the same phenomenon, called electromagnetism. You need to incorporate special

---

[1] Some authors distinguish between *Gaussian* and *CGS* by including a factor of $4\pi$ in Gauss's law.

relativity to see how electricity and magnetism are united, and there were some decades between Maxwell and Einstein. Consequently, it was quite some time after they were separately established, that electricity and magnetism were realized to be just different ways that electromagnetism can exert a force.

The starting place for an "electric" force is Coulomb's law. If some number of electrons is added to, or removed from, an object, then it acquires a "charge" $q$. A force appears between two charged objects separated by some distance. This force is proportional to the product of the charges, and inversely proportional to the square of the distance between them, that is,

$$F = k_E \frac{q_1 q_2}{d^2}. \tag{A.1}$$

Here $k_E$ is an arbitrary constant of proportionality; without describing what we mean by "charge," we can say no more about it.

A "magnetic" force appears between two wires, each of which carries something called a "current." For two long, parallel wires, the force per unit length is proportional to the product of the currents and inversely proportional to the perpendicular separation of the wires, that is,

$$\frac{F}{L} = k_M \frac{I_1 I_2}{d}. \tag{A.2}$$

As with the electric force, $k_M$ is a generic constant of proportionality which depends on what we mean by "current."

Today we understand that (A.1) and (A.2) are two different manifestations of "electromagnetism." A "current" is in fact a flow of "charge," and the theory of electromagnetism tells us that

$$2k_E = c^2 k_M. \tag{A.3}$$

In other words, if we make a choice for $k_E$, then (A.3) specifies $k_M$ and vice versa.

The essential point is that *the SI and Gaussian systems make different choices for $k_E$ or $k_M$*. Other choices will lead to other systems of units, but we will not be discussing them here. There are also variations on whether or not to combine factors of $4\pi$ into various quantities, but we are not going to distinguish between them in this appendix.

## The SI System: Inventing a Unit for Current

The *SI* system is based on (A.2). People learned how to make current, well before we understood it in terms of charge. Perhaps for these reasons, a new base unit, the Ampere (A), was created. One Ampere is the amount of current flowing in each of two long, parallel wires, separated by one meter, such that the force between the wires is $2 \times 10^{-7}$ N/m. We write

$$k_M = \frac{\mu_0}{2\pi} \qquad \textbf{SI} \tag{A.4}$$

where

$$\mu_0 \equiv 4\pi \times 10^{-7} \frac{N}{A^2}. \tag{A.5}$$

(The factor of $4\pi$ turns out to be handy to cancel out integrations over the unit sphere.)

This quantity $\mu_0$ turns out to describe the magnetic properties of the vacuum. It shows up, for example, in the inductance of a loop of wire surrounding empty space. This is all forced upon us by the invention of the Ampere. Some books refer to $\mu_0$ as the "permeability of free space."

Equations (A.3) and (A.4) tell us how to write Coulomb's law (A.1) in the *SI* system. We have

$$k_E = \frac{c^2}{2} k_M = \frac{\mu_0 c^2}{4\pi} \quad \textbf{SI} \tag{A.6}$$

$$= 8.99 \times 10^9 \frac{N \cdot m^2}{(A \cdot s)^2}. \tag{A.7}$$

This form of Coulomb's law shows that charge, in the *SI* system, has units Amperes×seconds ($A \cdot s$). This is defined to be the Coulomb (C). *SI* furthermore defines the quantity

$$\varepsilon_0 \equiv \frac{1}{\mu_0 c^2} \tag{A.8}$$

called the "permittivity of free space." It is another property of the vacuum, showing up, for example, as the capacitance of parallel plates separated by empty space. It is, as with $\mu_0$, forced upon us by choosing a new base unit for current. Thus, combining (A.1), (A.6), and (A.8), we write Coulomb's law in *SI* as

$$F = \frac{1}{4\pi\varepsilon_0} \frac{q_1 q_2}{d^2} \quad \textbf{SI} \tag{A.9}$$

which is the form presented in introductory physics textbooks that use the *SI* convention.

## The Gaussian System: No New Base Units

In the *Gaussian* system, we take the point of view that no new base units are necessary. We write

$$k_E = 1 \quad \textbf{Gaussian} \tag{A.10}$$

that is, a dimensionless number. In other words, Coloumb's law (A.1) is simply

$$F = \frac{q_1 q_2}{d^2} \quad \textbf{Gaussian}. \tag{A.11}$$

The unit of charge in the *Gaussian* system is *derived* in terms of centimeters, grams, and seconds. It is called the electrostatic unit (esu), or sometimes the statcoulomb, and is simply[2]

$$esu \equiv \sqrt{dyne \cdot cm^2} = g^{1/2} \cdot cm^{3/2}/s. \tag{A.12}$$

---

[2] Recall that the unit of force in *CGS* is called the dyne $\equiv g \cdot cm/s^2 = 10^{-5}$ N.

In this case, the magnetic force between wires is just (A.2) with (A.3), namely

$$\frac{F}{L} = \frac{2}{c^2} \frac{I_1 I_2}{d} \qquad \textbf{CGS}. \qquad (A.13)$$

The *Gaussian* system is not without its sources of confusion. Some authors use (A.13) to define a unit of current, the statampere, which gives 2 dyne/cm of force between two long parallel wires separated by 1 cm. Note that this *is not* the same as one esu/s, something called the "absolute ampere" or "abampere." The statampere and abampere differ by a factor of $2.998 \times 10^{10}$, although they have the same dimensions, namely those of the esu/s $= g^{1/2} \cdot cm^{3/2}/s^2$.

## Converting between SI and CGS

It should now be clear to you that the units of charge and current have different *dimensions* between *SI* and *Gaussian*, and this is why everyone encounters confusion when converting between one system and the other.

Of course, all of this boils down to experiment. You make a measurement, and use some equations (whether they are *Gaussian* or *SI*) to interpret the result. We will take the point of view of Coulomb's law as a starting point, and the classic work by Millikan[3] to measure the charge on a single electron, a (negative) quantity that we traditionally call $-e$. The modern best value for his measurement is $e = 4.8032042 \times 10^{-10}$ esu.

So let us start by giving ourselves the problem of expressing the charge on an electron in Coulombs. This is easy. Let $e_{esu}$ equal the dimensionless number $4.8032042 \times 10^{-10}$. The force between two electrons separated by one *meter* is $10^{-4} e_{esu}^2$ dyne $= 10^{-9} e_{esu}^2$ N. So, in *SI*

$$10^{-9} e_{esu}^2 \text{ N} = \frac{1}{4\pi\varepsilon_0} \frac{e^2}{(1 \text{ m})^2} = \frac{\mu_0}{4\pi} \frac{c^2 e^2}{(1 \text{ m})^2} - 10^{-7} c_{SI}^2 e_C^2 \text{ N} \qquad (A.14)$$

where $e_C$ is the electron charge in Coulombs, and $c_{SI} \equiv 2.998 \times 10^8$, is yet another dimensionless number. Then $e_C = e_{esu}/10 c_{SI} = 1.602 \times 10^{-19}$. This procedure is obviously valid regardless of the charge on an electron. We therefore write

$$q_C = q_{esu}/10 c_{SI} = q_{esu}/2.998 \times 10^9 \qquad (A.15)$$

as a general conversion between charge in the *CGS* system to that in *SI*. That is, one Coulomb represents a much larger amount of charge (i.e. very many more electrons) than one esu, by a factor of $10 c_{SI}$.

The trick here was to recognize that the *numerical* difference between Coulombs and esu is absorbed by the factor $c^2$ in (A.14). This *is not* to say that Coulombs and esu differ by the dimensions of velocity. One *cannot* equate Coulombs to esu without some conversion factor that explicitly cancels out the base unit Amperes.

---

[3] This experiment was a tour de force, which Millikan carried out systematically and carefully over two decades. For a culmination of this work, see his paper "The most probable 1930 values of the electron and related constants," *Phys. Rev.*, **35** (1930) 1231. He determined the value $e = (4.770 \pm 0.005) \times 10^{-10}$ esu.

We can extend from here. Consider the units of electric potential, defined by

$$1 \text{ Joule} = 1 \text{ Volt} \cdot \text{C} \qquad \textbf{SI}$$

$$1 \text{ erg} = 1 \text{ statvolt} \cdot \text{esu} \qquad \textbf{Gaussian}$$

therefore

$$1 \text{ Volt} \cdot \text{C} = 10^7 \text{ statvolt} \cdot \text{esu} \tag{A.16}$$

since one Joule equals $10^7$ ergs. (I like this way of writing things because it means I can use the "=" sign. Energy is energy, whether *Gaussian* or *SI*.) Now thinking again in terms of number of electrons, we know that one Coulomb corresponds to $10c_{SI}$ times as much charge as an esu. So we write, now having to abandon a strict equality,

$$1 \text{ Volt} \cdot 10c_{SI} \Longleftrightarrow 10^7 \text{ statvolt} \tag{A.17}$$

or

$$1 \text{ statvolt} \Longleftrightarrow 299.8 \text{ Volt}. \tag{A.18}$$

In other words, in practical terms, one statvolt is the same as 300 volts. Perhaps here is a reason that *SI* is more popular with electricians and engineers. We always prefer to use numbers on the order of unity when doing practical work. One volt is a reasonable potential difference from a human perspective, but 300 volts would give you a rather significant shock. So, if we worked in *Gaussian*, practical electronics would be discussed in terms of "millistatvolts," a somewhat unwieldy term.

The same thing works symbolically, of course. To go from *Gaussian* to *SI* we need to insert the factor $\mu_0 c^2 / 4\pi = 1/4\pi\varepsilon_0$ in front of Coulomb's law, and rederive things. So, wherever we encounter a value of charge $q$ in a *Gaussian* equation, we replace it with $q/\sqrt{4\pi\varepsilon_0}$. (The trivial conversions from centimeters and grams to meters and kilograms are irrelevant symbolically.) Similarly, any values of current $I$ are replaced by $cI\sqrt{\mu_0/4\pi}$.

You can easily check that these substitutions turn (A.11) into (A.9), and (A.13) into (A.2) w/(A.4). For a different example, the electric field $\mathbf{E} = \lim_{q_0 \to 0} \mathbf{F}/q_0$ is multiplied by $\sqrt{4\pi\varepsilon_0}$ leaving (of course) the force for a charge in an electric field $\mathbf{F} = q\mathbf{E}$ unchanged.

Now let us try going the other way, namely *SI* to *Gaussian*, with the Lorentz force law, namely

$$\mathbf{F} = q\mathbf{E} + q\mathbf{v} \times \mathbf{B} \qquad \textbf{SI}. \tag{A.19}$$

We know that the first term on the right is unchanged. In the second term, replace $q$ with $q\sqrt{4\pi\varepsilon_0} = q\sqrt{4\pi/\mu_0}/c$, but what about the $\mathbf{B}$ field? A study of electromagnetism leads to Ampère's law, which relates magnetic fields and currents. Indeed, steady currents give rise to static magnetic fields according to

$$\nabla \times \mathbf{B} = \frac{4\pi}{c}\mathbf{j} \qquad \textbf{Gaussian} \tag{A.20a}$$

$$\nabla \times \mathbf{B} = \mu_0\,\mathbf{j} \qquad \textbf{SI} \tag{A.20b}$$

where $\mathbf{j}$ is the current density. To go from *CGS* to *SI*, $\mathbf{j}$ is multiplied by $c\sqrt{\mu_0/4\pi}$, so equations (A.20) tell us to multiply $\mathbf{B}$ by $\sqrt{4\pi/\mu_0}$. So, getting back to (A.19) we multiply the second term by $\sqrt{4\pi/\mu_0}/c$ for $q$ and $\sqrt{\mu_0/4\pi}$ for $\mathbf{B}$. (Remember, we are going from *SI* to *CGS*.) We therefore arrive at the Lorentz force law

$$\mathbf{F} = q\mathbf{E} + q\frac{\mathbf{v}}{c} \times \mathbf{B} \qquad \textbf{Gaussian}. \tag{A.21}$$

An important lesson in this last example is that in going from *SI* to *Gaussian*, the magnetic field changes its dimensions differently than does the electric field. Indeed, (A.19) shows that in *SI*, the dimensions of $\mathbf{E}$ are the dimensions of $\mathbf{B}$ multiplied by the dimensions of velocity. On the other hand, (A.21) shows that in the *Gaussian* system, $\mathbf{E}$ and $\mathbf{B}$ have the *same* dimensions. Pedagogically, this is an advantage that the *Gaussian* system has over *SI*.

For a final example, let us express the magnetic moment in *Gaussian* units, which is the starting point for this book. In your introductory physics class, which most likely used *SI* units, you defined the magnetic moment $\mu = I\mathscr{A}$ for a current $I$ moving in a closed loop that enclosed an area $\mathscr{A}$. You also learned that the potential energy for such a current loop in a magnetic field $\mathbf{B}$ was $-\boldsymbol{\mu} \cdot \mathbf{B}$, which of course must be the same expression as in the *Gaussian* system. We already know that $I \to c\sqrt{\mu_0/4\pi}I$ and $\mathbf{B} \to \sqrt{4\pi/\mu_0}\mathbf{B}$. Therefore, in order to keep the same expression for potential energy, the definition of magnetic moment in *Gaussian* units must be $\mu = I\mathscr{A}/c$.

# Appendix B  Elementary Solutions to Schrödinger's Wave Equation

This appendix summarizes simple solutions of Schrödinger's wave equation corresponding to a variety of soluble potential energy functions. Most of these are derived and discussed in the body of the textbook, but all are included in nearly any undergraduate textbook on quantum mechanics. For some solutions, MATHEMATICA code is given for calculating the wave functions.

## B.1  Free Particles ($V = 0$)

The plane wave, or momentum, eigenfunction, from Section 1.7, is

$$\psi_{\mathbf{k}}(\mathbf{x},t) = \langle \mathbf{x}|\mathbf{k}\rangle e^{-iEt/\hbar} = \frac{1}{(2\pi)^{3/2}}\, e^{i\mathbf{k}\cdot\mathbf{x}-i\omega t}, \tag{B.1}$$

where

$$\mathbf{k} = \frac{\mathbf{p}}{\hbar} \quad \text{and} \quad \omega = \frac{E}{\hbar} = \frac{\mathbf{p}^2}{2m\hbar} = \frac{\hbar k^2}{2m}. \tag{B.2}$$

The normalization, as described in Section 1.6, is

$$\int \psi_{\mathbf{k}'}^* \psi_{\mathbf{k}}\, d^3x = \delta^{(3)}(\mathbf{k}-\mathbf{k}'). \tag{B.3}$$

The superposition of plane waves leads to the *wave packet* description. In one dimension,

$$\psi(x,t) = \frac{1}{\sqrt{2\pi}} \int_{-\infty}^{\infty} dk\, A(k)\, e^{i(kx-\omega t)} \quad \text{where} \quad \omega = \frac{\hbar k^2}{2m}. \tag{B.4}$$

For $|A(k)|$ sharply peaked near $k \simeq k_0$, the wave packet moves with a group velocity

$$v_g \simeq \left(\frac{d\omega}{dk}\right)_{k_0} = \frac{\hbar k_0}{m}. \tag{B.5}$$

The time evolution of a minimum wave packet can be described by

$$\psi(x,t) = \left[\frac{(\Delta x)_0^2}{2\pi^3}\right]^{1/4} \int_{-\infty}^{\infty} e^{-(\Delta x)_0^2(k-k_0)^2+ikx-i\omega(k)t}\, dk \quad \text{where} \quad \omega(k) = \frac{\hbar k^2}{2m}, \tag{B.6}$$

and

$$|\psi(x,t)|^2 = \left\{ \frac{1}{2\pi(\Delta x)_0^2 \left[1 + (\hbar^2 t^2/4m^2)(\Delta x)_0^{-4}\right]} \right\}^{1/2}$$

$$\times \exp\left\{ -\frac{(x - \hbar k_0 t/m)^2}{2(\Delta x)_0^2 \left[1 + (\hbar^2 t^2/4m^2)(\Delta x)_0^{-4}\right]} \right\}. \tag{B.7}$$

So the width of the wave packet expands as

$$(\Delta x)_0 \quad \text{at} \quad t = 0 \rightarrow (\Delta x)_0 \left[1 + \frac{\hbar^2 t^2}{4m^2}(\Delta x)_0^{-4}\right]^{1/2} \quad \text{at } t > 0. \tag{B.8}$$

## B.2 Piecewise Constant Potentials in One Dimension

For a constant potential $V = V_0$ and $E > V$, the solution is

$$\psi_E(x) = c_+ e^{ikx} + c_- e^{-ikx} \quad \text{where} \quad k = \sqrt{\frac{2m(E - V_0)}{\hbar^2}}. \tag{B.9}$$

For $E < V$, that is the classically forbidden region, the solution is

$$\psi_E(x) = c_+ e^{\kappa x} + c_- e^{-\kappa x} \quad \text{where} \quad \kappa = \sqrt{\frac{2m(V_0 - E)}{\hbar^2}}. \tag{B.10}$$

Note that $c_\pm$ must be set equal to 0 if $x = \pm \infty$ is included in the domain under discussion.

### B.2.1 Rigid-Wall Potential (One-Dimensional Box)

For a well with infinitely high walls at $x = 0$ and $x = L$, that is

$$V = \begin{cases} 0 & \text{for } 0 < x < L \\ \infty & \text{otherwise,} \end{cases} \tag{B.11}$$

the wave functions and energy eigenstates are

$$\psi_E(x) = \sqrt{\frac{2}{L}} \sin\left(\frac{n\pi x}{L}\right), \quad n = 1, 2, 3 \ldots,$$

$$E = \frac{\hbar^2 n^2 \pi^2}{2mL^2}. \tag{B.12}$$

### B.2.2 Square-Well Potential

For a well with finite walls at $x = \pm a$, that is

$$V = \begin{cases} 0 & \text{for } |x| > a \\ -V_0 & \text{for } |x| < a \quad (V_0 > 0), \end{cases} \tag{B.13}$$

the bound-state $(E < 0)$ solutions are:

$$
\psi_E \sim \begin{cases} e^{-\kappa|x|} & \text{for} \quad |x| > a \\ \left. \begin{array}{ll} \cos kx & \text{(even parity)} \\ \sin kx & \text{(odd parity)} \end{array} \right\} & \text{for} \quad |x| < a, \end{cases} \tag{B.14}
$$

where

$$
k = \sqrt{\frac{2m(-|E| + V_0)}{\hbar^2}} \quad \text{and} \quad \kappa = \sqrt{\frac{2m|E|}{\hbar^2}}. \tag{B.15}
$$

The allowed discrete values of energy $E = -\hbar^2 \kappa^2/2m$ are to be determined by solving

$$
\begin{array}{ll} ka \tan ka = \kappa a & \text{(even parity)} \\ ka \cot ka = -\kappa a & \text{(odd parity)}. \end{array} \tag{B.16}
$$

Note also that $\kappa$ and $k$ are related by

$$
\frac{2mV_0a^2}{\hbar^2} = (k^2 + \kappa^2)a^2. \tag{B.17}
$$

# B.3  Transmission Reflection Problems

In this discussion we define the transmission coefficient $T$ to be the ratio of the flux of the transmitted wave to that of the incident wave. We consider these simple examples.

## B.3.1  Square Well

For a wave incident on a finite square well defined by

$$
V = 0 \text{ for } |x| > a \quad \text{and} \quad V = -V_0 \text{ for } |x| < a,
$$

with $V_0 > 0$, the transmission coefficient is

$$
\begin{aligned}
T &= \frac{1}{\left\{ 1 + \left[ (k'^2 - k^2)^2/4k^2k'^2 \right] \sin^2 k'a \right\}} \\
&= \frac{1}{\left\{ 1 + \left[ V_0^2/4E(E + V_0) \right] \sin^2 \left( 2a\sqrt{2m(E + V_0)/\hbar^2} \right) \right\}}
\end{aligned} \tag{B.18}
$$

where

$$
k = \sqrt{\frac{2mE}{\hbar^2}} \quad \text{and} \quad k' = \sqrt{\frac{2m(E + V_0)}{\hbar^2}}. \tag{B.19}
$$

Note that resonances occur whenever

$$
2a\sqrt{\frac{2m(E + V_0)}{\hbar^2}} = n\pi, \quad n = 1, 2, 3, \ldots. \tag{B.20}
$$

### B.3.2  Square Barrier

This potential is the same as for the square well, but with $V = V_0 > 0$ for $|x| < a$. One finds for the case $E < V_0$,

$$T = \frac{1}{\left\{1 + \left[(k^2 + \kappa^2)^2/4k^2\kappa^2\right]\sinh^2 \kappa a\right\}}$$

$$= \frac{1}{\left\{1 + \left[V_0^2/4E(V_0 - E)\right]\sinh^2\left(2a\sqrt{2m(V_0 - E)/\hbar^2}\right)\right\}}. \tag{B.21}$$

For $E > V_0$, the result is the same as the square-well case with $V_0$ replaced by $-V_0$.

### B.3.3  Potential Step

The potential is $V = 0$ for $x < 0$ and $V = V_0$ for $x > 0$, and we take $E > V_0$. One finds for the transmission coefficient

$$T = \frac{4kk'}{(k+k')^2} = \frac{4\sqrt{(E - V_0)E}}{\left(\sqrt{E} + \sqrt{E - V_0}\right)^2} \tag{B.22}$$

with

$$k = \sqrt{\frac{2mE}{\hbar^2}} \quad \text{and} \quad k' = \sqrt{\frac{2m(E - V_0)}{\hbar^2}}. \tag{B.23}$$

### B.3.4  General Potential Barrier

For $E < V(x)$ in the range $a \leq x \leq b$ and $E > V(x)$ elsewhere, the approximate JWKB[1] solution for $T$ is

$$T \simeq \exp\left\{-2\int_a^b dx\sqrt{\frac{2m[V(x) - E]}{\hbar^2}}\right\}, \tag{B.24}$$

where $a$ and $b$ are the classical turning points.

## B.4  Simple Harmonic Oscillator

See Section 2.5.2. The potential energy function is

$$V(x) = \frac{1}{2}m\omega^2 x^2. \tag{B.25}$$

---

[1] JWKB stand for Jeffreys–Wentzel–Kramers–Brillouin.

We introduce a dimensionless variable

$$\xi = \sqrt{\frac{m\omega}{\hbar}}\, x, \tag{B.26}$$

and write the energy eigenfunctions as

$$\psi_E = (2^n n!)^{-1/2} \left(\frac{m\omega}{\pi\hbar}\right)^{1/4} e^{-\xi^2/2} H_n(\xi) \tag{B.27}$$

and the energy levels are

$$E = \hbar\omega\left(n + \frac{1}{2}\right), \quad n = 0, 1, 2, \ldots. \tag{B.28}$$

The Hermite polynomials have the following properties:

$$H_n(\xi) = (-1)^n e^{\xi^2} \frac{\partial^n}{\partial \xi^n} e^{-\xi^2}$$

$$\int_{-\infty}^{\infty} H_{n'}(\xi) H_n(\xi)\, e^{-\xi^2}\, d\xi = \pi^{1/2} 2^n n! \delta_{nn'}$$

$$\frac{d^2}{d\xi^2} H_n - 2\xi \frac{dH_n}{d\xi} + 2n H_n = 0$$

$$H_0(\xi) = 1, \quad H_1(\xi) = 2\xi, \tag{B.29}$$
$$H_2(\xi) = 4\xi^2 - 2,$$
$$H_3(\xi) = 8\xi^3 - 12\xi,$$
$$H_4(\xi) = 16\xi^4 - 48\xi^2 + 12.$$

The Hermite polynomials are calculated in MATHEMATICA with HermiteH[n, $\xi$]. The following code produces the first few wave functions as functions of $x$:

```
\[Psi]E = (2^{n} n!)^(-1/2) (m \[Omega]/(Pi \[HBar]))^(1/
    4) Exp[-\[Xi]^{2}/2] HermiteH[n, \[Xi]];
\[Psi]0 = \[Psi]E /. n -> 0 /. \[Xi] -> x Sqrt[m \[Omega]/\[HBar]];
\[Psi]1 = \[Psi]E /. n -> 1 /. \[Xi] -> x Sqrt[m \[Omega]/\[HBar]];
\[Psi]2 = \[Psi]E /. n -> 2 /. \[Xi] -> x Sqrt[m \[Omega]/\[HBar]];
\[Psi]3 = \[Psi]E /. n -> 3 /. \[Xi] -> x Sqrt[m \[Omega]/\[HBar]];
```

If you code these up into MATHEMATICA, or some other symbolic mathematics application, it is worth your while to check that your wave functions are orthonormal.

## B.5  The Central Force Problem

See Section 3.7. The basic time-independent Schrödinger equation is

$$-\frac{\hbar^2}{2m}\left[\frac{1}{r^2}\frac{\partial}{\partial r}\left(r^2 \frac{\partial \psi_E}{\partial r}\right) + \frac{1}{r^2 \sin\theta}\frac{\partial}{\partial \theta}\left(\sin\theta \frac{\partial \psi_E}{\partial \theta}\right) + \frac{1}{r^2 \sin^2\theta}\frac{\partial^2 \psi_E}{\partial \phi^2}\right] + V(r)\psi_E = E\psi_E \tag{B.30}$$

where our spherically symmetrical potential $V(r)$ satisfies

$$\lim_{r \to 0} r^2 V(r) \to 0. \tag{B.31}$$

The method of separation of variables,

$$\Psi_E(\mathbf{x}) = R(r) Y_l^m(\theta, \phi), \tag{B.32}$$

leads to the angular equation

$$-\left[\frac{1}{\sin \theta} \frac{\partial}{\partial \theta}\left(\sin \theta \frac{\partial}{\partial \theta}\right) + \frac{1}{\sin^2 \theta} \frac{\partial^2}{\partial \phi^2}\right] Y_l^m = l(l+1) Y_l^m, \tag{B.33}$$

where the spherical harmonics

$$Y_l^m(\theta, \phi), \quad l = 0, 1, 2, \ldots, \quad m = -l, -l+1, \ldots, +l \tag{B.34}$$

satisfy

$$-i \frac{\partial}{\partial \phi} Y_l^m = m Y_l^m \tag{B.35}$$

and the $Y_l^m(\theta, \phi)$ have the following properties:

$$Y_l^m(\theta, \phi) = (-1)^m \sqrt{\frac{2l+1}{4\pi} \frac{(l-m)!}{(l+m)!}} P_l^m(\cos \theta) e^{im\phi} \quad \text{for } m \geq 0,$$

$$Y_l^m(\theta, \phi) = (-1)^{|m|} Y_l^{|m|*}(\theta, \phi) \quad \text{for } m < 0,$$

$$P_l^m(\cos \theta) = (1 - \cos^2 \theta)^{m/2} \frac{d^m}{d(\cos \theta)^m} P_l(\cos \theta) \quad \text{for } m \geq 0,$$

$$P_l(\cos \theta) = \frac{(-1)^l}{2^l l!} \frac{d^l (1 - \cos^2 \theta)^l}{d(\cos \theta)^l},$$

$$Y_0^0 = \frac{1}{\sqrt{4\pi}}, \quad Y_1^0 = \sqrt{\frac{3}{4\pi}} \cos \theta, \tag{B.36}$$

$$Y_1^{\pm 1} = \mp \sqrt{\frac{3}{8\pi}} (\sin \theta) e^{\pm i\phi},$$

$$Y_2^0 = \sqrt{\frac{5}{16\pi}} (3 \cos^2 \theta - 1),$$

$$Y_2^{\pm 1} = \mp \sqrt{\frac{15}{8\pi}} (\sin \theta \cos \theta) e^{\pm i\phi},$$

$$Y_2^{\pm 2} = \sqrt{\frac{15}{32\pi}} (\sin^2 \theta) e^{\pm 2i\phi}$$

$$\int Y_{l'}^{m'*}(\theta, \phi) Y_l^m(\theta, \phi) d\Omega = \delta_{ll'} \delta_{mm'} \left[\int d\Omega = \int_0^{2\pi} d\phi \int_{-1}^{+1} d(\cos \theta)\right].$$

For the radial piece of (B.32), we define

$$u_E(r) = rR(r) \tag{B.37}$$

and then the radial equation is reduced to an equivalent one-dimensional problem, namely,

$$-\frac{\hbar^2}{2m}\frac{d^2 u_E}{dr^2} + \left[V(r) + \frac{l(l+1)\hbar^2}{2mr^2}\right] u_E = E u_E$$

subject to the boundary condition

$$u_E(r)|_{r=0} = 0. \tag{B.38}$$

For the case of *free* particles, that is $V(r) = 0$, in our spherical coordinates:

$$R(r) = c_1 j_l(\rho) + c_2 n_l(\rho) \quad (c_2 = 0 \text{ if the origin is included}) \tag{B.39}$$

where $\rho$ is a dimensionless variable

$$\rho \equiv kr, \quad \text{with} \quad k = \sqrt{\frac{2mE}{\hbar^2}}. \tag{B.40}$$

We need to list the commonly used properties of the Bessel functions and spherical Bessel and Hankel functions. The spherical Bessel functions are

$$j_l(\rho) = \left(\frac{\pi}{2\rho}\right)^{1/2} J_{l+1/2}(\rho),$$

$$n_l(\rho) = (-1)^{l+1} \left(\frac{\pi}{2\rho}\right)^{1/2} J_{-l-1/2}(\rho),$$

$$j_0(\rho) = \frac{\sin \rho}{\rho}, \qquad\qquad n_0(\rho) = -\frac{\cos \rho}{\rho},$$

$$j_1(\rho) = \frac{\sin \rho}{\rho^2} - \frac{\cos \rho}{\rho}, \qquad n_1(\rho) = -\frac{\cos \rho}{\rho^2} - \frac{\sin \rho}{\rho}, \tag{B.41}$$

$$j_2(\rho) = \left(\frac{3}{\rho^3} - \frac{1}{\rho}\right) \sin \rho - \frac{3}{\rho^2} \cos \rho,$$

$$n_2(\rho) = -\left(\frac{3}{\rho^3} - \frac{1}{\rho}\right) \cos \rho - \frac{3}{\rho^2} \sin \rho.$$

For $\rho \to 0$, the leading terms are

$$j_l(\rho) \xrightarrow[\rho \to 0]{} \frac{\rho^l}{(2l+1)!!}, \qquad n_l(\rho) \xrightarrow[\rho \to 0]{} -\frac{(2l-1)!!}{\rho^{l+1}}, \tag{B.42}$$

where

$$(2l+1)!! \equiv (2l+1)(2l-1)\cdots 5 \cdot 3 \cdot 1. \tag{B.43}$$

In the large $\rho$-asymptotic limit, we have

$$j_l(\rho) \xrightarrow[\rho \to \infty]{} \frac{1}{\rho} \cos\left[\rho - \frac{(l+1)\pi}{2}\right],$$

$$n_l(\rho) \xrightarrow[\rho \to \infty]{} \frac{1}{\rho} \sin\left[\rho - \frac{(l+1)\pi}{2}\right]. \tag{B.44}$$

Because of constraints (B.37) and (B.38), $R(r)$ must be finite at $r = 0$; hence, from (B.39) and (B.42) we see that the $n_l(\rho)$-term must be deleted because of its singular behavior as

$\rho \to 0$. Thus $R(r) = c_j j_l(\rho)$ [or, in the notation of Section 6.4, $A_l(r) = R(r) = c_j j_l(\rho)$]. For a three-dimensional square-well potential, $V = -V_0$ for $r < R$ (with $V_0 > 0$), the desired solution is

$$R(r) = A_l(r) = \text{constant} \, j_l(\alpha r), \tag{B.45}$$

where

$$\alpha = \left[ \frac{2m(V_0 - |E|)}{\hbar^2} \right]^{1/2} \quad \text{for} \quad r < R. \tag{B.46}$$

As discussed in (6.135), the exterior solution for $r > R$, where $V = 0$, can be written as a linear combination of spherical Hankel functions. These are defined as follows:

$$h_l^{(1)}(\rho) = j_l(\rho) + i n_l(\rho) \tag{B.47}$$

$$h_l^{(1)*}(\rho) = h_l^{(2)}(\rho) = j_l(\rho) - i n_l(\rho) \tag{B.48}$$

which, from (B.44), have the asymptotic forms for $\rho \to \infty$ as follows:

$$h_l^{(1)}(\rho) \xrightarrow[\rho \to \infty]{} \frac{1}{\rho} e^{i[\rho - (l+1)\pi/2]} \tag{B.49}$$

$$h_l^{(1)*}(\rho) = h_l^{(2)}(\rho) \xrightarrow[\rho \to \infty]{} \frac{1}{\rho} e^{-i[\rho - (l+1)\pi/2]}. \tag{B.50}$$

If we are interested in the bound-state energy levels of the three-dimensional square-well potential, where $V(r) = 0, r > R$, we have

$$u_l(r) = r A_l(r) = \text{constant} \, e^{-\kappa r} f\left( \frac{1}{\kappa r} \right)$$

$$\kappa = \left( \frac{2m|E|}{\hbar^2} \right)^{1/2}. \tag{B.51}$$

To the extent that the asymptotic expansions, of which (B.50) give the leading terms, do not contain terms with exponent of opposite sign to that given, we have, for $r > R$, the desired solution from (B.51):

$$A_l(r) = \text{constant} \, h_l^{(1)}(i\kappa r) = \text{constant} \, [j_l(i\kappa r) + i n_l(i\kappa r)], \tag{B.52}$$

where the first three of these functions are

$$h_0^{(1)}(i\kappa r) = -\frac{1}{\kappa r} e^{-\kappa r}$$

$$h_1^{(1)}(i\kappa r) = i\left( \frac{1}{\kappa r} + \frac{1}{\kappa^2 r^2} \right) e^{-\kappa r} \tag{B.53}$$

$$h_2^{(1)}(i\kappa r) = \left( \frac{1}{\kappa r} + \frac{3}{\kappa^2 r^2} + \frac{3}{\kappa^3 r^3} \right) e^{-\kappa r}.$$

Finally, we note that in considering the shift from free particles, i.e. $V(r) = 0$ to the case of the constant potential $V(r) = V_0$, we need only to replace the $E$ in the free-particle solution (B.39) and (B.40) by $E - V_0$. Note, though, that if $E < V_0$, then $h_l^{(1,2)}(i\kappa r)$ is to be used with $\kappa = \sqrt{2m(V_0 - E)/\hbar^2}$.

# B.6 Hydrogen Atom

See Section 3.7.4. The potential energy function for this problem is given by

$$V(r) = -\frac{Ze^2}{r}$$

(B.54)

and for bound states ($E < 0$) we introduce the dimensionless variable

$$\rho = \left(-\frac{2m_e E}{\hbar^2}\right)^{1/2} r = \frac{Zr}{na_0}$$

(B.55)

for electron mass $m_e$ and Bohr radius $a_0 = \hbar^2/m_e e^2$. The energy eigenfunctions and eigenvalues (energy levels) are

$$\psi_{nlm} = R_{nl}(r) Y_l^m(\theta, \phi)$$

$$R(r) = \frac{1}{(2l+1)!} \left(\frac{2Zr}{na_0}\right)^l e^{-Zr/na_0} \left[\left(\frac{2Z}{na_0}\right)^3 \frac{(n+l)!}{2n(n-l-1)!}\right]^{1/2} F(l+1-n; 2l+2; 2\rho)$$

$$E_n = -\frac{Z^2 e^2}{2a_0} \frac{1}{n^2}$$

(B.56)

where $n = 1, 2, 3, \ldots$ with $l = 0, 1, \ldots, n-1$, and $F(a; c; x)$ is the confluent hypergeometric function, which solves Kummer's equation and is normalized as $F(a; c; 0) = 1$.

One can instead write the radial wave function using associated Laguerre polynomials, but conventions and normalizations vary. Using the definition developed in Problem 3.30,

$$R(r) = \left(\frac{2Zr}{na_0}\right)^l e^{-Zr/na_0} \left[\left(\frac{2Z}{na_0}\right)^3 \frac{(n-l-1)!}{2n[(n+l)!]^3}\right]^{1/2} L_{n-l-1}^{2l+1}(2\rho).$$

(B.57)

The radial functions for low $n$ are:

$$R_{10}(r) = \left(\frac{Z}{a_0}\right)^{3/2} 2e^{-Zr/a_0}$$

$$R_{20}(r) = \left(\frac{Z}{2a_0}\right)^{3/2} (2 - Zr/a_0) e^{-Zr/2a_0}$$

(B.58)

$$R_{21}(r) = \left(\frac{Z}{2a_0}\right)^{3/2} \frac{Zr}{\sqrt{3}\, a_0} e^{-Zr/2a_0}.$$

Some convenient radial integrals are

$$\langle r^k \rangle \equiv \int_0^\infty dr \, r^{2+k} [R_{nl}(r)]^2,$$

$$\langle r \rangle = \left( \frac{a_0}{2Z} \right) [3n^2 - l(l+1)]$$

$$\langle r^2 \rangle = \left( \frac{a_0^2 n^2}{2Z^2} \right) [5n^2 + 1 - 3l(l+1)] \tag{B.59}$$

$$\left\langle \frac{1}{r} \right\rangle = \frac{Z}{n^2 a_0},$$

$$\left\langle \frac{1}{r^2} \right\rangle = \frac{Z^2}{n^3 a_0^2 \left( l + \frac{1}{2} \right)}.$$

The following MATHEMATICA code is useful for calculating these wave functions:

```
\[Psi]nlm[r_, \[Theta]_, \[Phi]_] = Nnl Rnl[r] Ylm[\[Theta], \[Phi]];
Ylm[\[Theta]_, \[Phi]_] = SphericalHarmonicY[l, m, \[Theta], \[Phi]];
Rnl[r_] = (2 Z r/(n a0))^{l}
     Exp[-Z r/(n a0)]
     Hypergeometric1F1[-n + l + 1, 2 l + 2, 2 Z r/(n a0)];
Nnl = 1/(2 l + 1)! Sqrt[(2 Z/(n a0))^{3} (n + l)!/(2 n (n - l - 1)!)];
```

Given these definitions, then, for example, the $1s$ and $2p$, $m = +1$ wave functions are

```
\[Psi]100 = \[Psi]nlm[r,\[Theta],\[Phi]] /. {n -> 1, l -> 0, m -> 0}
\[Psi]211 = \[Psi]nlm[r,\[Theta],\[Phi]] /. {n -> 2, l -> 1, m -> 1}
```

As always, it is wise to check that your code produces proper orthonormal wave functions.

# Appendix C  Hamiltonian for a Charge in an Electromagnetic Field

We will often need to know the Hamiltonian for a particle with mass $m$ and charge $q$ moving while in the presence of a static electric field $\mathbf{E}(\mathbf{x})$ and static magnetic field $\mathbf{B}(\mathbf{x})$. We will take a very utilitarian approach, namely to write down the equation of motion and then find the Lagrangian which, in classical physics, yields this equation. We then construct the Hamiltonian from the Lagrangian.

We know that the equation of motion is given by the Lorentz force law, namely

$$m\ddot{\mathbf{x}} = q\mathbf{E} + \frac{1}{c}q\dot{\mathbf{x}} \times \mathbf{B}. \tag{C.1}$$

We can derive the Hamiltonian from the Lagrangian, but it is not obvious how to build the Lagrangian when there is no obvious "potential energy" function for a charged particle in a magnetic field.

Our approach[1] will be to start with the conjecture that the correct Lagrangian is given by

$$\mathcal{L}(\mathbf{x}, \dot{\mathbf{x}}) = \frac{1}{2}m\dot{\mathbf{x}}^2 - q\phi(\mathbf{x}) + \frac{q}{c}\dot{\mathbf{x}} \cdot \mathbf{A}(\mathbf{x}) \tag{C.2}$$

where $\phi$ and $\mathbf{A}$ are the standard electrostatic and magnetic vector potentials, that is

$$\mathbf{E}(\mathbf{x}) = -\boldsymbol{\nabla}\phi(\mathbf{x}) \qquad \text{and} \qquad \mathbf{B}(\mathbf{x}) = \boldsymbol{\nabla} \times \mathbf{A}(\mathbf{x}), \tag{C.3}$$

and then show that Lagrange's equations lead us to (C.1). We have

$$\begin{aligned}
0 &= \frac{d}{dt}\frac{\partial \mathcal{L}}{\partial \dot{x}_i} - \frac{\partial \mathcal{L}}{\partial x_i} \\
&= \frac{d}{dt}\left[ m\dot{x}_i + \frac{q}{c}A_i(\mathbf{x}) \right] + q\frac{\partial \phi}{\partial x_i} - \frac{q}{c}\dot{\mathbf{x}} \cdot \frac{\partial \mathbf{A}}{\partial x_i} \\
&= m(\ddot{\mathbf{x}})_i + q(\boldsymbol{\nabla}\phi)_i + \frac{q}{c}\left[ \frac{d}{dt}A_i(\mathbf{x}) - \dot{\mathbf{x}} \cdot \frac{\partial \mathbf{A}}{\partial x_i} \right].
\end{aligned} \tag{C.4}$$

The first term on the right of (C.4) is just the left-hand side of (C.1), and the second term on the right of (C.4) is just the first term on the right-hand side of (C.1). It therefore remains to evaluate

$$\dot{\mathbf{x}} \cdot \frac{\partial \mathbf{A}}{\partial x_i} - \frac{d}{dt}A_i(\mathbf{x}) = \sum_j \dot{x}_j \frac{\partial A_j}{\partial x_i} - \sum_j \frac{\partial A_i}{\partial x_j}\dot{x}_j = \sum_j \dot{x}_j \left( \frac{\partial A_j}{\partial x_i} - \frac{\partial A_i}{\partial x_j} \right) \tag{C.5}$$

---

[1] One can derive this Lagrangian from first principles of relativity and electromagnetism, but we will leave that approach to an advanced course on classical or quantum field theory.

and show that it equals $(\dot{\mathbf{x}} \times \mathbf{B})_i = (\dot{\mathbf{x}} \times (\nabla \times \mathbf{A}))_i$. We can use the totally antisymmetric symbol $\varepsilon_{ijk}$ to write the cross product of two vectors as $(\mathbf{a} \times \mathbf{b})_i = \sum_j \sum_k \varepsilon_{ijk} a_k b_k$. Therefore

$$(\dot{\mathbf{x}} \times \mathbf{B})_i = \sum_j \sum_k \varepsilon_{ijk} \dot{x}_j \left( \sum_l \sum_m \varepsilon_{klm} \frac{\partial A_l}{\partial x_m} \right) = \sum_j \dot{x}_j \sum_l \sum_m \sum_k \varepsilon_{kij} \varepsilon_{klm} \frac{\partial A_l}{\partial x_m} \qquad (C.6)$$

where $\varepsilon_{ijk} = \varepsilon_{kij}$ because the indices are rearranged by an even number of exchanges. We then make use of the theorem $\sum_k \varepsilon_{kij} \varepsilon_{klm} = \delta_{il}\delta_{jm} - \delta_{im}\delta_{jl}$ to write

$$(\dot{\mathbf{x}} \times \mathbf{B})_i = \sum_j \dot{x}_j \sum_l \sum_m (\delta_{il}\delta_{jm} - \delta_{im}\delta_{jl}) \frac{\partial A_l}{\partial x_m} = \sum_j \dot{x}_j \left( \frac{\partial A_i}{\partial x_j} - \frac{\partial A_j}{\partial x_i} \right) \qquad (C.7)$$

which is the same as (C.5). This proves our conjecture that (C.2) is the correct Lagrangian.

We can now use definitions from classical mechanics to derive the Hamiltonian $\mathscr{H}(\mathbf{p}, \mathbf{x})$ from (C.2). First we determine the canonical momentum $\mathbf{p}$ from

$$p_i \equiv \frac{\partial \mathscr{L}}{\partial \dot{x}_i} = m\dot{x}_i + \frac{q}{c} A_i(\mathbf{x}). \qquad (C.8)$$

Then we construct the Hamiltonian using the Legendre transformation

$$\begin{aligned}
\mathscr{H} &= \sum_i \dot{x}_i p_i - \mathscr{L} \\
&= \frac{1}{m} \left( \mathbf{p} - \frac{q}{c}\mathbf{A} \right) \cdot \mathbf{p} - \frac{1}{2}m\frac{1}{m^2} \left( \mathbf{p} - \frac{q}{c}\mathbf{A} \right)^2 + q\phi - \frac{q}{c}\frac{1}{m} \left( \mathbf{p} - \frac{q}{c}\mathbf{A} \right) \cdot \mathbf{A} \\
&= \frac{1}{2m} \left( \mathbf{p} - \frac{q}{c}\mathbf{A} \right)^2 + q\phi. \qquad (C.9)
\end{aligned}$$

It is important to note that the canonical momentum $\mathbf{p}$ is not equal to $m\dot{\mathbf{x}}$ in this case.

# Appendix D   Proof of the Angular-Momentum Rule (3.358)

It will be instructive to discuss the angular-momentum addition rule from the quantum-mechanical point of view. Let us, for the moment, label our angular momenta, so that $j_1 \geq j_2$. This we can always do. From (3.355), the maximum value of $m$, $m^{\max}$, is

$$m^{\max} = m_1^{\max} + m_2^{\max} = j_1 + j_2. \tag{D.1}$$

There is only one ket that corresponds to the eigenvalue $m^{\max}$, whether the description is in terms of $|j_1 j_2; m_1 m_2\rangle$ or $|j_1 j_2; jm\rangle$. In other words, choosing the phase factor to be 1, we have

$$|j_1 j_2; j_1 j_2\rangle = |j_1 j_2; j_1 + j_2, j_1 + j_2\rangle. \tag{D.2}$$

In the $|j_1 j_2; m_1 m_2\rangle$ basis, there are two kets that correspond to the $m$ eigenvalue $m^{\max} - 1$, namely, one ket with $m_1 = m_1^{\max} - 1$ and $m_2 = m_2^{\max}$ and one ket with $m_1 = m_1^{\max}$ and $m_2 = m_2^{\max} - 1$. There is thus a twofold degeneracy in this basis; therefore, there must be a twofold degeneracy in the $|j_1 j_2; jm\rangle$ basis as well. From where could this come? Clearly, $m^{\max} - 1$ is a possible $m$-value for $j = j_1 + j_2$. It is also a possible $m$-value for $j = j_1 + j_2 - 1$, in fact, the maximum $m$-value for this $j$. So $j_1$, $j_2$ can add to $j$ of $j_1 + j_2$ and $j_1 + j_2 - 1$.

We can continue in this way, but it is clear that the degeneracy cannot increase indefinitely. Indeed, for $m^{\min} = -j_1 - j_2$, there is once again a single ket. The maximum degeneracy is $(2j_2 + 1)$-fold, as is apparent from Table D.1 constructed for two special examples: for $j_1 = 2, j_2 = 1$ and for $j_1 = 2, j_2 = \frac{1}{2}$. This $(2j_2 + 1)$-fold degeneracy must be associated with the $2j_2 + 1$ states $j$:

$$j_1 + j_2, j_1 + j_2 - 1, \ldots, j_1 - j_2. \tag{D.3}$$

If we lift the restriction $j_1 \geq j_2$, we obtain (3.358).

**Table D.1** Special Examples of Values of $m$, $m_1$, and $m_2$ for the Two Cases $j_1 = 2$, $j_2 = 1$ and $j_1 = 2$, $j_2 = \frac{1}{2}$, Respectively

| $j_1 = 2, j_2 = 1$ $m$ | 3 | 2 | 1 | 0 | $-1$ | $-2$ | $-3$ |
|---|---|---|---|---|---|---|---|
| $(m_1, m_2)$ | $(2,1)$ | $(1,1)$ | $(0,1)$ | $(-1,1)$ | $(-2,1)$ | | |
| | | $(2,0)$ | $(1,0)$ | $(0,0)$ | $(-1,0)$ | $(-2,0)$ | |
| | | | $(2,-1)$ | $(1,-1)$ | $(0,-1)$ | $(-1,-1)$ | $(-2,-1)$ |
| Numbers of States | 1 | 2 | 3 | 3 | 3 | 2 | 1 |
| $j_1 = 2, j_2 = \frac{1}{2}$ $m$ | $\frac{5}{2}$ | $\frac{3}{2}$ | $\frac{1}{2}$ | $-\frac{1}{2}$ | $-\frac{3}{2}$ | $-\frac{5}{2}$ | |
| $(m_1, m_2)$ | $(2,\frac{1}{2})$ | $(1,\frac{1}{2})$ | $(0,\frac{1}{2})$ | $(-1,\frac{1}{2})$ | $(-2,\frac{1}{2})$ | | |
| | | $(2,-\frac{1}{2})$ | $(1,-\frac{1}{2})$ | $(0,-\frac{1}{2})$ | $(-1,-\frac{1}{2})$ | $(-2,-\frac{1}{2})$ | |
| Numbers of States | 1 | 2 | 2 | 2 | 2 | 1 | |

Section 3.8 includes a discussion of how one goes about calculating Clebsch–Gordan coefficients. However, in practice, we generally use tables to look them up, or computer applications that carry out the calculation for us.

Of the many tables available, I am partial to the one available online from the Particle Data Group (PDG) (http://pdg.lbl.gov). A concise table for all of the Clebsch–Gordan coefficients is posted at

http://pdg.lbl.gov/2018/reviews/rpp2018-rev-clebsch-gordan-coefs.pdf

This page also lists the first several spherical harmonics and $d$-functions.

Programming languages such as PYTHON include libraries for calculating Clebsch–Gordan coefficients, as do symbolic manipulation applications such as MATHEMATICA. For example, in MATHEMATICA, the syntax is

ClebschGordan[{j1, m1}, {j2, m2}, {j, m}]

which returns the coefficient for adding $j_1$ and $j_2$ to get $j$, with $m_1 + m_2 = m$. If an unphysical combination of quantum numbers is given, then MATHEMATICA returns the value zero, along with a warning that the request was not physical. Purely symbolic function calls are of course possible, and return the conditions under which a nonzero answer is returned.

As a trivial example that compares the result from the PDG table with the output from MATHEMATICA, executing the cell

ClebschGordan[{2, 1}, {3/2, -1/2}, {7/2, 1/2}] == Sqrt[12/35]

returns the value True.

# Appendix F  **Notes on Complex Variables**

These notes are meant to accompany a graduate level physics course, to provide a basic introduction to the necessary concepts in complex analysis. They are not complete, nor are any of the proofs considered rigorous. The immediate goal is to carry through enough of the work needed to explain the Cauchy residue theorem.

## F.1  Complex Numbers and Complex Functions

A complex number $z$ can be written as

$$z = x + iy \qquad \text{or} \qquad z = re^{i\phi} \text{ with } r \geq 0$$

where $i = \sqrt{-1}$, and $x$, $y$, $r$, and $\phi$ are real numbers. Clearly, $x = r\cos\phi$ and $y = r\sin\phi$, leading to a description in terms of the "complex plane." The complex conjugate of $z$ is

$$z^* = x - iy \qquad \text{or} \qquad z^* = re^{-i\phi}.$$

The "modulus" of $z$ is $|z| \equiv \sqrt{z^*z} = r = \sqrt{x^2 + y^2}$, and $\phi$ is often called the "phase" of $z$.

A complex function $f(z)$ typically returns a complex number. Generically, we write

$$f(z) = u(x,y) + iv(x,y) \tag{F.1}$$

for purposes of proofs or illustrations. The behavior of the (real) functions $u(x,y)$ and $v(x,y)$ are critical for classifying complex functions, as seen when we consider taking derivatives.

## F.2  Differentiation and Analyticity

We define the derivative $f'(z) = df/dz$ of a complex function $f(z)$ in the same way as we do for the derivatives of real functions. That is, for $z_0 \equiv x_0 + iy_0$,

$$f'(z_0) = \left.\frac{df}{dz}\right|_{z=z_0} = \lim_{z \to z_0} \frac{f(z) - f(z_0)}{z - z_0}.$$

However, there is clearly an ambiguity, depending on whether we approach $z_0$ along the line $y = y_0$ or along $x = x_0$. (Of course, we could also say the ambiguity is along any line of constant $\phi = \phi_0$, but it is sufficient to consider just two orthogonal directions.) That is,

$$f'(z_0) = \lim_{x \to x_0} \frac{u(x,y_0) - u(x_0,y_0)}{x - x_0} + i \lim_{x \to x_0} \frac{v(x,y_0) - v(x_0,y_0)}{x - x_0} = \frac{\partial u}{\partial x} + i \frac{\partial v}{\partial x}$$

or

$$f'(z_0) = \lim_{y \to y_0} \frac{u(x_0,y) - u(x_0,y_0)}{iy - iy_0} + i \lim_{y \to y_0} \frac{v(x_0,y) - v(x_0,y_0)}{iy - iy_0} = -i \frac{\partial u}{\partial y} + \frac{\partial v}{\partial y}.$$

Therefore, in order to remove the ambiguity and have a consistent definition of the derivative,

$$\frac{\partial u}{\partial x} = \frac{\partial v}{\partial y} \qquad \text{and} \qquad \frac{\partial u}{\partial y} = -\frac{\partial v}{\partial x}. \tag{F.2}$$

These are called the Cauchy–Riemann conditions. A function $f(z)$ which satisfies these rather restrictive conditions is called **analytic**. Indeed, analytic functions have very many applications in physics, and we will merely scratch the surface here.

For example, the function $f(z) = e^z = e^x(\cos y + i \sin y)$ is analytic. This is easy to prove. Putting $u(x,y) = e^x \cos y$ and $v(x,y) = e^x \sin y$,

$$\frac{\partial u}{\partial x} = e^x \cos y = \frac{\partial v}{\partial y} \qquad \text{and} \qquad \frac{\partial u}{\partial y} = -e^x \sin y = -\frac{\partial v}{\partial x}$$

so the Cauchy–Riemann conditions (F.2) are satisfied.

It is simple to show that $f(z) = az$ is analytic, where $a$ is a complex constant. It is also not hard to show that the product of two analytic functions is analytic, so any function of the form $f(z) = a_n z^n$, where $n$ is a nonnegative integer, is also analytic. Of course, any sum of analytic functions is analytic, so we see that any polynomial in $z$ is analytic in the entire complex plane.

These examples beg the question: If a function $f(z)$ can be written explicitly in terms of $z$, is it analytic? The answer is "Yes." To see this, realize that instead of $x$ and $y$, we could always write a complex function in terms of $z$ and $z^*$ using $x = (z + z^*)/2$ and $y = (z - z^*)/2i$. Now consider

$$\frac{\partial f}{\partial z^*} = \frac{\partial f}{\partial x} \frac{\partial x}{\partial z^*} + \frac{\partial f}{\partial y} \frac{\partial y}{\partial z^*} = \left( \frac{\partial u}{\partial x} + i \frac{\partial v}{\partial x} \right) \left( \frac{1}{2} \right) + \left( \frac{\partial u}{\partial y} + i \frac{\partial v}{\partial y} \right) \left( -\frac{1}{2i} \right)$$

$$= \frac{1}{2} \left( \frac{\partial u}{\partial x} - \frac{\partial v}{\partial y} \right) + \frac{i}{2} \left( \frac{\partial u}{\partial y} + \frac{\partial v}{\partial x} \right) = 0$$

so long as the Cauchy–Riemann conditions (F.2) are satisfied. That is, if the expression for $f(z)$ contains only $z$ (and not $z^*$) then the function is analytic.

There is some common terminology. A function $f(z)$ need not be analytic in the entire complex plane. (If it is, we called the function "entire.") If it is analytic at a point $z_0$ then we call that a "regular point." Otherwise, $z_0$ is called a "singular point." Much of our discussion of complex integration will focus on the notion of singular points.

# F.3  Integration and Series Expansion

Similarly to differentiation, we approach integration of complex functions the same way as with real functions, but we need to be aware that there is now an arbitrariness of the "path" of integration. With $dz = dx + i\,dy$ and using using (F.1), we have

$$\int_{z_1}^{z_2} f(z)dz = \int_{z_1}^{z_2} (u\,dx - v\,dy) + i\int_{z_1}^{z_2} (v\,dx + u\,dy) = \int_{z_1}^{z_2} \mathbf{A}\cdot d\mathbf{x} + i\int_{z_1}^{z_2} \mathbf{B}\cdot d\mathbf{x} \qquad \text{(F.3)}$$

where $\mathbf{A} = u\hat{\mathbf{x}} - v\hat{\mathbf{y}}$ and $\mathbf{B} = v\hat{\mathbf{x}} + u\hat{\mathbf{y}}$. So, we can now think of the two integrals on the right as real integrals of vector functions over curves in the $xy$ plane. However, if we invoke Stokes's theorem, these become integrals of the curls, and using (F.2), we find

$$\nabla \times \mathbf{A} = \left(-\frac{\partial v}{\partial x} - \frac{\partial u}{\partial y}\right) = 0 \qquad \text{and} \qquad \nabla \times \mathbf{B} = \left(\frac{\partial u}{\partial x} - \frac{\partial v}{\partial y}\right) = 0 \qquad \text{(F.4)}$$

and each of the two integrals on the right in (F.3) is path independent. Hence, the integral of an *analytic* complex function $f(z)$ is path independent and can be unambiguously defined.

**From here on, we assume all functions to be analytic unless explicitly noted otherwise.** It is obvious from (F.3) that, when integrating around a closed path $C$,

$$\oint_C f(z)dz = 0$$

which is known as the Cauchy–Goursat theorem. We will be exploring circumstances where the integrand is explicitly singular at one or more points.

For the first example, we prove the Cauchy integral formula, namely

$$f(z_0) = \frac{1}{2\pi i}\oint_C \frac{f(z)}{z - z_0}dz \qquad \text{(F.5)}$$

where $C$ is a closed contour in the complex plane that contains the point $z_0$ and traversed in the counterclockwise direction. We can break up a contour $C$ into something that looks like Figure F.1. Notice that $C_0$ is a tiny *circular* contour around the singular point, but in the clockwise direction. That is, we replace $C$ with $\lim_{C_0 \to 0}(C + C_0)$. However, for $C_0 \neq 0$, the new contour $C$ does not include the singular point, so by (F.4) we write (F.5) as

$$f(z_0) = -\frac{1}{2\pi i}\oint_{C_0} \frac{f(z)}{z - z_0}dz. \qquad \text{(F.6)}$$

The shrinking contour $C_0$ is parameterized as $z - z_0 = re^{i\phi}$ for $r \to 0$ and $\phi = 2\pi \to 0$, so

$$-\frac{1}{2\pi i}\oint_{C_0} \frac{f(z)}{z - z_0}dz = -\frac{1}{2\pi i}f(z_0)\int_{2\pi}^{0} \frac{1}{re^{i\phi}}ire^{i\phi}d\phi = -\frac{1}{2\pi i}f(z_0)i(-2\pi) = f(z_0)$$

proving the Cauchy integral formula (F.5). A trivial, but suggestive, rewriting of (F.5) gives

$$f(z) = \frac{1}{2\pi i}\oint_C \frac{f(\xi)}{\xi - z}d\xi \qquad \text{(F.7)}$$

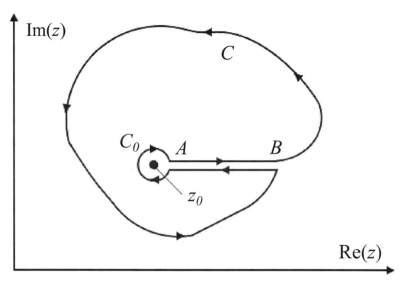

A contour in the complex plane for proving the Cauchy integral formula.

which leads to a convenient way to write the derivatives of a complex function, namely

$$f^{(n)}(z) = \frac{d^n f}{dz^n} = \frac{n!}{2\pi i} \oint_C \frac{f(\xi)}{(\xi - z)^{n+1}} d\xi. \tag{F.8}$$

Now consider the series expansion of an analytic function $f(z)$. We would naturally write

$$f(z) = f(z_0) + f'(z_0)(z - z_0) + \cdots = \sum_{n=0}^{\infty} a_n (z - z_0)^n \tag{F.9}$$

where

$$a_n \equiv \frac{1}{n!} f^{(n)}(z_0) = \frac{1}{2\pi i} \oint_C \frac{f(\xi)}{(\xi - z_0)^{n+1}} d\xi. \tag{F.10}$$

Such a Taylor series expansion works out as expected, but the curve $C$ specifies regions in which the series converges.

This idea can be expanded to include $-\infty \leq n \leq \infty$, still using the right side of (F.10) to define $a_n$, and with modified regions of convergence. Such an expansion is called a Laurent series. It clearly is not, in general, an analytic function because of poles that appear for $n < 0$. These, however, lead us to one of the most important theorems of complex analysis, so far as mathematical physics is concerned.

## F.4 The Cauchy Residue Theorem

Let $g(z)$ have an isolated singularity at $z = z_0$. If the Laurent expansion can be written as

$$g(z) = \sum_{n=-\infty}^{\infty} a_n (z - z_0)^n = \frac{b_1}{z - z_0} + \sum_{n=0}^{\infty} a_n (z - z_0)^n \tag{F.11}$$

then we say that $g(z)$ has a "simple pole" at $z = z_0$. Higher-order poles are possible, but we are not going to consider them here.

Consider a contour $C$ within the radius of convergence of $g(z)$. Separate the integral of $g(z)$ around this contour into two terms, one for each of the two terms on the right in (F.11). The second term is a polynomial in $z$; therefore it is analytic and the integral is zero. Recall that we reduced the contour to a small circle around the pole in order to prove the Cauchy integral formula. We can do the same thing here, and

$$\oint_C g(z)dz = b_1 \oint_C \frac{1}{z - z_0} = 2\pi i \, b_1. \tag{F.12}$$

We refer to $b_1$ as the "residue" of $g(z_0)$, sometimes written as $\mathrm{Res}[g(z_0)]$. We have

$$\mathrm{Res}[g(z_0)] = \lim_{z \to z_0} (z - z_0)g(z)$$

for a simple pole at $z_0$.

If there is more than one simple pole within the contour $C$, this result is easy to generalize. Instead of redrawing the contour with a small loop about the single pole, do it for all $N$ poles within the contour. The result is clearly

$$\oint_C g(z)dz = 2\pi i \sum_{k=1}^{N} \mathrm{Res}[g(z_k)]. \tag{F.13}$$

We refer to this as the Cauchy residue theorem. It is widely used in mathematical physics.

The usefulness of the residue theorem can be illustrated in many ways, but here is one important example. It is a warm-up to evaluating the integral in (6.44). The exercise is to evaluate the integral

$$I = \int_{-\infty}^{\infty} \frac{e^{ika}}{q^2 - k^2} \, k \, dk = \lim_{\varepsilon \to 0} \int_{-\infty}^{\infty} \frac{e^{ika}}{q^2 - k^2 + i\varepsilon} \, k \, dk \tag{F.14}$$

where $k$, $a$, $q$, and $\varepsilon > 0$ are all real variables. We use the second version above because this moves the singularities at $k = \pm q$ off the real axis. To be sure, we could have moved off the real axis by using $-i\varepsilon$ instead of $+i\varepsilon$, and in fact, this would give us a different answer. A physical rationale is needed to justify one sign or the other. Leave that for a physics course.

We can evaluate (F.14) using contour integration by first allowing $k$ to be complex and then noting that $e^{ikx} \to 0$ as $\mathrm{Im}(k) \to +\infty$. Therefore (F.14) can be rewritten as an integral over a semicircular contour $C$ that runs (counterclockwise) along the $\mathrm{Re}(k)$ axis and closes as a semicircle in the $\mathrm{Im}(k) > 0$ plane. Then for $\varepsilon \to 0$, the integrand in (F.14) has poles at

$$k = \pm\sqrt{q^2 - i\varepsilon} = \pm q\left(1 - i\frac{\varepsilon}{q^2}\right) \Rightarrow \pm q \mp i\varepsilon$$

where we redefine $\varepsilon$ (with $q > 0$) so that it is still small and has the same sign.

The pole at $k = k_0 \equiv +q - i\varepsilon$ does not matter to us, since it is outside the integration contour. However, the pole at $k = -k_0 = -q + i\varepsilon$ is inside, so we use the residue theorem to write

$$
\begin{aligned}
I &= \lim_{\varepsilon \to 0} \oint_C \frac{e^{ika}}{(k - k_0)(k + k_0)} \, k \, dk \\
&= \lim_{\varepsilon \to 0} 2\pi i \left. \frac{e^{ika}}{k - k_0} \, k \right|_{k = -k_0} = \pi i \lim_{\varepsilon \to 0} e^{-ik_0 a} = \pi i e^{-iqa}.
\end{aligned}
$$

# Bibliography

The following list includes textbooks and monographs that are, in general, cited earlier in this book. Some entries are not cited, but are included here to indicate general upper level textbooks on quantum mechanics that students and instructors may find useful for alternative explanations and additional exercises.

Arfken, G. B., and H. J. Weber (1995). *Mathematical Methods for Physicists*, 4th edn. Academic Press.

Arfken, G. B., H. J. Weber, and F. E. Harris (2013). *Mathematical Methods for Physicists: A Comprehensive Guide*, 7th edn. Academic Press.

Baym, G. (1969). *Lectures on Quantum Mechanics*. New York: W. A. Benjamin. Republished in paperback by CRC Press, 1974.

Bethe, H. A., and R. W. Jackiw (1968). *Intermediate Quantum Mechanics*, 2nd edn. New York: W. A. Benjamin.

Biedenharn, L. C., and H. Van Dam (editors) (1965). *Quantum Theory of Angular Momentum*. New York: Academic Press.

Bjorken, J. D., and S. D. Drell (1965). *Relativistic Quantum Fields*. New York: McGraw-Hill.

Blatt, J. M., and V. W. Weisskopf (1952). *Theoretical Nuclear Physics*. John Wiley & Sons.

Bohm, D. (1951). *Quantum Theory*. Englewood Cliffs, NJ: Prentice-Hall.

Bransden, B. H., and C. J. Joachain (2003). *Physics of Atoms and Molecules*. Pearson.

Byron, F. W., and R. W. Fuller (1992). *Mathematics of Classical and Quantum Physics*. Dover.

Cohen, M. L., and S. G. Louie (2016). *Fundamentals of Condensed Matter Physics*. Cambridge University Press.

Dicke, R. H., and J. P. Wittke (1960). *Introduction to Quantum Mechanics*. Reading, MA: Addison-Wesley.

Dirac, P. A. M. (1958). *Quantum Mechanics*, 4th edn. London: Oxford University Press.

Edmonds, A. R. (1974). *Angular Momentum in Quantum Mechanics*, 3rd printing. Princeton University Press.

Fermi, E. (1950). *Nuclear Physics*. Chicago, IL: University of Chicago Press.

Fetter, A. L., and J. D. Walecka (2003). *Quantum Theory of Many-Particle Systems*. Dover.

Feynman, R. P., and A. R. Hibbs (1964). *Quantum Mechanics and Path Integrals*. New York: McGraw-Hill.

Finkelstein, R. J. (1973). *Nonrelativistic Mechanics*. Reading, MA: W. A. Benjamin.

French, A. P., and E. F. Taylor (1978). *An Introduction to Quantum Physics*. New York: W. W. Norton.

Goldberger, M. L., and K. M. Watson (1964). *Collision Theory*. New York: Wiley.

Goldstein, H., C. Poole, and J. Safko (2002). *Classical Mechanics*, 3rd edn. Reading, MA: Addison-Wesley.

Gottfried, K. (1966). *Quantum Mechanics*, vol. I. New York: W. A. Benjamin.

Gottfried, K., and T.-M. Yan (2003). *Quantum Mechanics: Fundamentals*, 2nd edn. Springer-Verlag.

Griffiths, D. J. (2005). *Introduction to Quantum Mechanics*, 2nd edn. Pearson.

Heitler, W. (1954). *The Quantum Theory of Radiation*, 3rd edn. Oxford: Oxford University Press.

Henley, E. M., and A. Garcia (2007). *Subatomic Physics*, 3rd edn. World Scientific.

Holstein, B. R. (1992). *Topics in Advanced Quantum Mechanics*. Addison-Wesley.

Itzykson, C., and J.-B. Zuber (1980). *Quantum Field Theory*. McGraw-Hill.

Jackson, J. D. (1975). *Classical Electrodynamics*, 2nd edn. New York: Wiley.

Jackson, J. D. (1998). *Classical Electrodynamics*, 3rd edn. New York: Wiley.

Landau, R. H. (1996). *Quantum Mechanics II: A Second Course in Quantum Theory*. Wiley.

Loudon, R. (2000). *The Quantum Theory of Light*, 3rd edn. Oxford Science Publications.

Mandl, F. (1957). *Quantum Mechanics*. London: Butterworths Scientific Publications.

Marcus, G. E. (editor) (2000). *Zeroing in on the Year 2000: The Final Edition*. Chicago, IL: University of Chicago Press.

Matthews, P. T. (1974). *Introduction to Quantum Mechanics*, 3rd edn. London: McGraw-Hill.

Melissinos, A., and J. Napolitano (2003). *Modern Physics*. Academic Press.

Merzbacher, E. (1970). *Quantum Mechanics*, 2nd edn. New York: Wiley.

Merzbacher, E. (1998). *Quantum Mechanics*, 3rd edn. Wiley.

Messiah, A. (1961). *Quantum Mechanics*. New York: Interscience.

Morse, P. M., and H. Feshbach (1953). *Methods of Theoretical Physics*. New York: McGraw-Hill.

Mott, N. F. (1952). *Elements of Wave Mechanics*. London: Cambridge University Press.

Newton, R. G. (1982). *Scattering Theory of Waves and Particles*, 2nd. edn. New York: McGraw-Hill.

Park, D. (1974). *Introduction to the Quantum Theory*, 2nd. edn. New York: McGraw-Hill.

Pauling, L., and E. B. Wilson (1935). *Introduction to Quantum Mechanics*. New York: McGraw-Hill.

Preston, M. (1962). *Physics of the Nucleus*. Reading, MA: Addison-Wesley.

Rojansky, V. (1938). *Introductory Quantum Mechanics*. Englewood Cliffs, NJ: Prentice-Hall.

Sakurai, J. J. (1964). *Invariance Principles and Elementary Particles*. Princeton University Press. Republished 2016.

Sakurai, J. J. (1967). *Advanced Quantum Mechanics*. Reading, MA: Addison-Wesley.

Sargent III, M., M. O. Scully, and W. E. Lamb, Jr. (1974). *Laser Physics*. Reading, MA: Addison-Wesley.

Saxon, D. S. (1968). *Elementary Quantum Mechanics*. San Francisco, CA: Holden-Day.

Schiff, L. (1968). *Quantum Mechanics*, 3rd. edn. New York: McGraw-Hill.

Shankar, R. (1994). *Principles of Quantum Mechanics*, 2nd edn. Plenum.

Taylor, J. R. (2005). *Classical Mechanics*. University Science Books.

Tomanaga, S. (1962). *Quantum Mechanics I*. Amsterdam: North-Holland.

Townsend, J. S. (2000). *A Modern Approach to Quantum Mechanics*. University Science Books.

Van der Waerden, B. L. (editor) (1967). *Sources of Quantum Mechanics*. Dover.

Weinberg, S. (1995). *The Quantum Theory of Fields*. Cambridge University Press.

Zee, A. (2010). *Quantum Field Theory in a Nutshell*, 2nd edn. Princeton University Press.

# Index

Abelian group, 154
absorption of radiation, 347
adiabatic approximation, 328
Aharonov, Y., 133
Aharonov–Bohm effect, 120, 131, 335
Airy function, 102, 107
ammonia maser, 326
ammonia molecule, 262
Anderson, C., 495
angular momentum
    addition, 205
    addition of orbital plus spin $\frac{1}{2}$, 206
    addition of two spin $\frac{1}{2}$, 206
    barrier, 197
    commutation relations, 26, 154, 209
    Dirac equation, 497
    eigenvalues, 180
    orbital, 28, 188
    partial wave decomposition, 388
    quantization, 3
anticommutator, 26
anyons, 433
Argand diagram, 396
Aspect, A., 231
atomic fine structure, 307, 309, 310
atomic force microscope, 471
atomic polarizability, 299

Baker–Hausdorff lemma, 89
Balmer formula, 204, 507
Bell's inequality, 227
Bell, J. S., 227
Berry's phase, 138, 331
    Aharonov–Bohm effect, 336
    neutrons, 334
    photons, 333
Berry, M. V., 331
Bethe, H. A., 422
Bloch's theorem, 269
Bloch, F., 422
Bohm, D., 133, 226
Bohr model, 204
Bohr radius, 204, 318, 462
Bohr, N., 1, 69
Born approximation, 375, 384
    higher orders, 387

Born, M., 45, 83, 93, 95, 180
Bose–Einstein condensation, 434, 457
bosons, 433, 455
bra, 12
    representation as row matrix, 18
Brillouin zone, 269
Brillouin, L., 104
Burke, K., 451

canonical commutation relations, 45
Casimir effect, 468
Cayley–Klein parameters, 165
charge conjugation symmetry, 498
classical correspondence, 76, 79, 81, 90, 96, 116, 326
Clebsch–Gordan coefficients, 208, 210
    and tensor operators, 236
    recursion relations, 212
    rotation operator, 216
    unitary matrix, 211
Clifford algebra, 490
commutator, 26
compatibility, 27, 68
completeness, 17, 18
Compton effect, 1
Compton wavelength, 480
confluent hypergeometric function, 203
conservation laws, 249
    and symmetry operators, 250
    classical Hamiltonian, 250
constant of the motion, 250, 252
continuity equation, 94, 482, 492
Coulomb potential, 201
covariance, 481
CPT symmetry, 501
cross section
    photoelectric effect, 351
    radiation absorption, 348
    Rutherford, 386, 424
    scattering, 373, 381

Dalgarno, A., 300
Darwin term, 311
Davisson–Germer–Thompson experiment, 1
de Broglie, L., 1, 43, 62, 93
degeneracy, 27, 98, 201, 205
    and parity symmetry, 260

and perturbation theory, 300
  connection to symmetry, 251
  exchange, 430, 442
density matrix, 172
density of states, 98, 344, 350, 353, 373
density operator, 172
diagonalization, 28, 35
  in degenerate subspace, 302
Dirac delta function, 38
Dirac equation, 478, 491
Dirac picture, 322
Dirac, P. A. M., 1, 10, 15, 22, 45, 47, 78, 83, 115, 137, 338, 344, 490
Dyson, F. J., 67, 339

effective potential, 196
Ehrenberg, W., 133
Ehrenfest theorem, 81
Ehrenfest, P., 81
eigenket, 11, 16, 81
  energy, 67, 70
  position or momentum, 38, 53
  simultaneous, 28, 181
eigenvalue, 11, 16, 23
  angular momentum, 180
  continuous, 37
Einstein's locality principle, 226
Einstein, A., 226
Einstein–Debye theory, 1
Einstein–Podolsky–Rosen paradox, 226
electric dipole approximation, 348
energy bands, 268
energy-time uncertainty relation, 74
ensemble, 22, 169
  pure, 172, 177
  time evolution, 175
ensemble average, 171
Ernzerhof, M., 451
Euler angles, 166, 187, 194, 222
exchange degeneracy, 430, 442
exchange density, 436
exchange-correlation energy, 448, 450
expectation value, 23, 68
  under rotation, 155

Fermi energy, 457, 462
Fermi's golden rule, 344, 373
fermions, 433, 455
Fetter, A. L., 509
Feynman, R. P., 114, 115, 509
Fock space, 454
form factor, 421, 423, 428
Fortun, M., 469
Fourier transform, 51, 351
Franck–Hertz experiment, 1
free particle, 79, 98, 110, 146, 198, 479, 484, 491–493

gauge transformations, 120
  electromagnetism, 126
  gravitational effect, 122
generating function, 99, 101, 204
geometric phase, 331
Gerlach, W., 2
Glauber, R., 91, 142, 472
Green's function, 110
  Helmholtz equation, 379, 388
  scattering, 376, 424
Groups
  SO(3), 163
  SO(4), 252
  SU(2), 164
  SU(2)×SU(2), 255

Hamilton, W. R., 93
Hamilton–Jacobi equation, 96, 400
Hamiltonian, 63, 65
Heisenberg equation of motion, 68, 78
Heisenberg picture, 77, 321
Heisenberg uncertainty principle, 3, 33, 43, 88
Heisenberg, W., 1, 45, 180, 441
helium atom, 312, 319, 437, 451
Hermite polynomials, 99
hidden variables, 227
Hilbert space, 11
Hilbert, D., 11, 93
Hofstadter, R., 424
Hohenberg, P., 444, 447
Hohenberg–Kohn theorem, 447
hydrogenlike atoms, 201
  degeneracy, 205, 252, 255, 260
  Dirac equation solution, 505
  energy levels, 204
  generating function, 204, 244
  Klein–Gordon equation solution, 485
  linear Stark effect, 298, 303
  perturbation, 288
  principal quantum number, 204
  quadratic Stark effect, 298
  relativistic perturbation, 306
  spin-orbit perturbation, 308
  variational principle, 318
  versus Bohr model, 204
  wave function, 205, 528

inner product, 12
interaction picture, 321, 340, 355, 371

Jeffreys, H., 104
Jordan, P., 45, 180

ket, 8, 10
  representation as column matrix, 18
Klein–Gordon equation, 478, 480
Klein–Gordon field, 443, 485

Kohn, W., 443, 444, 447
Kohn–Sham potential, 449
Kramers degeneracy, 284
Kramers, A., 104
Kummer's equation, 203

Lamb shift, 305, 509
Lamoreaux, S., 469
Lande's formula, 313
Laporte's rule, 264
Lee, T. D., 264
Lewis, J. T., 300
Liouville's theorem, 175
Lipkin, H. J., 138
Lippmann–Schwinger equation, 375, 382, 387, 424, 427
long wavelength approximation, 353, 355
Lorentz force, 126
Lorentz transformation, 481

magnetic moment, 3, 4, 55, 307, 326
    from Dirac Equation, 496
    time evolution in magnetic field, 63, 66, 69, 147, 157, 158, 180, 333
magnetic monopole, 135, 337
matrix mechanics, 45
matrix representation, 18
    angular-momentum operators, 184
    diagonal matrix, 27
    eigenvalue problem, 35
    rotation operator, 185, 216
    unitary transformation, 34
Maxwell's equations, 464
Merzbacher, E., 509
muon $g - 2$, 71

neutrinos, 71, 138
neutrons
    Berry's phase, 334, 368
    bouncing energy eigenvalues, 103, 107
    gravitational phase, 123
    interferometry, 148, 158
    $np$ bound state, 410
normal ordering, 458
normalization, 13, 17, 22, 292
nuclear magnetic resonance, 326

observable, 11, 16, 22
    compatible, 27, 28
    incompatible, 29
    unitary equivalent, 36
operator, 13
    annihilation and creation, 84
    antiunitary, 273
    generator, 25, 42, 152, 189, 250
    Hermitian, 14, 16, 19
    ladder, 181

linear, 14
number, 84
pair distribution, 458
permutation, 430
projection, 18, 292, 301
representation as square matrix, 18
rotation, 152, 168, 216
symmetry, 250
tensor, 233
time evolution, 63, 65
trace of, 35
translation, 40
unitary, 34, 75, 210
vector, 232
optical theorem, 381
orthogonality, 13, 17
orthohelium, 440
outer product, 15

parabolic coordinates, 300
parahelium, 440
parity symmetry, 256
    Dirac equation, 497
    one-dimensional double well, 261
    perturbation theory, 297
    violation, 264
partition function, 112, 179
Paschen–Back effect, 313
path-integral formulation, 114
Pauli exclusion principle, 433, 456, 494
Pauli matrices, 160
Pauli, W., 21, 159
Perdew, J., 443, 451
Peshkin, M., 138
photoelectric effect, 98, 350, 464
photon, 464, 467
pion-proton scattering, 415
Planck radiation law, 1
Poisson brackets, 46
polarization
    circular, 8
    of light, 6
    spin, 170
Pople, J., 443
Positron, 492, 495
potential energy function
    $\delta$-function well, 144
    Coulomb potential, 201, 386, 505, 528
    double $\delta$-function, 145
    Gaussian potential, 403
    linear, 101
    local, 379
    one-dimensional finite well, 145, 521
    one-dimensional infinite well, 318, 521
    periodic, 266
    screened Coulomb, 309
    simple harmonic oscillator, 83, 99, 523

spherical well, 198, 385
symmetric double well, 261
three-dimensional harmonic oscillator, 199
Yukawa potential, 386, 387, 403
precession, 69, 157
projection theorem, 239
propagator, 109

quantum optics, 472
quarkonium, 103

Rabi, I. I., 323
Ramsauer–Townsend effect, 407
reduced matrix element, 239
representations, 18
resonance
  magnetic, 325
  scattering, 397, 412
  two-state system, 324
rotation, 149
  finite, 153
  infinitesimal, 150
  operator, 152
  orthogonal group, 163
  orthogonal matrix, 149
  two-component formalism, 161
  unitary unimodular group, 164
  vector, 231
rotational symmetry
  angular momentum conservation, 251
Runge–Lenz vector, 245, 252

$S$ matrix, 372
  poles as bound states, 410
scattering phase shift, 394
scattering shadow amplitude, 405
Schrödinger equation, 65, 82, 91
  energy quantization, 93
  probability density and flux, 94
  spherically symmetric potentials, 195
Schrödinger picture, 77, 321, 339
Schrödinger, E., 1, 62, 92
Schwinger, J., 23, 42, 218
second quantization, 443, 454, 485
selection rules
  parity, 263
Sham, L.J., 444, 447
Siday, R.E., 133
simple harmonic oscillator, 83
  as solution to the Schrödinger equation, 99
  coherent states, 91, 142, 473
  generating function, 99
  ground state, 86
  parity symmetry, 260
  propagator, 111
  quadratic perturbation, 296
  three-dimensional, 199

Slater determinant, 442
sodium atom, 309
sodium metal, 464
solid angle, 374
Sommerfeld, A., 106
space quantization, 3
spherical Bessel function, 199
spherical harmonics, 191, 217
  as spherical tensors, 233
spin-angular functions, 216, 246, 309, 502
spin magnetic resonance, 325
spin-orbit interaction, 307
spin-statistics theorem, 433
spinor, 493
spinor spherical harmonics, 216, 246, 309, 502
spontaneous emission of radiation, 352
squeezed states, 474
Stark effect, 264, 298, 329
stationary state, 69
statistical mechanics, 176
statistics
  Bose–Einstein, 433
  Fermi–Dirac, 433
  Maxwell–Boltzmann, 434
Stern, O., 1
Stern–Gerlach experiment, 1, 54
  sequential, 4, 8, 24
stimulated emission of radiation, 347
sudden approximation, 328
superfluid helium, 434, 474

$T$ matrix, 372, 374
tensor operator
  Cartesian, 233
  irreducible spherical, 234
  reducibility, 233
Thomas precession, 308
Thomas, L. H., 308
Thomas–Reiche–Kuhn sum rule, 349
time reversal, 14, 270
  Dirac equation, 499
totally antisymmetric symbol $\varepsilon_{ijk}$, 26
transition amplitude, 339, 372
transition probability, 340
translation, 40
  finite, 43
tritium atom, 328
two-state systems, 2, 289, 323

unitarity limit, 413–415
units, 43, 479, 510
Unsöld, A., 440

vacuum, 454, 468
van der Waals' force, 314
variational theorem, 316, 443, 447

vector, 149
    axial, 258
    operator definition, 232
    polari, 258
    rotations, 149, 152, 156
vector space, 10
    bra space, 12, 15
    dual space, 12
    ket space, 10, 15
virial theorem, 87, 306
von Neumann, J., 170

Walecka, J. D., 509
wave function, 39, 47, 82, 94
    orbital plus spin angular momentum, 206
    plane wave, 51, 97, 520
    renormalization, 295
    spherical well, 198, 397
    under parity, 258
    under time reversal, 279
wave mechanics, 92

wave packet, 51, 81, 111
    in scattering theory, 381
Weisberger, W. I., 138
Weisskopf, V. F., 358
Wentzel, G., 104
Weyl, H., 93, 131
Wiener, N., 83
Wigner, E. P., 185, 222, 227, 264, 270, 284, 358
Wigner–Eckart theorem, 218, 237, 298, 393
    examples, 238
    projection theorem, 239
Wigner–Seitz radius, 462
Wilson, W., 106
WKB approximation, 104
Wu, T. T., 137

Yang, C. N., 137, 264

Zee, A., 509
Zeeman effect, 311, 360, 362
    quadratic, 312, 364
zero point energy, 468, 487